Crystallization and Related Phenomena in Amorphous Materials

MATERIALS RESEARCH SOCIETY SYMPOSIUM PROCEEDINGS VOLUME 321

Crystallization and Related Phenomena in Amorphous Materials

Symposium held November 29-December 2, 1993, Boston, Massachusetts, U.S.A.

EDITORS:

Matthew Libera

Stevens Institute of Technology
Hoboken, New Jersey, U.S.A.

Tony E. Haynes

Oak Ridge National Laboratory
Oak Ridge, Tennessee, U.S.A.

Peggy Cebe

Massachusetts Institute of Technology
Cambridge, Massachusetts, U.S.A.

James E. Dickinson Jr.

Corning Incorporated
Corning, New York, U.S.A.

MATERIALS RESEARCH SOCIETY
Pittsburgh, Pennsylvania

CAMBRIDGE UNIVERSITY PRESS
Cambridge, New York, Melbourne, Madrid, Cape Town,
Singapore, São Paulo, Delhi, Mexico City

Cambridge University Press
32 Avenue of the Americas, New York NY 10013-2473, USA

Published in the United States of America by Cambridge University Press, New York

www.cambridge.org
Information on this title: www.cambridge.org/9781107409446

Materials Research Society
506 Keystone Drive, Warrendale, PA 15086
http://www.mrs.org

First published 1994
First paperback edition 2012

Single article reprints from this publication are available through
University Microfilms Inc., 300 North Zeeb Road, Ann Arbor, MI 48106

CODEN: MRSPDH

ISBN 978-1-107-40944-6 Hardback

Contents

PART I: THE STRUCTURE OF GLASSES

*Invited Paper

*Invited Paper

PART V: BEAM-ASSISTED CRYSTALLIZATION
AND AMORPHIZATION

*Invited Paper

viii

*Invited Paper

*Invited Paper

xi

Preface

This volume is a compilation of more than 100 papers presented during a symposium on crystallization held as part of the 1993 Fall Meeting of the Materials Research Society in Boston. This symposium was highly interdisciplinary. A main goal, largely achieved, was to include researchers working with different kinds of materials into a single forum. This symposium thus provided an opportunity for cross-fertilization of ideas and methods common to research on crystallization and related phenomena in amorphous ceramics, metals, polymers, and semiconductors. The publication of this proceedings volume will further that goal.

The symposium was organized to emphasize common themes, ranging from characterization of structure and relaxation in the amorphous state, to the fundamentals of crystal nucleation, to microstructural development in the nascent crystalline phase. The organization of this proceedings volume preserves this format; thus, papers describing research in each materials class will be found throughout all sections of this book. Additionally, there was in particular one topic that recurred repeatedly throughout the symposium and that serves as a unifying theme for this proceedings. A large number of papers, distributed throughout this volume, emphasize various properties of defects in the amorphous state, especially with regard to their structure and dynamics. This reflects a common recognition that defects in amorphous materials control nucleation and growth of the crystalline phase, in fact playing a role which is quite complementary to that of crystal defects in the inverse process of amorphization. Consequently, the contents of this proceedings volume document the substantial progress that is being made toward defining and controlling defects in the amorphous state, and thereby toward a better understanding of crystallization that ultimately will enhance our ability to control the final crystal microstructure for a range of advanced applications.

Matthew Libera
Tony Haynes
Peggy Cebe
Jim Dickinson

December 1993

Acknowledgments

This symposium was made possible through the financial assistance of the following sponsors:

> Oak Ridge National Laboratory
> Allied-Signal, Inc.
> Hoechst-Celanese

We would also like to thank the session chairs, who managed the refereeing of the papers contained in this volume in a timely manner:

K. Barmak
L.A. Boatner
L.A. Clevenger
J.S. Custer
H.J. Frost
A.L. Greer

J.S. Im
J. Kieffer
J.M. Poate
M.O. Thompson
B. Wunderlich

A special thanks also goes to Professor David Turnbull for opening the symposium with a commemorative lecture in honor of the Twentieth Anniversary of the founding of the Materials Research Society.

And finally, we wish to thank Shirley Cox of Oak Ridge National Laboratory and the MRS Publications staff for their capable assistance in the preparation of this volume.

PART I

The Structure of Glasses

HIGH TEMPERATURE CALORIMETRIC STUDY OF MIXING, PHASE SEPARATION, AND CRYSTALLIZATION IN SILICATE GLASSES

ALEXANDRA NAVROTSKY
Department of Geological and Geophysical Sciences and Princeton Materials Institute, Princeton University, Princeton, New Jersey 08544.

ABSTRACT

Several glass forming systems have been studied by in situ scanning calorimetry and by solution calorimetry in molten $2PbO \cdot B_2O_3$ at 973 K. In systems which show macroscopic glass-glass phase separation on annealing, namely K_2O-La_2O_3-SiO_2 and Li_2O-SiO_2, the enthalpies of solution of optically clear quenched glasses and milky annealed glasses are almost identical. This behavior, coupled with a linear variation of heat of solution with composition along joins which have a large change in overall degree of polymerization, support the contention from spectroscopy that short range clustering dominates even in nominally single phase materials, with exsolution observed as the initially present heterogeneous regions grow in size. Alkali titanosilicate melts show C_p anomalies consistent with breakdown of alkali titanate complexes or changes in Ti coordination. Calorimetry is a sensitive quantitative probe of crystallization as a function of time in both ceramic and geologic systems.

INTRODUCTION

Melting, crystallization, glass formation. and phase separation in silicate systems are closely linked and all reflect the structure of the noncrystalline phase at the microscopic level. The purpose of this paper is threefold: to review some recent advances in calorimetry and how they apply to glassy and molten silicates, to stress the role of short and mid range order (both ordering and clustering) in melt and glass energetics, and to show how calorimetry can be used to study kinetic as well as thermodynamic features of melting and crystallization.

CALORIMETRIC METHODOLOGY

Calorimetry at high temperature is essential to study enthalpies of formation, mixing, and phase transition for two reasons. First, silicates are generally sufficiently unreactive at room temperature to make determination of heats of formation difficult unless thermodynamic cycles involving chemical reactions near 1000 K are employed. Second, many structural changes are nonquenchable, and must be characterized in situ as a function of temperature. Melting and crystallization, especially complicated in glass-forming systems, often involves an intermingling of kinetic and thermodynamic factors. Fortunately, a variety of high temperature instrumentation is now available for quantitative calorimetry at 1000-1600 K. In the Princeton calorimetry laboratory, we have had extensive experience with several different instruments, and can now tailor calorimetric experiments specifically for different purposes.

Calvet twin calorimeters operating at 973-1073 K are used to determine heats of formation, mixing, and phase transition by solution calorimetry, mainly in molten lead borate solvent [1]. These calorimeters are custom built and consist of two calorimetric chambers surrounded by thermopile detectors. Samples can either be equilibrated at calorimeter temperature and then dissolved (solution calorimetry) or dropped from room temperature directly into the solvent (drop solution calorimetry). The former method has been used to study mixing properties in glasses and heats of vitrification [2]. The latter method is useful when the structural state or composition is hard to maintain at the temperature of calorimetry, such as in systems with variable oxidation states [3] or containing H_2O or CO_2 [4, 5]. This type of calorimeter is limited to 800-1100 K, and its temperature can not be varied on a daily basis, taking several weeks to

3

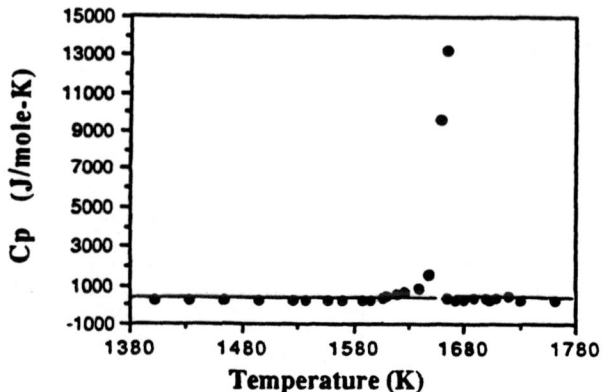

Fig. 1 Heat capacity of diopside, CaMgSi$_2$O$_6$, obtained by step scanning
calorimetry on heating (10 K steps). [10]

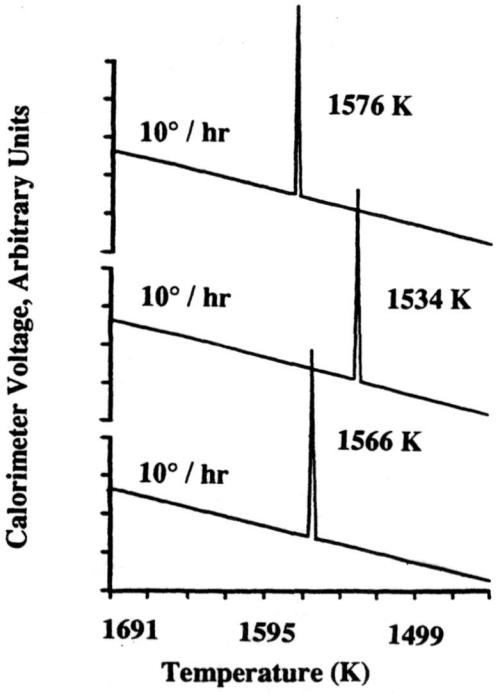

Fig. 2 Change in calorimetric signal during continuous cooling of diopside
melt. Exotherms show crystallization [10].

equilibrate. It is not suitable for in situ studies of most silicate melts. We are finishing construction of an ultrasensitive solution calorimeter for samples in the 1-2 mg range. Controlled atmosphere can be passed over the samples in solution calorimeters.

A commercial Setaram DSC 111 calorimeter, with miniaturized thermopile detectors, measures heat capacities and enthalpies of transition (including glass transitions), [6] at 300-1073 K. Though somewhat delicate, it is sensitive, accurate, and precise ($\pm 1\%$ in C_p). Calorimetric samples can be sealed in capsules and/or run in controlled atmosphere. A thermogravimetric apparatus (TGA), originally part of this instrument, now is run separately with a furnace capable of 1373 K. We found that running the calorimeter in a vertical position, necessary for TGA, compromised calorimetric accuracy severely.

A commercial Setaram HT 1500 calorimeter, whose temperature can be varied from day to day in the 900-1773 K range, can be used for transposed temperature drop calorimetry, in which a sample is dropped from room temperature to high temperature in the calorimeter. A home-built "hybrid calorimeter" combines features of the Setaram and Calvet designs [7]. These instruments have been the workhorses for determining heats of fusion of silicates and heats of mixing in situ in silicate melts [7-9]. A recent development has been to use these calorimeters in scanning and step-scanning mode for measurement of enthalpies of fusion [10], of heat capacities of silicate melts [6, 11], and kinetic studies of melting and crystallization [12]. The accuracy of C_p measurements appears to be 3 % at 900-1400 K, falling off to 5-8 % at 1500-1773 K. The experiments are time-consuming, and 1-3 g samples are required.

The heat capacity of diopside, obtained by heating in step scanning mode [10] is shown in Fig. 1. The increase in C_p at 1600-1680 K suggests incongruent melting and/or "premelting" phenomena in the solid. The major melting event at 1691 K is followed by a region of constant heat capacity in the liquid. Cooling results in a different sequence of events (see Fig. 2). At a continuous scan rate, diopside does not crystallize at its equilibrium melting point; rather, the liquid supercools, and crystallization to a complicated phase assemblage is seen [10]. This happens at variable temperatures in nominally identical experiments, suggesting that nucleation may be somewhat poorly controlled and presumably heterogeneous.

A recent advance is the availability of a more conventional differential scanning calorimeter, the Netzsch 404, for heat capacity measurements to 1673 K. Basically it is a quantitative DTA, since each sample pan has only one thermocouple, but the design of sample pan, platinum heat sink, and thermocouple make the quantitation very encouraging. Fig. 3 shows the heat capacity of silica glass (100 mg). The deviation from C_p measured accurately by drop calorimetry (several grams) is less than 1%. In addition to the advantages of smaller sample size

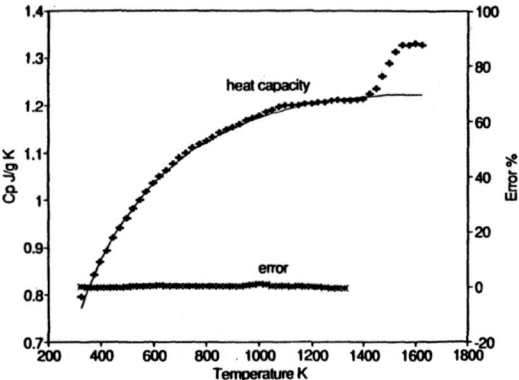

Fig. 3 Heat capacity of SiO_2 glass. Points represent recent measurements by DSC; curve represents equation derived from drop calorimetry [24]. The deviation between scanning and drop measurements is also shown as error. The jump in C_p at high T in the scanning experiments is strongly suggestive of a glass transition which, though hinted in the drop calorimetry, could not be resolved clearly.

and faster and more accurate C_p measurement, the rapid scan rates used by this type of instrument (typically 10 or 20 K/min) should let us explore supercooled liquids and glasses above T_g over a wider temperature range before crystallization sets in.

Development of calorimetric methodology for iron-bearing materials deserves comment. Oxidized (Fe^{3+}-bearing) glasses, melts, and crystals present no special problems in solution, drop-solution, transposed-temperature drop, or scanning calorimetry. Neither change in oxidation state nor Fe absorption by Pt are problems in air. Compounds containing Fe^{2+} have been successfully studied by solution calorimetry [13] and drop solution calorimetry [14]. Weight change experiments in the latter experiments confirm no change in redox state. For scanning calorimetry proper sample containment, presaturation of capsules with Fe, sealed capsules, and fO_2 control by gas mixing will need to be perfected.

RESULTS

Energetics of Titanium-Bearing Silicate Liquids

Titanium is an important component in natural systems, TiO_2 reaching 10 weight percent in alkali-rich silica-poor lavas such as ugandites and lamproites. In addition, because of its ability to take both tetrahedral and octahedral coordination, Ti may be an atmospheric pressure analogue for the high pressure behavior of Si. TiO_2 is also an important nucleating agent in glass ceramics.

The heat capacities of several TiO_2-bearing silicate glasses and liquids have been measured [6]. The results for liquids of M_2O-TiO_2-SiO_2 composition (M = Na, K, Cs) are compared to those for liquids of M_2O-$2SiO_2$ composition (see Fig. 4). The presence of TiO_2 has a profound influence on the heat capacity at 900-1300 K. Specifically, replacement of Si^{4+} by Ti^{4+} leads to doubling of the magnitude of the jump in C_p at the glass transition (T_g): this is followed by a progressive decrease in liquid C_p for over 400 K, until C_p eventually becomes constant and similar to that in Ti-free systems. The large heat capacity step at T_g in the TiO_2-bearing melts suggests significant configurational rearrangements in the liquid that are not available to TiO_2-free silicates. These "extra" configurational changes apparently saturate as temperature increases, implying the completion of whatever process is responsible for them, or the attainment of a random distribution of structural states. Above 1400 K where the heat capacities of TiO_2-bearing and TiO_2-free alkali silicate liquids are similar, their configurational entropies differ by 3.5 J/g.f.w.-K. Density and spectroscopic data on quenched glasses suggest that the anomalous configurational rearrangements may involve the breakdown of alkali and alkaline earth titanate complexes and/or changes in Ti coordination [15-17].

The C_p anomalies in Ti-bearing melts are absent in Ti-free melts, presumably because the silicate species do not undergo similar temperature-induced speciation changes. Ti may be right at the borderline of tetrahedral-octahedral coordination change. We predict that the Zr-bearing analogues would not show anomalies because only one coordination state is likely to be favorable for Zr.

Mixing, Phase Separation, Short Range Order, and Melt Models

For the titanosilicates Lange and Navrotsky [6] discussed restructuring in terms of the concept of strong and fragile liquids [18] and of relating configurational entropy and viscosity [19]. If exothermic complex formation (e.g TiOK species), dominant at lower temperature, becomes less pronounced with increasing temperature, Navrotsky [20] argued that the stage may be set for the onset of immiscibility as temperature increases, i.e. a lower critical consolute temperature (LCST) analogous to that seen in organics.

Joins along which the degree of polymerization changes with composition often tend toward clustering and a dome of immiscibility below a critical temperature (a UCST). In systems such as PbO-SiO_2 and $CaMgSi_2O_6$-$NaAlSi_3O_8$, calorimetry shows positive heats of

K2O-TiO2-SiO2

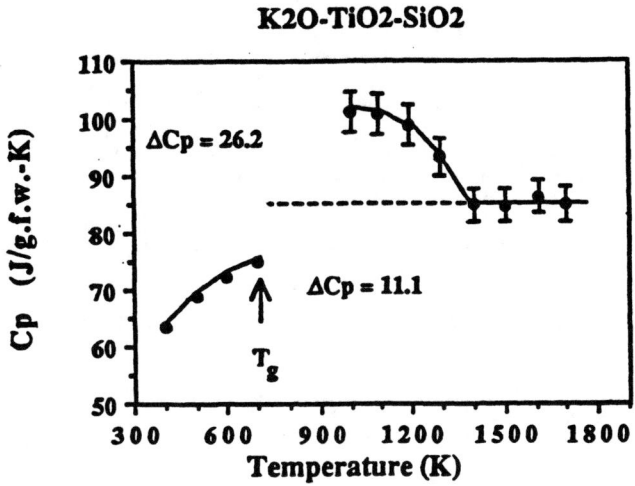

$\Delta Cp = 26.2$

$\Delta Cp = 11.1$

T_g

K2O-2SiO2

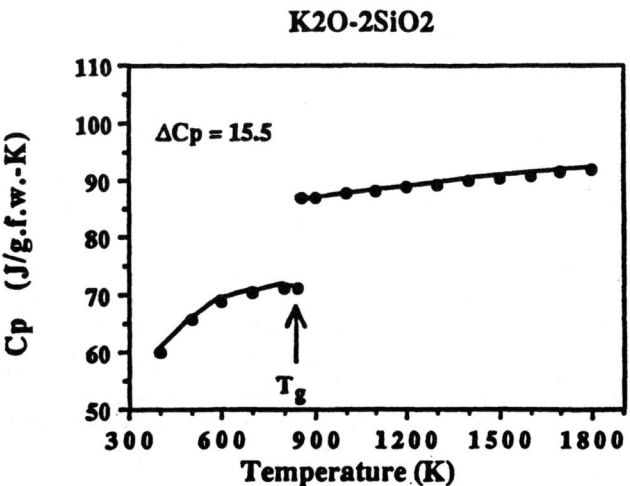

$\Delta Cp = 15.5$

T_g

Fig. 4 Heat capacities of glasses and liquids of K_2TiSiO_5 and $K_2Si_2O_5$ compositions [6].

mixing in optically homogeneous glasses [21]. However, if clustering in the glass is pervasive on a scale smaller than that which leads to macroscopic immiscibility, the energetics may show zero heats of mixing (e.g La, K silicates), [22] because the short range order along the join is not changing. Rather the mole fractions of two or more types of local environments change linearly with composition, resulting in behavior analogous to mechanical mixing. This implies that glasses that are optically clear and conventionally considered homogeneous and those that are milky and phase separated are energetically virtually the same, because only the size of the clusters has changed. This energetic equivalence has been seen in calorimetry of clear and phase separated La, K silicate [22] and $Li_2Si_4O_9$ glasses [23]. Such behavior implies that short range order is the dominant energetic factor, that entropies of mixing are much smaller than a simple polymerization model or homogeneous equilibrium model would imply, and that the thermodynamics of immiscibility do not follow classical formulations such as regular or subregular solution models. Such clustering is probably often important for systems containing non-framework cations of high charge and large size such as the rare earths, Ti, Zr, and Mo.

Calorimetry as a Probe of Melting and Crystallization in Model Silicate Melts and in Magmas: Thermodynamics and Kinetics

Understanding magmatic processes requires detailed knowledge of both kinetics and thermodynamics of melting and crystallization. A major focus of our research has been to explore the use of high temperature calorimetry to study kinetic as well as thermodynamic aspects of melting and crystallization.

In the lithium silicate system, chosen because it is a relatively simple model system with extensive previous work on crystallization kinetics, Sen et al., [23] determined the enthalpy of drop-solution as a function of annealing time at 773 or 793 K, and quantified the extent of crystallization by X-ray diffraction, density, and NMR for $Li_2Si_2O_5$ glass. The proportion of the total energy liberated as a function of measured crystallinity follows a mechanical mixture line joining the homogeneous glass and crystal as two end members. This indicates absence of any energetically significant short to medium range structural changes in the uncrystallized portion of this partially crystallized glass, and corroborates well with spectroscopic findings.

Phase separation in $Li_2Si_4O_9$ involves a small decrease in $Q_4 : Q_3$ ratio. The enthalpy of phase separation is very small compared to that of crystallization. ^{29}Si MAS NMR indicates that the optically homogeneous $Li_2Si_4O_9$ glass may actually represent a very early stage of unmixing, consisting of submicroscopic domains of two amorphous phases. Crystallization of $Li_2Si_2O_5$ in $Li_2Si_4O_9$ does not affect the SiO_2 regions structurally and silica starts to crystallize only after completion of disilicate crystallization.

In addition to the above ex situ experiments, calorimetry can be used to study the rates of crystallization and melting by appropriate in situ measurements. A major conclusion in these preliminary studies is that the admixture of even small amounts of a second component spreads out the enthalpy release over a large temperature interval.

This work was extended to natural basaltic systems by Lange et al., [12]. Step-scanning calorimetric measurements were performed between 1073 and 1673 K on a ugandite and an olivine basalt. The distribution of latent heat across the liquidus-solidus intervals of the two samples is distinctly different, (see Fig. 5), reflecting significant variation in the sequence and abundance of mineral phases during melting (clinopyroxene and leucite in the ugandite; olivine, clinopyroxene, and plagioclase in the basalt). The common assumption of a uniform distribution of latent heat across the liquidus-solidus interval of a magma is a reasonable approximation for the olivine basalt, but is grossly in error for the ugandite. This is due to cotectic precipitation of leucite and clinopyroxene, leading to a large, disproportionate release of latent heat early in the crystallization sequence. The implication for the thermal history of a crystallizing ugandite magma is that the rate of heat loss during conductive cooling will initially be more rapid than the average rate. The net result will be to produce lower magmatic temperatures after a given cooling interval relative to models assuming a uniform release of latent heat.

An additional series of scanning calorimetric experiments was performed at variable rates to evaluate the role of kinetics on the distribution of enthalpy during both melting and

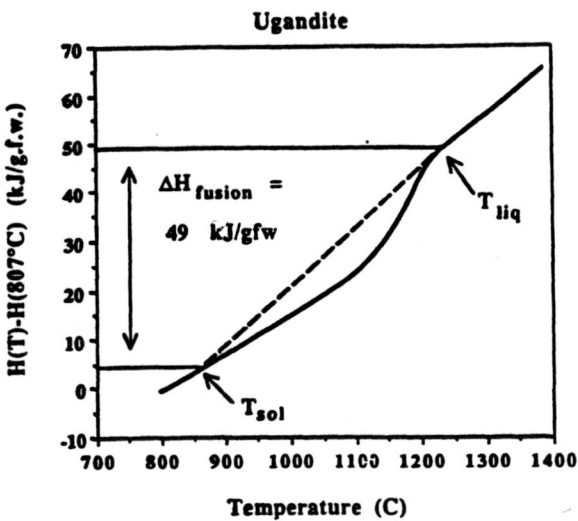

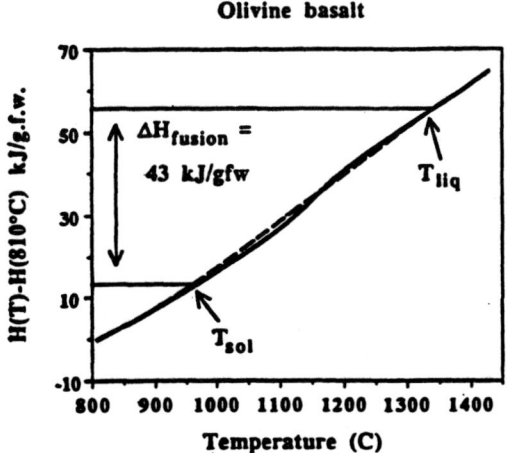

Fig. 5 Enthalpy as a function of temperature for ugandite and olivine basalt. Solid curve is measured value, dashed line shows constant release of enthalpy throughout the solidus-liquidus interval [12].

crystallization of the ugandite and olivine basalt. The enthalpy curve depends on direction (heating or cooling) and rate. Clinopyroxene is the most important mineral phase in controlling the shapes of the enthalpy profiles during cooling; this is due to its large enthalpy of fusion and its tendency for sluggish nucleation, followed by rapid crystallization at temperatures that vary with cooling rate.

CONCLUSIONS

High temperature calorimetry, combined with structural and spectroscopic studies, provides quantitative insight into both thermodynamics and kinetics of processes relating melts, glasses, and crystals. Many glasses contain considerable mesoscale order. When this is the result of exothermic interactions (complex formation), the liquid above T_g can be "fragile", the complexes break up with increasing temperature. This leads to a large C_p anomaly, less exothermic heats of mixing in the melt than in the glass and the possibility of the unmixing with increasing temperature (an LCST). When the interactions are positive (clustering), even an optically homogeneous glass may be effectively phase-separated in terms of local environments and energetics. Both crystallization and melting in multicomponent systems occur over a wide solidus-liquidus interval and the evolution of enthalpy is quite nonlinear in this temperature range. The release of enthalpy of crystallization (or fusion) can be used to study the kinetics of melting (crystallization).

ACKNOWLEDGMENTS

This work was supported by the National Science Foundation (Grant EAR 9104923) and by Corning, Inc.

REFERENCES

1. A. Navrotsky, Phys. Chem. Min. **2**, 89-104, (1977).

2. J.J. DeYoreo, A. Navrotsky and D.B. Dingwell, J. Am. Ceram. Soc. **73**, 2068-2072, (1990).

3. Z. Zhou and A. Navrotsky, J. Mater. Res. **7**, 2920-2935, (1992).

4. S. Circone and A. Navrotsky, Am. Min. **77**, 1191-1205, (1992).

5. L. Chai and A. Navrotsky, Cont. Min. Pet. **114**, 139-147, (1993).

6. R. Lange and A. Navrotsky, Geochim. Cosmochim. Acta **57**, 3001-3011, (1993).

7. L. Topor and A. Navrotsky, in High Pressure Research: Applications to Earth and Planetary Sciences, (Y. Syono and M.H. Manghnani, Eds.), Terra Publishing Co., Tokyo, Japan, p. 71-76, (1992).

8. A. Navrotsky, D. Ziegler, R. Oestrike and P. Maniar, Cont. Min. Pet. **101**, 122-130, (1989).

9. D. Ziegler and A. Navrotsky, Geochim. Cosmochim. Acta **50**, 2461-2466, (1986).

10. R. Lange, J.J. DeYoreo and A. Navrotsky, Am. Min. **76**, 904-912, (1991).

11. R. Lange and A. Navrotsky, Cont. Min. Pet. **110**, 311-320, (1992).

12. R. Lange, K.V. Cashman and A. Navrotsky, Cont. Min. Pet. (submitted).

13. M. Akaogi, E. Ito and A. Navrotsky, J. Geophys. Res. **94**, 15671-15685, (1989).

14. N.E. Brown and A. Navrotsky, Am. Min. (in press).

15. J.E. Dickinson and P.C. Hess, Geochim. Cosmochim. Acta **49**, 2289-2296, (1985).

16. T. Hanada and N. Saga, J. Noncryst. Solids **38-39**, 105-110, (1980).

17. R. Lange and I.S.E. Carmichael, Geochim. Cosmochim. Acta **51**, 2931-2946, (1987).

18. C.A. Angell, In Relaxation in Complex Systems (Eds. K. Ngai and G.B. Wright) pp. 3-11, Natl. Tech. Info. Serv. U. S. Dept. Commerce, Springfield, VA 22161, (1985).

19. P. Richet, Geochim. Cosmoshim. Acta **48**, 471-483, (1984).

20. A. Navrotsky, Nature **360**, 306, (1992).

21. A. Navrotsky, Min. Assoc. Canada Short Course in Silicate Melts, C.M. Scarfe, Ed., **12**, 130-153, (1986).

22. A.J.G. Ellison and A. Navrotsky, Scientific Basis for Nuclear Waste Management XIII, (V.M. Oversby and P.W. Brown, Eds.), Mat. Res. Soc. Symposium Proc. **176**, 193-207, (1990).

23. S. Sen, C. Gerardin, A. Navrotsky and J.E. Dickinson, J. Noncryst. Solids (in press).

24. P. Richet, Y. Bottinga, L. Denielou, J.P. Petitet and C. Tequi, Geochim. Cosmochim. Acta **46**, 2639-2658, (1982).

STRUCTURAL ANALYSIS OF AMORPHOUS PHOSPHATES USING HIGH PERFORMANCE LIQUID CHROMATOGRAPHY

B. C. SALES, L. A. BOATNER, B. C. CHAKOUMAKOS, J. C. McCALLUM, J. O RAMEY, AND R. A. ZUHR
Solid State Division, Oak Ridge National Laboratory, Oak Ridge, TN, 37830-6056

ABSTRACT

The technique of high-performance liquid chromatography (HPLC) provides unique detailed information regarding the structure of partially disordered or amorphous phosphate solids. Applications of the experimental technique of HPLC to phosphate solids are reviewed, and examples of the type of information that can be obtained with HPLC are presented.

INTRODUCTION

Determining the atomic-scale structure of amorphous solids has proven to be a formidable scientific and technological problem for the past 100 years. While significant advances have been made in understanding the structure-property relationships in crystalline solids, similar progress has not been achieved in the case of amorphous materials. For amorphous solids, the standard analytical techniques are primarily sensitive to short range order- normally defined as spatial correlations between an atom and neighboring atoms separated by no more than 5 or 10 Å. These techniques include: X-ray or neutron diffraction, nuclear magnetic resonance, extended X-ray absorption fine structure measurements, electron paramagnetic resonance, Mossbauer spectroscopy, Raman spectroscopy, infrared spectroscopy, and optical spectroscopy. The technique of high-performance liquid chromatography (HPLC) is, to our knowledge, the only technique currently capable of measuring the distribution of specific structural units in the amorphous state that are between 10 and 100 Å in size. The major disadvantage of the HPLC technique is that, at present, it can only be reliably applied to analyzing the structures of phosphates.

Inorganic phosphates encompass a large class of important materials whose applications include: catalysts, ion-exchange media, solid electrolytes for batteries, linear and nonlinear optical components, chelating agents, synthetic replacements for bone and teeth, phosphors, detergents, and fertilizers [1,2]. Phosphates may, in fact, represent the most interesting class of new inorganic materials because of the adaptability of the PO_4 tetrahedra in bonding to other diverse structural units. Phosphate ions also represent a unique link between living systems and the inorganic world [3]. For this reason, phosphate minerals play an important role in tracing the earliest life forms and in the reconstruction of biomolecular evolution.

*Research sponsored by the Division of Materials Sciences, U. S. Department of Energy under Contract No. DE-AC05-84OR21400 with Martin Marietta Energy Systems, Inc.

HPLC AND PHOSPHATES: EXPERIMENTAL TECHNIQUE

The metal-phosphates of interest here consist of chains of corner-linked PO_4 tetrahedra (Fig 1); which are, in turn , bonded to each other by the metal cations. A crystalline phosphate of this type will generally have a phosphate chain of only one specific length in its structure while an amorphous phosphate solid will be characterized by a distribution of "spaghetti-like" chains of different lengths. By applying the HPLC technique, the relative concentration and distribution of phosphate chains in an amorphous solid can be determined quantitatively. Selected crystalline-phosphate standards are used to calibrate the HPLC system so that a particular peak in the HPLC chromatogram can be unambiguously assigned to a known phosphate anion.

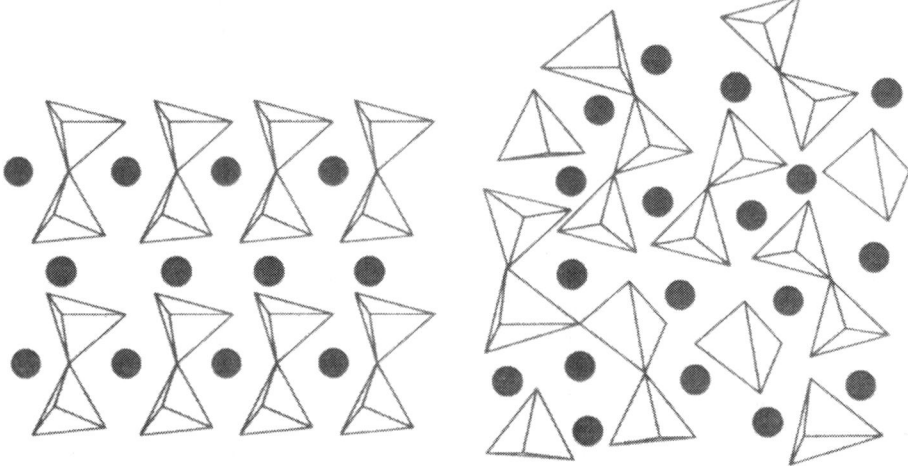

FIG. 1 Schematic drawing of (left) a crystalline metal-phosphate compound with only P_2 chains (two corner-linked PO_4 tetrahedra) and (right) an amorphous metal-phosphate with the same overall composition. In the amorphous phosphate, both longer and shorter phosphate chains are present. The metal cations are represented by the shaded spheres.

The HPLC technique is applicable to most metal-phosphate solids because of the manner in which these solids dissolve in high-pH aqueous solutions. In the present experiments the phosphate solid is placed in a pH 10 solution of 0.22 M NaCl and 5×10^{-3} M Na_4EDTA. Diffusion of water into the solid leads to hydration, disentanglement, and subsequent transport of entire polyphosphate anion chains into solution. Acid/base reactions between the H^+ and OH^- ions and the P-O-Metal groups affect the ionic interaction between the chains and assist in the disentanglement and transport of the chains into solution [4-7]. The presence of the EDTA in the solution is apparently critical when dealing with high-field-strength cations such as Mg^{+2} or Fe^{+3}. Once the "spaghetti-like" chains are in the chromatography solution, they are quite stable and do not exhibit any tendency toward either hydrolysis or polymerization for a period of at least 24 h [8]. The solution containing the dissolved phosphate solid is passed through a 0.45 micron filter and then injected into the HPLC system where the chains of linked PO_4 tetrahedra are bound on an ion-exchange column. An HPLC chromatogram is obtained as the sodium chloride molarity at the ion-exchange column is programmed to increase as a function of time, and thereby, sequentially release phosphate chains of various lengths beginning with phosphate anions consisting of only one PO_4 tetrahedron (P_1), then two PO_4 tetrahedra (P_2), three PO_4 tetrahedra (P_3) and so forth. The HPLC chromatogram, therefore, consists of a series of peaks whose position in time (i.e., the time at which the sodium chloride molarity has reached the value at which specific phosphate anions are released from the column) corresponds to the presence of PO_4 chains of a given length and where the area under these peaks is proportional to the amount of phosphorus contained in chains of that length. A detailed discussion of the HPLC system used in the present experiments has been given previously [9,10].

The strongest evidence that the information obtained from an HPLC chromatogram represents the phosphate anion distribution in the *amorphous* solid comes from the application of the same technique to *crystalline* solids. In a crystalline phosphate compound, the type of phosphate anion present in the solid can be determined independently using either X-ray or neutron diffraction, and this information can then be compared to the result obtained from an HPLC experiment on the same crystalline solid. For a large number of crystalline phosphate compounds, HPLC and diffraction results are in perfect agreement. In calibrating our HPLC system, a "standard solution" is used that is prepared by dissolving measured amounts of several crystalline phosphate compounds in the chromatography solution. The chromatogram obtained from this solution consists of a series of peaks (Fig 2), with each crystalline compound contributing only one peak whose area is proportional to the amount of that phosphate solid in solution.

FIG. 2. HPLC chromatogram of a "standard solution" prepared by dissolving several crystalline-phosphate compounds in the chromatography solution. Each of the phosphate compounds listed in the upper right of the figure contributes only one phosphate anion and is responsible for only one of the peaks in the chromatogram. The subscript m, for two of the compounds, indicates that these two solids contain cyclic (i.e. ringlike) phosphate anions (trimetaphosphate and tetrametaphosphate).

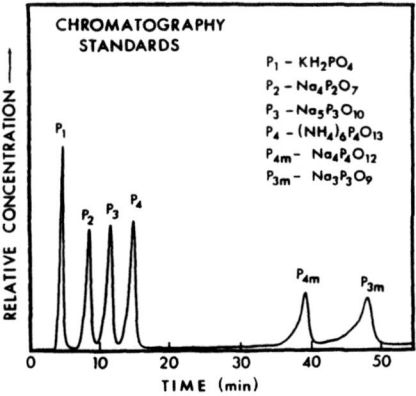

CHROMATOGRAPHY STANDARDS

$P_1 - KH_2PO_4$
$P_2 - Na_4P_2O_7$
$P_3 - Na_5P_3O_{10}$
$P_4 - (NH_4)_6P_4O_{13}$
$P_{4m} - Na_4P_4O_{12}$
$P_{3m} - Na_3P_3O_9$

HPLC AND PHOSPHATES: EXAMPLES

Dehydration of MgHPO₄•3H₂O

Synthetic crystals of the mineral newberyite (MgHPO₄•3H₂O) were grown from aqueous solution. Our interest in this material was stimulated by the previous work of Kanazawa and co-workers (11) who noted that as the waters of hydration were removed from newberyite during heating, the material underwent a crystalline-to-amorphous transition. Crystalline newberyite that is heated above about 150°C becomes amorphous and remains amorphous even when heated to temperatures as high as 600 °C. Above 600 °C, however, crystalline $Mg_2P_2O_7$ finally forms. This type of crystalline-to-amorphous transition upon heating is unusual and only occurs for a few materials.

The powder X-ray diffraction (XRD) results for MgHPO₄•3H₂O crystals heated for 1 h in flowing oxygen at the indicated temperatures are shown in Fig. 3 [12,13]. Specimens annealed at temperatures between room temperature and about 100 °C exhibited the characteristic diffraction pattern of crystalline newberyite as shown in the top trace of Fig 3. For crystals heated between 150 °C and 600 °C, however, no evidence of crystallinity could be detected by X-ray analysis, and as shown in Fig. 3, the X-ray patterns in this temperature region were all rather similar with only weak scattering evident at low scattering angles. Between 150 °C and 600 °C, the material is, therefore, amorphous by the standard definition of this state of matter.

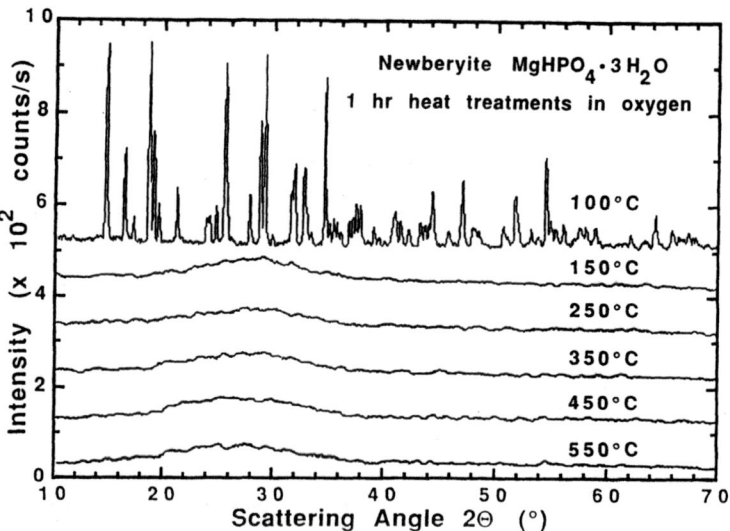

FIG. 3. Powder X-ray diffraction patterns (Cu K_α radiation) from MgHPO₄•3H₂O samples heated in flowing oxygen for 1 h at the indicated temperatures. For temperatures above 150 °C, the XRD results show that the material is amorphous.

The HPLC method was used to determine the distribution of chain lengths (phosphate anions) for the amorphous phosphates formed by heating newberyite over the same temperature range as that used to obtain the XRD results. The corresponding HPLC data for this set of samples are shown in Fig 4. The significant changes in the phosphate-anion distribution that occur when heating newberyite between 100 and 550 °C are immediately apparent in Fig 4. For a crystalline sample of $MgHPO_4 \cdot 3H_2O$ heated to 100 °C, the chromatogram consists of a single P_1 (i.e. orthophosphate anion) peak in agreement with the known structure of newberyite [14,15]. During the initial stages of the dehydration process (100-300 °C), the percentage of P_2 chains increases as one would expect for the transition from an orthophosphate to a pyrophosphate (Fig. 5). Between 300 and 500 °C, however, there is a surprising decrease in the P_2 concentration which may reflect the thermodynamic tendency of the material to form $MgHPO_4$, a phase that exists as a crystalline solid for all of the other alkaline earths, but has not been observed for Mg.

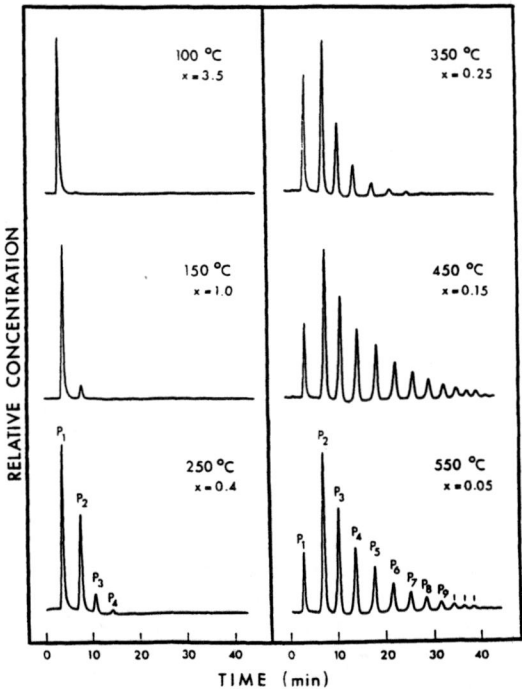

FIG. 4. HPLC chromatograms from $MgHPO_4 \cdot 3H_2O$ samples heated in flowing oxygen for 1 h at the indicated temperatures. The position in time (or NaCl molarity) and the area under each peak in the chromatogram indicate respectively the identity and quantity of a particular phosphate anion. A phosphate anion labeled P_n means that the anion consists of n corner-linked PO_4 tetrahedra. The "water content" x of each sample, as determined by thermogravimetric analysis, is also shown in the figure.

FIG. 5. Results from an analysis of the chromatograms for $MgHPO_4 \cdot 3H_2O$ samples annealed in flowing oxygen at temperatures between room temperature and 600 °C. (a) Percentage of phosphorus present in the P_3 and P_2 anions as a function of annealing temperature. (b) Percentage of phosphorus present in the P_5, P_7, and P_9 anions. (c) Average phosphate chain length at each annealing temperature.

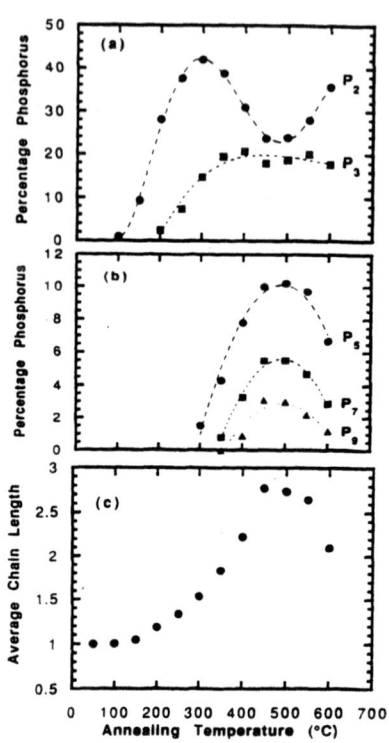

Samples heated at 500 °C exhibit the highest concentrations of longer chains (i.e. P_5 to P_{13}) and the maximum observed average phosphate chain length of about 2.75 (Fig. 5). At 500 °C, almost all of the water and hydrogen have been removed from the newberyite (as determined by thermogravimetric analysis). If a formal charge balance between the Mg cations and phosphate anions were maintained, the average chain length should be 2.0. Samples heated to temperatures near 500 °C would, therefore, be expected to have substantial internal stresses due to charge imbalance effects and accordingly, the amorphous phases may exhibit unusual dielectric and mechanical properties. The average chain length and concentration of the longer phosphate chains (i.e., >P_5) begin to decrease for samples annealed at temperatures >500 °C. These structural changes are precursors to the formation of crystalline magnesium pyrophosphate ($Mg_2P_2O_7$) that is only observed by XRD at temperatures above 600 °C. Similar local structural changes have been reported for other systems near the onset of devitrification [16].

18

Crystallization of $Pb_2P_2O_7$ Glass

Lead pyrophosphate was synthesized by melting and reacting appropriate amounts of PbO and $NH_4H_2PO_4$ for several hours in air at 950 °C. Large single-crystal plates of $Pb_2P_2O_7$ could be produced by slow cooling (1 to 2°C/h) of the melt in the range of 950 to 700 °C, and a homogeneous glass with the same composition could be prepared by quenching the melt between cold copper plates. X-ray diffraction and polarized-light analysis verified the absence of crystalline material in the $Pb_2P_2O_7$ glass. HPLC analysis of the $Pb_2P_2O_7$ single crystals showed that only the pyrophosphate anion (P_2) was present in agreement with the known crystal structure [17].

The chromatogram obtained for a lead pyrophosphate glass is shown at the top of Fig 6 [18]. This chromatogram shows that the glassy phase consists of a large peak corresponding to chains that are two PO_4 tetrahedra in length (P_2 chains) as well as both longer and shorter phosphate chains. The average chain length computed from this chromatogram is 2 ±0.02 indicating formal charge balance in the glass between the phosphate anions and the lead cations. Scanning-calorimetry measurements showed that the glass transition occurred at about 390 °C with rapid recrystallization beginning near 440 °C.

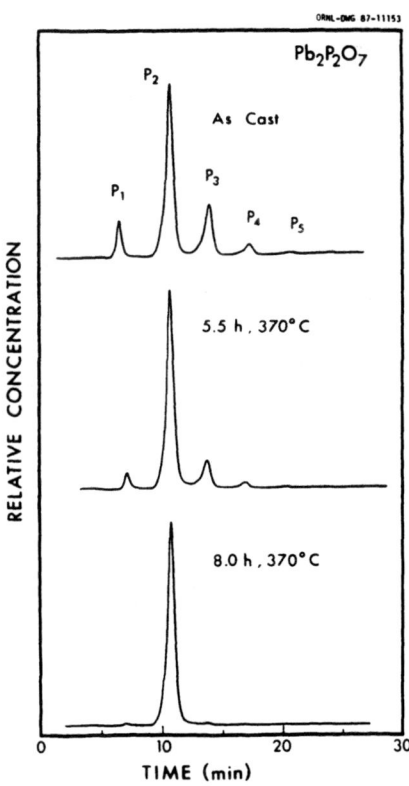

FIG. 6. HPLC chromatograms for (top) lead pyrophosphate glass as prepared; (middle) material at an intermediate state of the devitrification transition at 370 °C, and (bottom) almost fully recrystallized $Pb_2P_2O_7$ after 8 h at 370 °C.

19

Lead pyrophosphate glass samples were isothermally annealed at temperatures above, at, and below the glass-transition temperature. The samples were then analyzed using HPLC. The middle and lower chromatograms in Fig. 6 show the results obtained for a partially devitrified and a completely recrystallized lead pyrophosphate sample, respectively. By annealing the "as-cast" glass specimens for various time periods, the detailed structural evolution of material can be mapped as it tranforms to the fully crystalline state and finally exhibits a chromatogram that consists only of a P_2 peak.

The details of the structural evolution during the amorphous-to-crystalline transformation for annealing temperatures of 370, 390 and 410°C are illustrated in Fig. 7. The results illustrated in Fig. 7 show that the recrystallization process consists of an induction period during which the relative change in the percentage of the various phosphate chains is small. Following this induction period, there is a relatively rapid change in the ratios of the various phosphate anions until the P_2 anion associated with the crystalline state becomes predominant.

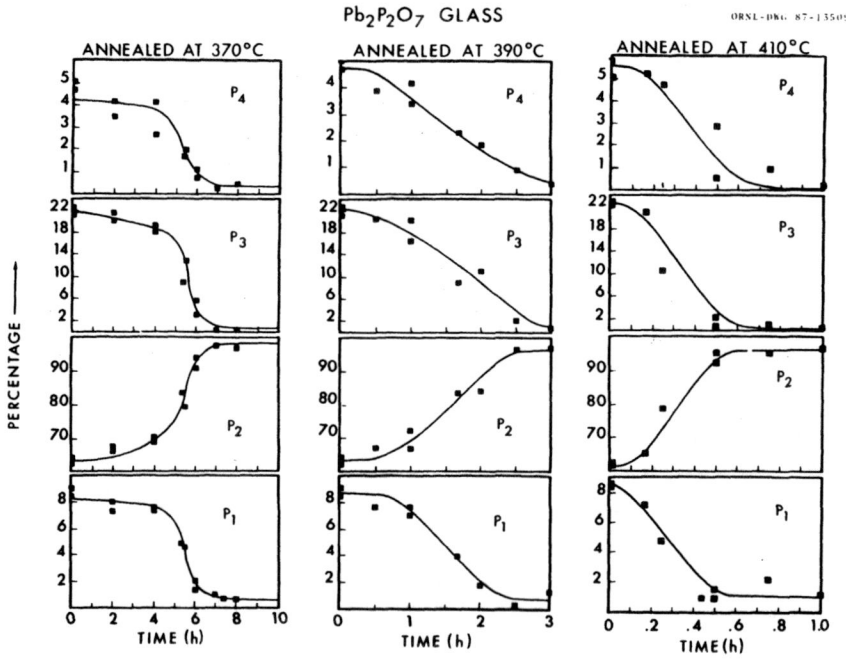

FIG. 7. The evolution with time of the relative concentration of the four major phosphate anions during the devitrification of $Pb_2P_2O_7$ at 370, 390, and 410 °C. The vertical axis corresponds to the percentage of the total phosphorus contained in chains of that length.

In a traditional devitrification experiment, only the volume fraction of crystallized material is monitored as a function of time and annealing temperature. Using the HPLC technique, however, more information about the devitrification process is available since the evolution of individual phosphate anions can be followed as well during the process. For example, devitrification data are often described by a modified Avrami equation [19] of the form $x(t) = 1 - \exp[-K(t-t_0)^n]$ where x is the volume fraction recrystallized at time t, K is a thermally activated rate constant for recrystallization, t_0 is the induction period for nucleation, and n is the Avrami coefficient that is used to determine the dimensionality of the controlling mechanism of recrystallization. For the data shown in Fig. 6, t_0 corresponds to about 3 h, and the value for n is close to 3 which corresponds to volume recrystallization either on preexisting nucleation sites or to interface-controlled growth. This type of analysis, however, does not utilize much of the detailed information that can be obtained with HPLC.

Amorphous $Pb_2P_2O_7$ Produced By Ion Damage

Single crystals of $Pb_2P_2O_7$ (typically $2x2x0.2$ cm^3) were grown from the melt as described in the previous section. In the crystal, the pyrophosphate groups (P_2 anions) are arranged in the form of sheets that are "stacked" to form a micaeous and easily cleaved crystal. One of the large surfaces of a freshly cleaved crystal (c axis) was implanted with various ions including Pb, O, or Ar. All of the remaining surfaces of the crystal were coated with silver paint so that these surfaces would be shielded when the implanted crystals were subsequently placed in the chromatography solutions. The implantation was performed at liquid nitrogen temperatures, and the energies of the implanted ions were adjusted so that the spatial extent of the resulting damage profiles were approximately the same (540 keV Pb^{3+}, 55 keV O^{2+} , or 120 keV Ar^{2+}). An ion-damaged crystal was placed in the chromatography solution and left until the outer 100 nm of the implanted-crystal face was dissolved. The solution was then injected into the HPLC system and the distribution of phosphate anions was measured.

A comparison of the chromatograms obtained for lead pyrophosphate glass prepared (as described in the previous section) by thermal quenching and the chromatograms from 100 nm thick amorphous layers produced by ion implantation of either Pb^{3+} (10^{15} ions/cm^2) or O^{2+} (10^{17} ions/cm^2) is shown in Fig. 8 [20]. Significant structural differences are apparent between the phosphate glass and the two amorphous layers produced via ion implantation. Relative to the glass, the percentage of P_2 chains is much lower for the ion-damaged layers which indicates that these layers are "more amorphous" or exhibit a higher degree of disorder than the thermally quenched glass of the same composition. In addition, the glass has a much larger percentage of P_3 anions whereas the amorphous layers have significantly higher concentrations of P_1, P_4, P_6, P_7, and P_8 anions. Figure 8 clearly illustrates that the structure of lead pyrophosphate glass is not the same as the amorphous solid of the same composition produced by ion implantation.

The HPLC results for a 100 nm layer of the implanted lead pyrophosphate crystal surface is shown in Fig. 9 for various O^{2+} implant doses. The HPLC results from 100 nm layers of crystals implanted with either Pb^{3+} or Ar^{2+}ions are qualitatively similar

FIG. 8. Comparison of the HPLC chromatograms from amorphous lead pyrophosphates produced by quenching the melt to for a conventional glass (top), by ion bombardment of the crystal surface with 10^{15} Pb^{3+} ions/cm^2 at 540 keV/ion (middle), and by ion bombardment of the crystal with 10^{17} O^{2+} ions/cm^2 at 55 keV/ion (bottom).

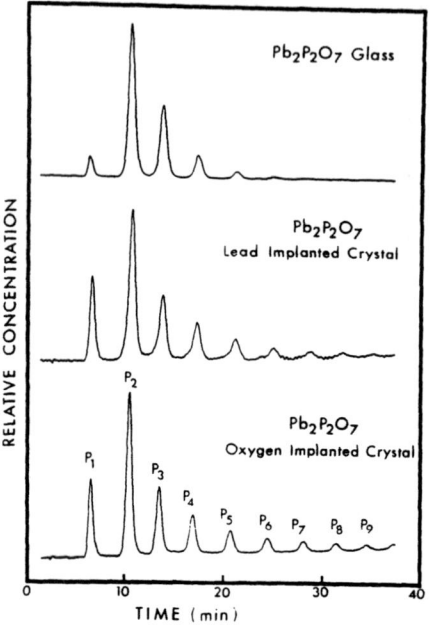

to the O^{2+} results and are not shown [20,21]. For the oxygen implanted surfaces, the first detectable alteration of the phosphate-chain structure occurs at a dose slightly less than 10^{13} O^{2+} ions/cm^2 . At doses near 10^{16} O^{2+}ions/cm^2, the distribution of phosphate chains begins to saturate, as illustrated in Fig. 10. Note that the lead pyrophosphate glass structure is not produced for any oxygen-implant dose. For lead pyrophosphate the amorphous state produced by ion implantation is fundamentally different from the glass state produced by thermally quenching the liquid.

CONCLUDING REMARKS

The technique of HPLC has been used to characterize the crystalline-to-amorphous transition that occurs during the dehydration of $MgHPO_4 \cdot 3H_2O$; the amorphous-to-crystalline transition that occurs during the devitrification of $Pb_2P_2O_7$; and the structure of amorphous layers produced on $Pb_2P_2O_7$ single crystals by ion bombardment. These experiments clearly demonstrate that the structure of an amorphous solid depends on how the solid was made amorphous. It was also shown

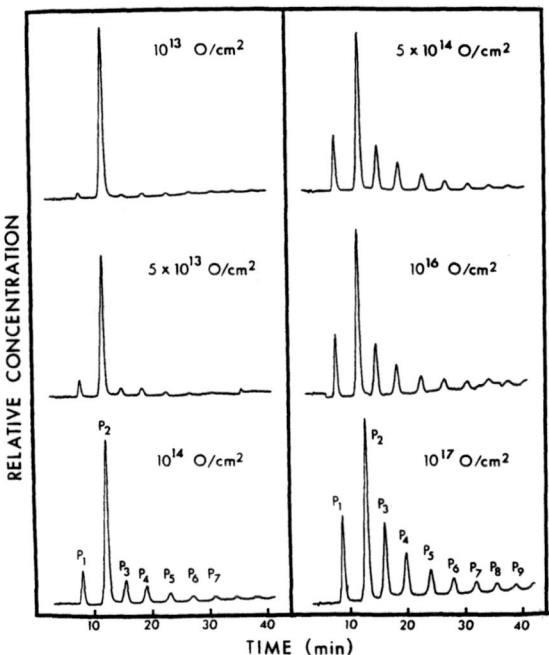

FIG. 9. HPLC results for 100nm thick layers of the surface of lead pyrophosphate single crystals that were implanted with various doses of oxygen ranging from 10^{13} to 10^{17} ions/cm^2 with an energy of 55 keV/O ion.

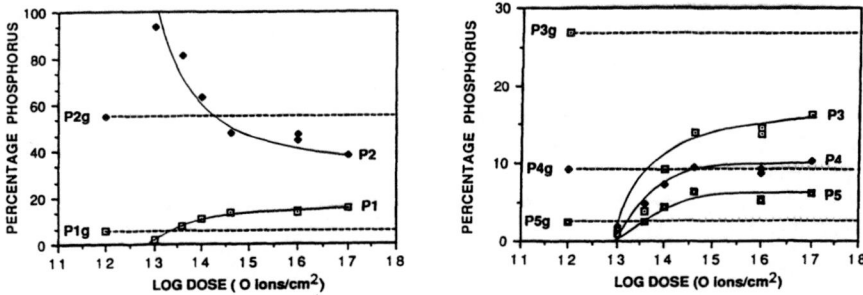

FIG. 10. The evolution with implant dose of the relative concentration of the five major phosphate anions in the ion-damaged layer. The corresponding concentrations of these same phosphate anions observed in a thermally quenched lead pyrophosphate glass are indicated by the dashed lines. The pyrophosphate glass structure is not produced at any oxygen implant dose.

that for some solids the HPLC technique can easily distinguish differences in the amorphous state that are not apparent from diffraction data. One general goal of the present research is to use HPLC in conjunction with more traditional techniques to generate an experimental data base for a particular amorphous solid that can then be used in a realistic test of sophisticated molecular-dynamics calculations.

REFERENCES

1. J. R. Van Wazer, Phosphorus and Its Compounds:Chemistry,Vol 1, (Interscience,New York, 1958).
2. T. Kanazawa, ed.,Inorganic Phosphate Materials (Elsevier, New York, 1958).
3. G. Arrhenius, B. Gedulin, and S. Mojzsis, in Chemical Evolution and the Origin of Life, edited by C. Ponamperuma and J. Chela-Flores, Trieste, Italy, Oct 1992 , in press (A. Deepak Publishing, Hampton VA, 1993).
4. E. Thilo, Angew. Chem. Int. Ed. 4, 1061 (1965).
5. B. C. Bunker, G. W. Arnold, and J. A. Wilder, J. Non-Cryst. Solids 64, 291 (1984).
6. A. E. R. Westman and P. A. Gartaganis, J. Am. Ceram. Soc. 40, 293 (1957).
7. T. R. Meadowcroft and F. D. Richardson, Trans. Faraday Soc. 61, 54 (1965).
8. R. S. Ramsey, Adv. Chromatogr. 25, 219 (1986).
9. R. S. Brazell, R. W. Holmberg, and J. H. Moneyhun, J. Chromatogr. 290, 163 (1984).
10. B. C. Sales, R. S. Ramsey, J. B. Bates, and L. A. Boatner, J. Non-Cryst. Solids 87, 137 (1986).
11. T. Kanazawa, Kagaku No Ryoiki 24, 222 (1970) (in Japanese).
12. B. C. Sales, B. C. Chakoumakos, L. A. Boatner, and J. O. Ramey, J. Non-Cryst. Solids 159, 121 (1993).
13. B. C. Sales, B. C. Chakoumakos, L. A. Boatner and J. O. Ramey, J. Mater. Res. 7, 2646 (1992).
14. F. Abbona, R. Boistelle, and R. Haser, Acta Crystallogr. B35, 2514 (1978).
15. H. Barth, M. Catti, W. Joswig, and G. Ferraris, Tscher. Mineral. Petrog. Mitt. 32, 187 (1983).
16. A. Chatelain and R. A. Weeks, J. Chem. Phys. 52, 3758 (1970).
17. D. F. Mullica, H. O. Perkins, D. A. Grossie, L. A. Boatner, and B. C. Sales, J. Solid State Chem. 62, 271 (1986).
18. B. C. Sales, J. O. Ramey, and L. A. Boatner, Phys. Rev. Letters 59, 1718 (1987).
19. M. Avrami, J. Chem. Phys. 7, 1103 (1939).
20. B. C. Sales, J. O. Ramey, L. A. Boatner, and J. C. McCallum, Phys. Rev. Letters 62, 1138 (1989).
21. B. C. Sales, J. O. Ramey, J. C. McCallum, and L. A. Boatner, J. Non-Cryst. Solids 126, 179 (1990).

CHARACTERIZATION OF METAMICT AND GLASSY LEAD PHOSPHATES

A. N. Sreeram and L. W. Hobbs

Department of Materials Science and Engineering
Massachusetts Institute of Technology
Cambridge, Massachusetts 02139, USA

ABSTRACT

Single crystals of $Pb_2P_2O_7$ (beam stable under 200 kV TEM electrons to a fluence $> 10^{27}$ e/m^2 and an ionizing dose $> 10^{14}$ Gy) were rendered metamict (amorphized) with ion-implantation (100 kV P$^+$ ions with several fluences between 5 x 10^{17}/m^2 - 2 x 10^{20}/m^2). $Pb_2P_2O_7$ and $PbO \cdot P_2O_5$ glasses were also ion implanted at identical fluences. Radial distribution functions for metamict and glassy phosphates generated using energy filtered electron diffraction (EFED) data collected on 100 kV field emission STEM (VG HB-5) indicate significant alterations in the medium range order.

INTRODUCTION

Electrons interact with matter far more strongly than X-rays or neutrons which enables them to be used as powerful characterization tools for the structural analysis of very small volumes of solids, such as those involved with the metamict regions generated using ion-beam irradiation. However, an electron diffraction pattern from an amorphous region contains information from both elastically scattered and inelastically scattered electrons. As the scattering intensity information from the elastically scattered electrons alone can be related to structure of the specimen, filtering out the inelastically scattered electrons becomes necessary. Using such a technique, Qin and Hobbs [1] have successfully generated the radial distribution functions from metamict silicas.

Lead phosphate glasses have been investigated for decades with respect to fundamental glass properties. In these glasses, [PO$_4$] tetrahedra, (like [SiO$_4$] tetrahedra in silicates) can link up together by corner sharing oxygen atoms to form polymerized structures up to and including three-dimensional glassy networks. Radial distribution functions have been generated for lime phosphate glasses [2] and amorphous aluminum phosphates [3] using X-ray diffraction techniques. More recently, Wright *et al.* [4] generated RDF for pure vitreous phosphorous pentoxide. Such techniques, though useful in generating RDFs from bulk amorphous materials, cannot prove useful in deriving such information from very small volumes; for example, ion-implanted metamict lead phosphate layers are 200 nm deep for pyrophosphate compositions and 225 nm for metaphosphate compositions [5, 6]. Sales *et al.* [7] have examined medium-range order in near-surfaces of single crystals of lead pyrophosphate ($Pb_2P_2O_7$), rendered aperiodic (metamict) by ion implantation as well as lead pyrophosphate glass, using high-performance liquid chromatography flow injection analysis (HPLC-FIA).

In this study, we report the RDFs generated for ion-irradiated lead pyrophosphate single crystals and for both ion irradiated and non-irradiated lead phosphate glasses using energy filtered electron diffraction (EFED) data collected on the 100 kV field emission STEM (VG HB-5).

EXPERIMENTAL PROCEDURE

Sample preparation techniques for single crystals of lead pyrophosphate ($Pb_2P_2O_7$) and various lead phosphate glasses and the ion-irradiation procedures have been discussed elsewhere [6]. The EFED data were collected on the 100 kV VG HB-5 STEM with a serial electron-energy loss spectrometer (EELS). The energy window used was about 1.5 eV. The selected-area diffraction mode was used to allow near-parallel illumination by electrons of the specimen. The selected area aperture (SAD) and the lead phosphate specimen (cross-section TEM specimen made from ion-irradiated samples) were relatively so positioned as to expose only the metamict portion of the sample. The EFED data for non-irradiated glasses were collected from parallel section TEM specimen thinned from bulk glasses and also from the ion-irradiated cross-section samples by moving the sample relative to the SAD aperture, so that only the non-irradiated portion was exposed to the electron beam. Data were aquired digitally and were used to reconstruct the RDFs using appropriate algorithms which are discussed briefly as follows.

The electron scattering intensity for a multicomponent isotropically arranged array of atoms (as found in amorphous and metamict materials) is given [8] as follows:

$$I(s) = N \sum_{uc} f_j^2 + N f_e^2 \sum_{uc} K_j \int_0^\infty 4\pi r^2 [\rho_j(r) - \rho_e] \frac{\sin(2\pi sr)}{2\pi sr} dr \quad (1)$$

where N is the number of units of composition uc (or the number of chemical molecular units), f_j is the atomic scattering amplitude with a mean value of f_e given by $f_j = K_j f_e$ (K_j a constant), $\rho_j(r)$ is the radial density function about an atom j with ρ_e being the average density of atoms, and s ($= \frac{2\sin(\frac{\Theta}{2})}{\lambda}$) is the scattering vector, with Θ being the scattering angle and λ being the electron wavelength. Introducing the reduced density function ($i(s) = (\frac{I(s)}{N} - \Sigma_{uc} f_j^2)/f_e^2$), we have

$$si(s) = 2 \int_0^\infty \sum_{uc} K_j r[\rho_j(r) - \rho_e] \sin(2\pi sr)dr \ . \tag{2}$$

The radial distribution function $G(r)$ is defined as

$$G(r) = 4\pi r^2 \sum_{uc} K_j[\rho_j(r) - \rho_e], \tag{3}$$

and by applying Fourier transforms, we can reconstruct the generalized RDF from the experimental intensity data $si(s)$ as

$$G(r) = 8\pi r \int_0^\infty si(s) \sin(2\pi sr)ds \ . \tag{4}$$

which is a linear superposition of partial RDFs. The maxima of the function correspond to various interatomic, intraatomic or composite interatomic distances.

A modification function is used to take into account the termination effect [8] applied on equation (4)

$$G(r) = 8\pi r \int_{s_{min}}^{s_{max}} si(s)M(s) \sin(2\pi sr)ds \ . \tag{5}$$

where $M(s) = \frac{\sin(\pi s/s_{max})}{\pi s/s_{max}}$; with $s_{max} = 16$ nm^{-1}. Equation (5) leads directly to the reconstruction of the RDF using the collected EFED data.

RESULTS AND DISCUSSION

Our earlier studies of cross-section (XTEM) samples of ion-irradiated lead phosphate specimen have indicated a strong decrease in the intensities of the first sharp diffraction peak (FSDP) [5, 6]. Using equation (5), the EFED data can further be used to reconstruct the RDFs. RDFs generated for $2\text{x}10^{20}$ 100 kV P$^+$ions/m^2 ion-irradiated, as well as non-irradiated, lead metaphosphate and lead pyrophosphate glasses are shown in Figs. 1a and 1b respectively. Fig. 2a shows the RDFs for three non-irradiated lead phosphate glasses of different compostions. Fig. 2b shows the RDFs for lead pyrophosphate single crystals which

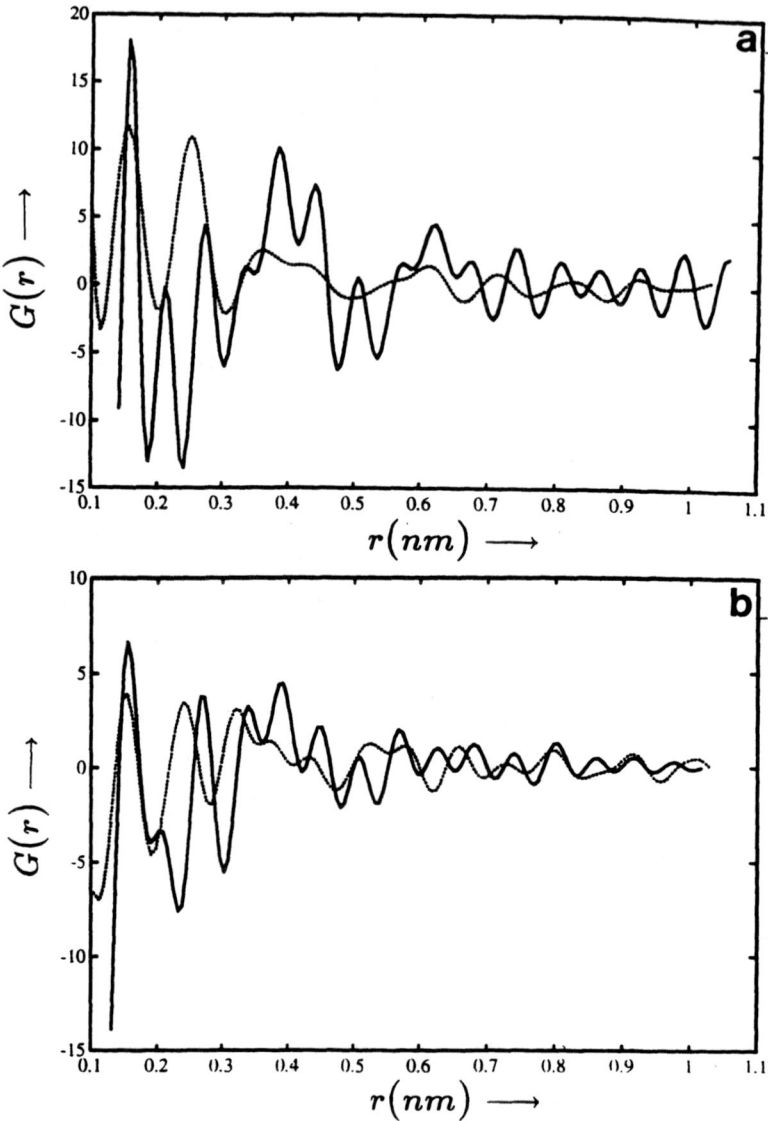

Figure 1: Reconstructed RDFs for non-irradiated and ion-irradiated lead metaphosphate glass (a) and lead pyrophosphate glass (b). The irradiation fluence of 100 kV P^+ ions was 2×10^{20} ions/m^2. The solid curves are for non-irradiated species and the dotted curves are for the irradiated ones.

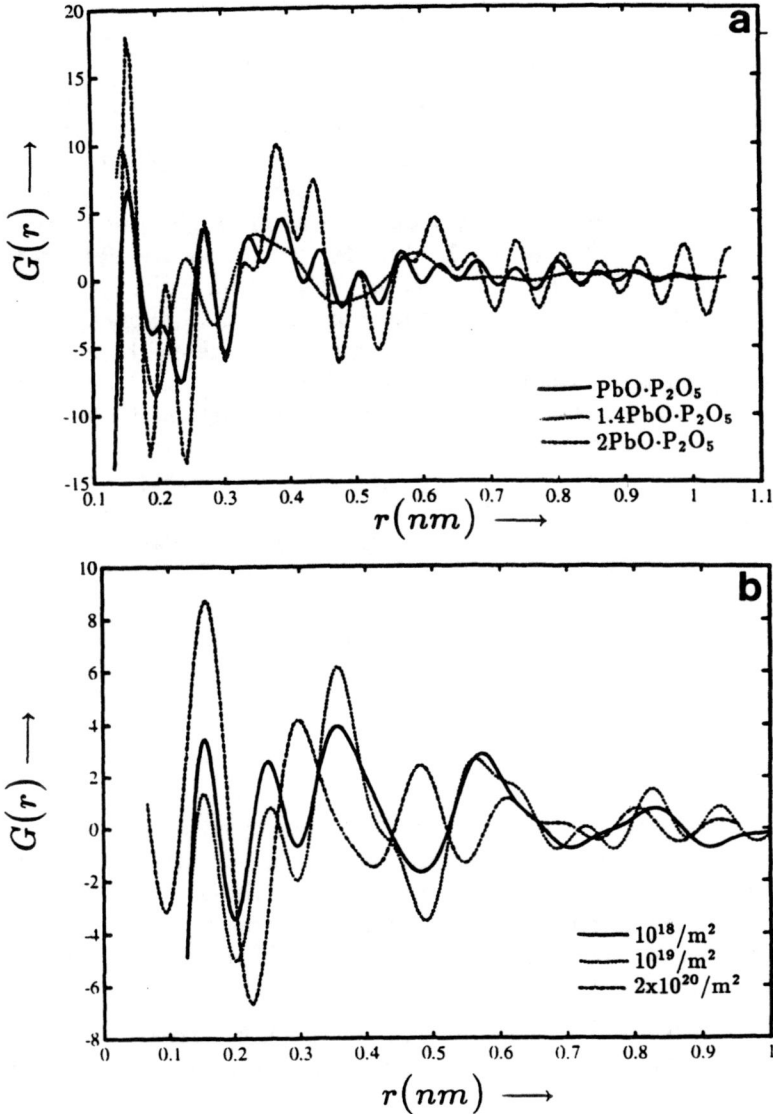

Figure 2: Reconstructed RDFs for (a) non-irradiated lead phosphate glasses of different compositions; (b) single crystals rendered metamict with various 100 kV P^+ ion fluences indicated in the plot.

were rendered metamict with three different fluences of 100 kV P$^+$ ion irradiation.

The first peak in all of the RDFs generated for the non-irradiated lead phosphate glasses represents the P-O bond distance (within the [PO$_4$] tetrahedra) of 1.57 Å. This is in fair agreement with the P-O bond distance reported for lime phosphate glasses by Biscoe et al. [2] who report it to be 1.57 Å and by Wingall et al. [3] who report it to be 1.6 Å. It is also quite clear from the RDFs for all of the metamict phosphates that the P-O bond distance remain unaltered, indicating that short-range order is preserved, even in the metamict state. For high dose P$^+$ ion implantation (2x10^{20}/m^2, Figs. 1a, 1b, 2b), the near-medium-range order is altered, indicated by a significant shift in the position of the second peak. Such an alteration of near-medium-range order (at high fluence) suggests that the metamict state so obtained is more 'disordered' than the conventionally preprared (thermally quenched) glasses. Also, by comparing Figs. 1b and 2b, for the same high fluence, the metamict state generated from the lead pyrophosphate glasses and single crystals are seen to differ significantly in near-medium-range and medium-range order. This is in agreement with the results obtained by Sales et al. [7] for high implant fluences of O^{2+} and Pb^{3+} on lead pyrophosphate glasses and single crystals, where the medium-range order was analyzed using HPLC-FIA. From Fig. 2b it is, however, quite apparent that for lower P$^+$ ion fluences the near-medium-range structures are very similar to each other, and only the medium-range structure is altered.

CONCLUSIONS

Radial distribution functions have been generated for aperiodic lead phosphates (both glassy and metamict states) using energy-filtered electron diffraction data collected on the STEM (VG HB-5). The short-range order in all of the aperiodic lead phosphates is preserved at a P-O bond distance of 1.57 Å. Near-medium-range order is also preserved, except at very high ion fluences.

Acknowledgement. The authors thank the U.S. Department of Energy, Office of Basic Energy Sciences, for support of this research under grant DE-FG02-89ER45396.

References

[1] L. C. Qin and L. W. Hobbs, 'Energy-Filtered Electron Diffraction From Silica Thin

Films', *Mater. Res. Soc. Symp. Proc.* **284** (1993) 331.

[2] J. Biscoe, A. G. Pincus, C. S. Smith and B. E. Warren, *J. Am. Ceram. Soc.* **24** (1941) 116.

[3] G. D. Wignall, R. N. Rothon, G. W. Longman and G. R. Woodward, *J. Mater. Sci.* **24** (1977) 1039.

[4] A. C. Wright, R. A. Hulme, D. I. Grimly, R. N. Sinclair, S. W. Martin, D. L. Price and F. L. Galeener, 'The Structure of Some Simple Amorphous Network Solids Revisited', *J. Non-Cryst. Solids* **129** (1991) 213.

[5] A. N. Sreeram, L. C. Qin, A. J. Garrat-Reed and L. W. Hobbs, 'Characterization of Metamict and Glassy Phosphates using Energy-Filtered Electron Diffraction', *Proc. 50th Ann. Meeting EMSA* (1992) 1230.

[6] A. N. Sreeram and L. W. Hobbs, 'Structure Characterization of Metamict and Glassy Lead Phosphates', *Mater. Res. Soc. Symp. Proc.* **279** 1993 559.

[7] B. C. Sales, J. O. Ramey, J. C. McCallum and L. A. Boatner, 'Structural Differences Between the Glass State and Ion-Beam-Amorphized States of Lead Pyrophosphates', *J. Non-Cryst. Solids* **126** (1990) 179.

[8] E. Lorch, 'Neutron Diffraction by Germania, Silica and Radiation-Damaged Silica Glasses', *J. Phys. C* **2** 1969 229.

HIGH TEMPERATURE DYNAMIC IN-SITU RAMAN SPECTRAL CHARACTERIZATION OF SODIUM BOROSILICATE GLASSES

AJIT JILLAVENKATESA AND ROBERT A. CONDRATE, Sr.
Institute of Glass Science and Engineering, NYS College of Ceramics, Alfred
University, Alfred, NY 14802

ABSTRACT

Dynamic Raman spectra of sodium borosilicate glasses containing large amounts of boron oxide have been studied as functions of temperature and holding time. Such characterization enabled measuring the changes in the nature and states of the molecular species that were present in the glasses and their melts. The data indicate that boroxol rings involving one or more four-coordinated boron ions in the sodium oxide-containing glasses decreased in concentration as they were heated through melt temperatures as was noted earlier for boroxol groups involving three-coordinated boron ions in pure borosilicate compositions.

INTRODUCTION

Structural changes can be noted in the boron oxide portion of the network of boron oxide-containing glasses and melts with changing temperature. Related glass structural investigations with temperature change have been conducted using Raman spectroscopy for glasses containing no alkali oxide additions. For instance, Walrafen *et al.* [1] have studied the high-temperature structural changes in vitreous boron oxide by Raman spectroscopy. While Furukawa and White [2] investigated high-temperature Raman spectra of B_2O_3 glasses that contained higher concentrations of boron oxide, Chakraborty and Condrate [3,4] have studied such spectra for both B_2O_3-GeO_2 and B_2O_3-GeO_2-SiO_2 glasses that contained large amonts of boron oxide. For all of these studies, the spectral data clearly indicated that structural changes occured in the borate portion of the glass networks as one approached and passed through their melt temperatures. The major structural change for each of the investigated glasses involved the breakdown of the boroxol rings possesing only three-coordinated boron ions into either borate chains or isolated borate (BO_3) groups. This current study will extend such spectral/structural investigations to glass compostions which possess boroxol rings containing one or more four-coordinated boron ions. The study will look at the Raman spectra of sodium borosilicate glasses containing large amounts of boron oxide at high temperatures.

33

EXPERIMENTAL PROCEDURES

Preparation of Glasses

Ten gram batches of the glass samples containing various compositions were prepared from reagent grade Na_2CO_3, H_3BO_3 and SiO_2. These batches were weighed out and melted in a covered platinum crucible. The glasses were melted for eight hours in a SiC globar furnace at temperatures varying from 800 to 1400°C, depending upon glass compostion. The resulting melts were poured into graphite molds to form 10x50 mm rods. Each glass sample was then annealed at 10°C above its glass transition temperature for 30 minutes. Flat glass specimens of dimensions 5x7x4 mm were cut from the rods. Table 1 lists the different glass compositions studied.

Wet chemical analysis of the glass samples showed the glass specimens to be within 1 mole percent of the calculated values. The glass transition temperature and softening point measurements were made using a Mettler DSC 30 Differential Scanning Calorimemter.

Table 1:

GLASS COMPOSITIONS STUDIED

0.15 SIO2	0.25 SIO2
0.10 Na2O - 0.75 B2O3	0.10 Na2O - 0.65 B2O3
0.20 Na2O - 0.65 B2O3	0.20 Na2O - 0.55 B2O3
0.30 Na2O - 0.55 B2O3	0.30 Na2O - 0.45 B2O3
	0.40 Na2O - 0.55 B2O3

0.35 SIO2	0.55 SIO2
0.10 Na2O - 0.55 B2O3	0.10 Na2O - 0.35 B2O3
0.20 Na2O - 0.45 B2O3	0.20 Na2O - 0.45 B2O3
0.30 Na2O - 0.35 B2O3	0.30 Na2O - 0.35 B2O3
0.40 Na2O - 0.25 B2O3	

Spectral Measurements

All Raman spectra were measured with an Instrument SA U1000 double grating spectrometer using a CR Innova 90 argon ion laser source. These spectra were measured in the 200-1500 cm^{-1} region with a 90° scattering geometry, and 400 μm slit widths, using the 514 and 488 nm lines of the laser. The laser power was 150-300 mW for these lines.

The hot-stage cell that is illustrated in Figure 1 was used for high temperature measurements. A platinum strip heater was used for the hot stage. The platinum strip was mounted in a water-cooled brass jacket with openings at suitable angles to enable entry of the incident laser beam onto the glass sample with collection of the scattered radiation at 90°. The strip was heated by passage of electrical current which was supplied through a suitable step-down transformer and variac. Two thermocouples (one on the surface of the strip, and the other on the top of the sample) were used to monitor the temperature of the sample.

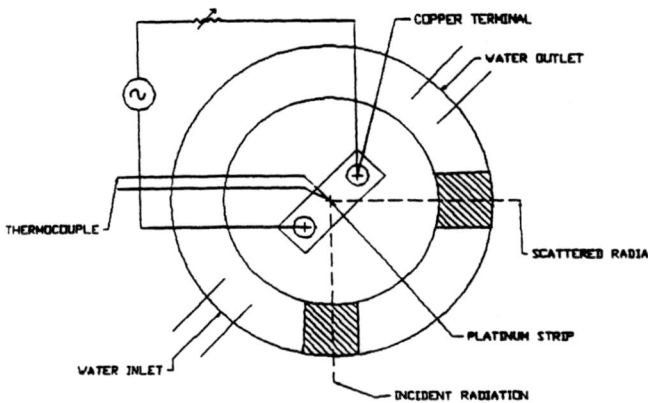

Figure 1: Schematic diagram of high temperature Raman cell.

RESULTS AND DISCUSSIONS

Band feature changes can be noted in-situ with changes in heat treatment temperature for the Raman bands at ca. 750-765 cm^{-1} that are associated with boroxol rings possessing one or more four-coordinated boron ions in sodium borosilicate glasses. Perusal of Figures 2 and 3 clearly indicate that these bands significantly decrease their band intensities relative to the silicate bending modes at ca. 480-490 cm^{-1} as they are heated to higher temperatures near melt conditions. A related Raman band at ca. 808 cm^{-1} was noted by earlier investigators [1-4] for pure boron oxide, borosilicate and borogermanate glasses which decreased its band intensity as the heat treatment temperature was increased. This Raman band was assigned to the ring breathing mode for boroxol rings possessing only three-coordinated boron ions. The spectral investigators argued that its intensity decreases indicated a breakdown of boroxol rings in the glass network as one increased the treatment temperature through melt conditions. In a similar manner, the Raman bands at 765-750 cm^{-1} can be assigned to ring breathing modes involving boroxol rings possessing one or more four-coordinated boron ions. Chakraborty and Condrate [5] have already made such Raman band assignments for ring breathing modes of boroxol rings with one or more four-coordinated boron ions in potassium borogermanate glasses containing larger amounts of boron oxide. The wavenumber value is lower for boroxol rings possessing one or more four-coordinated boron ions because the bond strengths (and therefore the force constants) are lower for B-O bonds involving four-coordinated boron ions than for those involving three-coordinated boron ions. The Raman spectral data clearly indicate that the concentration of boroxol rings possessing one or more four-coordination in sodium borosilicate glasses possessing larger amounts of boron oxide is also decreasing as one proceeds through melt temperatures.

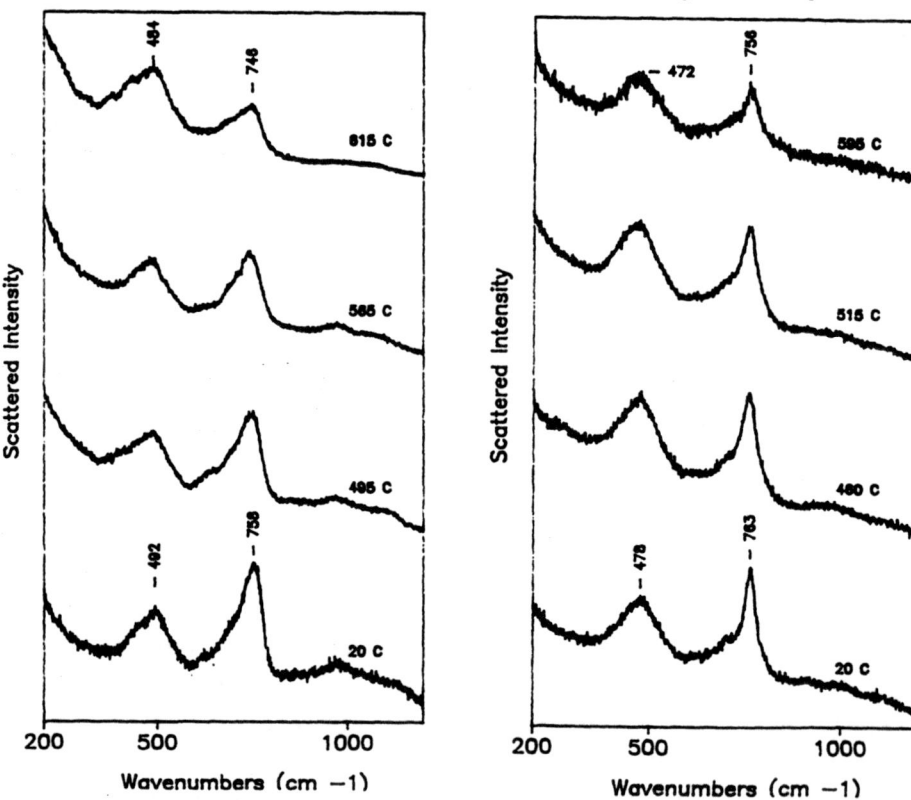

Figures 2,3 : Raman spectra of glasses containing 0.3 Na$_2$O - 0.55 B$_2$O$_3$ - 0.15 SiO$_2$ and 0.2 Na$_2$O - 0.55 B$_2$O$_3$ - 0.25 SiO$_2$ respectively, at different temperatures.

As one increases the sodium oxide content in the sodium borosilicate glass, the Raman band in the 750-765 cm^{-1} region lowers its wavenumber value. This wavenumber shift is due to three-coordinated boron ions being replaced by four-coordinated boron ions in the boroxol ring. The greater is the degree of replacement, the lower is the wavenumber shift. Further perusal of the spectral data indicates that the relative intensities of the bands decrease with treatment temperature increase even with the wavenumber shift. Both the ring breathing mode and the silicate bending mode decrease as one goes to higher treatment temperatures. These wavenumber decreases are probably due to expansion of the glass network with temperature increase, lowering the related force constants and therefore the related wavenumbers of the corresponding vibrational modes.

Figure 4 illustrates the Raman spectra for a sodium borosilicate glass in which the concentrations of boroxol rings containing three-coordinated boron ions and boroxol rings containing four-coordinated boron ions are high. Clearly, the spectral data does not indicate selective ring breakage for the investigated glass composition of either type of boroxol ring with increase in treatment temperature.

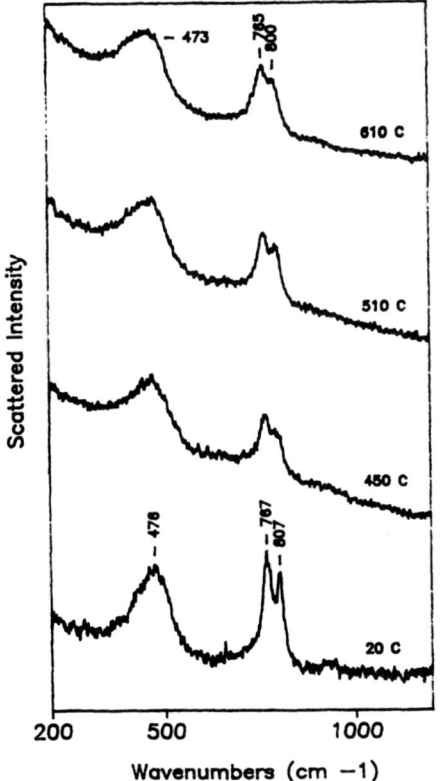

Figure 4: Raman spectra of glass containing 0.1 Na_2O - 0.55 B_2O_3 - 0.35 SiO_2, at different temperatures.

REFERENCES

[1] G. E. Walrafen, M. S. Hokmadi, P. N. Krishnan and S. Guha, J. Chem. Phys. 79, 3609 (1983).

[2] T. Furukawa and W. B. White, J. Amer. Ceram. Soc. 64, 443 (1981).

[3] I. N. Charaborty and R. A. Condrate, Sr., J. Non-Crys. Solids 81, 271 (1986).

[4] I. N. Chakraborty, and R. A. Condrate, Sr., J. Mater. Sci. Lett. 5, 361 (1986).

[5] I. N. Chakraborty and R. A. Condrate, Sr., J. Amer. Ceram. Soc. 66, C71 (1983).

ON THE DETERMINATION OF PARTIAL
RDFs FOR AMORPHOUS MATERIALS

L.C. Qin and L.W. Hobbs

Department of Materials Science and Engineering
Massachusetts Institute of Technology
Cambridge, Massachusetts 02139, USA

ABSTRACT

Radial distribution functions (RDFs) for vitreous silica (v-SiO$_2$) have been obtained from energy-filtered electron diffraction data obtained in the HB5 scanning transmission electron microscope. Results have been compared with those obtained from high-resolution neutron diffraction experiments, and are in good agreement within experimental errors. It was found to be impractical to obtain partial RDFs for this material from combined neutron, X-ray and electron diffraction data, because the similarities in characteristics of X-ray and electron scattering cause indeterminacies. A criterion equation has been given to determine feasibility.

INTRODUCTION

For disordered materials, the atomic arrangements cannot be determined uniquely as in the case of crystals, because the number of independent positional parameters is practically infinite. In crystals, the number reduces to a finite number due to the constraints imposed by symmetry as given by the characteristic space group of the material. For amorphous materials, radial distribution functions (RDFs) have instead been devised to describe the short range order of amorphous materials on a statistical basis, which nonetheless give the atomic environment surrounding an atom in ensemble average.

Though the RDF of a monoatomic material can be deduced from a single diffraction experiment, as was demonstrated long ago [1], there are some difficulties in determining the partial RDFs, which also specify the chemical order, for multiple-component systems since there is more than one unknown to be determined. Early attempts [2] using X-ray diffraction provided powerful insights, and the necessary approximation is still often employed in practice. In general there are $n(n + 1)/2$ partial structure factors to be determined for an n-component system, which in turn determine the partial RDFs. Since each diffraction experiment can provide such a linear equation containing the unknown partial structure factors, it is therefore quite reasonable to expect that combined electron, neutron, and X-ray diffraction experiments would be able to determine the partial RDFs unambiguously [3] for a binary system. Other methods involve the use of anomalous absorption [4] or the recently developed EXAFS technique [5] using very bright synchrotron sources.

In the present paper we report the results of our attempt to determine the partial RDFs for vitreous silica (v-SiO$_2$). The technique is to use three independent sets of diffraction intensity data obtained from neutron, X-ray and electron diffraction experiments, respectively.

EXPERIMENTAL

Energy-filtered electron diffraction data were collected in the VG HB5 STEM operating at 100 kV using an energy window of about 2 eV in width [6]. Neutron diffraction data were collected from the most readable recent published results [8].

THEORY

For a system of N atoms with n types of atoms, each type having N_α atoms, the partial atomic density functions are defined by

$$\rho_{\alpha\beta}(\vec{r}) = \frac{1}{N_\alpha} \sum_{j=1}^{N_\alpha} \sum_{k=1}^{N_\beta} \delta[\vec{r} - (\vec{r}_{\alpha j} - \vec{r}_{\beta k})] - \delta_{\alpha\beta}\delta(\vec{r}) , \qquad (1)$$

where $\alpha, \beta = 1, 2, ..., n$. For a chemically homogeneous system, the atomic density function $\rho_{\alpha\beta}$ gives the atomic distribution of β atoms with respect to an arbitrarily chosen α atom.

For systems with a random distribution of atoms, we can take an ensemble average over all orientations in both real and Fourier space. This will lead to the following equations in terms of the density functions $\rho_{\alpha\beta}(r)$ and they are related by Fourier transformation to the partial structure factors $S_{\alpha\beta}(q)$:

$$< r(\rho_{\alpha\beta}(r) - \bar{\rho}_\beta) >= 2 \int_0^\infty q S_{\alpha\beta}(q) \sin(2\pi q r) dq , \qquad (2)$$

where $\bar{\rho}_\beta$ is the average number density of atoms of type β and q is the scattering vector defined by

$$q = \frac{2\sin(\Theta/2)}{\lambda}$$

with Θ and λ being the scattering angle and the wave length, respectively. The partial structure factors are expressible in terms of the experimentally obtainable quantities by

$$i(q) = \sum_{\alpha=1}^{n} \sum_{\beta=1}^{n} c_\alpha c_\beta f_\alpha f_\beta S_{\alpha\beta} , \qquad (3)$$

where $i(q)$ is the reduced intensity

$$i(q) = \frac{I(q) - I(0)\delta_{q0}}{N} - \sum_{\alpha=1}^{n} c_\alpha f_\alpha^2 , \qquad (4)$$

in which c_α is the atomic fraction for atoms of type α

$$c_\alpha = \frac{N_\alpha}{N} , \qquad (5)$$

f_α is the atomic scattering amplitude for atom of type α, and $I(q)$ is the elastic scattering intensity which is experimentally measured.

Since $c_\alpha \rho_{\alpha\beta} = c_\beta \rho_{\beta\alpha}$, with the introduction of the weighted partial structure factors

$$S'_{\alpha\beta} = \frac{1}{2}(c_\alpha S_{\alpha\beta} + c_\beta S_{\beta\alpha}) , \qquad (6)$$

the number of unknown partial structure factors can therefore be reduced to $n(n+1)/2$, and the partial RDFs can be obtained by the following equation

$$< c_\alpha r(\rho_{\alpha\beta}(r) - \bar\rho_\beta) > \; = \; < c_\beta r(\rho_{\beta\alpha}(r) - \bar\rho_\alpha) >$$
$$= 2 \int_0^\infty q S'_{\alpha\beta}(q) \sin(2\pi qr) dq . \qquad (7)$$

The above equations are applicable to any multiple component system.

RDF FOR v-SiO$_2$ FROM ELECTRON DIFFRACTION

For the binary system of v-SiO$_2$ we used the following reduced structure factor modified from the reduced intensity with a sharpening function $\bar{f}(q)$, and the modified RDF is obtained as

$$r(\rho_e(r) - \bar\rho_e) \; = \; n_1 r [\frac{1}{r}\int_0^\infty q(\frac{f_1}{\bar f})^2 \sin(2\pi qr) M(q) dq] * [\rho_{11}(r) - \bar\rho_1] +$$

$$+ n_2 r [\frac{1}{r}\int_0^\infty q(\frac{f_2}{\bar f})^2 \sin(2\pi qr) M(q) dq] * [\rho_{22}(r) - \bar\rho_2] +$$

$$+ 2n_1 r [\frac{1}{r}\int_0^\infty q\frac{f_1}{\bar f}\frac{f_2}{\bar f} \sin(2\pi qr) M(q) dq] * [\rho_{12}(r) - \bar\rho_2]$$

$$= 2 \int_0^\infty q S(q) \sin(2\pi qr) M(q) dq , \qquad (8)$$

where the subscripts 1 and 2 refer to silicon and oxygen atoms, respectively; $M(q)$ is the modification function to account for the termination effect, and $S(q)$ is the dimensionless reduced structure factor

$$S(q) = \frac{I(q) - I(0)\delta_{q0}}{N_c \bar{f}^2} - \frac{1}{\bar{f}^2} \sum_{uc} f_j^2 \ ,$$ (9)

where uc refers to a molecular unit.

In the present study, the sharpening function used was

$$\bar{f} = f_1 + 2f_2 \ .$$ (10)

and the Lorch modification function [7] was used

$$M(q) = \frac{\sin(\pi q/q_{max})}{\pi q/q_{max}} \ ,$$ (11)

where q_{max} is the maximum value for q obtained in experiment.

Figure 1 shows an RDF for v-SiO$_2$ obtained by using energy-filtered electron diffraction data with $q_{max} = 16 \ nm^{-1}$ in the form of $g_e(r) = 4\pi r^2(\rho_e(r) - \bar{\rho}_e)$. Though it is a convoluted form, the sharp peaks due to the short range order in the structure are still well preserved. The peaks at 0.16 nm and 0.26 nm are due to the well-defined interatomic distances of Si-O and O-O in the constituent $[SiO_4]^{4+}$ tetrahedra. For comparison to high-resolution neutron diffraction studies [8], where $q_{max} = 72 \ nm^{-1}$, the deduced RDF is also plotted in the same figure. Since the scattering lengths for neutrons are constants, $(b_1 = 0.4149 \ pm; b_2 = 0.5804 \ pm)$, the RDF thus obtained is a linear combination of the partial RDFs

$$
\begin{aligned}
g_n(r) &= 4\pi r^2(\rho_n(r) - \bar{\rho}_n) \\
&= 8\pi r \int_0^\infty q i_n(q) \sin(2\pi q r) dq \\
&= 4\pi b_1^2\{r^2[\rho_{11}(r) - \bar{\rho}_1]\} + 8\pi b_2^2\{r^2[\rho_{22}(r) - \bar{\rho}_2]\} + \\
&\quad + 8\pi b_1 b_2\{r^2[\rho_{12}(r) - \bar{\rho}_2]\} \ .
\end{aligned}
$$ (12)

PARTIAL RDF FOR v-SiO$_2$

The partial structure factors are related to the experimentally measurable intensity data $i(q)$ in the following form:

$$i(q) = c_1 f_1^2 S_{11} + c_2 f_2^2 S_{22} + 2 f_1 f_2 S'_{12} \ .$$ (13)

There are three unknowns (S_{11}, S_{22} and S'_{12}) to be determined. However, each diffraction experiment will provide one such equation. So if we can

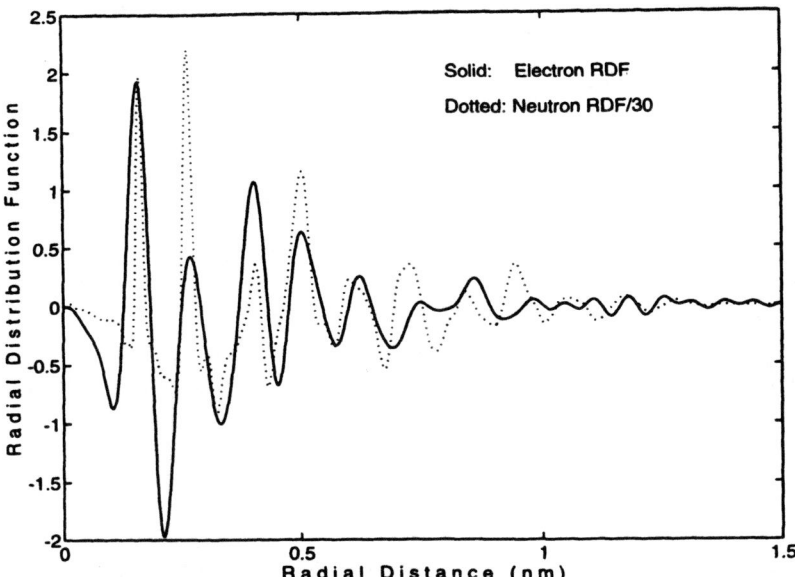

Figure 1. Radial distribution functions for vitreous silica obtained from electron diffraction data $[g_e(r) = 4\pi r^2(\rho_e(r) - \bar{\rho}_e)]$ and neutron diffraction data $[g_n(r) = 4\pi r^2(\rho_n(r) - \bar{\rho}_n)]$, respectively. The peaks indicate the short range order in the material as the interatomic distances.

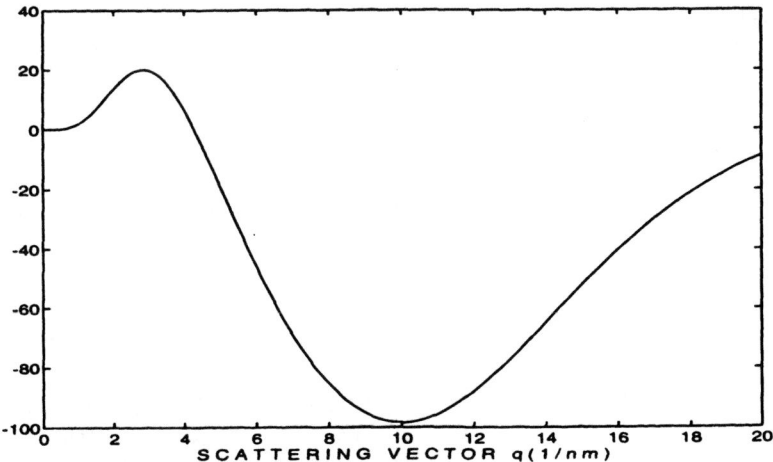

Figure 2. The determinant *vs.* scattering vector q for combined electron, neutron, and X-ray diffraction intensities from vitreous silica. A singularity occurs at $q = 4.2 \ nm^{-1}$ when the determinant equals zero.

have three different sets of experimental data obtained by using different radiation, such as neutrons, X-rays and electrons, then we can have a set of simultaneous linear equations

$$
\begin{cases}
c_1 b_1^2 S_{11} & + & c_2 b_2^2 S_{22} & + & 2 b_1 b_2 S'_{12} & = & i_n(q) \\
c_1 f_1^{(X)2} S_{11} & + & c_2 f_2^{(X)2} S_{22} & + & 2 f_1^{(X)} f_2^{(X)} S'_{12} & = & i_X(q) \\
c_1 f_1^{(e)2} S_{11} & + & c_2 f_2^{(e)2} S_{22} & + & 2 f_1^{(e)} f_2^{(e)} S'_{12} & = & i_e(q)
\end{cases}
\tag{14}
$$

where b_j, $f_j^{(X)}$ and $f_j^{(e)}$ $(j = 1, 2)$ are the atomic scattering lengths for neutrons, the atomic scattering amplitudes for X-rays, and the atomic scattering amplitudes for electrons, respectively.

However, the condition for having a unique solution to the above equations is that the determinant of the coefficient matrix is not equal to zero, namely,

$$
\begin{vmatrix}
b_1^2 & b_2^2 & b_1 b_2 \\
f_1^{(X)2} & f_2^{(X)2} & f_1^{(X)} f_2^{(X)} \\
f_1^{(e)2} & f_2^{(e)2} & f_1^{(e)} f_2^{(e)}
\end{vmatrix} \neq 0 .
\tag{15}
$$

Since the atomic scattering amplitudes for X-rays and electrons are closely related by the Mott formula

$$
f^{(e)}(q) = \frac{2me^2}{h} \frac{[Z - f^{(X)}(q)]}{q^2} ,
\tag{16}
$$

in which m is the relativistic mass of an electron, then the determinant condition (15) becomes

$$
\begin{vmatrix}
b_1^2 & b_2^2 & b_1 b_2 \\
f_1^{(X)2} & f_2^{(X)2} & f_1^{(X)} f_2^{(X)} \\
(Z - f_1^{(X)})^2 & (Z - f_2^{(X)})^2 & (Z - f_1^{(X)})(Z - f_2^{(X)})
\end{vmatrix} \neq 0 .
\tag{17}
$$

For v-SiO$_2$ this determinant is a function of the scattering vector q, and it is given in figure 2 for the range of q from 0 to 16 nm^{-1}. A singularity is seen to occur at about $q = 4.2$ nm^{-1}.

DISCUSSION

Comparing the neutron RDF and the electron RDF, we recognize that there are more details available in the neutron RDF. Most significant is loss of the peak at about 0.3 nm, which gives the Si-Si correlation. This is important in determining the mean value of the Si-O-Si bonding angle. The loss of it is caused by the termination in diffraction data, which is much more extensive in Fourier space for electron diffraction (data extending to only 16 nm^{-1}) than for neutron diffraction data (data up to 72 nm^{-1}).

It is interesting to note that the condition equations are independent of the relative atomic fractions c_1 or c_2, so the condition equations [eqn. (15) or (17)] are valid for any binary system. Given the elements of the system, equation (15) or equation (17) will determine whether the partial RDFs can be deduced from the combined neutron, X-ray and electron diffraction data without singularity. Unfortunately, the condition equations may not always be satisfied for certain values of q of interest. In this case, the partial RDFs may not be deduced from these three experiments with satisfactory accuracy.

Although these intrinsic indeterminacies may occur at only a few (often one) q values, it seems that in theory the partial structure factors can still be deduced except at these singularities. However, we should note that the existence of any such intrinsic singularity means that two of the linear equations are almost linearly related to each other; in particular this occurs for the electron and X-ray equations. In this case, the round-off error in the deduced partial structure factors may be above the limit to reconstruct any meaningful partial RDFs of the system.

Though it is possible for the singularity to occur for the combination of neutron diffraction data with either electron data or X-ray data, it is much more harmful when it occurs for the combination of electron and X-ray data. This is because of the close relationship in the atomic scattering amplitudes between electrons and X-rays via the Mott formula (16).

CONCLUSIONS

Radial distribution functions deduced from energy-filtered electron diffraction data are in a convolution with numerical windows involving relevant sharpening functions, but the basic feature reflecting the short range order is preserved. Termination in Fourier space causes the loss of fine details in the RDFs. A numerical criterion has been given to examine quickly the feasibility of deducing partial RDFs when combined data of electron, neutron and X-ray experiments are to be used. For vitreous silica, it is not feasible to deduce the partial RDFs by performing the three above-mentioned three diffraction experiments.

Acknowledgments: The authors wish to thank Dr. D.L. Griscom of the Naval Research Laboratory for the Supersil-W vitreous silica material used in this study. This research is supported by the Department of Energy, Office of Basic Energy Sciences, under Grant DE-FG02-89ER45396.

REFERENCES

1. F. Zernike and J.A. Prins, *Z. Phys.* **41**, 184 (1927).
2. B.E. Warren, H. Krutter and O. Morninger, *J. Am. Cer. Soc.* **19**, 202 (1936).
3. D.T. Keating, *J. Appl. Phys.* **34**, 923 (1963).
4. J. Krogh-Moe, *Acta Chem. Scand.* **20**, 2890 (1966).
5. B. K. Teo, *EXAFS: Basic Principles and Data Analysis*, (Springer-Verlag, Heidelberg, 1986).
6. L.C. Qin, A.J. Garratt-Reed and L.W. Hobbs, *Proc. 50th EMSA Ann. Meeting*, p.350, (1992).
7. E. Lorch, *J. Phys.* **C2**, 229 (1969).
8. A.C. Wright, *J. Non-Cryst. Solids* **123**, 129 (1990).

A POSITRON-ANNIHILATION-LIFETIME STUDY OF POLYMER BLENDS

J. LIU, AND Y. C. JEAN
Department of Chemistry, University of Missouri-Kansas City
Kansas City, MO 64110

H. YANG
Eastman Chemical Co. Kingsport, TN 37662-5001

ABSTRACT

Positron-annihilation-lifetime (PAL) spectroscopy has been utilized to investigate the free-volume properties of two types of polymer blends, a miscible blend of bisphenol-A polycarbonate (PC) and tetramethyl bisphenol-A polycarbonate (TMPC), and an immiscible blend of PC and polystyrene (PS). In the miscible blend, the free-volume hole size and its fraction follow a linear relationship with respect to the weight fraction while in the immiscible blend, the relationship is not linearly additive. The free-volume hole distributions in the immiscible blend are found to be significantly broader than those in the pure polymers. The difference is thought to be a result of the free volume formed and associated with the conformation and interchain packing between the dissimilar chains in incompatible polymers.

INTRODUCTION

In the last few decades, it has been realized that many physical and mechanical properties of polymers can be significantly improved by a process of blending.[1] The miscibility and the phase separation phenomena of polymer blends have received significant attention in polymer applications.[2] In order to predict and enhance the materials properties of blends, it is important to understand the nature and the underlying reasons of blending at a molecular level. One of rational approaches in this line of research is to investigate the free-volume properties on blending.[3]

There exist many physical probes for characterizing the structures and properties of polymer blends.[1,4] However, only a limited number of probes are available for characterizing the free-volume properties because of the very small size and the dynamic nature of the free volume.[4] In recent years, positron-annihilation-spectroscopy (PAS) has emerged as a unique probe for characterizing the free-volume properties in polymers.[5] In PAS, one employs the anti-particle of electrons, the positron, as a nuclear probe. Because of its positively charged nature, the positron is repelled by the ion cores and preferentially localized in atomic-size free-volume holes of a polymeric material. Therefore, the positron and positronium (a bound atom which consists of a positron and an electron) annihilation signals are found to be contributed mainly from the free-volume holes in a polymer.

Mat. Res. Soc. Symp. Proc. Vol. 321. ©1994 Materials Research Society

Currently, PAS has been mainly developed in monitoring the ortho-positronium (o-Ps) annihilation lifetimes (PAL) for polymeric applications. The results for o-Ps lifetime and its probability are related to the free-volume hole size, fraction and distribution. Free-volume hole sizes, fractions, and distributions in a variety of polymers have been reported using PAL methods. In polymer blends, some PAL results have been reported.[6,7] In this paper, we report a PAL study for two well-known polymer blends, namely PC-TMPC and PC-PS and the results of free-volume properties: size, fraction and distribution.

EXPERIMENTAL

Sample Preparations:

The initial chemical structures of the polymers under investigation, bisphenol-A polycarbonate (PC), and tetramethyl bisphenol-A polycarbonate (TMPC), and polystyrene (PS), are shown in Fig.1 below. These materials were purchased from Imperial Chemicals Inc. Physical properties of these polymers are listed in Table I below.

Table I. Physical properties of initial polymers in this study

Polymer	Mw	Mw/Mn	Density	T_g(DSC)	$V_f(\text{Å}^3)$	$f_v(\%)$
PC	83,000	3.9	1.200	150°	112±2	3.9±.2
TMPC	67,000	2.6	1.083	190°	141±2	7.6±.2
PS	105,000	1.06	1.058	100°	104±2	5.9±.2

Notes: V_f and f_v are the mean free volume and its fraction from this work. Other parameters were measured as reported in Ref.[8].

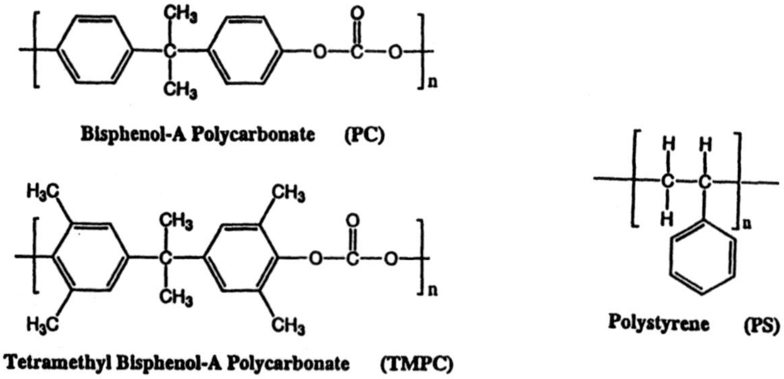

Bisphenol-A Polycarbonate   (PC)

Polystyrene   (PS)

Tetramethyl Bisphenol-A Polycarbonate   (TMPC)

Fig. 1 Molecular structures of pure polymers used in this study.

The appropriate wt % of polymer blends were prepared by dissolving them in dichloromethane solvents. The mixed polymer solutions were cast in a glass dish and dried at 80° in a vacuum oven for several days. The dried films were then used for the positron lifetime experiments.

Positron-Annihilation-Lifetime (PAL) Spectroscopy:

The positron-lifetime measurements were performed using a fast-fast coincident method which entails monitoring the signal (1.28 MeV γ-ray) from positron decay in the ^{22}Na (10 μCi) isotope as the start time and the signal (0.51 MeV γ-ray) from the positron annihilation in the materials as the end time. Two PAL spectra at different statistics (1, and 15 x 10^6 counts) were collected in each blend for a complete data analysis of finite lifetimes and lifetime distributions, respectively. Detailed descriptions of the PAL method can be found in Refs. [5,9].

Mean Free-Volume Size and Fraction:

All of the PAL spectra obtained were analyzed by two methods: (1) finite-term lifetime analysis, and (2) continuous-lifetime analysis. The former analysis employs the PATFIT program [10] and the later method uses the CONTIN program.[11,12] The finite-term lifetime decomposes a PAL spectrum into 2-5 terms of negative exponentials. In these polymers, we found that 3-lifetime results give the best χ^2 (<1.1) and most reasonable standard deviations. The shortest lifetime ($\tau_1 \approx 0.12$ns) is the lifetime of p-Ps (singlet Ps) and the intermediated lifetime ($\tau_2 \approx 0.40$ns) is the lifetime of the positron. The longest lifetime ($\tau_3 \approx 1-3$ ns) is due to o-Ps annihilation. In the current PAL method, we employ the results of o-Ps lifetime to obtain the mean free-volume hole radius by the following semi-empirical equation:[13]

$$\tau_3 = \frac{1}{\lambda_3} = \frac{1}{2} \, [1 - \frac{R}{R_o} + \frac{1}{2\pi} \sin(\frac{2\pi R}{R_o})]^{-1} \, , \qquad (1)$$

where τ_3 (o-Ps lifetime) and R (hole radius) are expressed in ns and Å respectively. R_o equals to $R + \Delta R$ where ΔR is the fitted empirical electron layer thickness (=1.66 Å).

The fractional free volume (%) is expressed as an empirically fitted equation as:[14]

$$f_v = AV_f I_3 \qquad (2)$$

where V_f (in Å^3) is the volume of free-volume holes calculated by using the spherical radius (R) of Eq. (1) from τ_3 (in ns), I_3 (in %) is its intensity, and A is empirically determined to be 0.0018 from the specific volume data.[14,15]

Free-Volume Hole Distributions:

Since the free-volume hole exists in a distribution, the lifetimes should be expressed more correctly as a distribution instead of discrete values. Hence, we employ the expression of positron-lifetime spectra in a form as:

$$N(t) = \int_0^\infty \lambda \alpha (\lambda) e^{-\lambda t} d\lambda + B \qquad \textbf{(3)}$$

where λ the annihilation rate is the reciprocal of lifetime τ with an annihilation probability-density function $\lambda\alpha$, and B is the background of the spectrum. A computer program CONTIN [11,12] was employed to provide $\lambda\alpha$ vs λ for a PAL spectrum by using the measured reference spectrum in Cu. Following the correlation Eq. (1) between τ_3 and the hole radius R and considering the difference of o-Ps capture probability in different hole size with a linear correction K(R) (=1+8R), and with a spherical approximation of free volume holes, the free-volume hole-volume-probability-density function, $V_f$pdf is expressed as:[16]

$$V_f pdf = -3.32\{\cos[2\pi R/(R+1.66)]-1\}\alpha(\lambda)/\{(R+1.66)^2 K(R) 4\pi R^2\} \qquad \textbf{(4)}$$

The fraction of hole volume between V_f and dV_f is $V_f$pdfdV_f. Detailed descriptions in this regard can be found elsewhere.[15,16]

RESULTS AND DISCUSSION

The results for the lifetime and intensity of o-Ps, τ_3, and I_3 as a function of wt% of TMPC and PS in PC blends are shown in Fig. 2. Free-volume hole volumes and fractions of free-volume holes are then calculated according to Eqs. (1), and (2) in a spherical-hole model and are shown in Fig. 3.

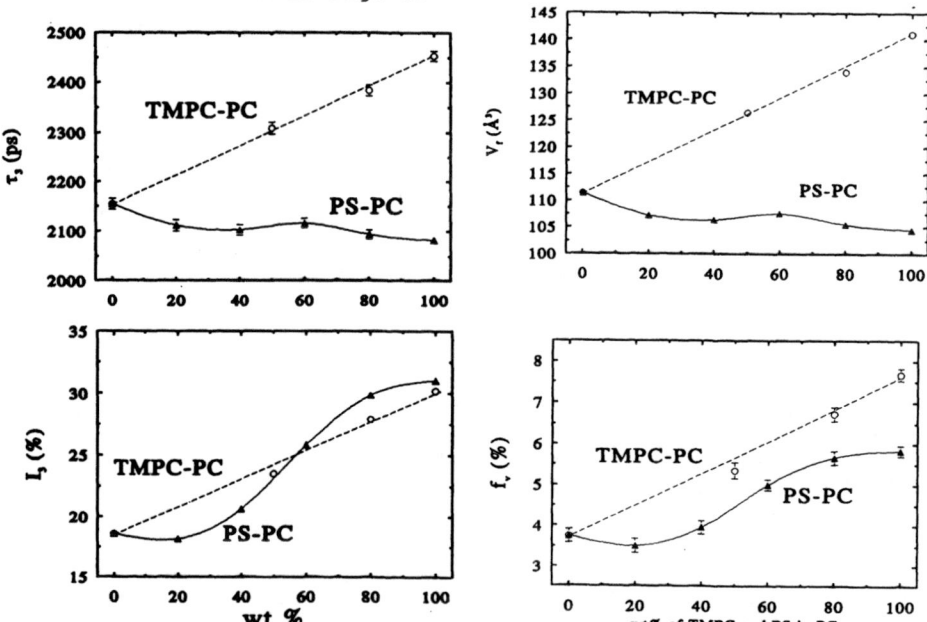

Fig.2 o-Ps lifetime and intensity vs wt% of TMPC or PS in PC.

Fig.3 Mean free volume and fraction vs wt%.

The τ_3 and I_3 in pure PC, TMPC, and PS are consistent with the existing results.[6,7,15] The results of V_f and f_v in pure polymers are also listed in Table 1. Larger values of V_f and f_v in TMPC than in PC can be explained from the substitution effect. Four methyl groups attached to the phenyl group of TMPC (see Fig. 1) can create more free volumes in molecular packing of polymer chains than in PC. In Fig. 3, we found that the variations of V_f, and f_v as a function of wt% are different between TMPC/PC and PS/PC blends: the variations are linearly additive for TMPC/PC but deviate from linear for PS/PC blends. There exists some free-volume theoretical interpretations on blending.[1,3] Here we simply consider a binary interchain interaction and express the mean-free volume f_v in a blend as:[3]

$$f_v = f_v^1 x + f_v^2 (1-x) + x(1-x) f_v^1 f_v^2 \beta \qquad (5)$$

where f_v^1, f_v^2 are the free-volume fractions in pure polymers 1, and 2, respectively. β is a parameter related to the interchain interaction between dissimilar chains, and x is the wt. fraction of polymer component 2. When we fit the results of f_v as shown in Fig. 3 to the above equation, we obtain $\beta \leq -.03$ and $\beta = -.18$ to .09 for TMPC/PC, and PS/PC, respectively. A very much smaller value of β in TMPC/PC indicates an insignificant change of free volume holes due to blending. It is known that TMPC/PC, and PS/PC are a miscible [17] and an immiscible blend [18], respectively. In TMPC/PC, the miscibility is shown to arise from the specific interaction between carbonyl groups of TMPC and PC.[8] A linearly additive relationship of the free volume ($\beta = 0$, the dashed line in Fig. 3) indicates that either no additional free volume or similar types of free volume are formed on blending. However, in the PS/PC immiscible blend, dissimilar chemical structures influence the segmental conformation and packing in the blend. The resulting free volumes of either a dilation at high fractions of PS (positive β value) or a contraction at low fractions of PS (negative β) due to blending are seen in Fig.3.

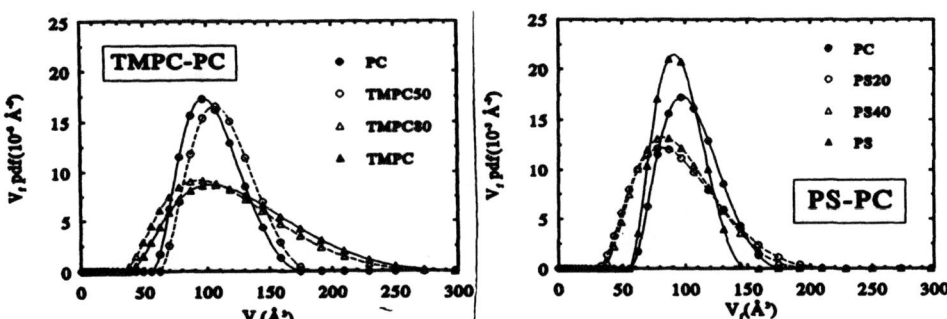

Fig.4. Probability of free-volume distribution in the TMPC/PC (miscible) and PS/PC (immiscible) polymer blends.

The difference in the free volume due to blending between two types can be seen more clearly in the distributions (Fig.4). A larger difference between the blends and the pure polymers in the distribution of free-volume holes is observed in the immiscible blend (PS/PC) than in the miscible blend (TMPC/PC). In the immiscible blend, the distribution is broader than the pure polymers due to the free volumes formed in the interfacial regions. Further systematic investigations on the free-volume effect using PAL on blends are in progress at our laboratory.

ACKNOWLEDGEMENT

This research has been supported by a grant from National Science Foundation (DMR-90040803) and Eastman Kodak Comp. Assistance from G.H. Dai Q. Deng and H. Shi is acknowledged.

REFERENCES
[1]   For examples, see D.R. Paul and S. Newman, Eds. "Polymer Blends", Vol. 1 and 2, Academic Press, N.Y. (1978); L.A. Utracki, "Polymer Alloys and Blends", Hanser Pub. N.Y. (1990)
[2]   For examples, see O. Olabisi, L.M. Robeson, and M.T. Shaw, "Polymer-Polymer Miscibility", Academic Press, N.Y. (1980)
[3]   For examples, see S. Wu, J. Polym. Sci. B, Polym. Phys. 25, 2511 (1987); R. Steller, D. Zuchowska, J. Appl. Polym. Sci. 8438, 1411 (1991)
[4]   For example, see J.D. Ferry, "Viscoelastic Properties of Polymers," 3rd ed. John Wiley & Sons, NY (1980)
[5]   For examples, see Y.C. Jean, J. Microchem. 42, 72 (1990)
[6]   J.E. Kluin, Z. Yu, S. Vleeshouwers, J.D. McGervey, A.M. Jamieson, R. Simha, and K. Sommer, Macromolecules, 26, 1853 (1993)
[7]   M.D. Zipper, G.P. Simon, P. Cherry, and A.J. Hill, J. Polym. Sci. B, Polym. Phys. (in press 1993)
[8]   H. Yang, and B. Ni, ACS Polymer Reprints 33(2), 573 (1992)
[9]   For examples, see "Positron Solid-State Physics," Eds. W. Brandt and A. Dupasquier, North Holland Pub. Amsterdam (1983); "Positron and Positronium Chemistry," Eds. D.M. Schrader and Y.C. Jean, Elsevier Pub. Amsterdam (1988)
[10]   PATFIT 88 package (1989 version) was purchased from Riso National Labs. Denmark.
[11]   S.W. Provencher, Comp. Phys. Comm. 27, 229 (1982)
[12]   R.B. Gregory, and Y. Zhu, Nucl. Inst. Methods in Phys. Res. A290, 172(1990)
[13]   H. Nakanishi and Y.C. Jean, in "Positron and Positronium Chemistry," ed. D.M. Schrader and Y.C. Jean, Elsevier Pub. Amsterdam (1988) Chapter 5, p. 159
[14]   Y.Y. Wang, H. Nakanishi, Y.C. Jean, and T.C. Sandreczki J. Polym. Sci. B, Polym. Phys. 28, 1431 (1990)
[15]   J. Liu, Q. Deng, and Y.C. Jean, Macromolecules (Dec. 1993)
[16]   Y.C. Jean, and Q. Deng, J. Polym. Sci. B, Polym. Phys. 29, 1359 (1992)
[17]   For example, see M.T. Shaw, J. Appl. Polym. Sci. 18, 449 (1974);C.K. Kim, M. Aguilar, and D.R. Paul, J. Polym. Sci. B, Polym. Phys. 30, 1131 (1992)
[18]   For examples, see G. Groeninckx, S. Chandra, H. Berghmans, and G. Smets, in "Multiphase Polymers", Adv. ACS Ser. 176, 337 (1979); M. Afeworki, and J. Schaefer, Macromolecules 25, 4097 (1992)

NONEQUILIBRIUM STATES OF SYMMETRIC TETRAALKYLAMMONIUM HALIDES

A. XENOPOULOS[1], M. MUCHA[2], AND B. WUNDERLICH

Department of Chemistry, University of Tennessee, Knoxville, TN 37996-1600
and Chemistry Division, Oak Ridge National Laboratory, Oak Ridge, TN 37831-6197.

ABSTRACT

The nonequilibrium behavior of symmetrical tetra-n-alkylammonium halides is described from a study of the transition behavior using thermal analysis and hot-stage optical microscopy. Results are discussed in terms of reversibility, supercooling, sharpness of transitions on thermal cycling, and glass transitions. The transitions of plastic to rigid crystals have the largest supercoolings. Gradual disordering on heating and incomplete ordering on cooling are found for some of the salts and represent interesting examples of order/disorder and glass transitions with reduced cooperativity.

INTRODUCTION

A large number of symmetric tetra-n-alkylammonium halides were analyzed with the goal to understand their mesophase transitions.[1] Symmetric cations were chosen to enhance the drive to crystallization, and simple halides, to avoid complications from anion rotation in the crystal. Plastic crystals, characterized by close-to-spherical motifs that can reorient without loss of positional order[2] were found for the small, rigid tetra-n-alkylammonium cations (methyl to propyl groups). For longer side chains (longer than butyl groups) only conformationally disordered mesophases (condis crystals) exist as high-temperature polymorphs. Condis crystals have some (or complete) conformational disorder and mobility with retention of positional and orientational molecular order.[3] It will be shown that on heating, the conformational disorder is sometimes introduced gradually. The rate of increase of disorder with temperature is in these cases accelerating with temperature and reaches ultimately the magnitude of the increase of disorder found when heating a liquid.[4] Details of the motion can be studied by solid state ^{13}C NMR.[5] Depending on the low-temperature state of order, this gradual change to a mobile, mesophase phase can be a disordering transition or a glass transition.

In addition to the general framework of equilibrium changes, a distinct non-equilibrium behavior was often noted. In this publication, we will also analyze the cases where ordering was impeded on cooling from the isotropic melt or mesophase.

EXPERIMENTAL DETAILS AND RESULTS

Commercial samples of many symmetric tetra-n-alkylammonium iodides, bromides, and chlorides with n from 1 to 18 carbon atoms were used for the measurements.[1,4] The

---

[1] Present address: Millipore Corporation, Bedford, MA 01730
[2] Permanent address: Polymer Institute of Technical University, 90-924 Lodź, Poland

purities were approximately 99%, except for 4Cl (92%). The abbreviated nomenclature used by us gives the number of carbon atoms in the alkyl group and designates the anion, so that 4Cl stands for tetra-*n*-alkylammonium chloride. For the DSC experiments a Perkin-Elmer DSC-7 was used, operating with a heating rate of 10 K/min and a constant flow of dry N_2. Temperature and heat-flow calibrations were carried out as before.[4] Measurements were made from 220 K to decomposition on 5–25 mg samples. A typical sequence of runs for the detection of nonequilibrium effects included an initial heating to the melt, followed by cooling, a second heating to the melt, quenching at nominally 500 K/min, and a final heating to the melt. Measurements were made during all, but the quenching step. The results from measurements on close to equilibrium samples were described earlier;[1,4] the nonequilibrium behavior is discussed here. Any weight loss during this sequence of thermal analyses (due to decomposition or evaporation) was kept to less than 3%.

For polarizing microscopy, an Olympus zoom stereo-microscope with up to 40× magnification was used. Temperature was controlled with a Mettler FP-82 hot stage, governed by a FP-80 processor. Microscopy was needed mainly to identify crystals and their transitions. Plastic crystals usually lack birefringence because of their cubic symmetry, while condis crystals often have higher birefringence than rigid crystals. The melt and plastic crystalline states, both without birefringence, can be distinguished by their appearance (droplets or crystals).

The results of the cooling experiments at 10 K/min are summarized in the Table by listing of the supercooling from the equilibrium transition temperatures established earlier.[1,4] Typical examples of heating and cooling sequences are shown in Figs. 1 and 4, below, heat capacity analyses of conformational disordering can be seen in Figs. 2 and 3.

DISCUSSION

A liquid sample can solidify either by ordering sufficiently so that the large-amplitude positional, orientational, and conformational motions are frozen (crystallization), or its large-amplitude motions can freeze at a glass transition without change in order. In case the ordering on crystallization is incomplete (mesophase

SALT	Supercooling (K)
	Bromides
C_2H_5	20, (PC→PC), 21 (PC→C)
C_3H_7	4 (PC→PC), 46 (PC→C)
C_4H_9	50 (M→CC)
C_5H_{11}	19 (M→C)
C_6H_{13}	49 (M→CC)
C_7H_{15}	24 (M→CC), 18 (CC→C?)
C_8H_{17}	17 (M→CC), 10 (CC→C?)
$C_{10}H_{21}$	14 (M→CC), 8 (CC→C)
$C_{12}H_{25}$	11 (M→CC), 31 (CC→C?)
$C_{16}H_{33}$	15 (M→CC), 12 (CC→C)
$C_{18}H_{37}$	12 (M→C)
	Iodides
C_2H_5	23 (PC→C)
C_3H_7	26 (PC→C)
C_4H_9	24 (M→PC), 28 (PC→C)
C_5H_{11}	17 (M→C)
C_6H_{13}	10 (M→CC), 4 (CC→CC), 7–17 (CC→C)
C_7H_{15}	11 (M→CC), 17 (CC→C?)
$C_{12}H_{25}$	11 (M→CC), 3 (CC→CC), 13 (CC→-C?)

Table *Supercooling for transitions of tetra-n-alkylammonium halides (10 K/min cooling rate). The phases are identified as C (crystal), PC (plastic crystal), and CC (condis crystal).*

formation), final solidification occurs at a lower temperature, either *via* further ordering or *via* a mesophase glass transition.[3] Although slow cooling and molecular symmetry facilitate crystallization, the tetra-*n*-alkylammonium salts studied demonstrate that it is not an *a priori* certainty that the gain and loss of crystalline order occurs in sharp transitions. In addition, there are many examples of nonequilibrium behavior on cooling melt and mesophases.

Orientational order: The transition of plastic crystals to crystals (PC→C) is observed only for short-chain salts. The supercoolings (at 10 K/min) range from 21 to 28 K for 2Br, 2I, 3I, and 4I, and reach 46 K for 3Br. These are the largest supercoolings in the Table. The PC→C-transition involves no positional ordering, and one might initially expect to have a lower supercooling relative to crystallization from the melt. The large molecule C_{60}, for example, has only about 10 K supercooling for its PC→C transition (T_d = 256 K).[6] In addition, melt-to-nematic liquid-crystal transitions, that also involve orientational ordering, are known to show practically no supercooling.[3] A good example for the study of the reason of this large supercooling is 4I, that has both a transition from the melt to a plastic crystal (positional ordering only) and a transition from the plastic crystal to a rigid crystal (orientational ordering and fixing of one conformational rotation per chain).[5] The slow step in the PC→C transition can not be the additional conformational ordering, since 2I, 3I, 2Br, and 3Br show no conformational mobility even in the melt and have similarly large supercoolings. A possible reason may be a cooperative deformation of the cation, needed to produce the nucleus for a rigid crystal in which the interaction with the anion supplies most of the cohesive energy. A deformed cation has, in fact, been reported for the crystal 2I.[7] Crystal structures of 3I and 4I are presently being investigated in our laboratory, and further insight into the balance between the coulombic interaction and the packing of nonpolar hydrocarbon chains is given below. The supercoolings in the Table for the PC→C transitions increase with cation size and are different for Br and I. Furthermore, bromides and chlorides also have a larger hindrances to full crystallization. Once the new phase is established, it can be noted from Fig. 1 that the PC→C transitions are sharp. This indicates that once the new phase is initiated (nucleated), a fast, cooperative stopping of the orientational motion ensues.

For 2Br and 3Br one finds in Fig. 1 a small PC→PC-transition, 15 K above their main C→PC disordering, not present in the iodide (see Table). Recent X-ray data indicate that 2Br transforms to *fcc* only at this second transition, with large changes in structure at both transitions (and similar supercoolings on cooling, see Table), while 3Br transforms to a simple cubic structure at the lower temperature, with little additional structural changes above. Microscopy shows a loss of birefringence above the second transition for 2Br and

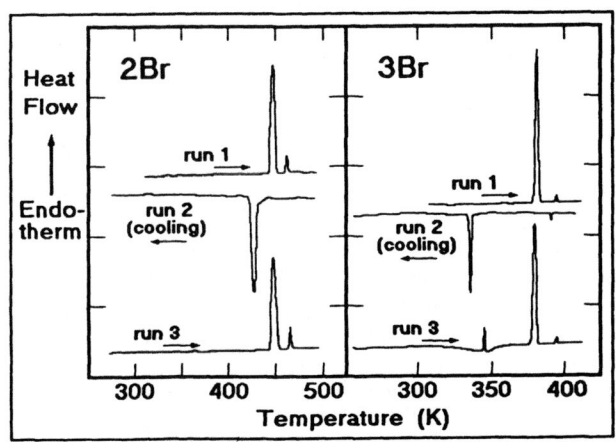

Fig. 1 *DSC traces for 2Br and 3Br showing supercooling.*

above the first for 3Br, in accord with the X-ray data. The DSC behavior in Fig. 1 is also different for the two salts. For 2Br, there is a similar supercooling for both transitions (the small transition occurs on cooling at the right shoulder of the large one) while for 3Br the main peak supercools by 46 K and the small peak by only 4K. In both transitions of 2Br, although they are largely different in enthalpy, similarly major structural rearrangements take place, i.e. in the upper transition large-scale compensation occurs of the gained coulombic lattice energy and the deformation energy of the nonpolar chains.

The heating trace of 3Br after cooling (run 3 of Fig.1) shows a broad exotherm with a superimposed, sharp endotherm, an effect even more pronounced after quenching. This can be explained by incomplete crystallization and formation of an intermediate, metastable phase. This phase is visible as highly-birefringent crystals that grow at room temperature on top of the crystals grown first on cooling (note the smaller exotherm in run 2 compared to runs 1 and 3). The crystallization of the short-chain tetra-n-alkylammonium salts contains thus many nonequilibrium steps, caused by packing and orientation difficulties.

Conformational order: The tetra-n-alkylammonium salts with chains longer than four carbon atoms do not show plastic crystalline phases, but may gain conformational mobility before isotropization. The simplest behavior is seen for 5I, 16Br and 18Br, where the disordering transition gives a condis crystal that melts a few kelvins higher, as shown in Fig. 2 for 5I.[4] On cooling, a single crystallization peak is obtained in these cases. Only one distinct transition can be seen with the microscope. At the disordering transition C→CC one

CH$_2$-group per chain becomes mobile with an entropy gain of 41 J/(K mol). At the about 7 K higher melting temperature, the other conformation disorders along with the loss of positional and orientational order.[5] The bonds close to the methyl group disorder first. The N–C bonds and the adjacent C–C-bond remain largely rigid, even in the melt.

No first-order transition to the condis state was seen for 7I, 7Br, and 8Br. The small transitions below isotropization are too small to account for a conformational

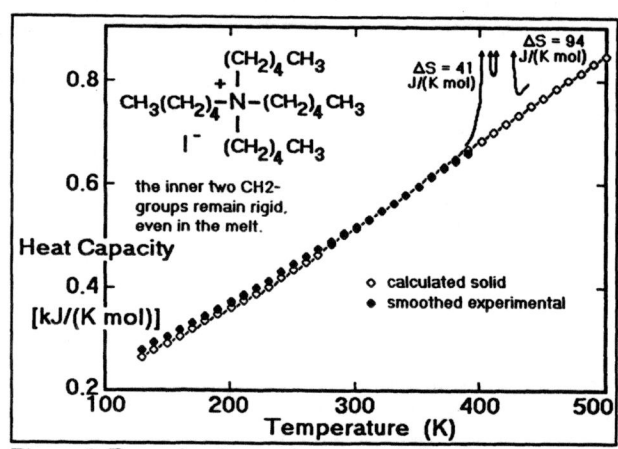

Figure 2 *Example of a conformational disorder transition.*

disordering, as is shown in Fig. 3 for the example of 8Br.[4] An analysis of the heat capacity shows, however, that considerable entropy is in these cases gained gradually (shaded area in Fig. 3). For 8Br, one expects three CH$_2$-groups per chain to gain mobility gradually (based on entropy estimates and solid state NMR results).[5] The entropies given in the figure account, however, for only two. It is thus likely that, instead of ordering only two CH$_2$-groups per chain, all three are partially ordered. It is known that an amorphous CH$_2$-group in the glassy state of polyethylene keeps a conformational entropy ΔS^o of about 3.0 J/(K mol), just enough to balance the total expected conformational entropy in the melt of $4 \times 5 \times 9 = 180$ J/(K mol) [per molecule: the gradual disordering (shaded) and small

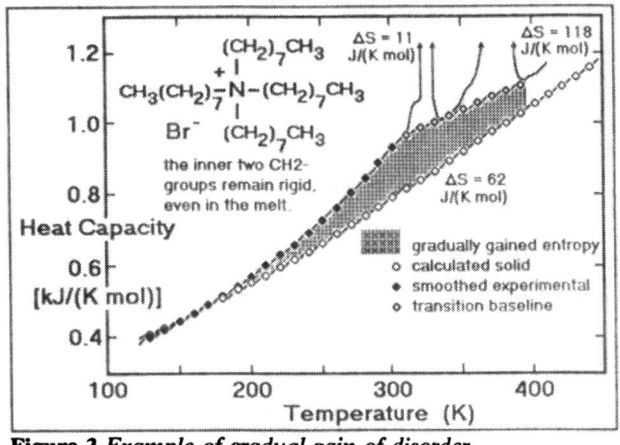

$(CH_2)_7CH_3$ $\Delta S = 11$ J/(K mol)

$CH_3(CH_2)_7 - N - (CH_2)_7CH_3$

Br^- $(CH_2)_7CH_3$

the inner two CH2- groups remain rigid, even in the melt.

$\Delta S = 118$ J/(K mol)

$\Delta S = 62$ J/(K mol)

gradually gained entropy
o calculated solid
• smoothed experimental
o transition baseline

Heat Capacity

1.2

1.0

0.8

0.6

[kJ/(K mol)]

0.4

100   200   300   400

Temperature (K)

Figure 3 *Example of gradual gain of disorder.*

transition give 73 J/(K mol), the isotropization entropy is 118 J/(K mol) of which 72 J/(K mol) are from conformational disordering of eight additional CH_2-groups, for a total of 145 J/(K mol), adding $4 \times 3 \times \Delta S^\circ$ yields 181 J/(K mol)]. The gradual transition in Fig. 3 is thus most likely a glass transition, broadened because of limited cooperativity. A similar condis-crystal glass transition has been seen before for N,N'-*bis*-(4-*n*-octyloxybenzal)-1,4-phenylene diamine.[8] In these cases of remaining disorder below the last ordering transition on cooling, the low temperature phase is marked as C? in the Table.

Also easy to understand is the broadening and 10–20% decrease in the transition-peak areas seen on cycling through the transitions for all compounds with side chains longer than octyl. It indicates increasingly incomplete ordering on cooling, obvious also from microscopy, where mixtures of rigid and condis crystals are seen at room temperature.

For 10Br, 12Br and 12I, disordering occurs in two widely separated steps, the second leading to isotropization, completing the melting. The condis crystals have the outer bonds disordered, while the central core remains largely rigid. On cooling 10Br and 12I, all bonds reorder reversibly. For 12Br the low-temperature peak becomes insignificantly small, both on cooling and on subsequent heating, showing that the outer bonds cannot order at the cooling rate employed. Only one transition is also seen by microscopy.

For 4Br there is only one transition on cooling, compared to three on heating, with a heat and entropy intermediate between the two major peaks on heating. A mixture of rigid and condis crystals remains at room temperature, as is also seen by microscopy. Ordering is completed on reheating, as evidenced by an exotherm and the exact reproducibility of the transition entropies. The transition on cooling is crystallization from the isotropic melt, and its supercooling of 50 K is extreme. It indicates that nucleation from the melt is also difficult. Similar results were obtained for 6Br, which is a condis crystal already at room temperature, and cannot give a rigid crystal even if cooled slowly (at 2 K/min) to 240 K. Cooling from the melt at 10 K/min gives one broad exotherm with a supercooling of 49 K. The change of birefringence on cooling is also small. Reheating gives a broad exotherm, followed by an endotherm with the same ΔH, and an additional endotherm.

Glass transition of amorphous salts: Although all samples were quenched in the course of the analyses, only for 4Cl was a normal glass transition observed. The DSC traces are shown in Fig. 4. The midpoint of the glass transition (T_g) is at 230 K with a ΔC_p of 140 J/(K mol). The glass was obtained after cooling from the melt at only 10 K/min, and the DSC trace was similar after cooling at 500 K/min. The glass transition is followed by a broad exotherm and a sharp endotherm, at 290 and 315 K, respectively. The exotherm and the endotherm have about the same area and correspond to an approximate entropy of 40 J/(K mol). Above 315 K, 4Cl is liquid, *i.e.* the endotherm includes only positional contributions for two ions and an orientational contribution, indicating that 4Cl is a condis crystal between exotherm and endotherm. On cooling at 10 K/min, no ordering can occur, as evidenced by the lack of any exotherm in the cooling run. The glass that is found on cooling must have all three types of disorder frozen. The large value of ΔC_p is in agreement with that expectation.

In order to check whether a faster cooling rate could produce the same behavior for the iodide, we

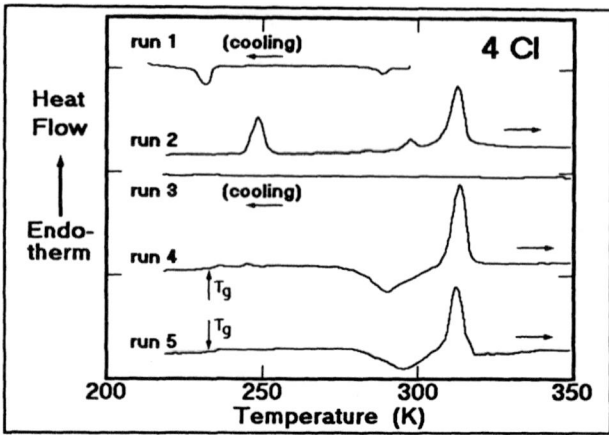

Figure 4 *Glass transition measured by DSC for 4Cl.*

quenched 4I both from the melt and from the plastic crystalline mesophase in liquid nitrogen. The two quenched samples were similar. There was a change in slope of C_p around 235 K that could indicate the onset of a glass transition, followed immediately by a broad exotherm. The area of the exotherm shows that, in both cases, about 40% of disorder was frozen-in by quenching, while the ensuing endotherms are similar for the two runs, and the run of the well crystallized sample. The heat capacities of 4I were measured in the regions below and above the transitions for the runs: slowly cooled, quenched from the melt, and quenched from the plastic crystal. Below the exotherm, the slowly cooled sample has, expectedly, the lowest heat capacity, while the sample quenched from the mesophase has a higher heat capacity than that quenched from the melt. Similar behavior has also been seen in the case of ethanol, where a metastable crystalline phase, identified as a plastic crystal, can be quenched into a "glassy crystal" that has a higher heat capacity than a glass obtained by quenching the melt.[9] Above the ordering exotherm, the heat capacities have opposite order, indicating gradual exothermic recrystallization, with larger latent heat for higher disorder.

Acknowlegements: This work was supported by the Div. of Mat. Res., Natl. Sci. Foundation, Polymers Program, DMR 90-00520 and the Div. of Mat. Sci., Office of Basic Energy Sci., U.S. Dept. of Energy, under Contract DE-AC05-84OR21400 with Martin Marietta Energy Systems, Inc.

References

1. A. Xenopoulos, J. Cheng, M. Yasuniwa, and B. Wunderlich, *Mol. Cryst. Liq. Cryst.* **214**, 63 (1992).
2. J. N. Sherwood (ed.), *The plastically crystalline state*, Wiley-Interscience, Chichester (1979).
3. B. Wunderlich, M. Möller, J. Grebowicz, and H. Baur, *Conformational motion and disorder in low and high molecular mass crystals*, Springer-Verlag, Berlin (1989) [*Adv. Polym. Sci.* **88**, 1].
4. A. Xenopoulos, J. Cheng, and B. Wunderlich, *Mol. Cryst. Liq. Cryst.* **226**, 87 (1993).
5. J. Cheng, A. Xenopoulos, and B. Wunderlich, *Mol. Cryst. Liq. Cryst.* **220**, 105; **220**, 127; **225**, 337 (1992).
6. Y. Jin, J. Cheng, M. Varma-Nair, G. Liang, Y. Fu, B. Wunderlich, X.-D. Xiang, R. Mostovoy, and A. K. Zettl, *J. Phys. Chem.* **96**, 5151 (1992).
7. E. Wait and H. M. Powell, *J. Chem. Soc.* 1872 (1958).
8. J. Cheng, Y. Yin, B. Wunderlich, and H. G. Wiedemann, *Mol. Cryst. Liq. Cryst.* **213**, 237 (1992).
9. O. Haida, H. Suga, and S. Seki, *J. Chem. Thermodynamics* **9**, 1133 (1977).

INTERDIFFUSION OF AMORPHOUS Si/Ge MULTILAYERS UNDER HYDROSTATIC PRESSURE

STEVEN D. THEISS, S. MITHA, F. SPAEPEN AND M.J. AZIZ
Division of Applied Sciences, Harvard University, Cambridge, MA 02138

ABSTRACT

We report initial results of an x-ray diffraction study of the pressure-dependence of the interdiffusion rate in amorphous Si/Ge multilayers. Anneals were performed in a diamond anvil cell at 700 K for various pressures and durations. Interdiffusion was measured by monitoring the rate of decay of the artificial Bragg peaks associated with the multilayer periodicity. A consistent increase in diffusivity was seen with pressure, characterized by an activation volume of -25±11 percent of the atomic volume of Si. An atomistic mechanism that might account for such behavior is discussed.

INTRODUCTION

The nature of the defects in amorphous semiconductors is not well understood. Given that a number of important properties such as diffusion rates [1] and electrical conductivity [2] are controlled by defects, it is highly desirable to develop a more complete and accurate picture of these defects. Unlike their crystalline counterparts where defects are well defined, it is likely that defects present in amorphous semiconductors exist in a continuum of states and configurations. However, it is still possible to conceptualize the existence of defects that have similar character to those present in crystalline materials, specifically the vacancy and the interstitial. While interstitial defects in amorphous materials might be quite similar in behavior to those in crystalline materials, the absence of periodicity would allow the unfulfilled bonds left over from the removal of an atom (i.e. a vacancy) to migrate and become individual "dangling" bonds. Although no direct evidence exists for the presence of interstitial defects, ESR experiments have demonstrated the existence of singly-occupied electronic states, which is widely regarded as evidence for the existence of neutral dangling bonds [3]. It has also been predicted that "floating" bonds, or over-coordination defects, exist in amorphous Si (a-Si) [4] The purpose of the present study is to help elucidate the role of defects in mass transport in amorphous semiconductors by looking at the effects of hydrostatic pressure on interdiffusion in a-Si/a-Ge multilayers. Given the chemical similarity of these two materials, it should eventually be possible to extend the understanding gained from our studies to the cases of pure a-Si and a-Ge.

As with other thermally activated processes, diffusivity is often written [5]:

$$D = D_0 \exp(-\Delta G^* / kT) \tag{1}$$

where D_0 is a prefactor containing a geometrical factor, a jump distance and a vibrational frequency and ΔG^* is the activation free energy. ΔG^* can be expressed as a combination of the activation energy, entropy and volume:

$$\Delta G^* = \Delta E^* - T\Delta S^* + P\Delta V^*,$$ (2)

so that

$$\Delta V^* = \left(\partial \Delta G^* \middle/ \partial P\right)_T.$$ (3)

As eq. (3) indicates, ΔG^* will either increase or decrease with pressure, depending on the sign of ΔV^*. Since the elements in the prefactor D_0 are expected to vary only slowly with pressure [6], combining eq. (1) and eq. (3) gives:

$$\Delta V^* \approx -kT\left(\partial \ln D \middle/ \partial P\right)_T.$$ (4)

Consequently the diffusivity, like ΔG^*, either increases or decreases with pressure, depending on the sign of ΔV^*. As is often done with ΔG^*, ΔV^* can be split into two components:

$$\Delta V^* = \Delta V_f^0 + \Delta V_m$$ (5)

where ΔV_f^0, the formation volume, is the volume change in the system upon formation of one defect in its standard state. ΔV_m, the migration volume, is the additional volume change when the defect reaches the saddle point in its migration path.

In crystalline materials, the formation volume of interstitials is negative and of vacancies is positive, with both having a magnitude of approximately one atomic volume. The correction for lattice relaxation around the defects is thought to be relatively small [7]. Thus an increase in pressure increases the equilibrium concentration of interstitials and decreases that of vacancies. In metals, the migration volume has been found to be smaller than the formation volume, indicating that their mobility is less strongly affected by pressure than is their concentration [8]. Consequently, because the diffusivity is proportional to the product of defect concentration and defect mobility, an increase in diffusivity with pressure would be an indication of interstitial-mediated diffusion, whereas a decrease in diffusivity with pressure would indicate vacancy-mediated diffusion. It should be noted that in the above analysis we have assumed that the defects are in thermodynamic equilibrium at the temperatures and pressures under which the experiment is conducted. If the time scale of the experiment is too short for the defects to equilibrate, then the volume of migration will dominate. For example, if the number of defects is constant over the duration of an experiment, the measured activation volume will be equal to the volume of migration.

In the case of amorphous materials, such clear-cut divisions in the types of defects cannot be made. It is difficult to predict *a priori* the magnitude of the formation volume of a dangling bond. It is also quite plausible that the migration volume of such a defect is a significant part of the overall activation volume. For example, the most plausible atomistic mechanism that has been developed to date to explain pressure-enhanced solid-phase epitaxial growth (SPEG) of amorphous Si and Ge is a dangling bond mechanism [9]. In the SPEG model of Spaepen and Turnbull [10], the dominant part of the activation volume for this process is the migration volume of a dangling bond residing at the interface between the crystalline and amorphous phases. Since this model applies to other covalent amorphous networks [11], it is reasonable to suggest that atomic transport in a-Si and a-Ge occurs by a similar process.

EXPERIMENTAL METHOD

The multilayer samples were prepared from elemental targets using Ar^+-ion sputter-deposition in a chamber with a rotating target block [12]. By controlling the length of time the ion beam strikes each target, multilayers of various periodicities can be obtained. The samples used in this experiment were designed to have approximately equal thickness layers of amorphous silicon and amorphous germanium, resulting in a 50/50 average composition for the entire film. The films are all from the same sputtering run and have a period of 46Å. The films were deposited onto small pieces of 20μm thick polished 100 Si (Virginia Semiconductor) which were

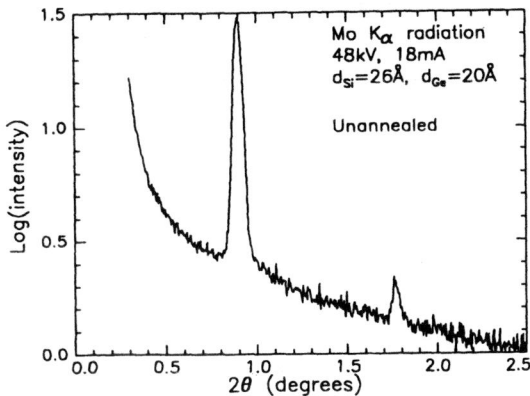

Fig. 1. θ-2θ x-ray scan of a-Si/a-Ge multilayer on 20 μm Si. Sample size is 250 μm X 150 μm.

affixed to a large piece of 0.35mm thick (100) Si backing wafer. Due to their large deposition stresses, the films on the 20 μm thick pieces had a noticeable curvature. In order to do proper x-ray reflectivity studies the samples needed to be flattened. This was accomplished by annealing them in an overpressure of He at 623 K for 3 hours. This was calculated to be sufficient to relieve the stresses as based on studies of relaxation in amorphous films [13]. After the anneal was completed no discernible curvature remained. No noticeable interdiffusion occurred at these temperatures. After this relaxation anneal, the sample was broken up into very small pieces suitable for use in a diamond anvil cell. Typical sample size is approximately 250 μm X 150 μm. These samples were then individually placed on a diamond x-ray mount and low-angle, θ-2θ x-ray scans were taken of the films using Mo K_α radiation in a fixed-anode x-ray system. The Mo radiation and the diamond x-ray mount allow diffraction peaks to be obtained from both the front (film-air) and back (film-Si wafer) sides of the samples. After the initial x-ray scan (Fig. 1), the samples were then annealed in a Merrill-Bassett diamond anvil cell under various pressures at 700 K. Other samples annealed at atmospheric pressure were also placed inside the cell (with Ar gas at one

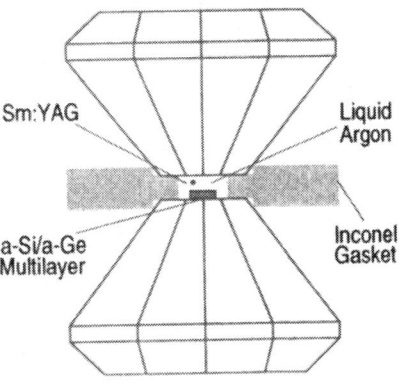

Fig. 2. Enlarged view of DAC sample area.

atmosphere) to ensure equivalent thermal environments during annealing. The annealing "chamber" consists of a 380 µm hole in an Inconel foil that had been pre-indented by the diamonds. The top and bottom of the "chamber" are the opposing diamond culets (1mm diameter). For the high pressure anneals, several small pieces of Sm doped YAG crystal were added to the chamber (see Fig. 2). When irradiated by the 488 nm line of an Ar-ion laser, the fluorescence lines of the Sm:YAG spectrum shift with pressure [14]. The advantage of Sm:YAG over ruby is that the Sm:YAG fluorescence lines are visible and relatively distinct at very high temperatures. The pressure medium used in this experiment is liquid argon, which is both inert and hydrostatic at high temperatures.

RESULTS

A direct comparison was made of the rates of interdiffusion for two samples: one annealed at atmospheric pressure and one annealed at high pressure. The diffusivity during each anneal was determined by measuring the drop in the integrated intensity of the first x-ray harmonic (at $2\theta = 0.9°$). If the composition modulation in the film is described as a Fourier series, the first Fourier component decays in time as [15]:

$$A(t) = A_0 \exp\left(-4\pi^2 \tilde{D}t \middle/ \lambda^2\right) \tag{6}$$

where A is the amplitude of the Fourier component, $\tilde{D}$ is the average interdiffusivity and λ is the multilayer repeat length. Since the intensity, I, of the x-ray harmonic is the square of the amplitude, the diffusivity is obtained by rearranging equation (6):

$$\tilde{D} = -\frac{\lambda^2}{8\pi^2} \frac{\partial \ln(I)}{\partial t} \quad . \tag{7}$$

where I is the intensity of the x-ray reflection.

To ensure that the samples began the experiment in the same structural configuration, they were both annealed together in the cell at atmospheric pressure for approximately 30 minutes at 700K (the temperature of all diffusion anneals). After this anneal, the samples were removed from the cell and their new multilayer peak intensities were measured. A comparison of the decreased intensities of the first harmonics of both films showed that diffusion had indeed occurred to the same degree (see Fig. 3). The samples were subsequently annealed separately, with sample 2 still at atmospheric pressure and sample 1 under high pressure. Following each anneal, the sample was removed from the cell and x-ray peak intensities were measured. In all cases the pressurized sample diffused at a higher rate than the unpressurized one. Pressurized anneals were done on sample 1 until the x-ray intensity dropped below a readily measured value. Pressure anneals were then performed on sample 2 after it had already undergone substantial diffusion at atmospheric pressure (see "d" in Fig. 3). At this stage, the rate of structural relaxation had significantly diminished. The purpose of these later measurements on sample 2 was to check the possible effect of structural relaxation on the activation volume. While the activation volumes in these more relaxed samples were larger than those obtained earlier, the difference is within the uncertainty of the earlier measurements.

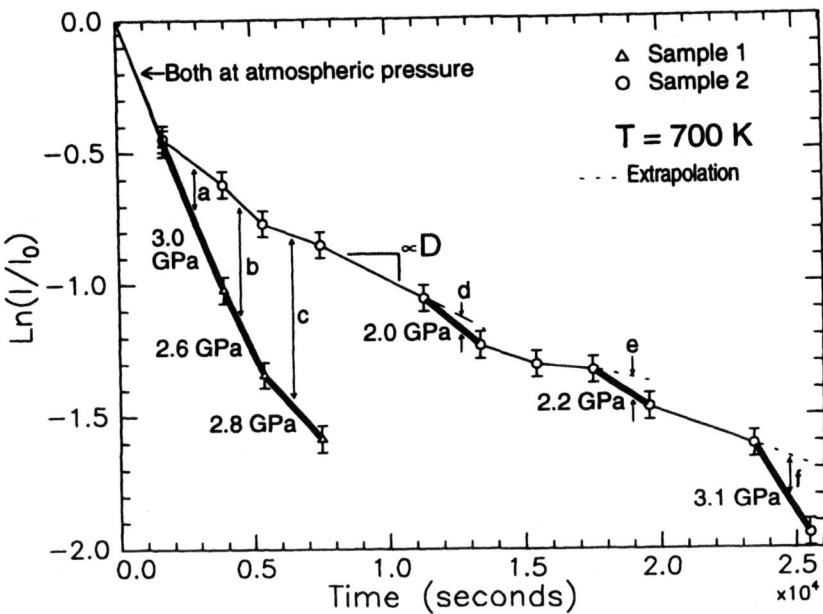

Figure 3. Change in first harmonic x-ray intensity vs. annealing time for various pressures at 700 K. Slope is proportional to diffusivity. Letters 'a' through 'f' indicate where comparisons of the diffusivity have been made to determine an activation volume.

The activation volumes are summarized in the table below. The activation volumes for sample 1 were obtained by comparing the diffusivity during each of the pressurized anneals with the corresponding diffusivity of sample 2 during an anneal of equal duration at atmospheric pressure (a,b,c). Those for sample 2 were obtained by comparing the diffusivity during each pressurized anneal with a linear extrapolation of the diffusivities measured over the previous 60 minutes (d,e,f).

Sample 1		Sample 2	
P (GPa)	ΔV^* (cm^3/mole)	P (GPa)	ΔV^* (cm^3/mole)
3.0	-2.5 ± 0.6 (a)	2.0	-3.0 ± 1.4 (d)
2.6	-1.4 ± 0.8 (b)	2.2	-3.4 ± 0.9 (e)
2.8	-3.9 ± 2.3 (c)	3.1	-3.9 ± 1.3 (f)
ΔV^* (avg.)	-2.6 ± 1.2	ΔV^* (avg.)	-3.4 ± 1.2

DISCUSSION

The average activation volume obtained from this experiment is -0.25±0.11 Ω_{Si}, where the atomic volumes of Si and Ge are 12.0 cm^3/mole (Ω_{Si}) and 13.6 cm^3/mole (Ω_{Ge}), respectively. Because as-made amorphous semiconductor films have large, non-equilibrium point defect concentrations [13], the activation volumes measured in this experiment probably are dominated by the *migration* volume of the defects. Pressure also enhances the rate of solid phase epitaxial growth (SPEG) during crystallization of amorphous Si and Ge [9]. The corresponding activation volumes are -0.28 Ω_{Si} and -0.46 Ω_{Ge}, respectively. Given the similarity of the activation volumes for interdiffusion and SPEG, it seems possible that a similar mechanism governs both processes. The Spaepen-Turnbull dangling bond mechanism for SPEG [10] suggests that the dominant contribution to the measured activation volume is a negative migration volume of dangling bonds located at the crystal-amorphous interface. Correspondingly, it is plausible that diffusion in the bulk of a-Si and a-Ge is controlled by dangling bond migration and that the activation volume that we have measured is that of dangling bond migration. More extensive experiments are necessary to definitively establish the mechanism.

CONCLUSIONS

X-ray diffraction measurements of the rate of decay of composition modulations in a-Si/a-Ge multilayers reveal a pressure-enhancement of the interdiffusivity, characterized by a negative activation volume of -0.25±0.11 Ω_{Si}. This quantity most likely represents the migration volume of the governing defects. Its value is close to those obtained for SPEG in a-Si and a-Ge, suggesting that similar mechanisms govern both processes.

ACKNOWLEDGMENTS

This work has been supported by the National Science Foundation through the Harvard Materials Research Laboratory under contract number DMR-89-20490. One of us (S.M.) has been supported by NSF-DMR-89-13268.

REFERENCES

1. B. Park, F. Spaepen, J.M. Poate, D.C. Jacobson and F. Priolo, J. Appl. Phys., **68**, 4556 (1990).
2. G. Müller and S. Kalbitzer, Non-Cryst. Solids 278 (1977).
3. P.A. Thomas, M.H. Brodsky, D. Kaplan, and D. Lepine, Phys. Rev. B, **18**, 3059 (1978).
4. S.T. Pantelides, Phys. Rev. Lett., **57**, 2979 (1986).
5. P.G. Shewmon, Diffusion in Solids, (J. Williams Book Co. Jenks, OK, 1983) p. 61.
6. M. Werner, H. Mehrer and H.D. Hochhiemer, Phys. Rev. B, **32**, 3930 (1985)
7. A. Antonelli, J. Bernholc, Phys. Rev. B, **40**, 10643 (1989).
8. R.M. Emrick, Phys. Rev., **122**, 1720 (1961).
9. G.Q. Lu, E. Nygren and M.J. Aziz, J. Appl. Phys., **70**, 5323 (1991).
10. F. Spaepen and D. Turnbull, AIP Conf. Proc. **50**, 73 (1979).
11. V.J. Fratello, J.F. Hays, F. Spaepen and D. Turnbull, J. Appl. Phys., **51**, 6160 (1980).
12. F. Spaepen, A.L. Greer, K.F. Kelton and J.L. Bell, Rev. Sci. Inst., **56**, 1340 (1985).
13. A. Witvrouw and F. Spaepen, J. Appl. Phys., **74**, 7154 (1993).
14. N. Hess and D. Schiferl, J. Appl. Phys., **71**, (1992).
15. S.M. Prokes and F. Spaepen Mater. Res. Soc. Symp. Proc. **77**, pp. 305-310 (1987).

ATOMIC MOTION IN AMORPHOUS $Ni_{81}B_{19}$ STUDIED BY REVERSE MONTE CARLO AND MOLECULAR DYNAMICS SIMULATION

BAREND J. THIJSSE, LEON VAN EE, AND JILT SIETSMA
Laboratory of Materials Science, Delft University of Technology, Rotterdamseweg 137,
2628 AL Delft, The Netherlands

ABSTRACT

Molecular dynamics simulations of glassy $Ni_{81}B_{19}$, starting with a configuration obtained by the Reverse Monte Carlo méthod, indicate a calorimetric glass transition at 960 K and point to a significant change in the atomic dynamics between 960 and 1200 K. Above this range, normal liquid-like behavior is found; at lower temperatures, we find a residual diffusivity and cooperative atomic motion. Atomic jumps are processes smeared out in time and space over continuous rather than discrete scales.

INTRODUCTION

Computer simulation studies of atomic motion are vital for understanding structural change in condensed matter, especially when the underlying microscopic processes are difficult to reduce to clear-cut mechanisms like vacancy jumping or hydrodynamic flow. Interesting phenomena of this kind are diffusion and structural relaxation in amorphous solids. Using results from Reverse Monte Carlo (RMC) and Molecular Dynamics (MD) simulation, we investigate in this paper the time and length scales over which atomic motions in an amorphous system correlate, and explore the behavior at elevated temperatures. Rather than working on a simple model system, we make a special effort to simulate an actual metallic glass, $Ni_{81}B_{19}$, to be able to test against experimental results as much as possible. Other authors have approached the same subject.[1-3]

SIMULATION PROCEDURES

In MD studies, one normally does not need a particular initial configuration, since the "correct" atomic-scale structure develops as the simulation proceeds. However, for amorphous systems well below the glass transition, where dynamics are slow, it is questionable whether the usual method of quenching a liquid to low temperatures leads to a realistic starting configuration. The quench rates attainable by MD are still orders of magnitude too high in comparison with laboratory conditions. For this reason we have used as the starting point for MD a room-temperature configuration of $Ni_{81}B_{19}$ obtained by RMC.[4] Since this is, in a certain respect, the most "objective" configuration, and since —by the very nature of the RMC technique— its partial Radial Distribution Functions (RDF's) agree extremely well with experimental results from neutron diffraction, the conditions for a rapid onset of realistic MD were expected to be most favorable. It remained to be seen, however, if the MD potentials would leave the RMC configuration essentially unaltered; this served as a cross-validation test of RMC and MD.

The MD runs were carried out using the constant-density velocity-Verlet algorithm with an integration time constant Δt of 3.5 fs; the temperature was kept constant to within 1% by velocity scaling, using a time constant of $5\Delta t$.[5] The system consisted of 1920 atoms (1555 Ni and 365 B) in a cubic box of 26.51 Å edge length under periodic boundary conditions (number density $\rho = 0.103$ Å^{-3}). As effective pair potentials, we used the ones recently calculated by Hausleitner and Hafner,[6] but they were modified in two ways: i) $V_{Ni-Ni}(r)$, $V_{Ni-B}(r)$ and $V_{B-B}(r)$

65

Table I
Equilibration and
production times

T^*	t_{eq}^*	t_{pr}^*
0.3	700	690
0.6	500	700
0.9	300	700
1.2	75	325
1.5	50	75
1.8	50	75
2.0	50	335

were multiplied by 1.61, 0.68, and 0.63, respectively, to obtain the well depths 0.14, 0.27, and 0.10 eV found by Li et al.,[7] and ii) the attractive parts of $V_{Ni-B}(r)$ and $V_{B-B}(r)$ were replaced by Lennard-Jones 6-12 tails. These modifications were necessary to avoid the almost instantaneous phase separation and unphysical diffusion behavior that were found when the original potentials were used.[8] Cut-off distances were taken as 6.7 Å for Ni-Ni and Ni-B, and 8.1 Å for B-B. Contrary to many other studies on the glass transition,[9-12] the pair potentials in this work are strongly non-additive (well positions 2.68 Å for Ni-Ni, 2.26 Å for Ni-B, and 3.34 Å for B-B — 1.84 Å for B-B would be "additive"), have different well depths, and have repulsive parts of different softness (the ratios of zero-crossing distance and well position are 0.87, 0.82, and 0.75, respectively). In addition, the density of the present system is high.

In this paper, reduced quantities, denoted by an asterisk, are expressed in units based on the original Ni-Ni pair potential: 1012 K for temperature, 0.087 eV for energy, 2.68 Å for length, and 0.71 ps for time. The simulation times are listed in Table I; all runs were started with the RMC configuration with zero velocities, which was immediately brought to the required temperature. Results shown here are averages over production time.

CONSISTENCY OF RMC AND MD RESULTS

Fig. 1 shows the effects of applying MD at room temperature ($T^* = 0.3$) to the RMC configuration. One sees that the agreement between the MD and the RMC results remains satisfying. There are a few imperfections: the nearest-neighbor peaks in the Ni-Ni and Ni-B partial RDF's

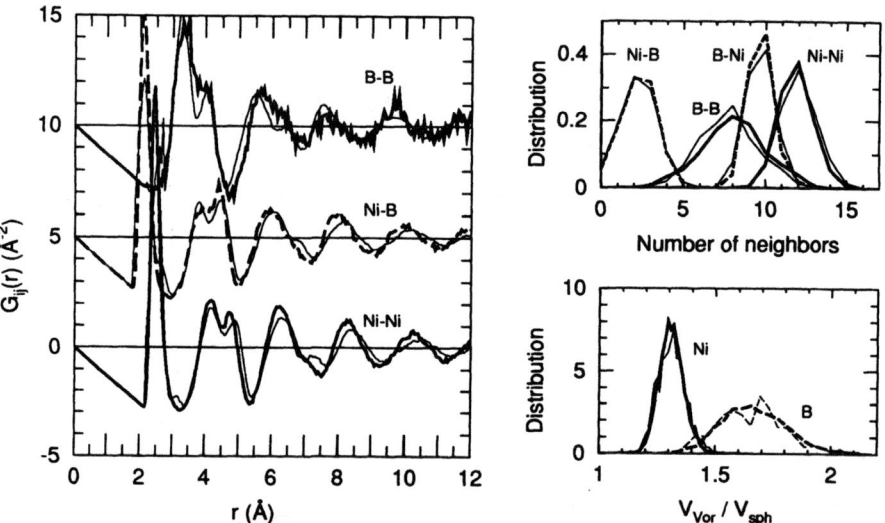

Figure 1. Partial RDF's, neighbor-number distributions, and normalized Voronoi volume distributions for the Molecular Dynamics system at T^*=0.3 (heavy curves) and for the starting configuration, obtained by Reverse Monte Carlo (light curves). Neighbors are defined as atoms within a distance of 3.3, 3.0, and 4.6 Å for Ni-Ni, Ni-B, and B-B pairs, respectively. Voronoi volumes were calculated using the radical-plane method with effective hard-sphere radii 1.26 Å and 0.85 Å for Ni and B; V_{sph} denotes the corresponding hard-sphere volume. Note that the boron atoms fit more loosely in their Voronoi volumes than Ni.

become sharper, the secondary maxima at ≈ 3 Å and ≈ 7 Å disappear, and a small mismatch in the medium-range order oscillations originates; however, the remaining characteristics are left almost entirely intact. It is particularly reassuring that the amplitude decay of the oscillations, the shapes of the split second peaks, and the absence of B-B nearest neighbors are well reproduced; normal temperature effects aside, the same level of coincidence between MD and RMC was observed at higher temperatures. This agreement between MD and RMC, which was also found in the structure factors (not shown), is crucial, since —as pointed out above— it implies that the MD results are in accordance with the experiment as well.[4] The correspondence between MD and RMC distributions of Voronoi volumes and neighbor numbers is also adequate. All this is substantial support both for the RMC method and for the pair potentials used. We consider the observed discrepancies in the partial RDF's a subject for further potential refinement, but not prohibitive for the current study of key aspects of microscopic behavior.

As an independent test, the system was melted at $T^* = 2.0$ and quenched to $T^* = 0.3$ at 1.4×10^{13} K/s. As a result, the partial B-B structure factor had changed considerably, the most significant feature being the appearance of a strong peak at 0.4 Å^{-1} which was absent in the RMC-started MD run. Also the velocity autocorrelation function of the boron atoms showed a markedly increased sustaining of the vibrational oscillations. These facts point towards a change of structure and dynamics as a result of the melt-and-solidify cycle, which indicates that at least in this case, the RMC configuration is a more reliable starting point for low-temperature MD than a quenched liquid.

In Fig. 2 it can be seen that the system is solid-like below $T^* \approx 0.95$ (960 K) and liquid-like above, as is evident from the change in the specific heat (from $2.9\,k$ to $3.7\,k$) and the T-dependence of the Wendt-Abraham ratio R. The transition can be identified as the "calorimetric" glass transition, because the quantities displayed reflect the structure rather than the dynamics of the system. The order-of-magnitude of the transition temperature is correct ($T_g \approx 680$ K experimentally). Interestingly enough, no clear transition point for R was reported for the simulation of Ni$_{80}$P$_{20}$.[1] The glass transition during the quench experiment was found to occur at a somewhat higher temperature ($T^* \approx 1.04$).

ATOMIC MOTION

Van Hove function and diffusion coefficient

Whereas the mean-square displacement of an atom inside the cage formed by its neighbors takes only a time interval $\tau^* \approx 0.25$ to reach its saturation value ($<\Delta r^{*2}>_{cage} \approx 0.023\ T^*$ for Ni and $\approx 0.033\ T^*$ for B), the escape rate out of this cage is extremely slow at low temperatures. This is evident from Fig. 3, in which the self part of the Van Hove probability function $P(\Delta r, \tau)$, i.e. the probability of finding at time τ an atom at a distance Δr away from its position at time 0, is shown for two temperatures. The gradual broadening of P illustrates that the system is not completely frozen below T_g; this is true also for $T^* = 0.3$. It should be stressed that at no temperature do we find any secondary maximum in P indicative of a preferred jump distance such as $\Delta r^* \approx 1$, even if we plot distributions of "jump"

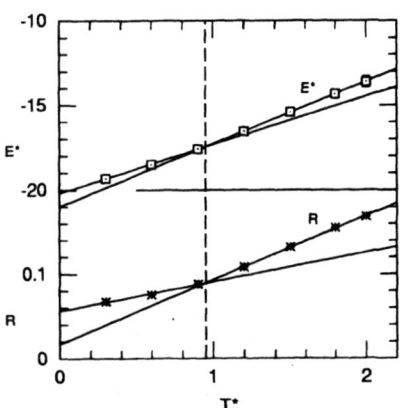

Figure 2. Total energy per atom E^* and Ni-Ni Wendt-Abraham ratio $R = g_{min}/g_{max}$, the ratio of the first minimum and first maximum of the pair correlation function, versus temperature. Lines are least squares fits. Dashed line shows calorimetric glass transition.

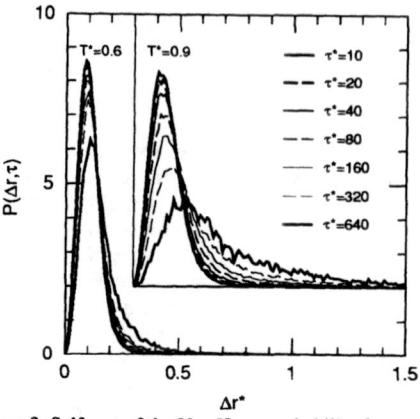

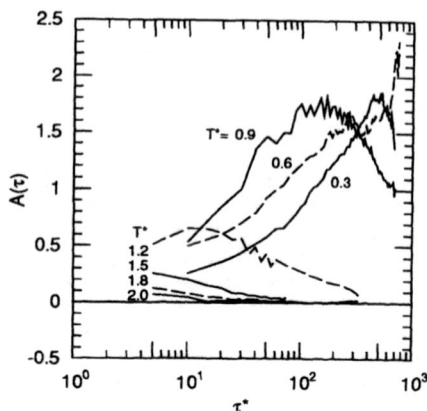

Figure 3. Self part of the Van Hove probability function $P(\Delta r,\tau)$ for temperatures $T^* = 0.6$ and 0.9. Ni atoms only.

Figure 4. Non-Gaussian parameter $A(\tau)$ for all temperatures investigated. Ni atoms only.

distances (see fig. 6 for a definition) instead of plain "displacement" distances. This suggests that at all temperatures there is a continuous spectrum of jump distances available to the atoms, rather than just one or a few single values such as in a crystal. The picture that emerges is that the atoms in a disordered medium have no well-defined "target positions" at their disposition, but can only "wriggle" their way out of their neighbor cage —with chance help from others. It is not clear to what extent differences in pair potentials and density are responsible for the fact that in some other work[1,3,9,10] a peak at $\Delta r^* \approx 1$ is reported.

Theoretically the Van Hove function has a Gaussian shape in two limiting cases: if the atoms are oscillating in harmonic potential wells, and in the case of hydrodynamic, purely uncorrelated diffusion. Hence the Non-Gaussian parameter $A(\tau) = \frac{3}{5} <\Delta r(\tau)^4> <\Delta r(\tau)^2>^{-2} -1$ can be used as an indicator to determine if an atom's history, beginning at an arbitrary moment, comprises a full transition from a pure, local vibration —i.e. $A = 0$ at very short times $\tau^* \ll 1$— to a state of total memory loss (again $A = 0$), and if so, at which time scale this happens.[11] Fig. 4 shows $A(\tau)$ for the temperatures indicated. At high temperatures this transition is rapidly completed, as can be expected for a liquid. At $T^* = 1.2$ the dynamics become visibly slower, while around and below T_g the system is very sluggish, and the atomic collisions remain correlated over the time span investigated.

This change in dynamics is also seen in Fig. 5, where the mean-square displacement is plotted versus time. For $T^* \geq 0.9$ we find regular diffusive behavior. The Ni diffusion constant $D = \frac{1}{6} \lim_{\tau \to \infty} d<\Delta r(\tau)^2>/d\tau$ has a non-Arrhenius T dependence; a Fulcher-Vogel fit[12] with $T_0^* = 0.30$ (300 K) and $B^* = 5.5$ (5600 K) yields the best results. At temperatures below T_g, the situation is less clear.

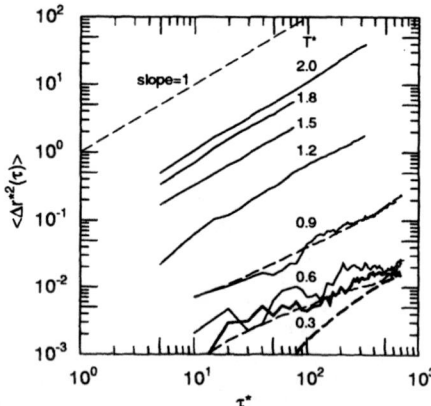

Figure 5. Mean-square displacement since start-of-production time for all temperatures investigated (Ni atoms only). The in-cage contribution $0.023\,T^*$ has been subtracted. Heavy lines are results for $T^* = 0.3$. Dashed lines denote averages over production time.

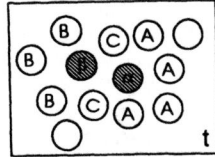 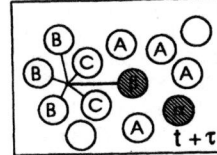

Figure 6. Detection of a two-atom jump-chain between times t and t+τ. Atom β is said to "follow" atom α and form a chain, if during time τ: (i) β has *gained* more than three original neighbors of α (i.e. A and C atoms) as its new neighbors; (ii) α and β have remained neighbors; and (iii) α and β have "jumped" more than a distance $\Delta r^* = 0.5$. A "jump" (shown for β as heavy line) is defined as the distance moved away from the center of the original neighbors (i.e. B and C atoms). An atom is the head of a chain when it doesn't "follow" any other atom, and the tail when it isn't "followed".

The difference between the production-time averages (dashed lines) and the plain mean square drift since the beginning of the production stage indicates that the system is relaxing. On the basis of the present data it is not easy to say whether this is caused by an artefact of the simulation —i.e. the equilibration stage still being unfinished— or by physically relevant structural changes. However, Fourier analysis of the time-dependence of the potential energy shows that the spectra are qualitatively very similar for all temperatures $T^* = 0.3$—1.2; this suggests that it is a real effect. In any case, a certain residual diffusivity is observed below T_g. As was also suggested by others,[3,13] cooperative atomic motion may be connected with this phenomenon. It is probably also related to the result in Fig. 4 that $A(\tau)$ is strongly positive around and below T_g, meaning that collisional-memory effects lead to a preference of larger atomic displacements at the cost of smaller ones.

Jump chains

To investigate whether a special kind of cooperative motion, namely atoms jumping as a chain,[3] occurs in our model of $Ni_{81}B_{19}$, series of configurations taken at $\tau^* = 10$ intervals were examined. We searched for chains using the method shown in Fig. 6. After identifying the atoms involved, the MD step was re-run with detailed output. An example for $T^* = 0.9$ is given in Fig. 7. It shows a 6-atom chain, of which the atoms advance almost simultaneously. The typical duration of the "jump" of an

Figure 7. Example of a chain of six atomic jumps taking place at $T^*= 0.9$. The vertical axis displays the positions of the six atoms projected on the head-to-tail axis at $\tau^*=0$. For comparison, the inset shows two successive vacancy jumps in a Ni fcc crystal ($\rho^* = 1.70$) at $T^*= 2.0$, calculated using the same Ni-Ni pair potential.

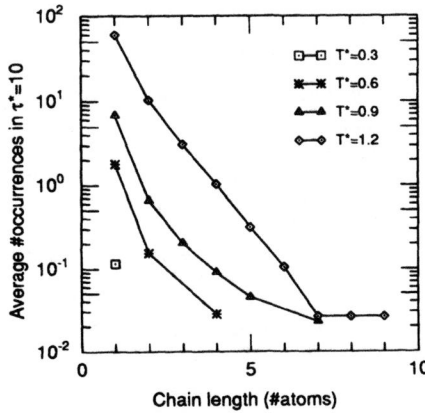

Figure 8. Chain length distribution for $\tau^*=10$ at temperatures $T^*= 0.3$—1.2.

69

atom is here $\tau^* \approx 2$, much longer than the mean vibration period, and much longer than a vacancy jump in a crystal (see inset). Although this example involves an almost linear chain with a short period of activity, folded chains and more gradually proceeding chains were also found. Chains are by no means exceptional vis-à-vis single atom jumps. As is shown in Fig. 8, their length distribution is a continuous function that includes isolated "jumps" (length=1). Since it also hardly changes shape with temperature, this kind of distribution appears to be an intrinsic property of the sub-T_g dynamics in amorphous solids.

CONCLUSIONS

The present results indicate a calorimetric glass transition at $T^* \approx 0.95$ and a transition region of the atomic dynamics from $T^* \approx 0.9$ to 1.2. We find cooperative atomic behavior in the amorphous solid state taking place at larger-than-atomic time and length scales. The distributions involved appear to be continuous rather than discrete. Finding out more about their properties and the conditions under which they emerge remains an ongoing challenge for theoreticians and computer physicists.

ACKNOWLEDGMENTS

Thanks are due to Peter Duine for stimulating discussions. This work is part of the research program of the Stichting voor Fundamenteel Onderzoek der Materie (Foundation for Fundamental Research of Matter), and was made possible by financial support from the Nederlandse Organisatie voor Wetenschappelijk Onderzoek NWO (Netherlands Organization for Scientific Research).

REFERENCES

1. L.J. Lewis, Phys. Rev. B **44**, 4245 (1991).
2. J. Hafner, M. Krajcí, and Ch. Hausleitner, in Proc. ILL/ESRF Workshop on Methods in the Determination of Partial Structure Factors, edited by J.B. Suck, P. Chieux, D. Raoux, and C. Riekel (World Scientific Publishing, Singapore, 1993), p. 179.
3. W. Frank, A. Hörner, P. Scharwaechter, and H. Kronmüller, to appear in Proc. 8th Int. Conf. on Rapidly Quenched Metals, Sendai, 1993; A. Hörner, PhD thesis, Max-Planck-Institut für Metallforschung, Stuttgart (Germany), 1993.
4. E.W. Iparraguirre, J. Sietsma, B.J. Thijsse, and L. Pusztai, Comp. Mat. Sci. **1**, 110 (1993).
5. M.P. Allen and D.J. Tildesley, <u>Computer Simulation in Liquids</u> (Oxford University Press, 1989).
6. Ch. Hausleitner and J. Hafner, Phys. Rev. B **47**, 5689 (1993).
7. J.C. Li, N. Cowlam, and F. He, J. Non-Cryst. Solids **112** (1989) 101.
8. B.J. Thijsse and J. Sietsma, to appear in Proc. 8th Int. Conf. on Rapidly Quenched Metals, Sendai, 1993.
9. J.-L Barrat, J.-N. Roux, and J.-P. Hansen, Chem. Phys. **149**, 197 (1990).
10. G. Wahnström, Phys. Rev. A **44**, 3752 (1991).
11. Y. Hiwatari, H. Miyagawa, and T. Odagaki, Solid State Ionics **47**, 179 (1991).
12. D. Thirumalai and R.D. Mountain, Phys. Rev. E **47**, 479 (1993).
13. G. Pastore, B. Bernu, J.-P. Hansen, and Y. Hiwatari, Phys. Rev. A **38**, 454 (1988).

SOLID STATE ORDERING OF WHOLLY AROMATIC COPOLYESTERS OF RANDOM MONOMER SEQUENCE

J. BLACKWELL, A.-I. SCHNEIDER AND C.M. McCULLAGH
Department of Macromolecular Science, Case Western Reserve University,
Cleveland OH 44106-7202

ABSTRACT

X-ray diffraction and computer molecular modeling methods are being used to investigate the ordering of random sequence copolyesters when cooled for the nematic melt. Data will be presented for the copolymers prepared from p-hydroxybenzoic acid (HBA) and 6-hydroxy-2-naphthoic acid. The non-periodic diffraction patterns arising from the random monomer sequences are also indicative of the presence of limited three dimensional order. For both polymer systems, we can generate the qualitative feature of the x-ray data using models in which short non-identical segments of 10-12 monomers on adjacent chains are approximately in register at their center. Molecular mechanics calculations show that these non-identical sequences can be packed on the observed hexagonal lattice with only small energy differences when compared to analogous homopolymer structures. Data will also be presented showing how x-ray diffraction can be used to follow transesterification of copoly(HBA/HNA) in the nematic melt.

Introduction

X-ray fiber patterns for thermotropic wholly aromatic copolyesters are particularly interesting because they demonstrate the existence of three dimensional order despite the fact that the chains have random monomer sequences. The most studied polymer in this class is the copolymer of p-hydroxybenzoic acid (HBA) and 2-hydroxy-6-naphthoic acid (HNA):

Various compositions have been studied in our laboratories, where in this paper we will deal only with the 75/25, 60/40 and 30/70 compositions. It has been previously shown that the meridional intensity maxima are not orders of a single repeat, and their positions shift continuously depending on the monomer ratio.[1-3] The meridional peak positions are reproduced with high accuracy by models consisting of parallel chains of completely random sequence,[4] which also means that the peak positions are a measure for the composition of these copolymers. A series of studies has been performed in our and other laboratories to analyze the structure within the polymer chains and the packing of these random chains,[4-7] and also the behavior when annealed and blended.[8-10]

In this paper we use molecular modeling to investigate the chain packing for the 75/25 composition, in comparison with the homopolymers. Our analyses have included the possibility of non-random monomer sequences in these copolymers. The main focus is on transesterification in blends of copoly(HBA/HNA), where two random copolymers undergo interchange and pass through non-random statistics to the random copolymer of intermediate composition. Changes in monomer sequence distribution and overall composition as the reaction progresses are detected by gradual shifting and merging of the first meridional maxima of the starting copolyesters.

Wide angle x-ray data for the as-spun fibers of the 75/25 copolymer show that extended chains are packed on an orthorhombic (pseudo-hexagonal) lattice with dimensions a = 9.18Å and

Mat. Res. Soc. Symp. Proc. Vol. 321. ©1994 Materials Research Society

b = 5.30Å.[4,8] Three dimensional order is present in the solid state, as indicated by the existence of an intense Bragg maximum on the first layer line. The intensity distribution on the layer lines is consistent with an approximately body-centered structure, in which successive aromatic units along the chain are inclined with respect to each other by ~60°, and adjacent chains are staggered by one monomer. Windle and coworkers[6,7] have suggested that several chains with identical non-repeating sequences segregate to form "non-periodic lattices", and that the Bragg scattering results from the average of a number of such arrays. The view in this laboratory has been that such a segregation is unlikely to account for the large lateral crystalline sizes and degree of crystallinity found in these materials. It was shown that the diffraction can be predicted by short, non-identical sequences approximately registered with their center on a layer. All such simulations require approximations and averaging. In the present work we study the intermolecular interaction of *actual* non-identical chains: we address the question whether or not non-identical chains can be packed according to the lattice dimensions defined by the x-ray data. This work follows on from previous molecular modeling studies of conformation,[11] stiffness[12] and packing[13,14] for LC polyesters.

Parameters determined previously from our x-ray data[1,15,16] were used to create arrays of chains. Strong equatorial maxima at d = 4.55, 2.63, and 2.29Å can be indexed by an hexagonal lattice with a = 5.30Å. The strong reflection on the first layer line at d = 3.3Å is indexed hk = 21 for a two chain orthorhombic (pseudohexagonal) unit cell with the base plane dimensions a = 9.18Å, and b = 5.30Å. This structure is similar to that of the high temperature form of the homopolymer poly(HBA),[8,17,18] which contains two monomer units in the c axis repeat of 12.5Å; the planes of successive aromatic units are mutually inclined by ~60°, and the monomers on adjacent chains are packed in a herring bone array. From the half width of the meridional maximum at d = 2.07Å[9] a correlation length of 70-80Å is determined depending on the polymer composition, corresponding to an extended chain conformation of 10-12 monomers. Line broadening applied to the first equatorial maximum at d = 4.55Å yields crystallite dimensions of ~70Å perpendicular to the chain axis, which indicates that the arrays consist of ~200 chains. However, in most of our calculations we have been limited to considering 19 chains of 12 monomers, due to memory and computing time limitations.

Transesterification occurs readily in the isotropic melt[19,20] and has also been reported for aromatic polyesters in the nematic melt.[21,22] Recently studies have examined transesterification between polyesters utilizing neutron scattering.[23-25] This method detects changes in block size, but does not address monomer statistics along the polymer chain as the reaction progresses. We present a study of transesterification between copolyesters of HBA/HNA of different compositions in which the reaction is monitored by x-ray diffraction.

Experimental

Atomic coordinates for the monomer residues HBA and HNA were derived from standard bond lengths and angles. The ester torsional angles (τ_1, τ_2 and θ; see below) taken from model compounds and energy minimizations indicate that the ester group is approximately coplanar with the aromatic plane attached to the carbonyl. In addition, the planes of successive aromatic moieties are inclined at about 60°, while alternate aromatic moieties are parallel to each other. The HNA residue was restricted to the shorter *cis* conformation, as suggested by the mechanical properties[26] and is consistent with the axial advance per residue as refined from the x-ray analysis.

Chains of 12 residues were constructed from the monomer models, with random sequences selected such that the overall composition in the array is 75/25 HBA/HNA. Two conformations for the chains were considered: conformation I, with symmetric disposition of the carbonyls to one side of the chain, where the axial projection can be described as a "butterfly conformation", and conformation II with random disposition of the carbonyls to both sides of the chain, described as a "distorted butterfly". In a particular array, all the chains had either conformation I or conformation II. The origin of the x,y projection was placed at a node of a hexagonal network with a = 5.3Å. The packing of the chains is described by an orthorhombic network with a = 9.2Å and b = 5.3Å, where each cell contains projections of two cells, the aromatic moieties in one plane being mutually inclined at 60°. The chains in an array were put in register by setting the ester oxygen of the central ester group at z = 0 (deviation from this register is represented by a standard deviation σ^4). Two different ways of packing were considered, which are specifically visible for the chains of conformation I: the disposition of the carbonyls to one side of the chain is the same within one ac layer, but opposite to the neighboring layer (type A packing), and the disposition of the carbonyls is the same for all chains in the array (type B). Likewise rotation of the type II chains results in different structures, although this is less obvious when viewed in projection. Arrays of 19 chains were submitted to energy minimization, using the Tripos SYBYL package loaded on a Silicon Graphics 4D/220. Further details of the minimization procedure are described in detail in reference 5. X-ray diffraction by the arrays was simulated by calculating the cylindrically averaged intensity transform I(R,Z) via

$$I(R,Z) = \sum_{n=-n'}^{n'} \left(\left| \sum_{j=1}^{N} f_j J_n \left(2\pi R r_j\right) \exp\left(2\pi i Z z_j - n\phi_j\right) \right| \right)^2$$

f_j is the atomic scattering factor of the jth atom in the array, with cylindrical polar coordinates r_j, θ_j, z_j; R, Z are the reciprocal space coordinates of the cylindrically averaged Fourier intensity transform I(R,Z); J_n represents the cylindrical Bessel functions of order up to n' = ±10. This expression was used to calculate the intensity on the equator, and on the first layer line at Z = 0.078Å^{-1}.

The x-ray scattering for the 19 chain arrays was predicted along the meridional direction, and along the equator and first layer lines. The arrays are small in the lateral direction, being only 5 chains wide across the diameter. Consequently the primary peaks are broader than those observed, and there are also significant subsidiary maxima. In the axial direction, the layer lines are shifted slightly due to the fact that the averaging is over small number of sequences. We also simulated the diffraction patterns for larger arrays of 127 non-identical chains of four monomers. These slices of the structure are sufficiently thick to generate the intensity on the first layer line, but are too large for optimization by energy minimization.

Specimens of 75/25 and 30/70 copoly(HBA/HNA) were provided by Hoechst Celanese Co., Summit, NJ, in the form of melt extruded pellets. Fibers of these copolyesters or their blends were drawn by hand from the nematic melt. The melt blend of 30/70 and 75/25 copoly(HBA/HNA) was obtained by first weighing the two copolymers in proportion to give an overall monomer mole ratio of 60/40. The mixture was melted at 310°C and stirred thoroughly by hand for 5 minutes. Fibers were drawn from the nematic melt and aligned parallel to each other in bundles for mounting on a diffractometer.

Compression molding of chopped fibers of the blend was performed using a Carver Laboratory Press. The platens were preheated to 315°C after which the sample was inserted and allowed to equilibrate without the application of pressure for 10 min. Specimens were subjected to 1800 psi pressure at 315°C for specified lengths of time and then cooled to room temperature within 20 min. The samples were then remelted, and fibers were drawn as described above.

θ/2θ diffractometer scans along the fiber axis direction were obtained using either a Philips P3100 diffractometer or a Rigaku Rotaflex diffractometer in transmission mode, with the incident and detector slits set at 0.15°. Scans over 2θ = 3-50° were performed at 0.01° increments using 10s counting intervals. Higher resolution scans of the 2θ = 7-17° region utilized a 0.02° increment and 160s counting interval. Resolution of the component peaks was accomplished using a Gaussian-Lorentzian curve fitting program.

Differential scanning calorimeter (DSC) scans were obtained using a Perkin Elmer DSC-7. Data were recorded in the temperature range of 150-350°C, with a heating/cooling rate of 20°C/min. The temperature of the solid-nematic transition, T_m, was determined from the second heating scan.

Results and Discussion

a) Energy Minimization

Twenty starting models, each of different chains of 12 monomers, with an overall 75/25 HBA/HNA monomer ratio were subjected to energy minimization. The models were selected such that mixtures of different sequences, different starting conformations (I or II) and different packing (A or B) could be considered. For all the starting models the calculated potential energy was very high (in excess of 10^4 kcals) due to overlap of atoms in different chains. The minimized models all had potential energies in the range of -1300 to -1450 kcals. The same calculations were performed for models of homopoly(HBA) using the lattice parameters for the analogous high temperature form. The minimized models for these structures had energies in the range -1200 to -1300 kcals, values which are comparable to those obtained for the copolymer. They are in fact higher energies, which is due simply to the different number of atoms in each of these models: 3021 for the homopolymer versus 3363 for the copolymer. A similar setup for homopoly(HNA) contained 4389 atoms, and had an energy of -1780 kcals after minimization. The important point is that the energies for the homopolymer and copolymer models are not greatly different, which shows that the random chains of the copolymer can be packed on the observed lattice with relatively little difficulty. The differences between

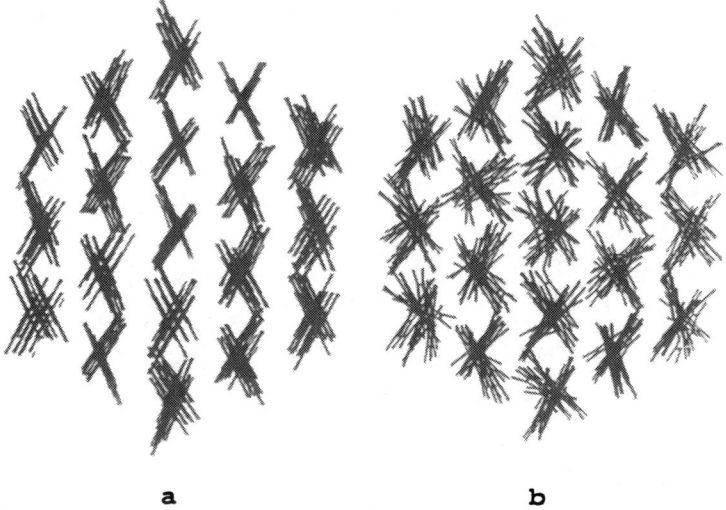

a   b

Fig. 1   c-axis projection of 19 chains of 12 residues with conformation I and type B packing:
a) starting array and b) energy minimized array.

the starting and final models are almost entirely in the torsion angles τ_1 and τ_2. The averages for these angles are close to the values found for model compounds. A survey of the torsion angles for the inner 7 chains of 20 arrays yields the following averages (and standard deviations):

$\tau_1 = 0°$ (10°), $|\theta| = 175°$ (4°) and $|\tau_2| = 70°$ (11°). Positive and negative values for τ_2 and θ are almost equally abundant. The conformational changes that occur on minimization are seen most easily in the c-axis projections (Fig. 1). The aromatic units, which appear as straight lines, are mutually inclined by ±60° to start with, but adjust to a distribution of orientations, resulting in an almost cylindrical projection. However, this is apparent only for projections involving several residues: the standard deviations indicate that the deviations from the one residue to the next are relatively small.

Starting models were arranged in register at their centers, such that the central ester oxygen was at $z = 0$. Minimization resulted in only small axial shifts, with a standard deviation of 0.2Å about the origin plane. This is not surprising since, even though better local packing might be achieved by shifting the chain by one monomer along the z axis, such a shift is unlikely to occur during static molecular mechanics modeling, given the cooperative conformational changes that would be necessary to move the chain in a screw-like manner. Also, registration is favored in a model consisting of finite chains of twelve monomers each, so as to avoid exposure of chain ends effects. However, it is unlikely that chain shifts can be simulated without dynamic modeling, which has not been performed with these models due to computational limits.

b) X-ray Diffraction Simulation

For comparison of the different types of conformation and packing, the results from the larger systems containing 127 tetramer chains will be shown here. Fig. 2 and 3 show the calculated intensity on the equator and on the first layer line, respectively. While the equatorial

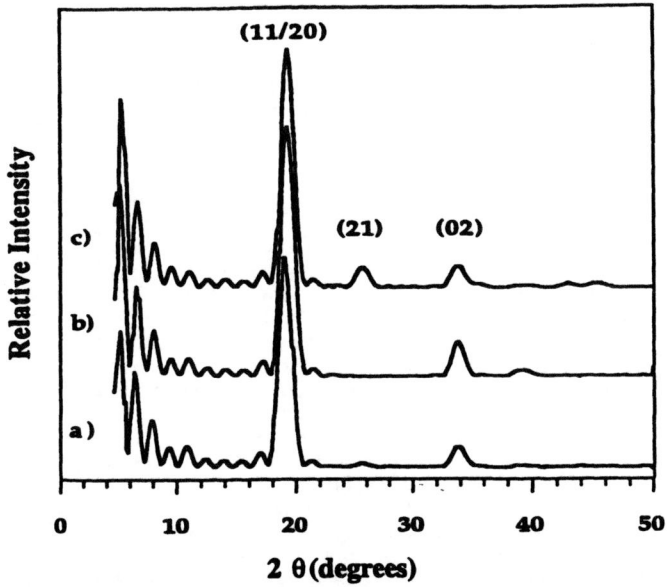

Fig. 2 Calculated data on the equator for unoptimized arrays of 127 tetramers for
a) type A packing of chains with conformation II, b) type A packing of
chains with conformation I, and c) type B packing of chains with
conformation I.

data are basically the same for the three types of arrays (conformation II with packing A, and conformation I with packing A and B), the main difference is observed in the scattering along the

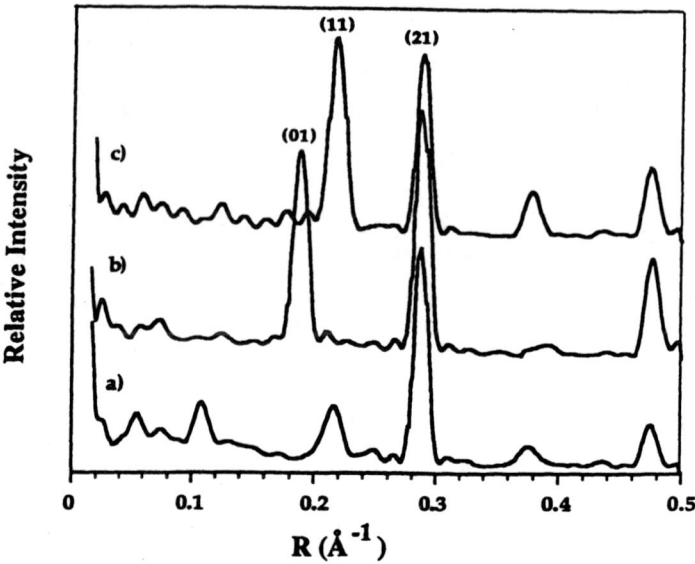

Fig. 3 Calculated data on the first layer line for unoptimized arrays of 127
tetramers for a) type A packing of chains with conformation II,
b) type A packing of chains with conformation I, and c) type B packing
of chains with conformation I.

first layer line. The best agreement is observed for the type II conformation, which shows a
random orientation of the carbonyl vectors. We predict high intensity only for the hk = 21
reflection on the first layer. The models containing chains of type I conformation gave this
reflection and also an appreciable hk = 01 reflection for type A packing, and hk = 11 reflection
for type B packing. These data therefore favor the more disordered type II conformation for the
quenched structure of this copolymer. We could not minimize these layer arrays. However, the
minimized 19 chain arrays also predicted the observed intensity on the first layer line.

The chains in the optimized arrays had small axial shifts of the origin ester oxygen about the
register plane, described by the standard deviation $\sigma = 0.2\text{Å}$. Earlier estimates from the x-ray
data set $\sigma \approx 2.0\text{Å}$, based on the distortions necessary to eliminate Bragg maxima on all but the
first layer line. Within the range of $\sigma = 0.2$-2.0Å, the main effect is to reduce the intensities of
the Bragg intensities on the first layer line; the broadening effects are secondary. Consequently,
the different σ values do not affect the above conclusions.

All the arrays of 19 chains reproduce approximately the positions of the main meridional
maxima. The deviations are within the range expected for use of a small sample of the possible
sequences. The intensities match reasonably well, except for the first maximum, which is
predicted to be much weaker than is detected by the diffractometer. However, the diffractometer
data is for a band on either side of the median, whereas the calculated data is I(0,Z). Crystallinity
has the effect of "concentrating" the layer line intensity towards the meridian, and this effect is
most appreciable for the first maximum, with the result that the observed peak in the
diffractometer scan is enhanced relative to I(0,Z). It is interesting that for the analogous
unoptimized array of antiparallel chains, the first maximum on the meridian is predicted to be
extremely weak and essentially unobservable. It is somewhat enhanced for the optimized
antiparallel structure, in which the axial registration is less perfect, but it is still very weak. To

some extent this observation argues against a regular alternating antiparallel array of chains, and presently we favor random packing of up and down chains.

c) Transesterification

The number and positions of the meridional maxima of HBA/HNA copolymers depends on the comonomer ratio. The first meridional maximum, corresponding approximately to the average monomer repeat, occurs at 7.91Å and 6.67Å for the 30/70 and 75/25 copolymers, respectively. In physical mixtures these two peaks can be clearly resolved (Fig. 4(a)). Upon

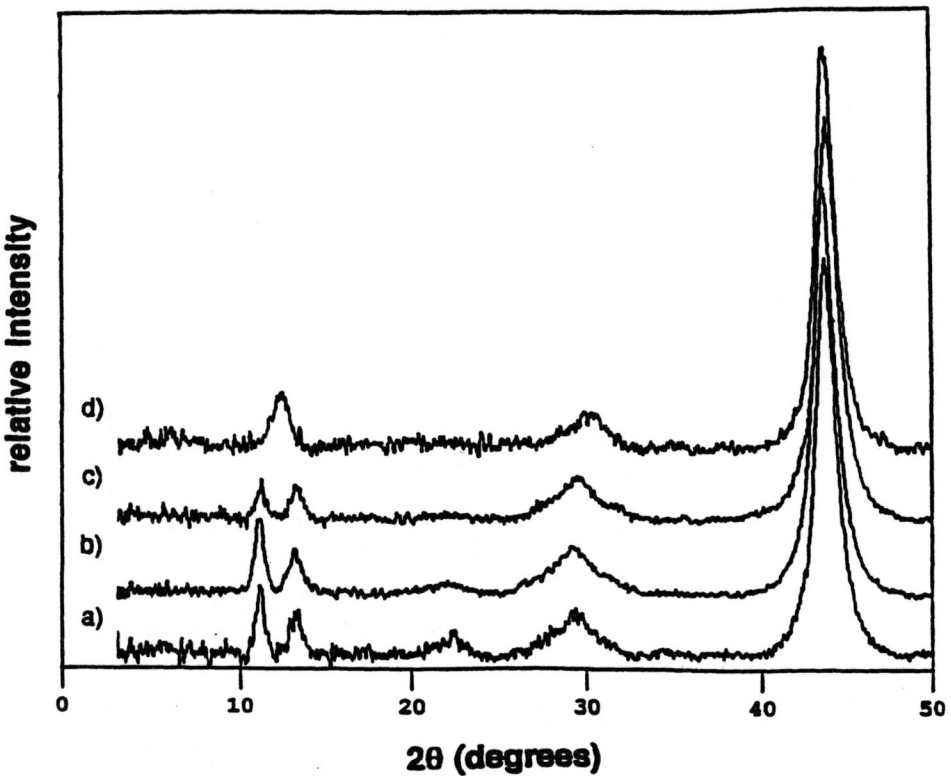

Fig. 4 θ/2θ scans along the fiber axis direction for a) a physical mixture of melt-spun fibers of 75/25 and 30/70 copoly(HBA/HNA) with overall monomer mole ratio of 60/40 b) the addition of scans for 75/25 and 30/70 copoly(HBA/HNA) in weight proportion to give a 60/40 monomer mole ratio c) melt blended 75/25 and 30/70 copoly(HBA/HNA) and d) melt blended 75/25 and 30/70 copoly(HBA/HNA) after compression molding at 315°C for 60 min.

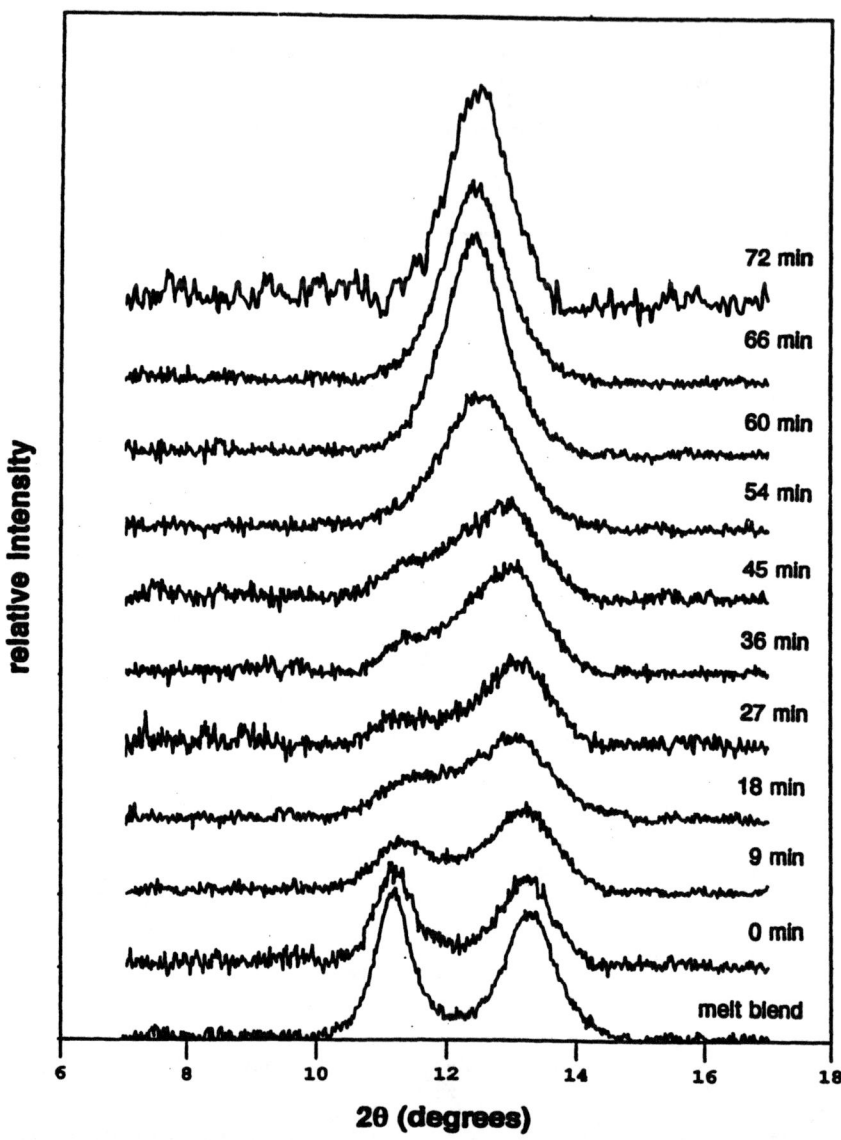

Fig. 5 θ/2θ meridional diffractometer scans of the 2θ = 7-17° region for the melt blend of 75/25 and 30/70 copoly(HBA/HNA) after compression molding at 315°C for the periods of time shown.

melt blending, T_m is greatly reduced from 303° and 288°C for the 30/70 and 75/25 copolymers, respectively, to 227°C. Despite physical co-mingling at the molecular level, the two initial maxima are still observed by x-ray diffraction (Fig. 4(c)).

Blending 75/25 and 30/70 copoly(HBA/HNA) appears to result in a compatible blend. The x-ray data contain the meridional maxima of both of the starting copolymers, indicating that the nearest neighbor statistics are unaffected. In addition, the blends shows a single solid state-to-nematic transition at 227°C, lower than the pure copolymers and significantly below the transition for the 60/40 intermediate composition (248°C). After compression molding at 315°C for 60 minutes, transesterification appeared to be complete: we observed the meridional maxima and solid state-to-nematic transition temperatures characteristic of the 60/40 copolymer (Fig. 4(d)). X-ray and DSC data for blends that were compression molded for shorter times show a slow transition with the progress of the transesterification reaction and allow us to follow the kinetics (Fig. 5). We have also been able to predict the observed changes based on changes in the monomer nearest neighbor statistics.[27]

Acknowledgment

This work was supported by NSF Materials Research Grant No. DMR 9122227 and the BP-CWRU Partners in Polymer Science Program.

References

1. R. A. Chivers, J. Blackwell, G. A. Gutierrez, Polymer 25, 435 (1984).

2. J. Blackwell, A. Biswas, G. A. Gutierrez, R. A. Chivers, Faraday Discuss. Chem. Soc. 79, 73 (1985).

3. J. Blackwell, R. A., Chivers, G. A. Gutierrez, A. Biswas, J. Macromol. Sci.-Phys. B24(1-4), 39 (1985).

4. A. Biswas and J. Blackwell, Macromolecules 21, 3146 (1988); ibid., 21, 3152 (1988); ibid., 21, 3158 (1988).

5. D. Hofmann, A.I. Schneider, J. Blackwell, Polymer, submitted for publication.

6. S. Hanna and A.H. Windle, Polymer 29, 207 (1988)

7. R. Golombok, S. Hanna, A.H. Windle, Mol. Cryst. Liq. Cryst. 155, 281 (1988).

8. Z. Sun, H.-M.Cheng, J. Blackwell, Macromolecules 24, 4162 (1991).

9. C.M. McCullagh, J. Blackwell, A.M. Jamieson, Macromolecules, submitted for publication.

10. M. T. DeMeuse and M. Jaffe, Mol. Cryst. Liq. Cryst. Inc. Nonlin. Opt. 157, 535 (1988).

11. D.J. Johnson, I. Karacan, J.G. Tomka, Polymer 31, 8 (1990).

12. B. Jung and B.L. Schurmann, Makromol. Chem., Rapid Commun.10, 419 (1989); Macromolecules 22, 477 (1989).

13. E.D.T. Atkins, E.L. Thomas, R.W. Lenz, Mol. Cryst. Liq. Cryst. 155, 263 (1988); ibid., 155, 271 (1988).

14. H.H. Chin, A. Azaroff, R.W. Lenz, J. Polym. Sci., Polym. Phys. Edn. **27**, 1993 (1989); ibid., **27**, 2001 (1989).

15. G.A. Gutierrez, R.A. Chivers, J. Blackwell, J.B. Stamatoff, H. Yoon, Polymer **24**, 937 (1983).

16. J. Blackwell, G.A. Gutierrez, R.A. Chivers, Macromolecules **17**, 1219 (1984).

17. D.Y. Yoon, N. Masciocchi, L.E. Depero, C. Viney, W. Parrish, Macromolecules **23**, 1793 (1990).

18. G. Lieser, J. of Polymer Sci. Phys. Ed. **21**, 1611 (1983).

19. A. M. Kotliar, J. Polym. Sci. **16**, 367 (1981).

20. R. W. Lenz, J. Jin, K. A. Feichtinger, Polymer **24**, 327 (1983).

21. J. Economy, R. D. Johnson, J. Lyerla, A. Muhlebach in "Liquid Crystalline Polymers," R. A. Weiss and C. K. Ober, eds., American Chemical Society, Washington, DC 1990. Chapter 10, "Synthesis and Microstructure of Aromatic Copolyesters"

22. A. Muhlebach, J. Economy, R. D. Johnson, T. Karis, J. Lyerla, Macromolecules **23**, 1803 (1990).

23. W. A. MacDonald, A. D. W. McLenaghan, G. McLean, R. W. Richards, S. M. King, Macromolecules **24**, 6164 (1991).

24. J. Kugler, J. W. Gilmer, D. Wiswe, H.-G. Zachmann, K. Hahn, E. W. Fischer, Macromolecules **20**, 116 (1987).

25. M. H. Li, A. Brulet, P. Keller, C. Strazielle, J. P. Cotton, Macromolecules **26**, 119 (1993).

26. M.J. Troughton, A.P. Unwin, G.R. Davis, I.M. Ward, Polymer **29**, 1389 (1988).

27. C.M. McCullagh, J. Blackwell, A.M. Jamieson, in preparation.

THE SENSITIVITY OF SMALL MOLECULE SORPTION TO ANNEALING IN GLASSY LIQUID CRYSTALLINE POLYMERS

ATSUSHI MORISATO*, N.R. MIRANDA*, J.T. WILLITS*, G.R. CANTRELL*, B.D. FREEMAN*, H.B. HOPFENBERG*, S. MAKHIJA**, I. HAIDER** AND M. JAFFE**
* Department of Chemical Engineering, North Carolina State University, Raleigh, NC 27695
** Hoechst Celanese Corp., Robert L. Mitchell Technical Center, 86 Morris Ave., Summit, NJ 07901

ABSTRACT

The sorption of organic penetrants is found to be sensitive to thermal annealing conditions in a series of glassy, nematic, thermotropic, random copolyesters. Controlled thermal annealing of two polymers in this series permitted a systematic variation of chain packing and, presumably, higher order molecular suprastructure, ranging from a disordered amorphous morphology to more ordered nematic liquid crystalline and semi-crystalline morphologies. The development of liquid crystalline order appears to reduce or preclude small molecule solubility in nematically ordered forms of these polymers.

INTRODUCTION

Liquid crystalline polymers (LCPs) such as poly(p-phenylene terephthalamide) and copolymers of p-hydroxybenzoic acid and 6-hydroxy-2-naphthoic acid have remarkably high barrier properties.[1] Oxygen permeability is at least an order of magnitude lower in LCPs than in common glassy polymers such as poly(ethylene terephthalate) and poly(vinylchloride) (PVC) and comparable to oxygen permeability through poly(acrylonitrile), a noted barrier polymer.[1]

While the molecular origin of high barrier properties in liquid crystalline polymers is not well-understood, barrier properties of conventional polymers are commonly understood to depend on individual polymer chain chemical structure and also on many-chain or hierarchical structure (*i.e.* morphology).[2] In polymers such as poly(ethylene) (PE), for example, conventional three dimensional crystalline order is commonly understood to preclude solubility of small penetrant molecules in the polymer.[3] This effect is illustrated in Figure 1, which presents the influence of amorphous phase content on gas solubility in semi-crystalline PE. The solubility of all three gases decrease linearly with decreasing amorphous phase content, and straight lines through the data extrapolated to 0% amorphous phase content suggest that crystalline regions effectively preclude solubility of gas molecules.[3] These data are consistent with the notion that sorption is limited to the amorphous regions of the polymer, and chains confined to a very efficiently packed three-dimensional crystalline lattice effectively preclude sorption of penetrant molecules. In this regard, the density of crystalline PE is 15% higher than the amorphous phase density.[3]

While the simple two-phase model adequately describes the effect of three dimensional crystallinity on penetrant solubility in many common polymers, poly(4-methyl-1-pentene) (P4MP) is an intriguing exception since it forms crystallites which are apparently accessible to small penetrant molecules.[3] Carbon dioxide and methane solubility in P4MP is presented in Figure 2 and, based on a simple linear extrapolation to 100% crystallinity, the crystalline phase seems to exhibit substantial solubility towards these penetrants. This result may be rationalized based on the unusual crystal structure of P4MP, where, in marked contrast to PE, the crystalline phase density is *lower* than the amorphous phase density.[3]

The current study is directed towards determining the influence of liquid crystalline order on penetrant solubility and diffusivity. The polyesters in this study include copolymers of 73 mole % p-hydroxybenzoic acid and 27 mole % 6-hydroxy-2-naphthoic acid (HBA/HNA), HIQ-40 (containing 40 mole % p-hydroxybenzoic acid and 30 mole % each of hydroquinone and isophthalic acid), and PICT (containing 25 mole % each of phenyl hydroquinone and t-butyl hydroquinone, 40 mole % chloroterephthalic acid, and 10 mole % isophthalic acid). These materials are randomly copolymerized, glassy, thermotropic, nematogenic LCPs. HIQ-40 and PICT have unusually useful properties since they may be dissolved in volatile solvents and cast into thin, uniform films which are completely disordered. The rapid evaporation of volatile solvent molecules, in comparison with the characteristic rate of mesogen ordering, results in the preparation of solvent-free, isotropic, and disordered amorphous films. Subsequent heating above the glass transition temperature, T_g, confers sufficient chain mobility to permit liquid crystalline order development, which persists upon cooling of the sample to below T_g. The sorption and transport

81

behavior of small molecules in the original, disordered, isotropic polymer can, therefore, be compared directly and unequivocally with corresponding behavior in the ordered, frozen liquid crystalline state.

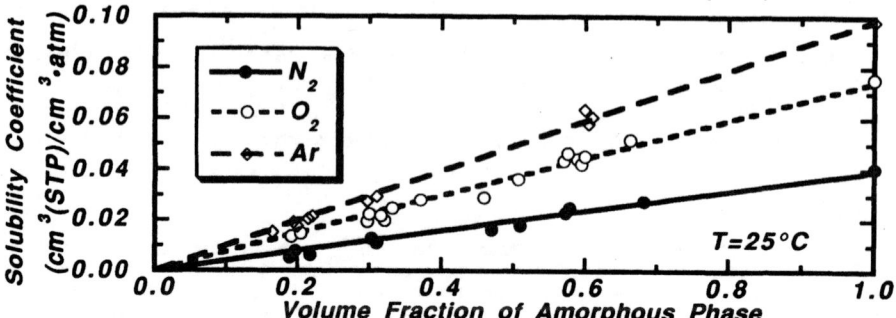

Figure 1. The effect of three dimensional crystallinity on gas solubility in polyethylene[3]

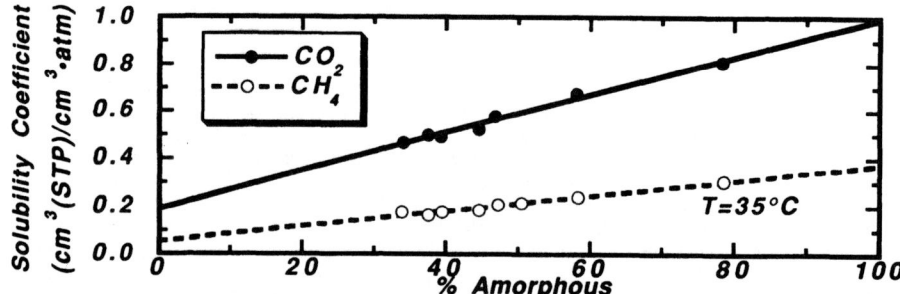

Figure 2. The effect of crystallinity on gas solubility in poly(4-methyl-1-pentene)[3]

In a nematic phase of a liquid crystalline polymer, rod-like polymer segments are locally aligned, somewhat imperfectly, in a common direction, described by a vector called the director.[4] Nematic microdomains formed by regions where the director value is constant or changes smoothly and continuously with position are separated by defect or disclination regions, where the director and, therefore, molecular orientation change abruptly. The characteristic width of these disclination regions may be on the size scale of microns in HBA/HNA.[5] While the molecular structure in disclination regions is not well-understood, polymer molecular weight appears to play an important role in the structure of disclinations and the related dynamics of disclination formation.[6, 7] Disclination regions have been reported to contain relatively high concentrations of chain ends.[4] Small angle light scattering patterns and crossed polarized light microscopy suggest, moreover, that thermal annealing can cause nematic microdomains to grow in size by disclination annihilation.[6, 7]

EXPERIMENTAL

Sample Preparation

HBA/HNA, which is insoluble in common solvents, was prepared for gravimetric sorption experiments by microtoming uniform microflakes of constant thickness (5 μm) at ambient conditions from the core of an extruded flex bar kindly supplied by the Hoechst Celanese Corp. Additional polymer morphological characterization information and sample preparation details are available elsewhere.[1] The nematic liquid crystalline domains in the microflake samples, taken from the core of this thick extruded bar, are presumed to be unoriented.[1] The glass transition temperature of HBA/HNA, determined by differential scanning calorimetry (DSC), was approximately 90°C.[1]

HIQ-40 films of uniform thickness of approximately 4-5 μm were prepared at Hoechst Celanese Corp. by casting in a clean room from a ten weight percent polymer solution. The solvent was a 75:25 mixture by weight of trifluoroacetic acid (TFA) and methylene chloride. Air drying and solvent extraction steps resulted in solvent-free, isotropic, non-crystalline films.[8]

To prepare HIQ-40 samples with varied morphology, amorphous films were heated, as described elsewhere,[8] to seven different temperatures between 100°C and 330°C. The annealing time was approximately one hour. The range of annealing temperatures provided samples annealed below T_g (approximately 130°C for the isotropic, unannealed polymer), between T_g and the melting point of three dimensional crystalline regions (approximately 310°C), and above the crystalline melting point. At 330°C, HIQ-40 is a nematic liquid crystalline fluid.[8] The content of three dimensional crystallites in the samples used in this study is, based on WAXD and DSC studies, believed to be small.[8]

PICT samples were kindly supplied by Dr. R.J. Kumpf of Miles, Inc. Thin uniform films (approximately 5 μm thick) of PICT were prepared by casting from 5% (w/v) solutions in methylene chloride under a nitrogen atmosphere at ambient conditions. The films were dried in nitrogen at ambient conditions for 24 hours and, afterwards, in air for at least 4 days to remove the solvent. PICT films prepared in this manner were isotropic and non-crystalline. Samples with a nematic morphology were prepared by annealing the isotropic films under vacuum at 160°C for 48 hours, followed by cooling under vacuum to ambient conditions. The T_g of PICT was 130°C as determined by DSC. No evidence of conventional three dimensional crystallinity was observed by DSC or WAXD in any of the PICT samples.

<u>Characterization of Sorption and Transport</u>

A McBain spring balance system and a Cahn electrobalance system were used to perform kinetic gravimetric sorption experiments as described previously.[8] A polymer sample was placed in the sample chamber of the sorption system and exposed to vacuum for at least 24 hours to remove air gases. Penetrant was then introduced into the sample chamber at a fixed pressure, and the subsequent increase in sample weight was recorded as a function of time. Afterwards, when sample weight was constant, the penetrant was removed from the sorption chamber by a vacuum pump, and the sample was kept under vacuum until all penetrant had been desorbed. Penetrant was then reintroduced into the sorption chamber to begin the next experiment. From these data, penetrant solubility may be determined as a function of penetrant pressure.[9] Penetrant activity is calculated as the ratio of the pressure of penetrant in contact with the polymer sample divided by the penetrant vapor pressure at the temperature of the experiment.

RESULTS AND DISCUSSION

<u>Acetone and Methylene Chloride Sorption in HBA/HNA</u>

Sorption isotherms of acetone and methylene chloride in HBA/HNA are presented in Figure 3. The uncertainty in the reported sorption values is approximately ±0.03 g penetrant/100 g polymer. To provide some perspective of the extraordinarily small sorption levels of these two highly condensable penetrants, these data are compared with acetone sorption in PVC, a conventional glassy polymer.

At high penetrant activity, sorption isotherms of vapors in typical glassy polymers are often convex to the pressure or activity axis.[10] This isotherm shape is associated with the swelling of the polymer to accommodate the penetrant at high penetrant activity. The shape of these sorption isotherms is often described by the Flory-Huggins sorption isotherm.[10] The Flory Huggins model was unable to describe, even qualitatively, the curvature in the sorption isotherms in HBA/HNA presented in Figure 3.[1] The Flory-Huggins model presumes that the penetrant is uniformly distributed throughout the polymer sample. Guided by previous studies[3], this model was modified to describe sorption in a material in which the penetrant is soluble in a volume fraction f of the material and completely insoluble in the remainder of the sample. The modified Flory-Huggins expression is given by:

$$\ln(a) = \ln(\phi_a) + (1 - \phi_a) + \chi(1 - \phi_a)^2 \tag{1}$$

where a is penetrant activity, ϕ_a is the volume fraction of penetrant in the soluble region, and χ is the polymer-solvent interaction parameter. The relationship between the amount of penetrant absorbed in the polymer, C (mass of penetrant per unit mass of polymer), and the penetrant volume fraction in the soluble region, ϕ_a, is given by:

$$\phi_a \equiv \left(C \frac{\rho_p}{\rho_s} \right) \Big/ \left(C \frac{\rho_p}{\rho_s} + f \right) \qquad (2)$$

where ρ_p and ρ_s are the polymer and penetrant densities, respectively.

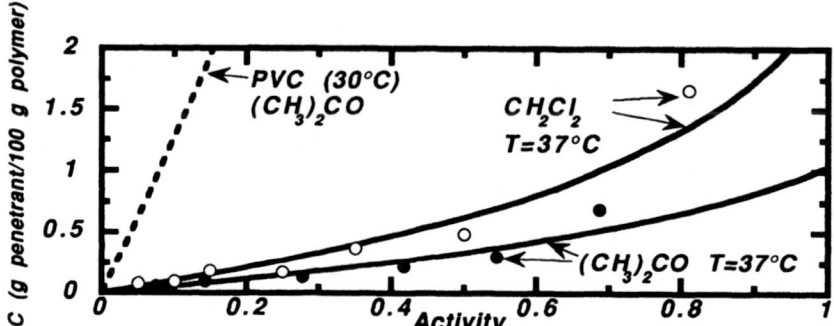

Figure 3. Sorption Isotherms of Acetone and Methylene Chloride in HBA/HNA (Solid Lines). The sorption isotherm of acetone in PVC (dashed line) is provided for comparison.[10]

A nonlinear least squares fit of this model to the HBA/HNA sorption data is represented by the solid lines in Figure 3. The model parameters which best fit the acetone data were: f=0.13 and χ=1.6; the parameters which best fit the methylene chloride data were: f=0.11 and χ=1.34.[1] Based on this model, the volume fraction of polymer available for penetrant sorption is quite small (on the order of 10% for both penetrants). The χ values are self-consistent and typical of χ values which would be observed for acetone and methylene chloride sorption in conventional, non-mesogenic polyesters.[1]

The presence of conventional crystallinity cannot explain the observed sorption behavior of acetone and methylene chloride sorption in HBA/HNA. DSC measurements of crystallinity in HBA/HNA microflakes suggest that perhaps only 5% of the material is crystalline,[1] clearly too low to account for the seemingly high fraction of inaccessible regions in HBA/HNA suggested by the modified Flory-Huggins model. If, however, the highly ordered liquid crystalline phase precludes solubility within the microdomains just as conventional crystallinity precludes sorption in PE, then penetrant molecules would be restricted to less ordered disclination regions between microdomains. The disclination regions account for a small fraction of the sample mass,[4] and, therefore, solubilities calculated based on the entire polymer mass would predictably be quite low compared to solubilities in largely amorphous polymers such as PVC.

Acetone Sorption in PICT

While HBA/HNA samples exhibit extraordinarily low penetrant solubilities, disordered amorphous samples of this polymer were not available to permit a direct comparison of the effect of liquid crystalline order on penetrant solubility. Amorphous, isotropic PICT films may be prepared, however, by casting from volatile solvents. When these disordered films are annealed above the glass transition temperature, cross-polarized optical microscopy reveals a birefringent texture consistent with a nematic mesophase. Acetone sorption in both the isotropic and liquid crystalline arrangements of PICT were determined and are presented in Figure 4. The solid lines in Figure 4 represent nonlinear least squares fits of the modified Flory-Huggins model accounting for the ordering in annealed PICT samples. The composite data are well represented by a single χ value and by confining acetone solubility differences between the two samples to f, the soluble volume fraction of the polymer. The model parameters are presented in the figure legend. The χ values are typical of those observed in the sorption of organic vapors in common, non-mesogenic glassy polymers.[10]

Within the context of this simple two-phase model, the decrease in f between the isotropic and nematic variants of this material, suggests that the accessible volume fraction of material is reduced by 27% in the nematic sample. However, between crossed polarizers in an optical microscope, annealed PICT, like HBA/HNA, exhibits a nematic liquid crystalline texture which spans the entire sample. These composite results are consistent with the notion that acetone is somewhat soluble in the nematic phase of PICT, much as small molecules are somewhat soluble in the crystalline phase of P4MP.

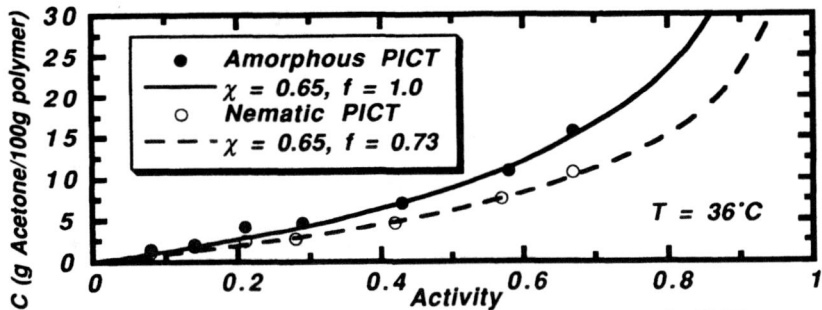

Figure 4. The Sorption of Acetone in Amorphous and Annealed (*i.e.* Nematic) PICT.

Density measurements reveal that the nematic film is only 0.8% more dense than the isotropic film. Moreover, PICT does not organize into conventional three dimensional crystallites. The inability to crystallize, in the conventional sense, is presumably related to the large, bulky, packing-inhibiting *t*-butyl, phenyl, and chlorine side groups. These groups may also contribute to the formation of a nematic phase which exhibits only slightly more efficient chain packing than an isotropic, amorphous arrangement of this polymer. The acetone sorption data are consistent with the notion that chain packing in the nematic mesophase of PICT may not be sufficiently ordered to preclude penetrant solubility.

Acetone Sorption in HIQ-40

HIQ-40 can be solvent-cast into thin uniform, isotropic, amorphous films.[8] However, unlike PICT, HIQ-40 is free from bulky side groups, and HIQ-40 chains pack efficiently into liquid crystalline microdomains which include low levels of conventional crystallinity.[8] Acetone solubility in HIQ-40 films was determined as a function of annealing temperature and is presented in Figure 5. Solubility determinations were executed at 35°C and an acetone activity of 0.15. Data related to the as-cast, unannealed, amorphous sample are reported at an annealing temperature of 25°C. Sorption in the as-cast film is the highest observed in the sample set. The solubility decreases monotonically as the treatment temperature is increased to 200°C, with most of the change occurring at annealing temperatures at or below the T_g of the isotropic sample, approximately 130°C. There is little change in solubility as the treatment temperature is raised to 300°C from 200°C. Finally, the solubility increases slightly as the annealing temperature in raised from 300°C to 330°C, which is above the melting point of the small amounts of conventional crystallinity in HIQ-40. Samples annealed at 200°C and 300°C exhibit solubilities which are approximately an order of magnitude lower than the acetone solubility observed in the unannealed sample.

WAXD and optical microscopy analysis of the as-cast film suggests that this sample is amorphous and does not contain measurable levels of conventional crystallinity.[8] Films annealed at 200°C and higher exhibited peaks in the WAXD spectrum consistent with the presence of very low levels of conventional crystallinity.[8] Films annealed below 140°C were isotropic as determined by cross-polarized optical microscopy. Films annealed at 170°C and 200°C exhibited birefringence but not a discernible nematic texture. Films annealed at 300°C and 330°C exhibited a strongly birefringent nematic texture.[8] No primary chemical structure changes were observed by NMR or FTIR at these temperatures, however, transesterification has been observed in samples annealed at high temperatures (≥300°C).[11]

The ten-fold solubility decrease is accompanied by a marked density increase (3.4%).[8] Most of the density change occurs in samples annealed at 200°C and lower. This density increase, which represents a substantial decrease in polymer free volume, is more than four times the density change observed when amorphous PICT is annealed to the nematic state. The accessible polymer fraction, f, decreases approximately tenfold as well, from 1 in the amorphous sample, to values near 0.1 in the sample annealed at 200°C. The lowest values are reminiscent of f values observed in HBA/HNA and suggest that liquid crystalline microdomains in HIQ-40 may be inaccessible to penetrant molecules. This hypothesis is supported by the density change upon annealing the sample to the fully-developed nematic state. The small changes in solubility observed at annealing temperatures above 200°C are consistent with the direction and magnitude of density changes in these samples,[8] and support the notion that penetrant solubility in HIQ-40 is sensitive to gross changes in chain packing.

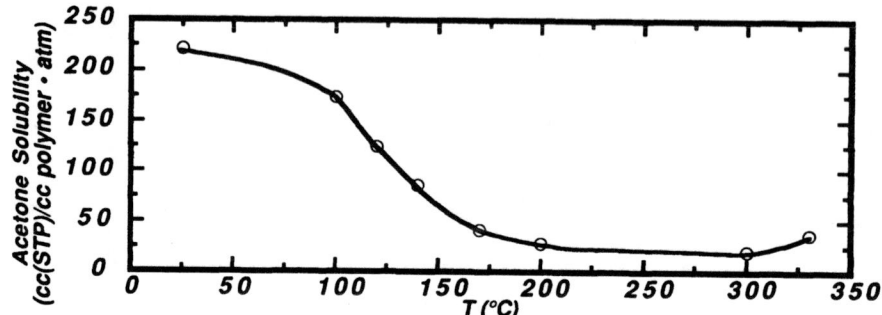

Figure 5. The Effect of Thermal Annealing Temperature on Acetone Solubility in HIQ-40.[8]

CONCLUSIONS

Small molecule solubility in HBA/HNA and HIQ-40, which exhibit very efficient chain packing, appears to be confined to the less ordered regions of the polymer sample, perhaps in the disclination regions between nematic liquid crystalline microdomains. In PICT, which contains bulky, packing disrupting side groups on the polymer backbone, small molecules appear to be somewhat soluble in the nematic phase of the material. Both PICT and HIQ-40 may be prepared as isotropic samples and annealed to the liquid crystalline state. In these materials penetrant solubility decreases (by as much as an order of magnitude in the case of HIQ-40) as ordering to the nematic mesophase becomes more complete. The composite data related to sorption of small molecules in these copolyesters suggest that liquid crystalline ordering, like conventional three dimensional crystallinity, may preclude or, at least, decrease penetrant solubility.

ACKNOWLEDGMENTS

This work was supported in part by the Electric Power Research Institute by EPRI Agreement RP8007-14 and the National Science Foundation by Grants CTS-9101587 and CTS-92579911. Moreover, we gratefully acknowledge the support of the Hoechst Celanese Corporation and ALCOA.

REFERENCES

1. N. R. Miranda, J. T. Willits, B. D. Freeman and H. B. Hopfenberg, J. Membrane Sci., in press.
2. D. H. Weinkauf and D. R. Paul, in *Barrier Polymers and Barrier Structures*, edited by W. J. Koros, (American Chemical Society, Washington, D.C., 1990), p. 60.
3. A. C. Puleo, D. R. Paul and P. K. Wong, Polymer, **30**, 1357 (1989).
4. A. M. Donald and A. H. Windle, *Liquid Crystalline Polymers*, (Cambridge University Press, Cambridge, 1992), p. 310.
5. A. Anwar and A. H. Windle, Polymer, **34**, 3347 (1993).
6. S. R. Rojstaczer and R. S. Stein, Macromolecules, **23**, 4863 (1990).
7. T. Shiwaku, A. Nakai, H. Hasegawa and T. Hashimoto, Polym. Comm., **28**, 174 (1987).
8. G. R. Cantrell, B. D. Freeman, H. B. Hopfenberg, S. Makhija, I. Haider and M. Jaffe, in *Liquid Crystalline Polymers*, edited by C. Carfagna, (Pergamon Press, Oxford, in press).
9. J. Crank and G. S. Park, in *Diffusion in Polymers*, edited by J. Crank and G. S. Park, (Academic Press, New York, 1968), p. 1.
10. A. R. Berens, J. Appl. Polym. Sci., **37**, 901 (1989).
11. J. J. Rafalko, M. Borzo, W. Choe and M. Jaffe, ACS Polymer Preprints, **34**, 770 (1993).

PHYSICAL STUDIES OF A SERIES OF MAIN-CHAIN LIQUID CRYSTALLINE POLYMERS WITH LONG SOFT SEGMENT

BING WU, C. PETER LILLYA AND RICHARD S. STEIN
University of Massachusetts, Chemistry Department, Amherst, MA 01003

ABSTRACT

A series of main-chain liquid crystalline polymers with long and polydisperse soft segments has been synthesized. The thermal properties of these polymers have been characterized by differential scanning calorimetry, optical microscopy and wide angle X-ray diffraction. Unexpectedly, these polymers have a strong tendency to form purely smectic mesophases. The degree of ordering of these polymeric mesophases is much higher than that of the mesophase formed by a low molecular weight model compound with the same mesogen. The soft segment length effects on thermal behavior of such segmented copolymers have been elucidated. A semi-quantitative theory has been formulated to interpret these observations. A phase diagram relating thermal properties to soft segment length has been proposed.

INTRODUCTION

Semiflexible main chain liquid crystalline polymers, which comprise alternating hard and soft segments, have received a lot of attention during the past decade. One major objective of these studies is to investigate the soft segment length effects on thermal behavior of liquid crystalline polymers. In most cases, the soft segments used are of the monodisperse polymethylene type and the number of methylene units in the soft segment main chain is from 4 to 18. As for many series of LCP's with such soft segments, the following effects are observed as the soft segment length increases [1]: (a) transition temperatures decrease; (b) the temperature range over which the mesophase is stable is decreased; and (c) the nature of the mesophase is changed from predominantly nematic to purely smectic. To interpret these observations, structural studies have been conducted [2,3]. By using various techniques, most researchers determined that in the mesophase the soft segments have highly, if not fully, extended conformations. However, a generally accepted structure has not been obtained.

Elucidation of the behavior of LCP's with long, polydisperse soft segments should enhance our understanding of how structure determines thermal phase behavior of semiflexible main chain LCP's. In this case, an open question is what kind of mesophase will be formed by such polymers. Since the soft segments with different lengths are randomly distributed along the polymer chains, formation of a regular smectic layer structure with fully extended soft segments is not possible. Therefore, these polymers might be expected to form nematic mesophases. However, the observations on semiflexible LCP's with monodisperse soft segments show that formation of smectic mesophases is favored by long soft segments.

Publications on liquid crystalline polymers with polydisperse long soft segments are very few and are principally on synthesis. Quite recently, Bilibin [4] and Kolb [5] have shown it is possible to make LCP's with increased hard and soft segment length. However, to our knowledge the nature of the mesophases formed by such LCP's has not been reported. Based on these considerations, we synthesized a series of segmented copolymers with polydisperse long soft segments and uniform hard segments. The thermal behavior of these polymers was characterized by differential scanning calorimetry, optical microscopy and X-ray diffraction. The observed effects were interpreted by a semi-quantitative thermodynamic theory which is based on Percec's work [6]. In this paper, we report our characterization results and the observed soft segment length effects on thermal behavior of these polymers. Some major results of our theoretical work will also be presented.

EXPERIMENTAL SECTION

We have studied series of segmented copolymers (1) and a low molecular weight model compound (2) have the following structures:

$$\left[\!\!\begin{array}{c}\overset{O}{\overset{\|}{C}}\!\!-\!\!\bigcirc\!\!-\!\!\overset{O}{\overset{\|}{OC}}\!\!-\!\!\bigcirc\!\!-\!\!N\!\!=\!\!N\!\!-\!\!\bigcirc\!\!-\!\!\overset{O}{\overset{\|}{CO}}\!\!-\!\!\bigcirc\!\!-\!\!\overset{O}{\overset{\|}{CO}}\!\!-\!\!R\!\!-\!\!O\end{array}\!\!\right]_n \tag{1}$$

$$n\text{-Bu}\!\!-\!\!O\overset{O}{\overset{\|}{C}}\!\!-\!\!\bigcirc\!\!-\!\!\overset{O}{\overset{\|}{OC}}\!\!-\!\!\bigcirc\!\!-\!\!N\!\!=\!\!N\!\!-\!\!\bigcirc\!\!-\!\!\overset{O}{\overset{\|}{CO}}\!\!-\!\!\bigcirc\!\!-\!\!\overset{O}{\overset{\|}{CO}}\!\!-\!\!n\text{-Bu} \tag{2}$$

where n-Bu is n-butyl group, and R is polydisperse poly(tetramethylene oxide) (PTHF). The number average molecular weight of R is ranged from 250 to 2900 and the M_w/M_n is about 1.9.

Thermal behavior was measured by a Du Pont 2000 Thermal Analyzer. Both heating and cooling scans were conducted for all samples, with the rate of 20°C/min. and 10°C/min. for heating and cooling scans respectively.

The transition characteristics were studied with a polarizing optical microscope (Leitz Laborlux 12 Pol Microscope), equipped with a hot stage, an Omega programmable temperature controller and a 35 mm camera. Textures were observed between two microscope cover glasses.

X-ray diffraction patterns of the unoriented samples were recorded on flat films using Cu K_α radiation. The sample was contained in a 1-mm capillary mounted in an electrically heated oven. A temperature controller was used to control the temperature within 0.5°C.

RESULTS AND DISCUSSION

The first three polymers in this series exhibit multiple phase transitions as evident in the DSC heating curve of A-2T250 in Fig. 1. Mesophase formation was confirmed by optical microscope observations, and the nature of mesophase was further identified by X-ray diffraction studies. Fig. 2 shows the focal conic texture of A-2T650 along with batonnets formation, which indicates the smectic nature of the mesophase. The thermal behavior and physical properties of these polymers and the low molecular model compound are summarized in Table I.

From these observations, the following effects were found for LCP's with polydisperse long soft segments:

(1) As soft segment length increases, the transition temperatures decrease and the temperature range of mesophase stability also decreases. LCP's with polydisperse long soft segments shows similar trends as LCP's with monodisperse short soft segments.

(2) As soft segment length increases, the nature of the mesophase changes from both nematic and smectic to purely smectic, despite the polydispersity of soft segments.

(3) The segmented copolymers form smectic mesophases while the corresponding low molecular weight model compound forms only a nematic mesophase.

(4) Increasing soft segment length causes clearing temperatures to drop faster than melting temperatures, resulting in loss of mesophase formation above a certain soft segment length.

(5) The melting temperatures can only be lowered to a limited value by increasing the soft segment length.

As to interpret these effects, a semi-quantitative thermodynamic scheme was formulated by modifying Percec's work [7]. One major result of this theory is a phase diagram which relates the thermal behavior of segmented copolymers to soft segment length as shown by Fig. 3.

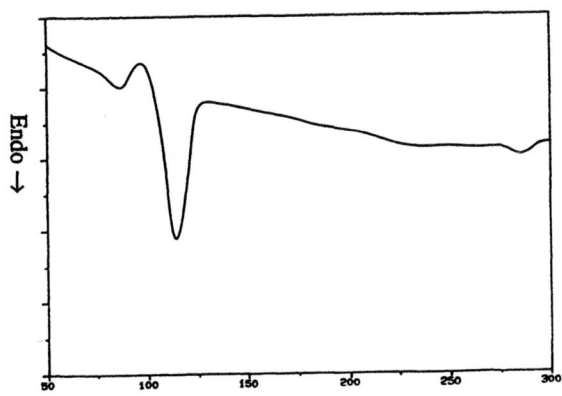

Fig. 1. The DSC heating curve of A-2T250

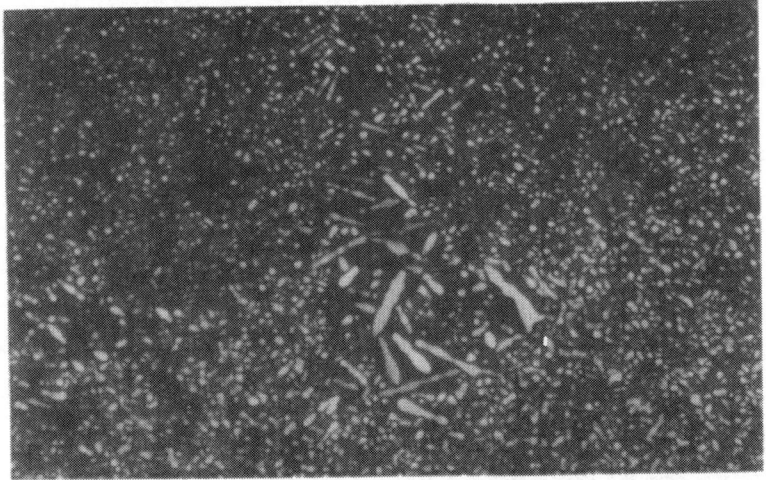

Fig. 2. Texture of A-2T650 as cooled form isotropic phase to mesophase

Table I. Thermal behavior and physical properties of studied samples

Sample	M_n of Soft Segment	Thermal Behavior (°C)	Physical Properties
A-2T250	250	K 117 S 229 N 286 I	nematic and smectic LCP
A-2T650	650	K 116 S 186 I	smectic LCP
A-2T1000	1000	K 116 S 140 I	smectic LCP
A-2T2000	2000	K 98 I	thermoplastic elastomer
A-2T2900	2900	K 78 I	thermoplastic elastomer
model compound	-----	K 163 N 320 dec.	nematic LC

The transition temperatures are measured on heating runs. K, S, N, I and dec. are for crystalline phase, smectic mesophase, nematic mesophase, isotropic liquid and decomposition, respectively.

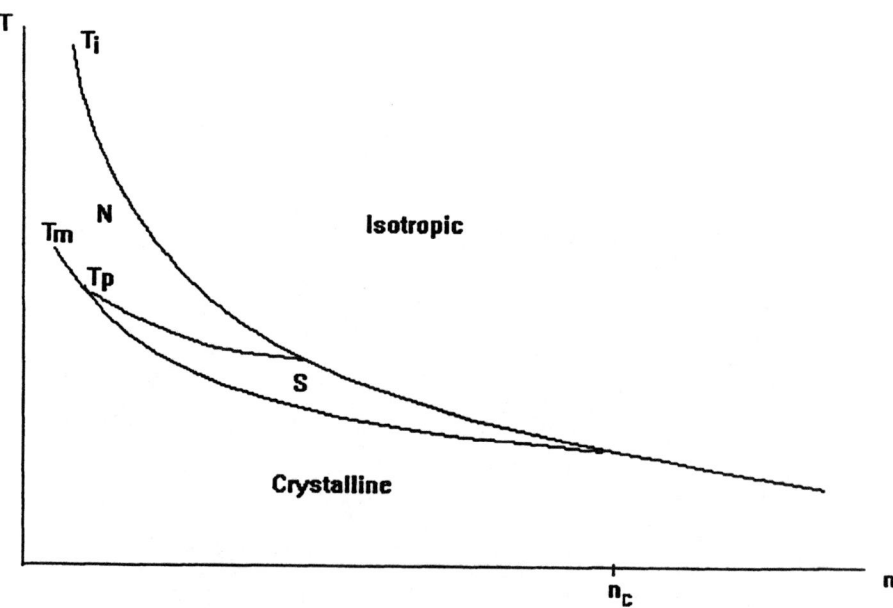

Fig. 3. Phase diagram of LCP's: Phase transition temperatures as a function of soft segment length. n designates the soft segment length, T_m, T_p and T_i are for melting, mesophase to mesophase and clearing transition temperatures, respectively.

ACKNOWLEDGMENT

Financial support for this work by CUMUP program in University of Massachusetts is gratefully acknowledged.

REFERENCE

[1] C. Noel and P. Navard, Prog. Polym. Sci., **16**, 55 (1991).

[2] Z. Jedlinski, J. Franee, A. Kulczycki, A. Sigiru and C. Carfagna, Macromol., **22**, 1600 (1989).

[3] P. P. Wu, S. L. Hsu, O. Thomas and A. Blumstein, J. Polym. Sci. Part B, Polym. Phys., **24**, 827 (1986).

[4] A. Y. Bilibin and O. N. Piraner, Makromol. Chem., **192**, 201 (1991).

[5] E. S. Kolb, PhD Dissertation, University of Massachusetts, Ameherst, 1991.

[6] V. Percec and A. Keller, Macromol., **23**, 4347 (1990).

[7] C. P. Lillya, B. Wu, E. S. Kolb and R. S. Stein, Macromol., will be submitted.

ATOMISTIC DETAILS OF DISORDERING PROCESSES
IN SUPERHEATED POLYMETHYLENE CRYSTALS
II. EFFECTS OF SURFACE CONSTRAINTS

B. WUNDERLICH, G. L. LIANG, B. G. SUMPTER, AND D. W. NOID

Department of Chemistry, University of Tennessee, Knoxville, TN 37996-1600 and
Chemistry Division, Oak Ridge National Laboratory, Oak Ridge, TN 37831-6197.

ABSTRACT

Atomistic details of disordering in superheated polymethylene crystals have been studied using full molecular dynamics simulations of crystals containing 9600 CH_2-groups. The crystal size was about 227 nm^3. Simulations were carried out for up to 100 ps, starting at temperatures about 100 K above the melting temperature. Typically 1.5 h of CPU time on a Cray X-MP were necessary per ps simulation. Superheating causes a quick development of large-scale disorder throughout the crystal, including reorientation, translation, and the destruction of crystal symmetry. This is followed ultimately by surface melting. Crystallization centers with hexagonal packing are found in superheated, unconstrained crystals. On cooling during the simulation, recrystallization processes compete with the disordering, resulting in a reorientation of the molecular chains and reorganization of the crystal. Neither the fully amorphous phase nor the ordered crystal are reached during these short-time simulations using an instantaneous temperature increase to above the melting temperature, followed by a slow cooling into the crystallization temperature region.

INTRODUCTION

Molecular dynamics simulations of the disordering process in unconstrained, superheated polymethylene crystals with sizes close to experimentally observed small crystals have been carried out earlier.[1] In these unconstrained crystals, the disordering developed instantaneously throughout the whole crystal. Surface melting followed by reorganization and recrystallization were observed when the temperature was reduced. We have now extended these simulations to surface-constrained crystals of the same size. In this paper we report new results from such molecular dynamics simulations and compare them with the prior results.

COMPUTATIONAL DETAIL

The simulations were carried out starting with $(CH_2)_{50}$-chains in perfect orthorhombic (ORTH) or monoclinic (MONO) arrangement. The crystals are illustrated in x-y-projection (along the chain axes) in Fig. 1. The following polyethylene crystal-lattice parameters were used: ORTH: a = 0.74, b = 0.49, c = 0.25 nm with c being the chain axis;[2] MONO: a = 0.81, b = 0.25, c = 0.48 nm, and β = 108° with b being the chain axis.[3] The double crosses mark the zigzag positions of the carbon atoms of the all-*trans* chains. In Fig. 1 each crystal contains 192 chains and each chain consists of 50 CH_2-groups, *i.e.* there are 9600 CH_2-groups

93

in each crystal with an initial dimension of about 6.0 × 6.0 × 6.3 ≈ 227 nm³. Each of the CH₂-groups was treated as a united atom with a mass of m = 14.03 dalton to reduce the computation time. All dynamic chains have stretching, bending, and torsional motion allowed. In the surface-constrained simulations, the outermost chains are frozen, (i.e. are not permitted to participate in any motion, but contribute to the van der Waals interaction with the dynamic chains). Three different cases of simulations were carried out: four-sided constraint (4-SC) in which layers of the surface chains of both, the x-z and y-z surface planes in Fig. 1 are static; three-sided constraint (3-SC) in which two surface layers of the y-z planes and only one of the x-z planes are static; and two-sided constraints (2-SC) in which two surface layers of the chains, one in a x-z-plane and one in y-z-plane are static. The chain ends of the dynamic chains are not constrained, i.e. the chains are free to diffuse out of the x-y-planes.

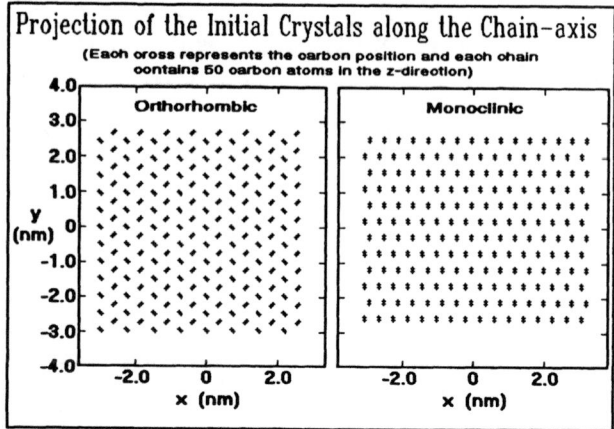

Figure 1 Representations of the initial crystal structures.

The Lagrangian operator $\mathcal{L}$ for the system is described as:

$$\mathcal{L} = \sum_{i=1}^{N} \frac{1}{2}(m\dot{x}_i^2 + m\dot{y}_i^2 + m\dot{z}_i^2)$$
$$- \sum_{1}^{M}\sum_{i=1}^{49} V_{2b}(r_{i,i+1}) - \sum_{1}^{M}\sum_{i=1}^{48} V_{3b}(\theta_{i,i+1,i+2}) \qquad (1)$$
$$- \sum_{1}^{M}\sum_{i=1}^{47} V_{4b}(\cos\tau_{i,i+1,i+2,i+3}) - \sum_{i+j\geq 4}^{9600} V_{NB}(r_{ij})$$

in which N is the number of dynamic CH₂-groups and M is the number of dynamic chains which depend on the type of the constraint. The subscripts of the potential energies V, 2b, 3b, 4b, and NB represent the two-body (stretching),[4] three-body (bending),[5] and four-body (torsional)[6] bonding interactions and the non-bonding interactions, respectively:[7]

$$V_{2b}(r_{i,i+1}) = D\{1 - \exp[-\alpha(r_{i,i+1} - r_e)]\}^2 \qquad (2)$$

$$V_{3b}(\theta_{i,i+1,i+2}) = \frac{1}{2}K_\theta(\cos\theta_{i,i+1,i+2} - \cos\theta_0)^2 \qquad (3)$$

$$V_{4b}(\cos\tau_{i,i+1,i+2,i+3}) = 8.37 - a\cos\tau_{i,i+1,i+2,i+3} + b\cos^3\tau_{i,i+1,i+2,i+3} \qquad (4)$$

$$V_{nb}(r_{ij}) = 4\varepsilon[(\sigma/r_{ij})^{12} - (\sigma/r_{ij})^{6}].$$ (5)

The parameters for Eqs. (2–5) are listed in Table I.

Table I Force field parameters.

Two-body interactions			Four-body interactions		
D	=	334.72 kJ/mol	*a*	=	18.41 kJ/mol
r_e	=	0.153 nm	*b*	=	26.78 kJ/mol
α	=	19.9 nm^{-1}			
Three-body interactions			Nonbonding interactions		
K_θ	=	130.22 kJ/mol	ε	=	0.477 kJ/mol
θ_0	=	113°	σ	=	0.398 nm

RESULTS

Six new simulations were carried out on ORTH and MONO crystals with different types of surface constraints, as is indicated in Table II. The D10-simulation is a reference-run, done earlier, with all chains being dynamic (unconstrained). Results of this run are shown in Fig. 2 for comparison, details of the data are described in Ref. 1. The S-series

Table II Time and temperature ranges for the superheated and constrained crystals.

Run Code	Crystal Type	Constraint Type	Time (ps)	Initial Temp (K)	Final Temp (K)
D10	MONO	NONE	100	456	337
S10	MONO	4-SC	20	480	446
S11	MONO	3-SC	100	442	255
S12	MONO	3-SC	100	222	144
S13	ORTH	3-SC	33	457	359
S14	ORTH	4-SC	66	340	227
S15	ORTH	2-SC	15	435	379

refers to simulations with various static surface-chain constraints. Of interest to the present discussion are the simulations done above the approximate melting temperature of pentacontane (365 K). After 0.1 ps of simulation (at which the first temperature is computed), the temperature decreases gradually due to computational energy-loss. The final temperature is the temperature at which the simulation was stopped (corresponding to the given simulation time). Selected simulations are graphically displayed in the figures of the discussion, many more have been analyzed to derive the crystal behavior.

DISCUSSION

A typical response of a crystal to an instantaneous heating to temperatures far above melting is shown in the x-y-projections of the unconstrained MONO crystal of Fig. 2.[1] The simulations reveal that fusion of a superheated crystal involves not only slow surface melting, which is known from experiment to take many orders of magnitude longer time for complete melting than available for simulation, time, but also a competing, fast, internal disordering process. As the temperature decreases below the melting point, the initial conformational disordering of the surface chains is reduced quickly, but the internal disorder, once formed, is slow in reaching the perfection expected for the lower temperatures. The final state in Fig. 2 is much more disordered than in the steady state achieved in analogous constant-temperature simulations at the final temperature.[8] Overall, the cooling through the melting process is, perhaps, an illustration of the Ostwald rule of stages, that states that, on crystallization, metastable states will be traversed before equilibrium is reached.[9]

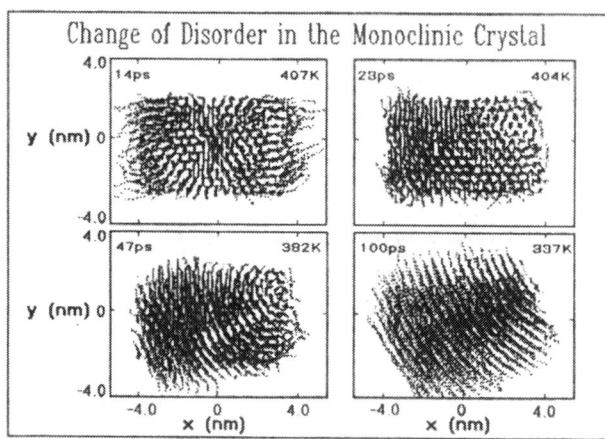

Figure 2 Unconstrained MONO (Run D10 in Table II).

The constrained crystals were observed to have a different dynamic behavior and crystal structure. The large-amplitude bending motion is restricted, particularly in the vicinity of the static chains. When all surface chains are constrained (4-SC runs S10 and S14, Table II), the x-y projections of the crystals shows, in contrast to the unconstrained crystals, that all chains remain in the position set by the orthorhombic or monoclinic crystal structure. No hexagonal domain-structure is set up, as was found for the unconstrained crystals. A check of the setting angle of the zigzag plane of the orthorhombic crystal shows, however, a deviation from the characteristic alternation (see Fig. 1) to parallel alignment, as in the monoclinic structure. Due to the initial choice of room temperature lattice parameters one can calculate from the known experimental expansivity and compressibility that the pressure would increase to about 500 MPa at 410 K. The unconstrained x-y-surfaces offer the only pressure relief, but require a diffusion in the chain direction. Figure 3 shows the results of such chain migration in the z-direction on the 8th layer of a 4-SC MONO crystal (see Fig. 1) at 10 ps ($\approx$465 K). The setting angle is still largely unchanged. The diffusion of the chains is largest in the center of the crystal, relieving the pressure close to the two open surfaces. The mechanism of diffusion does not seem to involve conforma-

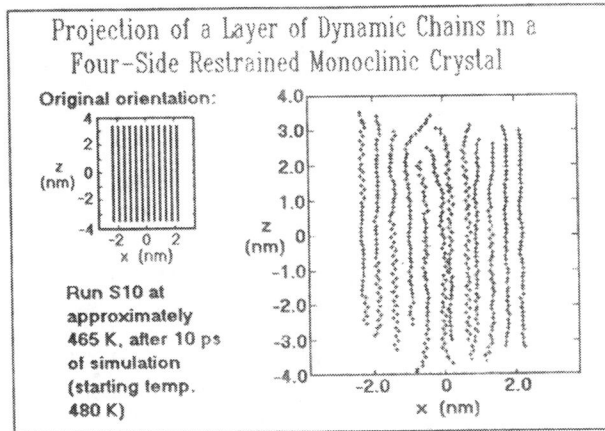

Projection of a Layer of Dynamic Chains in a
Four–Side Restrained Monoclinic Crystal

Original orientation:

Run S10 at
approximately
465 K, after 10 ps
of simulation
(starting temp.
480 K)

Figure 3 Longitudinal diffusion shown in x-z projection.

tional defects. It is thus most likely that the diffusion mechanism is coupled mainly to the longitudinal acoustic vibration (LAM) and the other skeletal vibrations. Similar diffusions were studied earlier by moving chains through the crystal by application of directed external forces.[10] Irregularities in the spacing of the chains can be seen in the center of the crystal, suggesting that a staggered packing of the chains, as seen for the orthorhombic packing of Fig. 1, but with largely parallel zigzag planes is more stable under the given conditions.

In the cases of a 3-SC crystal, as is shown by the x-y projections of a MONO crystal at 5 and 100 ps in Fig. 4 (Run S11 in Table II), some chains can be seen to move perpendicular to the chain axis toward the open side (unconstrained side). This occurs even at temperatures initially below the melting point (Run S12, Table II). The chains are squeezed laterally out of the enclosure, accompanied by surface melting, similar to the unconstrained case of Fig.2. In the center of the mobile chains a close-to-hexagonal arrangement with a distinct domain boundary developed. As the temperature decreases (at longer simulation time), the dynamic chains recrystallize within the enclosure, but are not able to fill the cavity smoothly. A residual nucleus of the hexagonal arrangement remains at the upper left

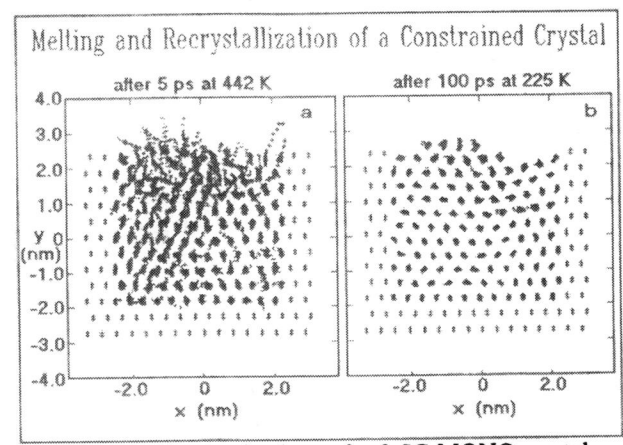

Melting and Recrystallization of a Constrained Crystal

after 5 ps at 442 K   after 100 ps at 225 K

Figure 4 The x-y projections of a 3-SC MONO crystal.

corner of Fig. 4b. In the bulk of the crystal, time was also not sufficient to heal the domain boundaries. The temperature has dropped after 100 ps of simulation to 255 K, *i.e.* below room temperature, and it can be seen that the crystal compensates the effect of the now too large enclosure by having bent x-z-crystal planes. Also note, that the top dynamic layer is not able to withdraw from the top right edge of the enclosure.

The simulation with a rigid wedge (2-SC) is illustrated in Fig. 5. In this short-time simulation, one can see the rate of formation of surface melting. Within about 5 ps major surface melting can be seen at the unrestrained corner. Even faster, within less than 1 ps one can observe the increase of the spacing of the dynamic chains relative to the fixed chains. As before, there seems to be a tendency to align the zigzag-chains parallel instead of alternating in setting angle as required for the orthorhombic packing. At the horizontal bottom layers a preference for parallel packing with the fixed chains can be noticed. The domains that develop, and can be best seen in Fig. 5b, are anchored at the

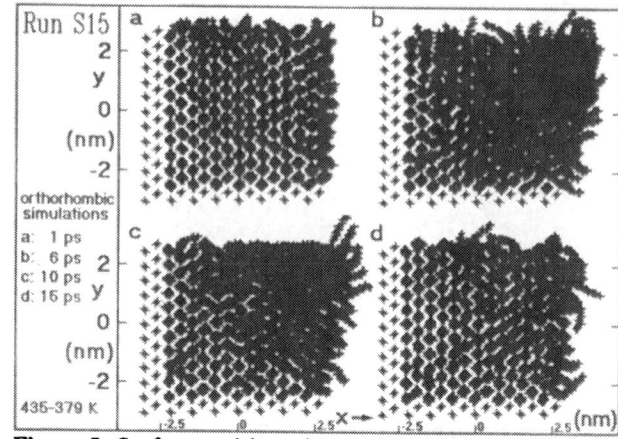

Figure 5 Surface melting of an ORTH crystal.

two solid surfaces. Most disorder is seen in the crystal after 6 – 10 ps. Although the melting temperature is not reached at 15 ps, a considerable internal improvement can be seen.

To summarize, as in the first part of this study of fusion after instantaneous heating,[1] volume expansion and internal disorder develop quickly (few ps). The amount of disorder produced in the more restrained crystals of Figs. 3 and 4 is less. The surface melting is shown to develop somewhat more slowly than the internal disorder (Fig. 5). Recrystallization of the interior disorder, that did not take place in the unrestrained crystals (Fig. 2), can be seen in Fig. 5 to be helped considerably by the presence of rigid surfaces.

Acknowledgements: Supported by the National Science Foundation, Polymers Program, Grant DMR 90-00520 and the Div. Mat. Sci., Office of Basic Energy Sciences, U.S. Dept. of Energy, under Contract DE-AC05-84OR21400 with Martin Marietta Energy Systems, Inc.

References

1. G. L. Liang, D. W. Noid, B. G. Sumpter, B. Wunderlich, *Acta Polym.*, **44**, 219 (1993).
2. C. W. Bunn, *Trans. Faraday. Soc.* **35**, 482 (1939).
3. T. Seto, T. Hara, K. Tanaka, *Japan. J. Appl. Phys.* **7**, 31 (1968).
4. R. P. Wool, R. S. Bretzlaff, B. Y. Li, C. H. Wang, R. H. Boyd, *J. Polym. Sci.* **24B**, 1039 (1986).
5. D. Brown, J. H. R. Clarke, *J. Chem. Phys.* **84** (5), 2858 (1986).
6. R. H. Boyd, S. M. Breitling, *Macromolecules* **7** (6), 855 (1974); R. A. Sorensen, W. B. Liam, R. H. Boyd, *Macromolecules* **21**, 194 (1988).
7. E. Neusy, S. Nosé, M. L. Klein, *Mol. Phys.* **52**, 269 (1984).
8. G. L. Liang, D. W. Noid, B. G. Sumpter, B. Wunderlich, *Makromol. Chem. Theory. Simul.* **2**, 245 (1993); *J. Polym. Sci.: Part B: Polym. Phys.*, to be published 1993.
9. B. Wunderlich, "Macromolecular Physics, Vol. 2," Academic Press, New York, 1976.
10. B. G. Sumpter, D. W. Noid, B. Wunderlich, *J. Chem. Phys.*, **93**, 6875 (1990); *Macromolecules*, **23**, 4671 (1992).

STRESS RELAXATION IN METALS AND POLYMERS: THEORY, EXPERIMENT AND COMPUTER SIMULATIONS

WITOLD BROSTOW *, JOSEF KUBÁT # and MICHAEL J. KUBÁT §
* University of North Texas, Center for Materials Characterization and Department of Physics, Denton, TX 76203-5308
# Chalmers University of Technology, Department of Polymeric Materials, 412-96 Gothenburg, Sweden
§ The Royal Institute of Technology, Department of Polymer Technology, 100-44 Stockholm, Sweden

ABSTRACT

There exist large amounts of experimental evidence on stress relaxation for metals and their alloys, synthetic and natural polymers, glasses and frozen non-polymeric organic liquids. The results, typically presented as curves $\sigma(\log t)$ of relaxation of stress σ as a function of logarithmic time t, exhibit common features, apparently independent of the type of material. All curves consist of three regions: initial, nearly horizontal, starting at σ_0; central, descending approximately linearly; and final, corresponding to the internal stress $\sigma_i = \sigma(\infty)$. We discuss briefly the experimental evidence as well as the main features of the cooperative theory which does not involve specific features of different classes of materials. The bulk of the paper deals with computer simulations. Simulation results obtained with the method of molecular dynamics are reported for ideal metal lattices, metal lattices with defects, and for polymeric systems. In agreement with both experiments and the cooperative theory, the simulated $\sigma(\log t)$ curves exhibit the same three regions. In agreement with the theory, the slope of the simulated central part is proportional to the initial effective stress $\sigma_0^* = \sigma_0 - \sigma_i$. The time range taken by the central part is strongly dependent on the defect concentration: the lower the defect concentration, the shorter the range. Imposition in the beginning of a high strain ε destroys largely the resistance of a material to deformation, resulting in low values of the internal stress σ_i. Since the cooperative theory assumes for particles (atoms, polymer chain segments) the existence of two states, unrelaxed and relaxed, and has a formal connection to the Bose-Einstein (B-E) distribution, we first simulate B-E systems, recording the formation of relaxed clusters of particles of different sizes. Differences in cluster sizes predicted from a B-E model and those obtained from the simulations are recorded and analyzed. On the joint basis of experimental, theoretical

Mat. Res. Soc. Symp. Proc. Vol. 321. ©1994 Materials Research Society

and simulation results, we explain the mechanism of stress relaxation in terms of deformations occurring in the immediate environment of the defects. These deformations, visible in simulations of both metals and polymers, correspond to cluster relaxations in the cooperative theory, and thus confirm a posteriori the assumptions made in developing the theory.

1. INTRODUCTION

Mechanical behavior of viscoelastic materials is most often characterized by peforming so-called transient experiments as a function of time: either the deformation under a constant load, that is creep, or the time decay of stress, $\sigma = \sigma(t)$ at a constant strain ε, that is stress relaxation [1, 2]. Large amounts of information on stress relaxation in polymers and polymer-based composites are available. The information is mostly but not only experimental. In this paper we shall consider experimental and theoretical information available, and add to it our own results of computer simulations. In distinction to more limited comprehension resulting from the use of a particular technique, we expect to acquire this way a better understanding of the phenomenon itself.

2. EXPERIMENTAL EVIDENCE

The experimental stress relaxation curves are plotted traditionally in the $\sigma = \sigma$ (log t) coordinates. Better comparison between curves for different materials can be achieved when σ_0 is taken into account as a reducing variable. Already in 1965 one of us [3] used σ^* / σ_0^* as the ordinate ($\sigma^* = \sigma - \sigma_i$; σ_0^* and σ_i are defined above in the abstract), with the usual abscissa of log t, and compared curves for a variety of materials: lead, polyisobutylene, cadmium, polyethylene, indium and rubber hydrochloride. It turned out that the curves for so vastly different materials have common features, and all exhibit three regions: initial, nearly horizontal; a long central region, descending approximately linearly; and final, again approximately horizontal. Subsequent plots for even more materials rendered similar results [4]; exemplary curves for three materials, each of a different kind, are shown in Fig. 1. Li [5] considered the final region of the diagram $\sigma = \sigma$ (log t) for *metals* and provided it with the name *internal stress*, and we have adopted his terminology for σ_i. Since it is well known that stress vs. strain curves for different types of materials show vastly different features (see for instance [6]), the objective of the present work was the explanation of the common character of stress relaxation curves.

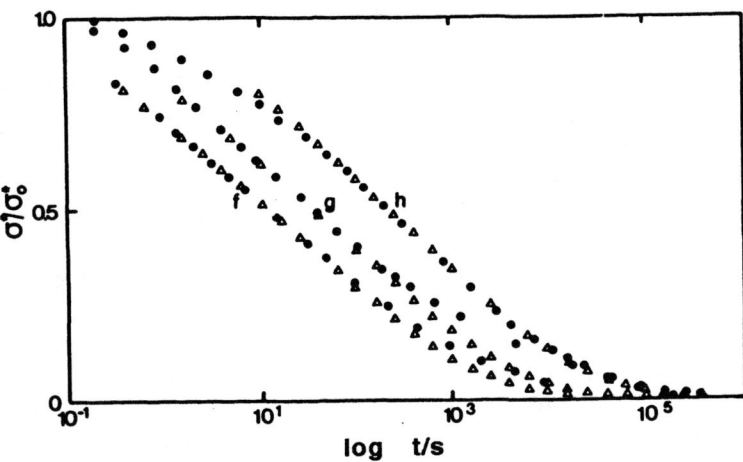

Fig. 1. Experimental stress relaxation curves for: f = Pyrex glass at 556°C; g = cetyl alkohol; and h = indium. • experimental; Δ theoretical. After [4].

3. THE THEORY

There exists a number of theoretical models aimed at predicting and explaining the stress relaxation curves. Such models have been reviewed in [2], and their various deficiencies pointed out. In general, models which can explain the behavior of metals serve poorly for polymers, as well as vice versa; see also [7]. For this reason, a material-independent *cooperative* model was developed [8, 7, 4, 2]. The model assumes a two-level system, with unrelaxed flow units in the upper level. The number n of such units at any given time serves as the measure of the difference $\sigma^* = \sigma - \sigma_i$. Details of the theory can be found in the papers quoted, in an application of the theory to the polymer aging problem [9], and a brief review is provided in [10]. We quote here only equations essential to the task at hand. The expected number of cooperatively relaxing particles (metal atoms, polymer chain segments) forming a cluster of size s is

$$\dot{n}_s = - s/[e^{s/\beta} - 1] \qquad (1)$$

the maximum cluster size at t = 0 is $\beta = (-6\dot{n}_0)^{1/2}/\pi$, while $\dot{n}_0$ is the total flow rate at t = 0. The clustering mechanism is formally equivalent to the interaction underlying the Bose-Einstein distribution and related to the phonon characteristics of the system disturbances produced by the elementary events. Given the shape of the experimental stress relaxation curves described above, we need an expression for the maximal slope

$$F = [d\sigma/d(\ln \ t)]_{max} \tag{2}$$

The theory leads eventually to the formula

$$F = c \ \sigma_0^* \tag{3}$$

Here c is a proportionality factor, found experimentally to have values near to 0.1 at ambient temperatures for metals, polymers as well as for a number of other solids [3].

4. MOLECULAR DYNAMICS SIMULATION OF METALS

Stress relaxation (an instantaneously applied strain, subsequently maintained as constant, typically $\varepsilon = 0.10$) was simulated by molecular dynamics (MD) for single-component systems of identical particles on the two-dimensional triangular lattice. An N-particle system was placed in a cell with periodic boundary conditions. Coupling between the simulation cell and an applied stress as well as an external thermal bath was used. The particles interacted via the Mie potential (often called the Lennard-Jones potential, but see [6]) with the most popular exponents, that is u(R) $= 4u_{min}[(r_o/r)^{12} - (r_o/r)^6]$, where r_o is the collision diameter and u_{min} is the depth of the well. The equations of motion used (in reduced coordinates) are defined in [10]. Calculated stress was averaged over 2000 time steps. Conversion for the reduced coordinates to those for a real metal such as copper can be accomplished by using multiplying factors [11]. The cut-off radius $r_{cut} = 2.5 \ R_{min}$ was applied, with $R_{min} = 2^{1/6}r_o$, so that both the first and the second nearest neighbors were included. It has been established in [10] that the size of the system, such as N = 8100 or N = 900, does not affect the stress relaxation curves, except for minor wiggles in the latter case.

We first performed simulations for ideal lattices without defects. The resulting curves were of the same qualitative shape as the experimental ones, but on the log t scale they extended for only slightly more than one decade, while the experimental curves typically span more than four decades. We have then tried defect generation by various procedures, and the resulting lattice with vacancies (typically about 0.03·N) were subjected to stress relaxation. The lattices *with defects* gave simulated stress relaxation curves similar in all respects to experimental ones, also in terms of the time span necessary to reach the internal stress σ_i level - a conclusion impossible to reach from experiments alone. Five examples of stress relaxation curves, all for the reduced temperature T^* = 0.2, are shown in Fig. 2. The melting point of the system studied is T^*_m

= 0.45 [12,13], with some premelting taking place below T^*_m, but the system was fully stable at $T^* = 0.4$ and below. Our simulations were typically conducted for round values of $T^* = 0.1$, 0.2, 0.3 and 0.4.

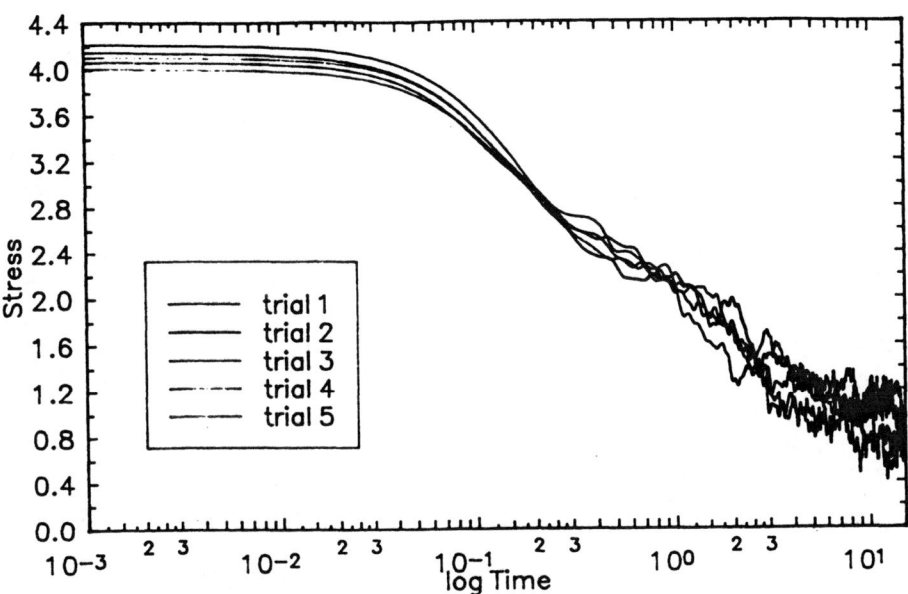

Fig. 2. Five stress relaxation curves for a lattice with vacancies at the reduced temperature $T^* = 0.2$. All three regions of the diagram known from experiments are visible in all cases, including somewhat larger fluctuations in the last segments of the curves corresponding to gradual establishment of the internal stress σ_i level.

While our simulations have been performed for two-dimensional lattices, we have clearly obtained curves analogous to experimental ones for three-dimensional materials. We infer that the dimensionality of the relaxing system is not important for the process, while the presence of defects is decisive for the time span of the relaxation.

5.   MOLECULAR DYNAMICS SIMULATION OF POLYMERS

We have conducted similar MD simulations for systems of chains on the same triangular lattice. One started with chains of length zero (single particles), and generated self-avoiding chains using a procedure due to Mom [14]. Particles *within* chains (chain segments) were interacting via a harmonic potential, while the same 6-12 potential was used for external interactions, that is for segments which are not nearest neighbors on the same chain. Stress relaxation simulations were performed in an analogous way as for metals.

Simulation results obtained for chain systems are very similar as those for metals, exhbiting the same three regions in the $\sigma = \sigma(\log t)$ curves and the same time span over at least four decades for the entire process. Hence, for brevity, we are not displaying here these results. As stressed in Section 2, experimental stress relaxation curves are essentially independent of the type of material, and our simulation confirm this fact as well. Given this indistiguishability, the question naturally arises, to what extent are our metals and polymers representative of the real materials. MD enables also simulation of tensile stress-strain experiments. Such simulations have been performed using the same hexagonal lattices and the same potentials [15], and stress *vs*. strain curves typical for metals and polymers obtained. The former exhibit maxima in the stress-strain curves with subsequent decline, the latter exhibit plateau regions with subsequent ascension. Hence our computer-generated systems are sufficiently representative for the real metals and polymers.

Eq. (3), a result of the cooperative theory, is also confirmed by experiments. Our simulations confirm this as well: for metals as well as for polymers, in both cases with vacancies, the values of c vary between 0.09 and 0.14, thus oscillating around the value of 0.10.

We have also investigated imposition of different values of the strain ϵ. It turns out that higher strains result in lower values of the internal stress σ_i. We infer from these results that the imposition of a high strain ϵ destroys largely the resistance of a material to deformation, as reflected in low values of the internal stress.

The most intriguing question is of course the similarity of stress relaxation curves for different classes of materials. We find consistently that stress relaxation is mainly caused by deformations which occur in the vicinity of the defects. There is a certain resemblance here between crystal nucleation from a melt and "nucleation" of cracks. In the ideal lattice, relatively large forces were needed to create a crack, but then the same forces were sufficient for quick propagation. In lattices with defects, relatively small forces cause atomic movements, but the movements are mainly of the ductile or flow type, rather than brittle crack propagation, extending the time span of the relaxation. The deformations we see resemble the regions of inhomogeneous atomic movement observed by Srolovitz, Vitek and Egami [16] - although they simulated amorphous metals to obtain stress-strain curves.   They observed nucleation of crack-like defects caused by loading, as well as localized viscous flow.

According to the cooperative theory, the flow units induce transitions of other unrelaxed flow units during the stress relaxation process. This is precisely what we see in simulations,   following the temporal change of the configurations of the system. We conclude that stress relaxation occurs via deformations in the immediate environment of the defects.   Hence it is the presence and concentration of the defects, rather than connectedness - or otherwise - of the particles that is decisive for the stress relaxation process. Needless to say, these deformations - visible in our simulations of both metals and polymers -correspond to cluster relaxations in the cooperative theory.   The necessary condition for stress relaxation curves such as in Figures 1 or 2 is a relatively uniform distribution of defects - as it occurs in real materials [3].

6.   BOSE-EINSTEIN CLUSTER RELAXATION SIMULATION

We have studied systems of particles such that each system had first all particles in their unrelaxed states.   We generated relaxing clusters of various sizes using Eq. (1).   However, with some sites already relaxed, competition for relaxed particle sites caused deviations shown in Fig. 3:

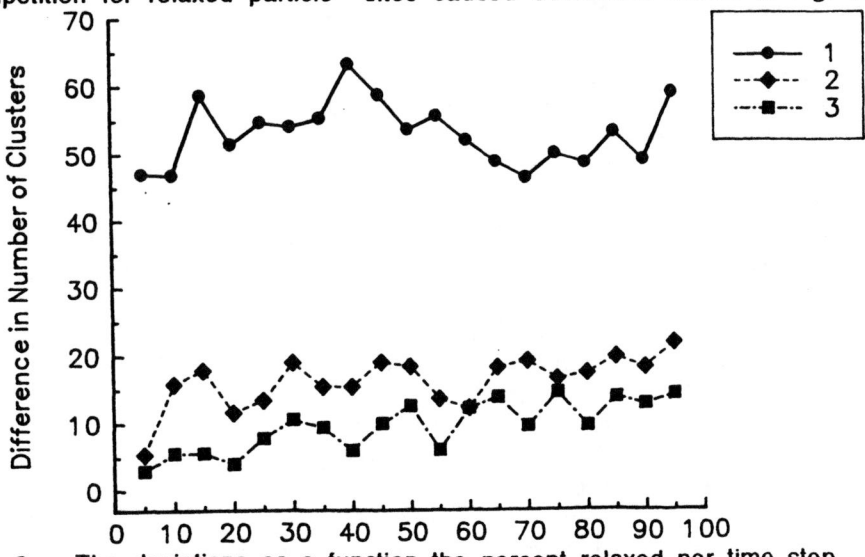

Fig. 3.   The deviations as a function the percent relaxed per time step.

Fig. 3 is an average of five runs for 900 particles. We infer that the competition for particles not yet relaxed increases the number of small clusters. This happens at the expense of clusters with more than six particles (not shown in the Figure) where negative deviations from Eq. (1) occur.

Acknowledgments

Dr. Slawomir Blonski, Mr. Wyatt A. Foard and Mr. James Palmer have participated in programming and in simulation runs. Computations were made on a Digital DecStation 3100 in Stockholm, on a Performance Optimization IBM RISC System/6000 Power Station in Denton, and on a Cray at the National Center for Computing Applications, Urbana, IL. We appreciate discussions with Prof. Robert Maksimov, Institute of Polymer Mechanics of the Latvian Academy of Sciences, Riga and with Prof. Nicholas W. Tschoegl at the California Institute of Technology, Pasadena.

References

1. V.H. Kenner in Failure of Plastics, edited by W. Brostow and R.D. Corneliussen (Hanser, Munich - Vienna - New York, 1986) Chap. 2.
2. J. Kubát and M. Rigdahl in Failure of Plastics, edited by W. Brostow and R.D. Corneliussen (Hanser, Munich - Vienna - New York, 1986) Chap. 4.
3. J. Kubát, Nature 204, 378 (1965).
4. J. Kubát, L.-Å. Nilsson and W. Rychwalski, Res Mechanica 5, 309 (1982).
5. J.C.M. Li, Can. J. Phys. 45, 493 (1967).
6. W. Brostow, Science of Materials (Wiley, New York, 1979); W. Brostow, Einstieg in die moderne Werkstoffwissenschaft (Carl Hanser, München - Wien, 1985).
7. J. Kubát, Phys. Status Solidi B 111, 599 (1982).
8. J. Kubát and M. Rigdahl, Mater. Sci. & Eng. 24, 223 (1976).
9. J.M. Kubát, J.-F. Jansson, M. Delin, J. Kubát, R.W. Rychwalski and S. Uggla, J. Appl. Phys. 72, 5179 (1992).
10. W. Brostow and J. Kubát, Phys. Rev. B 47, 7659 (1993).
11. Z.-G. Wang, U. Landman, R.L. Blumberg Selinger and W.M. Gelbart, Phys. Rev. B. 44, 378 (1991).
12. F.F. Abraham, Phys. Rev. Letters 44, 463 (1980).
13. F.F. Abraham, Phys. Rep. 80, 339 (1981).
14. V. Mom, J. Comput. Chem. 2, 446 (1981).
15. S. Blonski, W. Brostow and J. Kubát, Phys. Rev. B 49, in press (1994).
16. D. Srolowitz, V. Vitek and T. Egami, Acta Metall. 31, 335 (1983).

PHASE SEPARATION IN BLENDS
OF POLYSTYRENE AND POLY(P-METHYLSTYRENE) USING
THERMAL ANALYSIS AND SMALL-ANGLE NEUTRON SCATTERING

A. XENOPOULOS, J. D. LONDONO, G. D. WIGNALL, AND B. WUNDERLICH

Chemistry Division, Oak Ridge National Laboratory, Oak Ridge, TN 37831-6197
and Department of Chemistry, University of Tennessee, Knoxville, TN 37996-1600

ABSTRACT

Differential scanning calorimetry (DSC) was used to study blends of polystyrene (PS) and poly(p-methylstyrene) (PpMS). The presence of two glass transitions on heating after quenching was interpreted as evidence of phase separation at the temperature of the liquid before quenching. The small difference between the glass transitions of the homopolymers in the PS/PpMS system of ≈ 13 K limits the reliable detection of double glass transitions for blends to concentrations between 30 and 70%. The results of the DSC technique are supported by comparison with small angle neutron scattering (SANS) data.

INTRODUCTION

Studies of polymer blends are not only of importance for optimizing polymer properties, but are also of value for establishment of recycling protocols. The mechanical and melt-flow properties of polymer blends depend largely on their phase behavior which, in turn, is governed by the chain architecture. The scientific challenge is then to understand the relationship between structure, phase behavior and properties.

Miscibility in polymer blends, which requires a negative free energy of mixing, is limited by the small ideal entropy of mixing which is insufficient to overcome the positive enthalpy of mixing. Only if specific interactions are present does one expect a negative enthalpy of mixing. The classical description of polymer phase behavior was developed over 50 years ago by Flory and Huggins (F-H) [1]. The F-H theory suggests that the phase behavior is independent of chain geometry, and the interaction parameter, χ, was originally assumed to be dependent only on temperature. In order to make χ account for all non-ideal mixing effects, it must, however, also be made dependent on molecular mass, volume fraction, and temperature [2]. These effects are still poorly understood, despite some recent theoretical advances [3].

Differential scanning calorimetry (DSC) and small angle neutron scattering (SANS) are techniques that permit the study of the phase behavior of polymer blends. An unambiguous indication of miscibility is provided by DSC when the blend components have well-separated, sharp glass transitions with the corresponding increase of heat capacity, ΔC_p [4]. In temperature dependent SANS studies, phase separation is noted by determining the temperature at which the reciprocal of the scattering intensity at zero scattering angle vanishes [2]. The deuterium-labelling technique, which is used in SANS experiments to enhance the natural contrast of the blend components, has been shown, however, to lead sometimes to phase separation of its own [2]. An unambiguous indication of miscibility is

107

thus provided by SANS only when the isotopic effects on phase separation are negligible, a fact that must be established independently [5]. Calorimetry is the technique of choice for such analyses.

The study of polystyrene/poly(p-methylstyrene) blends (PS/PpMS) is of interest in the light of results from recent SANS experiments on isotopic polystyrene blends (PSH/PSD) [6], that showed a qualitatively different dependence of the isotopic interaction parameter, χ_{HD}, on concentration than the dependence found in isotopic blends of polyolefins [6,7]. Jung and Fischer [8] performed SANS experiments on PS/PpMS blends and block copolymers. Their results indicated a significant dependence of χ on temperature, composition, and molecular mass. The details of these dependencies were not further investigated. It is thus of interest to determine whether the anomalous concentration dependence of χ observed for isotopic blends of PS is also displayed by PS/PpMS.

EXPERIMENTAL PROCEDURE AND RESULTS

Table 1 shows details of the homopolymers from which the blends were made. The blend components were dissolved in toluene and mixed by volume. Methanol was then used

Table 1 Details of the Homopolymers

Polymer	Designation	M_w	N	M_w/M_n
d8-PS [deuterated polystyrene]	S1	32,700	291	1.05
	S2	25,300	226	1.06
	S3	55,000	490	1.04
PpMS [poly(p-methylstyrene)]	M1	58,800	498	1.04
	M2	131,000	1108	1.05

to precipitate the blends. The resultant white powder was dried at 443 K overnight. No controlled cooling procedure was followed. Table 2 includes details of the sample designa-

Table 2 Details of the Blends

Sample	ϕ_{PS}	T_s (K)
S1M1-20, S1M2-30, S1M1-85, S1M2-85, S2M2-85		$T_s < 400$ K
S1M2-40	0.4382	419
S1M1-50	0.5395	383
S2M2-70	0.7325	431
S3M2-85	0.8488	505

tion. An example is S1M1-50, where the number 50 refers to the approximate weight percent of the PS component, in this case S1. The volume fractions for the samples, ϕ_{PS}, are also listed in Table 2, along with T_g, to be discussed below.

For the DSC measurements a Perkin-Elmer DSC-7 with mechanical refrigeration unit and constant flow of dry nitrogen gas was used. Temperature and heat of fusion were calibrated with indium and are accurate to ±1 K and ±2%, respectively. About 10 to 15 mg of sample were enclosed in standard aluminum pans and sealed. All samples were heated to 400 K and then cooled at either 0.1 or 10 K/min through the glass transition region. A standard heating rate of 10 K/min was used for all heating runs. The DSC traces of four of the homopolymers are shown in Figure 1. The slight (1.5 K) difference in T_g may be a manifestation of the different molecular masses.

Figure 2 shows heating traces for the S1M2-40, S2M2-70 and S1M1-50 blends using cooling rates of 10 and 0.1 K/min. The traces for the S1M2-40 and S2M2-70 blends exhibit double glass transitions for both cooling rates. These blends are clearly phase separated. Cooling rates of 0.1 K/min result in pronounced hysteresis endotherms in the subsequent heating, facilitating the qualitative detection of double glass transitions [9]. The S1M1-50 blend displays a double T_g in the trace corresponding to the 0.1 K/min cooling rate, but only one T_g is apparent at the higher cooling rate. The reduced miscibility after slow cooling may well be caused by the longer time available for the phases to separate on slow cooling. Traces for other samples (S1M1-20, S1M2-30, S1M1-85, S2M2-85, S1M2-85; not shown in Fig. 2) displayed only one T_g, corresponding to miscible blends.

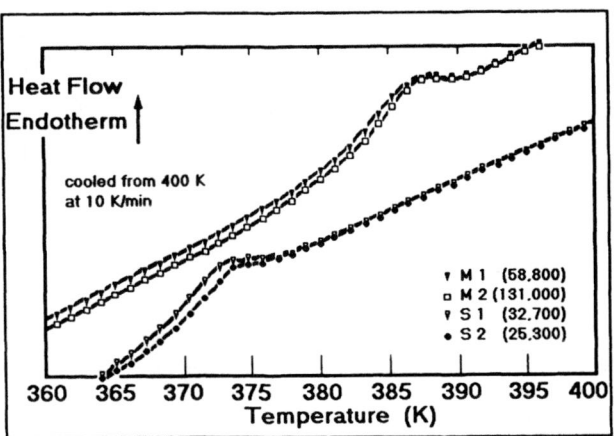

Figure 1 Heating Traces for Homopolymers

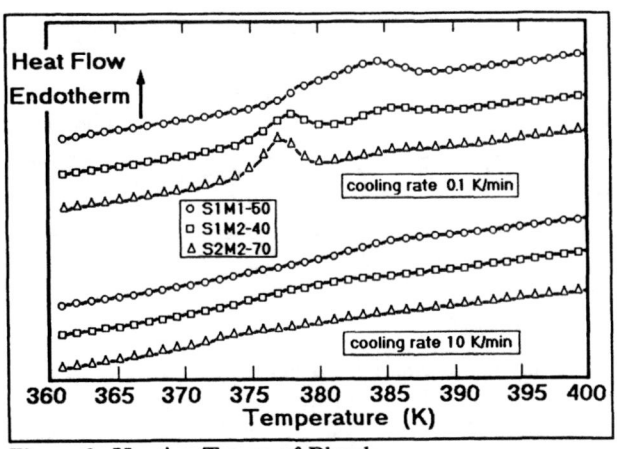

Figure 2 Heating Traces of Blends

In a second series of measurements, samples were annealed at temperatures higher than 400 K and then

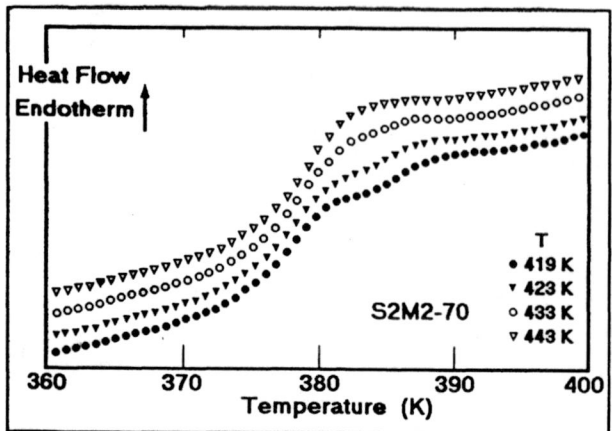

Figure 3 Heating Traces of S2M2-70 after Heating to the Indicated Temperatures

quenched. The purpose of these measurements was to assess the accuracy with which the location of the two-phase boundaries can be determined from the DSC measurements. Two sets of data are shown to illustrate the results. The first set is given in Fig. 3. The blend S2M2-70 was heated to temperatures between 413 and 443 K and quenched in the calorimeter with a rate of about 500 K/min. The instrument remained in control throughout the glass transition region. From Fig. 3 one can see that the phase boundary is located between 423 K and 443 K. At 423 K and below the quenched sample has two glass transitions, while at 443 K the blend is a homogeneous solution with one glass transition. The trace for the sample quenched from 433 K displays two glass transitions that are barely resolved, marking close proximity of the phase boundary. The second set of data are shown in Fig. 4. Sample S3M2-85 was annealed at temperatures between 423 K and 513 K and quenched by immersion into liquid N_2. Comparable results (not shown) could also be obtained by quenching inside the calorimeter, as done for sample S2M2-70. For S3M2-85 the low concentration of the M2 homopolymer makes it difficult to detect the two glass transitions. Careful examination of Fig. 4 reveals that S3M2-85 is a solution at 513 K, but phase-separated at 423 K. Detailed inspection of traces from intermediate annealing temperatures (not shown) do not reveal further details.

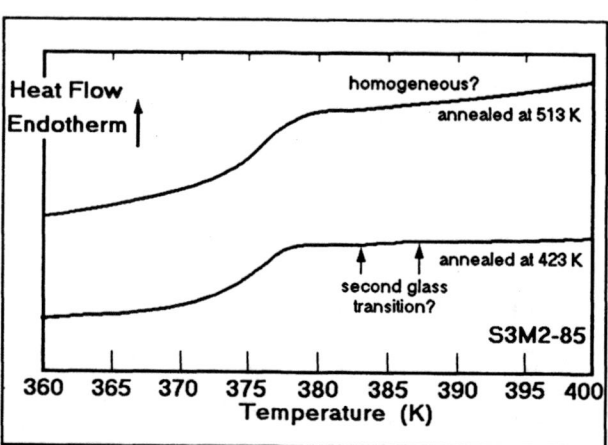

Figure 4 Heating Traces of S3M2-85 after Heating to the Indicated Temperatures

The small-angle neutron scattering experiments were performed at the W. C. Koehler Center for Small Angle Scattering Research at Oak Ridge National Laboratory. The SANS data were converted to differential scattering cross-sections, $d\Sigma/d\Omega(Q)$ (in units of cm^{-1}; with data collected between $0.05 < (Q = 4\pi\lambda^{-1}\sin\theta) < 0.4$ nm^{-1}; 2θ is the scattering angle;

and λ, the neutron wavelength). Corrections were made for effects of void coherent scattering and incoherent background scattering. The data collection and correction procedures have been described in numerous reports [5-7].

Only the SANS data that are useful for comparison with the DSC results will be given here. Further details will be reported elsewhere [10]. Figure 5 shows, that for sample S3M2-85, the scattering is relatively temperature insensitive down to 483 K, and that the intensity at 473 K increases at the smallest Q-values by almost one order of magnitude. This change may be interpreted as the result of a phase separation, that took place between 473 and 483 K. Ornstein-Zernike plots [of $(d\Sigma/d\Omega)^{-1}$ versus Q^2, not shown], are linear, and can be used for an extrapolation to zero scattering angle. They confirm the phase separation. The intensity at small Q for the pattern taken at 408 K is

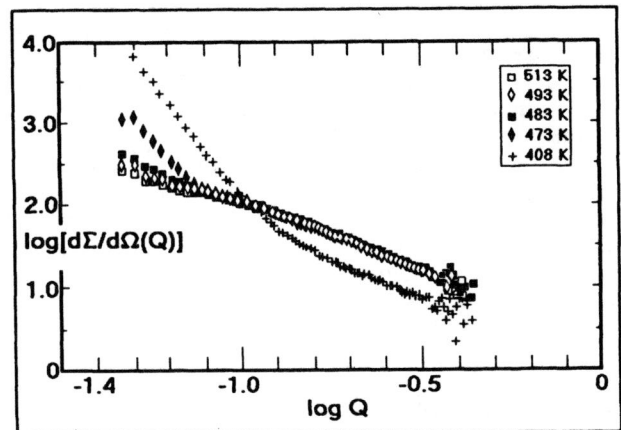

Figure 5 Small Angle Neutron Scattering Data for the Blend S3M2-85 at the Indicated Temperatures

three orders of magnitude larger than in the homogeneous phase, proving the phase-separation somewhere between 473 and 483 K, in agreement with Fig. 4.

DISCUSSION

From earlier SANS studies [8] on PS/PpMS [ϕ = 0.5, MW(PS) = 35,100, MW(PPMS) = 63,500], the dependence of χ_{SANS} on absolute temperature, T, was determined to be χ_{SANS} = −0.0002 + (2.16/T). This temperature dependence was combined with the F-H prediction of the value of χ at the spinodal, χ_s, to yield values of the phase-separation temperature, T_s, listed in Table 2.

$$\chi_s - \frac{v_o}{2} \left(\frac{1}{v_A \phi_A N_A} + \frac{1}{v_B \phi_B N_B} \right) \qquad (1)$$

In Eq. 1, v_A and v_B are the molar volumes of the homopolymers [11] and $v_o = (v_A v_B)^{1/2}$. The values of T_s calculated in this manner can be expected to be only an approximation, as χ is also a function of ϕ and N. Furthermore, care must be taken to interpret the SANS determined value of the interaction parameter, χ_{SANS}, as the interaction parameter χ, originally introduced by F-H. These two values of χ are identical only when χ has no ϕ-dependence [7]. Despite these approximations, the resulting values of T_s are useful in interpreting our results.

In Table 2 it is shown that, except for S1M2-40, S2M2-70, and S3M2-85, the blends have a calculated T_g lower than 400 K. This is thus the reason why S1M2-40 and S2M2-70 were found to be inhomogeneous after heating to 400 K. The value of T_g for sample S3M2-85 is listed in Table 2 as 505 K. This sample was observed, however, by SANS to phase-separate between 473 and 483 K (see Fig. 5). This discrepancy is likely to be a consequence of the approximations made in the calculation of T_g.

The results in Figs. 4 and 5 show that phase separation in the sample S3M2-85 was detectable by SANS, but the DSC data were difficult to interpret for the same sample because of the low pMS concentration. Figures 2 and 3 indicate that DSC can be used to detect phase separation unambiguously in this system, thus the detection limits are approximately $0.30 \leq \phi \leq 0.70$.

The combination of DSC and SANS can thus resolve the problem of phase-separation in the PS/PpMS system and clarify the influence of the isotope substitution. The DSC can, futhermore, provide information on the size of the phases in case of microphase separation and the symmetry of the broadening of the transition that may be connected with concentration fluctuations. A study of this type was completed for the system polystyrene/poly-α-methylstyrene mixed by blending and blockcopolymerization [9,12]. The full data on the PS/PpMS system and its link to the theory awaits completion [10].

ACKNOWLEDGMENTS

This work was supported by the Division of Materials Research, National Science Foundation, Polymers Program, Grant # DMR 90-00520 and the Division of Materials Sciences, Office of Basic Energy Sciences, U.S. Department of Energy, under Contract DE-AC05-84OR21400 with Martin Marietta Energy Systems, Inc.

REFERENCES

1. P. J. Flory, Principles of Polymer Chemistry (Cornell Univ. Press, Ithaca, NY, 1953). M. L. Huggins, Physical Chemistry of High Polymers (John Wiley and Sons, New York, NY, 1958).
2. G. D. Wignall in *Encyclopedia of Polym. Sci. Eng.*, **10**, 112 (1987).
3. K. S. Schweizer and J. Curro, *Phys. Rev. Lett.*, **60**, 809 (1988).
4. B. Wunderlich, Thermal Analysis (Academic Press, Boston, MA, 1990).
5. R. G. Alamo, J. D. Londono, L. Mandelkern, F. C. Stehling, and G. D. Wignall, *Macromolecules*, (submitted).
6. J. D. Londono, A. H. Narten, G. D. Wignall, K. G. Honnell, E. T. Hsieh, T. W. Johnson, and F. S. Bates, *Macromolecules*, (submitted).
7. F. S. Bates, M. Muthukumar, G. D. Wignall, and L. J. Fetters, *J. Chem. Phys.*, **89**, 535 (1988).
8. W. G. Jung, E. W. Fischer, *Makromol. Chemie, Makromol. Symp.*, **16**, 281 (1988).
9. S.-f. Lau, J. Pathak, and B. Wunderlich, *Macromolecules*, **15**, 1278 (1982).
10. J. D. Londono, G. D.Wignall (in preparation).
11. W. Patnode, W. J. Scheiber, *J. Am. Chem. Soc.*, **61**, 3449 (1939). R. Corneliussen, S. A. Rice and H. Yamakawa, *J. Chem. Phys.*, **38**, 1768 (1963).
12. U. Gaur and B. Wunderlich, *Macromolecules*, **13**, 1618 (1980).

LOCAL STRUCTURE SURROUNDING IMPLANTED As⁺ IONS IN POLYSULFONE FILMS

R.A. MAYANOVIC*, Y. FENG*, K.W. GROH*, Y. WANG*, R.E. GIEDD*, AND M.G. MOSS**
*Dept. of Physics and Astronomy, Southwest Missouri State University, Springfield, MO 65804
**Brewer Science Inc., Rolla, MO 65401

ABSTRACT

Results from As K-edge XAFS studies on polysulfone films implanted with 50 KeV As⁺ in the dose range of 10^{15} to 10^{16} ions/cm² indicate that As reacts chemically with O atoms to form As-(3)O based molecular structures having As-O bond lengths equal to 1.81±0.02 Å. In comparison to samples implanted in the dose range 10^{15} to 10^{16} ions/cm², the molecular environment surrounding As in polysulfone implanted at 10^{17} ions/cm² has additional structure beyond the nearest-neighbor O atoms.

INTRODUCTION

Polymers constitute one of the most widely used class of materials in modern society, second only to metals: Past trends indicate an increase in demand in the future of polymer-based materials.[1] In particular, modification of polymers leading to their improved physical properties, such as mechanical strength and electrical conductivity, is certain to have a significant role in the future. One way in which modification of polymers is being accomplished is by ion implantation. In general, ion implantation in polymers causes cross-linking between chains and formation of carbonized structures as a result of weakening and breakage of chemical bonds. Results of studies on modifications of physical properties of polymers resulting from ion-implantation have been reviewed elsewhere.[2, 3]

An interesting consideration regarding ion implantation in polymers is the nature of chemical activity of implanted ions after coming to rest in the material. The molecular structure in the immediate vicinity of the implanted ionic species is a direct consequence of its chemical activity in the polymer. Because the structural environment surrounding the implanted ionic species in polymers is expected to be highly disordered and short-ranged, experimental techniques based on long-ranged atomic or molecular order are inappropriate for investigative use. One technique which is ideally suited for such studies is x-ray absorption fine structure (XAFS). To our knowledge, there have been no prior direct studies on structure surrounding implanted ions in polymers. We report here on first direct investigations of this kind showing that As⁺ ions become chemically active upon implantation in polysulfone (Figure 1) as evidenced by bonding directly to oxygen.

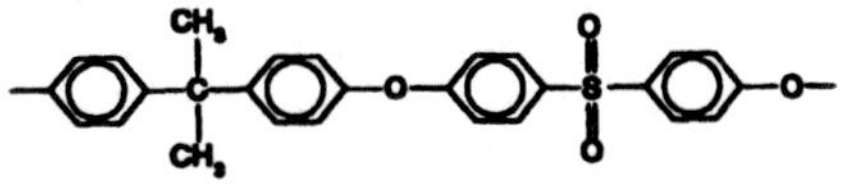

Figure 1. Polysulfone

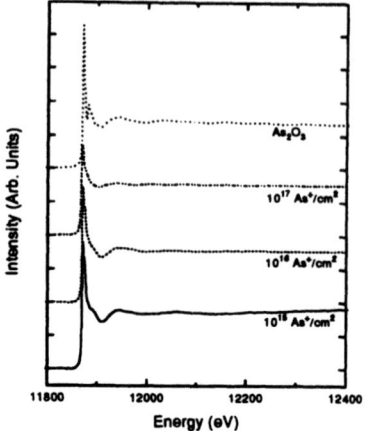

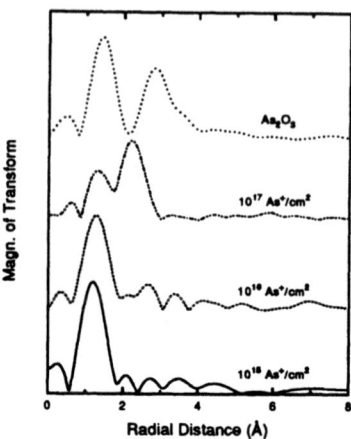

Figure 2. As K edge x-ray absorption fine structure spectra for 10^{15} (solid), 10^{16} (short-dash), 10^{17} (dot-dash) As$^+$/cm^2 implanted in polysulfone film samples, and As$_2$O$_3$ (long-dash) powder sample.

Figure 3. Fourier transform of (XAFS)·k^2, where k is the photoelectron wavenumber. The transform k-ranges used were 3.2-9.5, 3.2-9.5, 3.2-12.0, and 3.8-12.5 Å^{-1} for the 10^{15} (solid), 10^{16} (short-dash), 10^{17} (dot-dash) As$^+$/cm^2 implanted in polysulfone films, and As$_2$O$_3$ (long-dash) sample, respectively.

EXPERIMENTAL

The amorphous polysulfone films (~0.5 μm thickness) were formed by spin-coating from chemically synthesized solution on glass slides at Brewer Science. The implantation was performed at room temperature with 50 KeV As$^+$ ions, at 0.5 μA/cm^2 of beam current density, in the dose range of 10^{15} to 10^{17} ions/cm^2. Commercially available As$_2$O$_3$ was ground using ceramic tools, then sieved to fine powders to 400 mesh (each having particle size <38 μm in diameter) and rubbed onto Scotch Magic tape.

Room temperature XAFS measurements were performed on beamline X23A2, at the National Synchrotron Light Source, which is equipped with a double-crystal type monochromator having Si(220) crystals. A slit (~1 mm) was used to collimate the incident x-ray beam, providing a flux of ~5x10^9 photons/s with a maximum stored current of 220 mA. The XAFS spectra were obtained by measuring the fluorescence signal at the As K edge using a 5.08 cm diameter P.I.N. photodiode.[4] The signal flux was in the range 1-10x10^7 photons/s. The incident x-ray beam intensity was measured using nitrogen-filled ionization chambers. Calculations of concentration of 50 KeV As$^+$ implanted ions as a function of depth in polysulfone film using TRIM[5] indicate that the predominant contribution to each sample's fluorescence signal originated from a region ~300 Å thick, centered at a mean depth of ~550 Å.

As K edge XAFS spectra for the implanted polysulfone films and powdered As$_2$O$_3$ are shown in Figure 2. Further analysis involved isolating the above-edge absorption fine structure

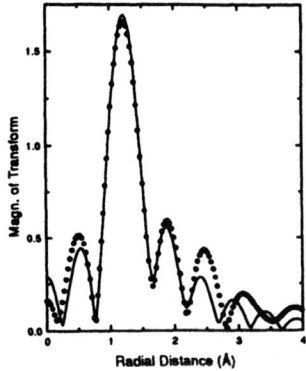

Figure 4. Comparison of Fourier transform of $XAFS \cdot k^2$ from experiment on 10^{15} As^+/cm^2-in-polysulfone (points) and fit (line) based on FEFF calculation.

and calculating Fourier transforms, shown in Figure 3, of the isolated XAFS: More detail on the analysis of XAFS data can be found elsewhere.[6] For unoriented samples, Fourier transforms such as those shown in Figure 3 reflect the radial distribution of atoms surrounding the photoexcited atom, in this case the As atom. For example, complete disorder of As relative to all other atomic species contained within the polysulfone would be reflected in Fourier transforms of XAFS having very little structure. We used non-linear least-squares fitting,[7] based on the Levenberg-Marquardt method, in order to obtain nearest-neighbor bond lengths and coordination numbers of As in polysulfone implanted at 10^{15} As^+/cm^2 dose. The fitting was done using theoretical XAFS, generated using FEFF,[8] a program employing a curved-wave and multiple-scattering formalism. A direct comparison of the fit and experimental Fourier transform for the polysulfone film sample implanted at 10^{15} As^+/cm^2 dose is shown in Figure 4.

RESULTS AND DISCUSSION

Figure 3 clearly indicates a local molecular structure surrounding the As atom in polysulfone dependent on implantation dose. Fourier transforms show a nearest-neighbor shell peak located at ~1.3-1.4 Å for all implanted polysulfone samples and a very prominent next-nearest-neighbor shell peak at ~2.2 Å for the 10^{17} As^+/cm^2-sample only. While this appears to be a clear indication of increasing structural complexity with dose, relative insensitivity of XAFS to hydrogen atoms and problems with shadowing or blocking effects make such conclusions somewhat tentative at present.

Based on distance and obvious similarity between the nearest-neighbor peaks of the transforms for the polysulfone samples and As_2O_3, the fitting represented in Figure 4 was performed assuming a distribution of oxygen atoms surrounding arsenic: Carbon was tried in place of oxygen but yielded substantially poorer agreement between fit and exerimental transforms. The results of the fitting are: $R_1 = 1.81 \pm 0.02$ Å, $N_1 = 3$, and $\sigma^2_1 = 0.008 \pm 0.004$ Å^2, where R_1, N_1, and σ^2_1 are the As-O nearest-neighbor distance, coordination number, and Debye-Waller factor, respectively. It is obvious from these results that at 10^{15} -10^{16} As^+/cm^2 dose, arsenic bonds to three oxygen atoms in a geometrical arrangement similar in nature to one found in As_2O_3, i.e., either planar-triangular or pyramidal. It is conceivable that the As-O bonding in polysulfone may occur shortly after the release of volatile oxygen resulting from irradiation damage.[2] There is high likelyhood that volatile hydrogen also participates in the reaction and possibly bonds on the exterior of the nearest-neighbor oxygen atoms. The relatively large σ^2_1 value, along with values of other fitting parameters, indicate significant structural disorder surrounding arsenic atoms in polysulfone. The additional structure in the Fourier transform data of the 10^{17} As^+/cm^2-polysulfone sample presents a substantial increase in degree of difficulty of determining the molecular structure surrounding the arsenic atom. Fitting of this data is presently underway.

CONCLUSIONS

We have shown for the first time direct evidence from As K edge XAFS results that implanted As^+ ions undergo bonding directly to oxygen atoms in polysulfone. The molecular structure is found to have arsenic bonded to three oxygen atoms, in either planar-triangular or pyramidal arrangement for the 10^{15} and 10^{16} As^+/cm^2-polysulfone samples, at bond lengths very close to those in As_2O_3. The structure for the 10^{17} As^+/cm^2-polysulfone sample appears to be more complicated as evidenced by having additional structure beyond the nearest-neighbor oxygen atoms.

ACKNOWLEDGMENTS

The X23-A2 beamline at Brookhaven National Laboratory is supported in part by the National Institute of Standards and Technology. The NSLS is supported by the Department of Energy under contract DE-AC02-76CH00016.

REFERENCES

1. M.F. Ashby, MRS Bull. **18**, 43 (1993).

2. T. Venkatesan, Nucl. Instrum. Meth. **B7/8**, 416 (1985).

3. G. Marietta, Nucl. Instrum. Meth. **B46**, 21 (1990).

4. S.M. Heald, in X-Ray Absorption: Principles, Applications, Techniques of EXAFS, SEXAFS, and XANES, edited by D.C. Koningsberger and R. Prins (John Wiley & Sons, New York, 1988), p. 87.

5. J.F Zeigler, J.P. Biersack, and U. Littmark, The Stopping and Ranges of Ions in Solids (Pergamon, New York, NY, 1985).

6. D. E. Sayers and B. A. Bunker, in X-Ray Absorption: Principles, Applications, Techniques of EXAFS, SEXAFS, and XANES, edited by D.C. Koningsberger and R. Prins (John Wiley & Sons, New York, 1988), p. 211.

7. M. Newville, P. Livins, Y. Yacoby, J.J. Rehr, and E.A. Stern, Phys. Rev. B **47**, 14126 (1993).

8. J. Mustre de Leon, J.J. Rehr, S.I. Zabinsky, and R.C. Albers, Phys. Rev. B **44**, 4146 (1991).

X-RAY AND ELECTRON DIFFRACTION STUDIES OF AS-DEPOSITED RF SPUTTERED THIN FILMS OF IrO_2

I. T. Penfold*, S. C. Moss*, J. Kulik** and K. G. Kreider***
*   Department of Physics, University of Houston, Houston, TX 77204-5506
** Texas Center for Superconductivity, University of Houston, Houston, TX 77204-5932
*** Process Measurement Division, NIST, Gaithersburg, MD 20899.

ABSTRACT

Iridium oxide films produced by reactive sputtering (SIROF's) have considerably higher densities than those made by other techniques such as the anodic reaction at metal surfaces which contain pores and microvoids. It has been previously reported that SIROF's deposited with substrate temperatures of 300K are amorphous. Here we report x-ray and electron diffraction measurements on SIROF's deposited on Al_2O_3, Si, NaCl and MgO substrates at 40°C. The x-ray diffraction patterns, I(Q), show "diffuse" structure that extends to beyond $Q(=4\pi\sin\theta/\lambda)=10Å^{-1}$. We show that it is possible to reproduce qualitatively the main features of the diffraction pattern by convoluting the crystalline rutile powder pattern with a Lorentzian profile. The width of this profile is compatible with that determined from the Scherrer equation although significant peak shifts and texture are observed. This analysis reveals that as-deposited SIROF's are crystalline with particle sizes in the range 25Å to 35Å rather than amorphous (for which no crystalline model would be appropriate). The electron microdiffraction data are also consistent with a crystallite size of a few nm and high resolution TEM reveals lattice fringes from crystallites few nm in size. The peak shifts, however, remain to be explained.

INTRODUCTION

Iridium oxide films, produced by anodic reaction of metal surfaces (AIROF's) or reactive sputtering of Ir targets (SIROF's) have been the subject of much interest due to the electrochromic [1],[2] and electrocatalitic [3] behaviour they exhibit. In particular, the chemical stability, low impedance and mechanical robustness of SIROF's have stimulated interest in their potential use as pH sensors [4].

Previous diffraction studies on SIROF's have shown that the structure of the as-deposited films is sensitive to the presence of water in the sputtering process as well as the temperature of the substrate. Films containing water in their bulk which were deposited at room temperature have been described as amorphous [2] while those deposited in dry environments have been considered as either amorphous or extremely small grained crystals (~30Å) [5][6]. The grain size of the films is observed to increase upon thermal annealing.

The crystalline films have the rutile structure found for bulk IrO_2. The structure is tetragonal with lattice parameters a=4.4983Å and c=3.1544Å [7]. Ir atoms are octrahedally coordinated by O atoms and the structure consists of both edge and corner sharing octahedra forming ordered chains along the c-axis.

The purpose of the present study has been to determine whether the as-deposited films are truly amorphous or, if crystalline, what structural characteristics are responsible for the details observed in the diffraction patterns.

EXPERIMENTAL

The Iridium oxide films were produced by d.c. magnetron reactive sputtering of an Ir target in an argon:oxygen atmosphere. Details of the sample preparation and characterization procedures have been reported elsewhere [4]. In addition, Electron Probe Microanalysis on a typical sample showed negligible contamination and, within the accuracy of the technique, gave the stoichiometry of IrO_2.

Single crystal Si, Al_2O_3, MgO and NaCl were used as substrates with the substrate temperature being held to approximately 40°C using a temperature controlled heat sink during the deposition process. As-deposited SIROF's made using similar methods have a density of ~$10g/cm^{-3}$ [2]. Film thicknesses (t) were found by stylus measurements and in the study reported here ranged from 1 to 10μm. All the films were strongly adhered to their substrates (excepting those deposited on NaCl substrates). The crystalline rutile powder sample was 99.99% of purity with respect to metallic impurities and was supplied by Aldrich. For the diffraction measurements it was pressed into a 0.5mm thick pellet.

The x-ray diffraction experiments were performed on a Rigaku 2-circle diffractometer with a sealed tube source and a graphite (002) monochromator. Measurements were taken using Mo Kα and Cu Kα radiation with wavelengths of λ=0.7098Å and 1.5418Å, respectively.

The Transmission Electron Mircoscopy (TEM) measurements were made with a JEOL 2000FX operated at 200keV. Diffraction patterns were recorded both in conventional selected area (SAD) mode and microdiffraction mode. Several high resolution images were also recorded. Suitable pieces of a film deposited on a NaCl substrate were chosen for these measurements.

RESULTS

The x-ray patterns for the films on all of the substrates were essential the same. Due to the presence of exposed substrate and a film thickness of 1μm thick for the sample deposited on the NaCl strong substrate scattering was observed. This made quantitative measurements from this sample difficult. The patterns presented here were taken from a 10μm thick film deposited on a sapphire (Al_2O_3) substrate using the Cu x-ray source. In this case the value of μt, where μ is absorption coefficient, was high enough to neglect substrate scattering. The data taken with Mo radiation showed continuous "diffuse" scattering to beyond $Q(=4\pi sin\theta/\lambda)=10Å^{-1}$.

The x-ray diffraction pattern for the sputtered film, shown in Fig. 1, has peaks at approximately the positions for the reported rutile powder pattern [7]. The profile of the first peak could be approximated to a Lorentzian with a full width at half maximum (FWHM) Δ $Q=0.17Å^{-1}$. The Scherrer equation may be used to get a particle size:

$$\beta(2\theta) = 0.94\lambda / Lcos\theta \qquad (1)$$

where $\beta(2\theta)$, the FWHM in the scattering angle, was used to estimate the particle size L=34Å. Grain sizes for the films on all of the substrates were in the range 25Å to 35Å. In the case of the films deposited on NaCl, the widths of the IrO_2 peaks between the substrate Bragg peaks was consistent with this small particle size (~30Å).

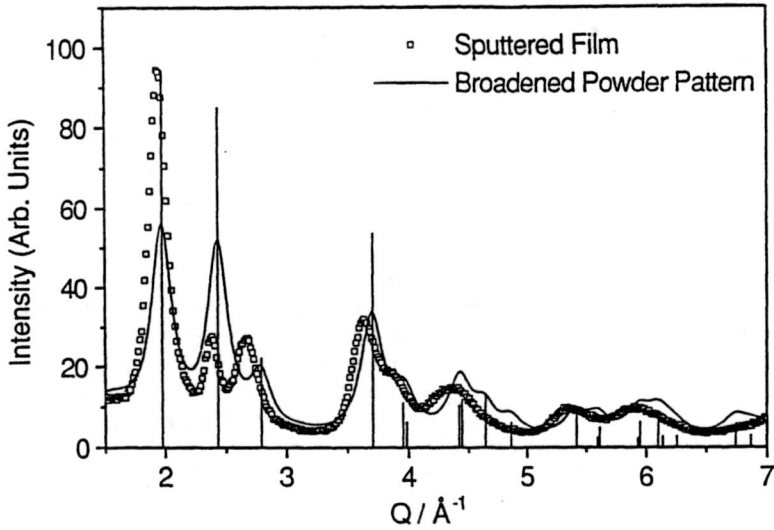

Fig. 1. Comparison of a 10μm IrO_2 film, sputtered at 40°C on sapphire, with a broadened powder pattern consistent with a particle size of 30Å. The vertical lines represent the positions and relative intensities of the crystalline IrO_2 Bragg peaks shown in Fig. 2.

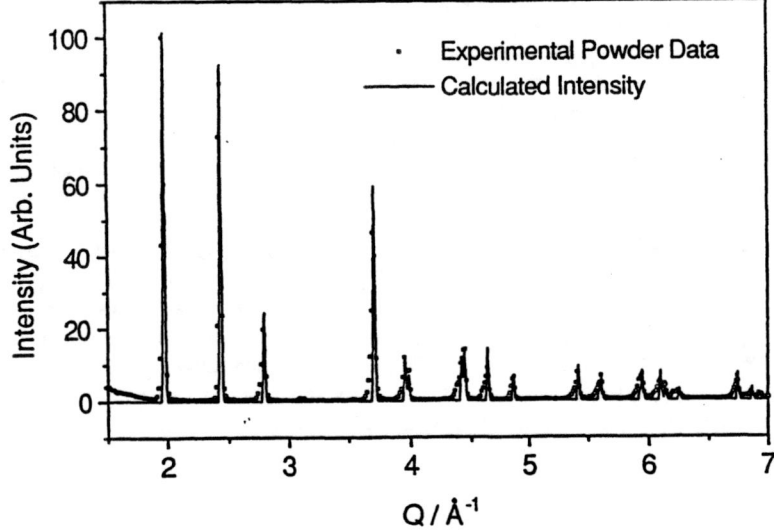

Fig. 2. Comparison of the powder pattern used in Fig. 1 and the calculated IrO_2 pattern.

Figure 2 shows the measured IrO_2 crystalline powder pattern with the calculated pattern from the reported structure [7]. The calculated intensities were convoluted by the instrumental resolution function which was determined from a Si powder pattern and dampened by an average temperature factor, e^{-2M}, estimated from difference between our calculated intensities and the experimental integrated intensities reported in [7]. The agreement between the measured and the calculated patterns is quite good. This confirms the reliability of the diffraction patterns of the sputtered films which were taken under the identical experimental conditions.

To summarize thus far, convoluting the experimental powder data with a Lorentzian profile for a particle size of 30Å, we were able to reproduce the main features of the pattern from the sputtered film (Fig. 1). It should be noted that using a particle size of 34Å, the broadened pattern exhibited more structure than the x-ray pattern from the sputtered film. This discrepancy may be due to the approximation that the peak shape is Lorentzian. The broadened powder pattern has been vertically scaled so that the agreement between the two data sets is closest at high Q. For $Q > 3Å^{-1}$ the broadened powder pattern has all the features of the sputtered film pattern including the shoulders on high Q side of the 4th and 5th peaks. The pattern was therefore indexed for the rutile structure for the discussion below.

There are clearly differences between the two patterns that require consideration. To begin with, the first peak (110) from the sputtered film has a far greater intensity, while the second peak (101) is smaller than that observed in the broadened powder. The effect of texture in the film was assessed by a pole figure analysis of the 110 peak. This revealed a higher pole density perpendicular to the surface of the film (5:1) indicating a strong preference for the 110 planes to align parallel to the substrate. This is a general feature of all the films examined here and is not unique to this particular substrate. The effect of the texture is less pronounced at higher Q where there are overlapping hkl peaks. This accounts for the better high Q agreement in the intensities of the broadened powder and sputtered films in Fig. 1.

The crystallinity of the films was confirmed by TEM. Figure 3a is a high resolution image from a film fragment obtained by crushing a small portion of one of the films. Several areas of the image exhibit lattice fringes consistent with the 110 spacing of the rutile structure. There is a large variation in grain size with many grains being significantly larger than 30Å (ranging from 30Å to 70Å in size). One particularly large grain, with a width of approximately 200Å is visible in the top right of the figure. Grains of this size must be uncommon because sharp Bragg peaks expected from such large grains are not observed in the x-ray diffraction data. Figures 3 b, c, and d are microdiffraction patterns obtained with probe sizes of 35, 100, and 250Å. In Fig 3b, the pattern arises almost entirely from a single grain. (There are some very weak reflections indicating that a small volume from one or two additional grains contributes to the pattern.) A progressively larger number of grains contribute to the patterns of Figs 3c and d. Note that the scale of all these patterns is the same, but the convergence angle is smaller in Fig. 3d. Finally, Fig. 3e is an SAD pattern obtained from an area of about $0.1\mu m^2$. It exhibits broad diffraction rings which have been indexed to the rutile pattern. The speckled nature of the rings illustrates that for the contributing volume ($< 0.01\mu m^3$) the grain orientation is not completely random and the number of grains is insufficient to yield a smooth Debye ring.

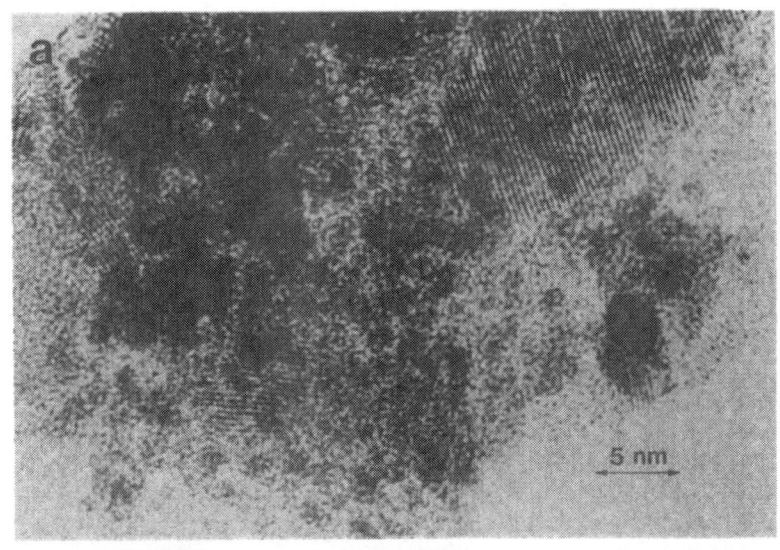

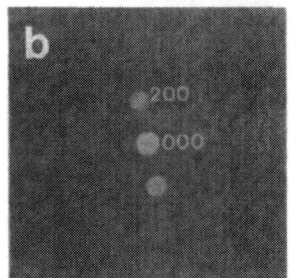
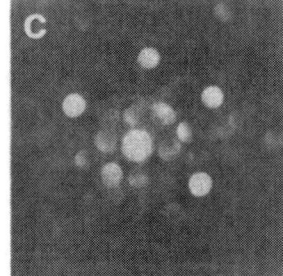
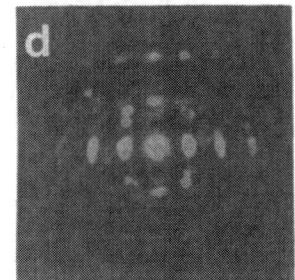

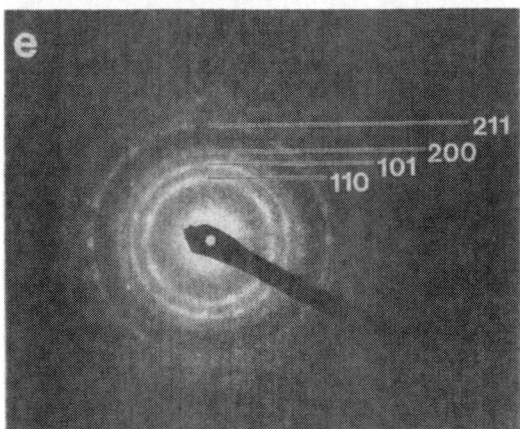

Fig. 3 (a) High resolution TEM image of a film fragment. Fig. 3 (b), (c) and (d) Microdiffraction patterns obtained with spot sizes of 40, 100 and 250 Å respectively. Fig 3 (e) SAD pattern.

CONCLUSIONS

The x-ray and electron diffraction data presented here show that the as-prepared IrO_2 films deposited at 40^oC are essentially crystalline. From the x-ray diffraction data, the structure of the films can best be interpreted as rutile with an average grain size on the order of 30Å. The films have a preference to align with the 110 planes parallel to the substrate.

The TEM data recorded from the sample deposited on a NaCl substrate suggests a larger size (30Å to 70Å) than the qualitative x-ray diffraction on the same sample. Further TEM measurements on the films for which more accurate diffraction data has been measured is underway to enable a more accurate comparison of grain size determined from the TEM and x-ray data.

The reason for the substantial shifts in the peak positions of the films relative to the calculated rutile powder lines is presently not clear. A strained or orthorhombic modification of the tetragonal lattice is unable to account for the observed shifts. The possibility that the presence of planar defects, such as stacking faults, are responsible for the observed shifts is currently being investigated. It should be noted that annealing an identical film at 420^oC for 4 hours produced a rutile diffraction pattern with peaks at the expected positions and the widths given by a particle size of ~55Å.

ACKNOWLEDGEMENTS

Research supported at the University of Houston by the DOE, Office of Basic Energy Sciences on DOE-FG05-87ER45325 and by the State of Texas at the Texas Center for Superconductivity.

REFERENCES

1. J. D. E. McIntyre, W. F Peck and S. Nakahara, J. Electrochem. Soc., 127, 1264 (1980).

2. S. Hackwood, A. H. Dayem and G. Beni, Phys. Rev. B 26 471 (1982).

3. S. Hackwood, L. M. Schiavone, W. C. Dautremont-Smith and G. Beni, J. Electrochem. Soc., 128, 2569 (1981).

4. M. J. Tarlov, S. Semancik and K. G. Kreider, Sensors and Actuators, B1 293 (1990).

5. K. G. Kreider, S. Semencik and J. W. Erickson in Proc. 4th Int. Conf. Solid State Sensors and Actuators (Transducers '87), Tokyo, Japan, June 2-5, 1987, 734.

6. K. G. Kreider, J. Vac. Technol., A4 626 (1986).

7. H. E. Swanson, M. C. Morris and E. H. Evans, NBS Monograph 25, Sec 4, 19 (1965).

DETERMINATION OF ALUMINUM COORDINATION ENVIRONMENTS IN AMORPHOUS Al₂O₃ BY SOLID STATE NMR SPECTROSCOPY

WILLIAM S. REES, JR., and LAMY J. CHOPIN
Department of Chemistry, Materials Research and Technology Center, and
The National High Magnetic Field Laboratory
The Florida State University, Tallahassee, FL 32306-3006

ABSTRACT

It previously has been difficult to probe the localized coordination environment of aluminum atoms in amorphous alumina samples by solid state NMR spectroscopy. Such an *in situ* technique has benefits in probing the structure of the material and monitoring the evolution of the microstructural development during thermal processing. The recent developments of more sophisticated NMR techniques by Pines, *et al.*, specifically devised for treatment of problems unique to the solid state spectroscopy of quadrupolar nuclei, have permitted a glimpse into these materials. As a result of applying these experimental techniques to new materials systems, a better understanding of the solid state structure of amorphous aluminas derived from polymer pyrolysis has emerged. Correlations between the onset and completion of crystallization and the localized structure may now be possible.

BACKGROUND

Alumina based materials have found use in a variety of technological applications from electronics to catalysis.[1] Since the microstructure of aluminas predominately controls the properties of these materials, it is desirable to obtain some means of controlling such microstructure. It has been found that certain aluminum-containing polymeric materials can be pyrolyzed into alumina ceramics, with the structure of the polymer determining the structure of the resulting ceramic.[2-5] Unfortunately, both the polymers and the ceramics are insoluble, amorphous materials, and have thus defied conventional attempts at bulk structural determination. Because of the amorphous qualities of the materials, XRPD fails to show the precise nature of the aluminum coordination (**Figure 1**); due to the quadrupolar nature of the ^{27}Al nucleus, and the insolubility of the materials, traditional NMR also is of limited value (**Figure 2**). A quantitative determination of the coordination at given aluminum sites in the polymers and the ceramics is desired in order to make correlations between the relatively easily controlled polymer structure and the relatively intractable ceramic structure.

Fortunately, improved technology in the field of NMR spectroscopy is allowing better resolution of signals from solid-state quadrupolar nuclei. Materials improvements have led both to higher fields and higher sample spinning speeds, each of which improves resolution under magic-angle spinning (MAS) conditions. MAS cannot completely eliminate quadrupolar broadening, but the technical improvements which Pines, *et al.*, have recently disclosed, provide the possibility for truly high-resolution solid-state NMR of quadrupolar nuclei.[6-9] Pines has shown that quadrupolar broadening in solids can be averaged out by allowing magnetization in the sample to evolve at a variety of angles in the magnetic field during the time of the experiment. Such a technique has been given the name DAS, for dynamic angle spinning. We have attempted to apply this technique

Mat. Res. Soc. Symp. Proc. Vol. 321. ©1994 Materials Research Society

to our samples in the hope of achieving a rapid, non-destructive, bulk analysis of the coordination of aluminum atoms within the samples.

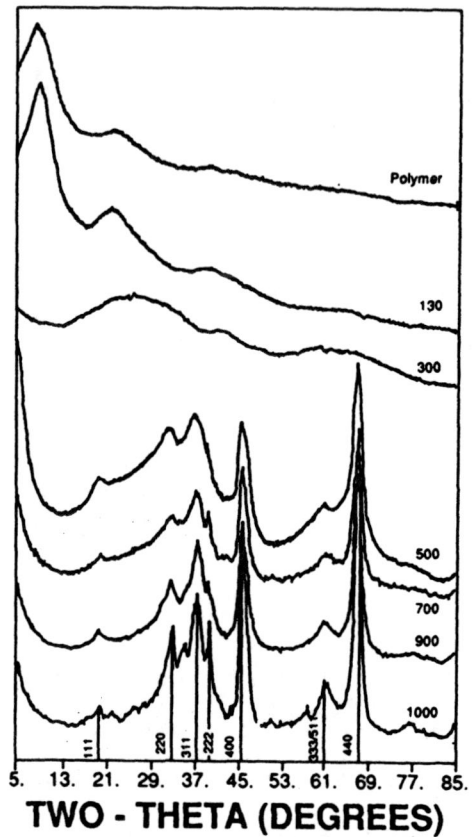

Figure 1: XRPD of polymer as it is pyrolyzed to alumina.

RESULTS AND DISCUSSION

NMR was performed using a Doty Scientific probe designed for a wide-bore Chemagnetics 400MHz magnet, with a home-built RF system. The experiment described by Pines involves applying a 90° pulse with the spinning (3-5kHz) sample oriented at one angle in the static field, allowing an evolution period of time t, then applying another 90° pulse to store the magnetization along the -z axis. While the evolved magnetization is stored, the sample quickly is reoriented in the field, and the magnetization is returned to the xy plane by a third pulse. After allowing another evolution period, t, the magnetization is detected. By varying t, a two-dimensional spectrum can be generated with the resolved isotropic spectrum on one dimension and the static powder pattern

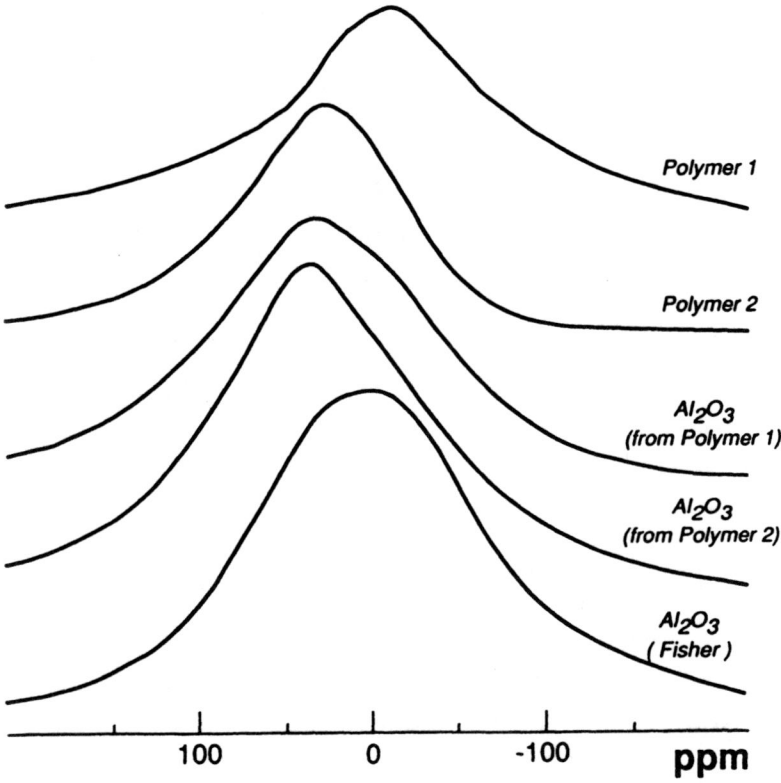

Polymer 1

Polymer 2

Al_2O_3
(from Polymer 1)

Al_2O_3
(from Polymer 2)

Al_2O_3
(Fisher)

100   0   -100   **ppm**

Figure 2: Solid state ^{27}Al NMR of polymer and ceramic samples. (static samples, 270MHz)

of the sample on the other. The simplest DAS experiment has angles of 37.38° and 79.19° relative to the static field axis, with equal evolution times at each position.[4] For a pure absorption experiment, a fourth pulse is added to achieve coherence transfer.[5] The sample is flipped by a servo motor (external to the probe) controlled by a computer, with the angles relative to a calibrated magic angle setting. The flip takes roughly 35ms, including stabilization.

The probe also is capable of high-speed MAS, and both ^{23}Na and ^{27}Al have been examined using both MAS and DAS (**Figures 3-5**). While a DAS signal is seen for the Na atom in NaC$_2$O$_4$, the resolution in the isotropic shift dimension remains poor. This is likely due to instabilities in the sample spinning speed and, possibly, to slight errors in the angle settings. Thus far, no signal for Al has been observed in a DAS experiment in our lab. This may be due to the rapid spin-lattice relaxation (T$_1$) for the sample (**Figure 6**). It may be necessary to install spin-control hardware

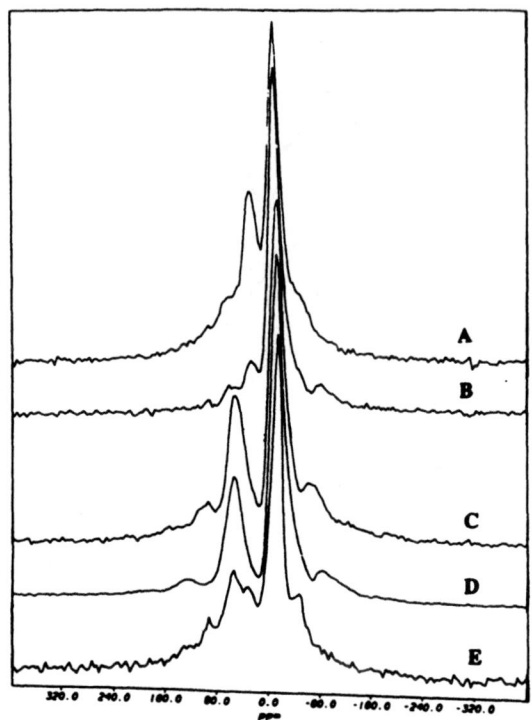

Figure 3: MAS at 5.5-6.0kHz of A) polymer 1, B) polymer 2, C) ceramic from polymer 2, D) ceramic from polymer 1, E) Al$_2$O$_3$ standard (Aldrich).

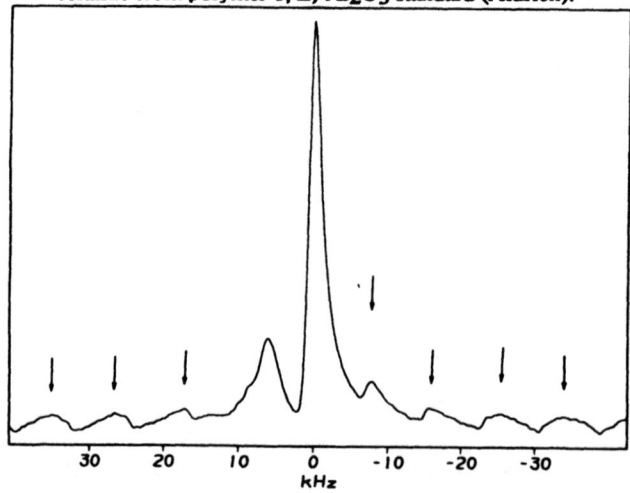

Figure 4: MAS of Al$_2$O$_3$ standard at 8kHz, showing improved resolution over the spectrum at 5.5kHz. Arrows indicate spinning side bands.

and to perform the experiments at lower temperatures. MAS provides some resolution enhancement (*c.f.*, **Figure 2**), and can be used to make a qualitative comparison of the sample structures.

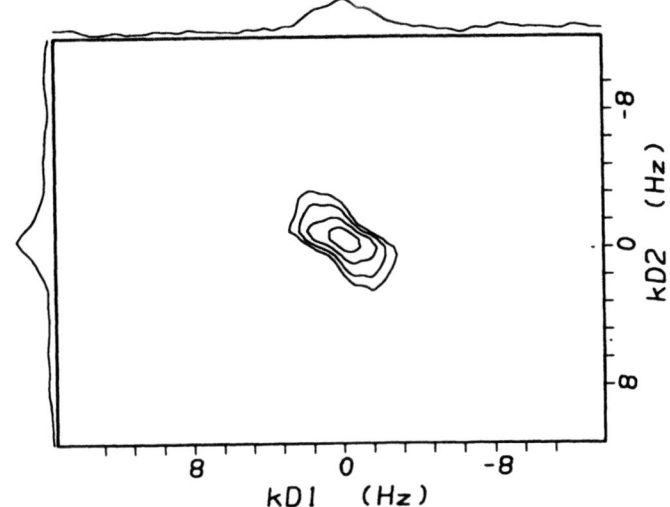

Figure 5: DAS NMR spectrum of sodium oxalate. Isotropic shift spectrum projection on the verticle axis, static powder pattern projection on the horizontal axis.

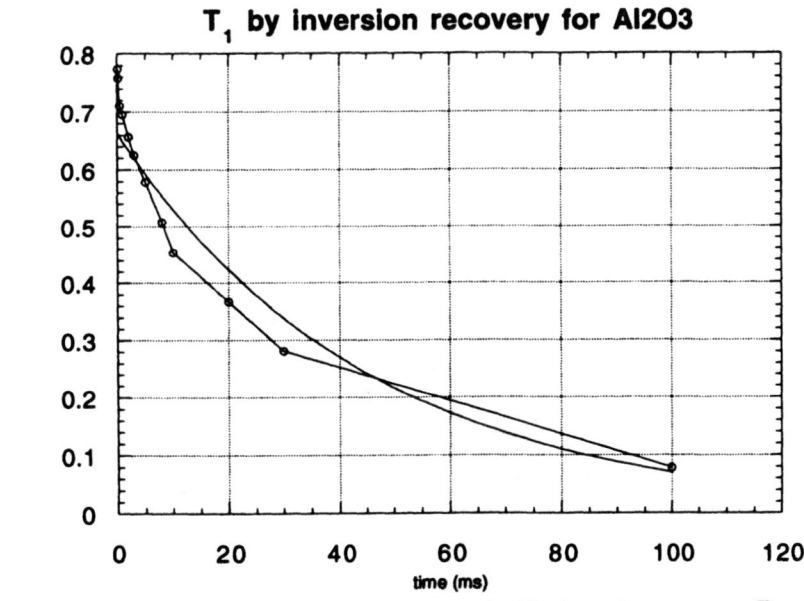

Figure 6: Determination of T_1 for Al_2O_3 standard by inversion recovery. Exponential fit (solid line, $y = 0.663 \, e^{(-0.0224x)}$, $R = 0.966$) implies T_1 of 44.5ms.

SUMMARY

The structures of some intractable, aluminum-containing polymers and aluminas have been examined by solid state NMR. While the promising technique of dynamic angle spinning NMR has so far yielded mixed results on structural data, work is continuing on understanding the intricacies of the technique, in order to make it a routine method for the determination of non-crystalline, solid state structures involving quadrupolar nuclei. High speed MAS has shown that the structures of the polymers and their corresponding ceramics are predominantly 6-coordinate at the aluminum sites, with the amount of 4- and 5-coordinate aluminum varying with the preparative route of the samples.

ACKNOWLEDGMENTS

We gratefully acknowledge Profs. Tim Cross and Terry Gullion for access to their facilities and willingly sharing ideas, and Dr. Gedris, Mr. Rosanskee, and Dr. Lazo for their assistance in acquiring these spectra.

REFERENCES

1. N. N. Greenwood, A. Earnshaw *Chemistry of the Elements*, (Pergamon, New York, 1984) pp. 272-284

2. W.S. Rees, Jr., W. Hesse, in *Chemical Perspectives of Microelectronic Materials II*, edited by L.V. Interrante, K. F. Jensen, L. H. DuBois, M. E. Gross (Mater. Res. Soc. Proc. **204**, Pittsburgh, PA, 1991) pp. 563-570.

3. W.S. Rees, Jr., W.Hesse, in *Synthesis and Processing of Ceramics: Scientific Issues*, edited by W. F. Rhine, T. S. Shaw, R. J. Gottschall, Y. Chen (Mater. Res. Soc. Proc. **249**, Pittsburgh, PA, 1991) pp. 51-57.

4. W.S. Rees, Jr., W. Hesse, Polymer Preprints, 32(3), 573 (1991).

5. W. S. Rees, Jr., W. Hesse, Polymer Preprints, **34** (in press) (1993).

6. A. Samoson, E. Lippmaa, A. Pines, Molec. Pys. **65** (4), 1013 (1988).

7. A. Llor, J. Virlet, Chem. Phys. Lett.,**152**, 248(1988).

8. R. Jelinek, B. F. Chmelka, Y. Wu, P. J. Grandinetti, A. Pines, P. J. Barrie, J. Kilnowski, J. Am. Chem. Soc. , **113**, 4097 (1991).

9. K. T. Mueller, B. Q. Sun, G. C. Chingas, J. W. Zwanger, T. Terao, A. Pines, J. Magn. Reson. **86**, 470 (1990).

SIMULATION STUDIES OF TITANIUM AND ZIRCONIUM WADEITE GLASSES

BEHNAM VESSAL* AND JAMES E. DICKINSON**
*BIOSYM Technologies, Inc., 9685 Scranton Rd., San Diego, CA 92121
**Corning, Inc., Corning, NY 14831

ABSTRACT

Molecular dynamics simulations of the structure of crystalline and glassy titanium and zirconium wadeites have been undertaken to study the microscopic origin of the striking difference between these two systems in terms of nucleation and surface crystallization. The results of simulation are in accord with Raman and EXAFS spectroscopy as well as neutron scattering and DTA scans on the same structures.

INTRODUCTION

The controlled nucleation and crystallization of glass to produce glass-ceramics is an extremely important technological process that is as much "art" as "science". That is, it is still difficult to predict "a priori" whether a given system will nucleate and crystallize homogeneously or heterogeneously. The system $K_2(Ti,Zr)Si_3O_9$ (Ti-and Zr-wadeite) provides an interesting test of hypotheses regarding nucleation and crystallization, because , TiO_2 and ZrO_2 are two of the most commonly used oxides used as nucleating agents. Since these oxides are major components of the wadeite system it would be expected that these compositions would be effectively nucleated and undergo bulk crystallization. However as shown by Dickinson [1] Ti-wadeite does not behave as expected. Instead, it surface crystallizes. Zr-wadeite, on the other hand, nucleates homogeneously and does not surface crystallize. The difference in the behavior of the two systems is effectively illustrated by the DTA scans shown in Figure 1. The crystallization exotherm for Ti-wadeite is very

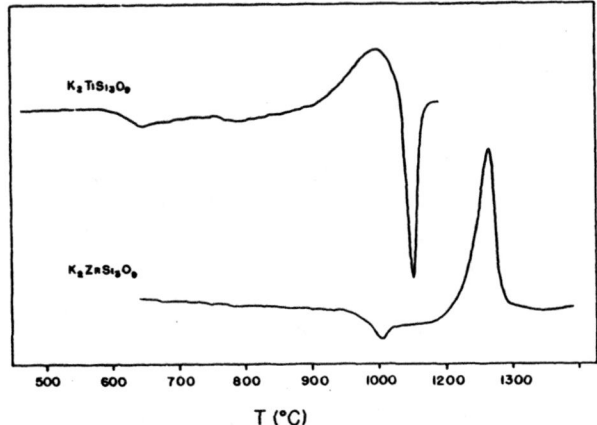

Figure 1 DTA scans of Ti- and Zr- wadeite.

broad whereas that for Zr-wadeite is sharply peaked. This type of behavior is typical of that expected for surface versus bulk crystallization. Based on Raman and EXAFS spectra of corresponding crystals and glasses, Dickinson [1,2] proposed that these differences are the result of structural dissimilarity between crystal and melt in the case of Ti-wadeite, but structural similarity in the case of Zr-wadeite. These differences/similarities influence the energetics of nucleation with the result that it is more difficult to form a critical nucleus of Ti-wadeite. The molecular dynamics simulations reported here have been undertaken to test this hypothesis.

COMPUTATIONAL DETAILS

Details of the Si-O, and O-O potential model used have been given elsewhere [3], but briefly the long-range Coulomb part of the potential is calculated using Ewald's [4] method, employing full ionic charges on both the silicon and the oxygen atoms. Both the direct lattice part and the reciprocal lattice part of the Ewald sum are computed. The short-range interactions are modeled using a four-range Buckingham potential. To model the covalent character of the Si-O bonds, three-body interactions between the O-Si-O triads are modeled using a form described elsewhere [3]. The short-range Ti-O and Zr-O interactions used are simple Buckingham interactions [5]. The K-O interaction used was derived by Parker [6].

In all of the simulations we started with a periodic hexagonal box containing 540 ions [108 Si^{4+}, 324 O^{2-}, 72 K^{+}, and 36 Ti^{4+} (or Zr^{4+})] with the Ti-wadeite [7] or the Zr-wadeite [8] crystal structure. Both constant volume and constant pressure simulations were undertaken. The same cyclic procedure described elsewhere [9] was used for the constant pressure simulations.

Figure 2 Molecular graphics representation of the structure of Ti-wadeite crystal. Titanium atoms have octahedral coordination, silicons have tetrahedral coordination, oxygens are bonded to either two silicons or a silicon and a titanium. Potassium atoms are shown as spheres.

RESULTS AND DISCUSSION

Constant Volume Simulations

Molecular graphics representations of the Ti-wadeite crystal and the corresponding glass simulated at constant volume are shown in Figures 2 and 3. These graphs are prepared using BIOSYM software. It is evident from Figure 2 that crystalline Ti-wadeite contains three-membered rings of SiO_4 tetrahedra connected together by TiO_6 octahedra, and the potassium atoms occupy interstitial positions. When the structure of the glass is analyzed (Figure 3) only one three-membered ring of SiO_4 tetrahedra can be found. This is in accord with the Raman experiments reported by Dickinson [1].

We have compared the first near neighbor peak in the radial distribution function (RDF) of both the Ti-O and Zr-O pairs for crystalline and glassy Ti- and Zr-wadeites in Figures 4 and 5. The Zr-O first neighbor distance for crystalline Zr-wadeite is at 2.03 Angstroms while that for glassy Zr-wadeite is at 1.98 Angstroms. However the Ti-O first neighbor distance shows remarkable difference between the crystalline and the glassy state. The Ti-O RDF for the crystalline phase of Ti-wadeite has a first sharp peak at 1.94 Angstroms while that of the corresponding glass shows evidence of a bimodal distribution of Ti-O bond lengths at 1.75 and 1.81 Angstroms.

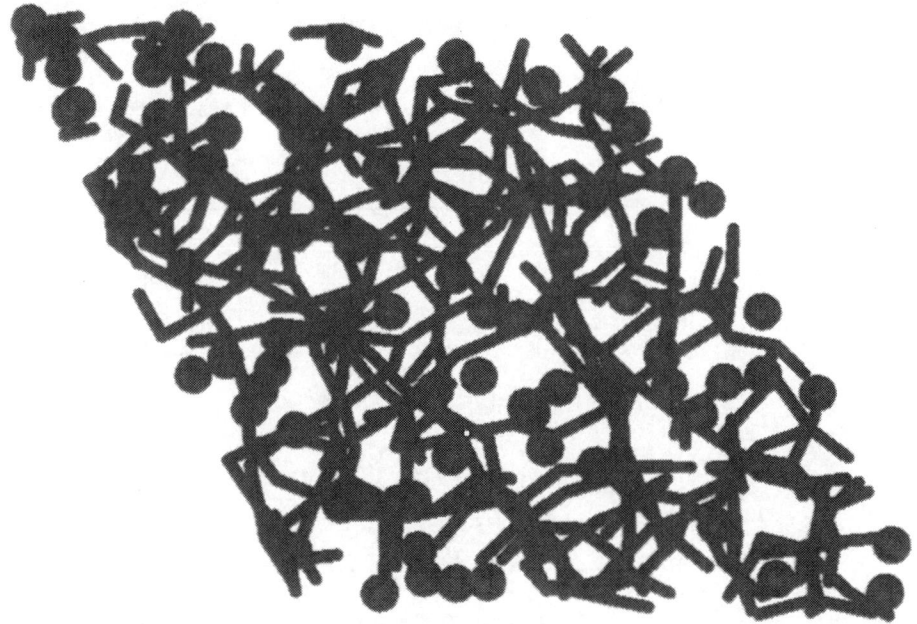

Figure 3 Molecular graphics representation of the structure of Ti-wadeite glass. Titanium atoms are shown in black, silicons in grey, and oxygens in dark grey. Potassium atoms are shown as spheres.

The Zr-wadeite crystal and glass RDF comparison is in excellent agreement with EXAFS measurements [2]. The MD results for the Ti-wadeite system, however, do not agree as well with EXAFS measurements which are best fit by two shells with one Ti-O distance at 1.67 Angstroms and the other at 1.97 Angstroms. Nevertheless, considering that the Ti-O potentials have not been optimized, we consider the results very encouraging. In any case, the simulations definitely are

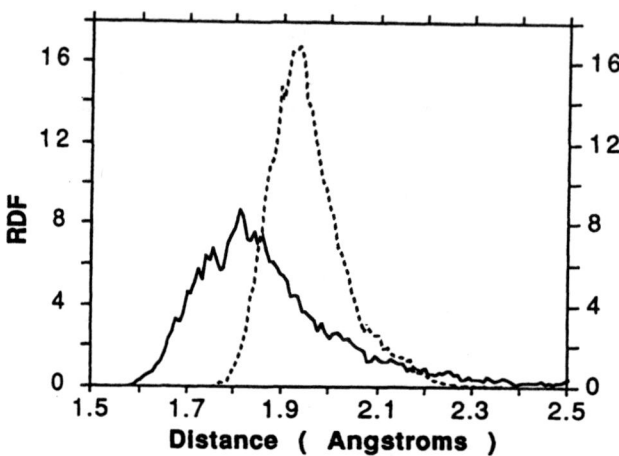

Figure 4 The Ti-O radial distribution function for Ti-wadeite crystal (dashed curve) and glass (full curve).

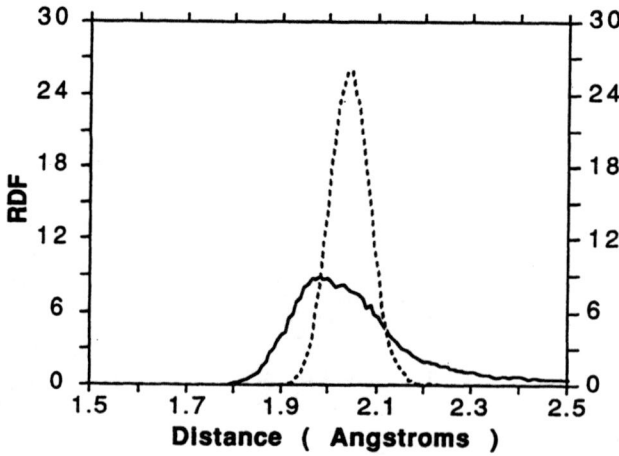

Figure 5 The Zr-O radial distribution function for Zr-wadeite crystal (dashed curve) and glass (full curve).

consistent overall with there being strong similarity between crystal and glass in the Zr-wadeite system and a very strong dissimilarity in the Ti-wadeite system.

Constant Pressure Simulations

We have also undertaken constant pressure melting of both the Ti- and Zr-wadeite. Figure 6 shows the plot of total energy versus temperature for the melting of Zr-wadeite. From such a plot we can estimate both the melting point and the glass transition temperature. Although the simulated melting point is high we have been able to observe the experimental trends in the melting points going from one system to the other. That is we observe the melting point of Ti-wadeite to be lower by about 600K in accordance with experimental results. To improve the agreement of the simulated melting points with the experiment, systems with defects should be built and free surfaces should be allowed. Also potential models with partial charges should be used where available and longer simulation times allowed. It should be noted that although the constant pressure simulations are CPU intensive, they can be analyzed to yield information on structure at the atomic level, radial distribution functions, bond angle distributions, coordination numbers, thermal expansion and heat capacity.

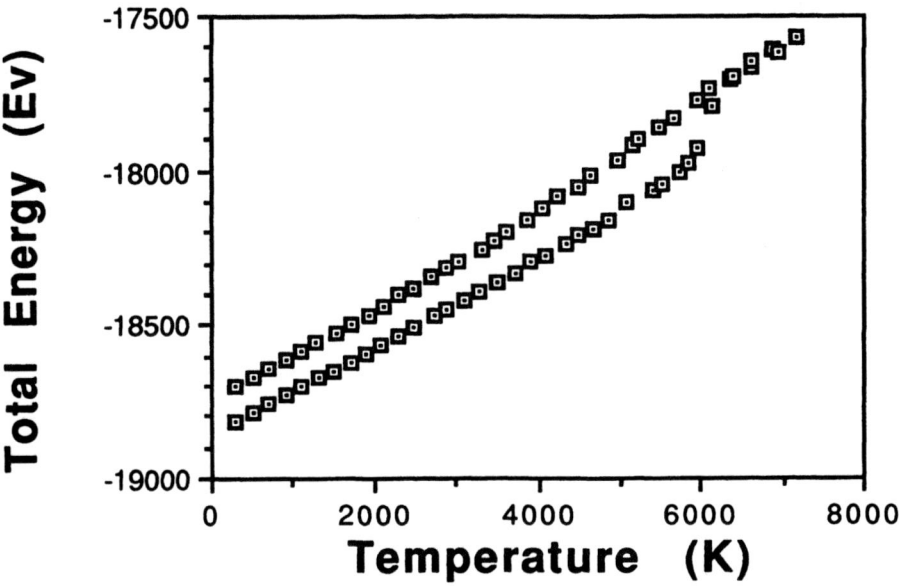

Figure 6 Total energy versus temperature for melting and glass formation in the Zr-wadeite system. The lower curve is the heating curve and the upper curve is the cooling curve.

CONCLUSIONS

Constant volume and constant pressure simulations of melting and glass formation in the Ti- and Zr-wadeite system show good agreement with DTA scans, neutron scattering, EXAFS and Raman spectroscopy which all point to the structural dissimilarity of crystalline and glassy Ti-wadeite and the structural similarity of crystalline and glassy Zr-wadeite.

REFERENCES

1. J.E. Dickinson, in Proceedings of the XV International Congress on Glass Leningrad, edited by O.V. Mazurin (Leningrad Nauka, Leningrad, 1989), pp. 192-197.

2. J.E. Dickinson, EOS Trans. Amer. Geophys. Union 69 (16), 501 (1988).

3. B. Vessal, M. Leslie, and C.R.A. Catlow, Molec. Simulation 3, 123 (1989).

4. P. Ewald, Annln. Phys. 64, 253 (1921).

5. G.V. Lewis, and C.R.A. Catlow, J. Phys. C: Solid State Phys. 18, 1149 (1985).

6. S.C. Parker, PhD thesis, University College, London, 1982.

7. N.G. Shumyatskaya, V.A. Blinov, A.A. Voronkov, V.V. Ilyukhin, and N.V. Belov, Sov. Phys. Doklady 18 (1), 17 (1973).

8. V.A. Blinov, N.G. Shumyatskaya, A.A. Voronkov, V.V. Ilyukhin, and N.V. Belov, Sov. Phys. Crystallogr. 22 (1), 31 (1977).

9. B. Vessal, M. Amini, D. Fincham, and C.R.A. Catlow, Phil. Mag. B, 60, 753 (1989).

PREPARATION, STRUCTURE, AND ELECTRONIC PROPERTIES OF AMORPHOUS GaAs BY TIGHT-BINDING MOLECULAR DYNAMICS

C. MOLTENI, L. COLOMBO AND L. MIGLIO
Dipartimento di Fisica - Universita di Milano, via Celoria 16, 20133 MILANO (Italy)

ABSTRACT

We investigate the short-range structural properties of a-GaAs as obtained in a computer experiment based on a tight-binding molecular dynamics simulation. The amorphous configuration is obtained by quenching a liquid sample well equilibrated at T=1600 K. A detailed characterization of the topology and defect distribution of the amorphous network is presented and discussed. The electronic structure of our sample is calculated as well. Finally, we discuss the reliability and transferability of the present computational scheme for large-scale simulations of compound semiconductor materials by comparing our results to first-principles calculations.

INTRODUCTION

Up to now the only reliable computer simulations at finite-temperature of III-V compounds have been obtained by first-principles molecular dynamics (MD) methods. In particular, careful investigations have been carried out on liquid (l) and amorphous (a) GaAs obtaining a good agreement with available experimental data.[1,2] More recently, a novel semi-empirical simulation scheme has been introduced,[3,4] where the forces governing the atomic motion are derived from the electronic structure of the sample as obtained by a semi-empirical tight-binding (TB) Hamiltonian. Such a TB molecular dynamics (TBMD) scheme has been successfully applied to both elemental[5-7] and compound materials[8,9] showing that it has both a large transferability and the required accuracy to describe complex semicondutor materials. On the other hand, the computational workload of TBMD is that one typical of a parametrized calculation and, consequently, large-scale simulations involving hundreds of atoms and/or long times (hundreds of picoseconds) are possible.

In the present paper, we will apply TBMD to study the structural properties of a-GaAs. The choice of such a material is mainly motivated by its fundamental importance as paradigmatic case of amorphous binary semiconductor whose short-range-order (SRO) features are still matter of discussion.[2,10] In particular, two main issues must be clarified: a full characterization of the topological and chemical disorder of the amorphous network and the identification of structural point defects (like dangling or wrong bonds) due to the presence of threefold- and fivefold-coordinated sites in the amorphous network

We have already demonstrated[7] that the SRO feature of an amorphous semiconductor, as obtained in a computer experiment, are affected by the sample preparation. Consequently, one of the key points of the present work is the procedure adopted to prepare the a-GaAs sample. Taking full advantage from the relatively low computational cost of the TBMD code, we have been able to cool slowly a good sample of l-GaAs from 1600 K down to 300 K (at an average cooling rate of 2.7 10^{12} K/s, much slower than that one achievable in a first-principles simulation[2]). The resulting amorphous network has global structural properties matching in a rather good way the experimental ones. Consistently, a structural model for the SRO properties of a-GaAs is here presented and discussed.

The reliability of the present simulation is also confirmed by the study of the electronic properties of a-GaAs. We will show that the liquid-to-solid transition has driven GaAs from a metallic liquid state to a solid semiconductor one, in good agreement with previous first-principles calculations.[2] Finally, we have calculated the vibrational properties of a-GaAs and compared them to the crystalline case.

COMPUTATIONAL FRAMEWORK

In our TBMD approach the time evolution of the atoms is obtained from the following hamiltonian:[9]

$$H = \sum_i \frac{p_i^2}{2m_i} + 2 \sum_n^{occupied} \varepsilon_n + U_{rep} \tag{1}$$

where the energies ε_n of the single-particle occupied states are obtained from a semi-empirical TB hamiltonian. Their sum is called band-structure energy E_{bs}.[12] We made use of a minimal sp^3s^* basis set where the two-center approximation[13] was adopted and interactions were restricted only to nearest-neighbouring particles. The TB hopping intergrals have been taken from Ref.[14] and their scaling upon the interatomic distance has been choosen according to the Harrison rule.[13]

The last term in eq.(1) can be suitably expanded in terms of two-body rapidly-decaying contributions that we cast in the following form:

$$U_{rep} = \sum_{i,j>i} \left[\Phi_1 \exp\left[-\frac{(r_{ij} - r_0)}{\alpha} \right] + \Phi_2 \frac{r_0}{r_{ij}} \right] \tag{2}$$

where r_{ij} is the relative distance between atom i and j, r_0 its equilibriom value and Φ_1 Φ_2 and α free parameters to be determined. The fitting procedure and the crucial role played by the last term in eq.(2) in describing charge-transfer effects have been discussed elsewhere.[8,9]

As for the like-atom interactions, which are necessary to deal with the disordered liquid and amorphous phases, we assumed that both the expression for the short-range interactions and the values of the parameters are the same as for Ga-As interactions, while the TB basis set was expanded to include the like-atom hopping parameters taken from Ref.[10].

The amorphous sample was obtained by quenching a metallic liquid sample[9] from 1600 K down to 300 K through direct rescaling of the atomic velocities. During the liquid-to-solid transition, the density of the sample was scaled linearly with the temperature from the experimental value for the liquid phase ($\rho = 5.71$ g/cm^3) to that one of the solid phase ($\rho = 5.32$ g/cm^3). The equations of motion for 64 atoms contained in the cubic simulation cell (with periodic boundary conditions) were integrated using a velocity-Verlet algorithm with a time step as large 10^{-14} s. The cooling process followed here is quite computer demanding. Nevertheless, we believe that this is an important feature of the present simulation. In fact, we have neither combined the MD simulation to Monte Carlo "moves" (as suggested in Ref.[11]), nor applied "educated guesses" for Ga-As switches (as described in Ref.[2]) in order to favour the formation of an amorphous network with a desired topology and point defect distribution.

136

After the cooling process, the sample was equilibrated during a constant-temperature constant-volume run for 70 ps and finally observed for 50 ps. The cutoff for the interactions was fixed to 3.185 A.

RESULTS AND DISCUSSION

The structure obtained by quenching from the melt has been found to closely reproduce a tetrahedral network. In fact, by using the value 2.8 A as the maximum first nearest-neigbour (1nn) distance, we found an average coordination number as large as 3.93 to be compared to the experimental 3.93 value.[17] The tetrahedral character of the a-GaAs network is confirmed by the total bond angle distribution function g(θ) shown in Fig.2(a). There, the main structure results in pretty good agreement to the experimental value.[17] The minor peak close to 60 degree is related to those bonds linking atoms whose relative distance falls in between the first maximum of j(r) (at 2.45 A) and our 1nn distance (2.8 A).

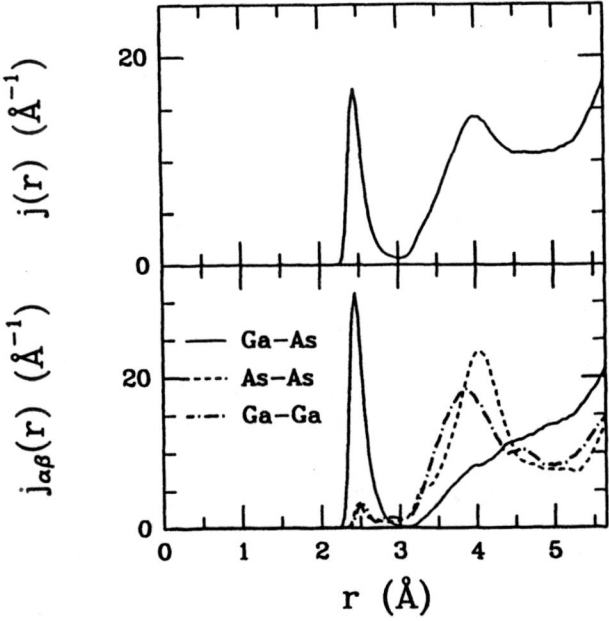

Fig.1 - Total (top) and partial (bottom) radial distribution function of a-GaAs at T=300 K

More direct information about the microscopic ordering of a-GaAs come out from the total and partial g(θ) shown in Fig.2 where minor differences can been observed between $g_{Ga-As-Ga}(\theta)$ and $g_{As-Ga-As}(\theta)$ (panel b and c, respectively). This clearly indicates that a-GaAs has a non-random chemical structure. Moreover, the presence of the shoulder peak at 60 degree is more evident in panels (b) and (c): this suggests that those bonds linking atoms lying at distances between 2.45 A and 2.8 A preferably require unlike atoms at the pivot site and at the vertex site of the triplet.

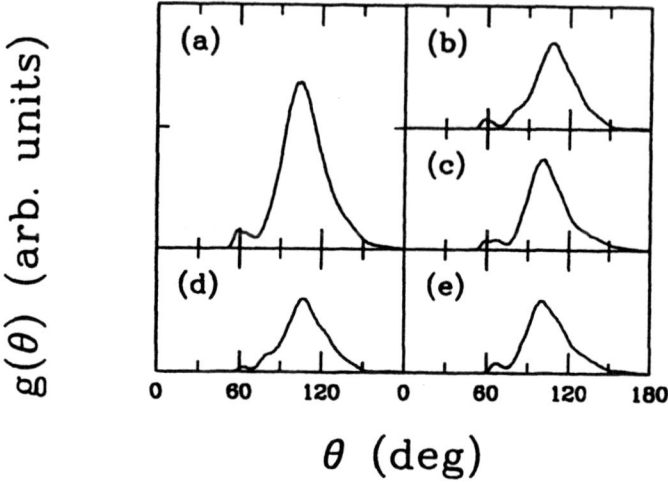

Fig. 2 - Bond angle distribution function of a-GaAs at T=300 K. Panel (a): total $g(\theta)$. Panels (b,c): $g_{X\text{-}Y\text{-}X}(\theta)$ with X,Y=Ga,As. Panels (d,e): unlike atom bond angle distribution.

As for the chemical order of a-GaAs, we found a sizeable amount of threefold-coordinated (C_3) sites with a concentration of 17%. The large amount of C_3 atoms, when compared to the case of a-Si where they have been found not to exceed the 4-5%,[7] is consistent with the chemical model by O'Really and Robertson.[10] There C_3 atoms are indicated as more stable in a-GaAs than in elemental amorphous semiconductors. A second important feature of our amorphous network is the presence of fivefold-coordinated (C_5) sites with a concentration of 10%. This result is in contrast with the first-principles simulation of Ref.[2] where no C_5 atoms were found. We believe that this large discrepancy is mainly related to the different procedure employed to obtain a-GaAs.

Another basic issue to characterize the amount of chemical order of a-GaAs is the fraction of wrong bonds, *i.e.* bonds linking like atoms. The qualitative picture derived from the available experimental data[17-22] indicates that such wrong bonds are indeed present in a-GaAs, but at a small concentration. Estimations range from few percent to 12%. This feature is confirmed by our simulation where the fraction of wrong bonds has been found to be 15% at 800 K and 12.9% at room temperature. This result is in good agreement with the available experimental data[17-22] and indicates that the atomic diffusion in the 1600-800 K temperature plays a major role in the process of ordering of the sample.

The cation and anion coordination number (3.92 and 3.93, respectively) are very similar thus confirming the picture of a chemically ordered amorphous network.

So far we have considered only the structural features of a-GaAs. We have also calculated the electronic density of states (DOS) in order to get a deeper insight into the present structural model. In Fig.3 we display the total electronic DOS (top panel) and the Ga-projected and As-projected DOS (middle and lower panel, respectively). They have been calculated by averaging over different configurations collected during the TBMD run. The main feature of the electronic DOS are in rather good agreement to the experimental valence x-ray photoelectron spectroscopy data[22] where three main structures at ~-2 eV, ~-6 eV and ~-11 eV were observed. In particular, we have reproduced the low-energy gap close to -8 eV separating the two occupied s-like bands of As and Ga (see middle and lower panels). Moreover, it is clearly seen that a-GaAs is a semiconductor.

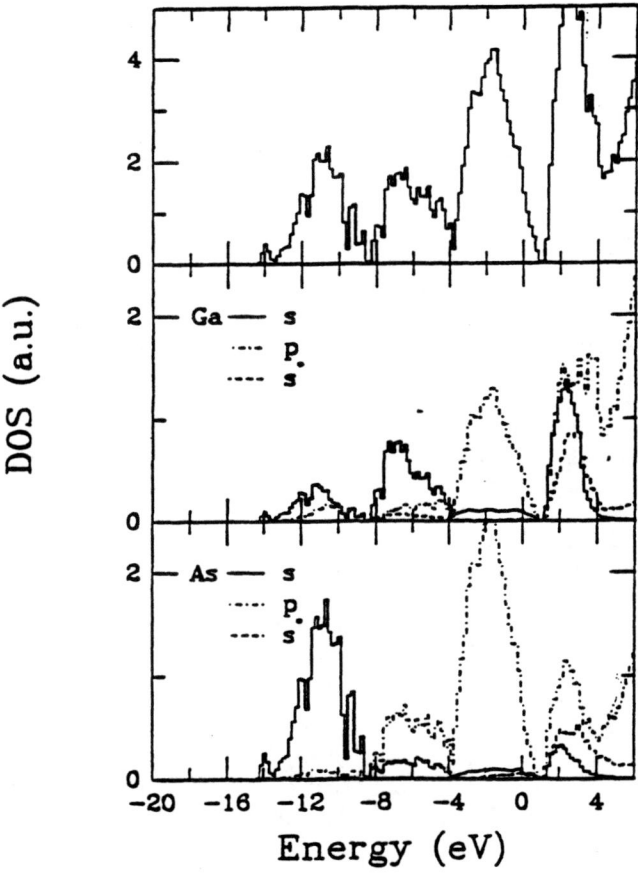

Fig.3 - Electronic density of states of a-GaAs at T=300 K. Panel (a): total DOS. Panels (b,c): Ga- and As-projected partial DOS, respectively. Full line: s-like band; dash-dotted line: p-like band; dashed line: s* band.

CONCLUSIONS

In summary, we have presented and discussed a structural model for a-GaAs as obtained by a tight-binding molecular dynamics simulation.- At variance with previous first-principle calculations, we have been able to produce a good amorphous sample without introducing ad *hoc* moves of atoms favouring a pre-selected distribution of point defects. Our results are in rather good agreement with the overall structural properties of a-GaAs as deduced from both experiments and first-principles simulations and provide a detailed microscopic characterization of such material. In particular, we have proved that C_3 sites are more stable than in a-Si and that C_5 sites are indeed present. A relatively high degree of chemical order has been deduced from the study of the coordination number and the bond angle distribution function. The investigation on the electronic density of states gives support the present structural model of a-GaAs.

REFERENCES

1. Q.M. Zhang, G. Chiarotti, A. Selloni, R. Car and M. Parrinello, Pys. Rev. B42, 5071 (1990)
2. E. Fois, A. Selloni, G. Pastore, Q.M. Zhang and R. Car, Phys. Rev.B45, 13378 (1992)
3. L. Goodwin, A.J. Skinner and D.G. Pettifor, Europhys. Lett. 9, 701 (1989)
4. C.Z. Wang, C.T. Chan and K.M. Ho, Phys. Rev. B39, 8586 (1989)
5. R. Virkkunen, K. Laasonen and R.M. Nieminen, J. Phys. Condens. Matter 3, 7455 (1991)
6. R. Virkkunen, K. Laasonen, R.M. Nieminen, J.Phys.Condens.Matter 2, 1537 (1990)
7. G. Servalli and L. Colombo, Europhys. Lett. 22, 107 (1993)
8. C. Molteni, L. Colombo and L. Miglio, Europhys. Lett. (1993), in press
9. C. Molteni, L. Colombo and L. Miglio, submitted for publication
10. E.P. O'Reilly and J. Robertson, Phys.Rev B34, 8684 (1986)
11. P.C. Kelires and J. Tersoff, Phys. Rev. Lett. 63, 1164 (1989)
12. D.J. Chadi, Phys. Rev. B29, 785 (1984)
13. W.A. Harrison, "*Electronic Structure and Properties of Solids*", W.H.Freeman Company, San Francisco (1980)
14. P. Vogl, H.P. Hjalmarson and J.D. Dow, J. Phys. Chem. Solids 44, 365 (1983)
15. J.A. Majewski and P. Vogl in "*The Structure of Binary Compounds*", North Holland, Amsterdam (1989), p. 287
16. S. Froyen and M. Cohen, Phys. Rev. B28, 3258 (1983)
17. N.J. Shevchik and W. Paul, J. Non-Cryst. Solids 13, 1 (1973)
18. A. Gheorghiu, K. Driss-Khodja, S. Fisson, M.L. Theye and J. Dixmier, J. Phys. (Paris) Colloq. 46, C8-545 (1985)
19. R.J. Temkin, Solid State Commun. 15, 1325 (1974)
20. M.L. Theye, A. Gheorghiu and H. Launois, J. Phys. C13, 6569 (1980)
21. N.J. Svevchik, J. Tejeda and M. Cardona, Phys. Rev. B9, 2627 (1974)
22. C. Senemaud, E. Belin, A. Gheorghiu and M.L. Theye, Solid State Commun. 55, 947 (1985)

PART II

Structural Relaxation

NUCLEATION AND MEDIUM RANGE ORDER IN SILICATE LIQUIDS: INFERENCES FROM NMR SPECTROSCOPY

JONATHAN F. STEBBINS AND SABYASACHI SEN
Department of Geological and Environmental Sciences, Stanford University, Stanford
CA 94305-2115

ABSTRACT

NMR spectroscopy is beginning to provide quantitative information on short– and medium–range structure in silicate glasses and liquids, and on molecular-scale dynamics in the liquids. Data on coordination numbers, second and higher neighbor connectivities, and larger scale heterogeneities, as determined by NMR at ambient and high temperatures, are potentially very useful in models of nucleation and growth of immiscible liquid and crystalline phases. New techniques such as dynamic angle spinning (DAS), high temperature magic angle spinning (MAS) are especially promising. Detailed studies of paramagnetic nuclear spin relaxation can characterize compositional heterogeneity in glasses to distance scales of at least several nanometers.

INTRODUCTION

Quantitative understanding of the kinetics of nucleation in silicate liquids remains an elusive, but important, goal for both earth and materials scientists. Similar experimental observations, of several types, are also needed in both fields. Information on local, short-range structure of single-phase liquids is crucial in understanding their thermodynamic properties, which enter into any model of nucleation of either crystalline or liquid domains. Results on microscopic mechanisms of viscous flow and diffusion are essential, as transport properties are also part of models of kinetics. A wide variety of spectroscopic and scattering techniques have contributed to our knowledge of the structure and dynamics of single-phase liquids. Here we will briefly describe only some recent results from NMR spectroscopy in these areas.

Much more difficult to obtain are experimental data on medium-range structure, which are at the heart of defining what a nucleus is and how it forms. Medium range order and compositional heterogeneity in liquids is of key importance, because liquid-liquid phase separation precedes crystal nucleation in many glass-ceramic systems [1]. Traditional models of nucleation have generally treated crystalline nuclei, and the surface energy that controls their size, in terms of bulk, macroscopic quantities such as free energy, density, and surface tension. However, it has become clear that more sophisticated approaches, involving a microscopic view of what a crystal-liquid (or liquid-liquid) interface might really look like [2, 3], and exactly how much internal ordering is present within nuclei, will need to be constrained by direct observation of nuclei themselves. Prospects for the use of NMR to study such medium range ordering in liquids and in nuclei, will be emphasized in this paper.

NMR is best known as a technique for the detection of relatively short-range structure, and it has thus become a good complement to diffraction methods, especially in

Mat. Res. Soc. Symp. Proc. Vol. 321. ©1994 Materials Research Society

amorphous materials. However, in organic molecules in solution, NMR has now become the method of choice for determining both bonding linkage topology and distances out to three, four, or even more bonds from a given carbon or hydrogen atom. The conformation of relatively large protein molecules (containing thousands of atoms) in solution can now be determined by NMR, using very high resolution, two–, three–, and even four–dimensional techniques. Organic liquids, with strong couplings to protons and rapid molecular motion to produce narrow NMR lines, are of course favorable cases. However, some of these same approaches (as well as techniques unique to the solid state) may begin to provide insights into the medium–range structure of inorganic materials as well.

SILICATE GLASS AND LIQUID STRUCTURE

Many studies of silicate glass structure have been done with NMR, beginning with now-classical work by Bray and colleagues on the distribution of borate species in borate and borosilicate glasses [4, 5]. More recently, most studies have used high resolution magic angle spinning (MAS) NMR of a variety of nuclides, especially ^{11}B, ^{29}Si, ^{27}Al, and ^{17}O. Several thorough reviews have been published recently [6, 7, 8]. Many of the most important findings of these studies have concerned short range structure, in particular the quantification of the distribution of coordination numbers for network forming cations B, Si, and Al. In situ, high temperature NMR studies are beginning to provide information on temperature effects on at least the average coordination numbers of Si, Al, Na, and Mg [9, 10, 11, 12, 13]. Such work is useful to constrain structure-based models of liquid and glass energetics and dynamics, which in turn are needed in any model of nucleation of liquid or crystalline phases. One type of model, for example, relates the thermodynamic activity of a crystallizing component, and thus the driving force for nucleation, to the abundance of a structurally similar species in the liquid. Although it is now clear for at least high-silica liquids that the lifetimes of large silicate "molecules" are probably too short to be of much consequence to structural relaxation and diffusion (see below), at the most local scale such "quasi-crystalline" approaches may have validity. For example, the mean coordination number of Al in high temperature aluminate, aluminosilicate, and fluoride melts has been estimated from *in situ* NMR studies [11, 12, 14], and is probably a key factor in controlling the crystallization behavior.

In many theories of nucleation (or any other rate process modeled using simple energy barriers), activation energies are the most readily predictable and testable quantities. Understanding the structure of hypothesized activated complexes or transition states is thus important. For example, the addition of single SiO_4 tetrahedra to a growing silicate crystal may be the fundamental growth step [15]. For this to occur, one or more oxygen ions must diffuse away from the surface after attachment has occurred. A commonly hypothesized intermediate is the SiO_5 group, which has also been postulated to play a similar role in the condensation of silicate polymers from aqueous solutions [16, 17]. Small concentrations of five-coordinated Si have been detected in both high– and ambient–pressure alkali silicate glasses by NMR [18, 19, 20] (figure 1). The abundance of the species increases with temperature as expected for a reactive intermediate [21].

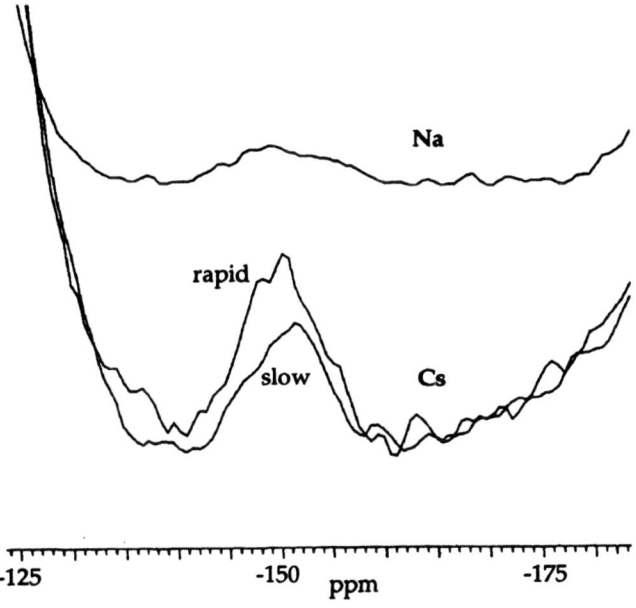

Fig. 1. ^{29}Si MAS spectra for $Na_2Si_4O_9$ (Na) and $Cs_2Si_4O_9$ (Cs) glasses, showing signal from five-coordinated Si near −150 ppm. Relative quench rates for $Cs_2Si_4O_9$ glass are marked [18].

Perhaps more directly relevant, however, are results that reveal the beginning of medium-range order, which is the distribution of second-neighbor atoms, most commonly the first cation neighbors to B, Al, or Si. For tetrahedrally coordinated Si, the number of oxygens bonded to other network formers is denoted by the superscript n in Q^n, where Q stands for "quaternary" and $0 \le n \le 4$. In at least simple binary alkali silicate glasses, MAS or static NMR can often give good quantitation of the relative abundances of various Q^n species (figure 2), although problems can arise with uniqueness of curve fitting. An obvious hypothesis to be tested by this approach relates the ease of crystal nucleation to the abundance of Q^0 species in the melt, because addition of this species to the surface requires the breaking of the fewest strong Si-O bonds, and because it is observed that lower-silica crystalline phases tend to form preferentially during extreme undercooling [15]. The abundance of this species is apparently too low to be readily detectable in glass-forming silicate compositions, but more studies on rapidly-quenched, low-silica glasses could allow this idea to be directly tested in the future.

A variety of studies have made it clear for some time that Q^n species distributions are much more ordered than predicted by a fully random model for the distribution of bridging and non-bridging oxygens, e.g. the continuous random network model [22, 23, 24], although ordering is far from complete. For example, in crystalline $Li_2Si_2O_5$, all Si sites are Q^3, but in the glass there are also about 10-12% Q^2 and 10-12% Q^4 sites, resulting in an ideal equilibrium constant of between about 0.01 and 0.02 for the disproportionation reaction $2Q^3 = Q^4 + Q^2$ [23, 25]. A fully random oxygen distribution

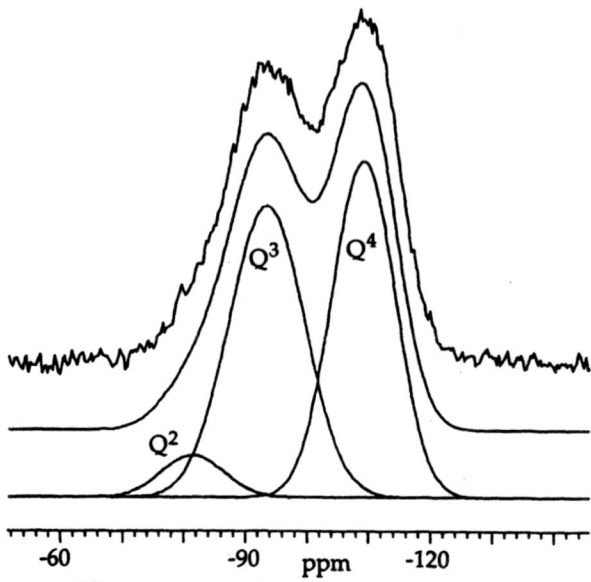

Fig. 2. ^{29}Si MAS spectrum for optically homogeneous, but partially phase-separated, $Li_2Si_4O_9$ glass. Spectrum has been fitted with three Gaussian peaks. The position of the Q^2 peak is constrained to be 12 ppm from the Q^3 peak, as is observed where these peaks are resolved in samples with lower silica contents.

would give about 30% Q^4 and Q^2 [25]. Studies of glasses with varying fictive temperature indicate that distributions do indeed become more random at higher T, as expected from entropic considerations [26]. Clustering of network modifiers will change this distribution. In particular, higher field-strength cations tend to concentrate non-bridging oxygens, increasing disproportionation and favoring nucleation of immiscible liquid phases. As in the case of the first coordination shell, species abundances may be related to thermodynamic activity of crystallizing components. In particular, the abundance of silica-like Q^4 species, and compositional and temperature effects on it, is clearly related to silica activity [27].

In general, however, Q species distributions provide only limited data on medium-range heterogeneity. The ^{29}Si MAS NMR spectrum of $Li_2Si_4O_9$ (figure 2), for example, is dominated by peaks for about 50% Q^3 and about 50% Q^4, as expected from stoichiometry alone. A wide range of medium or even long-range order is consistent with this distribution, ranging from the strict alternation of the two types of tetrahedra (as in the crystalline phase) to the presence of large domains each containing mostly Q^3 or Q^4 sites. The distinction between these alternatives lies in part in the abundances of minor (and more difficult to observe) species. The equilibrium constant derived from single-phase $Li_2Si_2O_5$ predicts (see above) only about 1% Q^2 species in a hypothetical homogeneous tetrasilicate, but spectra indicate that considerably more must be present in the real glass (about 5-10%). This does suggest that even an optically homogeneous sample

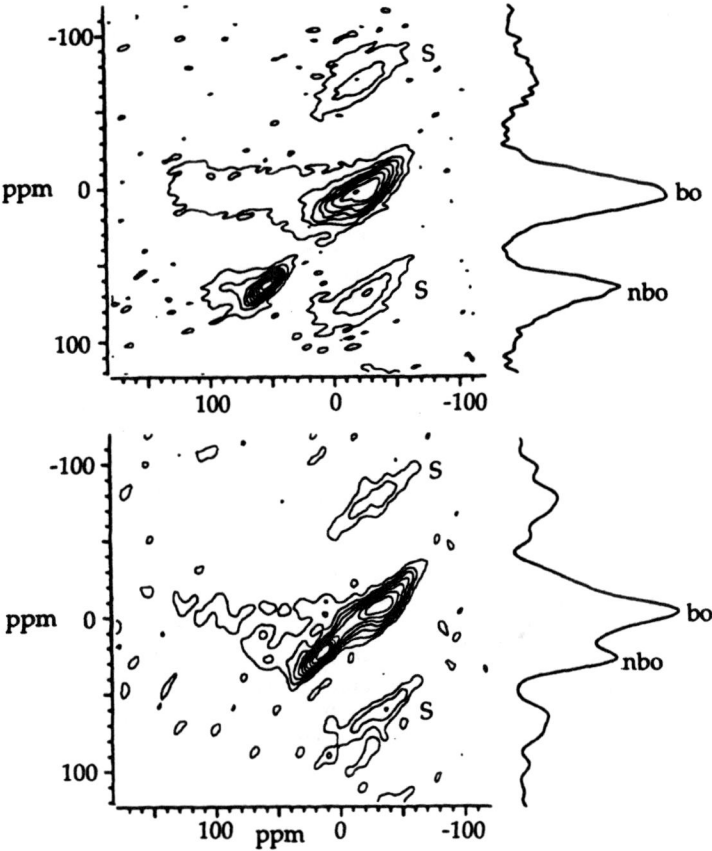

Fig. 3. Two-dimensional ^{17}O DAS NMR spectra and projections onto the isotropic (vertical) axis for (top) $K_2Si_4O_9$ and (bottom) $KMg_{0.5}Si_4O_9$ glasses. Peaks for bridging oxygens (bo), non-bridging oxygens (nbo) and spinning sidebands (S) are labeled [28].

has begun spinodal decomposition into lithium– (and Q^{2-}) rich regions, but still says little about the size of those regions.

Cation ordering can also be investigated from the "viewpoint" of the oxide anion using ^{17}O NMR. The new technique of dynamic angle spinning (DAS) NMR, in which the sample spinning axis is rapidly cycled between two or more special angles, has recently provided improved resolution among oxygen sites in glasses. In figure 3, for example, are DAS spectra for $K_2Si_4O_9$ and $KMg_{0.5}Si_4O_9$ glasses [28]. Peaks for bridging and non-bridging oxygens are clearly resolved. In this composition, each non-bridging oxygen has one Si and several network modifier neighbors. A disordered distribution of K and Mg in the mixed cation glass would require a variety of different non-bridging oxygen

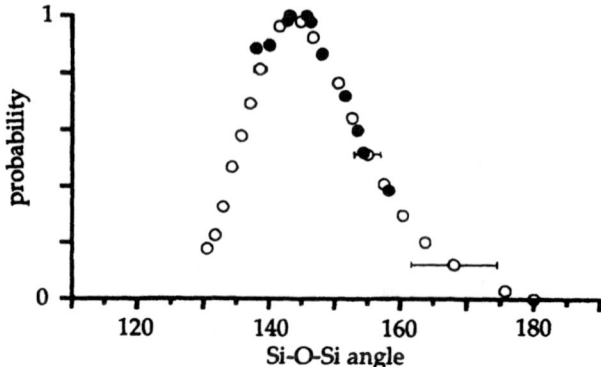

Fig. 4. Si–O–Si bond angle distribution for bridging oxygens in $K_2Si_4O_9$ glass, determined from two-dimensional ^{17}O DAS NMR spectra [28].

types, with different assortments of K and Mg neighbors. The fact that a single, narrow, non-bridging oxygen peak appears in the K-Mg glass, at a quite different position from the pure-K sample, indicates a high degree of cation ordering consistent with the relatively small amount of silicate species disorder in this material [22]. In this case, a *lack* of clustering or compositional heterogeneity is implied. When DAS NMR is combined with two-dimensional data acquisition and processing, correlations between parameters such as chemical shift, quadrupolar coupling constants, and quadrupolar asymmetry parameter can also be observed. New information about the distribution of Si-O-Si angles in glasses, which are directly related to inter-cation distances and hence to clustering, has been obtained with this approach (figure 4) [28].

LOCAL VS. LONG-RANGE ANISOTROPY

NMR chemical shifts, quadrupolar shifts, and dipole-dipole interactions are in general strongly dependent on the orientation of a local structural unit with respect to the external magnetic field. These anisotropies give rise to much of the breadth of NMR peaks in solids and have lead to the development of MAS, DAS, and other approaches that cause averaging and improve resolution. However, there is potentially a great deal of information (now generally discarded) about possible structural anisotropy in the spectra of non-spinning samples of glass quenched during rapid deformation of the liquid. For example, local structural alignment in organic and biopolymer films and fibers can readily be quantified. Anisotropies in the local environments of ^{11}B, ^{27}Al, and ^{29}Si have been sought for a variety of oxide glass films and fibers [29]. Results were definitively negative, indicating that local bond-breaking and randomization of structural alignment occurs all the way down in temperature to the glass transition. In itself, this result says nothing about intermediate range order. However, if optical anisotropy indeed exists as reported for some of these materials (such as fibers of e-glass), *something* must be deforming to give a non-uniform refractive index. If the anisotropy is not local, then the existence of a deformed domain structure is suggested.

Modeling of the deformation and diffusion processes in such an aggregate could lead to size constraints.

CONSTRAINTS FROM EXCHANGE DYNAMICS

NMR spectroscopy is particularly useful in studying dynamics at the "diffusive" time scales of seconds to nanoseconds. When structurally distinct species have distinguishable spectral features, exchange among the species can be quantified by its effect on spectra. If the exchange among two species is much slower than the separation in frequency of their NMR peaks, then no effect will be seen. If the exchange is much more rapid than the frequency separation, then complete averaging will be observed and a minimum for the exchange rate can be derived. In between these regimes, partial averaging occurs that in some cases can be accurately modeled to give quantitative exchange frequencies.

In silicate melts, this approach has been taken to determine the frequency of exchange between Q^3 and Q^4 species in $K_2Si_4O_9$ liquid [30, 31]. It was found that the exchange time, as determined from fitting partially averaged spectra, was similar to the shear relaxation time derived from the bulk viscosity. A simple Eyring model based on the jumping of Si or O ions from one site to another at the measured exchange rate predicted the bulk viscosity surprisingly well. Given that the strongest bonds in the system (Si–O) must break and re-form for this exchange to take place, and this composition must form a highly connected tetrahedral network, a close relationship between exchange and flow is sensible. These results have been quantitatively confirmed by ^{17}O NMR at high temperature, which allows observation of the exchange between bridging and non-bridging oxygens [32].

This work has been extended down to temperatures just above the glass transition by the application of two-dimensional exchange spectroscopy on static, and very recently, on spinning samples [33]. The latter technique permits much higher resolution and the detection of motion as slow as 1 Hz or slower. Even at viscosities as high as 10^{10} Pa·s, viscosity in this material seems to be controlled by local bond-breaking and species exchange. Bulk structural relaxation is thus not controlled by other possible mechanisms, such as chain entanglement or the sliding of domains relative to one another on intervening weak bonds. This again suggests a relatively homogeneous structure, which is not surprising in this material, in which both liquid-phase and solid-phase nucleation is very difficult.

Few dynamical studies have been done on other compositions, but all results so far show that when viscosity is low enough for local anisotropy to be averaged by rotation of silicate tetrahedra, all structural species are fully exchanged [12, 30]. This is true even in liquids as low in silica 40 mole per cent at temperatures not far above the liquidus. However, because of the experimental complications caused by crystallization, the quantitative correlation between bulk shear relaxation and microscopic exchange has not been established in these lower silica liquids. Indeed, it might be expected that flow could involve isolated but coherent groups of one or two tetrahedra bound only weakly through non-bridging oxygens to the rest of the structure. This question may be resolvable by using high temperature MAS NMR to further study exchange just above the glass transition temperature.

The observation of complete exchange places limits on the size of domains that may be present, if it is assumed that the compositions of coexisting domains or clusters are distinct enough to cause significant differences in NMR peak positions. Domains, if present, must be small enough so that diffusion of the exchanging species can take place in less than the observed exchange time, or in a time less than a minimum exchange time if that is all that can be determined. For example, in the two-dimensional high T MAS study on $K_2Si_4O_9$ mentioned above[33], the Q^3–Q^4 exchange time τ is about 4 s at 828 K. The diffusivity D of the oxide ion is likely to be the rate limiting step, and can be estimated from the Eyring relation $D = kT/\lambda\eta$, where k is the Boltzmann constant, λ the ion jump distance (about 0.3 nm), and η the viscosity [34, 35]. Assuming for the moment that the dimensionality of domains is 3 (see below), and taking the characteristic distance for diffusion as $x = (D\tau)^{1/2}$, and $\eta=10^{10}$ Pa·s, we obtain a distance of 0.12 nm, indicating that exchange is completely local and no domains are present. In a case where only a single, fully exchanged peak is observed, only a minimum exchange time and maximum diffusion distance can be estimated. In a study of isotropic (fully averaged) ^{29}Si chemical shifts, the lowest temperature observation for supercooled sodium disilicate liquid was at 1098 K (about 50 K below the liquidus) [12]. Here, complete exchange suggests an exchange time of less than about 0.0013 s, as this is the inverse of the frequency separation of typical Q^n peaks. With the above relations and η = 2200 Pa·s, we obtain a maximum domain size of about 5 nm. Note that if domains are "crystal-like," then they would have to be much smaller to be involved in exchange, given the much slower diffusion in a solid compared to that estimated from the Eyring relation in a liquid.

SPIN–LATTICE RELAXATION AND MEDIUM-RANGE HETEROGENEITY

One of the longer-range interactions that can be detected by NMR is the magnetic dipolar coupling between nuclear spins and those of unpaired electrons. Although the energy of this coupling decreases as a very strong function of distance ($1/r^6$), it can nonetheless potentially give information about structure to distances of at least several, if not 10's, of nm, and thus be used to explore heterogeneities containing dozens to 100's of atoms.

One approach to exploiting this coupling involves the effects of unpaired (paramagnetic) electrons on nuclear spin relaxation. Because of the low energies of nuclear spin transitions, spontaneous relaxation of excited states does not occur at a significant rate. Coupling to a magnetic field fluctuating at the NMR resonant (Larmor) frequency, or, for quadrupolar nuclides, to a fluctuating electric field gradient, is required for energy transfer. This "spin-lattice relaxation", characterized by a time constant T_1, is often not well-understood in solid materials, particularly for spin 1/2 nuclides in rigid structures. In high-resolution NMR structural studies of solids, relaxation effects are usually ignored or are considered only in how they either cause low signal intensity (if T_1 is inconveniently long) or, alternatively, signal broadening (if T_1 is too short). However, it is clear that in systems where internuclear dipolar couplings are relatively weak, relaxation of spin 1/2 nuclides is usually dominated by electron-nuclear dipolar interactions. The situation can often be simplified by doping a sample with enough paramagnetic cations (usually 100's to 1000's of ppm) to ensure that this mechanism

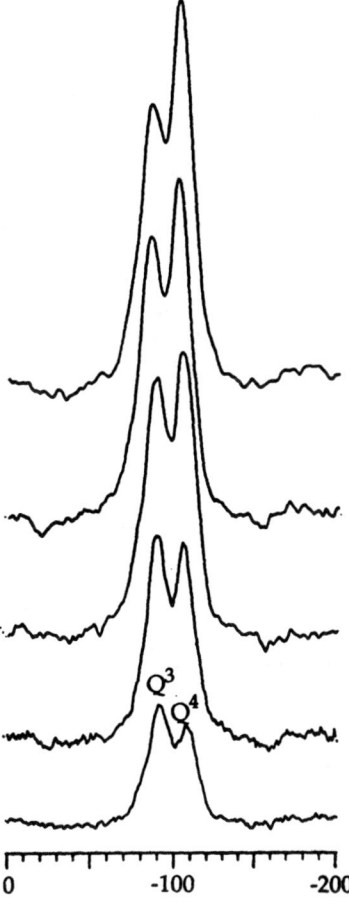

Fig. 5. ^{29}Si MAS spectra for optically homogeneous, but partially phase-separated, $Li_2Si_4O_9$ glass, collected with a range of delay times from 20 s (bottom) to 2000 (top) s. The sample was doped with 500 ppm Gd_2O_3 to insure that spin-lattice relaxation occurred via electron-nuclear dipolar interaction. Note that the Q^3 peak relaxes faster, and thus grows in intensity more rapidly, than the Q^4 peak [37].

dominates. Detailed measurements of the relaxation behavior can then provide information about both the spatial distribution of paramagnetics and that of the nuclide whose NMR signal is being observed. Because relaxation rate is a strong function of the distance from the nucleus to a paramagnetic center, distance information may be obtainable. In particular, non-random distributions caused by clustering or compositional domain formation may be detectable.

This approach has been used, for example, to confirm that the structure of silica gels is fractal in character [36], and to learn about the characteristic dimensions of the structure. We have recently applied this method to alkali silicate glasses to explore the effects of submicroscopic liquid-liquid phase separation [37, 38]. The composition $Li_2Si_4O_9$, for example, lies within a two-liquid spinodal, and thus should begin to phase-separate spontaneously. It is, however, possible to prepare optically-clear glasses of this composition, and then anneal them until macroscopic phase separation occurs. Even in rapidly-quenched samples, however, inspection of ^{29}Si MAS NMR spectra suggests that the average distance from Si to paramagnetics is different for Q^4 and Q^3 sites, as the former have considerably slower spin-lattice relaxation. Spectra collected with longer delays between radiofrequency pulses thus show increasingly larger Q^4 peaks (figure 5). This effect must be caused by compositional heterogeneity. The nature and scale of this heterogeneity can be explored by examining the power-law behavior of relaxation curves, with the magnetization M decaying with time t as $M(t) \sim t^\alpha$. These reveal a power-law exponent α of 0.35, consistent with the theoretical prediction for relaxation in a fractal percolation cluster. Q^3 peaks relax more rapidly because paramagnetics are concentrated in Q^3-rich regions. Annealing of samples causes domains to grow and the difference in Q^3 and Q^4 relaxation to be accentuated. Information on the characteristic sizes of the compositional heterogeneities can be obtained if some simple assumptions about the behavior of the electronic spins are made. In this case, this suggests that the species distribution is fractal to a length scale of at least 2 to 3 nm. The connection of these types of results to generally understood ideas of structural heterogeneity is confirmed by further work on $Na_2Si_2O_5$ and $Li_2Si_2O_5$ glasses, which are known to be outside the range of liquid-liquid immiscibility. These indeed show a power-law exponent of 0.51, a dimensionality of 3 (non-fractal behavior), and no differential relaxation of different Q species. This indicates, as expected, that the structures do not have major medium range compositional heterogeneities. We expect this approach to be very useful in the future for exploring fluctuations during nucleation of crystalline, as well as liquid, domains.

THE PROGNOSIS FOR DIRECT SPECTROSCOPIC DETECTION OF NUCLEI

NMR is generally considered to be a technique of low sensitivity, when compared to optical, infra-red, or x-ray spectroscopy. However, its ability to directly detect, quantify, and characterize a small amount of crystal-like silicate material in an amorphous matrix of similar composition may in some cases be unsurpassed. It has recently been shown that detection limits for amorphous minor species such as SiO_5 groups may be as small as about 0.01% in ^{29}Si MAS spectra at ambient T [18]. Sensitivity for the narrower peaks expected even for partially disordered crystalline nuclei should be even greater. With a 0.01 volume % detection limit, and nucleation densities of about $10^7/cm^3$ (as observed for homogeneous nucleation in $Li_2Si_2O_5$), the detectable size would be about 2.2 µm, which would probably constitute a more-or-less normal crystal and thus be relatively "uninteresting". However, for very high nucleation densities typical of glass ceramic precursors [1, 39], nuclei should be detectable at a much earlier stage. With a nucleation density of $10^{15}/cm^3$, for example, the minimum detectable size would be $(10^{-19})1/3 = 4.6 \times 10^{-7}$ cm or 4.5 nm, which would contain only a few hundred tetrahedra. Direct information on both abundance and structure of nuclei should eventually be obtainable.

ACKNOWLEDGMENTS

This work was supported by the National Science Foundation (EAR9204458). We thank our collaborator, Ian Farnan (CNRS, Orléans) for long and productive interactions in several of the studies described here, and J.E. Dickenson (Corning) and the M.R.S. for the opportunity to present this work.

REFERENCES

1. G. H. Beall, in *Glass: Current Issues*, edited by A. F. Wright and J. Dupuy (Martinus Nijhoff Publishers, Dordrecht, 1985), pp. 31-48.

2. E. Ruckenstein and B. Nowakowski, J. Coll. Interface Sci. **137**, 583-592 (1990).

3. G. Narasimhan and E. Ruckenstein, J. Coll. Interface Sci. **128**, 549-565 (1989).

4. P. J. Bray, J. Non-Cryst. Solids **73**, 19-45 (1985).

5. P. J. Bray, *et al.*, J. Non-Cryst. Solids **129**, 240-248 (1991).

6. H. Eckert, Ber. Bunsenges Phys. Chem. **94**, 1062-1085 (1990).

7. H. Eckert, Prog. Nucl. Mag. Reson. **24**, 159-293 (1992).

8. R. J. Kirkpatrick, *et al.*, in *Structure and Bonding in Noncrystalline Solids*, edited by G. E. Walrafen and A. G. Revesz (Plenum Press, New York, 1986), pp. 302-327.

9. D. Massiot, F. Taulelle and J. P. Coutures, Colloq. Phys. **51-C5**, 425-431 (1990).

10. B. T. Poe, *et al.*, Eos, Trans. Am. Geophys. Union **72**, 572 (1991).

11. B. T. Poe, P. F. McMillan, C. A. Angell and R. K. Sato, Chem. Geol. **96**, 241-266 (1992).

12. J. F. Stebbins and I. Farnan, Science **255**, 586-589 (1992).

13. P. S. Fiske, J. F. Stebbins and I. Farnan, Eos, Trans. Am. Geophys. Union **74**, 349 (1993).

14. J. F. Stebbins, I. Farnan, N. Dando and S. Y. Tzeng, J. Am. Ceram. Soc. **75**, 3001-3006 (1992).

15. R. J. Kirkpatrick, Am. Mineral. **68**, 66-77 (1983).

16. J. D. Kubicki and A. C. Lasaga, Geochim. Cosmochim. Acta **57**, 3847-3854 (1993).

17. F. Liebau, Inorgan. Chim. Acta **89**, 1-7 (1984).

18. J. F. Stebbins and P. McMillan, J. Non-Cryst. Solids **160**, 116-125 (1993).

19. J. F. Stebbins and P. McMillan, Am. Mineral. **74**, 965-968 (1989).

20. X. Xue, J. F. Stebbins, M. Kanzaki, P. F. McMillan and B. Poe, Am. Mineral. **76**, 8-26 (1991).

21. J. F. Stebbins, Nature **351**, 638-639 (1991).

22. H. Maekawa, T. Maekawa, K. Kawamura and T. Yokokawa, J. Non-Cryst. Solids **127**, 53-64 (1991).

23. R. Dupree, D. Holland and M. G. Mortuza, J. Non-Cryst. Solids **116**, 148-160 (1990).

24. R. Dupree, D. Holland and D. S. Williams, J. Non-Cryst. Solids **81**, 185-200 (1986).

25. J. F. Stebbins, J. Non-Cryst. Solids **106**, 359-369 (1988).

26. M. E. Brandriss and J. F. Stebbins, Geochim. Cosmochim. Acta **52**, 2659-2670 (1988).

27. J. F. Stebbins and I. Farnan, Science **245**, 257-262 (1989).

28. I. Farnan, *et al.*, Nature **358**, 31-35 (1992).

29. J. F. Stebbins, D. R. Spearing and I. Farnan, J. Non-Cryst. Solids **110**, 1-12 (1989).

30. I. Farnan and J. F. Stebbins, J. Am. Chem. Soc. **112**, 32-39. (1990).

31. I. Farnan and J. F. Stebbins, J. Non-Cryst. Solids **124**, 207-215 (1990).

32. J. F. Stebbins, I. Farnan and X. Xue, Chem. Geol. **96**, 371-386 (1992).

33. I. Farnan and J. F. Stebbins, Eos, Trans. Am. Geophys. Union **73**, 600 (1992).

34. D. B. Dingwell, Chem. Geol. **82**, 209-216 (1990).

35. D. B. Dingwell and S. L. Webb, Eur. J. Mineral. **2**, 427-449 (1990).

36. F. Devreux, J. P. Boilot, F. Chaput and B. Sapoval, Phys. Rev. Lett. **65**, 614-617 (1990).

37. S. Sen and J. F. Stebbins, Eos, Trans. Am. Geophys. Union, in press, (1993).

38. S. Sen and J. F. Stebbins, presented at 1993 MRS Spring Meeting, Boston, MA, 1993 (unpublished)

39. A. F. Wright, in *Glass: Current Issues* , edited by A. F. Wright and J. Dupuy (Martinus Nijhoff Publishers, Dordrecht, 1985), pp. 21-30.

DYNAMICS OF CRYSTALLINE AND AMORPHOUS POLYTETRAFLUOROETHYLENE STUDIED BY MULTIPLE QUANTUM NMR

DAVID A. LATHROP AND KAREN K. GLEASON
Department of Chemical Engineering, MIT, Cambridge, MA 02139.

ABSTRACT

We report a new technique for probing polymer dynamics through the refocussing of multiple quantum (MQ) nuclear magnetic resonance (NMR) coherences. The MQ-NMR experiment follows the correlated behavior of multiple spin-1/2 nuclei interacting through dipolar couplings. Motion which modulates the dipolar coupling strengths on the same time scale as the experiment (~1 to 20 kHz) alters the intensity of the observed coherences. Temperature dependent ^{19}F data are presented on polytetrafluoroethylene samples of varying crystallinity. For the as-polymerized 98% crystalline PTFE sample, a sharp increase in MQ coherence refocussing occurs, centered at ~298 K. The 64% crystalline melt-quenched sample shows a increase at the same temperature but which has a lower intensity. Thus, the ~298 K peak is most associated with motion in the crystalline phase. This temperature is intermediate between the two first order transition at 293 and 303 K. Oscillations in the refocussed fractions are observed from 208 to 230 K for the 98% crystalline sample, while this ratio is constant over the same temperature range for the 64% crystalline sample. These oscillations may be associated with paracrystalline defects found only in the first sample. Thus, the MQ refocussing experiment is able to clearly differentiate between polymer samples which have different thermal histories. The sharpness of the MQ refocussing features and their variations in magnitude, shape, and sign with temperature are signatures of the molecular level details of the underlying dynamics which produce them.

INTRODUCTION

The bulk properties of polymers are greatly influenced by molecular motion occurring within these solids. While a wide range of techniques including mechanical loss, dielectric loss, and NMR relaxation have been developed to study polymeric motion, complemenary information is sought via a new technique which relies on the refocusing of multiple quantum nuclear magnetic coherences (MQ NMR).[1] MQ coherences can involve anywhere from 2 to >100 nuclei, allowing the correlated motion for atomic groups of various sizes to be probed, rather than tracking the motion of a single site. Over a wide range of temperatures, this single technique is able to provide information which is complementary to the previously developed diagnostic techniques. In addition, unique information about the molecular level detail of motion is obtained.

Refocussing of MQ coherences can be carried out using any polymer containing hydrogen and/or fluorine since these nuclei give rise to strong homonuclear dipolar couplings. Polymeric motion modulates these dipolar couplings through variations in internuclear distances and/or the orientation of the internuclear vectors with respect to the applied magnetic field of the NMR spectrometer. Traditional MQ NMR experiments prepare these MQ coherences using a series of pulses for a fixed duration. These MQ coherences can not be detected directly. Thus, a mixing period, equal in duration to the preparation period, is used to refocus the coherence for detection, as shown below.

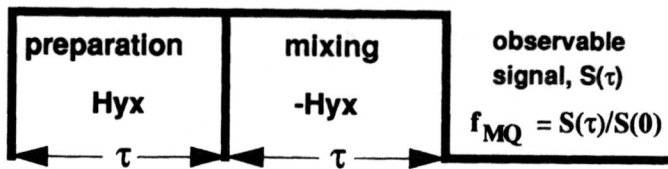

Ideally, if all dipolar couplings are identical during both the preparation and mixing periods, the fraction which is refocused, f_{MQ}, will equal one.

However, motion which modulates internuclear dipolar couplings on the time scale of the MQ experiment will cause f_{MQ} to vary. The duration of the MQ preparation period, τ, is between 30 μs and 1 ms, allowing polymeric frequencies between 1 and 33 kHz to be probed. Generally these motions will interfere with MQ refocussing, lowering the value of f_{MQ}. However, if the frequency of the molecular motion matches a half-integral multiple of the MQ experimental frequency, complete refocussing can occur.

As temperature is monotonically increased and the MQ preparation time is fixed, the frequency of motion can take on several of these special refocusing frequencies interspersed by defocusing frequencies. Thus, a series of oscillations in f_{MQ} with temperature is expected. In addition to identifying the frequency of motion, the shapes of these oscillations are signatures for the underlying mechanism of molecular motion, for instance, multi-site hops versus continuous rotations. Only motions which restore the relative positions of the nuclei after the MQ experiment will cause refocussing. Motions such as translational diffusion which do not restore the initial configuration of the nuclei will not produce refocussing under any condition.

BACKGROUND: POLYTETRAFLUOROETHYLENE.

Polytetrafluoroethylene (PTFE) is a partially crystalline polymer which has been extensively studied using x-ray diffraction[2,3], differential scanning calorimetry (DSC)[4], nuclear magnetic resonance (NMR)[5-7], mechanical relaxation[8,9] and dielectric loss[10] techniques. PTFE undergoes two first-order crystalline transitions at 292 and 303 K, as well as three viscoelastic relaxations termed α, β and γ.[11] The γ-relaxation occurs about 160 K in the amorphous chains and is believed to consist of torsional motions of polymeric backbone segments.[12] The activation energy of the g-relaxation has been determined to be about 18 kcal/mole.[5] Upon further increase in temperature to 292 K, PTFE undergoes the first of two first order crystalline transitions. This first transition corresponds to a slight uncoiling of the crystalline helix from 13 carbon atoms per 180° twist to 15 carbon atoms per 180° twist. The unit cell is triclinic below 292 K and hexagonal above 292 K. Between 292 and 303 K, small angle oscillations of chain segments about a preferred crystallographic direction are believed to occur.[3]

The second first order crystalline transition is seen at 303 K and involves a further uncoiling of the molecules in the crystalline phase. Above 303 K the oscillation of the molecular segments occurs with a random angular orientation in the lattice and the preferred crystallographic direction is lost.[3] Translational displacement of molecules in the crystalline region along the chain axis is believed to begin after this second transition.[6] The α-relaxation occurs above 400 K and involves both amorphous and crystalline regions. PTFE finally melts at about 600 K.

The β relaxation overlaps with the two phase transitions and is commonly associated with torsional oscillations and rotations of chain segments around the chain axes in the crystalline regions. It is believed that the motions associated with the β relaxation are linked to the uncoiling of the helix that occurs during the phase transitions.[6]

In virgin samples of as-polymerized PTFE there is evidence of the presence of two crystalline phases. Differential Thermal Analysis (DTA) has shown double peaks for the crystalline transitions in as-polymerized PTFE and a viscoelastic anamoly has been observed in samples of virgin PTFE at ca. 208 K that is not seen in melt-crystallized samples.[8] Likewise, Vega and English[6] have suggested the presence of more than one crystalline phase with differing mobilities to explain NMR $T_{1\rho}$ and T_{1xy}^{REV} results in as-polymerized PTFE. Starkweather et. al.[13] have estimated that approximately one quarter of the chains in as-polymerized PTFE belong to this more mobile phase. The motion responsible for the 208 K anamoly in the more mobile crystalline phase is not well understood, but, it has been attributed to segmental motions in paracrystalline regions that contain mobile conformational defects[14] and irregular interchain spacing[9].

EXPERIMENTAL

The MQ-NMR experiments were performed on a home-built NMR spectrometer operating at 270-MHz. Phase incrementation of the preparation period pulses with respect to the mixing period pulses was achieved with a 240-MHz Sciteq digital synthesizer with 8-bit phase resolution. The incremental preparation phase shift was set at 5.6°, allowing detection of coherences with $|n| \leq 16$. The length of the $\pi/2$ pulses used ranged from 1.6 to 2.0 μs. The basic MQ pulse sequence time was held fixed at 36 μs. The signal intensity experiments were performed at a phase shift of 90° (time reversal) and long relaxation delays ($> 5T_1$) for maximum signal strength.

The MQ intensity experiments were performed on DuPont dispersion polymerized T-60 in its as-polymerized form as well as after melting and slow cooling. The weight percent crystallinity of these two samples, as calculated from their densities, was determined to be 98% and 64%, respectively.

RESULTS AND DISCUSSION

Oscillations in f_{MQ} were observed over the range 206 K to 230 K for an as-polymerized, 98% crystalline polytetrafluoroethylene (PTFE) sample (solid lines in Fig. 1). Each line indicates a fixed value of the MQ preparation period, τ. The overall decrease in f_{MQ} with increasing τ indicates relaxation not associated with molecular motion. However, at each fixed τ, oscillations in f_{MQ} can be seen. A viscoelastic anomaly associated with the motion of defects through the intermediate paracrystalline regions has also been seen in as-polymerized samples over this same temperature range. In contrast, f_{MQ} is relatively constant in a melt-crystallized PTFE sample which is 64% crystalline and contains no paracrystalline phase (dotted lines, Fig. 1). The thermal processing received by the second sample is expected to eliminate the paracrystalline region which gives rise to the motion of point defects. Thus, no f_{MQ} oscillations are seen in this sample.

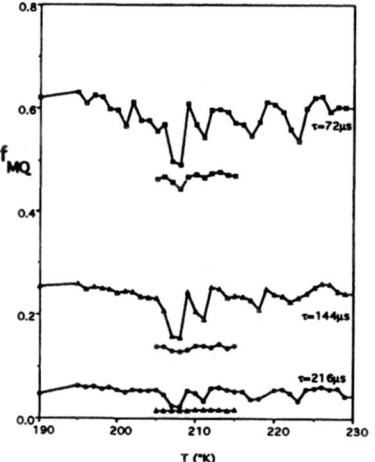

Fig. 1 Oscillations in the as-polymerized sample (solid symbols and lines correspond to the temperature region associated with motion of point defects. These defects are absent in the 64% crystalline sample (open symbols and dotted lines) and no oscillations are seen.

Differences in the behavior of f_{MQ} from the two PTFE samples are also seen at higher and lower temperatures, Fig. 2 and Fig. 3, respectively. Near room temperature (Fig. 2), the crystalline phase of PTFE undergoes two first order crystalline transitions and a corresponding maximum is seen in f_{MQ} of both samples. However, in the more crystalline as-polymerized sample this effect is larger. At much lower temperatures (Fig. 3), there is only motion in the amorphous phase corresponding to a decrease in f_{MQ} in the 68% crystalline sample while no change in f_{MQ} is noted for the 98% crystalline sample.

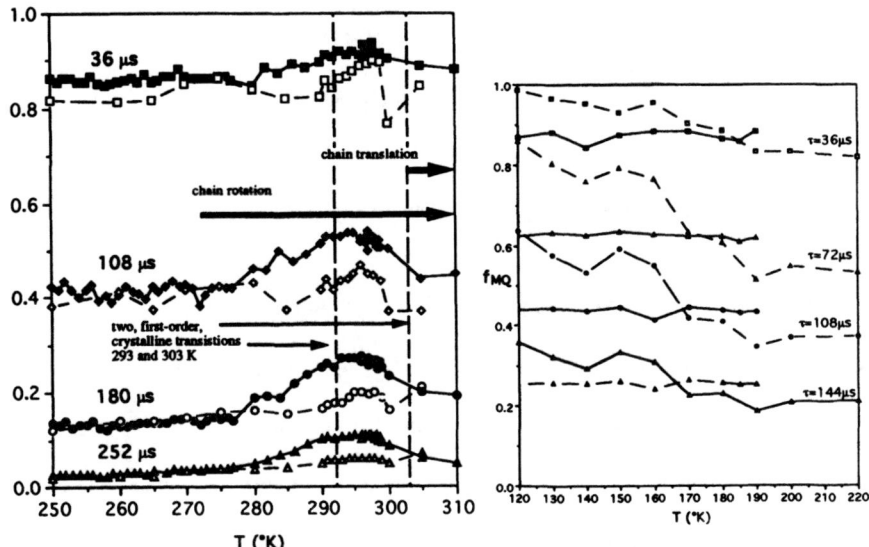

Fig. 2. fMQ for as-polymerized PTFE (98% crystalline, solid symbols and lines) and melt-crystallized PTFE (64% crystalline, open symbols and dotted lines). The rise in fMQ corresponds to onset of high speed crystalline chain rotations. Decrease near 303°K transition corresponds to onset of crystalline chain slippage.

Fig. 3. Decrease in fMQ with increase temperature in 64% crystalline sample (open symbols, dotted lines) is related to γ-relaxation in amorphous region. 98% crystalline sample (solid symbols and lines) is constant over this low temperature region.

CONCLUSIONS

These initial experiments on PTFE have demonstrated the potential of the MQ technique for measuring frequency specific motions and the ability to differentiate clearly between polymer samples which have different thermal histories and hence vary in crystallinity. In addition, the shape of the f_{MQ} features are signatures of the molecular level details of the underlying dynamics which produces them.

In the future, we seek to quantify the unique "signatures" of molecular motion present in the MQ NMR refocusing experiment. This understanding will allow valuable insight into the behavior of a wide variety of macromolecules in the solid state. Such information could be directly compared to the predictions of computational models which describe polymer dynamics.

ACKNOWLEDGEMENTS

The authors gratefully acknowledge the National Science Foundation, grant No CTS9057119. We would also like to thank Ted Treat and Paul Resnick at Dupont for the samples studied and Bruce Scruggs for valuable help.

REFERENCES

1. Y. Yen and A. Pines, J. Chem. Phys. **78**, 3579 (1983).

2. C. W.Bunn and E. R. Howells, Nature **174**, 549 (1954).

3. E.S. Clark and I.T. Muus, Z. Kristallogr Kristallgeom. Kristallphys. **117**, 119 (1962).

4. S.F. Lau, H. Suzuki, and B. Wunderlinch, J. Polym. Sci., Polym. Phys. Ed. **22**, 379 (1984).

5. R.K. Eby and K.M. Sinnott, J. Appl. Phys. **32**, 1765 (1961).

6. A.J. Vega and A.D. English, Macromolecules **13**, 1635 (1980).

7. D.W. McCall, D.C. Douglass, and D.R. Falcone, J. Phys. Chem. **71**, 998 (1967).

8. M. Takenaga, H. Obori, and K. Yamagata, J. Polym. Sci.Polym. Lett. Ed. **23**, 45 (1985).

9. Y. Ohzawa and Y. Wada, Japan. J. Appl. Phys. **3**, 436 (1964).

10. F. Krum and F. Müller, Kolloid Z. **164**, 81 (1959).

11. N.G. McCrum, B.E. Read, and G. Williams, Analastic and Dielectric Effects in Polymer Solids (Wiley, New York, 1967).

12. Y. Wada and R. Hayakawa, in Progress in Polymer Science, Japan, edited by Okamura and Takayanagi, (Kodansha, Tokyo, 1972), Vol. 3.

13. H. Starkweather, P. Zoller, G. Jones, and A. Vega, J. Polym. Sci. Polym. Phys. Ed. **20**, 751 (1982).

14. M. Takenaga, K. Yamagata, A. Kasai, and T. Ariyama, J. Appl. Polym. Sci. **39**, 1689 (1990).

RELAXATION AND DEFORMATION IN GLASSY POLYMERS

T. S. CHOW
Xerox Webster Research Center, 800 Phillips Road, 114-39D, Webster, NY 14580

ABSTRACT

A unified account of the structural relaxation and nonlinear deformation in glassy polymers is presented. The glassy state relaxation is derived from the local configurational rearrangements of molecular segments, and the dynamics of hole motion. We then apply this to predict the nonlinear stress-strain relationships.

INTRODUCTION

The physical properties of glassy polymers vary more strongly with time and temperature than those of other materials such as metal and ceramics, as a consequence the long time behavior and thermal history become a major concern. This important effect of the glassy state relaxation on the nonlinear deformation, which differs completely from metal plastic theory where material deforms without relaxation [1], forms the basis of this paper. The study of volume relaxation [2] below the glass transition temperature has indeed revealed that glassy states do undergo slow change processes. Furthermore, it has been recognized that volume relaxation, viscoelastic response and plastic yield share the same structural relaxation mechanism [3]. Constitutive equations [4] have been proposed to describe this nonlinear deformation, but they can not predict the properties of polymers as a function of physical aging, strain rate, temperature, and external stress fields. On the basis of our statistical dynamic theory of glasses [5], the deformation kinetics in the glassy state are considered as a result of the local configurational rearrangements of molecular segments, and the dynamics of holes (free volumes) provide a quantitative description of the segmental mobility. This new concept will be utilized in the discussion of the salient features of plastic yield and nonlinear stress-strain behavior of glassy polymers.

GLASSY STATE RELAXATION

The dynamics of the holes (free volumes) and bond rotations during vitrification and physical aging are analyzed on the basis of nonequilibrium statistical mechanics. We have reported that the conformational activation energy controlling the rotational relaxation of bonds is between 1 and 2 orders of magnitude lower than the hole activation energy [3]. As a result, the conformer relaxes much faster than the hole. Since the mechanical properties of glasses vary slowly in time, the dominant contribution to the structural relaxation and physical aging is from the holes. Consider a lattice consisting of n holes and n_x polymer molecules of x monomer segments each. The total number of lattice sites (N) and volume (V) are written as: $N(t) = n(t) + xn_x$, and $V = vN$ where t is time, and v is the volume of a single lattice cell. The hole number n(t) consists of both equilibrium and nonequilibrium contributions in the glassy state. The nonequilibrium part of n goes to zero above the glass transition temperature (T_g). The change below T_g defines the glassy state. Minimizing the excess Gibbs free energy due to hole introduction with respect to the hole number, the equilibrium hole fraction, $\bar{f} = \bar{n}/\bar{N}$, is given by

$$\bar{f}(T) = f_r exp\left[-\frac{\varepsilon}{R}\left(\frac{1}{T} - \frac{1}{T_r}\right)\right] \tag{1}$$

where ε is the mean energy of hole formation, R is the gas constant, and the subscript r refers to the reference condition at $T = T_r$, which is a fixed quantity near T_g. The ε characterizes the intermolecular interaction which affects the bonding between chain segments.

We have analyzed the hole dynamics and fluctuations on a fractal lattice [5], and have obtained the solution of nonequilibrium hole fraction, $\delta(t) = f(t) - \bar{f}$, for a system started from equilibrium

$$\delta(T,t) = -\frac{\varepsilon}{R}\int_0^t \frac{q\bar{f}}{T^2}\Psi(t-t')dt' \tag{2}$$

where $q = dT/dt < 0$ is the cooling rate, and Ψ is the relaxation function. Different paths of time integration in Eq.(3) describe different thermal history behavior of the glassy state relaxation and recovery kinetics -- the memory effect. The hole configurational space in a quenched and annealed glass is divided into regions separated by barriers. The size of the region is proportional to the time needed for a hole to penetrate an energy barrier, which is affected by large stress fields. From such time-length scaling, we have derived [5] the relaxation function

$$\Psi(t) = exp\left[-\left(\frac{t}{\tau}\right)^\beta\right], \qquad 0 < \beta \leq 1 \tag{3}$$

where β defines the shape of hole energy or the relaxation spectrum, and $\tau = \tau_r a$ is the relaxation time, and the shift factor

$$a(T,\delta) = \left(\frac{\bar{f}+\delta}{f_r}\right)^{-\frac{1}{\beta(\bar{f}+\delta)}} \tag{4a}$$

At $T = T_r$, we have $\delta = 0$, $\bar{f} = f_r$, and $\tau = \tau_r$ which requires $a = 1$. In the vicinity of T_g, Eq. (4a) becomes

$$\ln a(T,\delta) \simeq \frac{1}{\beta}\left(\frac{1}{\bar{f}+\delta} - \frac{1}{f_r}\right), \qquad as \quad \frac{\bar{f}+\delta}{f_r} \to 1 \tag{4b}$$

When $T \geq T_g$, we have $\delta = 0$ and Eq. (4b) reduces to the exact form of Doolittle's equation [6].

We have determined [3] the hole parameters for polystyrene (PS): $\varepsilon = 3.58$ kcal/mol, $\beta = 0.48$, $\tau_r = 30$ min and $f_r = 0.032$. These parameters will also be used to calculate the stress-strain behavior later. The departure from a Doolittle-WLF dependence to Eq. (4) type of shift factor is calculated in Figure 1. There are also important differences between Eqs. (3-4) and the Kohlrausch-Williams-Watts (KWW) equation [7], which has the same form as Eq. (3) except by replacing β and τ with β_w and τ_w, respectively. The β_w and τ_w are treated as empirical parameters which must alter continuously through the glass transition region in order to fit the relaxation data. Since our τ is a time dependent quantity for $T < T_g$, the accurate way of determining β from relaxation data is by treating t/τ as an independent variable plotted in the transient or dynamic master curves. Using the above mentioned hole parameters for PS, we have obtained [8] $\beta_w = 0.0785$ for $T < T_r - 20$ K and $\beta_w = 0.48$ for $T > T_r$, which

162

reveals the drastic change in β_w with temperature in the vicinity of the glass transition, but β remains to be a *constant*.

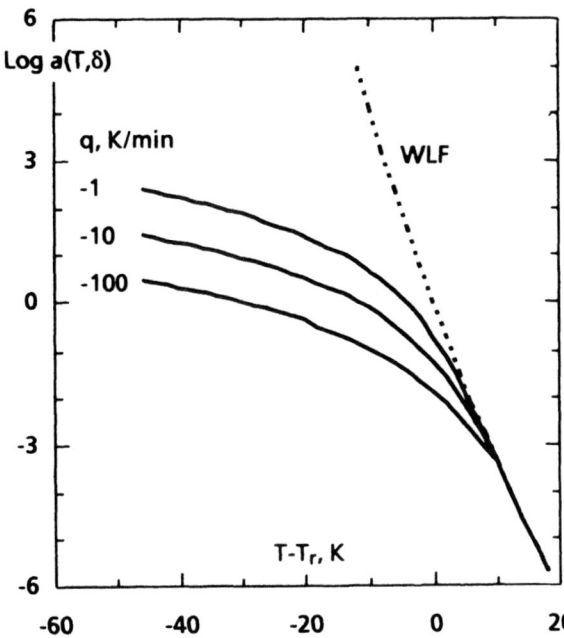

Figure 1. Relaxation time scale (a) as a function of temperature and cooling rate, when PS is cooled through the glass transition.

STRESS-STRAIN BEHAVIOR

For simplicity, we shall limit the scope of discussion to the uniaxial compression except in the case of differing stress fields. We have mentioned earlier: both the viscoelastic and volume relaxations share the same relaxation mechanism [3]. Therefore, the relaxation modulus (E) of physically aging polymers can in general be written in the following form: $E(t) = E_0\Psi(t/\tau)$ where E_0 is the Young's modulus and Ψ is given by Eq. (3). At high stress levels, the contribution from the external work done on a lattice cell has to be included in the structural relaxation. By taking into account the long range cooperative interaction, the external work done by a compressive stress (σ) acting on a hole cell during yielding is $\Delta w = -\sigma\Omega N/n = -\sigma\Omega/f$, where Ω is the compressive activation volume. The ratio Ω/f represents the volume of the polymer segment which has to move as a whole in order for plastic deformation to occur. The nonlinear relaxation time takes the form [3]

$$\tau_\sigma(T,\delta;\sigma) = \tau(T,\delta)\exp\left(-\frac{\sigma\Omega}{2f\beta kT}\right) \tag{5}$$

Combining Eqs. (3-5), and putting $e = \dot{e}t$, we have the stress dependent relaxation modulus written in the form

$$E(T,\delta;\sigma) = E_o\Psi = E_o(T,\delta)\exp\left\{-\left[\frac{e}{\dot{e}\,\tau_\sigma(T,\delta;\sigma)}\right]^\beta\right\} \tag{6}$$

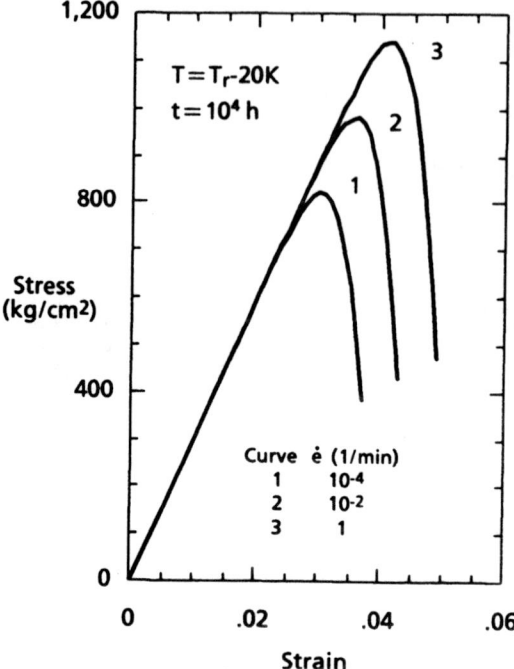

Figure 2. Effect of strain rate [8] on the compressive stress-strain curves of PS.

where e is the compressive strain and $\dot{e}$ is the strain rate. The δ in Eqs. (2-6) is the same parameter. Eq. (6) reveals that plastic yielding occurs at

$$\dot{e}\,\tau_o(T,\delta;\sigma_y) \sim constant \tag{7}$$

Eq. (7) is the well known Eyring's yield criterion [9]. Using Eqs (5) and (7), we obtain the compressive yield stress

$$\sigma_y = A + K\left[log\ \dot{e} + log\ a(T,\delta)\right] \tag{8}$$

where A and $K \sim 1/\Omega$ are constants. For nonlinear deformation, the dependence of E on σ in Eq. (6) results in the continuous changing of E, which defines the slope of the σ-e curve with other parameters held constant. Thus, integrating Eq. (6), we get a nonlinear integral equation:

$$\sigma(e) = E_o(T,\delta)\int_0^e exp\left\{-\left[\frac{e'exp\,(2.303\,\sigma(e')/K)}{\dot{e}\,\tau(T,\delta)}\right]^\beta\right\}de' \tag{9}$$

Eqs. (8-9) together with Eqs. (1), (2), and (4) will be applied to calculate many different effects on the nonlinear stress-strain behavior of PS, including plastic yielding, with a single set of hole parameters. The shape of stress-strain curves depends on the time scales in which the solid polymer is relaxed and measured.

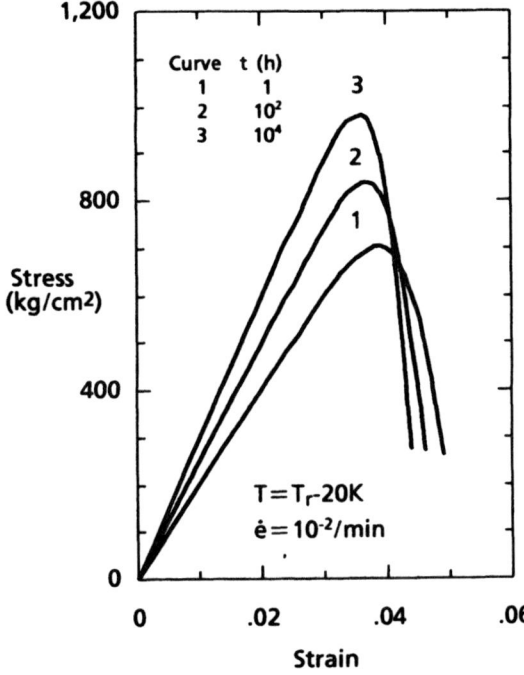

Figure 3. Effect of physical aging [8] on the compressive stress-strain curves of PS.

Using Eq. (8) and σ_y-e data under uniaxial compression at room temperature, one obtains $K = 90$ kg/cm^2. In addition to K and β, Young's modulus $E_0 = 29.6 \times 10^3$ kg/cm^2 and $\tau = 2 \times 10^{14}$ min at 23°C for a well aged PS are determined separately from experimental data. The calculated strain rate effect is shown in Figure 2. Consider that the polymer is quenched isobarically from liquid to glass for annealing before it is quenched again to room temperature for assessing the stress-strain behavior. By using Eqs. (9) and (4) together with the hole parameters, the effect of physical aging is calculated in Figure 3.

In order to discuss the influence of external stresses, Δw has to be generalized. The external work done on a lattice cell is now written in the form: $\Delta w = -\sigma_{ij}\Omega_{ij}/f$ where σ_{ij} and Ω_{ij} are the stress and activation volume tensors, respectively. For isotropic glasses, the activation volume tensor has only two independent components. They are the bulk activation volume, which is equal to the lattice volume v, and the shear activation volume Ω_{12}. The activation volumes in uniaxial tension, $\Omega(+)$, and compression, $\Omega(-)$, can be expressed in term of v and Ω_{12}. For PS, we determined [10]

$$\left[\Omega_{11}(+), \Omega_{11}(-), \Omega_{12}, v\right] = \left[111.7, 98.8, 158.6, 19.2\right]\mathring{A}^3 \tag{10}$$

which is used to calculate the effect of stress fields in Figure 4. Except in the case of hydrostatic compression, Eq.(10) reveals that the volume of molecular segments which has to move as a whole at yielding is much larger than the volume of a single lattice site. The motion of the polymer segments is no longer local, which is in contrast to linear deformation. This reinforces our earlier assumption that yielding is

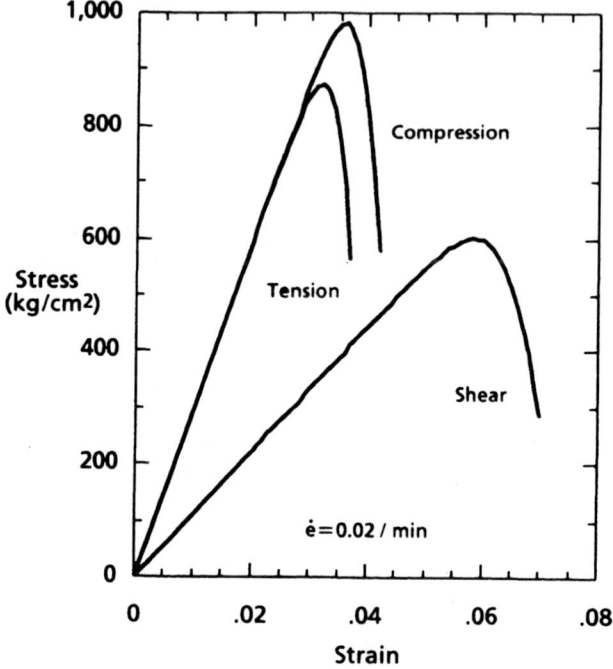

Figure 4. Effect of stress fields [10] on the nonlinear stress-strain curves of PS at 23°C.

a cooperative phenomenon, which is consistent with recent experimental investigation [11].

REFERENCES

1   P. Polukhin, S. Gorelik and V. Vorontsov, Physical Principles of Plastic Deformation, (Mir, Moscow 1983).
2   A. J. Kovacs, Adv. Polym. Sci., 3, 394 (1963); A. J. Kovacs, J. J. Aklonis, J. M. Huchinson and A. R. Ramos, J. Polym. Sci., Polym. Phys. Ed., 17, 1097 (1979).
3   T. S. Chow, Adv. Polym. Sci., 103, 149 (1992).
4   R. F. Landel and S. T. J. Peng, J. Rheology, 30, 741 (1986).
5   T. S. Chow, Macromolecules, 24, 440 (1992).
6   A. K. Doolittle, J. Appl. Phys., 22, 1471 (1951); M. L. Williams, R. F. Landel and J. D. Ferry, J. Am. Chem. Soc., 77, 3701 (1955).
7   R. Kohlrausch, Ann. Phys. (Leipzig), 21, 393 (1847); G, Williams and D. C. Watts, Trans. Faraday Soc., 66, 80 (1970).
8   T. S. Chow, Polymer, 34, 541 (1993).
9   H. Eyring, J. Chem. Phys., 4, 283 (1936).
10 T. S. Chow, J. Rheology, 36, 1707 (1992).
11 C. Xiao and A. F. Yee, Macromolecules, 25, 6800 (1992)

GENERALIZED ACTIVATION ENERGY SPECTRUM THEORY: A NEW APPROACH FOR MODELING STRUCTURAL RELAXATION IN AMORPHOUS SOLIDS

JUNG H. SHIN AND HARRY A. ATWATER, Thomas Watson Laboratory of Applied Physics, California Institute of Technology, Pasadena CA 91125

ABSTRACT

A general approach to the dynamics of structural relaxation in amorphous solids is developed. A form of the recombination kinetics of defects is chosen which removes the *ad hoc* assumption made in previous theories that defects recombine only with others of identical activation energy. The generalized theory is tested quantitatively by modelling the structural relaxation of amorphous silicon, and comparing the results with the experimental data on structural relaxation. It is found that the generalized theory is necessary in order to accurately describe the time-resolved relaxation data. The generalized theory is also applied to estimate the effect of irradiation on the nucleation kinetics of crystal silicon, and is found to agree well with experimental data.

INTRODUCTION

The activation energy spectrum (AES) theory for defect annihilation [1] has been very valuable in describing structural relaxation of many amorphous solids [3, 4]. The main point of the AES theory is that the activation energies of thermal processes responsible for structural relaxation form a continuous, broad spectrum rather than sharply defined peaks at one or several activation energies. Therefore, the defect population is described by $N(Q)$, such that $N(Q)dQ$ is the density of defects in a given experimental sample with activation energy between Q and $Q + dQ$. Another important concept in the AES theory is that of the density of relaxation states, $D(Q)$. Physically, it is the saturation value of $N(Q)$, as could be achieved by a full unrelaxation to the saturation value of defect density at 0 K.

The activation energy spectrum theory as it has been used so far (hereafter termed the simple theory), assumes that defects with different activation energies are independent of each other. Furthermore, the annihilation kinetics of defects are approximated by a step function. These approximations considerably simplify the analysis of experimental data, and the simple theory has been successful in at least qualitative analysis. However, it is unphysical to assume that two defects somehow do not interact with each other if their activation energies differ by dQ, especially if the defects undergo bimolecular recombination.

We present here a new, more physically reasonable theory, hereafter referred to as the generalized activation energy spectrum theory, which takes into account the possibility that defects may recombine with each other regardless of their activation energies. We will use amorphous silicon (a-Si) as the model system, and calculate the dynamics of defects in a-Si both under irradiation and during thermal anneal using both the simple and the generalized activation energy spectrum theory. Amorphous silicon is chosen to be the model system because there has been a great interest in studying the structural relaxation of amorphous silicon [3, 2, 5, 6]. Furthermore, using electrical conductivity as a probe, the density of relaxation states $D(Q)$ of a-Si has previously been measured [3, 7].

MODELLING DEFECT DYNAMICS IN AMORPHOUS SILICON

To model the evolution of defect population in a-Si during ion irradiation and following anneals, we make the following assumptions. First, we assume that an impinging ion creates defects with different activation energies with equal probability. Second, we assume that annihilation occurs via bimolecular recombination [5, 8]. Finally, we assume that the reaction cross-section is independent of defect activation energy. If we use the simple theory, we can write

$$\frac{dN(Q)}{dt} \;=\; g\dot{\phi}\left(1 - \frac{N(Q)}{D(Q)}\right) - \nu_o \exp\left(\frac{-Q}{kT}\right) N(Q)^2 \tag{1}$$

where g, $\dot{\phi}$, and ν_o are defects/flux, ion flux, respectively. ν_o is the reaction constant. In the case of bimolecular recombination, it would be the product of the attempt frequency, lattice constant, and the reaction cross section. The first term on the right side of Eq. (1) is the defect generation term. It ensures that the actual defect population, $N(Q)$, saturates at its maximum value, $D(Q)$. A similar form for generation of defects has been used successfully before [9].

The generalized activation energy spectrum theory takes into account the possibility that a defect may recombine with defects with different activation energies. This is achieved by writing

$$\frac{dN(Q)}{dt} \;=\; g\dot{\phi}\left(1 - \frac{N(Q)}{D(Q)}\right) - \frac{\nu_o N(Q)}{2}\left[\exp\left(\frac{-Q}{kT}\right)\int N(Q')dQ' \;+ \right.$$
$$\left. \int \exp\left(\frac{-Q'}{kT}\right) N(Q')dQ'\right], \tag{2}$$

where $\nu_o N(Q)\exp(-Q/kT)\int N(Q')dQ'$ is the rate at which a defect with activation energy Q annihilates defect with different activation energies, and $\nu_o N(Q)\int \exp(-Q'/kT)N(Q')dQ'$ is the rate at which a defect with activation energy Q is annihilated by other defects with different activation energies. The factor of $1/2$ is needed because we are counting each defect recombination event twice. In both cases, we have g and ν_o as fitting parameters to be determined.

The underlying assumption of the interaction of defects with different activation energies is that the defects are of similar kinds (presumably point defects), and differ only in their activation energies. The activation energies are unlikely to be that of self-diffusion, since the activation energy for self-diffusion in a-Si is between 0.13 and 0.22 eV [10]. Most likely, the activation energies are that of some rate-limiting step to defect migration, possibly that of the transition states necessary for the defects to become mobile.

Both Eq. (1) and Eq. (2) were solved by computer simulations such that $N(Q, t+\delta t) = N(Q,t) + (dN/dt)\delta t$, where δt is the timestep. For simulations, the measured value of $D(Q)$ from Ref. [3] was extended to 0.6 eV, which was the low limit used for activation energy. The results of calculations of defect dynamics in a-Si using the simple theory will be hereafter referred to as the modified simple theory, since we no longer make the step function approximation. For Eq. (1), the results of simulations were compared with analytical solution, and found to agree very well, thus confirming the accuracy of simulation.

EXPERIMENTAL CONDITIONS

In all experiments, a-Si resistors defined as mesas from a-Si films were used. Irradiations were performed at room temperature, with energies chosen to confine all the damage within a-Si film. All anneals were performed in a high vacuum furnace for 15 min. A more detailed description of these experiments can be found in Ref. [11].

The quantity actually measured in experiments was the electrical conductivity of a-Si. However, based on existing experimental data, we have shown previously that the electrical conductivity can be used as a sensitive probe of defect dynamics of a-Si [3]. Therefore, we will hereafter discuss structural relaxation of a-Si in terms of defect density without explicit reference to the measured conductivity.

RESULTS AND DISCUSSION

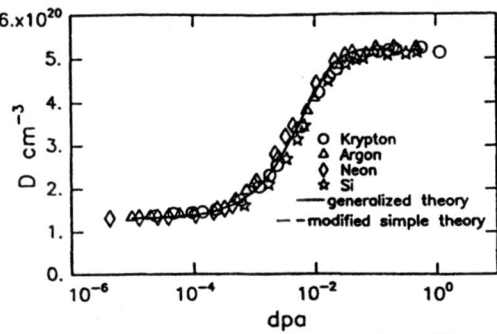

Figure 1: Calculated and measured defect density in amorphous silicon 15 min after irradiation to various doses using indicated ions. The solid and dashed lines are fit using indicated theories.

Figure 1 shows the defect density of a-Si, measured 15 minutes after termination of irradiation, as a function of irradiation dose. Prior to irradiation, samples were relaxed by an anneal at 900 K. The symbols are experimental data. The ion doses are normalized to displacement per atom (dpa) as calculated by TRIM simulation program [12]. Since the change in defect density scales very well with dpa, we will discuss unrelaxation in terms of dpa without making specific reference to the irradiating ion. The solid curve is the result of calculations using the generalized theory, and the dashed curve is the result of calculations using the modified simple theory. The values for g and ν_o used to fit the data are 4.1×10^{22} cm^{-3} dpa^{-1} eV^{-1} and 1.7×10^{-10} cm^3 sec^{-1}, respectively, for the generalized theory, and 4.1×10^{22} cm^{-3} dpa^{-1} eV^{-1} and 5.1×10^{-8} cm^3 sec^{-1} eV, respectively, for the modified simple theory. Both theories fit the experimental data well. It must be noted that the value of ν_o for the generalized theory depends on Q_{min}, the lower limit on the activation energy used in simulations. However, the conclusions that can be drawn from the results do not change if the same two values of ν_o and Q_{min} are used consistently. The same values of g and ν_o are used in all the following figures. *Since there are no more adjustable parameters, the results which follow are results of calculation, not a fit to the data.*

Figure 2 shows the calculated defect density $N(Q)$ of a-Si, originally irradiated at room temperature to saturation, after isochronal anneals. Also shown is the traditional step function approximation made in the simple theory. It is clear that the step function approximation is inaccurate, especially for high temperature anneals. Comparing the modified simple theory with the generalized theory, we find that within the modified simple theory, high temperatures are needed to anneal out defects with high activation energies. Within the generalized theory, however, some of the high activation energy defects are annihilated even after a low temperature anneal, because the low activation energy defects recombine not only with themselves, but also with some of the defects with higher activation energies. Therefore, the total defect density decreases much more steeply with initial increase in anneal temperature than is predicted by either the simple or the modified simple theories.

Such a steep initial decrease in the total defect density, however, is observed experimentally, as Fig. 3 shows. The lack of exact agreement between theories and prediction is expected, since theoretical results are predictions, and not an attempt at a fit. However, the generalized theory predicts values that are in better agreement with the experimental data, including the observed curvature in the decrease of total defect density with increasing anneal temperature. Furthermore, as the inset in Fig. 3 shows, the generalized theory predicts the experimentally observed "crossover" effect. That is, for a certain combination of irradiation dose and anneal temperature, samples irradiated to a higher dose have a lower defect density after an anneal, since they initially contain more mobile defects to annihilate defects that are immobile. This "cross-over" effect cannot occur in the simple theory. Its prediction by the generalized theory confirms that the generalized theory produces new physics, not just better quantitative agreement.

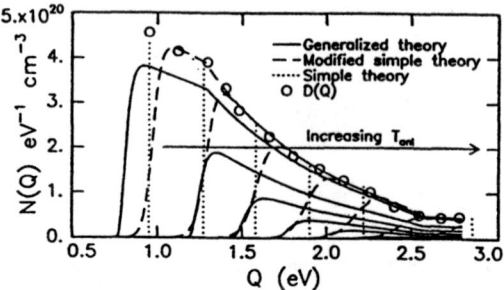

Figure 2: Calculated values of $N(Q)$ after isochronal anneals. The symbols are the density of defect states $D(Q)$. The solid, dashed, and dotted lines are results of calculations using indicated theories.

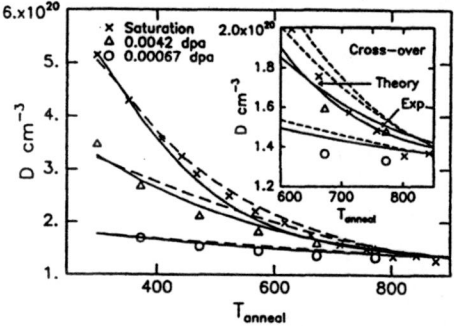

Figure 3: Calculated and measured defect density after 15 min anneals for samples irradiated to different doses. Solid lines are results of the generalized theory, and the dashed lines are results of modified simple theory. Symbols are experimental data. Inset shows the "cross-over" effect, marked by arrows.

Figure 4 (a) shows the defect density immediately following termination of irradiation for different fluxes, plotted against $\ln(t)$. Solid lines are results of the generalized theory, and the dashed lines are results of the modified simple theory. Symbols are experimental data measured *in situ*. Again, the results of the generalized theory agree better with the experimental data, converging quickly to same values within experimental resolution. More importantly, the generalized theory again predicts the observed "cross-over" effect, shown more clearly in the Fig. 4 (b). Again, the reason for this effect is that with higher ion flux, more mobile defects are available to annihilate defects that are immobile. This "cross-over" effect, however, need not always be present even within the generalized theory. Previous investigation [7] has failed to observe this effect, and it was verified by calculation that under experimental conditions in Ref. [7], the defect generation rate is too low for the "cross-over" effect to occur. The "cross-over" effect will never occur, however, if defects of different activation energies are independent of each other, which would be the case for unimolecular recombination of defects at fixed sinks. Since evidence suggests that structural relaxation in a-Si occurs via annihilation of point defects [5, 13], the existence of a "cross-over" effect may be interpreted as a strong evidence for bimolecular recombination of defects in a-Si.

Recently, the role of ion irradiation in enhancing the nucleation of crystal silicon in amorphous

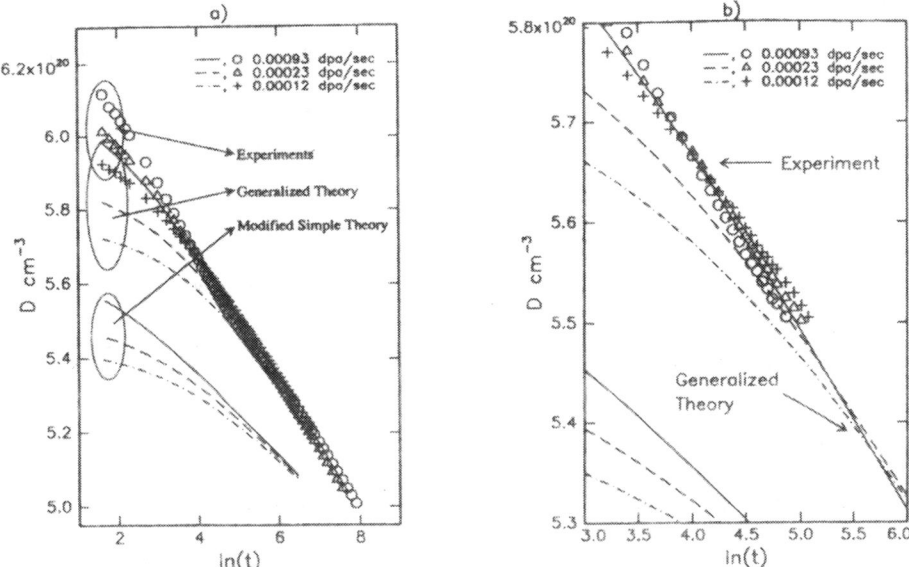

Figure 4: Calculated and measured defect decay transient immediately following irradiation to saturation with different fluxes. Symbols are experimental data, and lines are results of calculation. Figure (b) shows the "cross-over" effect, marked by arrows, in detail.

matrix was investigated [14]. Using *in situ* measurement of nucleation rate at high temperature and low ion flux, we have demonstrated that nucleation rate is substantially enhanced under these conditions although the growth rate is unaffected [14]. Furthermore, by interrupting the ion irradiation before the transformation was completed, and then observing the resulting dynamic changes in the nucleation rate, we have shown that the free energy barrier which control the nucleation is affected by irradiation. Using the classical theory of nucleation, and assuming that the change in the free barrier is due to increase of the free energy difference between amorphous and crystalline silicon, we have measured an increase in the free energy difference of 0.002 eV/atom [14].

With the generalized theory, it is possible to estimate the defect population under irradiation conditions where the nucleation rate is affected, and thus estimate the change in the free energy difference between amorphous and crystal silicon. Excess defects in amorphous silicon are known to increase the stored enthalpy of a-Si [5], and could account for the observed increase in the free energy difference. Unfortunately, it is not practical to simulate the defect dynamics directly, since the time step required for an accurate simulation at high temperatures is exceedingly short. However, it is possible to put upper and lower bounds on the total defect density. The lower limit on the defect concentration was determined by simulating the evolution of defect population with time steps that are longer than the estimated defect lifetime. This is equivalent to letting defects continue to annihilate even after they are already annihilated, and results in an underestimate of the defect population. The upper limit on defect concentration during irradiation was determined by simulating the evolution of defect population with a lower limit of activation energy that is greater than 0.6 eV. The defects with low activation energies exert controlling influence on the total defect density. Therefore, using a lower limit of activation energy that is higher than 0.6 eV is equivalent to neglecting their dominant contribution to defect annihilation, and results in an overestimate of the defect population.

The change in the free energy difference between amorphous and crystalline silicon was estimated by assuming that each defect contributes equally to the stored enthalpy, and scaling the excess defect density to the known value of stored enthalpy of fully unrelaxed a-Si. The estimated change in the free energy difference between a-Si and c-Si, $\delta(\Delta G)$ is

$$0.0015 \text{ eV} < \delta(\Delta G) < 0.0031 \text{ eV}$$

This is in excellent agreement with 0.002 eV, the experimentally estimated value of $\delta(\Delta G)$, and further supports the validity of the generalized theory.

CONCLUSION

In conclusion, a generalized theory is presented to describe the structural relaxation of amorphous solids. Applied to structural relaxation of a-Si, only the generalized theory is sufficient for accurate description of the relaxation dynamics. Applied to irradiation enhanced nucleation of crystal silicon, the generalized theory predicts a change in the free energy difference between a-Si matrix and crystal silicon under irradiation that is in good agreement with experimentally derived value.

ACKNOWLDGEMENT

We thank Michael Easterbrook and Chih M. Yang for technical assistance. This work was supported by U. S. Department of Energy under grant DE -FG03-89ER45395

References

[1] M. R. J. Gibbs, J. E. Evetts, and T.A. Leake, J. Mater. Sci. **18**, 278 (1983).

[2] J H. Shin, J. S. Im, and H. A. Atwater, in *Phase Formation and Modification by Beam-Solid Interactions,* edited by G. S. Was, D. M. Follstaedt, and L. E. Rehn, Mat. Res. Soc. Symp. Proc. **235**, 21 (1992).

[3] J. H. Shin and H. A. Atwater, Phys. Rev. B **48** 5964 (1993).

[4] B. J. Bartholomeusz, P. M. Kopalidis, and D. G. Stinson, J. Appl. Phys. **71** 4595 (1992).

[5] S. Roorda, W. C. Sinke, J. M. Poate, D. C. Jacobson, S. Dierker, B. S. Dennis, D. J. Eaglesham, F. Spaepen, and P. Fuoss, Phys. Rev. B **44** 3702 1991).

[6] S. Coffa, J. M. Poate, D. C. Jacobson, W. Frank, and W. Gustin, Phys. Rev. B **45**, 15 8355 (1992)

[7] S. Coffa, F. Priolo, and A. Battaglia, Phys. Rev. Lett. **70** 3756 (1993).

[8] C. A. Volkert, J. Appl. Phys. **70**, 3521 (1991).

[9] P.A. Stolk, L. Calcagnile, S. Roorda, W. C. Sinke, A. J. M. Berntsen, Appl. Phys. Lett. **60** 1688 (1992).

[10] B. Park and F. Spaepen, J. Appl. Phys. **68** 4556 (1990).

[11] J. H. Shin and H. A. Atwater, in *Phase Transformations in Thin Films–Thermodynamics and Kinetics* edited by M. Atzmon, A. Greer, J. Harper, and M. Libera, Mat. Res. Soc. Symp. Proc. **311**. Mat. Res. Soc. Symp. Proc. **311** (accepted for publication).

[12] J. F. Ziegler, J. P. Biersack, and U. Littmark, *The Stopping and Range of Ions in Solids* (Pergammon, New York, 1985); Copyright J. F. Ziegler, 1991.

[13] G. N Vandehoven, A. N. Liang, L. Niesen, and J. S. Custer, Phys. Rev. Lett. **68** 3714 (1992).

[14] J. H. Shin and H. A. Atwater; Nucl. Inst. Meth. B **80** 973 (1993).

Crystal to Glass Transition and Melting in Two Dimensions

M. LI[1,2], W. L. JOHNSON[1] AND W. A. GODDARD III[2]
[1] W. M. Keck Laboratory, 138-78, California Institute of Technology, Pasadena, California 91125
[2] Molecular & Materials Simulation Center, Beckman Institute, 139-74, California Institute of Technology, Pasadena, California 91125

ABSTRACT

Thermodynamic properties, structures, defects and their configurations of a two-dimensional Lennard-Jones (LJ) system are investigated close to crystal to glass transition (CGT) via molecular dynamics simulations. The CGT is achieved by saturating the LJ binary arrays below glass transition temperature with one type of the atoms which has different atomic size from that of the host atoms. It was found that for a given atomic size difference larger than a critical value, the CGT proceeds with increasing solute concentrations in three stages, each of which is characterized by distinct behaviors of translational and bond-orientational order correlation functions. An intermediate phase which has a quasi-long range orientational order but short range translational order has been found to exist prior to the formation of the amorphous phase. The destabilization of crystallinity is observed to be directly related to defects. We examine these results in the context of two dimensional (2D) melting theory. Finite size effects on these results, in particular on the intermediate phase formation, are discussed.

INTRODUCTION

Both crystal to glass transition and melting are topological order to disorder transitions. The similarities of the starting and end phase and the disappearance of long range translational symmetries at the transitions suggest that they are the same [1]. In general the thermodynamics and in particular, the kinetics of these transitions resemble each other remarkably [1]. Although the mechanism of melting is still in debate [2], the theories and models developed for melting have been extensively used in the understanding of the CCT [1, 3-10]. However, these analogous criteria suffer from the same problem as they do in melting, namely that they are not able to provide a detailed microscopic picture of how these transitions proceed. In this paper, 2D LJ binary arrays under the polymorphic constraint are studied. By keeping the array at a sufficiently low temperature and thus eliminating chemical inhomogeneity, we expect to see certain intrinsic mechanisms of the CGT and to establish relationships between microscopic properties (such as atomic interactions and atomic size differences) with the thermodynamic and kinetic properties during the CGT.

The crystal to glass transitions can be obtained by a number of methods [1,11]. There are rarely cases where a pure system can be amorphized unless some extreme kinetic constraints are imposed. As in rapid quenching of melts, the majority of systems that can be amorphized usually contain several components. These multicomponent amorphizable systems invariably become metastable prior to the transition with much increased free energy. The composition-induced crystal to glass transition seems to be present in most of these systems and thus can be regarded as a general characteristic of solid state amorphization. Even mechanical deformation- and irradiation-induced amorphization can be regarded as composition-induced. The defects produced in these processes such as dislocations, antiphase and grain boundaries, antisite defects and interstitials directly contribute to the rise of free energy of the systems as well as the solute solubility. In particular, one can regard the chemical disordering such as the formation of antisite defects and interstitials in intermetallics caused either by irradiation or by mechanical deformation as a process producing metastable solid solutions. This motivated us to use solid solutions as a model system. In our early work in a binary LJ solid solution [6] characterized by two parameters, the solute concentration and solute-solvent atomic size difference, we found that a crystal saturated with solute atoms of sufficiently large atomic size mismatch can transform into a glass. Dramatic softening of elastic constants, especially shear elastic constants, prior to the CGT led us to believe that it is caused by mechanical instability. Furthermore, the molar volume and enthalpy of the crystalline and the amorphous phases at the transition differ only slightly. Compared with abrupt changes in both volume and enthalpy at thermal melting, the CGT seemed to be a much weaker transition. However, the finite system size, short simulation time and possible insufficient relaxation in the system made it difficult to draw a firm conclusion about the nature of the transition. In addition, three-dimensional microscopic configurations of the system could not be observed easily. In 2D arrays, however, we can minimize these problems by using larger systems and longer simulation times. The atomic configurations and defects and their evolution can also be visualized directly.

CRYSTAL TO GLASS TRANSITIONS IN BINARY LJ ARRAYS

To simulate the CGT, we used a model system of binary arrays of atoms, A and B types, interacting with a Lennard-Jones potential. Arrays of solid solutions were generated by dispersing small solute atoms B on a hexagonal lattice occupied by solvent atoms A. Structural and thermodynamic properties were calculated using MD methods. Further computational details can be found in ref. 12 and 13. The most relevant parameters that can be varied are solute and solvent atomic size ratio, $\alpha = R_{BB}/R_{AA}$, where R_{AA} and R_{BB} are A and B atomic radii, LJ potential well depths of three types of interactions, $\epsilon_{AA}, \epsilon_{AB}$ and ϵ_{BB}, and solute concentrations, $X_B = N_B/N_{tot}$, where N_B and N_{tot} are number of B atoms and the total number of atoms used in simulations. A large potential depth between solute and solute atoms can lead to a large negative heat of mixing and a large atomic size difference can cause large local lattice distortions and fast solute atom diffusion, both of which, for instance, have been found to be essential for solid state amorphization. In our simulations, however, the potential depths of all three types of interactions were set equal to minimize the possibility of cluster formation, phase separation or any other chemical inhomogeneity. So we are left with only two parameters, the atomic size ratio and solute composition.

Since the system is finite, possible wavelengths of fluctuations of composition, structure, thermodynamic and defect properties, are restricted to this size. The finite size can affect the simulation results significantly, especially close to the CGT. The quality of the simulation results can also be affected by inadequate sampling of the system because of the randomness of the solute distributions in a finite sample. To make a finite system representative of that of an infinite system, several initial configurations with different random solute distributions were used. However, $N_{tot} = 256$ atoms used in this simulation are still relatively small. A more detailed work which checks the finite size effects will be presented elsewhere [13]. In the remainder of this paper, we will present these preliminary simulation results close to the CGT in the binary arrays with $\alpha = 0.75$, temperature $T = 0.25$ (in reduced LJ units), which is slightly below the glass transition temperature, and pressure $P = 0.0$.

Structure evolution of the binary LJ solid solution can be observed through radial distribution functions (RDF) shown in Fig. 1. As the solute concentration increases, the peaks of RDF's become broadened but can still be well resolved. The double peaks corresponding to the second and third nearest neighbor positions remain distinguishable even when the amorphous phase forms at $X_B \geq 17.2\%$. To have a more quantitative measurement of atomic order and symmetry of the system, we calculated the the translational and bond-orientational order correlation functions [14]. The translational order correlation function is defined by $\rho_G(r) = <\rho_G(r)\rho_G(0)>$, where $\rho_G(r) = \exp(-i\vec{G} \cdot \vec{r})$ is the translational order parameter with the shortest reciprocal lattice vector G in a perfect hexagonal lattice. The orientational order correlation function is defined by $\Psi(r) =< \psi(r)\psi(0) >$, where $\psi(r)$ is the orientational order parameter that measures the angular order of the nearest neighbor atom bonds with respect to a reference axis.

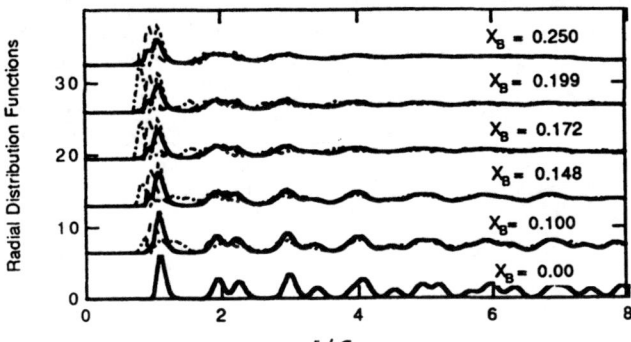

Fig. 1 Solid line: total RDF; dotted line: partial RDF for AA atoms; dashed line: partial RDF for AB atoms; dot-dashed line: partial RDF for BB atoms

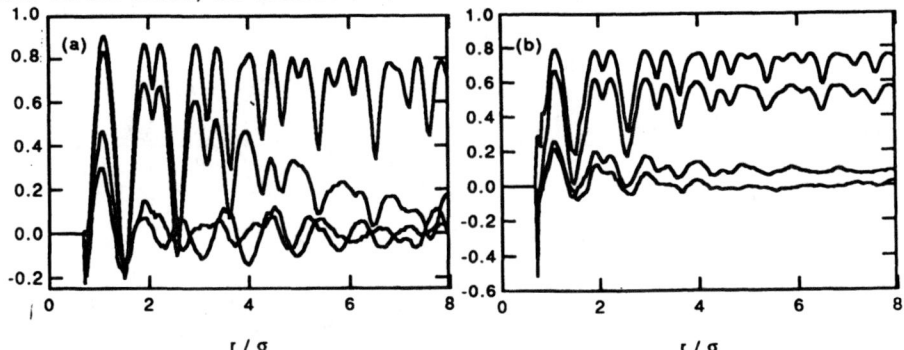

Fig. 2 (a) Translational order correlation functions; (b) Bond-orientational order correlation functions. From top to bottom: $X_B = 0.10, 0.148, 0.172, 0.25$.

In 2D crystalline phases, the quasi-long range translational symmetry decays algebraically, while the orientational correlation function remains a constant over a long range [14]. In liquid or amorphous phases, both correlation functions are short-range and decay exponentially over distance. More interesting is a possible intermediate phase, called the hexatic phase, existing between these ordered and disordered phases, which has quasi-long-range orientational order but short-range translational order. Nelson, Rubinstein and Spaepen [15] demonstrated the existence of such phase in an experiment using hard sphere binary arrays. We will show in this paper that such a phase could also exist in binary arrays with atoms interacting with soft LJ potentials.

Fig. 2 shows the calculated correlation functions at different solute concentrations in the binary arrays. It can be seen clearly that the correlation functions behave remarkably similarly to those predicted by the theory of two dimensional melting [14]. For $X_B < 0.148$, $\Psi(r)$ remains a constant while $\rho_G(r)$ decays only slightly at large distances; the translational correlation function decays fast to almost zero for $X_B \geq 0.148$ over the distance of the sample size, but the orientational correlation function shows a slight decay in the same manner as $\rho_G(r)$ does at $X_B < 0.148$. Finally both of them exponentially decay to zero at roughly the third or fourth nearest neighbor distances when $X_B \geq 0.199$.

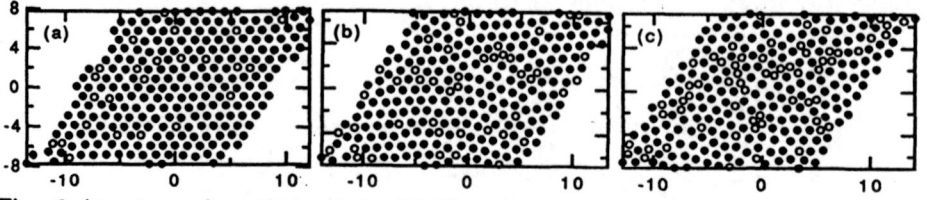

Fig. 3 Atomic configurations of the LJ Binary arrays at (a) $X_B = 0.10$ (crystal); (b) $X_B = 0.172$ (Hexatic ?); (c) $X_B = 0.25$ (amorphous). •: A atom; o: B atom.

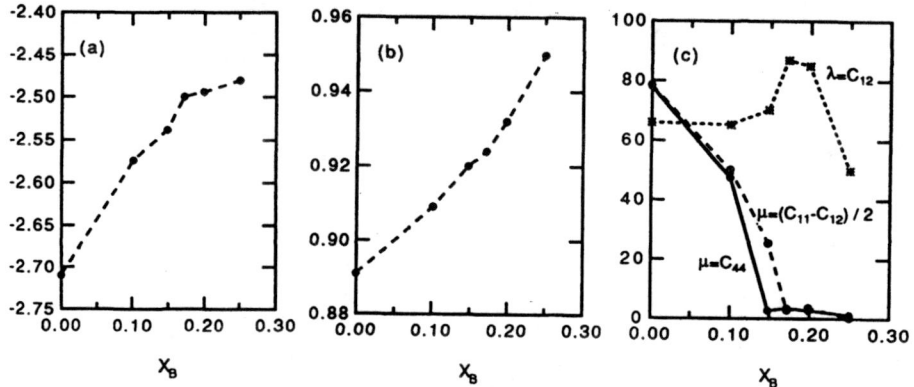

Fig. 4 (a) Molar enthalpy; (b) density and (c) elastic constants.

We also plotted the atomic configurations from which these correlation functions were calculated. An edge dislocation is defined by an extra row of atoms in two dimensions. The translational order defined above will be interrupted across such a row of atoms defining a dislocation. At $X_B = 0.148$ the distance over which the translational correlation function tends to vanish is approximately the same as the mean distance between two dislocation aggregates. Fig. 3b shows the atomic configuration at $X_B = 0.172$ where translational order is absent beyond the third or fourth nearest neighbor, but the long-range orientational order is still present at large distances. The dislocations in this system become more dense and some well defined crystalline patches start showing different orientations. The amorphous structure shown in Fig. 3c corresponds to the binary arrays at $X_B = 0.25$ with both short-ranged $\rho_G(r)$ and $\Psi(r)$.

Fig. 4a and 4b show the changes of density and enthalpy during the CGT as X_B increases. Interestingly no abrupt changes have been observed for the density at $X_B = 0.172$ where the long range translational order disappears and at the composition $X_B = 0.199$ where long range orientational order vanishes. The enthalpy has a slight jump ($< 2\%$) when long-range translational order disappears at $X_B = 0.172$. In addition, we have not observed any hysteresis.

Fig. 4c shows the variations of the two independent elastic constants in the two dimensional binary arrays with X_B. The shear constants μ decreases and λ increases at $X_B < 0.199$, leaving the bulk modulus almost a constant. μ decays sharply and reaches an extremely small value at $X_B \geq 0.148$. It goes to zero only if stress fluctuations in the system are large enough to spontaneously activate dislocations and drive them sliding across the system. λ starts decaying only after the system enters the amorphous region. As the binary array gets close to the CGT, it develops an elastic anisotropy. The relation $\mu = C_{44} = C' = (C_{11} - C_{12})/2$ in an isotropic media such as the hexagonal lattice is no longer obeyed, particularly in the region of "hexatic phase". This may be attributed to the preferred orientation of dislocations with the shortest Burgers vectors parallel to the closed packed atomic directions and thus breaking the isotropic symmetry. Finite size can also enhance this effect significantly. One presumably expects this effect to get smaller as the size gets larger [13].

IS THE CGT MEDIATED BY DEFECTS UNBINDING?

A dislocation and a disclination in 2D can be defined conveniently by nearest neighbor numbers [16]. A disclination occurs when an atom has a number of nearest neighbors different from six, and usually two disclinations with typically 5 and 7 nearest neighbors bind together to form a dislocation. Thus dislocations and other more complicated defect configurations can be visualized by simply mapping the coordination numbers. Fig. 5 shows the defect configurations at the different stages of the CGT. The increasing number of dislocations are responsible for the destruction of long-range translational order at $X_B < 0.172$ (Fig. 5a), but further increase of dislocations will not significantly change the long-range orientational order. At $0.172 \leq X_B < 0.199$ (Fig. 5b) orientational order still persists over a long distance but translational order vanishes over not more than the fourth nearest neighbor distance.

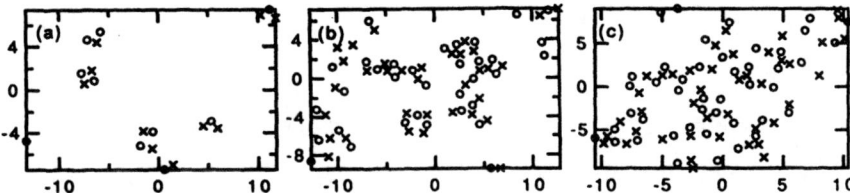

Fig. 5 Defect configurations of the LJ binary arrays at (a) $X_B = 0.148$ (crystal); (b) $X_B = 0.172$ (Hexatic ?); (c) $X_B = 0.25$ (amorphous). X denotes a disclination with 5 nearest neighbors; O a disclination with with 7 nearest neighbors.

Above results of the CGT suggest that its mechanism is different from the defect-unbinding theory of melting [14] in which dislocation-pairs need to unbind to single dislocations to destroy long-range translational symmetry at T_m and these single dislocations need to unbind again at $T_2(\leq T_m)$ into disclinations to destroy long-range orientational symmetry. In the LJ binary arrays, the increasing number of dislocations and their clusters can be sufficient to disrupt long-range translational order. If dislocation density increases further to form "dislocation cell structure", then the long-range orientational order will be destroyed. The correlation length of the orientational correlation function defines the domain over which certain crystalline features are preserved (Fig. 5b). Boundaries with such large dislocation density are full of solute atoms. If the domain size is large enough, one expects to see the formation of a so called nanocrystalline phase [17]. If dislocation densities increase further and these boundaries start proliferating, an amorphous phase forms. Another convincing evidence against the defect unbinding mechanism for the CGT is provided by the dislocation coupling constant, or Kosterlitz-Thouless constant K, which, at $X_B = 0.172$, is still much larger than the value 16π as predicted if dislocation unbinding presumably occurs [14]. However, we do not rule out the possibility of the defect unbinding mechanism in certain systems in which a fairly large repulsive interaction between solute atoms may raise the barriers of dislocation clustering and thus forces them to unbind [13].

SUMMARY

We summarize some general features of the CGT in 2D from our simulation results. First, like its 3D counterpart, the 2D binary arrays undergo the CGT for sufficiently large atomic size differences with increasing solute concentrations. This is in agreement with experimental observations in both liquid to glass transitions and solid state amorphizations where atomic size difference was found to be necessary (but not sufficient) in determining the glass-forming ability [1, 18]. We also found that it is more difficult to make a glass in 2D than in 3D. Second, decreasing of the crystallinity of the binary arrays as more and more solute atoms are saturated is directly caused by defects created around them. At low temperatures, thermal vibrations and resulted anharmonicities are relatively small. So density as well as configurations of these defects, most of which are dislocations, play a dominant role in destabilizing the crystalline order and inducing the CGT. Third, the CGT proceeds in three stages at each of which an unique microstructure exists. An intermediate phase consisting of ensembles of crystalline-like clusters was found to exist prior to the formation of the amorphous phase. This phase is characterized by a quite long range orientational order but short-range translational order. However, finite sample sizes used in our simulation can also lead to similar results. This topic is currently under investigation [13].

ACKOWNLEDGEMENT

The financial support of this work is provided by grants from the DOE under contract No. DEFG038GER45242 and NSF under contract No. DMR-8811795 and CHE-91-100284.

REFERENCES:

1. W. L. Johnson, Prog. Mater. Sci., **30**, 81, (1986); R. W. Cahn and W. L. Johnson, J. Mater. Sci., **1**, 724, (1986).
2. K. J. Strandburg, Rev. Mod. Phys., **60**, 161, (1988); F.F. Abraham, Phys. Rept., **80**, 339, (1981).
3. D. Wolf, P. R. Okamoto, S. Yip, et al., J. Mater. Res., **5**, 286, (1990); P. R. Okmoto, and M. Meshii, *Science of Advanced Materials*, edited by H. Wiedersich and M. Meshii (ASM International, Materials Park, OH, 1990), pp. 33.
4. M. Born, J. Chem. Phys., **7**, 591, (1939).
5. J. L. Tallon and W. H. Robinson, Phil. Mag. A, **36**, 741, (1977); J. L. Tallon, Phil. Mag. A., **39**, 151, (1979).
6. M. Li and W. L. Johnson, Phys. Rev. Lett., **70**, 1120, (1993); W. L. Johnson, M. Li and C. E. Krill, J. Non-crystal. Sol., **156**, 481, (1993).
7. C. E. Krill, J. Li, C. Ettl, K. Samwer and W. B. Yellon, J. Non-cryst. Sol., **156**, 506, (1993).
8. J. Koike, Phys. Rev. B, **47**, 7700, (1993).
9. R. Devanathan, N. Q. Lam, P. R. Okamoto and M. Meshii, MRS Symposia Proceeding No. 291. pp. 653 (MRS, Pittsburgh, 1992).
10. H. Fecht and W. L. Johnson, Nature, **334**, 50, (1988).
11. D. E. Luzzi (ed.), J. Alloys and Compounds, **194**, (1993)
12. M. Li and W. L. Johnson, Phys. Rev. B, **46**, 5237, (1992).
13. M. Li, W. L. Johnson and W. A. Goddard, unpublished results.
14. D. R. Nelson and B. I. Halperin, Phys. Rev. B, **19**, 2457, (1979).
15. D. R. Nelson, M. Rubinstein and F. Spaepen, Phil. Mag. A, **46**, 105, (1982).
16. M. P. Allen, D. Frekel and W. Gignac, J. Chem. Phys., **78**, 4206, (1983).
17. W. L. Johnson, unpublished results.
18. T. Egami and Y. Waseda, J. Non-Cryst. Sol., **64**, 113, (1984);B. C. Giesson, *Proc. 4th Int. Conf. on Rapidly Quenched Metals*, edited by T. Masumoto and K. Suzuki, (Japan Institute of Metals, Sendai, 1982), Vol. 1, pp. 213.

THERMAL STABILITY AND RELAXATION PROCESSES
OF THE Ti-Cu- AND Ti-Ni-BASED METALLIC GLASSES

NIKOLAY G. BABICH, NIKOLAY N. DASHEVSKY, OLESJA I. NAKONECHNA,
SERGEY L. REVO AND NIKOLAY I. ZAKHARENKO
Taras Shevchenko University, Department of Physics, Volodimirskaya st. 64, 252017 Kiev,
Ukraine

ABSTRACT

Temperature dependences of magnetic susceptibility for Ti-Cu- and Ti-Ni-based amorphous alloys have been obtained in the temperature range 300-850 K. All investigated alloys are shown to be Pauli paramagnets. Fe and Co impurity atoms possess zero localized moment due to effect of spin moment quenching. Relaxation processes were studied using thermopower measurements. It was shown the thermal stability of investigated alloys is determined predominantly by Nagel-Tauc electronic criterion.

INTRODUCTION

Amorphous metallic alloys by their nature are metastable. Different authors note a number of criteria defining the stability of the amorphous state. These are the structural criterion based on the Bernal model, the electronic criterion, the size factor and the Donald-Davies thermodynamic criterion [1]. The question of using the Nagel-Tauc electronic criterion for "metal-metal" type amorphous alloys is still discussed, because the peculiarities of electronic structure of components in amorphous metallic alloys (in particular - the problem of electron transfer between components at amorphyzation) has not been well studied.

The correlation between electronic structure and thermal stability intervals on the basis of magnetic susceptibility data and the processes of structural relaxation of the Ti-Cu- and Ti-Ni-based amorphous metallic alloys were investigated in this work.

EXPERIMENTAL PROCEDURE

The amorphous Ti-Cu- and Ti-Ni-based alloys were prepared with the melt spinning technique. Amorphous structure was verified by X-ray diffraction. Temperature dependences of magnetic susceptibility $\chi(T)$ were obtained using the Faraday-type technique with automatic microbalance (the sensitivity was equal to 10^{-11} cm^3/g and accuracy was better than 1.5%). Thermopower E_T as a function of annealing time t at different annealing temperatures $T_a < T_x$ was measured in a couple containing annealed and as-quenched alloys of a same composition.

Mat. Res. Soc. Symp. Proc. Vol. 321. ©1994 Materials Research Society

RESULTS AND DISCUSSION

The temperature dependences of magnetic susceptibility $\chi(T)$ of the Ti-Cu- and Ti-Ni-based alloys while heating and cooling were obtained using Faraday technique. The compositions of investigated samples are listed in Table I. All alloys may be separated to three groups. The first group (a) includes nonmetalloid-type alloys $Ti_{50}Cu_{45}Me_5$ (Me = Fe, Co, Ni, Cu) and $Ti_{64}Cu_{36}$ alloy. The second group (b) is formed of $Ti_{50}Cu_{45}Ni_5$ basic alloy and those for which the part of Ti atoms are substituted by Si, P metalloid atoms. The third group (c) consists of $Ti_{64}Ni_{36}$-based alloys with metalloid Si, P additions and $Ti_{70}Ni_{25}Si_5$ alloy.

Table I.

Magnetic susceptibility in amorphous (χ_a) and crystalline (χ_k) state and crystallization temperature T_x of Ti-Cu- and Ti-Ni-based alloys.

Alloy	$\chi_a \cdot 10^6$, cm³/g	$\chi_k \cdot 10^6$, cm³/g	T_x, K
(a)			
$Ti_{64}Cu_{36}$	2.30	1.80	610
$Ti_{50}Cu_{50}$	2.18	-	637
$Ti_{50}Cu_{45}Ni_5$	2.02	1.32	647
$Ti_{50}Cu_{45}Co_5$	2.20	1.91	680
$Ti_{50}Cu_{45}Fe_5$	2.29	1.56	642
(b)			
$Ti_{48}Cu_{45}Ni_5Si_1P_1$	1.94	1.23	683
$Ti_{48}Cu_{45}Ni_5P_2$	1.91	1.46	697
$Ti_{47}Cu_{45}Ni_5Si_3$	2.07	1.31	690
$Ti_{46}Cu_{45}Ni_5Si_4$	2.09	1.12	675
$Ti_{46}Cu_{45}Ni_5Si_2P_2$	1.88	1.24	705
$Ti_{47}Cu_{45}Ni_5P_3$	2.24	1.46	645
$Ti_{48}Cu_{45}Fe_5Si_2$	2.23	1.44	650
$Ti_{48}Cu_{45}Co_5Si_2$	2.31	1.66	740
$Ti_{62}Cu_{33}P_5$	2.05	1.53	687
(c)			
$Ti_{60}Ni_{36}P_4$	2.65	3.14	630
$Ti_{60}Ni_{36}Si_2P_2$	2.76	3.22	655
$Ti_{70}Ni_{25}Si_5$	2.98	3.23	650
$Ti_{47}Ni_{34}Cu_{17}Si_6$	-	-	640

The most typical χ(T) curves are shown in Fig. 1. The analysis of susceptibility polytherms for a-group alloys demonstrates that magnetic susceptibility is almost temperature independent (dχ/dT ≈ 0) in the amorphous state. In the 640-680 K temperature region χ(t) curves undergo an uneven change and stay invariant with further heating. We identified the temperature of magnetic susceptibility change with the crystallization temperature T_x that is confirmed by X-ray diffraction data. χ is temperature independent while cooling the samples. The χ measurements in cryogenic temperature region for some selected alloys did not show any essential temperature dependence, indicating the absence of Curie-Weiss paramagnetism. Taking into account such behaviour of magnetic susceptibility one can assert the total χ consists of Pauli susceptibility and ion-core diamagnetism. In our case the latter one contributes no more than 10% to total susceptibility [2], so the investigated alloys belong to the Pauli paramagnets class. Magnetic susceptibility of such paramagnets is known to be determined mainly by the itinerant electrons and is proportional to electron density of states at the Fermi level $N(E_F)$. The fact that magnetic susceptibility is temperature independent indicates the absence of localized magnetic moments on impurity Me atoms both in amorphous and crystalline states. The electron transfer between component atoms in amorphous metallic alloys is the problem to discuss. On different opinion [3] the rigid band model isn't suitable for these objects. Therefore it is necessary to find the other reason for localized magnetic moment absence on the impurity atoms. To our mind it

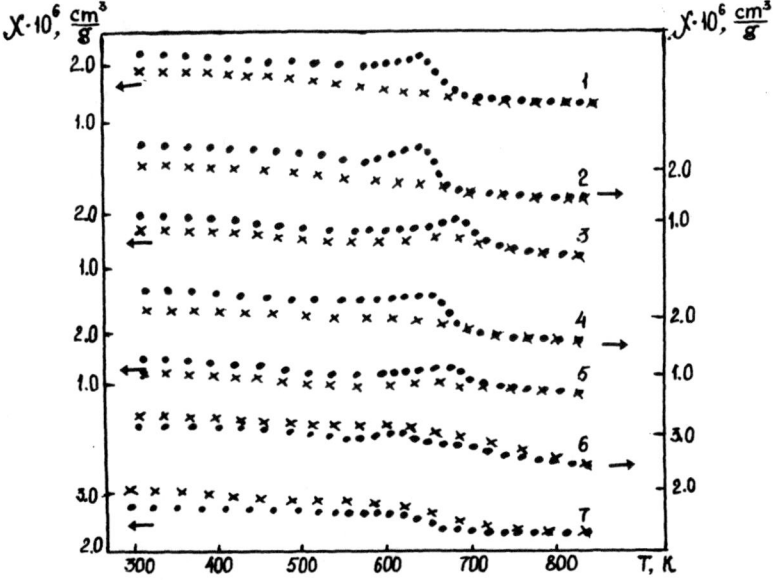

Fig.1. Temperature dependences of magnetic susceptibility of $Ti_{50}Cu_{45}Fe_5$ (1), $Ti_{50}Cu_{45}Ni_5$ (2), $Ti_{48}Cu_{45}Ni_5P_2$ (3), $Ti_{47}Cu_{45}Ni_5Si_3$ (4), $Ti_{62}Cu_{33}P_5$ (5), $Ti_{60}Ni_{36}P_4$ (6), $Ti_{60}Ni_{36}Si_2P_2$ (7) at heating (•) and cooling (×).

connects with the quenching of spin moments of impurity atoms by an internal field. This phenomenon takes place due to d-electrons interaction peculiarities between components of the alloy and is known for crystalline alloys too [4]. In the case of amorphous metallic alloys this phenomenon has to displayed more clearly due to influence of two factors that lead to essential weakening of exchange interatomic interaction. Firstly, it is the absence of the long-range order in amorphous alloys and, secondly, the presence of Si, P metalloid atoms for which d-states are empty. The above mentioned reasons lead to essential decreasing of d-d bond formation probability between metallic atoms. Under these conditions intratomic interaction becomes preferable. Therefore, when the orbital moment is quenched, the spin moment is quenched also.

Let us consider correlation between T_x and χ, which as noted above, is proportional to $N(E_F)$. The qualitative dependence between these values follows from obtained results (Table I). The alloys having smaller value of χ have higher crystallization temperatures, T_x. This fact confirms that the thermal stability of the alloys is defined preferably by the Nagel-Tauc electronic criterion. The amorphous alloys containing Fe and Co impurities are the exception in general picture: the correlation between T_x and χ is not as good. In our opinion, in order to connect χ and T_x it is necessary to take into account the following. According to the Stoner's theory [5] susceptibility has to increase because of Stoner enhancement effect in alloys with high density of electronic states. Consequently, under these conditions the Pauli susceptibility does not reflect directly $N(E_F)$ values.

Now we consider the properties of b-group alloys. T_x is increased in comparison with those of a-group. It is possible to explain the increasing of T_x probably by strengthening of interatomic bonds in these amorphous alloys and, as a result, by increasing of thresholds for diffusion transfer which accompanies the crystallization. This is confirmed by thermopower data (Fig. 2).

The dependences of thermoelectric power E_T on annealing time t, obtained for different annealing temperatures T_a were analysed and it was shown that E_T values were changed with T_a according to Arrenius law:

$$E_T = E_{T0} \times \exp(-Q/kT_a) \tag{1}$$

where Q is the activation energy of relaxation processes. The experimental Q values are listed in Table II. One can see the alloys having smaller T_x have larger χ and Q values, hence the Nagel-Tauc criterion may be used for this group of alloys.

Let us use for obtained results the Mott theory. According to [6] thermopower is given by:

$$E_T/T \sim \partial \ln N(E)/\partial E \big|_{E=E_F} \tag{2}$$

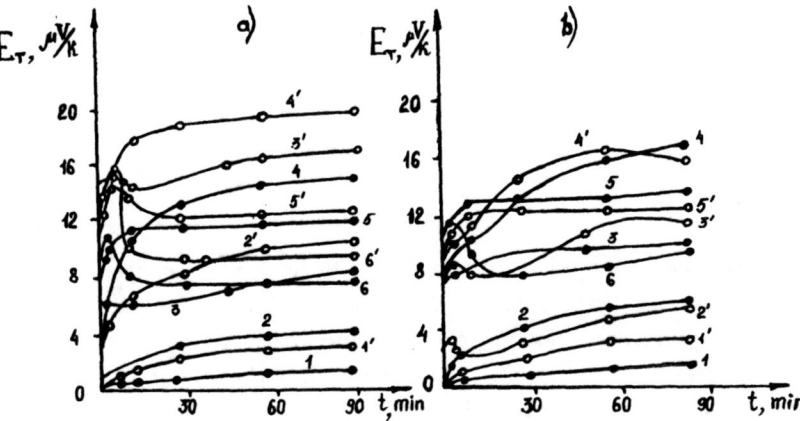

Fig 2. Thermopower vs. annealing time curves at T_a = 150 (1,1'), 200 (2,2'), 250 (3,3'), 300 (4,4'), 370 (5,5') and 350°C (6,6') for $Ti_{48}Cu_{45}Ni_5P_2$ (a,1-6), $Ti_{46}Cu_{45}Ni_5Si_2P_2$ (a,1'-6'), $Ti_{60}Ni_{36}Si_2P_2$ (b, 1-6) and $Ti_{62}Cu_{33}P_5$ (b, 1'-6') amorphous alloys.

Table II.

Activation energy of relaxation (Q) calculated from thermopower data.

Alloy	Q, eV	$\chi_a \cdot 10^6$, cm³/g	T_x, K
$Ti_{48}Cu_{45}Ni_5P_2$	1.19	1.91	697
$Ti_{46}Cu_{45}Ni_5Si_2P_2$	1.58	1.88	705
$Ti_{60}Ni_{36}Si_2P_2$	1.45	2.65	655
$Ti_{62}Cu_{33}P_5$	1.68	2.05	687

So, the $E_T(T)$ dependence is linear and the inclination coefficient is defined by the density of electronic states at the Fermi level. Analysis of the obtained data permits us to make a conclusion that the Fermi level is shifted to higher N(E) values as a result of relaxation after heat treatment. The increasing of magnetic susceptibility observed before the crystallization of samples (Fig. 1) confirms the above conclusion. On the base of optical experiments [7] it is possible to make assumption that Si and P atoms formed covalent-like bonds with metal ones. These bonds are characterized by high binding energy, so they make the diffusion motion of atoms difficult which is resulted in T_x rising. Therefore the addition of metalloid atoms leads to increasing of thermal stability and to infringement of the correlation between T_x and χ. Alloys of the c-group have weak $\chi(T)$ dependences, consequently the Ni atoms have zero localized moment. It is obviously that the reason of this is the quenching of spin moments that is noted above. However, only for these alloys the increasing of χ at crystallization takes place due to magnetic properties of crystalline phases.

CONCLUSIONS

As a result of our investigation Ti-Cu- and Ti-Ni-based alloys in amorphous and crystalline states were shown to be Pauli paramagnets. Fe, Co and Ni impurity atoms possess zero localized moment due to spin-moment quenching effect. Thermal stability of the investigated alloys is in agreement in general with the Nagel-Tauc electronic criterion. However, the addition of metalloid atoms (Si, P) leads to T_x rising probably due to formation of covalent-like bonds between metallic and metalloid atoms.

REFERENCES

1. K.Suzuki, H.Fujimori and K.Hasimoto, Amorphous Metals, (Metallurgiya Publishers, Moscow, 1987) p.328.
2. M.M.Schieber, Experimental Magnetochemistry (North-Holland Publishers, 1967), p.573.
3. E.P.Wohlfarth, in Amorphous Metallic Alloys, edited by F.E.Luborsky (Metallurgiya Publishers, Moscow, 1987), pp.283-300.
4. N.I.Zakharenko, PhD thesis, Taras Shevchenko University, 1983.
5. R.White, Quantum Theory of Magnetism (Mir Publishers, Moscow, 1985), p.350.
6. B.L.Gallagher, A.B.Kaiser and D.Greig, J.Non-Cryst.Solids, **61/62**, 1231 (1984).
7. L.V.Poperenko (private communication).

Amorphous-to-crystalline transition in $Tb_{33}Fe_{66}$ films studied via EXAFS

V.G. Harris, B.N. Das, W.T. Elam and N.C. Koon
US Naval Research Laboratory, Washington, D.C. 20375

Abstract: The amorphous-to-crystalline transition in sputter-deposited amorphous $Tb_{33}Fe_{66}$ films has been studied using a conversion-electron extended X-ray absorption (EXAFS) technique. Modeling of the EXAFS data using theoretical and empirical standards allowed the quantitative measurement of coordination, radial distance, and disorder of local atom shells around the Fe sites after varying degrees of heat treatments.

Introduction:

In this paper we present EXAFS studies of amorphous $Tb_{33}Fe_{66}$ films before and after heat treatments at temperatures approaching and exceeding the crystallization temperature. We focus on the qualitative changes in the Fourier transformed Fe EXAFS data and the results of modeling the near neighbor environments using empirical and theoretical standards. A multiple shell model is found to provide the best fit to the local environment around Fe sites. The evolution of structure with heat treatments is found to occur in three distinct stages; relaxation, short-range-ordering, and long-range-ordering. Fits obtained using empirical standards allow the measurement of the fractional amounts of amorphous and bcc crystalline phase after various heat treatments illustrating the progressive crystallization of the sample.

Studies investigating the atomic structure have been performed on amorphous $R_{33}Fe_{66}$ (R: rare earth) alloys; among them Cargil used x-ray scattering to probe $Gd_{33}Fe_{66}$ [1] and D'Antonio et al. used neutron scattering to probe the structure of $Tb_{33}Fe_{66}$ and $Y_{33}Fe_{66}$ [2]. Both of these studies report that the local environment of the amorphous $R_{33}Fe_{66}$ samples bare no resemblance to the local structure of the crystalline Laves phase (RFe_2). D'Antonio and co-workers report that the Fe tetrahedral coordination of the Laves phase is preserved in the amorphous analogs while the Tb tetrahedral coordination is not. EXAFS studies of amorphous TbFe films have been performed on TbFe alloys in the compositional range of Tb_xFe_{1-x}, $x \leq 0.25$, by Robinson et al. [3], and Harris et al [4], with Harris et al. using a multiple shell model to incorporate trapped voids in the atomic structure which manifest asymmetric Gaussian distribution of atoms.

Experimental:

A series of 2500Å thick Fe-rich, TbFe films were ion beam sputter-deposited onto glass substrates under ambient conditions. Films were capped with a 150Å layer of Ge to retard oxidation after removal from the deposition chamber. X-ray fluorescence and absorption measurements of the deposited films and empirical standards $(TbFe_2$ and $TbFe_3)$ were used to establish the composition of the films. In this paper we will focus our discussion on the results of the $Tb_{33}Fe_{66}$ alloy film which is representative of the amorphous-to-crystalline transition in other Fe-rich Tb-Fe films.

Samples were annealed under vacuum at temperatures ranging from 200 to 500°C for a period of 60 minutes. The heating apparatus employed a dual furnace configuration to heat the film samples while simultaneously maintaining a Ti sponge material at 1000°C; the Ti charge served as a gettering source to improve the cleanliness of the annealing environment.

X-ray absorption measurements were collected using a conversion electron technique [5] at the Naval Research Laboratory's materials analysis beamline, X23B, at the National Synchrotron Light Source (Brookhaven National Laboratory, Upton, NY). [The optical and X-ray properties of this beamline are presented in Ref. 6.] EXAFS analysis and modeling procedures follow those outline in Ref. 7 resulting in Fourier transforms of the Fe EXAFS data and structure parameters describing the local environment of the Fe atoms. In this paper we present the Fe EXAFS data which has superior signal-to-noise to that of the Tb EXAFS.

Mat. Res. Soc. Symp. Proc. Vol. 321. ©1994 Materials Research Society

Specifically, quantitative information of the near neighbor (NN) environment around the Fe atom was obtained by fitting Fourier-filtered EXAFS data of an r-space range encompassing the NN Fourier peak (r=1-3Å) with theoretical EXAFS data generated using the FEFF codes developed by Rehr et al [8]. Prior to the modeling exercise the validity of the theoretical data were fit to EXAFS data collected from empirical standards of α-Fe, ε-Tb, TbFe₂ and TbFe₃ to determine phase shifts and amplitude reduction factors.

A second modeling analyses involved the fitting of the FF EXAFS data with empirical standards of α-Fe and the TbFe amorphous structure. The Fe EXAFS data collected from the as-deposited film served as the standard for the amorphous structure. Because the primary crystallization product of this TbFe alloy is α-Fe this analysis method serves to define the relative atom fraction of Fe in the amorphous and bcc structures.

Results and Discussion:
Heat treatment induced changes in Fourier transformed Fe EXAFS

Figure 1 is a plot of Fourier transformed Fe EXAFS data collected from the as-deposited and annealed $Tb_{33}Fe_{66}$ film samples. Also shown in the plot are similar data collected from an α-Fe film and a powder sample of the $TbFe_2$ Laves phase. These standards are used as empirical standards of the local environment of Fe in the bcc structure and the crystalline analog of the amorphous sample, respectively. The data collected from both standards have been subjected to the same analysis procedures as the $Tb_{33}Fe_{66}$ film samples but normalized to a near neighbor peak amplitude of 2.0 to clarify comparisons. The inset to Figure 1 is an expanded view of an r-space range of 3-5Å. This region of the Fourier transform best illustrates the onset of long-range-order (LRO) by the appearance of the higher order atomic shells.

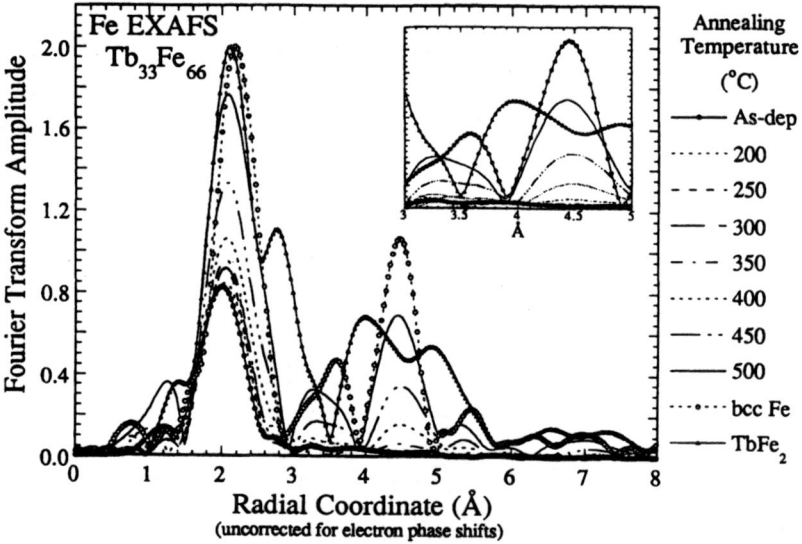

Figure 1. Fourier transformed Fe EXAFS data collected from as-deposited and annealed samples, with data collected from α-Fe and crystalline $TbFe_2$. Annealing temperatures are listed in the figure. The inset plot illustrates the ordering of higher order atomic shells with increases in annealing temperature. The data have been Fourier transformed using a k-range of 2.3-9.7Å^{-1} with a k^2-weighting. Data are uncorrected for electron phase shifts.

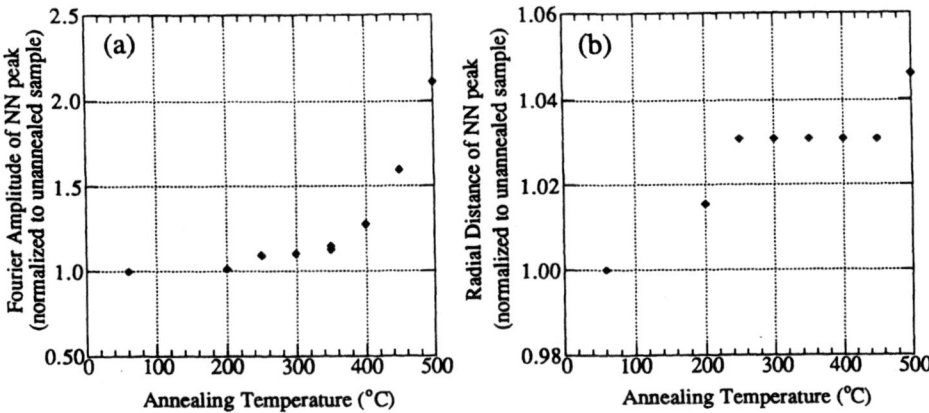

Figure 2. (a) Change in Fourier amplitude of the near neighbor peak normalized to the as-deposited data; (b) radial distance of near neighbor peak normalized to as-deposited data.

The as-deposited amorphous sample features a large broad asymmetric Fourier peak centered near 2Å (uncorrected for electron phase shift) and has no structure appearing above the background signal at $r \geq 3$Å signifying the lack of LRO in the sample. With heat treatments at increasing temperatures the NN peak is observed to increase in amplitude and shift to higher radial distances. Figure 2 (a) and (b) are plots of the NN peak amplitude and radial distance normalized to the data collected from the as-deposited sample. The initial increase in amplitude after an anneal at 250°C is likely due to an increase in ordering of the local atom shells corresponding with a stress relief in the film. The shift to higher radial distances signals the transition from an amorphous dense-random, close-packed structure to an intermediate structure having rigid bonds.

The structure appearing at $r \geq 3$Å changes very little with annealing over a temperature range of $200°C \leq T_{ann} < 350°C$. Subtle changes in the structure over this region are difficult to interpret because of the weak amplitude of the features compared with the background signal. However, after an anneal at 350°C a broad peak appears above the background signal at a radial distance near 4.5Å. This peak corresponds in radial distance with the 4th and 5th neighbors of the bcc unit cell; a photoelectron diffraction event provides additional amplitude to this peak. With anneals at $400°C \leq T_{ann} \leq 500°C$ this peak increases in amplitude and other peaks appear above the background signal near 3.5Å; these peaks also correspond to atomic shells in the bcc unit cell. The radial distance and relative amplitude of these higher order peaks signifying that a fraction of these samples exists as a bcc crystalline phase, likely α-Fe. X-ray diffraction measurements using a $Cu_{k\alpha}$ rotating anode source reveals only a small broad peak appearing in the θ-2θ scan of the sample annealed at 450°C. After a heat treatment at 500°C the diffraction scan reveals weak diffraction features corresponding with the (110), (200), and (220) planes of α-Fe. After an anneal at 500°C the Fourier profile matches closely with the bcc Fe profile with the exception of the radial distance of the NN peak which is shifted to a lower distance; this can be attributed to the relative fraction of the amorphous phase remaining in the annealed samples.

Being that the near neighbor Fourier peak of the bcc Fe standard consists of contributions from both the first and second neighbors, at 2.482Å (8 neighbors) and 2.867Å (6 neighbors), respectively, the presence of a bcc crystalline phase in the annealed samples would both increase the near neighbor peak amplitude, due to the increase in the number of near neighbors contributing to the NN peak, and shift the peak to higher radial distances. It is unclear if the changes in the NN peak after anneals at $T_{ann} \leq 350^\circ C$ are due to the presence of a small amount of the bcc phase or the ordering of some other local symmetry. For example, the formation of $TbFe_2$ clusters would also shift the NN peak to higher radial distances but increase its amplitude only slightly in comparison to the ordering of the bcc structure as seen in Fig. 2(a).

Modeling of the near neighbor Fourier peak using theoretical standards

Modeling of the FF NN peak with theoretical EXAFS standards indicates that the environment around the Fe atom is best fit using a two Fe atomic shells at $2.475 \pm 0.01 \text{Å}$ and at $2.65 \pm 0.02 \text{Å}$, and a third shell of Tb atoms at $2.99 \pm 0.03 \text{Å}$. A similar model for the atomic structure around both Fe and Tb atoms in amorphous TbFe was presented in Ref. 4. For this composition the occupation of the first Fe shell is 5 ± 0.7 atoms, the second Fe shell is 3 ± 1 atoms, and the Tb shell is 2.5 ± 1 atoms. Error bars reflect a decrease of 20% in the goodness of fit parameter. These findings are in qualitative agreement with Refs. 1 and 2 when one takes into account the multiple shell model.

EXAFS modeling indicates a near linear decrease in the Debye-Waller coefficients of the local shells with increasing annealing temperature over the range $200^\circ C \leq T_{ann} < 350^\circ C$. This result indicates the progressive relaxation and ordering of the local environment around the Fe sites. Corresponding with anneals at higher temperatures, $400^\circ C \leq T_{ann} \leq 500^\circ C$, the Debye-Waller coefficients are measured to decrease more rapidly (albeit linearly). This corresponds with the appearance of the higher order bcc peaks in Fig. 1. This behavior to heat treatments can be examined in terms of three phases of structural evolution: (I) initial relaxation, (II) ordering of the local environment, and (III) formation of LRO. The initial relaxation (phase I) of metallic glasses have been the subject of much study and is now fairly well understood to result from the annihilation of trapped voids in the disordered structure. However, no direct evidence has been presented for the existence of a preferred symmetry to the relaxed, disordered structure. Similarly, the formation of LRO in these materials (phase III) has also been studied using structure sensitive probes (i.e. electron, neutron, and x-ray scattering). These studies have proven effective in identifying both the onset of LRO and the products themselves. For example, using EXAFS we illustrate unambiguously that the primary crystallization product of this sample is α-Fe. Furthermore we identify an annealing temperature of $350^\circ C$ as the temperature at which LRO is first observed in the local structure. However, the intermediate stage (phase II), were the relaxed, disordered structure transitions to LRO is not well understood. We show here direct evidence that ordering takes place and that rigid bonds are formed after annealing over the range of $250^\circ C \leq T_{ann} < 350^\circ C$. It is not clear what bond symmetry, if one exists at all, is present in these annealed samples.

Modeling of the near neighbor Fourier peak using empirical standards

In order to determine the relative amounts of amorphous and crystalline phases contributing to the NN environment of annealed samples one might attempt to model the data using theoretical EXAFS data corresponding with all the atomic shells of each phase contributing to the local environment. For a sample having just two phase, say bcc Fe and an amorphous TbFe phase, this would consist of Fe-Fe correlations at 2.48Å, 2.65Å, 2.86Å, and a Fe-Tb correlation near 3.0Å; all having varying occupations and disorder parameters. Unfortunately, this would be a difficult and likely an untrustworthy approach due to the complications encountered with the large number of floating parameters and the coupling of the atomic shells close to one another. Alternatively, we have adopted a technique were we model the data using empirical standards of bcc-Fe and Fe in an amorphous structure (simulated by the Fe EXAFS of the as-deposited sample). Using this technique we float only two parameters, the atom fraction of each phase, and this we constrain to sum to 1.0. The results indicate that this technique is capable of measuring to a high degree of precision the fractional amounts of Fe sites in the bcc crystalline phase and the amorphous phase. Figure 3 is a plot of the results of this analysis illustrating the fractional amount of the two phases in all the samples studied here. The use of only two phases is a crude approximation of reality as it is likely that the TbFe intermetallic phases form as secondary crystallization products during high temperature anneals. The increasing error bars measured in samples annealed at $T_{ann} > 400^oC$ is likely the result of the ordering of TbFe intermetallic phases. In Fig. 3 the onset of LRO around the Fe sites appear near 350°C where a significant increase in the bcc fraction is measured. Anneals at increasing temperatures foster the growth of the bcc fraction at the expense of the amorphous phase. The inset plot is a comparison between the FF Fe EXAFS data collected from the sample annealed at 450°C and a linear combination of 60 at. % amorphous phase and 40 at. % of the bcc phase. The fit is excellent over the entire k-space range available.

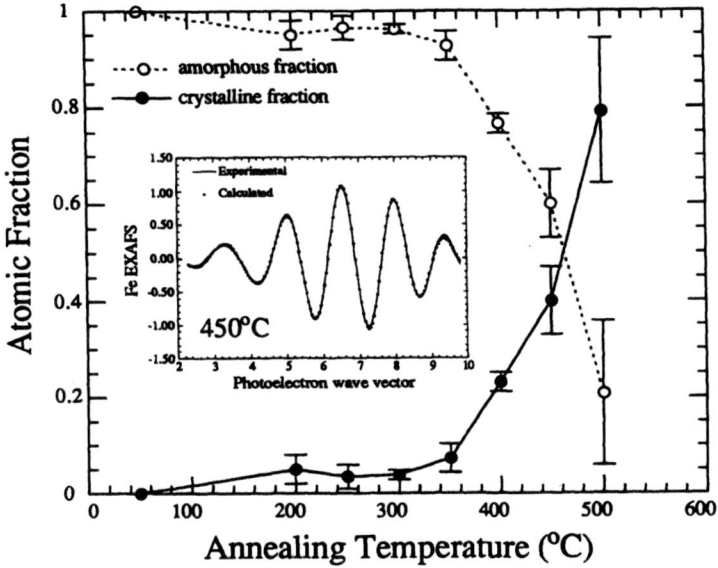

Figure 3 . Results of the fitting of annealed FF Fe EXAFS data using empirical standards of α-Fe and an the Fe EXAFS collected from the as-deposited sample representing the amorphous structure. Data illustrate the atomic fraction of Fe sites in the amorphous and crystalline phases. The inset plot contains the FF Fe EXAFS data for the sample annealed at 450°C and the best fit using a linear combination of empirical standards (2 curves are presented, the experimental data as a solid curve, the calculated data as symbols). Error bars presented in the main body of the figure represent a 100% decrease in the goodness of fit parameter.

In summary, we have performed EXAFS measurements on amorphous $Tb_{33}Fe_{66}$ film samples after heat treatments over a temperature range of 200-500°C. We find that the structural evolution around the Fe sites occur in three phases: initial relaxation (still structurally disordered), short-range-ordering of the local environment (< unit cell dimensions), and long-range-ordering (> unit cell dimensions). The amorphous structure is described in terms of a multiple shell model having two Fe-Fe correlations at 2.475Å and 2.65Å, and a Fe-Tb correlation at 2.99Å. The two shells of Fe are used to simulate a near neighbor shell of Fe having an asymmetric distribution due to trapped voids. From the systematic study of the local structure and its response to heat treatments we identify the primary crystallization product as α-Fe and its ordering temperature as 350°C; x-ray diffraction does not detect LRO until 450°C. Using empirical standards in linear combination to fit the local environment of annealed samples we illustrate that the sample annealed at 350°C consists of 8 at. % of bcc Fe and 92 at. % amorphous phase. After an anneal at 500°C the sample consists of 80 at. % bcc Fe. Evidence is presented in support of that local symmetry exists in samples annealed at low temperatures, 200°C≤Tann<350°C.

This research was carried out in part at the National Synchrotron Light Source (Brookhaven National Laboratories, Upton, NY), which is sponsored by the U.S. Department of Energy (Division of Material Science and Division of Chemical Sciences of the Office of Basic Energy Sciences).

References:

[1] G.S. Cargill III and T. Mizoguchi, J. Appl. Phys. **49**, 1753 (1978).

[2] P. D'Antonio, J.H. Konnert, J.J. Rhyne, and C.R. Hubbard, J. Appl. Cryst. **15**, 452 (1982).

[3] C.J. Robinson, M.G. Samant, and E.E. Marinero, Appl. Phys. A **49**, 619 (1989).

[4] V.G. Harris, K.D. Aylesworth, W.T. Elam, B.N. Das, and N.C. Koon, Phys. Rev. Lett., **69**, 1939 (1992).

[5] W.T. Elam, J.P. Kirkland, R.A. Neiser and P.D. Wolf, Phys. Rev. B, **38**, 26 (1988).

[6] R.A. Neiser, J.P. Kirkland, W.T. Elam, and S. Sampath, Nucl. Instrm. Methods Phys. Res. Sect. A**266**, 220, (1988).

[7] D.E. Sayers and B.A. Bunker, in *X-ray Absorption: Basic Principles of EXAFS, SEXAFS, and XANES*, D.C. Koningsberger and R. Prins (editors), (Wiley, New York, 1988).

[8] J.J. Rehr, J. Mustre de Leon, S.I. Zabinsky, and R.C. Albers, J. Amer. Chem. Soc., **113**, 5135 (1991).

IONIC MOTION AND STRUCTURE OF ION CONDUCTIVE GLASSES

T. Akai, M. Yamashita, H, Yamanaka, and H. Wakabayashi.
Optical Material Department , Osaka National Research Institute, AIST*
1-8-31, Midorigaoka, Ikeda, 563, Japan.

ABSTRACT

The dynamic structure of xLi_2S-Ga_2S_3-$6GeS_2$ (x=4 and 6) glasses has been investigated by ^7Li nuclear magnetic resonance. In two samples similar values of spin-lattice relaxation time (T_1) were obtained. The relaxation mechanism at 20MHz and 78MHz is therefore attributed to the local motion of lithium ions. In the glass corresponding to x=6, which shows higher conductivity, the slow motion of ions showing an activation energy of 24.3kJ/mol has been detected by the spin-lattice relaxation time in the rotating frame ($T_{1\rho}$). This value is comparable to the activation energy determined by the conductivity. The existence of this mode is supported by the motional narrowing of the line width which is sensitive to the motion less than 10kHz.

INTRODUCTION

In the last decade, much attention has been concentrated on the use of solid state lithium ion conductors as an appropriate element for microbatteries [1]. Since Ribes et al. proved the high conductivity of alkali-containing sulfide glasses in 1978[2], numerous sulfide glass systems such as Li_2S- SiS_2[3], Li_2S-B_2S_3[4], Li_2S-P_2S_5[5] have been prepared. The characteristic conductivity of these sulfide glasses is as high as $10^{-3}Scm^{-1}$ - $10^{-4}Scm^{-1}$ at room temperature. The xLi_2S - $(1-x)GeS_2$ system is also known to form a glass over the range $0<x<0.5$[6] and shows a conductivity of 10^{-4} S cm^{-1}. It is thought that the addition of Ga_2S_3 may extend the glass-forming region of the system, which will lead to higher conductivity. Hence we are now investigating the glass forming region of the system Li_2S - Ga_2S_3 - GeS_2 . We have successfully prepared glass which shows a conductivity as high as in the Li_2S-GeS_2 glass system, taking into account of lithium content.

Despite the extensive work on the preparation of ion conducting sulfide glasses described above, little is known about the structure and ionic conduction mechanism. Nuclear magnetic resonance is a powerful method to study the static and dynamic structure of sulfide glasses. The spin-lattice relaxation time (T_1) is useful in detecting motion within the glass. Some work has appeared concerning T_1 in several sulfide systems [7,8], but the measurements were performed only at a single resonant frequency. Ions in ionic conductors, both in crystals and in glasses, may have several motional modes with different frequencies and activation energies. Hence in examining the dynamic structure of the sulfide glasses, the measurements should be performed over a wide range of frequencies. The only sulfide glass system which has been examined so far over a wide frequency range is the Li_2S-SiS_2 system[9]. To understand the ion conduction mechanism wide frequency range measurement should, however, be performed over a number of systems.

In this work, we examined the lithium motion in Li_2S-Ga_2S_3-GeS_2 ranging from 10^3Hz to 10^8Hz, changing several nuclear magnetic resonance parameters. We start from the discussion of T_1 at 20MHz and 78MHz, proceed to the discussion of the spin-lattice relaxation time in the rotating frame ($T_{1\rho}$), and then discuss the motional narrowing analysis which is sensitive to the motion in the several kHz region.

* Institution renamed from the Government Industrial Research Institute.

EXPERIMENTAL

The composition of the glass prepared in this work was xLi_2S - Ga_2S_3 - $6GeS_2$. The ratio of Ga_2S_3 to GeS_2 is set at 1/6, under which a glass easily forms. The starting materials for the xLi_2S-Ga_2S_3-$6GeS_2$ glass system were $Li_2S(99\%)$, $Ga_2S_3(99.999\%)$, and $GeS_2(99.999\%)$. They were mixed in a glove box, sealed in a silica glass tube coated with silicon and melted at 1050°C for 15 minutes. The molten mixture was then quenched by immersing the glass tube in water.

The impedance of the samples was measured using an HP4192A impedance analyzer in the frequency range 5 Hz to 13MHz from room temperature to 250°C. The data was analyzed using a complex impedance plot in order to obtain the conductivity. The conductivity at 25°C and activation energy of the xLi_2S - Ga_2S_3 -$6GeS_2$ (x=4 and x=6) are listed in table 1.

For NMR measurement the samples were ground, degassed and sealed in a pyrex glass tube. The ^7Li spectra and spin-lattice relaxation time at 78MHz was measured on a Bruker MSL 200 system. The spectra were obtained by pulse FT method. T_1 was measured by both train-90° pulse sequence and inversion recovery methods.

The measurement of $T_{1\rho}$ was performed on a Chemmagnetics 200 spectrometer by the spin-lock method with the rotating field as large as 150kHz. The spin lattice relaxation time at 20 MHz was measured on a JEOL spin echo instrument.

Table 1. Conductivity at 25°C (σ_{25}) and activation energy (E_σ).

Sample No.	x	$\sigma_{25°C}$/Scm^{-1}	E_σ/kJmol^{-1}
1	4	2.0×10^{-6}	54.4
2	6	2.0×10^{-5}	38.8

RESULTS and DISCUSSION

Spin-lattice relaxation time (T_1)

The spin - lattice relaxation time, T_1, can be expressed as [10],

$$1/T_1 = C(J(\omega_L) + 4J(2\omega_L)), \tag{1}$$

where C is the coupling constant of the lattice and the spin system, $J(\omega_L)$ is the spectral density and ω_L is the Larmor frequency. As is easily seen from eq.(1), T_1 is sensitive to motion as fast as the Larmor frequencies. In the case of glasses, it is well known that $\log T_1$ vs. $1/T$ plots do not show a symmetric shape on both sides of the T_1 minimum because the correlation function is different from a single exponential [11]. Several types of correlation function have been proposed, which used the Cole-Davidson correlation function[12] and Kohlrausch-Williams-Watt (KWW) equation[13] . It is, however, more appropriate to use the KWW function, $g(t) = \exp((-t/\tau_c)^\beta)$, which is widely used in the analysis of relaxation phenomena in glassy state materials.

The numerical calculation of spectral density is a difficult task, but the following relations were found by several authors[14,15].

$$\omega_L\tau_c \ll 1; \quad T_1 \propto \tau_c^{-1} \tag{2}$$

$$\omega_L\tau_c \gg 1; \quad T_1 \propto \omega^{(1+\beta)}\tau_c^\beta \tag{3}$$

where ω_L is the Larmor frequency. From eqs.(2) and (3), a $\log T_1$ vs. $1/T$ plot gives an asymmetric V shape with a slope equal to the activation energy (E_a) in the higher temperature range and βE_a in the lower temperature range. The frequency dependence of T_1 is expected to be proportional to $\omega^{(1+\beta)}$.

Spin-lattice relaxation times of samples 1 and 2 are plotted against the reciprocal temperature in figure 1. One remarkable feature in figure 1 is that the value of T_1 of sample 1 is nearly the same as that of the sample 2 despite the significant difference of composition and conductivity. The spin relaxation mechanism is considered as the fluctuation of nuclear quadrupole moment[16]. Therefore, the same relaxation time means a similar rate of motion and nuclear quadrupole moment. It is unlikely that this common motion in two samples is related to the ionic conduction, so we have attributed this to the local motion of the lithium ion.

Only T_1 on the lower temperature side was observed because τ_c tends to be much longer than $\omega_L{}^{-1}$. It is desirable to observe both sides of the T_1 minimum for accurate analysis, however, it is possible to analyze the data to some extent using (2) and (3) if the measurement is performed at more than one frequency.

The apparent activation energy equal to the value of βE_a is 14kJ/mol and is common for both samples at each frequency. The absolute value of T_1 is proportional to $\omega^{1.08}$. The obtained parameters are $E_a = 170$kJ/mol and $\beta = 0.08$. The problem, then, arises whether the value of E_a and β obtained here is the true value or does another higher frequency motional mode lower the relaxation time at 78MHz. To solve this, the measurement should be performed on much higher frequency region. The anomalies of T_1 below 170K is characteristic of glassy states and is usually explained as a two level system of local motion[17].

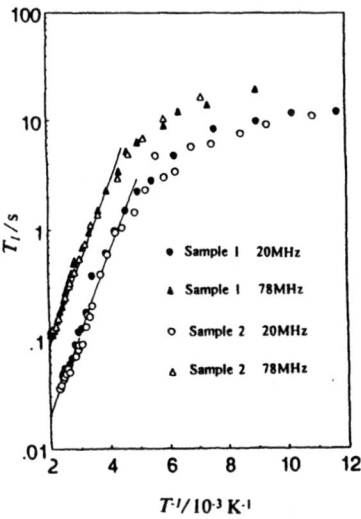

Figure 1. Temperature dependence of T_1. Close and open circles show the T_1 of sample1 and sample 2 at 20MHz. Close and open triangles show the T_1 of sample 1 and sample 2 at 78MHz, respectively.

Spin - lattice relaxation time in the rotating frame ($T_{1\rho}$)

In comparing NMR with conductivity data, it is important to consider the motion in the kHz range. Spin-lattice relaxation times in the rotating frame are useful in detecting the slow motion ranging from 10kHz to 10^2 kHz. Recently, Kuchler et al. have reported that the activation energy of $T_{1\rho}$ is in accordance with the conductivity data[18].

$T_{1\rho}$ can be expressed as follows[19],

$$T_{1\rho} = 1/4 \ C'J \ (2\omega_l) \qquad (4)$$
$$\omega_l = \gamma H_l$$

where γ is the gyromagnetic ratio of the nuclei and H_l is the strength of the rotating field. Since $J(\omega_L)$ in (1) is replaced by $J(2\omega_l)$, $T_{1\rho}$ can be explained in the same way as eqs.(2) and (3). The only difference is that the limiting condition is $\omega_l\tau_c$ instead of $\omega_L\tau_c$.

Figure 2 shows the temperature dependence of

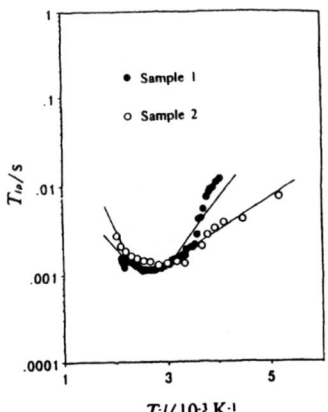

Figure 2. Temperature dependence of $T_{1\rho}$. Close circles show the $T_{1\rho}$ of sample 1 and open circles show that of sample 2.

$T_{1\rho}$ in the rotating field at $\omega_1 = 150 \mathrm{kHz}$. The obtained parameters from samples 1 and 2 are $E_a = 12.5$ kJ/mol, $\beta = 1.0$ and $Ea = 24.3$ kJ/mol, $\beta = 0.31$, respectively. The motion detected in sample 2 must be related to the ionic conduction, as the value of E_a is close to that of E_σ. On the other hand, there is a great deviation between the values of E_a and E_σ in sample 1. It is, thus, revealed that only in sample 2, which shows higher conductivity, that E_a comparable to E_σ is obtained at this frequency. In the next section, we will examine much slower lithium motion by the motional narrowing of the line width.

Motional narrowing of ^{7}Li line width

The ^{7}Li line width is narrowed by the onset of lithium motion which occurs above 170K in sulfide glasses. This so-called motional narrowing contains information on the lithium motion which is on the order of the rigid line width (several kHz in this case). Figure 3 shows the motional narrowing of samples 1(x=4) and 2 (x=6). The analysis of motional narrowing using the KWW equation has been already derived[20], but it is very difficult to obtain the parameters correctly if more than one motional mode exist. We, therefore, have adopted the Hendrickson-Bray equation[21] to analyze the motional narrowing to overcome this problem. Hendrickson and Bray assumed that the motional narrowing is caused by the increasing number of activated ions (highly mobile ions), the fraction of which is determined by the Boltzmann distribution. From this assumption, T_2, inversely proportional to the line width, can be expressed as the weighted average of the rigid lattice and the activated state. Thus the equation is derived as follows,

$$\ln(1/W - 1/A) = -Ea/RT + \ln(1/B - 1/A) \qquad (5)$$

where W is the observed line width, R is the gas constant, A the rigid lattice line width, B the high temperature line width and Ea the activation energy of the lithium motional mode responsible for motional narrowing. According to eq.(5), a semilogarithmic plot of $(W^{-1} - A^1)$ gives a straight line with a slope equal to E_a. It is well known that several linear regions are observed if eq.(5) is applied to a glass. Usually the highest activation energies are attributed to long range motion of the ions and the lower activation energy attributed to local motion. Equation(5) has been successfully applied to many oxide glass systems[21]. The model is very

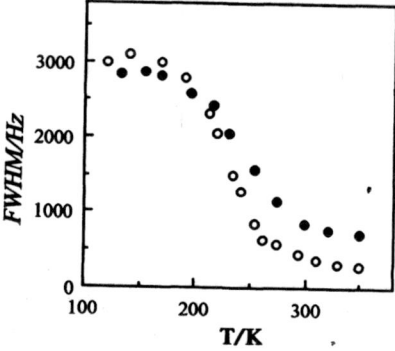

Figure 3. ^{7}Li line width is plotted against the temperature. Close circles show FWHM of sample 1 and the open circles show that of sample 2.

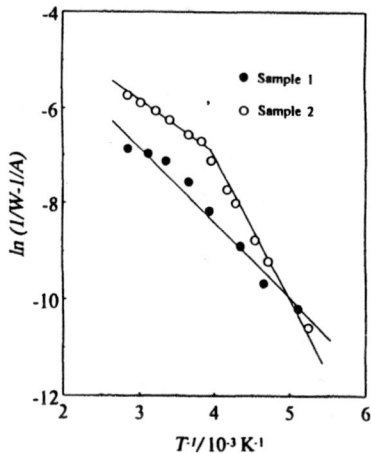

Figure 4. Semilogarithmic plots of $(W^{-1} - A^{-1})$ vs. T^{-1} for sample 1 (close circle) and sample 2 (open circle).

approximate but has the advantage that the distribution of T_2 is compensated by the distribution of the fraction of activated ions.

We applied eq.(5) to the line width as shown in figure 3. The result is shown in figure 4. The plot for sample 2 shows two linear regions while sample 1 shows only one. The derived parameters from figure 4 are listed in table 2. Since the activation energy in the low temperature region in sample 2 is close to $E\sigma$, this can

Table 2. Parameters obtained from the Hendrickson-Bray analysis.

Sample No.		Ea / kJmol^{-1}	B / Hz
1		13.4	7.4
2	low temp.	26.6	3.18×10^{-3}
	high temp.	8.3	18.3

be related to the long range motion. The activation energy over the high temperature region is related to short range motion. A similar phenomenon is also observed in other sulfide glasses such as Li_2S - P_2S_5-LiI and Li_2S-As_2S_3-LiI [23].

The result obtained from motional narrowing is similar to that of the $T_{1\rho}$. Some workers claim that the conductivity of the glasses deviates from the dc value above 10^5Hz far below their glass transition temperature so that the motion less than that is important in considering the dc conductivity[11]. This work has shown that there exist motional modes related to ionic conduction in higher conductive glass. This fact may suggests that another migration path of ions is created as the lithium content increases, but greater information on the local structure is needed for further discussion.

CONCLUSION

Two sulfide glasses (xLi_2S-Ga_2S_3-$6GeS_2$, x=4 and 6), which exhibit relatively high conductivity, have been prepared and the dynamic structure of lithium within these glasses was investigated by three NMR parameters, T_1, $T_{1\rho}$ and the line width. The value of T_1 at high frequency (20MHz, 78MHz) show similar values for the two samples, which implies relaxation is caused by local motion. $T_{1\rho}$ at 150kHz shows significant difference between the two glasses in that the higher conductive glass shows a higher activation energy comparable to $E\sigma$, while the E_a of the lower conductive glass deviates from $E\sigma$. Analogous results were obtained by a Hendrickson-Bray analysis of the motional narrowing of the line width, which is sensitive to the motion in several kHz range. These facts assure the motion less than 10^5 Hz is essential for higher ionic conduction and also suggest that the migration path of the ions varies with lithium content.

REFERENCES
[1] C. Julien, NATO ASI Ser. B, 199, 335 (1989).
[2] M. Ribes, D. Ravaine, J. L. Souquet and M. Maurin, Revue de Chimie Minerale, 16, 339 (1979).
[3] J. H. Kennedy and Y. Yang, J. Electrochem. Soc., 133, 2437 (1986).
[4] A. Levasseur, R. Olazcuaga, M. Kbala, M. Zahi and P. Hagenmuller, C. R. Acad. Sci. Paris, C, 563 (1981).
[5] R. Mercier, J. P. Malugani, B. Fahys and G. Robert, Solid State Ionics, 5, 663 (1981).
[6] J. L. Souquet, E. Robinel, B. Barrau and M. Ribes, Solid State Ionics, 3/4 317(1981).
[7] J. Senegas and J. Olivier-Fouracade, J. Phys. Chem. Solids., 44, 1033 (1983).
[8] S. Visco, P. Spellane and J. H. Kennedy, J. Electrochem. Soc., 132, 1766 (1985).

[9] F. Borsa, D. R. Torgeson, S. W. Martin and H. K. Patel, Phys. Rev. B, **46**, 795 (1992).

[10] A. Abragam, <u>Principle of Nuclear Magnetism</u>, (Oxford University Press, New York, 1961). p291.

[11] S. W. Martin, Mat. Chem. Phys., **23**, 225 (1989).

[12] D. W. Davidson and R. H. Cole, J. Chem. Phys., **19**, 1484 (1951).

[13] R. Kohlraush, Ann. Phys.(Leipzig), **12**, 393 (1847).

[14] J. L. Bjorkstam, J. Listerud and M. Villa, Solid State Ionics., **18/19**, 117 (1986).

[15] J. L. Bjorkstam, J. Listerud and M. Villa, Phys. Rev. B, **22**, 5025 (1980).

[16] E. Göbel, W. Müller-Warmuth and H. Olyschläger, J. Magn. Reson., **36**, 371 (1979).

[17] T. L. Reinecke and K. L. Ngai, Phys. Rev. B, **12**, 3476 (1975).

[18] R. Kuchler, O. Kanret and S. Rückstein and H. Jain, J. Non-Cryst. Solids, **128**, 328 (1991).

[19] D. C. Look and I. J. Lowe, J. Chem. Phys., **44**, 2995 (1966).

[20] J. L. Bjorkstam, J. Listerud, M. Villa and C. I. Massara, J. Magn. Reson., **65**, 383 (1985).

[21] J. R. Hendrickson and P. J. Bray, J. Magn. Reson., **9**, 341 (1973).

[22] J. R. Hendrickson and P. J. Bray, J. Chem. Phys., **61**, 2754 (1974).

[23] S. Visco, P. Spellane and J. H. Kennedy, J. Electrochem. Soc., **132**, 1766 (1985).

RELAXATION OF NEW-TPI THERMOPLASTIC POLYIMIDE STUDIED BY THERMALLY STIMULATED DEPOLARIZATION CURRENT

SHARON X. LU, WENDY C. RUSSELL AND PEGGY CEBE
Department of Materials Science and Engineering, Massachusetts Institute of Technology,
Cambridge, MA 02139

INTRODUCTION

Semicrystalline polyimide, Regulus[TM] NEW-TPI, is a thermoplastic polyimide produced by Mitsui Toatsu. This aromatic polyimide possesses outstanding heat resistance, high tensile strength, excellent electrical properties, and strong solvent resistance. It has become the subject of recent study in our group [1-6] and others [7-13]. The flexible ether and meta linkages in the monomer unit lower its glass transition temperature (T_g) to 250°C in quenched amorphous material. The chemical repeat unit has been reported previously [1,13] and is shown below:

NEW-TPI is a melt-processible semicrystalline polymer with a melting temperature (T_m) of 385°C from differential scanning calorimetry (DSC). Our previous studies [1,5] have shown that T_g shifts only slightly to higher temperature for semicrystalline polymer and the crystallinity is low (less than 30%). A very small amount of tightly bound, or rigid, amorphous phase exists and relaxes completely about 30°C above T_g. Further investigation of molecular relaxation behavior of semicrystalline NEW-TPI is carried out in the present work using a more powerful technique, thermally stimulated depolarization current (TSDC) along with DSC and wide angle x-ray scattering (WAXS).

EXPERIMENTAL

Regulus[TM] NEW-TPI film was kindly supplied by Mitsui Toatsu Chemical Co. The as-received film was transparent and 100μm thick. It is considered amorphous as no peaks were observed from Bragg scattering in the wide angle x-ray diffractogram and equal heats for crystallization and melting were observed in DSC. No orientation was seen in the plane of the film using polarizing optical microscopy. Crystallinity of NEW-TPI is obtained from the ratio of the heat of fusion of the semicrystalline sample to that of the perfect crystal, which is reported as 139 J/g [9]. The film was dried in a vacuum oven at 100°C for 20 hrs prior to any further treatment. Some of the samples were annealed at 260°C for 1hr. As no crystallization was found, these samples will be referred to as 'amorphous' in our later discussion. Some of the samples were cold crystallized at 300°C for 1hr, and will be referred to as 'semicrystalline'.

Crystallization and melting behavior of Regulus[TM] were studied using a Perkin Elmer DSC-4 with Indium calibration for temperature and heat flow. Sample mass around 8 mg and a scan rate of 10°C/min were used for all the samples. WAXS data were obtained at room temperature using a Rigaku RU-300 x-ray diffractometer which was operated at 50kV and 200mA with Cu-Kα radiation and a diffracted beam graphite monochromator. The measurement range was from $2\theta = 5°-55°$ with a step scan interval of 0.02 degrees and a 2θ scan rate of 1°/min.

197

Mat. Res. Soc. Symp. Proc. Vol. 321. ©1994 Materials Research Society

TSDC measurement was carried out on our self-designed and self-assembled apparatus. The major part is the sample cell which contains two layers. The first layer is for liquid nitrogen cooling. The second layer is for heating and Faraday shielding. The triaxially shielded electrical conduction path is triaxially concentric with the two layers of the cell. A polymer film sample is loaded onto a spring loaded disc-shaped electrode. Helium gas is used as a heat transfer agent inside the cell. To obtain a global TSDC spectrum, Regulus™ was first heated to a temperature T_p where the poling field is applied ($E_p=1\times10^6$ V/m). After holding at T_p for 10 min, the cell was quickly quenched with liquid nitrogen at a cooling rate of about -15°C/min to a temperature T_0 where the poling field was reduced to zero and the two high voltage electrodes were short-circuited. The current vs. temperature was then recorded continuously as the cell was heated up slowly at a heating rate of 2°C/min. For all our TSDC experiments (unless otherwise mentioned), gold was evaporated on the samples to avoid the spurious charges that might come from the poor contact between the sample and the electrodes. Details about TSDC techniques have been discussed in some excellent reviews [14,15].

RESULTS AND DISCUSSION

DSC AND WAXS

The DSC scans are presented in Figure 1, curve 1 and curve 2 for both amorphous and semicrystalline Regulus™ respectively. The amorphous sample shows a sharper glass transition compared to the semicrystalline sample, and its T_g occurs around 250°C. A crystallization exotherm and a melting endotherm are observed at 313°C and 382°C, respectively, with equal heats. The semicrystalline sample has a broader glass transition with a T_g around 255°C. A small endothermic peak appears at 314°C followed by a larger endothermic peak at 382°C. This double melting peak phenomenon in polymers has been extensively discussed and is considered here to be due to crystal reorganization [16,17]. The crystallinity χ_c, is found to be 0.27, which is consistent with our previous study [1,5].

Figure 2, curve 2 and curve 1, shows the WAXS intensity vs. scattering angle for both amorphous and semicrystalline sample. The crystallinity is also calculated by subtracting the amorphous halo from the WAXS scan of the semicrystalline sample. The result is $\chi_c = 0.30$ which is also in good agreement with the crystallinity determined by DSC, considering the large error inherent in the WAXS area subtraction.

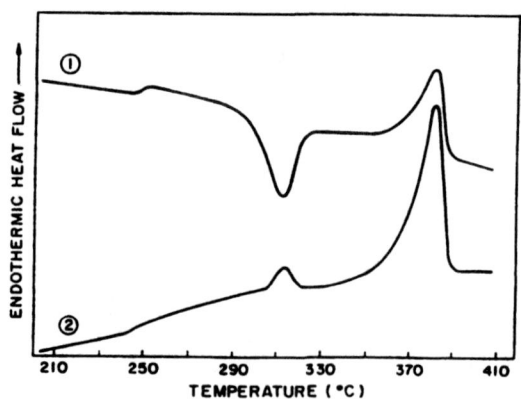

Figure 1 DSC scans for amorphous (curve 1) and semicrystalline (curve 2) Regulus NEW-TPI

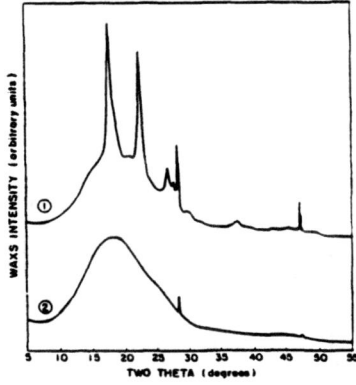

Figure 2   WAXS for amorphous
(curve 2) and semicrystalline (curve
1) Regulus NEW-TPI

Global TSDC Spectrum

The global TSDC spectrum for amorphous NEW-TPI poled at 260°C followed by rapid quenching is shown in Figure 3 [6]. Four peak regions have been identified and labeled. Peak 1 is the small peak showing at the high temperature end at 277°C. Peak 2 is the strongest peak with a maximum peak position at 255°C. Peak 3 has the second highest current peak at 178°C covering quite a broad temperature range. Around 0°C Peak 4 occurs with the smallest height and relatively narrow width. The assignment of amorphous peak origins has been reported in our previous study [6]. Peak 1 is assigned to the crystallization of the amorphous sample. Peak 2 is from the cooperative motion of the long polymer chains, i.e., glass transition relaxation. Peak 3 is the result of space charge and/or ionizable species inside the polymer bulk. Peak 4, which shifts its position and sometimes even disappears, is disregarded because it is a spurious charge effect.

We adopted the same experimental conditions and same method of peak assignment for semicrystalline Regulus™ film cold crystallized at 300°C for 1 hr. The resulting two global TSDC spectra are both shown in Figure 4, sample A and sample B for two independent trials of similarly treated films. Fresh samples were used for both cases and the sample treatment and experimental conditions were identical. Two strong peaks are seen in sample A and three peaks are seen in sample B. No peak is observed in region 1 in sample B. Testing for sample A was halted at too low a temperature due to experimental difficulties. The experimental temperature of the TSDC scan is below the sample treatment temperature for semicrystalline sample, therefore no new crystals formed during the TSDC test. Thus no peak is seen for semicrystalline sample which corresponds to Peak 1 in the amorphous sample. This observation confirms our previous assignment of Peak 1 in the amorphous polymer as due to additional crystals forming. The glass transition, Peak 2, is seen in both samples at the same location and is discussed in the following paragraph. Peak 3 appears in both sample A and B, however it shifts its position. While Peak 3 occurred in all the semicrystalline and amorphous samples we tested, its peak height and location were widely variable. We have not yet determined which stage of our sample treatment affects its location. As in the amorphous sample, Peak 4 comes from spurious charges and will not be considered further.

After carefully examining the glass transition relaxation peak, Peak 2 in Figure 4, we report the following observations which are very consistent with previous DSC results. First, T_g of the semicrystalline polymer occurs at 260°C, which is 5°C above that of the amorphous sample. The crystals act as thermo-reversible crosslinks and constrain amorphous phase mobility. This leads to higher T_g in semicrystalline films. Second, this peak has a greater full width at half maximum than the amorphous NEW-TPI indicating a broader distribution of relaxation times in semicrystalline film. Third, the height of Peak 2 is also reduced in semicrystalline film. The relative peak ratio of

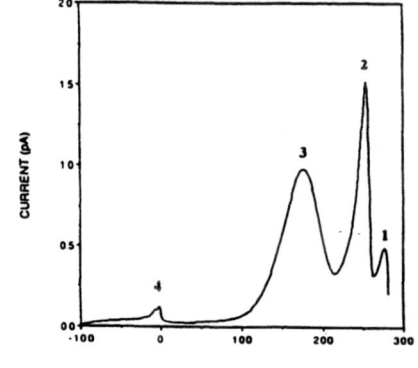

Figure 3   TSDC scan for Au-coated amorphous NEW-TPI with fast quenching

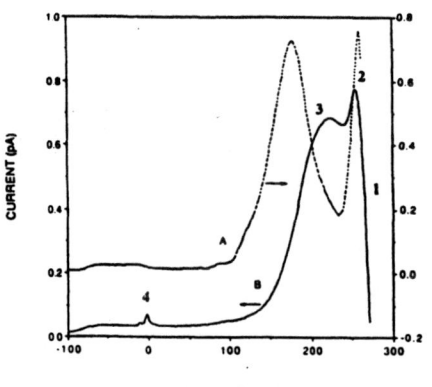

Figure 4   TSDC scan for Au-coated semicrystalline NEW-TPI with fast quenching

semicrystalline Peak 2 to amorphous Peak 2 is calculated to be 0.65. This may be related to the relative quantity of amorphous phase dipoles in the semicrystalline sample compared to the amorphous sample. Using DSC, our previous calculation of amorphous moieties from the step height of the heat capacity change at T_g gave an amorphous fraction of 0.63 by weight [5].

The origin of Peak 3, which in the amorphous sample we attributed to space charge and/or ionizable species inside the polymer bulk, appears to be even more complex in the semicrystalline sample. Peak 3 for semicrystalline Regulus™ (Figure 4) sample A has a maximum at 178°C while sample B shows up at 225°C. Generally speaking, there exist at least two phases inside the semicrystalline polymer bulk, i.e., amorphous phase and crystalline phase, which provide current pathways of different conductivities. For Regulus™ NEW-TPI, in addition to the crystal phase, both mobile and rigid amorphous phases are observed [1,5], and different conductivity between them could be safely assumed. When the electric field is applied, homocharges will build up at the phase boundaries due to the differences in conductivity in addition to the space charge and/or ionizable species inside the semicrystalline bulk. These factors come into play to give Peak 3. To further test this idea, we have performed two tests which are described below.

When a polymer sample is quickly quenched from the poling temperature, it is in an non-equilibrium state where excess free volume exists inside the polymer bulk. As the sample is heated slowly, it tends to resume as much as possible its equilibrium state. This relaxation is similar to physical aging where the glassy polymer is held for a period of time at or below its glass transition temperature. On the other hand, if we cool the sample slowly, the polymer has enough time to approach to its equilibrium state. Slow cooling will allow the free volume to be reduced, and in addition, the free charges will also have time to separate for a longer distance in a temperature high enough to give them a rather high mobility. Figure 5 shows the result of TSDC measurement on semicrystalline sample slowly cooled after being poled at 260°C for 10 min. Again, Peak 1 was not observed because the polymer has already crystallized completely. The glass transition, Peak 2, is at the same location and about the same peak current as the rapidly cooled sample of Figure 4. No significant narrowing is observed, which suggests that amorphous phase is constrained in between the crystalline phase and does not relax as it does in the pure amorphous polymer. Peak 3 broadens significantly in the slowly cooled sample. This is probably due to the longer separating distance of the charges and the impeded ability of the free carriers to move through the crystalline and amorphous interphase region.

Figure 6 shows the TSDC result for semicrystalline sample coated with aluminum electrodes, poled at 260°C for 10 min and cooled rapidly. Similar to what we have observed for

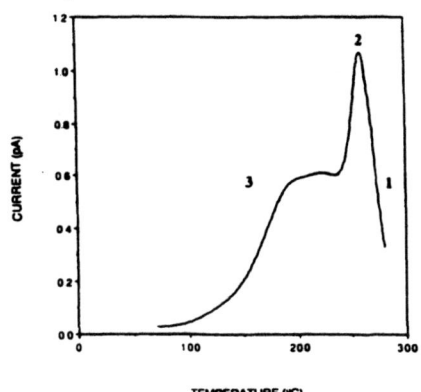

Figure 5 TSDC scan for Au-coated semicrystalline NEW-TPI with slow cooling

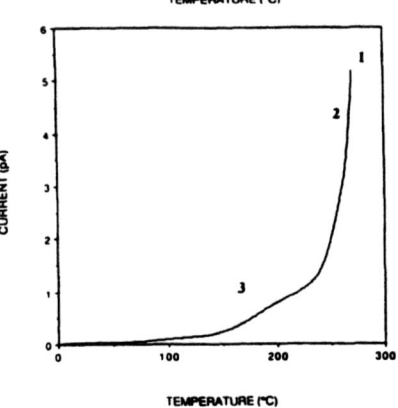

Figure 6 TSDC scan for Al-coated semicrystalline NEW-TPI with fast quenching

the amorphous sample [6], the sample coated with Al shows an ionic conduction [18] at high temperature, which suggests a different contact mechanism for Au and Al. This strong current starts to occur just below T_g and masks Peak 2. Peak 3 can be observed around 180°C but it is difficult to determine the peak shape. Combining these observations, it is clear that Peak 3 is not dipolar relaxation. Its origin is quite complicated, and space charge, electrode type, ionizable species and interfacial charges can all contribute to its shape and position.

CONCLUSIONS

Regulus™ NEW-TPI polymer has been studied by DSC, WAXS and TSDC. The glass transition relaxation of semicrystalline polymer is quite strong and shifts to a slightly higher temperature compared to that of the amorphous polymer. Thus, the crystals appear to have only a minor effect on the relaxation time distribution of the amorphous chains. These results are very consistent with our prior studies using DSC and dynamic mechanical relaxation [1,5]. Space charge, electrode type, ionizable species and interfacial charges may all affect the TSDC spectrum.

ACKNOWLEDGEMENT

This research is supported by the Electric Power Research Institute, contract RP: 8007-13. The authors thank Mitsui Toatsu for film samples of Regulus™ and for support. The authors thank Prof. Lacabanne for helpful discussions about TSDC cell design.

REFERENCES

1. P.P. Huo and P. Cebe, Polymer 34, 696 (1993).
2. P.P. Huo and P. Cebe, Colloid Polym. Sci. 270, 840 (1992).
3. J.B. Friler and P. Cebe, Polym. Eng. & Sci. 33, 587 (1993).
4. Y. Aihara and P. Cebe, Polym. Eng. Sci., in press (1993).
5. P.P. Huo, J.B. Friler and P. Cebe, Polymer 34, 4387 (1993).
6. S.X. Lu and P. Cebe, Polym. Prepr., (American Chem. Soc., 1994), in press.
7. T.H. Hou and R.M. Reddy, SAMPE Q. 38, January (1991).
8. B. Hsiao, B. Sauer and A. Biswas, J. Poly Sci.: Poly. Phys., in press (1993).
9. Mitsui-Toatsu Chem., Inc. Tokyo, Japan. Technical Data SheetA-00.
10. H. Sakaitani, K. Okuyama and H. Arikawa, Poly. Prepr. Jpn. 40, 478 (1991).
11. S. Yuasa, M. Truji and T. Takahashi, Polym. Prepr. Jpn 40(1-4), 491 (1991).
12. T. Sasuga, Polymer 32, 1539 (1991).
13. P.M. Hergenrother, SPE Conference on High Temperature Polymers and Their Uses, (Case Western Reserve University, 1989).
14. P. Braunlich, Thermally Stimulated Relaxation in Solids, 1st ed. (Springer-Verlag, Berlin, 1979).
15. J.P. Ibar, P. Denning, T. Thomas, A. Bernes, C. de Goys, J.R. Saffell, P. Jones and C. Lacabanne, Advances in Chemistry Series No. 227, Polym. Characterization: Phys. Prop., Spectroscopic, and Chromatographic Methods, American Chemical Society 227, 167 (1990).
16. J.S. Chung and P. Cebe, Polymer 33, 2312 (1992).
17. J.S. Chung and P. Cebe, Polymer 33, 2325 (1992).
18. J. van Turnhout, Polym. J. 2, 173 (1971).

VISCO-ELASTIC RELAXATIONS IN ALKALI BORATES AND ALKALI SILICATES

J. KIEFFER*, J.E. MASNIK*, B.J. REARDON*, AND J.D. BASS**
*Department of Materials Science and Engineering, University of Illinois at Urbana-Champaign, Urbana, IL 61801
**Department of Geology, University of Illinois at Urbana-Champaign

ABSTRACT

The Brillouin light scattering technique was used to study short time relaxations in glass forming oxide melts. The line shape analysis of Brillouin spectra yields a complex mechanical modulus. The temperature dependence of this modulus provides insight into the structural disintegration above the glass transition, as well as the thermally activated mechanisms which facilitate the momentum transport in the process of viscous dissipation.

A series of binary and ternary alkali borates and alkali silicates have been investigated. The results show fundamental differences in the way alkali cations affect the network structure in borates as compared to silicates. While these cations move relatively freely within the rigid silicate network, their motions are strongly coupled to the network relaxation in borates. The activation energies for cation motion are in good agreement with those found by other methods.

INTRODUCTION

The successful design of glassy materials that exhibit high ionic conductivities, for application as electrolytes, depends on a thorough understanding of the relationship between a structure and its dynamic behavior.[1,2] High ionic conductivity is typically achieved when combining small cations with network forming elements that assemble into a backbone structure with large openings.[3] On the other hand, larger network openings reduce the rigidity of this backbone. In a mechanically soft network the coupling between the motion of the mobile cations and that of the network elements is more pronounced, which causes an increase in the collision rate and the dissipative momentum transfer. The correlations between structural relaxations and the motion of highly mobile cations are still poorly understood.

Dielectric relaxation spectroscopy is by far the most common technique for the investigation of fast ion conductors.[4] Mechanical actuation can be realized up to MHz frequencies by means of ultrasonic transducers. At GHz frequencies, inelastic light scattering provides a unique way to probe the propagation and dissipation of acoustic phonons. Being close to the first Brillouin zone boundary (or equivalent construct for amorphous substances), this frequency range is crucial for diffusive jumps of charged species, due to the proximity between acoustic and optical branches in the dispersion relation. In this paper we report our results on the determination of the complex mechanical modulus in alkali silicates and borates, using Brillouin scattering.

THEORETICAL BACKGROUND

Light scatters as a result of the local change in the dielectric constant due to density fluctuations, and therefore acts as a gauge capable of evidencing the propagation and attenuation of acoustic phonons. Elastically scattered light produces a central Rayleigh line. Sustainable density fluctuations, propagating with the velocity of sound, v_0, give rise to Brillouin scattering

Mat. Res. Soc. Symp. Proc. Vol. 321. ©1994 Materials Research Society

at frequencies that are shifted with respect to that of the incident light by $\pm q \cdot v_0$, where q is the wavevector of the elastic wave. The dissipative exchange of energy and momentum attenuates the propagating elastic deformations and causes broadening of the Brillouin lines.

The theoretical shape of the Brillouin spectrum is given by the dynamic structure factor, $S(q,\omega)$. For fluids, $S(q,\omega)$ can be derived using either the generalized hydrodynamic formalism[5] or mode coupling theory.[67] The normalized scattering intensity is

$$\frac{S(q,\omega)}{S(q)} = \frac{2(\gamma-1)}{\gamma} \cdot \frac{q^2 \cdot \kappa/\rho_0 c_p}{\omega^2 + \left(q^2 \cdot \kappa/\rho_0 c_p\right)^2} + \frac{1}{\gamma}\left[\frac{q^2 \cdot \Gamma}{(\omega+v q)^2 + \left(q^2 \Gamma\right)^2} + \frac{q^2 \cdot \Gamma}{(\omega - v q)^2 + \left(q^2 \Gamma\right)^2}\right], \quad (1)$$

where $\gamma = \dfrac{c_p}{c_v}$, $\Gamma = \dfrac{1}{2}\left[\dfrac{\eta(\omega)}{\rho_0} + \left(\dfrac{\kappa}{\rho_0 c_p}\right)\right] \cdot (\gamma - 1)$, ρ_0 is the average density, κ is the thermal conductivity, c_v is the heat capacity at constant volume, and c_p that at constant pressure. The coefficient $\eta'(\omega)$ is the dynamic viscosity and the ratio $\left(\kappa/\rho_0 c_p\right)$ is the thermal diffusivity. By fitting eq. (1) to the Brillouin spectra, it is possible to extract the complex mechanical modulus of the scattering medium,[8]

$$M^*(\omega) = M' + iM'' = M_0 + M_2 \frac{\omega^2 \tau^2}{1+\omega^2\tau^2} + iM_2 \frac{\omega\tau}{1+\omega^2\tau^2}. \quad (2)$$

The real part of the modulus represents the capability of the structure to store energy elastically. It can be related to the frequency shift between the Brillouin and Rayleigh lines, $\omega = |\omega_s - \omega_R|$. This elastic modulus comprises a static and a dynamic component,

$$\left(M_0 + M_2 \frac{\omega^2 \tau^2}{1+\omega^2\tau^2}\right) = \frac{\rho_0 \omega^2}{q^2}. \quad (3)$$

The imaginary part, on the other hand, reflects the amount of energy dissipated in the probed structure by friction. From eq. (1) follows that the width of the Brillouin peak is predominantly controlled by the dynamic viscosity $\eta'(\omega)$, which is related to the loss modulus as

$$M_2 \frac{\tau}{1+\omega^2\tau^2} = \eta'(\omega). \quad (4)$$

τ is the characteristic relaxation time of a mechanism. If the mechanisms involved in structural relaxation are thermally activated, we can write $\tau = \tau_0 e^{E_A/k_B T}$, where τ_0 is the fundamental time constant of the mechanism. Substitution in eq. (2) yields

$$M^*(\omega,T) = M_0(T) + M_2 \frac{\omega^2 \tau_0^2 e^{2E_A/k_B T}}{1+\omega^2\tau_0^2 e^{2E_A/k_B T}} + iM_2 \frac{\omega\tau_0 e^{E_A/k_B T}}{1+\omega^2\tau_0^2 e^{2E_A/k_B T}}. \quad (5)$$

While the real part of this expression shows an inflection, at which the elastic modulus changes from its low-temperature to its high-temperature limit, the imaginary part exhibits a maximum near the same temperature. If several mechanisms can be activated within a certain temperature range, their effects superimpose. Due to the high frequency of the Brillouin probe, the product $\omega \cdot \tau_0$ assumes values between 10^{-4} and 10^{-2}. The influence of τ_0 on the location of the maximum (or inflection point) is intensified, which makes the technique particularly sensitive towards differentiating between individual mechanisms by their pre-exponential term.

EXPERIMENTAL

A detailed description of the experimental setup has been given elsewhere.[9,10] Here we will only summarize the most important points. As ingredients for the sample preparation we used Cabo-Sil fumed silica (99.8 % pure), anhydrous boron oxide powder (99.2 % pure), and analytical grade anhydrous lithium-, potassium-, and sodium carbonates with a purity of 99.985 %. The powders were melted and homogenized before the sample holders, which consisted of a platinum wire wound into a double loop, were filled. Upon immersion, the wire loop filled due to surface forces. The so contained melt assumed the shape of a slightly bulged cylinder.

The samples were then mounted in a small furnace located in the optical path of the Brillouin setup. Before measurement the samples were held for twelve hours at their highest temperature under a dried air atmosphere, to remove any water that was introduced during prior handling. Chemical analysis before and after the measurements revealed that the alkali loss from the samples during this period was less than 1 mol%. The laser beam enters from the bottom and the scattered light is collected from the horizontal orifice over a solid angle of 15°. The furnace orifices are capped with silica windows. The scattered light is collimated and analyzed with a six-pass tandem interferometer. Light from a single-mode argon laser, with a wavelength of 0.5145 μm was used. To eliminate the effect of the finite collection angle, and any spectral broadening due to imperfections in the optical components or interferometer, the raw spectra were deconvoluted before being fit with eq. (1).

RESULTS AND DISCUSSION

In the analysis of our data, we compare the changes in both parts of the complex mechanical modulus to gain information about the structural developments during glass formation. With the elastic part of the mechanical modulus we can monitor the connectivity in the network. The larger the connectivity, the more rigid the network and the larger the elastic modulus. In fig. 1 we show the longitudinal elastic moduli of various potassium silicates, as a function of the temperature. In binary silicates, the modulus changes only slightly up to T_g. In terms of network connectivity, the glass structure does not change in this temperature range. The glass transition is marked by a discontinuity in the temperature dependence of the elastic modulus. Above T_g, the modulus decreases with increasing temperature, reflecting the thermally activated network disintegration. The rate of decrease is higher, the larger the alkali

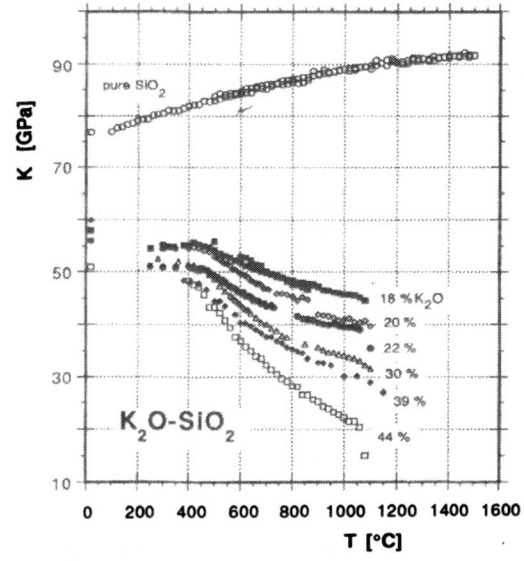

Fig. 1 Longitudinal elastic moduli of potassium silicate systems with various compositions, as a function of the temperature.

concentration. In terms of Angell's *strong – fragile* classification of glass forming liquids,[11] a fragile liquid exhibits a rapid network depolymerization with increasing temperatures.

With increasing alkali concentrations, the elastic moduli of the silicates decrease. At temperatures far above T_g, this decrease is significantly more pronounced than at room temperature, which suggests that at high temperatures the effect of the alkali cations on the ability of the network to support strain is stronger than at room temperature. The oxygen which is brought into the system with the alkali oxide, mediates the breakdown of the network by reacting with a Si-O-Si link, to form two non-bridging oxygens. The number of network interruptions depends on the amount of alkali oxide introduced, a balance which is unlikely to shift noticeably at higher temperatures. However, at room temperature the alkali cations are located near non-bridging oxygens. They are relatively immobile and participate in the transmission of elastic deformation. At high temperatures, on the other hand, alkali cations are very mobile. When impacted by a propagating phonon, the cations can use the transmitted energy to overcome the activation barrier for a diffusional jump, a process which reduces the structural rigidity.

In fig. 2 the longitudinal elastic moduli for sodium borates with various compositions are shown as a function of the temperature. As for the silicates, the temperature dependence of the modulus exhibits a discontinuity at T_g for the borates. The rate of decrease of the modulus with temperature, again, reflects increasing fragility of the liquids with increasing alkali content. A significant difference between borates and silicates becomes evident in the concentration dependence of the room-temperature and high-temperature moduli. In borates, the room-temperature elastic modulus increases with increasing alkali concentration. This behavior is attributed to the formation of tetrahedrally coordinated borons. The additional oxygen, introduced by the alkali oxide, is shared between two borons and establishes cross-links between originally planar boron-oxygen sheet structures. The cross-linking enhances the structural rigidity.

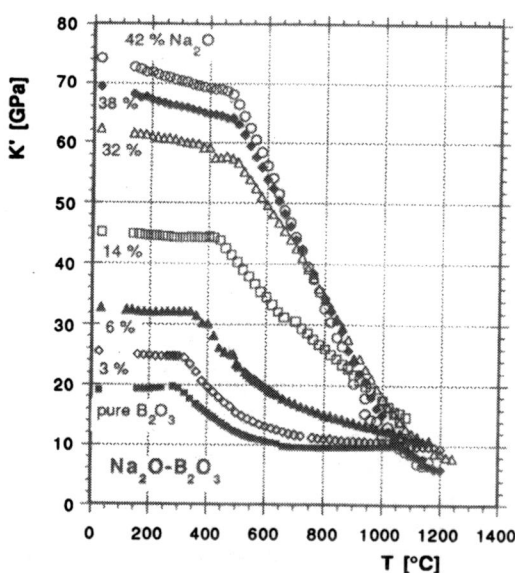

Fig. 2   Longitudinal elastic moduli of sodium borate systems with various compositions, as a function of the temperature.

At high temperatures, the elastic properties of the liquid borate structures are virtually unaffected by the alkali cation concentration. Although the formation of four coordinated boron seems to initially facilitate the network breakdown, once sufficient thermal activation is provided, the network has disintegrated to the smallest possible structural units, independent of the amount of modifier cations. In other words, all cations present in the system continuously exchange their oxygen neighbors. For boron this may involve the passage through an intermediate state of tetrahedral coordination. This transition is possible any time a fourth oxygen approaches the boron atom from outside of the BO_3 plane, whether this additional oxygen belongs to another BO_3 unit or whether it stems from the alkali oxide.

From the comparison between the temperature dependence of the elastic moduli of silicates and borates it can be concluded that a network structure persists to much higher temperatures in the silicates.

A stronger network allows for the alkali cations to move more freely within this structure. This can be verified by examining the dissipative component of the complex modulus. In figs. 3 and 4 the dissipation coefficients, $\zeta = v' \cdot \omega$, (where $v' = \eta'/\rho_0$ is the kinematic viscosity), are shown for alkali silicates and alkali borates respectively. ζ corresponds to the mechanical loss modulus per unit mass. At the high probing frequencies employed here, the mechanisms that can contribute to viscous dissipation include atomic diffusion, rotational motions of small structural segments and the anharmonic response of structural defects that constitute mechanical discontinuities.

The data shown in figs. 3 and 4 can be fit by a linear combination of terms given as the imaginary part of eq. (5). These fits allow one to determine the activation energies and pre-exponential time constants of the various mechanisms reflected in the dissipation coefficient spectrum. In all binary silicates a distinct me-

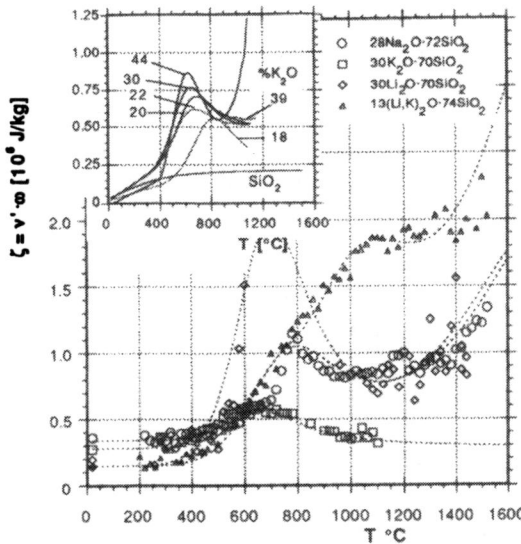

Fig.3 Dissipation coefficients of various binary and ternary alkali silicate systems, as a function of the temperature. Insert: Dissipation coefficients of potassium silicates with various alkali concentrations.

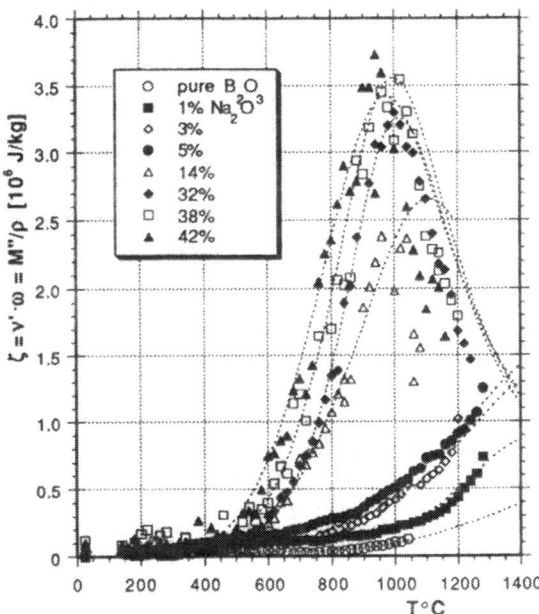

chanism characterized by an activation energy of approximately 65 kJ/mol was found. This dissipation can be attributed to the momentum exchange during 'collisions' of diffusing alkali cations with the network. The activation energy of this process agrees well with that found by Day and Shelby for the alkali peak in their internal friction spectra.[12] The motion of network elements becomes activated only at much higher temperatures. From the distinct appearance of alkali dissipation maxima in case of binary silicates, we can conclude that the motion of these cations does not involve a strong correlation with network relaxations.

Fig.4 Dissipation coefficients of various sodium borate systems, as a function of temperature. Insert: Magnification of the low activation energy hump in B_2O_3-rich compositions.

For the ternary $(Li_2O,K_2O)\cdot 3SiO_2$ system, on the other hand, we found a wider spread of activation energies for cation diffusion mechanisms. This supports the concept that in mixed alkali silicates specific sites exist for each type of cation. When the cations jump to a vacant site which has previously been occupied by a cation of the other kind, the network will relax to accommodate the new cation. This process consumes an additional amount of energy, and causes a stronger coupling between cations and network.

In borates the motion of alkali cations is less distinguishable from that of network elements. In pure B_2O_3, the motion of fully disintegrated network components is characterized by an activation energy of 76 to 80 kJ/mol. In binary alkali borates, the corresponding dissipation peak shifts downwards, while another peak, which has an activation energy of 56 to 60 kJ/mol and which we attribute to the alkali motion, shifts upwards with increasing alkali concentration. As seen with the elastic behavior of the borates, the impact of the alkali cations on the network backbone is stronger. Above T_g the network disintegrates more rapidly with increasing alkali content. As a consequence, the alkali cations and boron atoms find themselves in comparable, rapidly fluctuating oxygen coordinations. Therefore, the motions of alkali cations and boron atoms require a similar amount of activation energy at these high temperatures.

CONCLUSION

Brillouin light scattering was used to monitor the structural developments in glass forming liquids. The high-frequency complex mechanical modulus measured by this method provides insight into the degree of networking within a structure, and allows one to detect molecular scale mechanisms involved in structural relaxations. In silicates the network structure prevails to high temperatures, allowing for the alkali cations to move more or less decoupled from network relaxations. In comparison, the borate network disintegrates more rapidly with temperature. At high temperatures, the mobilities of boron and alkali cations are equally constrained by the oxygens.

Acknowledgment: The National Science Foundation, for the support under MSS 90-08918.

References:

1   C.A. Angell, Materials for Solid State Batteries, B.V.R. Chowdari and S. Radhakrishna, eds., pp. 31
2   L. Borjesson, L.M. Torell, S.W. Martin, C. Liu and C.A. Angell, *Phys. Letters* 125, 330 (1987)
3   T. Minami and N. Nachida, *Mat. Chem. Phys.* 23, 63 (1989)
4   P.B. Macedo, C.T. Moynihan and R. Bose, *Phys. Chem. Glasses* 13, 171 (1972)
5   R.D. Mountain, *Journal of Research of the NBS* 70A, 207 (1966)
6   N.J. Tao, G. Li and H.Z. Cummins, *Phys. Rev.* B43, 5815 (1991)
7   W. Götze and M. Lücke, *Phys. Rev.* A11, 2173 (1975)
8   K.F. Herzfeld and T.A. Litovitz, Absorption and Dispersion of Ultrasonic Waves, Academic Press, N.Y. 1959
9   J.E. Masnik, J. Kieffer and J.D. Bass, *Mat. Res. Soc. Symp. Proc.* 248, 505 (1992)
10 J.E. Masnik, J. Kieffer and J.D. Bass, *J. Am. Ceram. Soc.* 72, (1993) accepted
11 C.A. Angell, "Strong and Fragile Liquids", in Relaxations in Complex Systems, K. Ngai and G.B. Wright, eds., National Information Service, U.S. Department of Commerce, Springfield, VA 22161 (1985)
12 J.E. Shelby, Jr., and D.E. Day, *J. Am. Ceram. Soc.* 52, 169 (1969)

AN INVESTIGATION OF THE VISCOUS FLOW PROBLEM ASSOCIATED WITH THE HEATING OF THE GLASS PREFORM DURING OPTICAL FIBER PROCESSING

SARA E. ROSENBERG, HARIS PAPAMICHAEL, AND IOANNIS N. MIAOULIS*
Tufts University, Thermal Analysis of Materials Processing Laboratory, Mechanical Engineering Department, Medford, MA 02155

ABSTRACT

A thermal study of the optical fiber manufacturing process involves some of the most challenging, and in some cases, unsolved fundamental problems in heat transfer and fluid mechanics. The heating stage of the process, where the glass cylinder (preform) is heated radiantly by the cylindrical muffle furnace, greatly influences the resultant quality of the fiber. During the process a neckdown region is formed that is characterized by the stretching of the glass. The two-dimensional transient equations of motion and mass conservation, with viscosity dependent on temperature, were solved to obtain the velocity profiles in the glass and the shape of the neckdown region. Axial velocity contours and the neckdown profiles were examined for various drawing conditions. Differences between the new method presented here and existing one-dimensional method were examined. The analysis can be used for drawing of other materials such as metals and polymers, and the modeling is applicable to other viscous liquids.

INTRODUCTION

For a typical drawing process, a glass preform is placed in a cylindrical furnace to be heated peripherally to a temperature where the glass softens. It is then pulled downward to form a glass fiber, as shown in Figure 1. It is essential that the diameter of the exiting fiber is kept constant in order to avoid deficiencies in the fiber that might cause it to fail. A thermal analysis of the heating region can give us essential information on the quality of the final product. The drawing force on the fiber, as it is pulled, and the decreasing viscosity as the glass heats up causes the radius of the preform to decrease, thus forming a neck down.

The uniformity of the fiber diameter can be affected by temperature fluctuations caused by convection-induced perturbations in the furnace. In cases where a non-uniformity of the diameter is observed, the operator manually adjusts the setup until the desired diameter is achieved. A better understanding and accurate modeling of the fiber drawing process is the first step in process optimization and automation, which should help to avoid the fluctuations and error due to manual adjustments.

Investigators examining the optical fiber process in the past have focused mainly on the cooling region. Papamichael and Miaoulis,[1] Glicksman,[2] and other investigators[3-9] considered the cooling of glass fiber of a given constant diameter as it exits the furnace. Others have investigated subjects related to the heating region of the drawing process, most of which refer to the problem of a stretching surface.[10-21] Geyling and Homsy[22] performed a one-dimensional analysis to examine the instabilities of the drawing of a preform into a fiber for various glasses and drawing conditions. Jaeger et al,[23,24] and Nagel[25] included experimental methods to measure the diameter of the fiber and the instabilities that can occur during the process, and they discussed types of heating and fiber drawing. Smithgall[26] presented a method that reduces diameter variations in the fiber

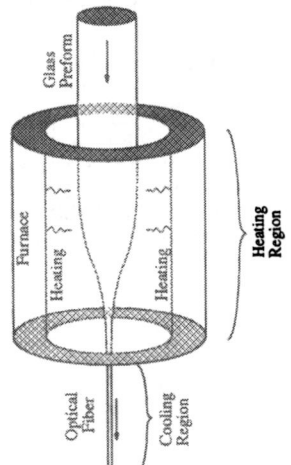

Figure 1. Fiber drawing process

---

* Author to whom correspondence should be addressed.

drawing process by implementing a feedback control loop. A particle tracking method to model axisymmetric die swell for flow through a jet was presented by Caswell and Viriyayuthakorn.[27]

Manfré,[28] Brown,[29] Geyling,[30] Glicksman,[2] and Kase and Matsuo[31] performed a one-dimensional analysis referring to the fiber formation from glass flowing through a nozzle rather than from a preform in a furnace. Manfré[28] concentrated on the forces affecting this fiber drawing process in one dimension. Geyling[30] started his study with a two-dimensional model that he later reduced to a one-dimensional one by simplifying with the small slope assumption that $|dR/dz| \ll 1$ and $|dV_r/dV_z| \ll 1$ through the neckdown region, where R is the radius, z is the axial location, V_r is the radial velocity, and V_z is the axial velocity. Glicksman[2] acknowledged the limitations of the one-dimensional assumption, but indicated that the results were qualitatively accurate. Kase and Matsuo[31] examined the neckdown profile and considered the material and heat transfer properties of the glass in the formation region. Paek and Runk[32] investigated the neckdown profile and temperature distribution in the fiber using a one-dimensional model. Myers[33] concentrated more specifically on the disturbances that may occur in the fiber drawing process, ignoring inertia, gravity, and surface tension in the solution. He predicted the radius, velocity, and temperature in one dimension. Myers concluded that the numerical and experimental results agreed well considering the limitations of a one-dimensional model.

Vasiljev et al, [34] studied, in one dimension, the drawing process of glass fibers from a preform. An extensive study of the forces involved in the fiber formation process showed that a numerical model of such a process must include all forces due to gravity, inertia, surface tension, and viscous friction. These forces have a significant effect on the neckdown shape.

The one-dimensional models can be used to simplify more complicated solution techniques and to provide a qualitative understanding of the drawing process. In an actual drawing process, the glass flows in both axial and radial directions; therefore, the problem should be approached as a two-dimensional one. Clark and Viriyayuthakorn[35] examined a horizontal model of the fiber drawing process to determine the two-dimensional flow field and neckdown profile, excluding the gravity effects that Vasiljev et al showed to be essential.

This paper presents a complete two-dimensional analysis used to determine the axial and radial velocities, and the radial neckdown profile, given a previously determined axial temperature distribution. The model includes the effects of gravity and refers to the axisymmetric, vertical neckdown of a preform pulled into a glass fiber.

ANALYSIS

As the preform is pulled through the furnace vertically, its temperature increases; therefore, the viscosity within the glass changes. This change in viscosity, combined with the pulling and gravity effects, cause the neckdown of the fiber. An optical fiber pulled down from a glass preform can be modeled as an axisymmetric cylinder of varying radius and fixed length. The equations that describe the motion of the fiber are the continuity, r-momentum, and z-momentum, respectively, as given below:

$$\frac{\partial r}{\partial t} + \frac{1}{r}\frac{\partial}{\partial r}(rRV_r) + \frac{\partial}{\partial z}(rV_z) = 0 \tag{1}$$

$$\rho\left(\frac{\partial V_r}{\partial t} + V_r\frac{\partial V_r}{\partial r} + V_z\frac{\partial V_r}{\partial z}\right) = -\frac{\partial p}{\partial r} + 2\left[\frac{\partial}{\partial r}\left(\mu\frac{\partial V_r}{\partial r}\right) + \frac{\mu}{r}\left(\frac{\partial V_r}{\partial r} - \frac{V_r}{r}\right)\right] + \frac{\partial}{\partial z}\mu\left(\frac{\partial V_z}{\partial r} + \frac{\partial V_r}{\partial z}\right) \tag{2}$$

$$\rho\left(\frac{\partial V_z}{\partial t} + V_r\frac{\partial V_z}{\partial r} + V_z\frac{\partial V_z}{\partial z}\right) = -\frac{\partial p}{\partial z} + 2\frac{\partial}{\partial z}\left(\mu\frac{\partial V_z}{\partial r}\right) + \frac{\partial}{\partial r}\left[\mu\left(\frac{\partial V_z}{\partial r} + \frac{\partial V_r}{\partial z}\right)\right] + \frac{\mu}{r}\left(\frac{\partial V_z}{\partial r} + \frac{\partial V_r}{\partial z}\right) \tag{3}$$

where r is the radial coordinate measured from the axis of the fiber, z is the axial coordinate along fiber measured from the furnace entrance, R is the outer radius of the fiber, μ is the viscosity of the fiber, ρ is the fiber density, V_r is the radial velocity, V_z is the axial velocity, and p is the pressure within the fiber. A solution to these equations would give the axial and radial velocities in the preform as well as its neckdown profile.

The axial velocity at the furnace entrance and exit are assumed to be constant across the entire radius,

$$V_z\,(0, r\,) = V_i\,,\quad V_z\,(l, r\,) = V_e \tag{4}$$

where V_i is the preform velocity at the entrance of the furnace, V_e is the fiber exit velocity, and l is the length of the furnace. No radial velocity exists at the furnace entrance or exit, and the radial velocity at the axis of the preform ($r = 0$) is also zero,

$$V_r\,(0, r\,) = V_r\,(l, r\,) = V_r\,(z\,, 0) = 0, \tag{5}$$

where V_r is the radial velocity. The outer radius of the fiber, R, is prescribed condition at the furnace entrance,

$$R\,(0) = R_0 \tag{6}$$

where R_0 is the initial preform radius. The viscosity profile is prescribed by the following equation[36]

$$\log(\mu) = 1.456 \times 10^7 (T)^{-2.1993} \tag{7}$$

where μ is the viscosity and T is the temperature ($^\circ$C) of the fiber at any particular point. For this analysis, the temperature, and thus viscosity, are assumed to vary axially along the fiber.

SOLUTION TECHNIQUE

A two step, iterative method was used to solve for the velocity field and radial profile. First, equations (1) - (3) were solved using Patankar's[37] SIMPLER pressure correction algorithm to obtain the velocities of a given radial profile. His method generates a pressure field using an initial guess for the velocity field and radial profile, which are then iteratively recalculated to satisfy the laws of conservation of mass and momentum. Once converged, the newly found velocities are then used to calculate a revised radial profile from the continuity equation (1). The entire procedure repeats to obtain a new velocity field with the pressure correction method, continuing until the radial profile has converged.

The fiber was divided into a normalized grid in both the radial and axial directions, using 60 axial divisions along an axial length of 0.14 m, and 10 proportionally spaced radial divisions. Trial calculations determined that the stability of the solution or the results obtained are not significantly affected by altering the radial or axial grid size. The radial divisions were normalized after each calculation of the radius to account for the adjusted radial profile.

A staggered grid used in this analysis, as seen in Figure 2 and specified by the SIMPLER algorithm, reduces difficulties in calculating the mass flow across the

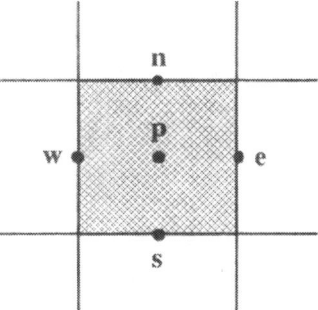

Figure 2. Staggered grid (n=north, s=south, e=east, w=west, p=center)

faces and in calculating values for all variables at the same grid point. This grid allows for the radial velocity calculations at the east and west control volume faces, the axial velocities at the north and south faces, and the pressures at the center. As the outer radius is recalculated, the radial grid size and radius at each node also changes for the next iteration. The radial profile was considered to be converged when the change in radius between iterations was less than a required precision (< 0.01).

RESULTS AND DISCUSSION

Vasiljev's one-dimensional and our two-dimensional model were compared with the same boundary conditions in order to verify the validity of the two-dimensional model. The preform radius was 0.02 m, the preform velocity at the furnace entrance was 0.0005 m/s, the furnace length was 0.14 m, and for this test, the fiber exit velocity was 0.8 m/s. Figure 3 shows that the fiber neckdown predicted by the two-dimensional model begins sooner than the one-dimensional case. As expected, the radial velocity contributes to the inward flow of the glass, which has not been accounted for in the one-dimensional case. The two models showed a maximum difference of approximately 10% in radial profile.

Our model was used to conduct a study for a variety of fiber drawing conditions, but with the same preform furnace entrance conditions as in those of Figure 3. The density of the glass fiber, ρ, was 2500 kg/m^3, and all cases used a linear viscosity profile along the length of the fiber. This is a legitimate assumption since a single glass was used and viscosity linearity would not significantly alter the results. Using a prescribed exit velocity, the exit radius was calculated by mass conservation, and the previously described method predicted the intermediate fiber profile. Figures 4-6 show the radial shape of the fiber as it is formed and the axial velocity contours for three exit velocities, 0.2 m/s, 0.8 m/s, and 3.2 m/s, respectively. As the fiber exit velocity increases, it is observed that the slope of the neckdown increases.

For all three cases, the axial velocity near the preform entrance is slower at the outer radius of the fiber, and higher at the central axis. This is a result of the pulling force distribution in the upper region of the fiber before the viscosity changes have an effect. However, as the viscosity increases, the velocity at the outer radius of the fiber also increases, overtaking the flow of the

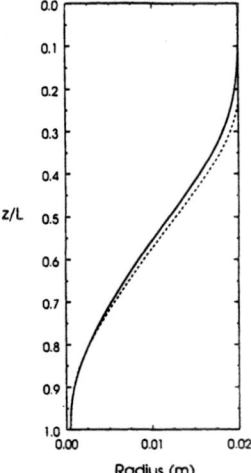

Figure 3. Neckdown profile for a glass preform; fiber drawn at 0.8 m/s; ---- Vasiljev et al,[34] (1-d model); —— present 2-d model

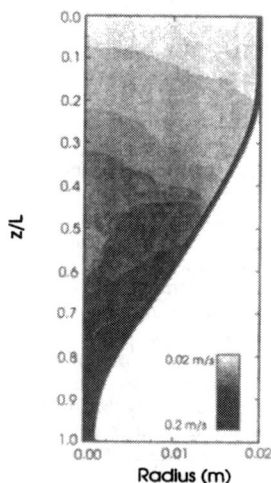

Figure 4. Neckdown profile and axial velocity contour for a glass preform; fiber drawn at 0.2 m/s

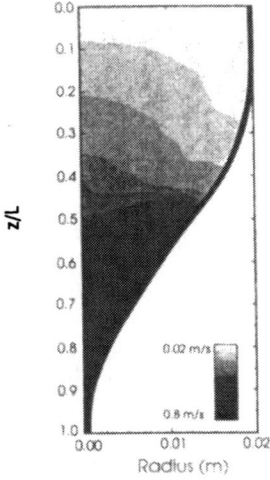

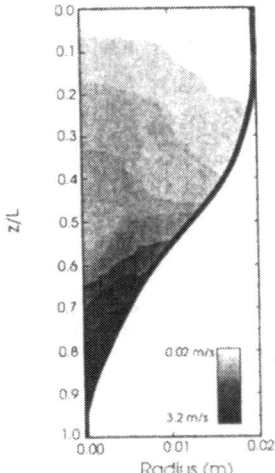

Figure 5. Neckdown profile and axial velocity contour for a glass preform; fiber drawn at 0.8 m/s

Figure 6. Neckdown profile and axial velocity contour for a glass preform; fiber drawn at 3.2 m/s

inner fiber as the fiber begins to narrow. This causes the axial velocity at the central axis to decelerate in order to conserve mass. At the furnace exit, the flow is again constant across the width of the fiber, as the fiber is too narrow to allow a velocity gradient.

The axial velocity gradient across the radius becomes steeper as the fiber exit velocity increases from 0.2 m/s to 3.2 m/s, as seen in Figure 4 and Figure 6, respectively. As this exit, speed increases, the center axis of the preform is affected by the motion of the exiting fiber before the surface. Thus, the fiber neckdown begins at a point closer to the furnace exit and has a steeper slope. When the fiber is drawn slower, the exiting fiber motion is more evenly distributed across the radius, causing a more gradual neckdown.

The axial velocity changes significantly with the fiber radius, as seen in Figures 4-6, making the one-dimensional assumption that the axial velocity is constant across the radius inaccurate for our purposes. Including this velocity gradient in an analysis affects the predicted neckdown profile since the laws of mass and momentum conservation link the values of radius with the axial and radial velocities. Because of this variation of axial velocity with radius, it can be concluded that the most accurate prediction of the neckdown profile must be considered in two dimensions. The one-dimensional analyses do provide qualitative insight into the fiber drawing process. However, for the most exact knowledge of the procedure, the neckdown process necessitates an entire two-dimensional model.

CONCLUDING REMARKS

The velocities and neckdown profile within an optical fiber during the drawing process were examined. An analytical model that included gravity effects was established to solve the mass and momentum conservation laws for a two-dimensional axisymmetric cylinder of varying radius. The study concluded that higher drawing speeds cause a steeper neckdown region, and the two-dimensional model predicted a more accurate neckdown profile. Furthermore, an investigation of the axial velocity contours in the fiber showed a large gradient across the radius, determining that the one-dimensional assumptions of previous models are not applicable when exact information on the glass fiber neckdown process is necessary.

ACKNOWLEDGMENTS

This research was supported by the National Science Foundation under grant CTS-9157278.

REFERENCES

1. H. Papamichael and I. N. Miaoulis, Glass Tech. **33** (4), 136-140 (1992); **32** (3), 102-108 (1991); J .Mater. Res. **6**, 159-167 (1991); ASME paper EEP-11, American Society of Mechanical Engineers Winter Annual Meeting, Dallas Texas, (1990).
2. L. R. Glicksman, Glass Tech. **9** (5), 131-138 (1968); Journal of Basic Engineering, Transactions of the ASME , 343-354 (1968).
3. R. Maddison, and P. W. McMillan, Glass Tech. **19**, 127-129 (1978).
4. S. M. Oh, Am. Cer. Soc. Bull. **58**, 1108-1110 (1979).
5. O. L. Anderson, J. Appl. Phys. **29**, 9-12 (1958).
6. S. Bateson, J. Appl. Phys. **29**, 13-21 (1958).
7. D. E. Bourne, and D. G. Elliston, Int. J. Heat Mass Transfer **13**, 583-593 (1970).
8. W. M. Kays, Convective heat and mass transfer (McGraw-Hill, New York, 1966).
9. D. E. Bourne and H. Dixon, Int. J. Heat Mass Transfer **16**, 985-990 (1971).
10. L. O. J. Crane, Zeitschrift fur angewandte Mathematic und Physik **21**, 645-647 (1970).
11. C. Y. Wang, Physics of Fluids **27**, 1915-1917 (1984).
12. C. D. S. Devi, H. S. Takhar and G. Nath, Int. J. Heat Mass Transfer **29**, 1996-1999 (1986).
13. K. N. Lakshmisha, S. Venkateswaran and G. Nath, J. of Heat Transfer **110**, 590-595 (1988).
14. B. K. Dutta, P. Roy and A. S. Gupta, Int. Comm. Heat and Mass Transfer **12**, 89-94 (1985).
15. M. Char and C. Chen, Int. J. Heat Mass Transfer **31**, 917-921 (1988).
16. S. N. Bhattacharrya and A. S. Gupta, Quarterly of Appl. Mathematics **XLIII**, 359-367 (1985).
17. B. K. Dutta and A. S. Gupta, Ind. Eng. Chem. Res. **26**, 333-336 (1987).
18. I. T. Drummond and W. Munch,J. Fluid Mech. **215**, 45-59 (1990).
19. P. S. Gupta and A. S. Gupta, The Canadian J. of Chemical Engineering **55**, 744-746 (1977).
20. B. K. Dutta, Zeitschrift fur angewandte Mathematic Mech. **68**, 231-236 (1988).
21. C. Chen and M. Char, J. Mathematical Analysis and Applications **135**, 568-580 (1988).
22. F. T. Geyling and G. M. Homsy, Glass Tech. **21** (2), 95-102 (1980).
23. R. E. Jaeger, A. D. Pearson, J. C. Williams, and H. M. Presby, Optical Fiber Telecommunications (ed. by Miller, S. E. & Chynoweth, A. G., Academic Press, NY, 263-298, 1979).
24. R. E. Jaeger, Fiber Optics: Advances in Research and Development (ed. by Bendow, B. & Mitra, S. S., Plenum Press, 33-53, 1979).
25. S. R. Nagel, Optical Fiber Telecommunications II, 121-215 (1988).
26. D. H. Smithgall, Tech. Digest, Top. Digest, Opt. Fiber Transmission, WF4 , 70-71 (1979).
27. B. Caswell and M. Viriyayuthakorn, J. of Non-Newtonian Fluid Mech. **12**, 13-29 (1983).
28. G. Manfré, Glass Tech. **10** (4), 99-106 (1969).
29. G. A. Brown, Fiber Optics: Advances in Research and Development (ed. by Bendow, B. & Mitra, S. S., Plenum Press, 55-76, 1979).
30. F. T. Geyling, The Bell System Technical Journal **55** (8), 1011-1056 (1976).
31. S. Kase and T. Matsuo, Journal of Polymer Science: Part A **3**, 2541-2554 (1965); Journal of Polymer Science **11**, 251-287 (1967).
32. U. C. Paek and R. B. Runk, Journal of Applied Physics **49** (8), 4417-4422 (1978).
33. M. R. Myers, AIChE Journal **35** (4), 592-602 (1989).
34. V. N. Vasiljev, G. N. Dulnev, and V. D. Naumchic, Glass Tech. **30**, (2), 83-90 (1989).
35. H. R. Clark, and M. Viriyayuthakorn, J. of Lightwave Tech. **LT-4** (8), 1039-1047 (1986).
36. H. Papamichael, Thermal Analysis of the Drawing Process of Optical Fibers: the Cooling Region, Master's Thesis, Tufts University (1990).
37. S. V. Patankar Numerical Heat Transfer and Fluid Flow, (McGraw-Hill, 1980).

STRAIN RELAXATION AND OXIDE FORMATION ON ANNEALED W/C MULTILAYERS

J.F. Geisz*, Y.H. Phang**, T.F. Kuech*, M.G. Lagally**, F. Cardone***, and R.M. Potemski***
*Department of Chemical Engineering, **Material Science Program, University of Wisconsin, Madison, WI 53706, ***T.J. Watson Research Center, IBM, Yorktown Heights, NY 10598

ABSTRACT

Tungsten-carbon (W/C) multilayer structures are used as X-ray mirrors and other optical elements. The optical properties of such elements are highly sensitive to changes in strain due to thermal processing. Sensitive curvature measurements were performed on 40Å period W/C multilayer structures on Si substrates using a two beam laser reflection technique. A compressive stress of approximately 1530 MPa was measured in these sputtered multilayer films. Thermal annealing to 500°C in air and under vacuum resulted in very little strain relaxation in the multilayers but X-ray diffraction data show a slight increase of the multilayer period. Significant strain relaxation, though, was observed when a 400Å W buffer layer was included. Thermal annealing of these samples to 400-500°C resulted in large strain relaxation due to the formation of α-W crystals in the buffer layer. Moderate oxide formation on air annealed samples as measured by SIMS was shown not to be a dominant mechanism of strain relaxation.

INTRODUCTION

Thin multilayer films of alternating high and low optical density materials are being used as artificial Bragg diffractors in optical elements such as X-ray mirrors. W/C multilayers on polished Si substrates are commonly used for such X-ray mirrors. Macroscopic bending of these structures due to stress in the thin films may have a considerable effect on their optical properties. Stability of these mirrors under common thermal processing is therefore of considerable interest.

Stress in sputtered films has been found to depend on sputtering conditions [1]. Sputtered tungsten films in particular have been observed with 500 MPa tensile to 1600 MPa compressive stress depending on substrate temperature [2]. In this study, we report measurements of the stress in sputtered W/C multilayers by a wafer curvature measurement technique. The relaxation of the as-grown strain upon thermal annealing to 400-500°C is also studied both with and without oxide formation. The effects of oxidation, crystal growth, and compound formation in multilayers and in various thick buffer layers are considered to explain the mechanism of strain relaxation.

EXPERIMENTAL

Multilayer films were deposited by dc magnetron sputtering with 5 mTorr Ar sputtering gas. The substrates were not intentionally heated. The base pressure in the vacuum system was 3×10^{-7} Torr. In most samples, a 400Å buffer layer of W or C was first deposited on a 20 mil thick 3 inch diameter (100) Si substrate. A multilayer consisting of 20 layers of 40Å nominal period (20Å C and 20Å W) was then deposited on the buffer layer. The high quality of these multilayers has previously been exhibited by their high reflectivity and low interfacial roughness

215

[3, 4]. Annealing studies were carried out on samples that were cleaved from the same wafer.

Curvature measurements were made by a two beam laser reflection technique [5] referred to as SSIOD (surface-stress-induced optical deflection.) The apparatus is shown in figure 1. An optically chopped HeNe laser beam was split into two parallel beams 7mm apart and reflected off the sample normal. The samples were not clamped down to prevent added strain. The divergence of the two reflected beams was measured by their separation at a distance of 1.93m away using a quadrant photodiode and lockin amplifier [6]. The curvature, $1/R$, is given by

$$\frac{1}{R} = \frac{\Delta x - \Delta x_{\text{flat}}}{2Dl} \tag{1}$$

where l is the separation of the center of the beams at the sample, Δx is the separation of the center of the beams at the detector, and D is the distance from the sample to the detector. An optical mirror on a thick quartz substrate was used as an absolute reference of curvature to find Δx_{flat}.

The average stress in the multilayer film may than be calculated by Stoney's formula [7],

$$\sigma_f = \frac{Y_s t_s^2}{6 t_f}\left(\frac{1}{R_1} - \frac{1}{R_2}\right) \tag{2}$$

where t_f and t_s are the film and substrate layer thicknesses, Y_s is the biaxial modulus given by $Y_s = C_{11} + C_{12} - 2C_{12}^2/C_{11}$ in (100) oriented cubic single crystal substrate, and $1/R_1$ and $1/R_2$ are the sample curvatures before and after multilayer deposition respectively.

Several curvature measurements were made of the samples before and after annealing using the apparatus pictured in figure 1. In one case, the measurements were also made before multilayer deposition. Curvature measurements were made consistently at the same spot on a sample since the wafers showed significant macroscopic non-uniformities in bowing.

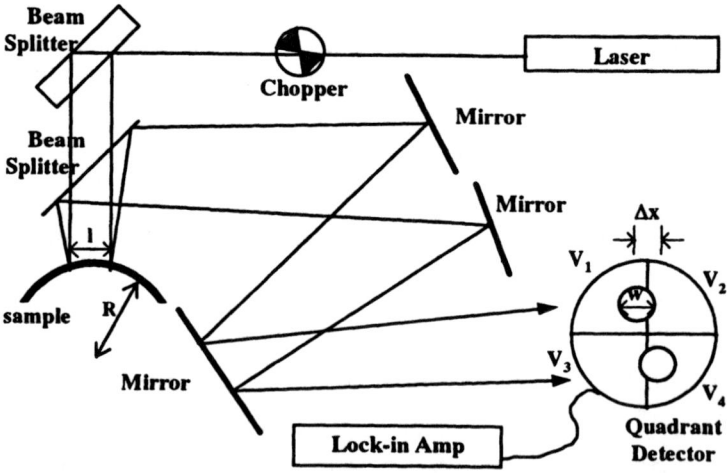

Figure 1: SSIOD two laser beam deflection curvature measurement apparatus

Samples were annealed in either vacuum or air environments for 1 hour. Resistive heating was used in a vacuum chamber at about 4×10^{-5} Torr. Partial pressures of water at 1×10^{-5} Torr and oxygen at 7×10^{-7} Torr were found to be present by a mass spectrometer. A small quartz tube furnace was used to anneal samples under a flow of nitrogen found to contain approximately 20% oxygen.

X-ray data was collected using a Nicolet Polycrystalline X-ray diffractometer. Cu K-α radiation was used with 0.05 mm entrance and exit slits. Thin Ni filters and a graphite monochromator were used at the detector for maximum resolution. A 2Θ resolution of 0.008° was achieved. Intensities were normalized by the full undiffracted intensity.

Composition depth profiling was performed using Secondary Ion Mass Spectroscopy (SIMS.) An oxygen sputter beam was used to analyze for C, W, and Si, while a Cs sputter beam was used for C, W, and O.

RESULTS AND DISCUSSION

Orthogonal strain measurements taken across a 3 inch Si wafer before multilayer deposition showed considerable non-uniformity in curvature. The wafer was found to have an average convex curvature, $1/R$, of 5.5 ± 15.1 km^{-1}. Measurements taken after the deposition of the W/C multilayer without a buffer layer revealed a fairly uniform increase in convex curvature of 15.8 ± 2.8 km^{-1} corresponding to an average compressive stress of 1530 MPa by equation 2. Tungsten films sputtered under similar conditions were also found to be under compressive stress on the order of 1000 MPa [2].

Curvature measurements were taken on multilayer samples with tungsten, carbon, and no buffer layers both before and after annealing to 400 or 500°C. Some samples were annealed in air while others were annealed under vacuum. The changes in curvature, $\Delta 1/R$, tabulated in table I indicate significant stress relaxation in all the samples with a W buffer layer, while stress relaxation in samples with C and no buffer layers was within the error of the measurement.

Table I: Effects of Annealing W/C Multilayers

		X-ray			SIMS	SSIOD	
buffer	anneal	d (Å)	σ (Å)	t_{oxide} (Å)	t_{oxide} (Å)	$1/R_{initial}$ (km^{-1})	$\Delta 1/R$ (km^{-1})
W	none	40.1	3.5	20	18	-	-
W	400°C air	40.9	3.5	100	-	-29.8	7.4
W	400°C air	40.9	3.2	20	-	-39.7	20.7
W	500°C vac	40.9	3.2	20	41	-24.3	17.7
W	500°C air	36.7	3.2	300	285	-27.0	19.3
C	none	39.5	3.6	20	-	-	-
C	500°C vac	40.2	3.6	20	-	-19.2	0.7
C	400°C air	37.0	3.7	80	-	-31.0	1.6
none	500°C vac	-	-	-	-	-17.4	0.7

Unannealed W/C multilayers were found to contain a thin oxide region (20 Å) by SIMS data. Annealing W/C multilayers in air has previously been found to result in a loss of composition modulation in a surface oxide region containing crystalline WO_3 and segregated microcrystalline graphite particles [8]. Most of the carbon is believed to effuse out of this oxide in the form of CO. The SIMS profile pictured in figure 2 shows the increased oxygen content and decreased carbon content within the oxide region in the sample annealed in air to 500°C. The thickness of the oxide region in annealed samples was measured by SIMS and by modeling low angle X-ray data using the theory of Peterson *et al.* [9]. These thicknesses are tabulated in table I. The modeled structure consisted of a W/C multilayer with period, d, and average interfacial roughness, σ, below a WO_3 oxide of thickness, t_{oxide}. The number of periods remaining in the multilayer was calculated by conservation of W mass. The shape of the oxide peak marked in figure 3 was highly sensitive to the thickness of the oxide.

Significant oxidation of W/C multilayers was observed only for samples annealed in air, where the oxide thickness increased with anneal temperature. Changes in sample curvature, though, in no way were correlated to this oxide thickness. The multilayer structure below the surface oxide did not roughen upon annealing to 400-500°C but the multilayer period increased slightly. This increase in multilayer period has been attributed to expansion of the amorphous carbon layers [10-12]. The absence of interface roughening indicated that interlayer diffusion and tungsten-carbide formation was minimal at these annealing temperatures. Growth of α-W and WC crystalline phases has been observed in W/C multilayers with carbon layers > 10Å thick only at much higher annealing temperatures [13].

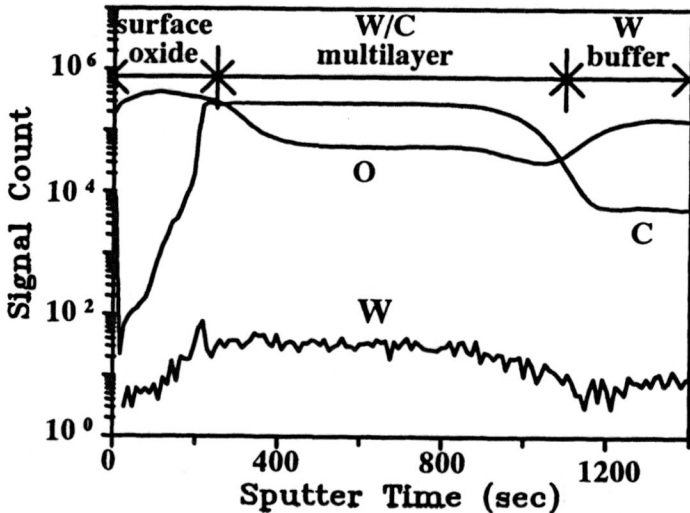

Figure 2: SIMS data for sample with a W buffer layer annealed to 500°C in air under a Cs sputter beam

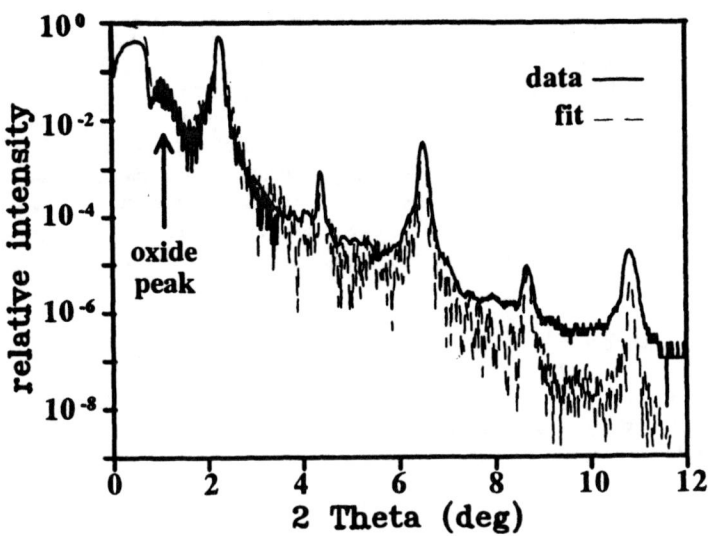

Figure 3: Low angle Θ-2Θ X-ray data from a sample with a W buffer layer annealed to 400°C in air. The theory of Peterson *et al.* [9] was used to fit the data.

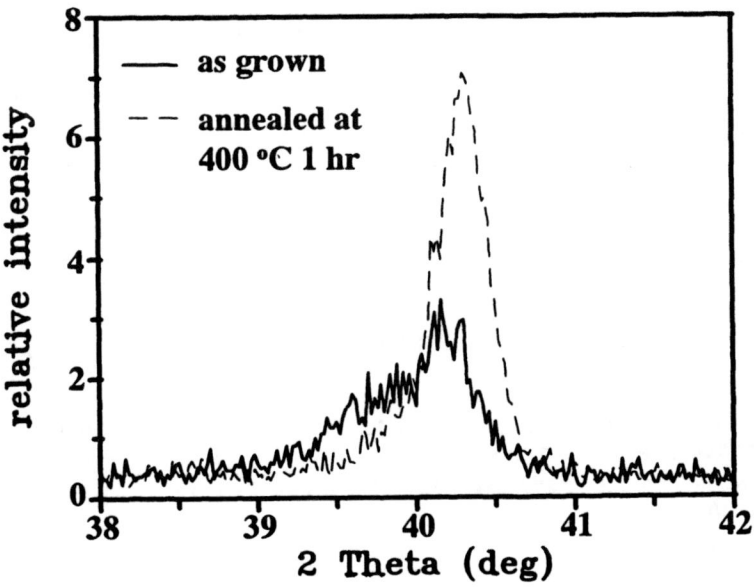

Figure 4: Θ-2Θ X-ray data showing the (110) α-W diffraction peak in a multilayer sample with a W buffer layer.

The actual mechanism of the stress relaxation observed appears to be crystal growth in the 400Å W buffer layer. The intensity of the (110) α-W diffraction peak in the Θ-2Θ X-ray data shown in figure 4 increased significantly upon annealing to only 400°C in samples with a W buffer layer. Such crystal growth in sputtered W layers upon annealing has been noted previously [14], but only in films above a critical thickness of about 40 Å [15]. The stress in sputtered W films has been found to be highly dependent on grain size [2]. The large compressive stress in the sputtered W buffer layer appears to be relieved by the contraction due to BCC α-W crystal growth. While these crystal grains may easily grow in the thick W buffer layer, extensive crystal growth in the thin W regions is not expected.

ACKNOWLEDGMENTS

The authors would like to thank Jim MacKay and Don Savage for helpful discussions. Generous equipment donations were provided by IBM and additional funding provided by the NSF through the Materials Research Group (DMR-9121074)

REFERENCES

1. H. Windischmann, J. Vac. Sci. Tech. A **9**, 2431 (1991).
2. R. Sun, T. Tisone, and P. Cruzan, J. Appl. Phys. **46**, 112 (1975).
3. D. Savage, J. Kleiner, N. Schimke, Y. Phang, T. Jankowski, J. Jacobs, R. Kariotis, and M. Lagally, J. Appl. Phys. **69**, 1411 (1991).
4. D. Savage, N. Schimke, Y. Phang, and M. Lagally, J. Appl. Phys. **71**, 3283 (1992).
5. A.J. Schell-Sorokin and R.M. Tromp, Phys. Rev. Lett. **64**, 1039 (1990).
6. J.F. Geisz, T.F. Kuech, M.G. Lagally, F. Cardone, and R.M. Potenski, J. Appl. Phys., scheduled February, 1994
7   R.R. Kola, D.L. Windt, W.K. Waskiewicz, B.E. Weir, R. Hull, G.K. Celler, and C.A. Volkert, Appl. Phys. Lett. **60**, 3120 (1992).
8. S. Chao, D. Pawlik, J. Gonzalez-Hernandez, Q. Wang, and D. Allred, Solid St. Comm. **79**, 205 (1991).
9. B.G. Peterson, L.V. Knight, and H.B. Pew, SPIE **563**, 328 (1985).
10. Z. Jiang, X. Jiang, W. Liu, and Z. Wu, J. Appl. Phys. **65**, 196 (1989).
11. X. Jiang and D. Xian, Appl. Phys. Lett. **57**, 2549 (1990).
12. V. Dupuis, M. Ravet, C. Tete, and M. Piecuch, J. Appl. Phys. **68**, 5146 (1990).
13. J. Gonzalez-Hernandez, B. Chao, D. Pawlik, D. Allred, and Q. Wang, J. Vac. Sci. Technol. A **10**, 145 (1992).
14. Y. Takagi, S.A. Flessa, K.L. Hart, D.A. Pawlik, A.M. Kadin, J.L. Wood, J.E. Keem, and J.E. Tyler, SPIE **563**, 66 (1985).
15. G.M. Lamble, S.M. Heald, D.E. Sayers, E. Ziegler, and Viccaro, J. Appl. Phys **65**, 4250 (1989).

PART III

Crystal Nucleation

TRANSIENT NUCLEATION IN DEVITRIFICATION

A.L. GREER
University of Cambridge, Department of Materials Science and Metallurgy, Pembroke Street,
Cambridge CB2 3QZ, U.K.

ABSTRACT

A review is given of transient crystal nucleation in glassy or amorphous solids. The types of behaviour are surveyed. It is shown that the kinetics can be quantitatively modelled and that the matching of experiment and theory provides an important test of the classical theory. Examples are considered of homogeneous nucleation (affecting glass formation), heterogeneous nucleation, and nucleation at an interphase interface. While the emphasis is on transient effects of the kind implicit in the classical theory, it is shown that transients can arise for other reasons as well, thus potentially complicating the interpretation of experiments.

1. INTRODUCTION

In the classical theory of nucleation attention is often focused on the critical clusters of the nucleating phase, i.e., those clusters which are in unstable equilibrium with the parent phase. Clusters larger than the critical size can grow spontaneously, and the nucleation frequency at any instant can be taken to be proportional to the population of critical clusters. But a further important element of the theory is that all clusters of the nucleating phase (pre-critical, critical and post-critical) are assembled molecule-by-molecule, and do not arise by single large fluctuations. During nucleation, then, there exists a distribution of cluster sizes; the nucleation frequency can be in steady state only when the steady state size distribution can be established. Attaining the steady state cluster size distribution cannot be instantaneous, and for this reason transient effects arise in nucleation.

In many cases, however, such transient effects are negligible, either because transformation of the sample occurs from only a few nucleation events or because the nucleation kinetics themselves are so rapid. A good example is crystal nucleation in the solidification of liquid metals. This typically occurs at low undercooling on a few heterogeneous sites, which cause transformation of the entire sample. Even if nucleation kinetics were important in the overall transformation, it is estimated that steady state nucleation (homogeneous or heterogeneous) would be established after (negligible) times of the order of 1 μs given the large atomic mobilities near the melting point.

In devitrification, on the other hand, transient nucleation effects are particularly important. Slow crystal growth means that nucleation kinetics are important in the transformation. Large numbers of nuclei and the ability to interrupt the transformation at any degree of transformation mean that the nucleation kinetics can be accurately determined. Furthermore, the kinetics can be determined under isothermal conditions. Low atomic mobility in the glassy or amorphous state can lead to very substantial times before steady state nucleation is established. The simplest type of transient nucleation effect to consider is that of an effective time-lag before any nucleation is observed. If the glassy or amorphous state were to be produced with no (or a negligible population of) clusters of any kind, then on subsequent annealing the initial nucleation frequency would be zero. Nucleation would then develop and eventually the steady state frequency would be established. At long times the population of clusters greater than the critical size (i.e., the number of nuclei, χ) is asymptotic to:

$$\chi = I^s(t - \theta), \qquad (1)$$

where I^s is the steady state nucleation frequency, t is the total annealing time at the nucleation temperature and θ is the effective time lag. It is unlikely that a glassy or amorphous sample would have no clusters initially. However, the initial population of clusters can be negligible in its effect on transformation kinetics if, for example, the glass was produced by quenching at a rate substantially greater (by ~2 orders of magnitude) than the critical quench rate for glass formation. For all the main categories of amorphous or glassy solids, transient nucleation effects of the kind considered can be observed. Good examples are found for a silicate glass [1], an amorphous semiconductor [2] and a metallic glass [3]. Of course, an effective time lag not of the maximum extent may still be found even when there is a significant population of initial clusters; this would complicate a quantitative interpretation of the time lag.

In what follows, various aspects will be considered of the transient nucleation (both homogeneous and heterogeneous) which arises from the evolution of the cluster size distribution. It will also be pointed out, however, that there are additional possible origins of transient effects.

2. TRANSIENT NUCLEATION KINETICS

2.1 Numerical Modelling

The basic kinetic model within the classical theory is illustrated schematically by Figure 1. The cluster size distribution develops through bimolecular reactions in which clusters gain or lose one molecule at a time, thereby being converted to clusters of neighbouring size. The development can be modelled by a series of coupled differential equations. The populations $N_{n,t}$ of clusters with n molecules at time t vary according to:

$$\frac{dN_{n,t}}{dt} = k_{n-1}^+ N_{n-1,t} - \left(k_n^- + k_n^+\right)N_{n,t} + k_{n+1}^- N_{n+1,t} \qquad (2)$$

where the rate constants k are dependent on molecular mobility and on the size dependence of the work of cluster formation (details, for homogeneous nucleation, in [4]). A variety of methods have been proposed for the solution of these linked equations [4,5]. Perhaps the least elegant method, but certainly the most powerful and versatile, is the numerical simulation of the evolution of the cluster populations. In this approach, discrete time intervals are used to transform the equations into:

$$N_{n,t+\delta t} = N_{n,t} + \delta t\left[-\left(k_n^+ + k_n^-\right)N_{n,t} + k_{n-1}^+ N_{n-1,t} + k_{n+1}^- N_{n+1,t}\right] \qquad (3)$$

and details of how to select a suitable time interval δt are given in [4]. Figure 2 shows the ideal type of transient nucleation behaviour given by the numerical modelling for the isothermal case in which there are initially no clusters. The transient nucleation observations in [1], [2] and [3] are of this form.

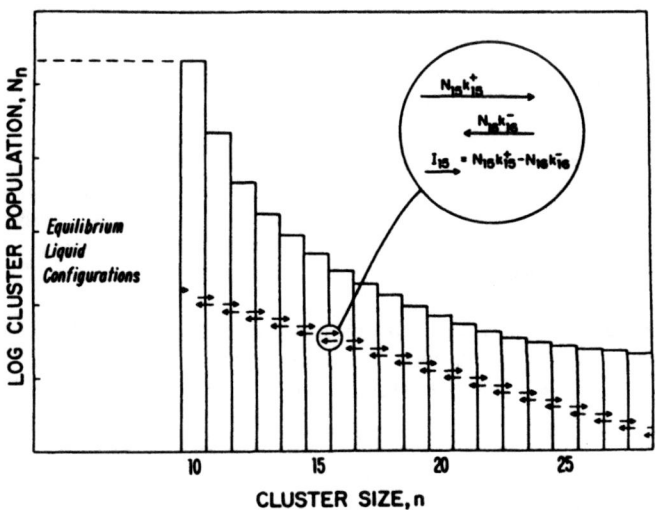

Figure 1: A schematic illustration of the cluster size distribution during nucleation. Apart from very small clusters, a steady state distribution will eventually be established through reactions in which clusters gain or lose single molecules.

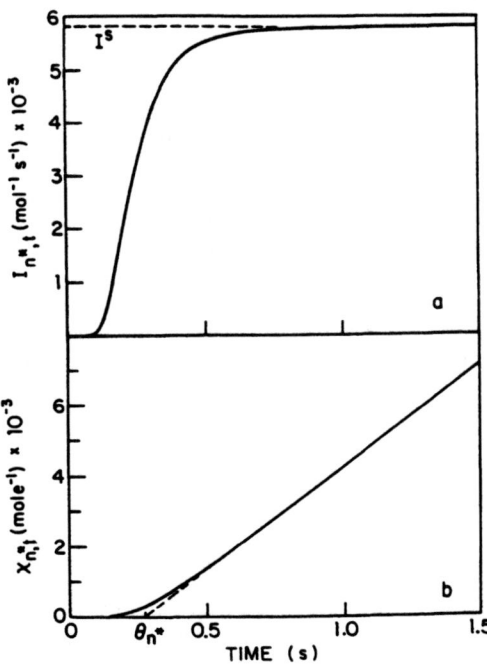

Figure 2: An example of transient nucleation behaviour as calculated using a numerical simulation, as described in [4]. The evolution with time of nucleation frequency is shown in (a) and that of the total number of nuclei (clusters greater than the critical size) is shown in (b). Isothermal conditions are assumed, with no clusters present at the start of the anneal. (After [4])

2.2 Analytical Approximations

For this simplest case (i.e., isothermal with no clusters initially), there have been a number of attempts to find good analytical descriptions of the nucleation kinetics. Early approaches are reviewed in [4]. The general form of the behaviour shown in Figure 2 is well fitted by a number of expressions of similar form, the most widely quoted being that of Kashchiev [6]. Kelton et al. [4] made a comparison of the analytical expressions with a numerical simulation conducted as in §2.1. They concluded that Kashchiev's expression gave the best fit. It seems, however, that this agreement may have been fortuitous, arising for the particular materials parameters used in the comparison. The most detailed, and recent, treatments of this problem have been by Wu [5,7] and Shneidman [8].

Although acceptable analytical approximations can be found for this simple case, more complex cases (for example, non-isothermal kinetics) clearly require a numerical approach. There can also be problems relating the measured nucleation data to the kinetics at the critical size [9].

3. VERIFICATION OF THE KINETIC MODEL

The existence of transient effects of the general form predicted by the classical theory itself supports the basic kinetic model. In this section, two further examples of analysis are considered which can be used to test the classical theory.

3.1 Ion Irradiation Effects on Crystal Nucleation in Amorphous Silicon

Amorphous silicon is tetrahedrally bonded, and is believed to form from liquid silicon (which has metallic bonding) by a first-order transformation. It is therefore not a glass, but an amorphous solid. However it can show crystallization behaviour analogous to that of conventional glasses. In a sample which is partially crystalline, it is known that ion irradiation causes amorphization at the crystal-amorphous interface. The interface migrates at a well defined rate dependent on the temperature and ion flux. If this behaviour were to extend to small crystalline clusters, then the clusters would shrink under irradiation and their size distribution would be altered. Im, Shin and Atwater [10] have shown that this is the case. As would be expected, irradiation leads to an increased time lag for nucleation. Further work [11] has shown that quantitative modelling of the effects is possible. Indeed, the effects of the irradiation can be used to control crystallization kinetics, and this may be useful in obtaining a grain size which is not only larger, but also more uniform. The ability to interpret the experimental results in terms of the alteration of the cluster size distribution provides strong evidence for the kinetic basis of the classical theory.

3.2 Multi-Stage Annealing in Lithium Disilicate

Lithium disilicate forms a glass which is believed to undergo polymorphic crystallization (i.e., with crystal and glass having the same composition throughout) initiated by homogeneous nucleation. The nucleation kinetics have been extensively analysed, with the first detailed examination of transient effects by James [1]. Kelton and Greer [12,13] used the classical theory with the numerical simulation outlined in §2.1 to fit data on (temperature-dependent) steady state nucleation frequencies and effective time lags collected from several authors. In this fitting there are in principle three main materials parameters, all of which are temperature-dependent. They are: the free energy driving force for crystallization, the crystal-glass interfacial energy, and the molecular mobility. The crystallization free energy is known from independent measurements and so was not an adjustable parameter. The molecular mobility is not known but was presumed to scale inversely with the viscosity, which is known. Thus there was just one (temperature-independent) adjustable parameter (the constant of proportionality) connected with the molecular mobility. The interfacial energy is completely unknown, and was treated as an independently adjustable parameter at each temperature. As detailed in [13], a consistent fitting to almost all of the data could be achieved, with reasonable values being obtained for both the constant of proportionality for mobility and the temperature-dependent interfacial energy. The interfacial energy σ (in J m^{-2}) was found to vary linearly with absolute temperature (in K), according to:

$$\sigma = 0.094 + (7 \times 10^{-5})T. \tag{4}$$

This type of variation reflects the enthalpic contribution (first term on right-hand side of equation) and the entropic contribution (second term) to the interfacial free energy. The data which could not be fitted with interfacial energies given by eq. (4) were for anneals below 700 K. At such low temperatures the critical nucleus would have less than 20 formula units, and there could be a number of reasons for the breakdown of the theory in this region. Using the materials parameters determined from the fitting it was possible to estimate the populations of clusters which would be produced during the production of the glassy samples. It was verified that these populations would have a negligible effect on the experimentally observed transient nucleation behaviour. It was thus valid to interpret the observations as corresponding to a complete absence of initial clusters, and the fitting was internally consistent.

The fitting suggests that the classical theory can provide a reasonable interpretation of the observed data. However, the theory can be tested more comprehensively. Once the fitting parameters have been established, they can be used (without further adjustment) to predict the results of different (and in particular, more complicated) anneals. The predictions can be tested against data on multistage anneals from the work of Fokin et al. [14]. An example from [13] is shown in Figure 3. This shows the effect of annealing first at a lower temperature and then measuring the nucleation kinetics during an anneal at a higher temperature. Note that for the data in Figure 3, application of eq. (1) would give an effective time-lag which is strongly negative. This arises because the pre-anneal at a low temperature generates a large population of clusters which can grow on subsequent annealing. Such effects clearly show that pre-existing clusters can have a large effect on transient nucleation kinetics, and that there is not a single type of relaxation, for example of the kind shown in Figure 2. In some cases, the pre-existing clusters can give rise to a marked maximum in the nucleation frequency during annealing. As detailed in [13], this behaviour, and indeed all the effects of multistage anneals (involving anneals at T > 700 K) can be quantitatively fitted using the numerical simulation with the materials parameters as determined from single isothermal annealing treatments. This fitting provides clear evidence of the usefulness of the classical theory.

4. HOMOGENEOUS NUCLEATION AND GLASS FORMATION

In some cases it can be the presence of heterogeneous nuclei which inhibits glass formation. However in general for glass formation, homogeneous nucleation is important in determining the critical cooling rate. Slow crystal growth limits the effect of heterogeneous nuclei. The application of non-isothermal numerical modelling relevant during cooling to form the glass is reported in detail in [15]. It is clear that transient effects lead to a very strong dependence on quench rate of the number of nuclei produced during the quench. However, the effects on the critical cooling rate for glass formation are not so straightforward. An arbitrary criterion for glass formation is often adopted, that a fraction of less than one part per million should crystallize during the quench. For systems such as lithium disilicate for which the critical cooling rate for glass formation is modelled to be of the order of 1 K s^{-1}, the volume fraction transformed during the quench (for rates close to the critical rate) is not at all dependent on transient nucleation. This is because the fraction transformed is dominated by crystals which nucleated at an early stage of the quench (i.e., at a high temperature at which molecular mobility is high and transient effects are negligible). On the other hand, materials which require a very high quench rate for glass formation (such as most metallic glasses) are expected to have their critical cooling rate substantially lessened by transient nucleation effects. For example, in the modelling in [15], it is estimated that the critical cooling rate for glass formation in $Au_{81}Si_{19}$ (at.%) would be ~10^8 K s^{-1} if the nucleation were in steady state throughout, but only 10^5 K s^{-1} when transient effects are considered. This means that glass formation in such a system is in fact made possible by normal processing routes only by the inhibition of crystal nucleation through transient effects. The procedure for estimating whether or not transient effects are significant in glass formability is described in [15].

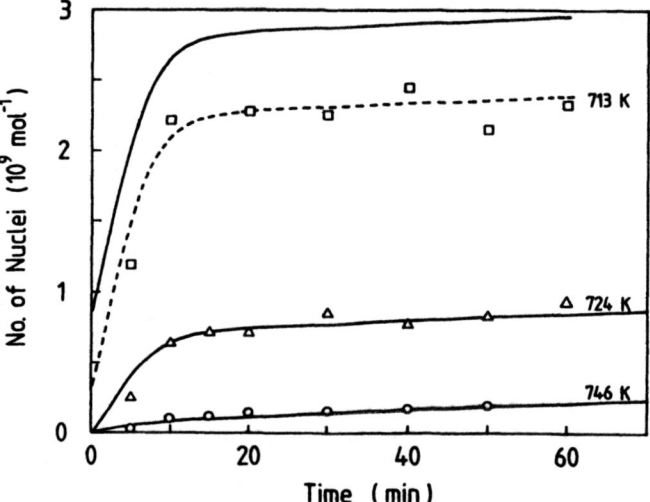

Figure 3: Number of nuclei produced in lithium disilicate glass as a function of annealing time at 758 K after a pre-anneal at 713, 724 or 746 K. The data points are from Fokin et al. [14], and the solid lines from a numerical simulation using materials parameters obtained by fitting simple nucleation data [13]. The dashed line shows a modification in which the number of nuclei at the end of the 713 K pre-anneal is altered to match the experimental value. This modification does not involve the values of the materials parameters obtained by fitting, but was necessary because of a deviation in the nucleation rate for the particular set of samples [13]. (After [13])

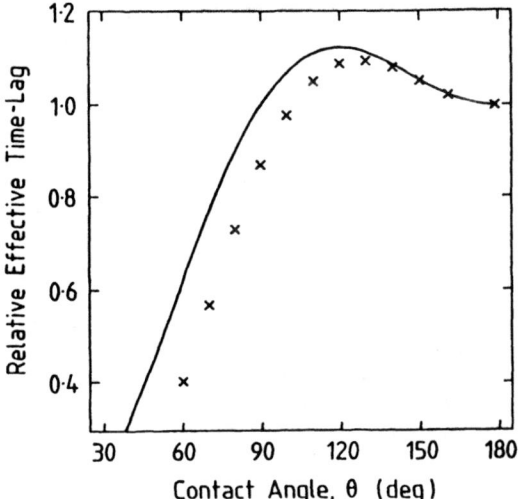

Figure 4: The effective time-lag for transient heterogeneous nucleation (×) as a function of contact angle θ, evaluated using a numerical simulation [16]. The values are relative to the value at θ = 180°. The solid line shows the ratio of cluster volume to crystal-liquid interfacial area as a function of θ, with values again relative to that at θ = 180°. (After [16])

5. HETEROGENEOUS NUCLEATION

So far consideration has been of homogeneous nucleation. In this section some consideration is given to heterogeneous nucleation. The adaptation of the kinetic model is considered in §5.1, and an example of a recent study relevant to glasses is given in §5.2.

5.1 Nucleation Kinetics in the Spherical Cap Model

Analogous to the consideration of a spherical cluster in the simplest version of homogeneous nucleation theory, the heterogeneous case uses a spherical cap. The contact angle θ is determined by the interfacial energies, $\theta = 0°$ corresponding to perfect wetting and the absence of a nucleation barrier, and $\theta = 180°$ being essentially homogeneous nucleation. Under particular conditions (i.e., at a particular temperature) the crystallization free energy and the crystal-glass interfacial energy will each be the same in the homogeneous and heterogeneous cases. It follows then that the radius of curvature of the critical nucleus will also be the same in each case. However, the volume of the equivalent heterogeneous nucleus is clearly less and consequently the work of formation of the critical nucleus is less. As a result heterogeneous nucleation can generally become significant at smaller undercooling than homogeneous nucleation. With appropriate adaptation of geometrical terms, the numerical simulation in §2.1 can be applied to the heterogeneous case [16]. Figure 4 shows how the effective time lag is a function of θ. It is found that θ scales quite closely with the ratio of the volume of the spherical cap to its curved interface area [17]. (The reason why the fit is not perfect is discussed in [16].) This scaling has an interesting consequence. Although the work of formation is much less for a heterogeneous nucleus than for a homogeneous nucleus, the effective time lags are similar in the two cases for much of the range of contact angle, and can even be greater in the heterogeneous case. When the value of θ is low and the driving force for crystallization is high, i.e. when the number of molecules in the critical heterogeneous nucleus is low) the spherical cap shape cannot be a good description of the nucleus. This may be one case in which an alternative to the classical theory is required, as recent experimental observations also imply [18].

5.2 Heterogeneous Nucleation on Added Particles

The addition to glasses of particles which can act as substrates for heterogeneous nucleation is of considerable direct interest. For example, in the production of glass ceramics it is common to add nucleating agents to the glasses to control the final microstructure. It is also possible to use the glass as a high-viscosity analogue of the liquid in order to study nucleation on particles added to a liquid. An example of a commercial process in which particles are added to a liquid is the grain-refining of aluminium, and this has recently been studied using an Al-based metallic glass [19]. Particles of TiB_2 were incorporated in the melt-spun metallic glass $Al_{85}Y_8Ni_5Co_2$ (at.%). On annealing, crystals of α-Al nucleated and grew on the (0001) faces of the added particles. Transmission electron microscopy of the transformation at a early stage is particularly useful in enabling the nucleation sites to be identified. This shows one advantage of studying the nucleation in the glass rather than the liquid is in being able to interrupt the transformation. Detailed observation shows that between the nucleating α-Al and the TiB_2 particles is a thin layer of Al_3Ti. Crystallography is important in the nucleation; the close-packed planes and close-packed directions are parallel in all three phases. The presence of the Al_3Ti layer is significant in elucidating the mechanism of grain refinement in conventional processing. Sometimes a contact time with the melt is required before added grain-refining particles become active, and this could be associated with the time taken to form the layer. The observation that sometimes particles become less effective after long contact times could be associated with dissolution of the layer. The mechanisms of grain refinement are not of direct interest here; however, it is important to note that the intermediate layer could give rise to transient effects with an origin quite different from that considered so far. Particularly in heterogeneous nucleation, transient effects can arise from causes other than the evolution of the cluster size distribution.

5.3 Nucleation in Interfacial Reactions

Interfacial reactions have attracted much attention recently. In the formation of metal silicides at metal-silicon interfaces relevant for electronic devices, it is notable that in thin films often only one silicide forms; it is important to understand the phase selection. At metal-metal interfaces in thin films in some cases the product phase formed is amorphous; an example is the formation of amorphous Ni-Zr [20]. Such reactions are not directly relevant to devitrification. However, they do illustrate another possible origin of transient effects at interphase interfaces, and this could be significant for devitrification by reaction at an interface.

In many of the cases of interest in thin-film interfacial reactions, the driving force for reaction is very high. A simple application of classical nucleation theory often leads to the prediction that there should be no nucleation barrier. Yet there is ample evidence that nucleation barriers do exist. This apparent paradox is considered in [21] and [22]. The two phases which will undergo the reaction may start not in diffusional equilibrium with each other. On annealing, strong interdiffusional fluxes arise, and these inhibit nucleation until the reacting phases approach their equilibrium compositions in contact with each other. After the compositions have changed in this way the driving force for formation of the product phase may be much reduced. Not only may there be ultimately a significant nucleation barrier, but also there is a transient period during which there is interdiffusion but no nucleation. This, then, is a further example of a transient effect which has its origin other than in the basic cluster dynamics of classical nucleation theory.

6. NUCLEATION IN TWO-COMPONENT SYSTEMS

The numerical modelling in §2.1 is for a one-component system. The clusters can be described by a single number — the number of molecules they contain. Only very rarely have devitrification kinetics been studied in true one-component systems (the studies of crystallization of amorphous silicon, §3.1, are an example). Nonetheless the one-component analysis can be reasonably applied in multicomponent systems in which the crystallization is polymorphic. However, there are many cases in which composition changes during crystallization are significant. As yet there has been very little work on transient nucleation effects in such cases. Preliminary numerical studies have been reported for partitioning binary systems in which the composition of the clusters is considered as well as their total size [16, 23]. The difference equations of the type in eq. (3) become more complicated as each cluster can now gain or lose an atom of either type. As yet, composition changes in the remaining untransformed glass have not been accounted for; the analysis may therefore be most appropriate for the very early stages of transformation. It is found that when the two components are assigned equal mobilities at the crystal-glass interface, the overall nucleation behaviour is closely similar to that which would be obtained if only the most probable clusters (according to their energies of formation, any cluster size has a most probable composition) were considered. In such a case, the analytical treatment of Reiss [24] yields good values for the steady state nucleation frequency. However, when the two components have very different mobilities (and such an asymmetry is common in glass-forming systems, at least in alloys [25]) the Reiss analysis is not valid. Wu [26] has given an analysis of the multicomponent case, identifying the conditions under which an analytical approach is tractable. In the general case, even for two-components, a numerical treatment is necessary. Clearly much remains to be done in this area.

7. CONCLUSIONS

Transient nucleation effects can arise from the evolution of the cluster size distribution. In devitrification such effects are particularly significant because of low atomic (or molecular) mobility. When a glass is formed at a quench rate greatly exceeding the critical quench rate for glass formation, the population of crystal clusters in the initial glass may be negligible; in such a case there is a simple form of transient behaviour in which the isothermal nucleation rate rises sigmoidally from zero up to its steady state value. In this case there exist good analytical approximations for the nucleation kinetics. Transient nucleation reduces the population of clusters quenched in a glass and reduces the critical cooling rate for glass formation. The latter effect, however, is significant only if the critical cooling rate is high. A wide variety of transient nucleation behaviour is observable in the presence of quenched-in clusters. The behaviour does not show a single relaxation time and in the general case cannot be well described analytically; in such cases a numerical approach is necessary. Transient effects are important in providing a good test of classical nucleation theory, in particular of the cluster dynamics basic to the theory. Quantitative agreement between theory and experiment, even in complex cases (such as nucleation under irradiation, or multistage anneals), strongly supports the theory. Heterogeneous nucleation can be modelled in a similar way to homogeneous. Although the work of formation of critical nuclei can be much reduced, effective time lags for nucleation can be comparable with the homogeneous case. When the number of atoms in the critical nucleus (homogeneous or heterogeneous) becomes small (for example <20), it becomes difficult to apply the classical theory quantitatively. In heterogeneous nucleation there are potential sources of transient nucleation unrelated to the evolution of the cluster size distribution; these are the formation of intermediate layers and diffusional fluxes in the adjacent phases. Very limited work has been done on the quantitative modelling of nucleation kinetics in partitioning two-component systems.

REFERENCES

1. P.F. James, *Phys. Chem. Glasses* 15, 95 (1974).
2. J.S. Im and H.A. Atwater, *Appl. Phys. Lett.* 57, 1766 (1990).
3. M. Buchwitz, R. Adlwarth-Dieball and P.L. Ryder, *Acta Metall. Mater.* 41, 1885 (1993).
4. K.F. Kelton, A.L. Greer and C.V. Thompson, *J. Chem. Phys.* 79, 6261 (1983).
5. D.T. Wu, *J. Chem. Phys.* 97, 2644 (1992).
6. D. Kashchiev, *Surf. Sci.* 14, 209 (1969).
7. D.T. Wu, in Kinetics of Phase Transformations, edited by M.O. Thompson, M.J. Aziz and G.B. Stephenson (Mater. Res. Soc. Proc. 205, Pittsburgh, PA 1992) pp. 411-416.
8. V.A. Shneidman, *Sov. Phys. Tech. Phys.* 33, 1338 (1989).
9. K.F. Kelton and A.L. Greer, in Rapidly Quenched Metals, edited by S. Steeb and H. Warlimont (Elsevier, Amsterdam, 1985), p. 223.
10. J.S. Im, J.H. Shin and H.A. Atwater, *Appl. Phys. Lett.* 59, 2314 (1991).
11. C.M. Yang and H.A. Atwater, in Phase Transformations in Thin Films — Thermodynamics and Kinetics, edited by M. Atzmon, A.L. Greer, J.M.E. Harper and M.R. Libera (Mater. Res. Soc. Proc. 311, Pittsburgh, PA 1993) pp. 185-190.
12. K.F. Kelton and A.L. Greer, *Phys. Rev. B* 38, 10089 (1988).
13. A.L. Greer and K.F. Kelton, *J. Am. Ceram. Soc.* 74, 1015 (1991).
14. V.M. Fokin, V.N. Filipovich and A.M. Kalinina, *Fiz. Khim. Stekla* 2, 129 (1976), (in Russian).
15. K.F. Kelton and A.L. Greer, *J. Non-Cryst. Solids* 79, 295 (1986).
16. A.L. Greer, P.V. Evans, R.G. Hamerton, D.K. Shangguan and K.F. Kelton, *J. Cryst. Growth* 99, 38 (1990).
17. S. Toschev and I. Gutzow, *Phys. Stat. Sol.* 21, 683 (1967).
18. W.T. Kim and B. Cantor, *Acta Metall. Mater.* 40, 3339 (1992).
19. P. Schumacher and A.L. Greer, *Mater. Sci. Eng. A* (1994) in press.
20. B.M. Clemens, W.L. Johnson and R.B. Schwarz, *J. Non-Cryst. Solids* 61&62, 817 (1984).
21. C.V. Thompson, *J. Mater. Res.* 7, 367 (1992).
22. A.L. Greer, *J. Magn. Magn. Mater.* 126, 89 (1993).
23. P.V. Evans and A.L. Greer, in Principles of Solidification and Materials Processing, edited by R. Trivedi, J.A. Sekhar and J. Mazumdar (Oxford & IBH Pub. Co., New Delhi, 1989), Vol. 2, p. 741.
24. H. Reiss, *J. Chem. Phys.* 18, 840 (1950).
25. A.L. Greer, N. Karpe and J. Bøttiger, *J. Alloys Compounds* 194, 199 (1993).
26. D.T. Wu, *J. Chem. Phys.* 99, 1990 (1993).

CRYSTAL NUCLEATION IN SUBMICRON DROPLETS OF PURE ELEMENTS

LOUIS M. HOLZMAN*, THOMAS F. KELLY*,**, AND W.N.G. HITCHON*,***
*Materials Science Program, **Materials Science and Engr., ***Electrical and Computer Engr.,
University of Wisconsin, Madison, WI 53706

ABSTRACT

Liquid-to-crystal nucleation has been studied extensively through droplet experiments to locate examples of homogeneous nucleation. However, prior to this work very few examples have been found, which implies that the experiments have not been able to isolate heterogeneous nucleants in a small percentage of the droplets as is required. In this research, electrohydrodynamic atomization (EHD) is used to produce sub-micron droplets of pure elements that are largely free of heterogeneous nucleants.

Diffraction patterns of individual EHD-produced droplets are viewed to determine the fraction of crystalline droplets produced as a function of droplet radius. These results are compared to theories for surface and volume heterogeneous nucleation and for homophase nucleation. It is found that Si and Ge nucleate through either homogeneous nucleation or nucleation by homophase impurities. Nucleation results for vanadium and iron were not conclusive.

INTRODUCTION

Previous experimental research on nucleation has shown that it is very difficult to produce homogeneous nucleation. Turnbull [1, 2] established the droplet experiment as a fundamental method of studying homogeneous nucleation. The material to be studied would be broken into a large number of small droplets in order to isolate the heterogeneous nucleants that were present in a small fraction of the droplets. Only homogeneous nucleation would remain as a possible mechanism to cause the crystallization of the droplets. Turnbull studied the crystallization of micron-sized droplets suspended in a variety of solutions. He found that those in mercury benzoate or mercury laurate solutions exhibited nucleation dependent on the droplet volume [2], which is consistent with homogeneous nucleation. Miyazawa and Pound found a volume dependence for nucleation of gallium from the melt [3]. However, experiments by Perepezko, et al. [4] demonstrated that greater supercoolings than those obtained in the research above were possible, showing that conditions allowing homogeneous nucleation had probably not been met.

In the experiments discussed above, it is not possible to differentiate qualitatively between volume heterogeneous and homogeneous nucleation. Homogeneous nucleation depends on the droplet cooling rate as well as the volume of the droplet, but the cooling rate of droplets of all sizes was the same. However, if the cooling rate is dependent upon droplet radius, homogeneous nucleation will result in an $R^3 * R^n$ dependence, where n represents the dependence of cooling time on radius. In such cases, homogeneous nucleation and volume

Mat. Res. Soc. Symp. Proc. Vol. 321. ©1994 Materials Research Society

heterogeneous nucleation can be qualitatively differentiated. Drehman and Turnbull [5] studied a case in which the cooling time was proportional to $R^{1.6}$, so homogeneous nucleation was expected to show an $R^{4.6}$ dependence, but surface heterogeneous nucleation was found in this case.

EHD atomization has been used to produce amorphous droplets of several pure elements [6]. A droplet experiment using EHD atomization [7, 8] offers advantages over previous nucleation experiments. First, the droplets produced through EHD atomization are very small, which isolates the heterogeneous nucleants to a smaller fraction of the droplets. EHD droplets are in the 10 - 100 nm range, which isolates the nucleants over a thousand times better than most previous experiments. The processing takes place in high vacuum, so surface nucleants are also minimized. Finally, EHD droplets cool radiatively, so the cooling rate is inversely proportional to droplet radius. Therefore, (steady-state) homogeneous nucleation will increase as R^4 so homogeneous and volume heterogeneous nucleation can be differentiated.

Turnbull discussed the possibility that the nucleation in his experiments with mercury could have been caused by homophase impurity nucleation [9], in which single impurity atoms, molecules or clusters of atoms are perfectly wet and form the cores on which nucleation takes place. This process is very similar to pure homogeneous nucleation, but the critical free energy of nucleus formation is lowered and the number of potential nucleation sites is reduced from that of homogeneous nucleation. Nucleation caused by a homophase impurity will also result in a nucleation rate dependent on the cooling rate, so it is not possible to differentiate qualitatively between homogeneous and homophase impurity nucleation in the EHD experiment. A schematic diagram of the different nucleation mechanisms is shown in Figure 1.

EXPERIMENTAL METHOD

Droplet Formation

The droplets for this study were produced by EHD atomization, a process developed by Phrasor Scientific, Inc. of Duarte, CA [7, 8] which uses strong electric fields to spray liquids as submicron droplets in a vacuum system (10^{-7} Torr). A schematic illustration of the EHD process is shown in Figure 2. An electron beam is used to melt a consumable electrode that may be a metal wire or ceramic rod. The droplets that are produced are typically smaller than 100 nm in diameter. At this size, the radiative cooling rate is quite high, on the order of 10^6 to 10^7

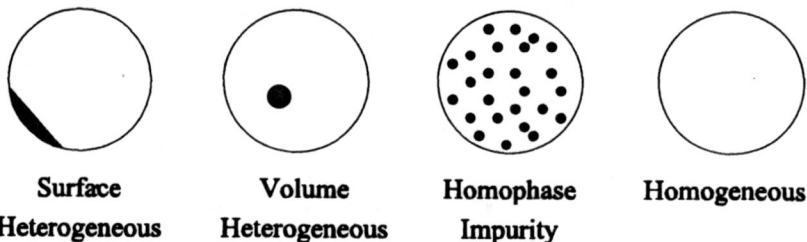

| Surface | Volume | Homophase | Homogeneous |
| Heterogeneous | Heterogeneous | Impurity | |

Figure 1   Schematic diagram of different nucleation mechanisms.   Heterogeneous and homophase impurity nucleants are represented by solid objects.

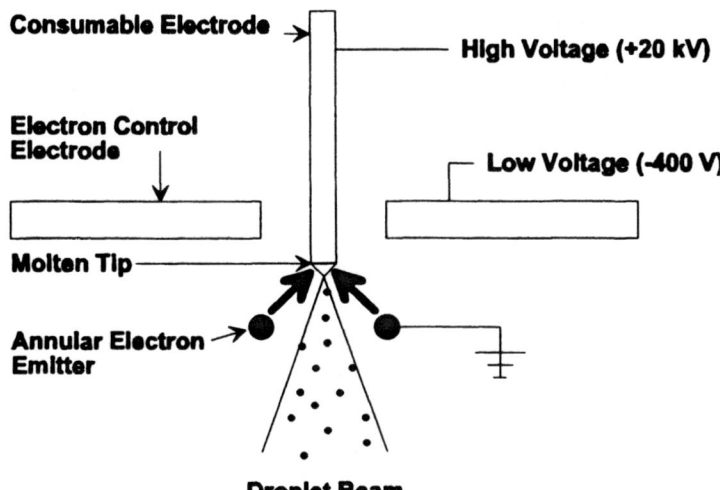

Consumable Electrode

High Voltage (+20 kV)

Electron Control
Electrode

Low Voltage (-400 V)

Molten Tip

Annular Electron
Emitter

Droplet Beam

Figure 2   Schematic illustration of the EHD atomization process. Droplets with sub-micron diameters are produced.

K/s.

The availability of heterogeneous nucleants is minimized through this technique. The high vacuum environment and high cooling rates reduce to a very low level the number of impurity atoms that strike a droplet. It is expected that less than one gas molecule on average will collide with the droplet during cooling [6]. High purity elemental semiconductor and metallic rods were used to form the EHD droplets. The rods were etched prior to insertion in the vacuum system to remove any impurities on the surface of the samples. It is expected that very few droplets would contain an impurity that is large enough to cause heterogeneous nucleation.

<u>Droplet Characterization</u>

To determine the active nucleation mechanism, the microstructure of a large number of droplets must be determined. This was done by observing the diffraction patterns of individual droplets. The EHD droplets are sprayed onto a thin carbon film (5 - 10 nm thick), which is then placed on a TEM finder grid. A STEM micrograph of Ge EHD spheres on a thin carbon film is shown in Figure 3. The finder grid allows individual droplets to be observed

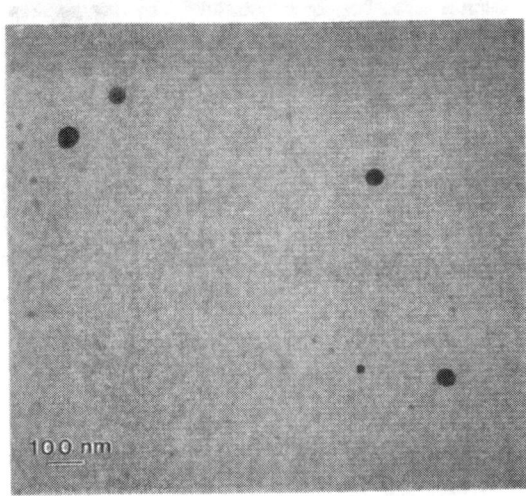

Figure 3   STEM image of Ge EHD droplets.

in a systematic manner so that double counting can be avoided. The solidified droplets are small enough that they are directly electron transparent. Thus, the structure of the solidified droplet can be determined using scanning transmission electron microscopy (STEM). The droplets are examined with a Vacuum Generators HB-501 STEM in normal image mode. A measurement of droplet diameter is obtained with nanometer resolution and the diffraction pattern from the same droplet is observed.

DETERMINATION OF NUCLEATION MECHANISM

The type of nucleation mechanism that is active was determined by examining the variation of the fraction of crystalline droplets with droplet diameter (i.e. - the dependence of the probability of nucleation on radius). Droplets with diameters ranging from 10 - 150 nm were examined. The results were tabulated in ten nm diameter ranges to exhibit graphically the nucleation dependence on radius. These findings were compared to the best fit for surface and volume heterogeneous and homogeneous nucleation mechanisms in order to determine which mechanism is active.

If the nucleants or nucleation events are distributed uniformly within the droplets, the fraction of droplets remaining nucleation-free can be expressed using a Poisson distribution:

$$X = \exp\left[-\left(d / d_i\right)^i\right] \qquad (1)$$

where X is the fraction of amorphous droplets (nucleation-free), d is the droplet diameter, and i=2 for surface heterogeneous nucleation, i=3 for volume heterogeneous nucleation, and i=4 for homogeneous nucleation with radiative cooling (volume and time-dependence). A single parameter, d_i, is varied to find the best fit with the experimental results for each nucleation mechanism. The best fit is determined by using a least-squares test on the value of X where each of the ten radii ranges is used as an individual trial in the test. This procedure does not yield directly any physical parameters of the element being studied, but does indicate the likelihood of each nucleation mechanism.

RESULTS

Droplet Characterization

The size and structure of nearly 1000 Si droplets and over 8000 Ge droplets were recorded. Most of these droplets were found to be either fully crystalline or fully amorphous. Iron and vanadium were also studied, but the amorphous droplets were too small (<20 nm) to determine the functional form of the nucleation dependence on droplet diameter.

A comparison between the experimental results and the different nucleation theories is shown graphically in Figure 4. The results for both Ge and Si are seen to exhibit behavior most similar to homogeneous nucleation (or nucleation by homophase impurities). For both Ge and Si, the

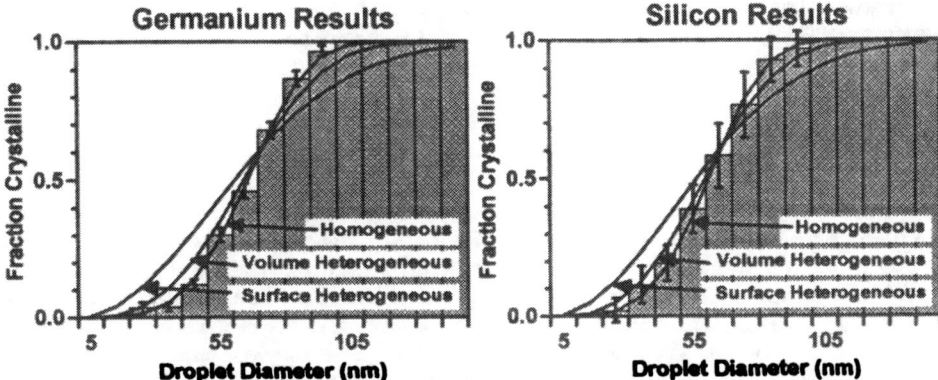

Figure 4   Comparison between the experimental nucleation results (histograms) and the best fit curves for the different nucleation mechanisms. Two-σ error bars are indicated. The value of d for all nucleation mechanisms is approximately 73 nm for Ge and 67 nm for Si.

sum-of-squared-errors value for the fitted homogeneous nucleation function is approximately one order of magnitude smaller than the value for volume heterogeneous and two orders of magnitude smaller than the value for surface heterogeneous nucleation. The results for Ge are quite conclusive, owing to the large number of Ge droplets analyzed.

Analysis Using Classical Nucleation Theory

Classical nucleation theory is used to compare the experimental results with theory. The equations used in this analysis have been described previously [10, 11]. It should be noted that the results of the theoretical calculations are highly dependent on properties of the supercooled liquids. The values of these parameters at high supercoolings have not been reported and their temperature dependence is not known.

The nucleation rate is integrated over the cooling time to find the expected number of nucleation events. The nucleation rate is proportional to droplet volume and the radiative cooling time is proportional to droplet diameter, so the number of nucleation events will be proportional to the fourth power of droplet diameter. The theoretical value of d_4 in Equation 1 is calculated for comparison with the experimental results.

The value of the glass temperature in the equations was varied to fit the theory with the experimental d_4 results. The values that were determined for the glass temperature were $0.33*T_m$ for Si and $0.34*T_m$ for Ge.

This analysis supports the case that homogeneous nucleation is occurring in EHD droplets, but the parameters can be changed to other reasonable values that support a homophase impurity nucleation mechanism [9]. For example, if it is assumed that single-atom homophase impurities are present at one part per 100,000, the glass temperature calculated in a similar analysis for homophase impurity nucleation is found to be $0.25*T_m$. Because the quantitative accuracy of the theory is not established and values of the parameters in the analysis are not known, it is not possible to conclude which homophase nucleation mechanism (homogeneous or homophase impurity) is exhibited in EHD droplets. It may be possible to establish this experimentally by repeating the study with elements of different purity.

CONCLUSIONS

Sub-micron droplets of molten Si and Ge were produced by EHD atomization. The solidified droplets were found to be either fully crystalline or fully amorphous. A function of the form $X=\exp[-(d/d_4)^4]$ was found to have the best fit to the experimental results of the possible nucleation functions. This result shows that nucleation took place through a mechanism that is both volume and cooling time dependent, which signifies that nucleation took place through either homogeneous nucleation or through nucleation about a homophase impurity. Calculations using classical nucleation theory of homogeneous nucleation were found to be in good agreement with the experimental results. However, the calculated results are quite sensitive to the parameters that are used, so it is also possible that nucleation may have occurred through a homophase impurity mechanism.

BIBLIOGRAPHY

1. D. Turnbull, J. Appl. Phys. **21**, 1022 (1950).
2. D. Turnbull, J. Chem. Phys. **20**, 411 (1952).
3. Y. Miyazawa and G. M. Pound, J. Cryst. Growth **23**, 45 (1974).
4. J. H. Perepezko, Mat. Sci. and Engr. **65**, 125 (1984).
5. A. J. Drehman and D. Turnbull, Scripta Metall. **15**, 543 (1981).
6. Y.-W. Kim, H.-M. Lin, and T. F. Kelly, Acta Metall. **37**, 247 (1989).
7. J. Perel, J. F. Mahoney, B. E. Kalensher, K. E. Vickers and R. Mehrabian, in Rapid Solidification Processing Principles and Technologies, R. Mehrabian, B. H. Kear and M. Cohen (Claitor's Publishing Division, Baton Rouge, LA, 1978) 258.
8. J. Perel, J. F. Mahoney, P. Duwez and B. E. Kalensher, in Rapid Solidification Processing Principles and Technologies II, edited by R. Mehrabian, B. H. Kear and M. Cohen (Claitor's Publishing Division, Baton Rouge, LA, 1980) 287.
9. D. Turnbull, Progress in Mat. Sci. - Chalmers Anniversary Volume, 269, (1981).
10. T. F. Kelly, M. Cohen, and J. B. Vander Sande, Met. Trans. **15A**, 819 (1984).
11. T. F. Kelly and J. B. Vander Sande, Int. J. of Rapid Solid. **3**, 51 (1987).

ACKNOWLEDGMENTS

This work was supported in part by a contract from the Department of Energy, contract number DE-FG02-85ER45215. We would like to thank Dr. P. P. Camus for helpful discussions.

EVIDENCE OF SILICA POLYMERIZATION DURING INTERNAL NUCLEATION OF GLASSY LITHIUM DISILICATE

JANE W. ADAMS* AND BERNARD H.W.S. DE JONG**
*Center for Ceramic Research, Rutgers University, Piscataway, NJ 08855-0909
** Institute for Earth Sciences and Vening Meinesz Institute for Geodynamic Research, Utrecht University, 3508TA Utrecht, the Netherlands

ABSTRACT

The principal problem in developing a theory of internal nucleation lies in the constitution of the nascent nuclei. To get at this constitution in an internally nucleating glass, oxygen and lithium core level x-ray photoelectron spectra have been measured on glassy and nucleated lithium disilicate and on the three stable crystalline compounds in the Li_2O-SiO_2 system. Our results indicate that nucleated lithium disilicate glass consists of completely polymerized silica in the formation of which non-bridging oxygens are consumed. During this silica polymerization lithium atoms move away from presumably four-fold coordinated sites towards higher coordinated ones.

INTRODUCTION

The principal problem in classical nucleation theory of glasses is the absence of a firm experimental basis. This lack is due to the inability to resolve nanometer-size nuclei even using modern microscopic techniques, and due to severe limitations in observing nuclei as formation can only be ascertained after some crystal growth. The connection of bulk thermodynamic and microstructural concepts is a recurrent problem in any nucleation theory as a consequence of this.

This paper reports our attempt to provide an experimental basis for glass structure changes due to nucleation using photoelectron spectroscopy of internally nucleated lithium disilicate glass. We chose this glass system because it has been the principal focus since the recognition of its peculiar nucleation behavior by Jaccodine [1] for the study of internal nucleation [2-7]. In addition, the glasses are readily made, they are not hygroscopic and the peak nucleation and growth temperatures, at 450 and 610°C respectively, are well known and experimentally accessible. Moreover, structural changes in nucleated lithium disilicate glasses can be "quenched in" from the melt or nucleation temperature to room temperature without phase change, thereby permitting observation at ambient conditions [7].

Our results indicate that the principal process taking place during nucleation is coupled movement of lithium and oxygen ions. The spectral variations can be interpreted as being due to nearly complete silica polymerization during nucleation and concomitant migration of lithium atoms from 4-fold to 5-fold oxygen coordinated sites. A secondary, but important inference is an increase in Si-O-Si angle and contraction of the Si-O bond distance in glassy lithium disilicate as inferred from binding energy shifts of the O 1s photoelectron spectroscopy [8].

EXPERIMENTAL

Glasses of the proper molar composition were prepared by conventional means using silicic acid and lithium carbonate as starting materials. Charges were melted in air in Pt crucibles at ~150°C above the liquidus temperature of the given composition. The disilicate liquid was rapidly quenched to a glass, the estimated quench rate being ~100°C/s. The resulting optically clear glass samples were not annealed. Lithium disilicate glasses were heat-treated at 450(±1)°C at times varying between 0.5 and 200 hours. Crystalline lithium di-, meta- and orthosilicate were made either by glass devitrification or from the melt. Resulting phases were checked for composition and crystallinity by ICP, XPS and X-ray diffraction.

XPS spectra were collected on both PHI 5500 and Kratos XM-800 spectrometers using 400W $MgK_{\alpha1,2}$ radiation at a constant pass energy of 23.5 eV. Base pressure of the system during

239

experiments was better than 5×10^{-9} torr. The problem of charging effects due to a loss of expelled electrons was alleviated by using thin samples and by flooding the sample surface with an electron gun. Absolute peak position calibration was carried out by referencing to the C 1s signal (284.8 eV) from adsorbed CO_2 on the sample. Samples were slightly argon etched for a few minutes to remove surface contaminants after the C 1s reference signal was detected. No changes were observed in the O 1s and Li 1s peak positions during sputtering. Gaussian curves were fitted according to a standard fitting routine to deconvolute the photoelectron peaks. The estimated error in peak position is less than 10 percent.

RESULTS

The binding energies and relative intensities of the O 1s and Li 1s spectra of all phases studied are collected in Tables I and II.

Table I. O 1s XPS binding energies (eV), FWHM (eV) and relative intensities (%) for crystalline and glassy lithium silicates and nucleated lithium disilicate.

Compound	Nucleation, h at 450°C	O(br), eV	I,%	FWHM, eV	O(nbr), eV	I,%	FWHM, eV
Glass							
$Li_2Si_2O_5$	0	531.2	75.6	1.9	529.2	24.2	2.2
	0.5	532.0	93.8	2.1	529.7	6.2	1.9
	1.5	531.7	96.4	2.0	529.7	3.6	1.8
	3.0	531.8	97.3	1.8	529.7	2.7	1.5
	6.0	531.8	93.5	1.9	529.9	6.5	1.5
Crystalline							
$Li_2Si_2O_5$	----	532.5	74.6	1.8	530.7	25.4	1.7
Li_2SiO_3	----	532.4	56.2	1.9	530.8	43.8	1.9
Li_4SiO_4	----	---	----	----	531.4	100	2.1

Table II. Li 1s XPS binding energies (eV), FWHM (eV), and relative intensities (%) for crystalline and glassy lithium silicates and nucleated lithium disilicate.

Compound	Nucleation, h at 450°C	Li, eV	I,%	FWHM,eV	Li, eV	I,%	FWHM,eV
Glass							
$Li_2Si_2O_5$	0	55.11	100	2.0	—	---	----
	0.5	55.2	65.3	1.9	51.7	34.7	2.0
	1.0	55.2	44.5	2.0	51.6	55.5	2.0
	1.5	55.2	50.2	2.0	51.7	49.8	2.0
	2.0	55.2	49.5	2.0	51.6	50.1	2.0
	6.0	55.2	73.5	2.2	51.7	26.5	2.2
	200	55.3	88.6	2.2	51.5	11.4	2.2
Crystalline							
$Li_2Si_2O_5$	----	55.3	100	1.8	—	----	----
Li_2SiO_3	----	55.2	100	1.8	—	----	----
Li_4SiO_4	----	55.1	78.4	2.0	51.8	21.6	2.0

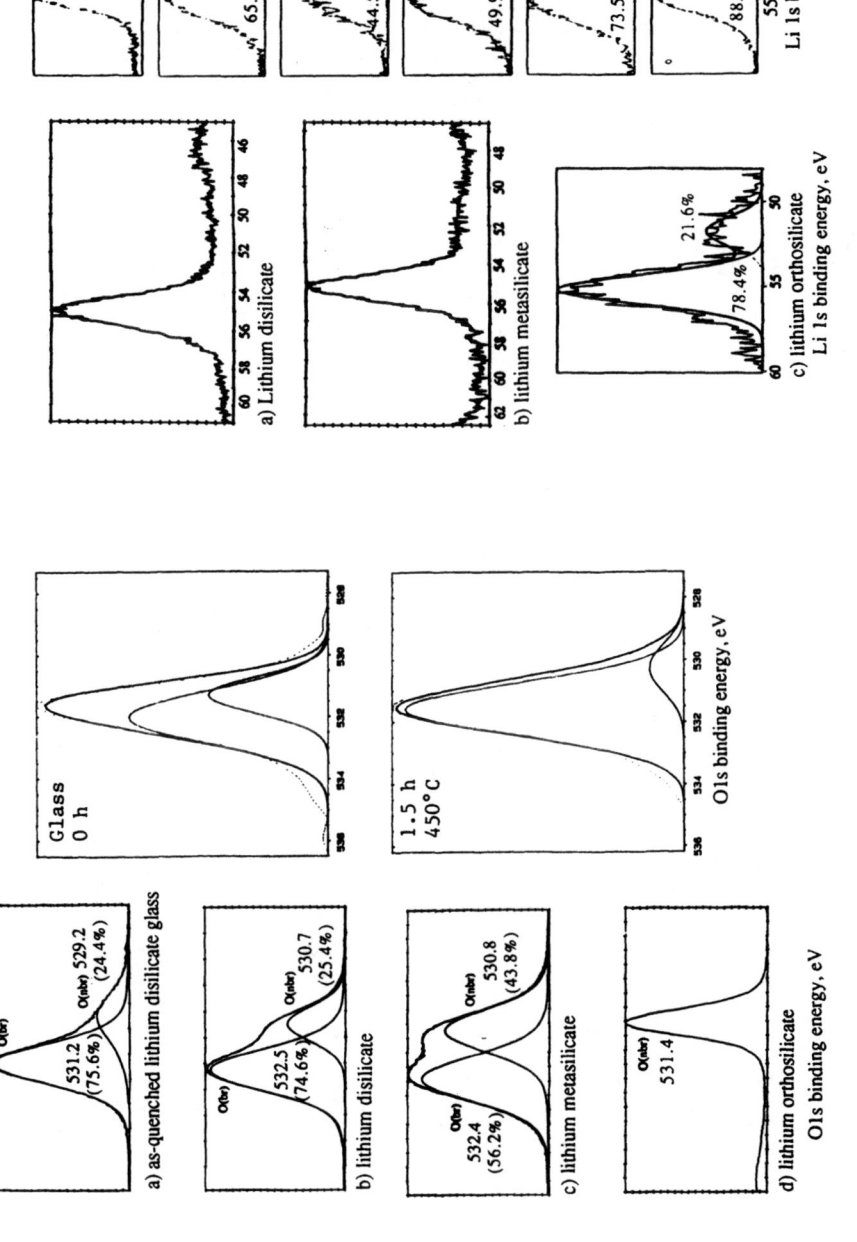

Figure 1. O 1s photoelectron spectra of glassy (a) and crystalline lithium silicates (b-d), and glassy and nucleated lithium disilicate.

Figure 2. Li 1s photoelectron spectra of crystalline lithium silicates (a-c) and glassy and nucleated lithium disilicates.

241

Figures 1 and 2 show the observed spectra from which the data in Table I and II were taken. Inspection of these tables and figures show that the O 1s spectra for crystalline and glassy lithium disilicate are similar in shape but shifted in position with respect to one another. The crystalline di- and metasilicate O 1s peak positions are similar, whereas the O 1s peak position for the orthosilicate occurs at a position between the two other crystalline lithium silicates. Notice the O 1s and Li 1s spectra of nucleated Li disilicate glass. In the O 1s spectrum of glassy lithium disilicate two peaks in the proper 3 to 1 ratio are present for O(br) and O(nbr). After nucleation the intensity of the O(nbr) peak decreases substantially whereas the O(br) peak shifts about 0.5 eV to smaller binding energy. In the Li 1s spectrum the glass spectrum before nucleation shows one peak. After heat treatment for 30 minutes, a second peak appears which increases and subsequently decreases in intensity depending on the residence time of the sample at 450°C. The principal change in glass structure during nucleation occurs in the first hour.

DISCUSSION

XPS chemical shifts correlate with atomic charge in that the absolute value of the core level binding energy increases with increasing positive charge on the atom [9, 10]. The core level serves as a probe of the valence system, since lines appearing at different core chemical shifts for the same element reflect different chemical valence states. For anions the level which is used as a basis is the ion with a complete octet. O^{2-} has therefore the least negative O 1s core level binding energy, whereas O^{1-} and O being more positive, will have increasingly larger O 1s binding energies. This is qualitatively reasonable. The difficulty for a core level electron to escape increases with increased electron density in the outer shells, and decreases with decreasing outer shell electron density. For an atom stripped of its shielding valence electrons, it is easier for a core level electron to escape to the continuum, and this gives an effectively less negative binding energy. Thus the 1s level functions as some sort of a fulcrum balancing the positive nuclear and valence charge densities.

O 1s shifts in crystalline and glassy lithium disilicate

It has already been shown by Bruckner et al. [11] and confirmed by others [7, 12-14] that the O 1s photoelectron spectrum at the silica rich end of binary alkali-silicate systems quantitatively distinguishes between bridging O(br) and non-bridging(O(nbr)) oxygen atoms and that non-bonding oxygen atoms, i.e. atoms connected to alkali atoms only, do not occur. Thus all oxygen atoms in a disilicate glass are connected to at least one silicon atom with as consequence that the ratio of O(nbr) to O(br) is set by the stoichiometry of the sample. Inspection of Table I and Figure 1 bears this out, the intensity ratios for the O 1s peaks for glassy and crystalline $Li_2Si_2O_5$ being about 3 to 1. Similarly the ratio for crystalline metasilicate is roughly 1 to 1. In crystalline lithium disilicate there are two types of O(br) atoms, one of which is connected to two silicon atoms only, the other to two silicon and one lithium atom, and one O(nbr) atom which has in its coordination sphere one Si and three Li atoms [15]. It is therefore somewhat unexpected that the FWHM of the O(br) and O(nbr) peak are similar. Furthermore, the O(br) to O(nbr) ratios being the same in as-quenched glassy and crystalline lithium disilicate, it came as a big surprise that during nucleation the O 1s spectral intensity of O(nbr) substantially decreases, as inspection of Table I indicates.

From the positions of the O 1s peaks an XPS charge, Q_O, can be derived utilizing the results of Bagus and Bauschlicher [16] according to the following formula:

$$O \ 1s(\text{binding energy}) = 545.509 + 19.622(Q_O) + 2.244(Q_O)^2 \qquad (1)$$

The oxygen charges for lithium silicates using this formula are collected in Table III. It should be kept in mind that assignment of charges is a partitioning problem and that the charges reported here, albeit internally consistent and close to values obtained from Mulliken overlap population analysis on ab initio molecular orbital results on related molecules [17-20], do not have any absolute meaning.

Table III. O 1s XPS derived charges

Compound	Nucleation, h at 450°C	O(br) charge	O(nbr) charge
Crystalline			
SiO2*	---	-0.74	
Li2Si2O5	---	-0.72	-0.83
Li2SiO3	---	-0.73	-0.83
Li4SiO4	---		-0.79
Glass			
Li2Si2O5	0	-0.79	-0.93
	0.5	-0.75	-0.90
	1.5	-0.77	-0.90
	3.0	-0.74	-0.90
	6.0	-0.74	-0.88

*quartz, cristobalite

The XPS derived charges for O(br) and O(nbr) charges vary between 531.2 and 529.2 eV, respectively, for glass and crystal, corresponding to oxygen charges of -0.79 and -0.93, cnf. a O(br) charge of -0.74 for quartz. During nucleation the charge on O(br) becomes slightly less negative at ~ -0.74, similar to the O(br) charge on crystalline SiO2. The change in O(nbr) concentration coupled with the shift in charge in O(br) towards SiO2 (refer to Table III) strongly suggests that internal nucleation in lithium disilicate involves silica polymerization.

Li 1s shifts in crystalline and glassy lithium disilicate

Li 1s spectra also indicate dramatic changes in local lithium environments during nucleation. The single Li 1s peak at 55.2 eV in as-quenched glass is similar to that observed for crystalline Li di- and metasilicates with only one peak at 55.2 eV, whereas spectra of crystalline Li4SiO4 and the nucleated lithium disilicate glasses contain two peaks indicative of two lithium sites. This peak at less negative binding energy indicates Li atoms with less positive charge.

Unfortunately not much is known about lithium 1s photoelectron spectroscopy. Previous XPS of lithium compounds have been fairly limited and data are not very consistent. Povey and Sherwood [21] measured a Li 1s shift for lithium metal, which has a bcc structure with each lithium atom connected to 8 other ones. Their value, normalized to the CO2 peak, is 49.5 eV [8]. They give the Li 1s shift for Li2O, which has the antifluorite structure with each lithium atom coordinated to 4 oxygen atoms and each oxygen atom to eight lithium atoms, as 51.2 eV. Li4SiO4 is known to consist of 4-, 5- and 6-fold coordinated lithium [22], the charges on the 5- and 6-fold sites being less positively charged. Lithium orthosilicate is the only crystalline compound known to exhibit two Li 1s peaks, at 54.6 and 51.1 eV [7]. These two peaks are similar in position to those observed in internally nucleated glassy lithium disilicate. With these structures in mind, the XPS results indicate that in glassy lithium disilicate all lithium atoms occur in 4-fold oxygen coordination as indicated by the peak around 55 eV, and that during nucleation significant migration of lithium atoms occurs towards 5-fold oxygen coordinated sites.

SUMMARY AND CONCLUSIONS

O 1s and Li 1s photoelectron spectra show that large atomic rearrangement occurs during internal nucleation in lithium disilicate glass. The nucleation process exhibits a sudden burst of charge transfers in which in a short time the principal rearrangement of the glass is accomplished. During nucleation silica polymerizes, expelling lithium atoms in the process and annealing defect O(nbr) atoms.

ACKNOWLEDGMENTS

We thank W. Wildenberg of Utrecht University for preparing the thin samples, M. Onyiriuka of Corning, Inc. for making the initial XPS evaluations and D.Hensley of Rutgers University for training J. Adams to do this spectroscopy. A portion of J. Adams' work was funded by an Earnest Oppenheimer Fellowship for Surface and Colloid Research from Cambridge University.

REFERENCES

1. R.J. Jaccodine, J. Am. Ceram. Soc. **44**, 472 (1961).
2. K. Matusita and M. Tashiro, J. Non-Cryst. Sol. **11**, 471 (1973).
3. G.F. Neilson and M.C. Weinberg, J. Non-Cryst. Sol. **34**, 137 (1979).
4. B.H.W.S. de Jong, C.M. Schramm and V.E. Parziale, J. Am. Chem. Soc. **106**, 4396 (1984).
5. B.H.W.S. de Jong, in *Ullmann's Encyclopedia of Industrial Chemistry* **A12**, 365 (1989).
6. M.C. Weinberg and E.D. Zanotto, J. Non-Cryst. Sol. **108**, 99 (1989).
7. J.W. Adams, PhD thesis, Cambridge University, 1993.
8. J.W. Adams and B.H.W.S. de Jong, in preparation.
9. H.S. Gotts and A.J. Glick, Phys. Rev. **B27**, 4729 (1983).
10. K. Siegbahn et al., *ESCA; atomic, molecular and solid state structure by means of electron spectroscopy*, (Almquist and Wiksells, Uppsala, 1967).
11. R. Brueckner, H.U. Chun, H. Goretzki and M. Samet, J. Non-Cryst. Sol. **42**, 49 (1980).
12. H. Nasu, J. Heo and J.D. Mackensie, J. Non-Cryst. Sol. **99**, 140 (1988).
13. D. Sprenger, H. Bach, W. Meisel and P. Gütlich, J. Non-Cryst. Sol. **126**, 111 (1990).
14. D. Sprenger, H. Bach, W. Meisel and P. Gütlich, in *The physics of non-crystalline solids*, edited by L.D.Pye, W.C. LaCourse and H.J. Stevens, (Taylor and Francis, London, 1992) pp. 42-47.
15. F. Liebau, Acta Cryst. **14**, 389 (1961).
16. P.S. Bagus and C.W. Bauschlicher, J. Electr. Spec. Rel. Phenom. **20**, 183 (1980).
17. D.J.M. Burkhard, B.H.W.S. de Jong, A.J.H.M. Meyer and J.H. van Lenthe, Geochim. Cosmochim. Acta **55**, 3453 (1991).
18. K.A. Van Genechten, W.J. Mortier, P. Geerlings, J. Chem. Phys. **86**, 5063 (1987).
19. J. Sauer, H. Haberlandt and W. Schirmer, *Structure and reactivity of modified zeolites*, edited by P. A. Jacobs et al.(Elsevier, Amsterdam, 1984) pp. 313-320.
20. J.D. Kubicki and D. Sykes, Am. Mineral. **78**, 253 (1993).
21. A.F. Povey and P.M.A. Sherwood, J. Chem. Soc. Faraday Trans.II **70**, 1240 (1974).
22. B.H.W.S. de Jong, D. Ellerbroek and A.L. Spek, Acta Cryst., submitted.

FIRST STAGES OF NANO-CRYSTALLIZATION
OF AMORPHOUS $Fe_{75.5}Cu_1Nb_3Si_{12.5}B_8$
STUDIED BY THE POSITRON ANNIHILATION LIFETIME TECHNIQUE

A.J. KRUK[*,**], H. SCHUT[**], J. SIETSMA[*] AND A. VAN VEEN[**]
[*]Laboratory of Materials Science, Delft University of Technology, Rotterdamseweg 137,
2628 AL Delft, The Netherlands
[**]Interfaculty Reactor Institute, Delft University of Technology, Mekelweg 15, 2629 JB Delft,
The Netherlands

ABSTRACT

The first stages of the nano-crystallization process of amorphous $Fe_{75.5}Cu_1Nb_3Si_{12.5}B_8$ into a nano-crystalline structure are investigated by the positron annihilation lifetime technique. Samples have been isothermally annealed at 643 K for times varying between 600 and 10^5 seconds. The positron lifetime spectra have been analyzed allowing for three lifetimes. The shortest and the longest lifetime, $\tau_1 = 150 \pm 2$ ps and $\tau_3 = 1500\text{-}2000$ ps respectively, are attributed to annihilation of positrons in the amorphous phase and to the formation and annihilation of ortho-positronium at the surface of the stacked foils and did not change significantly upon the annealing. The intermediate positron lifetime τ_2 increased from 324 ps to 387 ps. The intensity of this component increased from 5 to 15%. Comparison with resistivity measurements indicates that the change of this lifetime component occurs at an early stage in the crystallisation process, i.e. when the fraction of crystalline material is on the order of 10^{-3}. The increase of τ_2 is attributed to positrons annihilating in a region with lower average density surrounding the small crystallite.

INTRODUCTION

The FeCuNbSiB alloy is a typical example of a material with a nanometer-sized grain structure showing extremely good magnetic properties such as a high permeability and a high saturation magnetization [1]. For this reason the crystallization from the amorphous phase into the nano-crystalline structure, which takes place at a temperature around 800 K, has intensivily been studied. Examples of experimental techniques commonly used to this aim are X-ray diffraction, isothermal resistivity measurements [2] and positron annihilation. Positron annihilation techniques have shown to be very powerful for the study of open-volume defects in solids [3]. In particular the positron annihilation lifetime (PAL) technique is capable of detecting small changes in the structure and density of this type of defects. This has been demonstrated by Schaefer et al. [4] who attributed observed positron lifetimes and intensities to selected positron trapping sites, such as vacancy-sized free volumes and microvoids, at the intersections of the crystallites produced by compacting techniques. The application of PAL for annealing studies starting from the amorphous phase has been demonstrated by Sui et al. [5], Kristiaková et al. [6], and Novotny and Zemcík [7]. All measurements reported so far, however, concern the formation of a considerable fraction of the crystalline phase. In this article we present the results of PAL experiments on $Fe_{75.5}Cu_1Nb_3Si_{12.5}B_8$ during annealing at a relatively low temperature (643 K), by which the very early stages of the development of nano-crystals can be studied.

245

EXPERIMENTAL

Sample preparation

The amorphous alloy used in these experiments was produced in ribbon of 20 mm width and 24 μm thickness by the melt-spinning technique. The composition of the alloy is $Fe_{75.5}Cu_1Nb_3Si_{12.5}B_8$(atom %). From the ribbon two sets of each 10 foils with dimensions 20x20 mm^2 were cut. On one of the foils of each set small droplets of a ^{22}NaCl solution were deposited to a total activity of approximately 1.8 MBq (50 μCi). After drying at room temperature a source-sample sandwich was obtained by stacking five foils facing the deposited side and stacking the remaining four foils at the opposite side. The diameter of the active source area was about 4 mm. One set was used for the annealing experiments, the other acted as a reference. The reason for using a sandwich with the radioactive material deposited directly on the sample is that a positron source deposited and sealed between two kapton foils has been shown to introduce a lifetime component of 385 ps. As will be shown later, this value just lies in the range of the lifetime component τ_2 and therefore hampers the analysis.

The annealing of the sample at 643 K has been performed in a tube-furnace with a long-term temperature stability better than 5 K. Annealing times were varied from about 600 to 10^5 seconds.

Positron lifetime spectrometer

For the positron lifetime measurements a fast-fast coincidence setup was used. The detectors consisted of BaF$_2$ scintillators coupled to XP2020Q photomultiplier tubes. In order to increase the timing resolution the output signal from the tubes was taken from the 9th dynode [8]. The Time to Amplitude Converted signals (ORTEC 467) were converted into a spectrum using a PC-based PHA/MCA data aquisition board. In order to monitor the stability of the spectrometer each measurement was divided into seven sub-measurements of two hours. No significant time drifts have been observed.

The time resolution of the spectrometer was experimentally determined by measuring the positron lifetime for a sandwich of well annealed tungsten single crystals. Analyzing the data with the computer code RESOLUTION [9] yielded a resolution function with a FWHM of 190 ps. The lifetime spectra measured for both the annealed and the reference alloy were analyzed with the code POSITRONFIT [10].

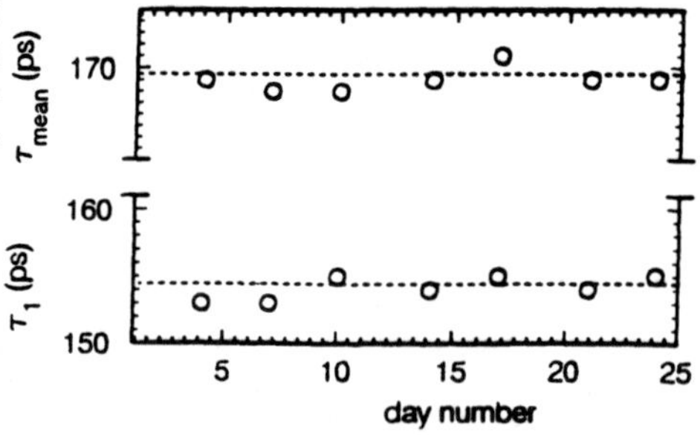

Fig. 1. The short lifetime component and the mean lifetime in the as quenched reference sample over a period of 24 days.

RESULTS

During the four weeks the experiments lasted the stability of the lifetime spectrometer has been monitored by measuring the lifetime spectra for the as quenched reference alloy after every two annealing measurements. Figure 1 shows the fitted short and mean positron lifetime components as a function of day number. From this figure it is concluded that the stability is good enough to reproduce mean lifetimes within ± 1 ps. The average value of the mean lifetime in the reference sample is 169 ps. The average value of the short lifetime component in this case is 154 ps.

The fitting results of the PAL-measurements for the annealed sample are given in figure 2. Shown are the mean positron lifetime, the intermediate lifetime τ_2, the short lifetime τ_1 and the intensity ratio of τ_1 and τ_2 as a function of annealing time. The long lifetime component (not shown) varied between 1500 and 2000 ps with an intensity of about 1%. This contribution is believed to result from the formation and annihilation of ortho-positronium at the surfaces of the stacked foils. The results are obtained by fitting with a three lifetime model without constraints on lifetime and intensity. In the analysis the source contribution to the spectra is estimated to have a lifetime of 430 ps with an intensity of 8%. These values are close to those of 470 ps and 5.5%, respectively, reported by de Vries [8] for ^{22}NaCl deposited on aluminium samples. From figure 2 it clearly follows that the increase of the mean lifetime is caused by an increase of the intermediate lifetime accompanied by an increasing intensity of this component. For the as quenched state a ratio $I_1/I_2 = 20$ is obtained. The difference in lifetimes for the two samples in the as quenched state (fig.1 and fig. 2 at annealing time zero) is larger than the statistical error for each of the samples. The main cause for this uncertainty in the absolute value of the lifetimes is the source contribution, which cannot be established with a great precision.

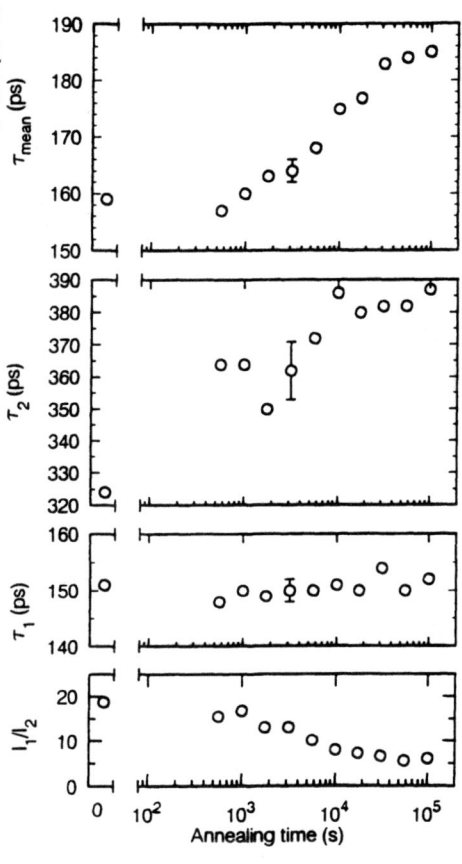

Fig. 2. The mean lifetime, the intermediate component τ_2, the short component τ_1 and the intensity ratio of τ_1 and τ_2 as a function of the annealing time at 643 K.

DISCUSSION

The annealing temperature in this study, 643 K, can be considered as very low with respect to normally used temperatures to induce nano-crystallization in this alloy (e.g. [1]). From measurements at higher temperatures (e.g. [2]) it can be estimated that even after 10^5 s at 643 K the crystalline fraction will not exceed 10^{-3}. Positron lifetimes can therefore supply information on the formation and initial growth of nuclei in the amorphous phase. It has been reported before that the positron lifetime technique detects the crystallization process in

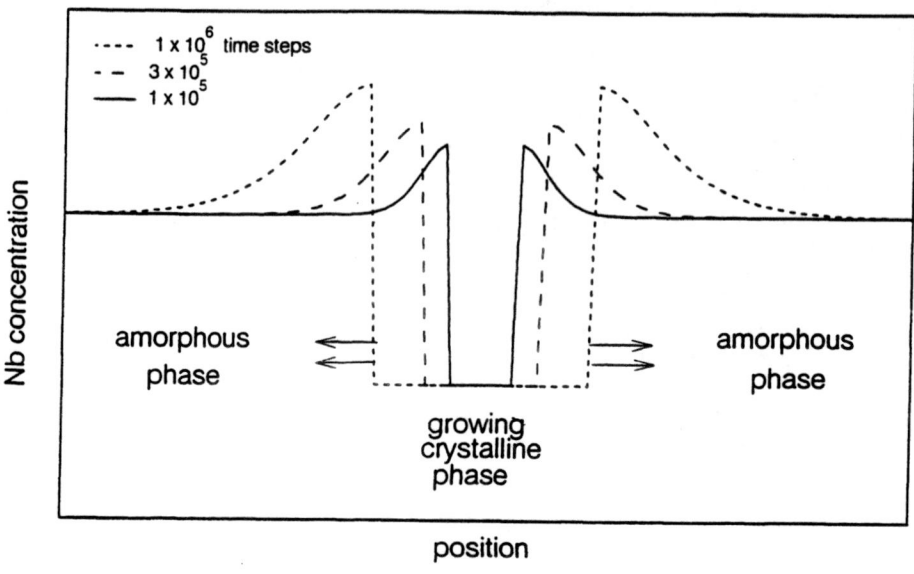

Fig. 3. *Schematic composition profile around crystalline grains. The evolution of the Nb concentration profile with time is governed by diffusion processes, resulting in a higher Nb concentration in the amorphous phase surrounding the crystallite. Similar curves describe the density profile, for which case the figure should be flipped over vertically.*

amorphous metals at an earlier stage than most other techniques [11].

The shortest lifetime, 150 ps, is attributed to the amorphous phase [7], and does not change significantly. The second lifetime, 324 ps for the as quenched state, increases continuously with increasing annealing time, and is therefore likely to be connected to the very early stages of nano-crystallization. The value of this lifetime indicates that it stems from a region with a density and/or composition that differs considerably from that of the "normal" amorphous phase. Such regions can very well be understood to form around (very small) crystallites. In addition, for iron crystals the range of τ_2 values is typical for positrons annihilating in vacancy clusters of about 5 to 15 vacancies [12]. Composition differences between the nano-crystalline and the amorphous phase have been experimentally observed by Hono et al. [13] and by van Bouwelen et al. [14]. Also a density difference must exist, the crystalline phase having a higher density than the amorphous phase. Through diffusion processes a density and composition gradient will develop that can easily be calculated schematically. The gradients shown in figure 3 result from diffusion calculations for a growing crystallite grain that has a lower content of the element under consideration (e.g. Nb) than the overall average, initially present in the amorphous phase. The growth rate of the grain is related to the concentration in the immediate surroundings, based on the increase of the crystallization temperature with increasing Nb-content in amorphous FeCuNbSiB [15]. The gradient in the local density should have a very similar appearance, albeit flipped upside down. This schematic picture can be used to explain the characteristics of τ_2 that have been observed in our experiments. Upon growth of the nucleus, the maximum in the composition profile becomes higher, the minimum in the density profile gets deeper. This explains the continuous trend in the value of the lifetime. The large difference between τ_1 and τ_2 cannot be expected to be caused by composition changes only. The value of τ_2, and its sensitivity on the the annealing time, is therefore a reflection of the density profiles around growing crystalline nuclei. The increase in the relative intensity of this lifetime component is connected to the increase of total volume of the "shells" around the developing crystallites. From these one-temperature data no information about continuing nucleation can be extracted. The intensity of about 5% for the as-quenched sample (see fig. 2) is a strong indication for the presence of quenched-in nuclei in this alloy.

CONCLUSIONS

At the very low temperature of 643 K positron lifetime measurements provide information on the presence and the development of nuclei in amorphous FeCuNbSiB. A lifetime component of 150 ps, connected to the amorphous phase, remains constant with increasing annealing time, the lifetime component of 324 ps continuously increases to 387 ps. This lifetime shows the development of nuclei, since it can be related to density or composition gradients in the shells surrounding the nuclei.

ACKNOWLEDGEMENTS

Dr. G. Herzer of Vakuumschmelze GmbH (Hanau, Germany) kindly supplied the amorphous $Fe_{75.5}Cu_1Nb_3Si_{12.5}B_8$ ribbon.

REFERENCES

1. G. Herzer, Mat. Sci. Eng. A133, 1 (1991).
2. F.van Bouwelen, J.Sietsma, and A. van den Beukel, J.Non-Cryst.Solids 156-158, 56 (1992).
3. P.J. Schultz and K.G.Lynn, Rev. Mod. Phys. 60, 701 (1988) and references therein.
4. H.-E.Schaefer, R.Würschum, R.Birringer and H.Gleiter, Phys.Rev.B 38 (14), 9545 (1988)
5. M.L.Sui, K.Y.Le, L.Y.Xiong, Y.Liu and J.Zhu, to be published in Conf. Proc. RQ8, Sendai 1993
6. K.Kristiaková, J.Kristiak, O.Sausa and P.Bandzuch, Key Eng. Mat. 81-83, 215 (1993).
7. I.Novotny and T.Zemcík, Key Eng. Mat. 81-83, 229 (1993).
8. J.de Vries, PhD thesis, Delft University of Technology, 1987
9. P.Kirkgaard, M.Eldrup, O.E.Mogensen and D.J.Pedersen, Comput.Phys.Com. 23, 975 (1981).
10. P.Kirkgaard and M.Eldrup, Comput.Phys.Com. 7, 401 (1974).
11. J. de Vries, G.W. Koebrugge and A.van den Beukel, Scr. Met.22, 637 (1988).
12. M.J. Puska and R.M.Nieminen, J.Phys.F 13, 333 (1983)
13. K. Hono, K. Hiraga, Q.Wang, A. Inoue and T. Sakurai, Act. Metall. Mat. 40 (9), 2137 (1992).
14. F.M.van Bouwelen, J.Sietsma, C.D.de Haan, and A.van den Beukel, App.Phys.Lett. 61, 2536 (1992)
15. Y.Yoshizawa and K.Yamouchi, Mat.Trans.Jpn.Inst.Met. 31, 307 (1990)

PLASTIC DEFORMATION-INDUCED NANOCRYSTALLINE ALUMINUM IN AL-BASED AMORPHOUS ALLOYS

[1]H. CHEN*, [2]Y. HE, [1]G.J. SHIFLET and [2]S.J. POON
[1]Department of Materials Science and Engineering, and [2]Department of Physics, University of Virginia, Charlottesville, VA 22903, U.S.A.

ABSTRACT

We report the first direct observation of crystallization induced in the slipped planes of aluminum based amorphous alloys by bending the amorphous ribbons. Nanometer-sized crystalline precipitates are found exclusively within a thin layer (shear band) in the slipped planes extending across the deformed amorphous alloy ribbons. It is also found that the nanocrystalline aluminum can be produced by ball-milling. It is likely that local atomic rearrangements within the shear bands create the nanocrystals which appear after plastic deformation.

INTRODUCTION

Amorphous alloys or metallic glasses produced by rapid solidification are thermodynamically unstable. There is always a thermodynamic tendency for the amorphous state to transform to the crystalline state. The crystallization of amorphous alloys has been studied for many years[1-2]. Crystallization was usually found to occur homogeneously in the bulk of amorphous alloys, and the initiation of crystallization is usually due to thermal treatment or a change of cooling rate in melt-spinning. There are several reports of surface crystallization induced by various techniques, both chemical and mechanical, e.g., polishing, scratching, bending, cold rolling, beating, and deposition of other elements on the amorphous alloys surface. Surface crystallization was also induced by isothermal annealing of amorphous alloys at a temperature well below the normal crystallization temperature prior to or followed by these surface treatments[3-7]. Some of the mechanisms proposed to explain surface crystallization include atom segregation, oxidation, surface diffusion, plastic deformation, decrease of total surface energy and the stress relaxation on the surface.

Aluminum-based amorphous alloys were first found in 1988[8-17]. These amorphous alloys, containing up to 90% aluminum and in the form of Al-RE-TM where RE=lanthanides and Yttrium, TM=Fe, Co, Ni and Rh, are made by rapid solidification. They show unusual mechanical properties, with many of the original alloys having fracture tensile strength over 800 MPa, which greatly exceeds that of the conventional high strength aluminum alloys. The fracture mode is ductile. The combination of those characteristics -- low density, high strength and ductility -- gives these alloys great potential as engineering materials. Their excellent mechanical properties can even be improved sometimes by partial crystallization[14-16]. The microstructure of the partially crystallized aluminum based amorphous alloys is that of tiny crystalline precipitates (7-10 nm in diameter) homogeneously embedded in an amorphous matrix[13].

EXPERIMENTAL PROCEDURES

Several aluminum-based amorphous alloys, $Al_{90}Fe_5Gd_5$, $Al_{90}Fe_5Ce_5$ and $Al_{87}Ni_{8.7}Y_{3.4}$, were studied. All showed similar characteristics. Alloy ingots were prepared by melting nominal amounts of high purity elements in an arc furnace under an argon atmosphere. Amorphous alloy ribbons were obtained by using a single roller melt-spinner in a partial helium atmosphere. Typical circumferential speed of the copper wheel was 45 m/s. The sample dimensions were 1-2 mm wide, 15-20 μm thick and up to several meters long. Small pieces were cut from the ribbon and were bent through 180° at room temperature. The bent samples were thinned in a twin-jet electropolisher. Shear bands were observed in a scanning electron microscope (SEM).

Microstructures were examined with a conventional transmission electron microscope (TEM) and a high resolution transmission electron microscope (HRTEM).

RESULTS AND DISCUSSION

Aluminum-based amorphous alloys have very good ductility. They can be bent through 180° at room temperature without cracking. Heavy plastic deformation is induced in the bending tip and its vicinity, where shear bands are concentrated (Figure 1). Further from the bending tip, the distance between shear bands increases. These shear bands are imaged as straight lines in the ribbon surface and are roughly parallel to the bending axis, which indicates that, at this scale, they are flat thin layers inside amorphous alloy ribbons.

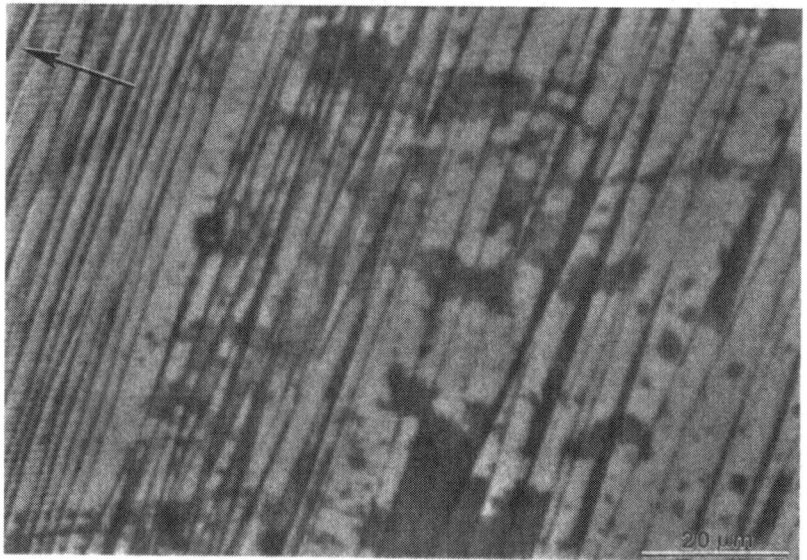

Fig. 1 SEM micrograph taken from the bent tip of an $Al_{90}Fe_5Gd_5$ amorphous alloy.

Possible microstructural changes associated with plastic deformation (bending, in this case) were investigated using TEM. Interest was primarily on the shear band itself, however, the region between shear bands was also examined. Figure 2 is a TEM micrograph taken near the bending tip of a bent-deformed $Al_{90}Fe_5Gd_5$ amorphous alloy ribbon. Running through the micrograph is a shear band which appears lighter than the rest of the micrograph. Within the shear band, nanometer-sized crystalline precipitates are homogeneously distributed. Regions outside the shear band are completely amorphous. Figure 3 is another TEM micrograph at lower magnification. This micrograph reveals waviness in the shear band whereas at much lower resolution (Figure 1) the bands appear straight. Therefore, the slip planes of the aluminum based amorphous alloys are actually rough when viewed on a scale of sub-micrometer. This contradicts the general view that slip planes of amorphous alloys are smooth[1,2].

Further investigations employing HRTEM showed that the shear band is not completely occupied by nanostructured crystallites, but contains amorphous regions in which the crystallites are embedded (Figure 4). The diameter of the crystallites varies between 7-10 nm. From the structure image and its corresponding optical diffraction pattern (Figure 4), the crystal structure of the precipitates is determined to be face centered cubic (f.c.c.) with a lattice parameter of 0.4 nm. Therefore, these nanocrystals are essentially pure aluminum since the solubility of Fe and Gd in

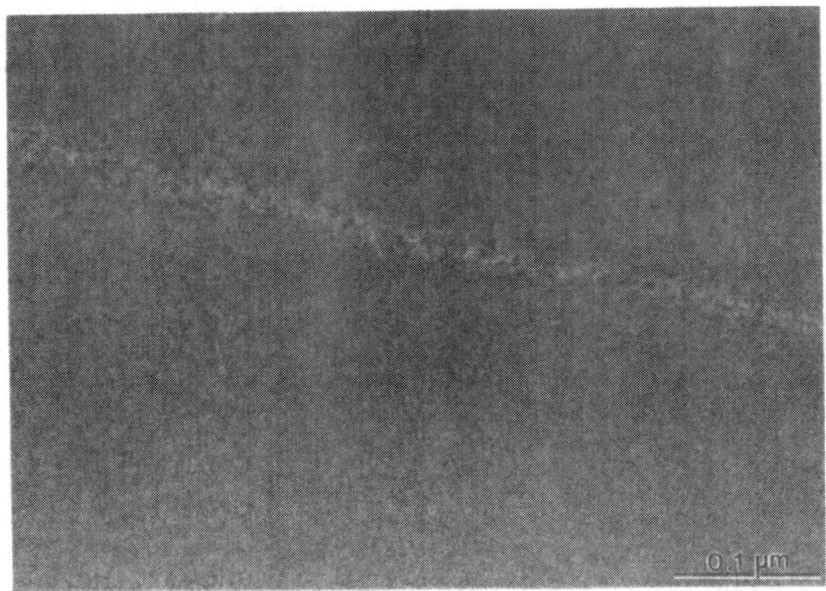

Fig. 2 TEM micrograph of an $Al_{90}Fe_5Gd_5$ amorphous alloy near its bent tip.

Fig. 3 TEM micrograph of a bent $Al_{90}Fe_5Gd_5$ amorphous alloy ribbon.

crystalline aluminum is rather small. The structure and morphology of the crystallites induced after bending are the same as those produced by isothermal annealing or lowering of the quenching speed[13-15].

To further understand the influence of plastic deformation on structural changes in aluminum-based amorphous alloys, these alloys were subjected to high-energy ball-milling. The experiments were performed using a standard SPEX-8000 ball mill. The glass ribbons were cut to 5 cm in length and sealed under argon atmosphere in a hardened tool steel vial. High hardness, good wear-resistant chrome steel balls were used. The diameter of the balls was 3/8 inch and the ball-to-sample weight ratio was about 15:1. After ball-milling for 1 minute, nanocrystals were detected in $Al_{90}Fe_5Gd_5$ glass by TEM dark-field imaging. After milling for 1 hour, both $Al_{90}Fe_5Gd_5$ and $Al_{90}Fe_5Ce_5$ showed traces of f.c.c. aluminum peaks with an amorphous background in the X-ray diffraction pattern. These results are consistent with the results of structural change by bending. The details of the structural changes induced by ball-milling will be published elsewhere.

For the aluminum-based amorphous alloys studied, varying the percentage of the TM or RE element led to different crystallization temperatures. The crystallization temperature for some of the alloys is around 300°C, while for the majority it is around 150°C. An $Al_{85}Ce_5Ni_{10}$ amorphous alloy, which crystallizes around 300°C, was used to study the shear deformation-induced nanocrystal within the shear band. After bending at room temperature, TEM observation revealed no crystalline precipitates in the shear band. Then the bent samples were heated from room temperature to their *normal* crystallization temperature (around 300°C). The aluminum nanocrystals begin to precipitate in shear bands at 150°C. Therefore, the $Al_{85}Ce_5Ni_{10}$ amorphous alloy, which has no shear deformation-induced crystallization at room temperature, does have it when heated after bending to a temperature higher than room temperature but well below its *normal* crystallization temperature.

The possibility of crystallization caused by localized heating arising from plastic deformation was considered. First, the heat generated during plastic deformation in a single shear band was estimated. Then the temperature increment of the shear band was obtained by assuming that the

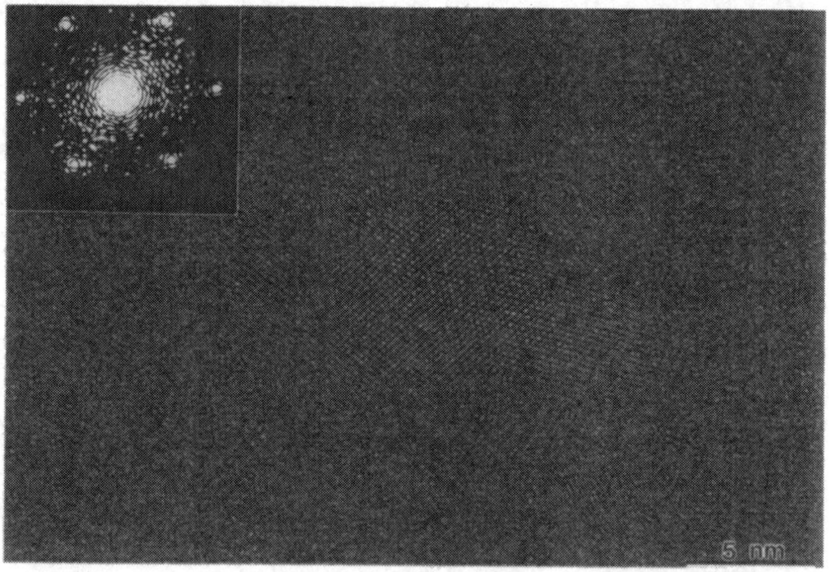

Fig. 4 HRTEM atomic image of a nanometer-sized crystalline precipitate formed within a shear band of an $Al_{90}Fe_5Gd_5$ amorphous alloy after bending. The crystallite is embedded in a matrix which remains amorphous. The inset is the optical diffraction of the atomic image.

work done by the applied stress during deformation of a single slip is transferred to heat and confined within the shear band. It shows a temperature increment of 3000K which is much higher than the crystallization and melting temperatures of the alloy. But the heat transfer from the shear band to the matrix is very fast. The temperature drops from 3000K to 40K within 1 nanosecond. The cooling rate is on the order of 10^{12} K/s which is much higher than the required cooling rate for the formation of amorphous alloys (10^6 K/s). Also, the heat generated in the shear band is not established instantaneously, especially in bending, where shear instability can be avoided. The detailed analysis of the localized heating effect can be found in [15]. The crystallization induced by bending-then-heating in the $Al_{85}Ce_5Ni_{10}$ amorphous alloy proves that the crystallization cannot be caused by localized heating.

Thus the observed bending-induced crystallization is likely due to the large permanent strain within the shear bands. In other words, it is caused mainly by a mechanical driving force. Although the atomic mechanism of such structural change has not been clarified, long range atomic diffusion is probably not involved in the process since temperature is not the main factor. On the other hand, the shear strain within the shear bands can be as large as 100%-1000% in the case of bending, and a large portion of atoms within the shear bands are subject to local displacements under the stress. These local atomic displacements can result in short range changes of topological and chemical order of the amorphous state. It is possible that such atomic rearrangements will lead to shifting of Al atoms towards more stable positions and will cause the formation of Al nanocrystals, since the environment around an Al atom in the amorphous state is not very different from that in the crystalline state. Once the size of an Al cluster reaches a critical radius (1.5-2 nm), thermodynamics favor further growth, thus giving rise to the nano-crystallites observed inside the shear bands. Naturally, such shear strain-induced instability of the amorphous state is highly dependent on the atomic structure and the nature of chemical bonding of the amorphous alloy.

CONCLUSIONS

For the first time, deformation-induced crystallization within the shear bands of amorphous alloys was directly observed. Nanometer-sized crystalline precipitates were found exclusively within the shear band extending across the deformed amorphous alloy ribbons. The crystal structure of these nanocrystal is f.c.c., which is an aluminum structure. The nanocrystal can be produced by either bending or ball-milling. The nature of such shear deformation-induced structural change is likely due to local atomic rearrangements within the shear bands which exhibit enormous plastic strain. Such reshuffling of atoms depends strongly on the topological order and chemical bonding strength of the amorphous state.

This work has been supported by the US Army Research Office under Contract No. DAAL03-91-G-0009.

* H. Chen is now with the Center for Advanced Materials, University of Massachusetts, Lowell, MA.

REFERENCES

1. F.E. Luborsky, Amorphous Metallic Alloys (Butterworth, Boston 1983).

2. H.J. Guntheroldt and H. Beck, Glassy Metals (Springer-Verlag, New York, 1981).

3. G. Wei and B. Cantor, Acta Met., **37**, 3409 (1989).

4. U. Koster and B.P. Witteler, MRS Symp. Proc., **80**, 355 (1987).

5. K. Muller and M.V. Heimendahl, J. Mat. Sci., **17**, 2525 (1982).

6. K. Miyoshi and D.H. Buckley, Thin Solid Films, **118**, 363 (1984).

7. U. Koster, Mat. Sci. Eng., **97**, 233 (1988).

8. Y. He, S.J. Poon and G.J. Shiflet, Science, **241**, 1640 (1988).

9. G.J. Shiflet, Y. He and S.J. Poon, Scripta Met., **22**, 1661 (1988).

10. G.J. Shiflet, Y. He and S.J. Poon, J. Appl. Phys., **64**, 6863 (1988).

11. Y. He, S.J. Poon and G.J. Shiflet, Scripta Met., **22**, 1813 (1988).

12. A. Inoue, K. Ohtera, A.P. Tsai and T. Masumoto, Jpn. J. Appl. Phys., **27**, L479 (1988).

13. Y. He, H. Chen, G.J. Shiflet and S.J. Poon, Phil. Mag. Lett., **61**, 297 (1990).

14. H. Chen, Y. He, G.J. Shiflet and S.J. Poon, Scripta Met., **25**, 1421 (1991).

15. H. Chen, PhD Dissertation (University of Virginia, Charlottesville, Virginia 1992).

16. A. Inoue, K. Ohtera, Z. Tao and T. Masumoto, Jpn. J. Appl. Phys., **27**, L1853 (1988).

17. Y.H. Kim, A. Inoue and T. Masumoto, Mater. Trans. JIM, **31**, 747 (1990).

INTERMEDIATE PHASE NUCLEATION AT DIFFUSION COUPLE BOUNDARIES

J. J. HOYT
Department of Mechanical and Materials Engineering, Washington State University, Pullman, WA 99164-2920

ABSTRACT

The nucleation of intermediate phases at the interphase boundary in diffusion couples is examined theoretically. A variational procedure is described which allows for a self consistent determination of the critical nucleus size and shape, the work of formation of the critical nucleus and the nucleation rate.

INTRODUCTION

The formation of an intermediate phase at the boundary of two other dissimilar phases is a phenomenon which is important in many areas of materials processing. Nucleation of intermediate phase precipitates at semiconductor/metal interfaces has important consequences in the metallization of electronic devices and, as a result, nucleation rates and product phases have been determined in a large number of systems [1]. The fabrication of reinforced ceramics by displacement reaction [2] requires the nucleation and growth of the final phases at the original interphase boundaries. In addition, phase formation at the reinforcement/matrix interface plays a critical role in the processing and mechanical properties of metal matrix composites [3].

Early work in the study of intermediate phase formation at interfaces was performed by Walser and Bene' [4] and Bene' [5] who established an empirical rule for predicting the first phase to form out of the many possible phases dictated by the equilibrium diagram. The Bene' rule holds for many systems, but there are notable exceptions. The early work did not address the rate of nucleation and it was not recognized until somewhat later that solute interdiffusion between the two original phases is a necessary prerequisite to the nucleation of a third phase at the boundary. Subsequent discussions of the intermediate phase formation problem were contained in the studies by d'Heurle [6], Thompson [7] and Desre' [8]. In each of these treatments a specific shape for the critical boundary nucleus was assumed. In a recent paper by Hoyt and Brush [9] it was shown that ad-hoc assumptions concerning precipitate shape need not be invoked; the critical nucleus size and shape is uniquely determined by the mathematics of the variational procedure applied to the system free energy change.

The purpose of the present note is to extend the work of Hoyt and Brush, hereafter referred to as paper I, to more complicated cases. A full numerical solution to the second order differential equation describing the critical nucleus shape will be employed rather than the perturbation technique developed in I. A numerical solution implies that the shape and work of formation of the critical nucleus can be found at all interdiffusion times and not just in the limit of late times. The next section briefly reviews the necessary mathematics and derives a nucleation rate for an intermediate phase. The results section examines the critical nucleus shape and the nucleation rate as a function of the interdiffusion time.

257

THEORETICAL BACKGROUND

Consider a diffusion couple whose terminal phases are denoted by α and β. Furthermore let an intermediate phase γ - that is a phase which appears between the α and β phase fields on the binary phase diagram - possess a free energy vs. composition curve which is sufficiently sharp such that the concentration of solute in the γ nucleus is constant. To simplify further, assume, as was done in I, that the diffusion coefficient is the same in α and β, that the α/γ and β/γ surface energies are equivalent and that the free energy vs. concentration curves for α and β are completely symmetric about the γ phase composition. Although somewhat unrealistic, the latter set of conditions reduce the mathematics to a one sided problem.

With the assumptions outlined above, one can write the work of formation of an interphase boundary precipitate as:

$$\frac{W}{W_-^*} = \frac{12}{P(\theta)}\int_0^a xy\,dx - \frac{12}{P(\theta)}s\int_0^a xy\operatorname{erf}\left(\frac{y}{\sqrt{\tau}}\right)dx - \frac{6}{P(\theta)}\int_0^a x(1+y'^2)^{1/2}\,dx - \frac{6}{P(\theta)}\cos\theta\int_0^a x\,dx \quad (1)$$

where τ is the interdiffusion time, $P(\theta)$ is defined by:

$$P(\theta) = 2 - 3\cos\theta + \cos^3\theta \quad (2)$$

and θ is the contact angle between nucleus and phase boundary plane; the latter quantity is related to various surface energies present [10]. The quantity s contains the thermodynamics of the problem and is given by:

$$s = -\frac{c^i - c^\infty}{c^i - c^o} \quad (3)$$

Here c^i is the α phase composition at the α/β interphase boundary and is derived from the metastable equilibrium tie line between the α and β phases. The quantities c^∞ and c^o are the initial and equilibrium α phase concentrations respectively.

In eq. 1 dimensionless quantities have been employed. As the interdiffusion time becomes very long the problem reduces to the well known case of grain boundary nucleation in a single phase of homogeneous concentration equal to c^i. As time approaches infinity the critical nucleus shape becomes a double spherical cap with a radius of curvature given by [10]:

$$R = \frac{2\sigma}{\Delta F_v^\infty} \quad (4)$$

where σ is the precipitate matrix surface energy and ΔF_v^∞ is the volume free energy change for the constant composition case. The above definition establishes a convenient length scale for the problem. In eq. 1, the distance along the interphase boundary, x, as well as the distance y measured perpendicular to the boundary have been normalized by R. Also the dimensionless time is defined as:

$$\tau = \frac{(\Delta F_v^\infty)^2 D}{\sigma} t \qquad (5)$$

where D is the solute diffusion coefficient and t is the dimensional time. Finally the work of formation has been normalized by the critical work of formation for the grain boundary problem; meaning the ratio on the left hand side of eq. 1 has limits of zero to unity.

To determine the critical work of formation for an interphase boundary nucleus, one must obtain the function $y(x)$ such that the variation of W is zero. The Euler-Lagrange equation describing the nucleus shape [11] is readily found to be:

$$\frac{x}{2} \frac{(y')^2 y''}{[1+(y')^2]^{3/2}} - \frac{x}{2} \frac{y''}{[1+(y')^2]^{1/2}} - \frac{1}{2} \frac{y'}{[1+(y')^2]^{1/2}} - x - sx\mathrm{erf}\left(\frac{y}{\sqrt{\tau}}\right) - \frac{2}{\sqrt{\pi}} \frac{sxy}{\sqrt{\tau}} \exp\left(-\frac{y^2}{\tau}\right) = 0 \qquad (6)$$

where the primes denote spatial derivatives. The boundary conditions appropriate to the problem are $y'(0)=0$ and $y(a)=0$. The value of a is found by maximizing the value of W subsequent to obtaining $y(x)$ via. eq. 6.

In keeping with the dimensional analysis described above, the quantity of interest here is the nucleation rate at any given interdiffusion time divided by the nucleation rate at $t \to \infty$. In general, the nucleation rate can be written as:

$$J = J_o \exp\left(-W^*/kT\right) \qquad (6)$$

where kT has the usual meaning. The pre-exponential factor is not well defined for the interphase boundary nucleation case [10], but one can assume that it should be proportional to the surface area of the precipitate A and a Zeldovich factor Z. In homogeneous nucleation one has:

$$Z = \left(-\frac{1}{2\pi kT} \frac{\partial^2 W}{\partial g^2}\right)^{1/2} \qquad (7)$$

where g is the number of atoms in the nucleus and the second derivative is evaluated at the critical size. In the present case an estimate for Z was found by numerically evaluating the second derivative with respect to the aforementioned quantity a. Thus, the ratio of nucleation rates sought is:

$$\frac{J(\tau)}{J_\infty} = \frac{AZ}{A_\infty Z_\infty} \exp\left[\frac{W_\infty^*}{kT}\left(1 - \frac{W^*}{W_\infty^*}\right)\right] \qquad (8)$$

A typical numerical solution of eq. 1 and 6 reveals that the rate ratio is much more sensitive to the exponential term than the pre-exponential factor. Thus, in the results to follow the term containing the surface areas and Zeldovich factors was taken to be unity.

RESULTS

Figure 1 shows the normalized nucleation rate vs. interdiffusion time found by numerically solving the differential eq. 6 and the work of formation expression 1. In this example, the quantity W_∞ $^*/kT$ was taken to be 45 and the contact angle was 20°. The increase in the nucleation rate with increased interdiffusion demonstrates that some exchange of solute between the two initial phases must take place prior to nucleation. The method outlined here represents a quantitative means of predicting the first intermediate phase to form and one need not rely on the empirical prediction of ref. 4. In other words, by generating a result similar to fig.1 for any possible intermediate phase, one can immediately determine the first phase to nucleate.

The critical nucleus shapes at the interdiffusion times labeled 1 and 2 in fig.1 are shown in figs. 2 and 3 respectively. For early times when the concentration profile in the vicinity of the interphase boundary is sharp the nucleus is larger and occupies a greater area of boundary than for later times. As time approaches infinity the double spherical cap (dotted line) shape is reproduced.

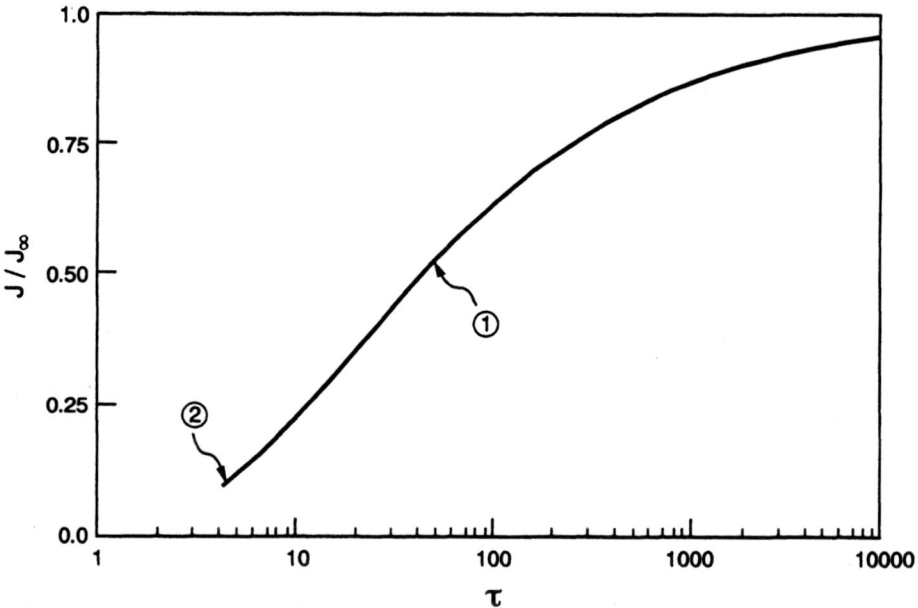

Figure 1. The normalized nucleation rate as a function of the interdiffusion time for a contact angle θ equal to $20°$ and W_∞^* / kT equal to 45.

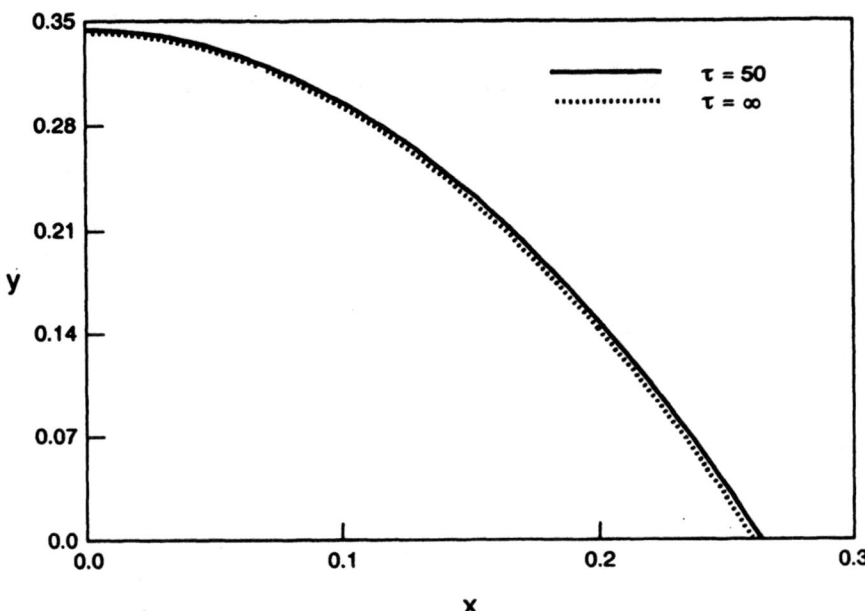

Figure 2. The critical nucleus shape $y(x)$ for the interdiffusion time labeled 1 in fig. 1. The dashed curve is the double spherical cap shape obtained at $\tau \to \infty$.

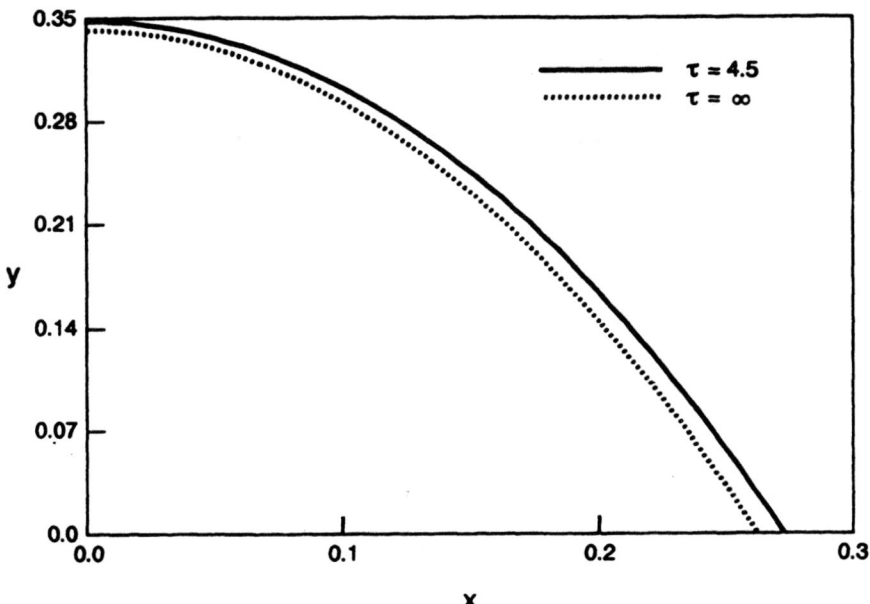

Figure 3. The critical nucleus shape for the interdiffusion time labeled 2 in fig.1.

CONCLUSIONS

The critical nucleus size and shape, the work of formation of the critical nucleus and the nucleation rate for intermediate phase nucleation at interphase boundaries can be determined self consistently by a variational technique. With the method, one can determine which intermediate phase will form first without resorting to empirical rules and one need not assume a given critical nucleus geometry.

ACKNOWLEDGMENTS

This work was supported by the National Science Foundation under contract # DMR 8919193.

REFERENCES

1.   K. N. Tu and J. W. Mayer, in Thin Films - Interdiffusion and Reactions, edited by J. M. Poate, K. N. Tu and J. W. Mayer (Wiley, New York, 1978).
2.   C. H. Henager, J. L. Brimhall and J. P. Hirth, Mater. Sci. Eng., A155, 109 (1992).
3.   J. H. Perepezko, to be published Comp. Interfaces, (1993).
4.   R. M. Walser and R. W. Bene', Appl. Phys. Lett., 28, 624 (1976).
5.   R. W. Bene', Appl. Phys. Lett., 41, 529 (1982).
6.   F. M. d'Huerle, J. Mater. Res., 3, 167 (1988).
7.   C. V. Thompson, J. Mater. Res., 7, 367 (1992).
8.   P. J. Desre', Acta Metall., 39, 2309 (1991).
9.   J. J. Hoyt and L. N. Brush, to be published.
10. J. W. Christian, Transformations in Metals and Alloys, (Pergamon Press, Oxford, 1975) p. 452.
11. C. Fox, An Introduction to the Calculus of Variations, (Dover Publications Inc.,New York, 1987).

THE ELASTIC STRAIN ENERGY OF COHERENT ELLIPSOIDAL PRECIPITATES IN ANISOTROPIC CRYSTALLINE SOLIDS: APPLICATIONS TO THE ARAGONITE–CALCITE TRANSFORMATION

MING LIU, Department of Geological Sciences, Brown University, Providence, RI 02912

ABSTRACT

Eshelby's theory has been applied to the aragonite–calcite transformation in order to calculate the elastic strain energy of coherent ellipsoidal precipitates in anisotropic solids. Using a quasi-Newton's method and a finite–difference gradient, the elastic strain energy was incorporated in order to calculate the activation energy ΔG^* for homogeneous nucleation under hydrostatic and non-hydrostatic conditions.

INTRODUCTION

The elastic strain energy associated with a coherent ellipsoidal precipitate has received much attention in the literature. As first shown by Eshelby[1], assuming isotropic elasticity, the stress field inside an ellipsoidal precipitate is constant when the stress–free transformation strain is constant. Eshelby's theory has since been extended to the anisotropic case[2-6]. However, most previous studies on the elastic strain energy either focused on the isotropic case[7-9] or assumed that the precipitate and matrix have the same elastic stiffnesses[9]. Recent studies have incorporated the anisotropy and elastic inhomogeneity in applications to some metal systems and have broadened the applicability of Eshelby's theory[6]. Metals generally have high symmetries; hence, the calculation of elastic strain energy has been restricted to an ellipsoid of revolution. The tri–axial case is very pertinent to phase transformations in ceramics and geological materials because these materials generally have lower symmetries.

Recently, workers have considered the effect of elastic strain energy on the activation energy for nucleation ΔG^*[6,9,10]. Lee et al.[6] have treated the case of an ellipsoid of revolution and have demonstrated the important role of elastic strain energy in determining ΔG^* and the shape of a critical nucleus. However, an ellipsoid of revolution is not a general case, and the effect of an external load on the elastic strain energy and the activation energy for nucleation has not been considered, although Eshelby[1] discussed the total energy change for a homogeneous system.

$CaCO_3$ is a common geologic material which has two polymorphs: aragonite (orthorhombic) and calcite (rhombhedral). Preserved evidence for the aragonite to calcite transformation is often used to infer the conditions of geological processes[10,11]. This phase transformation will be used as an example to first calculate the elastic strain energy for coherent ellipsoidal precipitates as a function of the ratios of the lengths of the three axes of an ellipsoid, following the procedure of Lee et al.[6] Then, using a quasi–Newton's method and a finite–difference gradient, the elastic strain energy will be incorporated into the minimization process to calculate the activation energy ΔG^* for the simple case of homogeneous nucleation and orientation–independent interfacial energy.

THEORY

For an arbitrary anisotropy, it has been shown[5,6] that if an ellipsoidal precipitate undergoes a uniform 'stress–free transformation strain', e_{ij}^T, the constrained strain e_{ij}^c is given as

$$e_{ij}^c = S_{ijkl}\, C_{klmn}\, e_{mn}^T, \tag{1}$$

where C_{klmn} are the elastic stiffnesses of the matrix phase, and in spherical coordinates

$$S_{ijkl} = \frac{\alpha\,\beta}{8\pi} \int_0^{2\pi} d\theta \int_0^{\pi} \sin\phi\, d\phi\; \frac{z_i\, z_l\, M_{jk}^{-1} + z_j\, z_l\, M_{ik}^{-1}}{(z_1^2 + \alpha^2\, z_2^2 + \beta^2\, z_3^2)^{3/2}}, \tag{2}$$

263

where

$$M_{ik} = C_{ijkl}\, z_j\, z_l,$$

and

$$z = (\sin \phi \cos \theta,\ \sin \phi \sin \theta,\ \cos \phi),$$

where $\alpha = a_2 / a_1$, $\beta = a_3 / a_1$ and a_1, a_2, and a_3 are the three semi-axes of an ellipsoid. The elastic strain energy per unit volume of a precipitate is given by[6]

$$W = \frac{1}{2}\ (e_{kl}^{T} - e_{kl}^{c})\, C_{ijkl}\, e_{ij}^{T}. \tag{3}$$

In the presence of an applied load where a precipitate forms in a system with strain e_{ij}^{A} due to an external stress P_{ij}^{A}, the mechanical energy change ΔE for this system due to the formation of a unit volume of a precipitate would be[1]

$$\Delta E = W + W_{int}, \tag{4}$$

where W_{int} represents the interaction energy between the precipitate and the applied load and is given by[1]

$$W_{int} = -\, C_{ijkl}\, e_{kl}^{A}\, e_{ij}^{T}. \tag{5}$$

In the case of an inhomogeneous system, where the elastic stiffnesses of the precipitate and matrix are not the same, an 'equivalent' stress–free transformation strain e_{ij}^{T} can be obtained by solving[6]

$$C_{ijkl}^{*}\ (e_{kl}^{c} - e_{kl}^{T*}) = C_{ijkl}\ (e_{kl}^{c} - e_{kl}^{T}), \tag{6}$$

where C_{ijkl}^{*} are the elastic stiffnesses of the precipitate and e_{ij}^{T*} is the actual stress–free transformation strain suffered by the precipitate. Therefore, the strain energy per unit volume of such a precipitate is given by[6]

$$W = \frac{1}{2}\ (e_{kl}^{T*} - e_{kl}^{c})\, C_{ijkl}^{*}\, e_{ij}^{T*}. \tag{7}$$

Analogous to the homogeneous case, if a system is under an externally applied load which produces stress P_{ij}^{A} and strain e_{ij}^{A}, then the mechanical energy change ΔE for the system after formation of a unit volume of such a precipitate is also given by Eq. (4), and W is given by Eq. (7). The interaction energy is given by

$$W_{int} = -\frac{1}{2}\ (C_{ijkl}^{*}\, e_{kl}^{A}\, e_{ij}^{T*} + C_{ijkl}\, e_{kl}^{A}\, e_{ij}^{T}). \tag{8}$$

For the case of an applied load, an 'equivalent' stress–free transformation strain e_{ij}^{T} is obtained by solving[1]

$$C_{ijkl}^{*}\ (e_{kl}^{c} - e_{kl}^{T*} + e_{kl}^{A}) = C_{ijkl}\ (e_{kl}^{c} - e_{kl}^{T} + e_{kl}^{A}). \tag{9}$$

COMPUTATIONAL PROCEDURE

For the given values of α and β, the S_{ijkl} in Eq. (2) can be calculated by numerical integration if the orientation relationship between the precipitate and matrix is known. Because $S_{ijkl} = S_{jikl}$ in Eq. (1), Eqs. (1) and (6) [or Eqs. (1) and (9) in the case of externally applied load] can be combined as a group of 12 linear equations with the constrained strain e_{ij}^{c} and the 'equivalent' stress–free transformation strain e_{ij}^T as the independent variables. The actual stress–free transformation strain e_{ij}^{T*} can be evaluated using the lattice parameters of both precipitate and matrix phases and the orientation relationship between the two phases.

The aragonite to calcite transformation has been extensively studied[10,11]. Although there are several possible orientation relations between the two phases during the transformation, the predominant relation for calcite nucleating inside single crystal aragonite is $(001)_A = (0001)_C$ and $[100]_A = [10\bar{1}0]_C$ [11]. This orientation has been used in the present study and the Cartesian coordinates taken parallel to the crystal axes of aragonite. Hence, the stress–free transformation strain e_{ij}^{T*} can be calculated from the lattice parameters[12] and the corresponding thermal expansion data of both phases[13]. The elastic stiffnesses of the two phases[13] are listed in Table 1. In order to clearly show the effect of applied load, large values of applied stress P_{ij}^A and strain e_{ij}^A have been chosen (Table 1), although aragonite would probably fail at such a high stress[14].

Table 1. Elastic stiffnesses of aragonite and calcite[13] and applied stress P_{ij}^A and strain e_{ij}^A

Phase	C_{ij} (Mbar)	11	22	33	44	55	66	12	13	23	14
Aragonite		1.60	0.87	0.85	0.41	0.26	0.43	0.37	0.02	0.168	0
Calcite		1.445	1.445	0.831	0.327	0.327	0.437	0.571	0.534	0.534	–0.205

$$P_{ij}^A \text{ (kbar)} = \begin{bmatrix} 5 & & 0 \\ & 5 & \\ 0 & & -10 \end{bmatrix} \qquad e_{ij}^A = \begin{bmatrix} 0.24\% & & 0 \\ & 0.75\% & \\ 0 & & -1.36\% \end{bmatrix}$$

RESULTS OF CALCULATED STRAIN ENERGY

The strain energies W, as a function of the ratios of the three semi–axes, α and β, calculated at 400°C, 1 atm are normalized to the strain energy per unit volume of a spherically shaped (i.e., $\alpha = 1$ and $\beta = 1$) transformed phase W_s and are plotted in Figs. 1a and 1b.

Similarly, the normalized values for the mechanical energy changes, $\Delta E/W_s$ at 400°C under applied stress P_{ij}^A (Table 1), are plotted on Figs. 1c and 1d as a function of α and β. The overall trend of $\Delta E/W_s$ as a function of α and β is similar to that of W/W_s. However, the value of ΔE can be negative because ΔE is the total mechanical energy change which consists of the interaction energy term W_{int} [Eqs. (4) and (8)] which can be either positive or negative depending on the value of P_{ij}^A.

INCORPORATION OF STRAIN ENERGY INTO ΔG^* CALCULATION

Theory and Computational Procedure

The standard free energy change, ΔG°, associated with the formation of an ellipsoidal nucleus under hydrostatic condition is given as[6]

$$\Delta G^\circ = (\Delta G_v + W) V + \sigma S, \qquad (10)$$

where ΔG_v is the volume free energy change associated with nucleation, W is the elastic strain energy per unit volume which is a function of α and β, σ is the interfacial energy between the nucleus and matrix, V and S are the volume and interfacial area of the ellipsoidal nucleus, and S can be expressed as

$$S = 8 a_1^2 S'(\alpha, \beta), \qquad (11)$$

where

$$S'(\alpha, \beta) = \int_0^{\pi/2} d\theta \int_0^1 \left[\frac{\beta^2 t^2 (\alpha^2 \cos^2\theta + \sin^2\theta)}{1 - t^2} + \alpha^2 \right]^{1/2} t \, dt . \qquad (12)$$

265

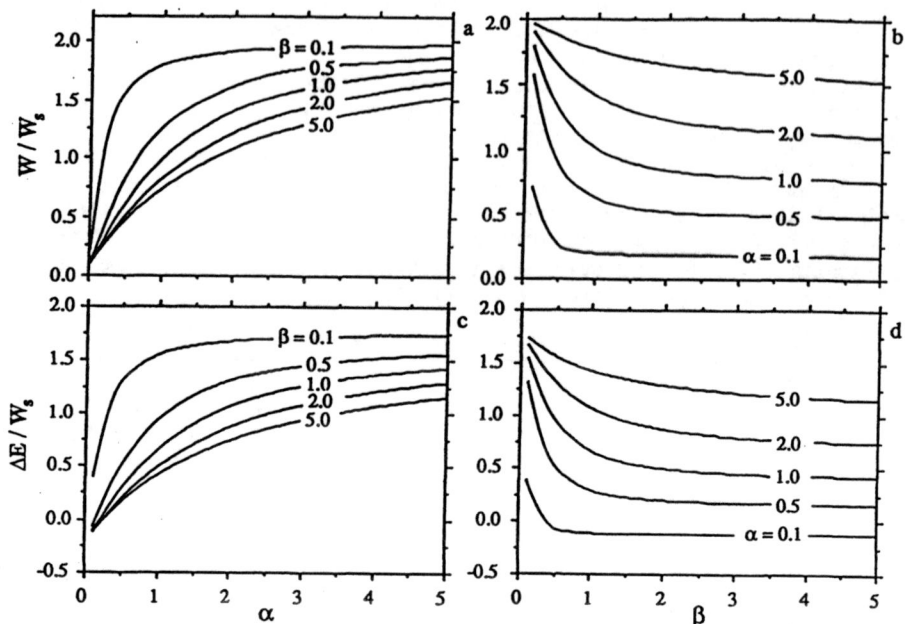

Fig. 1. Normalized strain energies W / W_s (a and b at 1 atm) and mechanical energy changes $\Delta E / W_s$ (c and d for an applied stress P_{ij}^A listed in Table 1) as a function of the ratios of the lengths of three semi-axes of an ellipsoid, α and β, for the aragonite–calcite transformation at 400°C. W_s (α, $\beta = 1$) = 2.15×10^5 kJ m^{-3}.

Although the interfacial energy σ is probably a function of orientation in a solid–solid transformation, it is assumed to be orientation independent in order to simplify the calculation and because of the large uncertainties in estimating σ.

From Eq. (10), the free energy change associated with formation of the critical nucleus, i.e., the activation energy for nucleation, can be written as:

$$\Delta G^* = f(a_1^*, \alpha^*, \beta^*) \tag{13}$$

where a_1^*, α^*, and β^* simultaneously satisfy the following equations:

$$[\partial(\Delta G^\circ) / \partial a_1]_{\alpha, \beta} = 0, \tag{14}$$

$$[\partial(\Delta G^\circ) / \partial \alpha]_{a_1, \beta} = 0, \tag{15}$$

$$[\partial(\Delta G^\circ) / \partial \beta]_{a_1, \alpha} = 0. \tag{16}$$

The differentiation of Eq. (14) gives:

$$a_1^* = -\frac{4 \sigma S'(\alpha, \beta)}{\pi \alpha \beta (\Delta G_v + W)}. \tag{17}$$

The differentiation of Eqs. (15) and (16) can not be performed analytically due to the complicated functional dependence of W and S' on α and β. However, a numerical method can be used to determine α^* and β^*. Substituting Eq. (17) into Eq. (10) gives:

$$\Delta G^* = \frac{4}{3}\, \pi\, a_1^{*3}\, \alpha\, \beta\, (\Delta G_v + W) + 8\, a_1^{*2}\, \sigma\, S'(\alpha, \beta). \qquad (18)$$

Using a quasi–Newton's method and a finite–difference gradient, α^* and β^* can be determined by minimizing the function ΔG^*. During the minimization process, $S'(\alpha, \beta)$ can be evaluated by numerical integration.

It should be emphasized that the above discussion is suitable to any hydrostatic case provided the strain energy W is calculated using the lattice parameters corresponding to that specific hydrostatic condition. In the non–hydrostatic case, the calculation procedure is similar except that W is replaced by ΔE, and ΔG_v is calculated at the corresponding temperature and mean normal stress $[(\sigma_1 + \sigma_2 + \sigma_3)/3]$ condition while taking the applied stress P_{ij}^A as a deviatoric stress.

Input Data and Calculated Results

In order to perform the calculation discussed above, the interfacial energy σ as well as ΔG_v must be known. ΔG_v can be calculated from thermodynamic data[15]. However, due to the unavailability of experimental data, the value of σ must be estimated.

Girifalco and Good[16] have related the free energies of cohesion of separate phases to the free energy of adhesion to yield the following relation:

$$\sigma_{ab} = \gamma_a + \gamma_b - 2\Phi(\gamma_a\, \gamma_b), \qquad (19)$$

where γ_a and γ_b are the surface energies of phases a and b, respectively, while σ_{ab} is the interfacial energy between the a and b phases. The value of Φ can be approximated as

$$\Phi = \frac{4\, V_a^{1/3} V_b^{1/3}}{(V_a^{1/3} + V_b^{1/3})^2}, \qquad (20)$$

where V is the molar volume.

The surface energy γ in Eq. (19) can be estimated using the Obreimoff–Gilman method[7,17]:

$$\gamma = E\, r^2 / d_0\, \pi^2, \qquad (21)$$

where E is Young's modulus normal to the plane of interest, which is readily obtainable from elastic compliances, and r is the atomic radius of atoms lying in that plane. The value of d_0 is equal to the interplanar spacing (see Brace and Walsh[17] for details). The calculated surface energies based on this theory agree very well with those of the experimental measurements for a number of materials, including calcite[17]. The calculated surface energies range from 270 to 840 mJ/m^2 for aragonite and from 190 to 480 mJ/m^2 for calcite[17].

Once the surface energy γ is approximated, the interfacial energy σ_{ab} between contacting faces of transformed phase and matrix can be calculated using Eqs. (19) and (20). Because the uncertainty associated with Eqs. (19) and (20) is unknown and may be large, the average interfacial energy σ is calculated to be ~20 mJ/m^2 by averaging the interfacial energies corresponding to the three faces perpendicular to the three axes of the ellipsoid.

The calculated activation energies ΔG^* are plotted as $\log_{10}(\Delta G^*/k_B T)$, where k_B is Boltzmann's constant, against temperature T

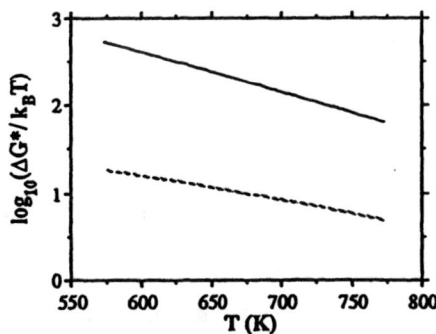

Fig. 2. Logarithm of $\Delta G^*/k_B T$ as a function of T for the homogeneous nucleation of calcite in aragonite. The solid and dashed lines are for 1 atm and an applied stress P_{ij}^A (Table 1), respectively.

on Fig. 2, and the solid and dashed lines on Fig 2 represent the conditions at 1 atm and the applied stress P_{ij}^A listed in Table 1, respectively. Fig. 2 indicates that ΔG^* for both cases decreases significantly with increasing temperature because of the increasing driving force for nucleation. The dashed line on Fig. 2 also indicates that with the assumed applied stress P_{ij}^A (Table 1), ΔG^* can be substantially decreased. However, in order to have a significant effect on ΔG^*, a very large differential stress may be required. This high differential stress may only be achievable at a high confining pressure because at ambient conditions most materials will undergo brittle failure.

CONCLUSIONS

Using Eshelby's method, the elastic strain energy for coherent ellipsoidal precipitates in anisotropic solids has been calculated for the aragonite to calcite transformation. The normalized strain energy W/W_s (hydrostatic condition) and mechanical energy change $\Delta E/W_s$ (non-hydrostatic condition) are reported as a function of the ratios of three semi-axes of an ellipsoid (Fig. 1). The results indicate that the strain energy of an ellipsoid is markedly different from that of a sphere.

A quasi-Newton's method and a finite-difference gradient were used to calculate the activation energy ΔG^* for homogeneous nucleation by incorporating the elastic strain energy. For the calculation, an orientation-independent interfacial energy was assumed and estimated using Girifalco and Good's method[16], and the surface energy was estimated using the Obreimoff–Gilman method[7,17]. The calculations indicate that ΔG^* decreases markedly with increasing chemical driving force, and an applied stress P_{ij}^A with proper values may substantially decrease ΔG^* compared to that at hydrostatic condition.

ACKNOWLEDGEMENTS

The author wishes to thank R. A. Yund, J. Tullis and J. R. Farver for their helpful comments. This research was supported by grant EAR–8904530 from the National Science Foundation to RAY.

REFERENCES

1. J.D. Eshelby, Proc. Roy. Soc. A241, 376 (1957); A252, 561 (1959); Progress in Solid Mechanics 2, 89 (1961).
2. L.J. Walpole, Proc. Roy. Soc. A300, 270 (1967).
3. J.R. Willis, Adams Prize Essay, University of Cambridge, 1970 (unpublished).
4. N. Kinoshita and T. Mura, Phys. Status Solidi A5, 759 (1971).
5. R.J. Asaro and D.M. Barnett, J. Mech. Phys. Solids 23, 77, (1975).
6. J.K. Lee, D.M. Barnett, H.I. Aaronson, Metall. Trans. A 8, 963 (1977); S. Satoh and W.C. Johnson, ibid., 23, 2761 (1992).
7. K. Robinson, J. Appl. Phys. 22, 1045 (1951). J.J. Gilman, ibid., 31, 2208 (1960).
8. D.M. Barnett, J.K. Lee, H.I. Aaronson, K.C. Russell, Scr. Metall. 8, 1447 (1974); K.C. Russell, ibid., 3, 313 (1969).
9. M. Shibata, and K. Ono, Acta Metall. 23, 587 (1975); J.A. Wert, ibid., 24, 65 (1976); M. Hayakawa and M. Oka, ibid., 32, 1415 (1984); F.K. LeGoues, H.I. Aaronson, Y.W. Lee, ibid., 32, 1845 (1984).
10. M. Liu and R.A. Yund, Contrib. Mineral. Petrol. 114, 465 (1993).
11. W.D. Carlson and J.L. Rosenfeld, J. Geol. 89, 615 (1981).
12. W.A. Deer F.R.S., R.A. Howie, J. Zussman, An Introduction to the Rock–Forming Minerals, (Longman Group Limited, London, 1975), p. 477, pp. 497–498.
13. B.J. Skinner, in Handbook of Physical Constants, edited by S.P. Clark, Jr. (Geo. Soc. Am. Memoir 97, New York, NY, 1966) pp. 75–96; F. Birch, ibid., pp. 97–173.
14. M. Liu, R.A. Yund, J. Tullis, EOS, Trans. Am. Geophys. Union 74, 162 (1993).
15. T.J.B. Holland and R. Powell, J. Metamorphic. Geol. 3, 343 (1985).
16. L.A. Girifalco and R.J. Good, J. Phys. Chem. 61, 904 (1957).
17. W.F. Brace and J.B. Walsh, Am. Mineral. 47, 1111 (1962).

PART IV

Kinetics of Crystallization

ANALYSIS OF CRYSTAL NUCLEATION AND GROWTH IN AMORPHOUS COBALT DISILICIDE.

D.A.Smith*, P.V.Evans** and S.R.Koppikar*
* Stevens Institute of Technology, Hoboken NJ 07030
** Alcan Research Center, Banbury, Oxon, UK

ABSTRACT

Extensive *in situ* investigations of the crystallization of amorphous cobalt disilicide have been conducted using a transmission electron microscope with a hot stage. Thermodynamic and kinetic parameters describing the heterogeneous transformation have been evaluated. The angle of contact betwen crystalline and amorphous material was determined from tilting experiments to be 76° . Nucleation rates in samples 40 nm thick were evaluated at various temperatures and compared with thermodynamic models to deduce an interfacial energy between amorphous and crystalline $CoSi_2$ (σ_{ca}) of 121 mJ/m^2 and an activation energy for crystallization of 1.27 eV. Johnson-Mehl-Avrami analysis of the observed continuous nucleation and steady state isotropic growth in 100 nm thick samples points to a gradual transition of the crystallization mode from 3-dimensional (n=4) to 2-dimensional (n=3) growth as might be expected. Comparison of the nucleation and growth rates in 40 nm and 100 nm thick samples demonstrated the influence of surfaces on crystallization phenomena in thin films.

INTRODUCTION

Silicides for application as contacts in integrated circuits can be formed by chemical reactions between a deposited metal film and a silicon substrate. This method requires interdiffusion which requires time at temperatures undesirably high for semiconductor device fabrication; a further difficulty is that precursors may crystallize before the formation of the desired phase, e.g., Co_2Si and $CoSi$ before $CoSi_2$.[1] Another method is the crystallization of a co-deposited amorphous film. This requires no long range diffusion; since the stoichiometry is easily controlled by the co-deposition process the formation of precursors is also precluded. Of all the "self-aligned" silicides available $TiSi_2$ and $CoSi_2$ seem the most suitable choices. Although $CoSi_2$ is not as susceptible to impurity and dopant interactions as $TiSi_2$, exposure to high temperatures induces grain coarsening. The interface between silicon and silicide subsequently roughens, leading to penetration of underlying shallow junctions. It is therefore important to elucidate the kinetics and thermodynamics of $CoSi_2$ formation and crystallization.

Crystallization of amorphous $CoSi_2$ was previously studied by resistometry[1]. This approach provides a measure of the total fraction transformed but does not permit the

271

nucleation and growth kinetics to be evaluated separately. In situ studies of crystallization allow the determination of the fraction of crystallized material, the nucleation rate, the growth rate and the influences of temperature, time or irradiation (ion or electron). [1,2,3]

EXPERIMENTAL PROCEDURE

Amorphous films of $CoSi_2$, 40 and 100 nm in thickness, were prepared by dual e-gun co-deposition at room temperature onto amorphous silicon nitride window substrates. A combined deposition rate of approximately 3 nm s^{-1} was maintained at a base pressure of $\approx 10^{-8}$ Torr. No further sample preparation was necessary for TEM observation. The stoichiometry of the deposited $CoSi_x$ was verified by Rutherford backscattering. Both cobalt rich and silicon rich alloys were studied, but in this paper we concentrate on the stoichiometric compound. Thermal crystallization was carried out in a Philips EM430 operating at 300kV under defocussed illumination. The heating holder was stable enough to maintain a selected field of view throughout the process. For experimental convenience a region near a corner of a window was selected for observation in each case. TEM images obtained at regular intervals during crystallization of the 100 nm films were digitized using a scanner; crystallized area fraction was then computed from each micrograph using public domain imaging software from the NIH. This data was used in the Avrami analysis reported later.

Angle of contact

It was observed that small crystals (i.e. those where the diameter was smaller than the film thickness) projected as circles when the beam was normal to the specimen, but as ellipses upon tilting, suggesting that the crystals were not spheres in a matrix but spherical caps at the surface or discs. Measurement of the projected ellipse minor axis enabled us to

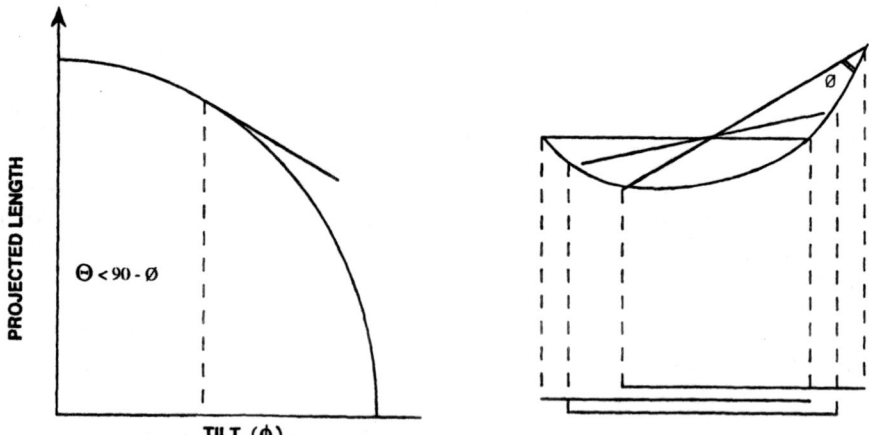

Fig. 1 Illustrates the variation of the projected length with tilting, for a spherical cap.

estimate the contact angle of the cap and to establish that the crystals were spherical caps. Consider a spherical cap, of contact angle θ, radius of curvature r and which projects as a circle of radius given by $a = r \sin \theta$ when viewed normally (Fig. 1).

On tilting through ϕ, provided $\phi > \pi/2 - \theta$, it may easily be shown that the projected minor axis of the ellipse will have a length l such that

$$l = \frac{a}{\sin \theta} + a \tan (\theta - \phi)(\sin \phi + \frac{\cos \phi}{\tan \theta}) \qquad (1)$$

For small tilts, the projected axis is simply $l = 2a \cos \phi$. Using equation (1), we have estimated that $\theta = 76°$ (± 7). The tilt angle ϕ was determined to an accuracy of $\pm 0.5°$ from the change in separation of two particles widely spaced compared to their diameters.

RESULTS

Nucleation analysis

(1) Surface nucleation kinetics

The TEM observations indicate that nuclei initially grow as spherical caps with a contact angle given by $\theta = 76°$. Ignoring transient effects, the steady state nucleation rate, I_s, for such spherical caps per unit surface area is written most generally as:

$$I_s \Sigma \left[1 / N_e(n) \, k_n{}^+ \right] = 1 \qquad (2)$$

where n is the number of atoms in a given cap, $N_e(n)$ is the equilibrium number of such caps per unit area of surface, $k_n{}^+$ is the rate constant for the reaction whereby a cluster of size n is promoted to $n+1$. The summation is carried out between limits $n = u$ and $n = v$ such that $u << n^* << v$ where n^* is the number of atoms in the critical nucleus (that size of cap which is in unstable equilibrium with the parent phase)[2]. I_s is determined by measuring $N(l)$, the population of crystals of diameter l within the area under observation, from the diameter statistics and converting this to $N(t)$ knowing the growth rate. A representative micrograph and a histogram are shown in figs. 2 and 3.

The equilibrium distribution of clusters is given by $N_e(n) = N_s \exp(-\Delta G_n / kT)$, where N_s is the number of atoms of the initial phase per unit area in the monolayer at the surface, and ΔG_n is the work of formation of a cap containing n atoms. The requirement for static equilibrium of the interfacial energies allows ΔG_n to be written

$$\Delta G_n = -n\Delta G_c + \sigma_{ca} (36 \pi v^2 f(\theta) n^2)^{1/3} \qquad (3)$$

where σ_{ca} is the crystal - amorphous interfacial energy, v is the atomic volume and $f(\theta) = (2 + \cos\theta) (1 - \cos^2\theta) / 4$.

The rate constant, k_n^+ can be decomposed to the product of the number of surface sites available on a cap of size n, $O(n)$, and a jump frequency, $\gamma\exp(-\delta g_n / 2kT)$, where γ is the unbiased atomic jump frequency, and δg_n is the free energy by which a cap of size $n+1$ exceeds a cap of size n Consideration of the geometry of a spherical cap yields $O(n) = 2n^{2/3} (1 - \cos\theta)[f(\theta)]^{-2/3}$. We assume the unbiased jump frequency is thermally activated and write $\gamma = \nu_0 \exp(-Q/kT)$, where ν_0 is the jump attempt frequency.

(2) The free energy of crystallization

 The Gibbs free energy of crystallization, that is the difference in free energy between the amorphous and crystalline phases can be expressed as a sum of enthalpic and entropic contributions , $\Delta G_c = \Delta H_c - T\Delta S_c$. The enthalpy of crystallization, ΔH_c was determined from differential scanning calorimetry to have a value of 0.313 kJ/mol [3]. In glass-forming systems it is usually found that the large entropy difference between liquid and crystal at the melting temperature, ΔS_f , is progressively reduced as the temperature is lowered, because of the excess specific heat of the glass relative to the crystal, ΔC_p. At a temperature T below T_m , the residual excess entropy may be expressed

$$\Delta S_c = \Delta S_f - \int_T^{T_m} \frac{\Delta C_p}{T} \, dT \qquad (4)$$

Indeed, an 'ideal' glass transition can be defined as the temperature at which undercooled liquid and crystal are isentropic, (*i.e.* $\Delta S \rightarrow 0$, of necessity at a temperature lower than the physically measured glass transition). If we neglect the residual entropy of crystallization($\Delta S_c \approx 0$), then ΔH_c represents an upper bound for the free energy of crystallization, ΔG_c .

(3) The interfacial jump frequency

 The jump frequency can be determined from crystal growth data, if it is accepted that the process of transporting an atom across the amorphous-crystal interface is similar during both nucleation and growth. Smith, Tu and Weiss [4] reported the size-independent crystal growth velocities for $CoSi_2$ at temperatures 427, 440 and 453 K. Assuming an Arrhenius type behavior for growth rates of the form $u=u_0 \, exp(-Q/kT)$ yields an activation energy for growth of Q = 1.09 eV per atom in 40 nm thick films. Writing the growth rate in conventional form as u = $f\nu_0 \, exp(-Q / kT)\lambda[1-exp(-\Delta G_c / kT)$]; λ is an interatomic distance, here taken as an average atom diameter of 3 Å, ΔG_c is the free energy of

Fig. 2 Bright field image of partially crystallized $CoSi_2$ in a 40 nm thick film.

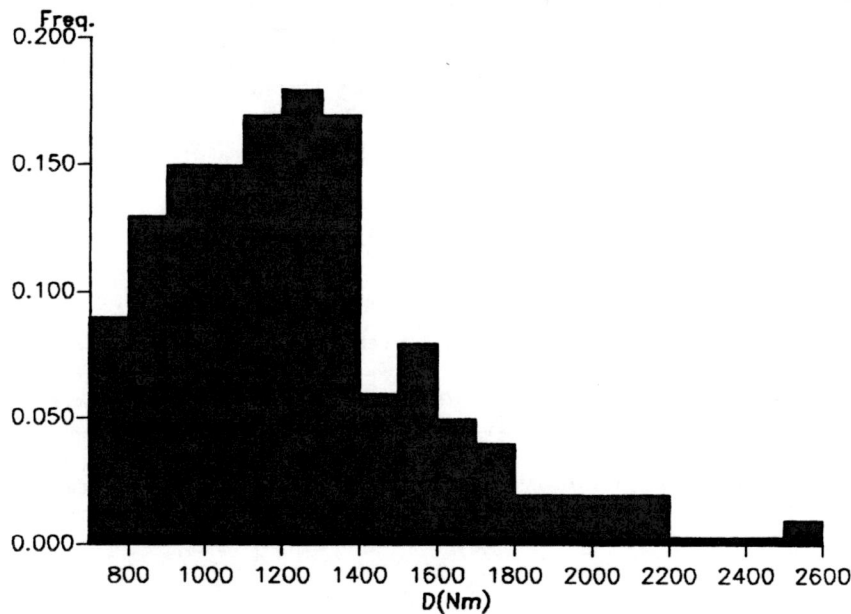

Fig. 3 A histogram showing the typical size distribution of crystals at a particular time for a 40 nm thick film.

crystallization and f is the fraction of interface sites at which growth occurs. It is seen that over the range in which the measurements were taken, the thermodynamic factor in square brackets (which accounts for the reverse flux of atoms from crystal to amorphous phase) varies very little, having an average value of around 0.61 . Thus the attempt frequency, $f\nu_0$ has a value of the order of 10^{13} s^{-1} , which is a reasonable atomic vibration frequency (with f=1). One would then expect a diffusion coefficient $D = (1/6)\lambda^2\nu_0 \, exp(-Q/kT)$ in the bulk amorphous phase; substitution of the above values and the melting temperature (T_m =1325 °C) yields $D(T_m) \approx 10^{-10}$ m^2 s^{-1} which is a plausible value for diffusivity in a liquid at its melting point, suggesting that the analysis is physically reasonable.

(4) The amorphous-crystal interface energy

We can now attempt to fit the nucleation measurements conducted on 40 nm films, using equation 2. Having made independent measurements and estimates of the cap contact angle, θ, the attachment kinetic parameters embodied in ν and the enthalpy of crystallization, ΔH_c, we are left with the evaluation of two thermodynamic quantities, namely the interfacial free energy σ between the amorphous and crystalline phases, and the entropy of crystallization, ΔS_c .

It has long been known that 'ideal' glasses are not achieved in practice [5]. A glass, viewed as a configurationally frozen liquid, can be characterized by the amount of excess entropy frozen in (measured relative to the entropy of the equilibrium crystal)[6]. Hence even an ideal glass would be expected to exhibit a non-zero ΔS_c. Thus far, we have assumed this residual excess entropy to be negligibly small. Clearly, the excess entropy may not exceed the entropy of fusion, ΔS_f . At temperatures below T_g we assume that the heat capacities of glass and crystal are very similar, consequently the enthalpy and entropy of crystallization do not vary with temperature. The free energy of crystallization may then be written:

$$\Delta G_c = \Delta H_c - T\Delta S_c \qquad (5)$$

As we already know ΔH_c from calorimetry, we are left with the evaluation of ΔS_c and σ. Fig.4 shows the results of two fits of experimental data on nucleation in a 40 nm film with these equations. The predicted curve with an initial value of $\sigma = 0.139$ Jm^{-2} has a slope different from that of the data. We find that the slope of the predicted curve can indeed be varied by varying σ. The combination of ΔS_c and σ giving the correct magnitudes of nucleation rates and the correct temperature dependence are $\Delta S_c = 2.16$ x 10^{-5} K^{-1}atom^{-1} and $\sigma = 0.121$ J m^{-2}, as shown in Fig.4. We deduce that the codeposition process produces a glass characterized by an excess entropy of about 10% of the entropy of fusion.

Another approach towards fitting the curve in fig.4 was explored. The summation in eqn. (2) can be approximated as:

$$I_s = I_0 \, exp(-Q/kT) \, exp(-\Delta W^*/kT) \qquad (6)$$

where ΔW^* is the work of formation of a critical cluster, Q is determined from growth rates as mentioned earlier and I_0 is weakly temperature dependent. It follows that the effective activation energy for the nucleation process, $(Q+\Delta W^*) = -[d \, \ln I_s/d(1/T) \,]/k$ with the result that the nucleation rate is proportional to $exp[-(Q+\Delta W^*)/kT]$. It is possible that the activation energy for atom transport across the crystal/amorphous interface is not the same for the processes of nucleation and growth. However, agreement cannot be improved by simply varying Q. This is because we have assumed that the driving free energy for the transformation, $\Delta G_c \approx \Delta H_c$ which is invariant with temperature. Consequently for a given σ, we are also assuming that the work of formation of a cluster is also temperature independent. A change in Q alone would indeed change the slope in fig. 4, but would produce orders of magnitude discrepancy between the predicted and measured nucleation rates. This inconsistency can be avoided by adjusting Q and ΔW^* such that their sum is constant, but it is unlikely that Q is different for nucleation and growth. This method was therefore not pursued further.

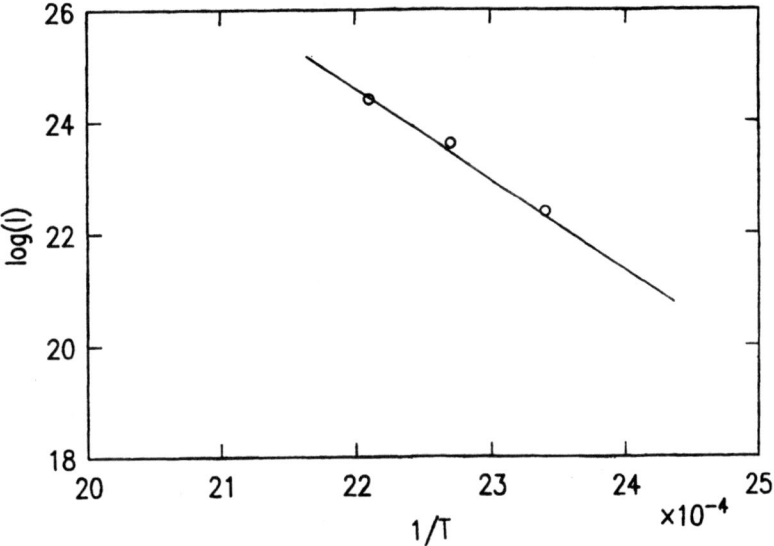

Fig. 4 The fit between experimental measurements of nucleation rate (circle) and the predicted rate in 40 nm films. Good fit is achieved with σ_{ac} =121mJ/m^2 and ΔS_c = 2.16X 10-5 eV/K.atom

Thickness effects

Transient heterogeneous nucleation has been reported to be a typical feature of crystallization of glasses[2]. This means that a devitrifying glass will invariably take some time at the crystallization temperature to attain the steady state distribution of embryos, or in other words a constant nucleation rate. Fig.5 was plotted to determine the variation of the crystal population within the available amorphous volume (for the area under observation). All three 100 nm thick samples approached a steady state nucleation rate that increased with temperature, after a lag time that decreased with temperature. The lag time is fundamentally related to the transient time that is a characteristic of any nucleation and growth mechanism. At a thickness of 40 nm, however, the material had a transient time of only a few seconds and a considerably higher nucleation rate than the thicker films, possibly an effect of the proximity of the two surfaces. Consequently it was not possible to suggest the occurrence of any bulk nucleation in the 100 nm foils. Growth rates in the 100 nm thick films were quite rapid ($\approx$ 100, 120 and 160 nm/min at 132 °C, 137 °C and 146 °C respectively). Growth in the 40 nm films occurred at slower rates ($\approx$ 130, 370 and 740 nm/min at 154°C, 167°C and 180°C respectively). This appears to contradict the earlier assumption that atom transport across the amorphous-crystalline interface is similar during nucleation and growth, which implies that growth rates should scale with nucleation rates. The activation energy for growth was determined from Arrhenius plots to be 0.5 eV for the 100 nm thick films (however this number was deduced from measurements over a temperature range of only 14 °C and may therefore not be very accurate) and 1.09 eV in the 40 nm films [4]. All the samples were processed under similar

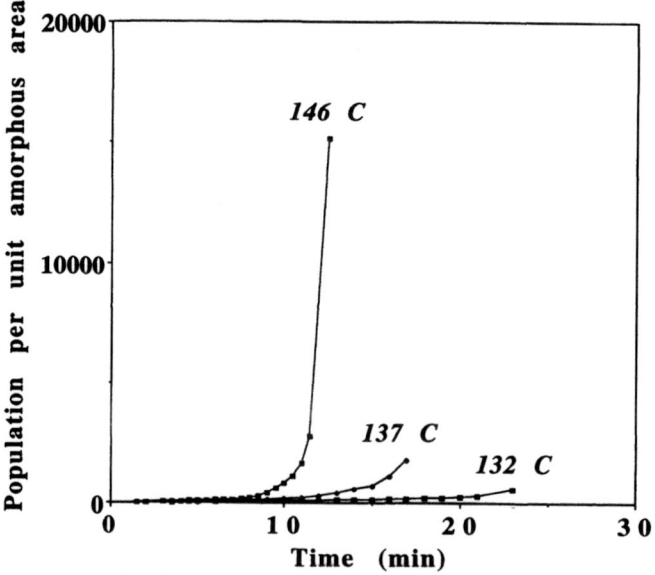

Fig.5 A steady nucleation rate is approached after a lag time of over 10 min in 100 nm films.

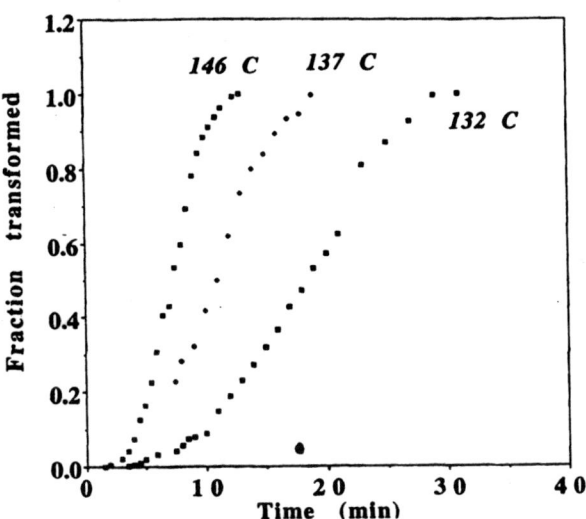

Fig.6 The characteristic sigmoidal variation of volume crystallized as a function of time, in a 100 nm thick film.

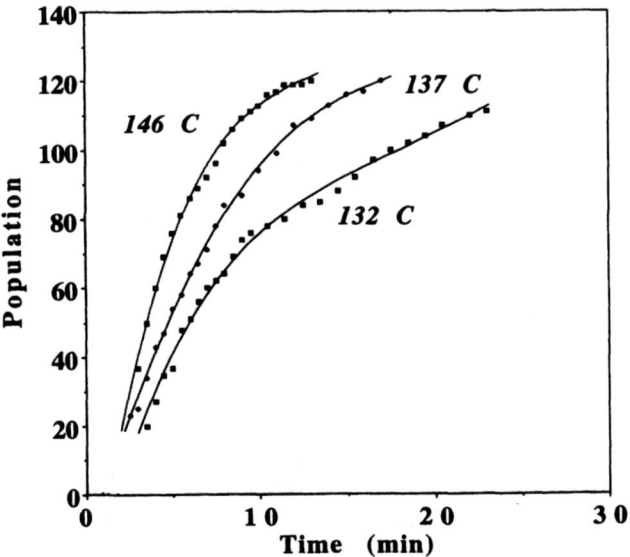

Fig.7 The time dependence of nucleation rate for a 100 nm film.

279

conditions with Co and Si deposited in stoichiometric proportions. The difference in growth activation energies must then be an effect of thickness. The situation certainly calls for a closer look at growth mechanisms in these films. Figures 6 and 7 show the nucleation and growth curves for $CoSi_2$ to be of the form expected of continuous nucleation with a lag time, at randomly distributed sites but with a constant growth rate.

Johnson-Mehl-Avrami analysis

The Avrami Equation is: $f(crystal) = 1-exp(-K\ t^n)$, where n is the mode parameter for crystallization and K is a temperature dependent coefficient, given by: $K=K_0\ exp(-E/kT)$, where E is the activation energy for crystallization driven by a nucleation and growth mechanism. Although f(crystal) is the *volume* fraction crystallized at time t, we were constrained to using *area* fractions computed from a sequence of TEM micrographs of the 100 nm fthick films. In our experiments this gives a small overestimate of volume fraction for crystals of diameter less than the film thickness. Since the crystals grew from a smallest visible diameter around 40 nm at ≈ 300 nm/min this approximation quickly becomes valid. Linear regression was used to determine slopes of Avrami plots with confidence levels in the range of 95-99%. It was determined that $n \approx 4$ at the beginning of transformation and gradually decreased to $n \approx 3$ towards completion, pointing to a mode change from 3-dimensional to 2-dimensional (planar) growth for all 100 nm films as shown in fig.8. Similar behavior was observed in the earlier resistometric studies[1].

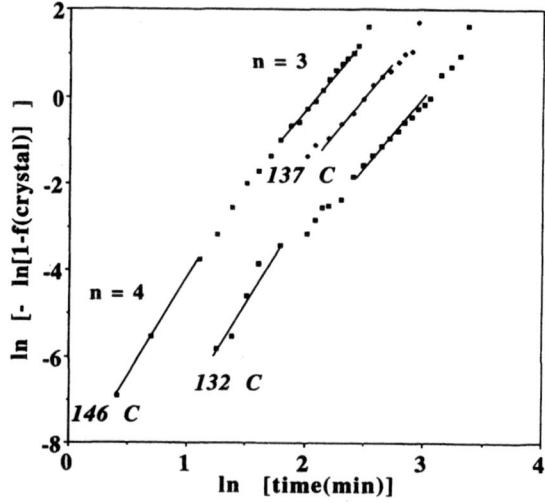

Fig.8 Avrami plots showing the change in growth mode from 3- to 2-dimensional; nucleation rate is constant after an initial transient.

DISCUSSION

That the Avrami curves change as the transformation proceeds may initially be attributed to the transition from three- to two-dimensional growth of randomly distributed nuclei. However it must be noted that an underlying presumption of Avrami's analysis[8] is that the nucleation rate I_s must be constant or decreasing. It is therefore quite possible that these JMA curves may only be valid for times exceeding the lag time. The other assumption of constant growth rate was certainly validated by the experiments. Hong et al[3] studied the crystallization of a range of thin-film amorphous alloys with compositions between 42 and 80 at-% silicon. They correlated DSC, transmission electron microscopy, and in situ stress and resistivity measurements. The activation energy for crystallization of stoichiometric $CoSi_2$ was found to be 1.4 ± 0.2 eV for the stoichiometric and the mode parameter was $3\cdot1$. The incubation time was of the order of 10 minutes. These data are substantially in agreement with our findings.

The effect of beam heating in the TEM might be expected to be more important in thicker films due to the greater absorption of energy from the electron beam . However a greater nucleation rate in the 40 nm films is proof that surface effects are stronger than any beam heating that may have occurred. It has been reported in an earlier analysis[7] that combining electron beam irradiation with ion beam irradiation in various ways can drastically affect nucleation. To check for possible heating effects due to inelastic scattering a pattern of tungsten dots each a few nanometers across was deposited on a film (Fig.9). Heating of the dots was expected to induce early nucleation on them, but did not do so. Beam heating may therefore be excluded from the analysis under the prevailing conditions. That our results agree with those of the earlier resistometric studies[1] supports this conclusion.

The microstructure of the crystals and their interfaces have been analysed in greater detail elsewhere[9]; no visible differences were detected in the microstructures of 40 and 100 nm foils.

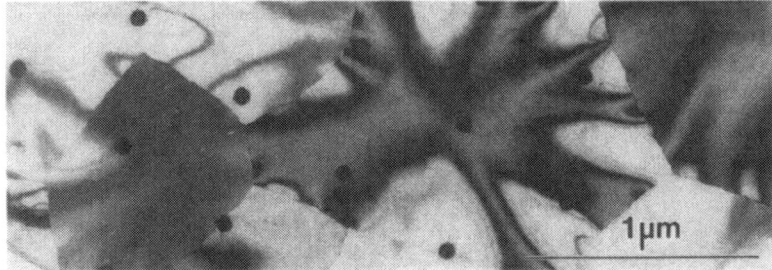

Fig. 9 An electron micrograph showing a crystallized $CoSi_2$ film with an evaporated hexagonal array of tungsten dots which scatter electrons strongly. Note the absence of an effect on nucleation.

CONCLUSION

The fundamental problem in phase transformations is to account for the stable states and to identify the mechanism by which a system transforms between states. Hot-stage crystallization in a TEM enabled was conducted to enable *in situ* observation of the progress of crystallization. It was possible to quantify an effect of thickness on nucleation and growth. The thicker films exhibited lower nucleation rates but higher growth rates. For the crystallization of $CoSi_2$ an amorphous-crystalline contact angle of $76°$ was determined from tilting studies in the TEM; a fit between theory and experimental data on nucleation rates in 40 nm films yielded an interface energy of 121 mJm^{-2}. The activation energy for crystallization was found to be 1.27 eV. Growth rate studies measured an activation energy for growth of 0.5 eV in 100 nm thick films, lower than the 1.09 eV measured earlier in 40 nm thick films. JMA analysis indicated that crystallization transforms from a 3 dimensional mode (n=4) to 2 dimensional (n=3). It is evident from this work that *in situ* TEM studies of crystallization offer a tool uniquely capable of analysing the two stages of nucleation and growth from kinetic and thermodynamic standpoints.

REFERENCES

1. A.Cros, K.N.Tu, D.A.Smith and B.Z.Weiss, Appl.Phys.Lett., 52 (16), 1311 (1988)
2. K.F.Kelton in Solid State Physics,vol.45, edited by H.Ehrenreich and D.Turnbull (Academic Press, San Diego, CA, 1991)
3. Q.Z.Hong, K.Barmak and L.A.Clevenger, J.App.Phys., **72**, 3423 (1992)
4. D.A.Smith, K.N.Tu and B.Z.Weiss, Ultramicroscopy, **30**, 90 (1989)
5. W.Kauzmann, Chem. Rev.,**42**, 219(1948)
6. H.Kanno, J. Non. Cryst. Solids, **37**, 203 (1980)
7. C.W.Allen and D.A.Smith in Mater.Res.Soc.Proc **201**, 405-410 (1991)
8. J.W.Christian, The Theory of Transformations in Metals and Alloys, 1st ed. (Pergamon Press, New York,1965)
9. D.A.Smith and C.W.Allen, Ultramicroscopy,37,279 (1991)

KINETIC STUDY OF CRYSTALLISATION
IN AMORPHOUS THIN LPCVD SI FILMS

B. PIERAGGI*, J. P. GUILLEMET**, B. de MAUDUIT**
*LMP, URA 445 du CNRS, ENSCT, 118 route de Narbonne, 31077 Toulouse Cedex, France
**CEMES-LOE, UPR 8011 du CNRS, BP 4347, 31055 Toulouse Cedex, France

ABSTRACT

The crystallisation behaviour of LPCVD silicon films has been investigated by TEM from
in situ isothermal annealing of undoped a-Si films deposited from disilane (Si_2H_6) at
temperatures 450, 465 and 480 °C and at gas pressure of 200 mTorr. Nucleation kinetics, grain
growth rates and crystallisation kinetics were determined for temperatures ranging from 600 to
675 °C. Nucleation kinetics have been experimentally determined in the early first stages of
annealing : they do not show any steady-state rate and are fitted according to a power law.
Experimental data for crystallisation kinetics are fitted by an Avrami law without introducing any
incubation time.

INTRODUCTION

The crystallisation of amorphous silicon (a-Si) films is an important and popular research topic
because of its potential applications in microelectronics such as thin film transistors (TFTs).
Most of previous studies have been performed on silicon films deposited by low pressure
chemical vapor deposition (LPCVD) from silane (SiH_4) or by physical deposition such as
electron-beam evaporation [1,2,3]. Nucleation kinetics are usually derived by counting the
number of Si crystallites after isothermal annealings for different times assuming a constant
nucleation rate after some transient period [4]. Crystallite growth kinetics are extracted, through a
kinetic model, from the nucleation rate and global transformation kinetics [5]. The activation
energies for nucleation and growth are then deduced from the activation energies of both the grain
size and a characteristic crystallisation time [1,2,6].
This paper concerns LPCVD Si films deposited at low temperature from disilane (Si_2H_6). Far
fewer kinetic and thermodynamic data have been previously published on the crystallisation of
such films [6,7,8,9,10] than on films deposited from SiH_4. They showed that a large grain size
could be obtained from crystallisation at a temperature as low as 600°C.
In this study, the crystallite growth kinetics, the nucleation and the global crystallisation
kinetics were separately determined without referring to a kinetic model. This was possible
through an initial determination of the crystallite growth rate by *in situ* transmission electron
microscopy (TEM) annealing experiments [11]. The knowledge of the crystallite growth kinetics
permitted us to accurately determine the nucleation kinetics [12]. The crystallisation kinetics were
then experimentally determined and compared to the crystallisation kinetics calculated through a
classical nucleation and growth model.

EXPERIMENTAL PROCEDURES

The investigated materials were undoped a-Si thin films, 150 nm thick, grown from disilane by
LPCVD, at deposition temperatures T_d of 450, 465 and 480 °C and total gas pressure p_d of 200
mTorr, onto thermally oxidized (111) silicon wafers. Plan-view samples were prepared by
chemical dissolution of the silicon oxide (SiO_2) interlayer and by subsequent cleaning. The *in situ*
annealing experiments were performed in a heating JEOL specimen stage permitting the control of
specimen temperature with an accuracy of ± 5 °C.

RESULTS

Grain growth rate.

Growth rates were measured by video recordings of several annealing sequences of about 20 minutes for crystalline silicon (c-Si) grains of different sizes. Previous results on films deposited at 480 °C and 75 mTorr [11] have shown that the grain growth kinetics are linear with an apparent activation energy of 2.4 ± 0.1 eV. Linear grain growth kinetics were observed in the entire time-temperature range of the investigated annealing conditions. The grain growth rate was also estimated from the gradient of the dependence of the maximum grain size on the annealing time as done in [7] for conventionally furnace annealed specimens. This method is less accurate than the direct measurement of grain growth rate by *in situ* annealing, particularly for crystallized fractions exceeding 30%. However, growth rates measured by these two methods are of the same order of magnitude.

The grain growth rates V_g measured along the major axis of crystallites are reported in Table 1.

[°C] T_a T_d	600	625	650	675
450	0.4	1.0	2.3	5.0
465	0.9	2.0	5.0	
480	0.4	1.0	2.4	5.4

Table 1 : Grain growth rate [nm.s^{-1}] as a function of the
deposition T_d and annealing T_a temperatures ($p_d = 200$ mTorr).

These results show that the a-Si / c-Si interface velocity strongly depends on the deposition conditions. The highest grain growth rate is measured on films deposited at 465 °C. Such an evolution of growth rate with the deposition temperature is difficult to explain on the basis of the present data because the growth process strongly depends on the impurity level and stress state of the deposited films.

By contrast, the apparent activation energy of the growth process does not appear to depend on the deposition parameters and always equals 2.4 ± 0.1 eV. Similar activation energies have been obtained by studying the solid phase epitaxy (SPE) of silicon through Rutherford backscattering experiments [13] or *in situ* TEM [14], as well as the solid phase crystallisation of amorphous films deposited by electron beam evaporation [15].

Nucleation Kinetics.

The determination of nucleation kinetics, deduced from crystallite sizes and growth rates, have been previously shown to be in good agreement with the kinetics obtained from the examination of specimen conventionnaly annealed in a furnace [12].

However, *in situ* high resolution electron microscopy (HREM) examination of cross sectionnal specimens has shown that nuclei are formed at the a-Si / SiO$_2$ interface [16], and thus that the nucleation process is heterogeneous. In the present experiments, the substrate was removed, but we have carefully checked that *in situ* experiments are indeed representative of actual bulk behaviour. It is difficult to determine, from plan-view examinations only, whether or not the nucleation process starts from the a-Si / SiO$_2$ interface. During the film deposition, the first adatoms arriving on the SiO$_2$ surface have more time to rearrange - and so to initiate potential nucleation sites - than the last adatoms deposited on the top of the Si film. This fact may explain the observed nucleation at the a-Si / SiO2 interface.

Figure 1(a) shows the nucleus density as a function of annealing time for a-Si films grown at 480 °C and 200 mTorr and annealed at various temperatures T_a. The corresponding nucleation kinetics can be expressed by a power law,

$$N_n = (k_n t)^q \tag{1}$$

where N_n is the nucleus density (expressed as the number of nuclei per unit area) and k_n is the rate constant for nucleation.

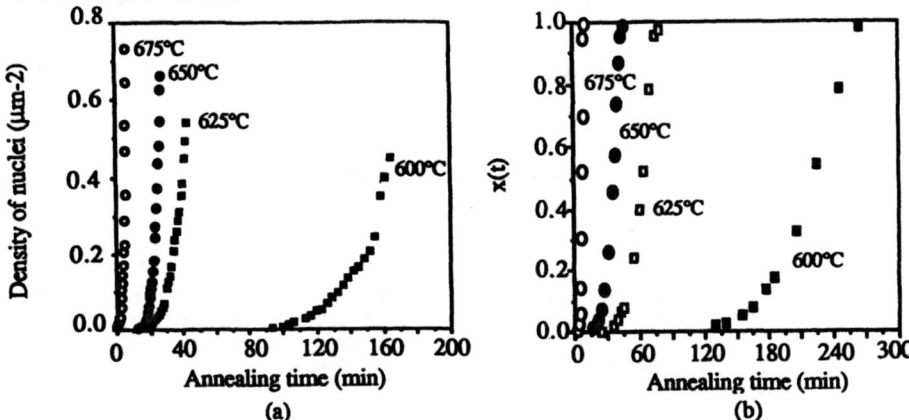

Figure 1 : Nucleation kinetics (a) and Crystallisation kinetics (b)
(films grown at 480 °C and 200 mTorr).

The exponent q was found to vary with deposition and annealing temperatures. Its value was fixed in such a way that it was consistent with the crystallisation kinetics. The difference between the nucleation kinetics at 600 °C and at higher temperature is apparent in Fig. 1(a). Table 2 gives the values of k_n and q as a function of annealing temperature for T_d = 480 and 465 °C.

Table 2 : Values of k_n [(nuclei.μm^{-2})$^{1/q}$.min^{-1}] and q of Eq. (1)
and values of k_c [min^{-1}] and n of Eq. (2) (p_d = 200 mTorr).

[°C] Td	Ta	600	625	635	650	675
465	q	11	4	4	4	
	k_n	5.8 10^{-3}	2.8 10^{-2}	3.6 10^{-2}	1.4 10^{-1}	
	n	13	6	6	6	
	k_c	5.2 10^{-3}	2.2 10^{-2}	3.0 10^{-2}	9.2 10^{-2}	
480	q	5	4		4	4
	k_n	5.2 10^{-3}	2.0 10^{-2}		2.9 10^{-2}	1.6 10^{-1}
	n	7	6		6	6
	k_c	4.4 10^{-3}	1.5 10^{-2}		2.7 10^{-2}	1.3 10^{-1}

Crystallisation Kinetics.

The crystallisation kinetics were determined from *in situ* experiments and subsequent image analysis. The area fraction of crystallites was measured on several TEM micrographs taken at the same magnification (5000x). A good correlation was observed between *in situ* and *ex situ* annealings. The assumption that this area fraction is equal to the crystal volume fraction is justified by the relative magnitudes of the specimen thickness and the grain growth rate. Indeed, at the annealing temperature of 600 °C, a nucleus formed at the specimen mid-plane reaches the specimen surfaces after an annealing time (specimen thickness / growth rate) varying from about 3 to 6 minutes, i.e., a very short time compared to the entire annealing time. Figure 1(b) shows the crystallised fraction X(t) vs. time for a-Si films grown at 480 °C and 200 mTorr. The crystallisation kinetics can be expressed by an Avrami law,

$$X(t) = 1 - \exp(-(k_C t)^n) \qquad (2)$$

The values of the constant k_C and exponent n are reported in Table 2.

DISCUSSION

The measurements of grain growth show that the process is linear. This could mean that the reaction is interface-controlled. Identical results have been previously obtained in the case of SPE of silicon [13]. However, *in situ* HREM experiments show that the a-Si/c-Si interface moves by microbursts [16]. This observation implies that the measured growth rates are average values that do not correspond to the actual local growth process.

The nucleation kinetics that we have experimentally determined in the first stages of annealing, where the crystalline fraction is low (approximately 10 or 15%), are usually referred to as a "transient nucleation". In figure 2 (a), the calculated nucleation kinetics are reported for the entire annealing time range. This curve presents a linear part resulting from the reduction of the amorphous area during annealing. One third of the grains nucleates during this transient period that represents a large fraction of the total annealing time. So, crystallites which have nucleated during this transient period correspond to a great part of the total crystalline fraction (approximately 60%).

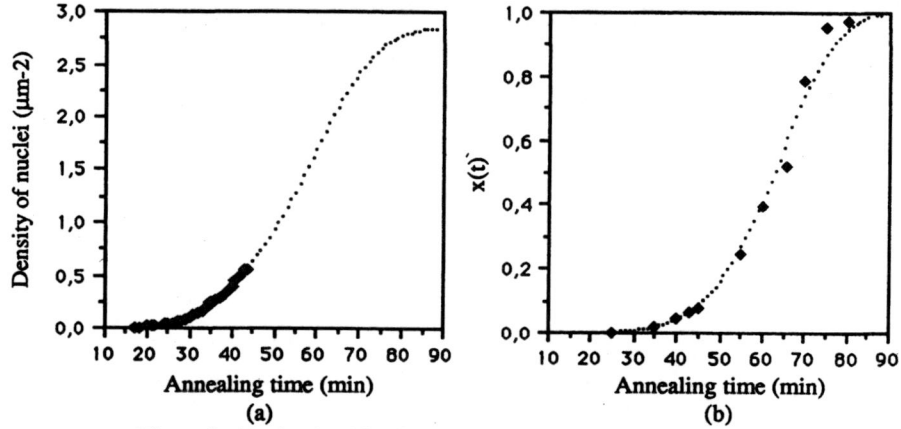

Figure 2 : Nucleation kinetics (a) and Crystallisation kinetics (b) at 625 °C
(experimental data (♦), calculated data (.))
(Films grown at 480 °C and 200 mTorr)

The power law is usually interpreted as resulting from a nucleation process that involves several steps to form a stable nucleus. For the amorphous to crystalline transformation of silicon, it can be assumed that the critical nucleus is formed through several identical steps. The exponent q hence corresponds to the number of sequential steps having the same rate constant k_n and therefore the same apparent activation energy. This apparent activation energy depends on the deposition temperature and is equal to 4.3 eV and 3.0 eV for films grown at 465 and 480 °C, respectively [12].

The large q value at 600 °C cannot be explained on the basis of the present data. The time-dependence of k_n does not depend on the value of the exponent q (as shown by Eq. (1)) and all the k_n values can be fitted to the same Arrhenius relationship. Indeed, the change in exponent q might result from a change in the number of steps involved in the formation of a critical nucleus.

Experimental data for crystallisation kinetics are fitted by an Avrami law (Fig. 2(b)) without introducing any incubation time as nucleation kinetics are fairly well fitted by a power law.

The constant k_C in (Eq.2) is not a real rate constant because it depends on both grain growth and nucleation kinetics. The values of exponents n and q are the best couple of integers that permits a simultaneous description of the observed nucleation and crystallisation kinetics. Indeed, from the usual relation between nucleation, growth rate and crystallisation kinetics, the exponents q and n, and the growth geometry term or "crystallite dimension" d, must be such that,

$$n = q + d \qquad (3)$$

From Eq (3), the dimensions of k_C depend on the dimensions of k_n for the nucleation and d for the growth. Because of the small specimen thickness, d is taken equal to 2. Thus, the variation of k_C with the annealing temperature would depend on the activation energies for growth and nucleation. Within the investigated temperature range, an apparent activation energy E_{cryst} may be deduced from the k_C values (3.8 eV and 3.0 eV for films grown at 465 and 480 °C, respectively [12]). E_{cryst} is related to the apparent activation energies E_{growth} and E_{nucl} for growth and nucleation, respectively,

$$nE_{crist} = qE_{nucl} + 2E_{growth} \qquad (4)$$

Equation (4) is fairly well verified for the investigated annealing temperature range. Futhermore, the large exponents found for nucleation and crystallisation processes are likely related to a continuous and heterogeneous process occuring at the a-Si / c-Si interface.

CONCLUSIONS

The present investigations have confirmed the usefulness of *in situ* TEM annealing experiments for studying thermal crystallisation of thin amorphous silicon films. Grain growth rates and the corresponding activation energies have been determined for films produced by thermal decomposition of disilane for various deposition conditions.

The average linear growth of Si crystallites permits an accurate determination of nucleation kinetics. The transient period of nucleation process is fitted by a power law. This period is followed by a constant nucleation rate due to the reduction of the amorphous area with annealing time. The global transformation kinetics can fitted to an Avrami law with a high value of the time exponent, which is representative of the strong influence of nucleation.

One crucial point is now to correlate the crystallisation behaviour and the deposition conditions, by determining the influence of the deposition parameters on the a-Si structure and also the influence of the a-Si structure on the nucleation and growth mechanisms. The change in the apparent activation energy of nucleation with the deposition parameters illustrates the importance of both the structure of amorphous Si and the nucleation processes in controlling the crystallisation of silicon thin films.

ACKNOWLEDGEMENT

The specimens were provided by Dr. E. Scheid and Dr. D. Bielle-Daspet from LAAS-CNRS (Laboratoire d'Analyse et d'Automatisme des Systèmes, Toulouse, France) whose helpful comments and suggestions throughout this work are gratefully acknowledged.

REFERENCES

1. R. B. Iverson and R. Reif, J. Appl. Phys. **62**, 1675 (1987).
2. M.K. Hatalis and D.W Greve, J. Appl. Phys. **63**, 2260 (1988).
3. K. Zellama, P. Germain, S. Squelard and J. C. Bourgoin, J Appl Phys. **50**, 6995 (1979).
4. J. L. Batstone, Phil. Mag. **A67**, 51 (1993).
5. L. Haji, P. Joubert, M. Guendouz, N. Duhamel and B. Loisel, Mat. Res. Symp. Proc. **230**, 177 (1992).
6. T. Kretz, R. Stroh, P. Legagneux, O. Huet, M. Magis and D. Pribat, presented at POLYSE '93, Saint-Malo, France, September 1993 (to be published by Trans. Tech. Publications).
7. K. Nakasawa, J. Appl. Phys. **69**, 1703 (1991).
8. E. Scheid, B. de Mauduit, P. Taurines and D. Bielle-Daspet, Jpn. J. Appl. Phys. **29**, L 2105 (1990).
9. C. H. Hong, C. Y. Park and H. J. Kim, J. Appl. Phys. **71**, 5427 (1992).
10. A. T. Voutsas and M. K. Hatalis, J. Electrochem. Soc. **140**, 871 (1993).
11. J. P. Guillemet, B. de Mauduit, B. Pieraggi, E. Campo and E. Scheid, J. Mater. Sci. Lett. **12**, 910 (1993).
12. J. P. Guillemet, B. de Mauduit, B. Pieraggi, D. Bielle-Daspet and E. Scheid, Proc. E- MRS Spring Meeting, Strasbourg, France, May 1993, to appear in Mater. Sci. and Eng. A (1993) in press.
13. L. Csepregi, E. F. Kennedy, J. W. Mayer and T. W. Sigmon, J. Appl. Phys **49**, 3906 (1978).
14. R. Sinclair, T. Yamashita, M. A. Parker, K. B. Kim, K. Holloway and A. F. Schwartzman, Acta. Cryst. **A44**, 965 (1988).
15. U. Köster, Phys. Stat. Sol. (a) **48**, 313 (1978).
16. R. Sinclair, J. Morgiel, A.S. Kirtikar, I. W. Wu and A. Chiang, Ultramicroscopy **51**,41 (1993).
17. J. P. Guillemet, B. Pieraggi, B. de Mauduit and A. Claverie, presented at POLYSE '93, Saint-Malo, France, 1993 (to be published by Trans. Tech. Publications).

CRYSTALLIZATION KINETICS IN A $Ni_{24}Zr_{76}$ AMORPHOUS ALLOY

G. Ghosh*, and F.-R. Chen**
*Dept. of Materials Science and Engineering, Northwestern University, Evanston, IL 60208-3108
**Materials Science Center, National Tsing Hua University, Kuang Fu Road, Hsinchu, Taiwan

ABSTRACT

Isothermal crystallization kinetics of $Ni_{24}Zr_{76}$ amorphous alloy has been studied by differential scanning calorimetry (DSC). Theoretically estimated transient times are found to be consistent with the experimentally observed incubation periods. The crystallization kinetics has been analyzed in terms of Kolmogorov-Johnson-Mehl-Avrami (KJMA) model and significant departure from the linear KJMA behavior was observed. Such non-linearity is due to the surface-induced crystallization, anisotropic growth of the crystals, impingement effects and variation of the nucleation rate during crystallization etc. By applying high-resolution electron microscopy (HREM), it has been shown that the nucleation and growth of the crystals do not take place at a constant state of disorder.

INTRODUCTION

It is very important to understand the atomic processes involved during amorphous to crystal transformation both from technological and fundamental perspectives. Depending on the alloy composition, crystallization may proceed by the formation of one or more than one crystalline phases which may be chemically ordered or disordered in nature. In earlier investigations [1-3], the crystallization kinetics of Ni_xZr_{1-x} amorphous alloys has been studied by continuously heating the specimens in a differential scanning calorimeter (DSC). Isothermal crystallization kinetics of a limited number of Ni-Zr metallic glasses have also been performed [4-6], and they have been analyzed in terms of the classical kinetic relation developed by Kolmogorov [7], Johnson and Mehl [8], and Avrami [9] (KJMA). Amorphous $Ni_{24}Zr_{76}$ alloy transforms into (α-Zr) and $NiZr_2$ phases, upon crystallization. Very recently, it has been reported that the crystallization kinetics of $Ni_{24}Zr_{76}$ amorphous alloy depart significantly from the linear KJMA behavior [5, 6]. In this investigation we are interested in studying the isothermal crystallization kinetics and the microstructural evolution of an $Ni_{24}Zr_{76}$ amorphous alloy, in order to identify the reasons for the departure of the isothermal kinetics from KJMA formalism. Apart from studying the crystallization kinetics by conventional techniques, such as DSC and transmission electron microscopy (TEM), high-resolution electron microscopy (HREM) has also been applied in this investigation in order to monitor the atomic processes involved during the nucleation and growth of the crystalline phases from the amorphous matrix.

EXPERIMENTAL

Amorphous ribbons of 2-3 mm wide and 25-30 μm thick were produced by "chill-block melt-spinning" technique. Dynamic and isothermal crystallization kinetics were studied in a Du-Pont DSC 910 cell coupled with computer controlled thermal analyzer (TA9900). During isothermal tests high purity Ar was continuously purged (3 liters/hr) into the DSC cell and the temperature was maintained within $\pm 0.15^\circ C$. Isothermal annealing of the amorphous alloys was also carried out in salt-bath after wrapping the specimens with Zr-foils and encapsulating them in quartz tubes partially filled with Ar. The temperature of the salt bath was controlled within $\pm 1.5^\circ C$.

The amorphous nature of the melt-spun ribbons and the phases formed after crystallization were characterized by X-ray Diffraction as well as by TEM. The thin foils, prepared by dual jet electropolishing using an electrolyte of methanol (80%) and perchloric acid (20%) at 223K, were examined in JEOL200CX TEM at 200kV or HITACHI9000 HREM at 300kV.

Mat. Res. Soc. Symp. Proc. Vol. 321. ©1994 Materials Research Society

RESULTS

The isothermal crystallization kinetic curves were obtained by integrating the DSC thermograms at different test temperatures, and as shown in Fig. 1 they are typically sigmoidal in nature. The presence of a significant incubation period, τ, can also be noticed. The incubation period was evaluated by measuring the time from the instant the DSC cell had reached the required temperature until the onset of the crystallization. This satisfies the operational definition of τ given by Christian [10]. The conventional KJMA plots, i.e., $\ln[-\ln(1-X)]$ vs $\ln(t-\tau)$, over the full range of volume fraction transformed are shown in Fig. 2. It is obvious that instead of being straight lines, such plots exhibit significant departure from linearity from the beginning of crystallization.

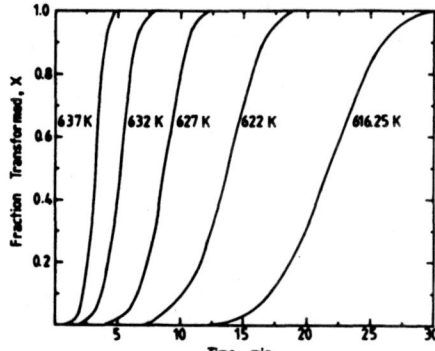

Fig. 1 Isothermal transformation curves of $Ni_{24}Zr_{76}$ amorphous alloy at different test temperatures.

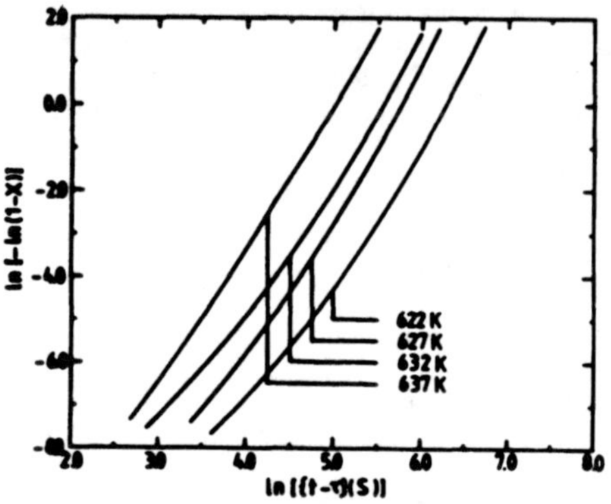

Fig. 2 KJMA plot of $Ni_{24}Zr_{76}$ amorphous alloy at different isothermal test temperatures.

The activation energy for crystallization, E_c, was evaluated from the data of isothermal crystallization and also from dynamic crystallization experiment. Figure 3 shows the Kissinger's plot i.e. $\ln[\dot{T}.T_p^{-2}]$ (where $\dot{T}$ is the heating rate and T_p is the peak temperature) vs T_p^{-1} the slope of which gives the activation energy for crystallization. The latter was also evaluated by applying Arrhenius-type of relation to the time required for 50% transformation at different temperatures, and this is also shown in Fig 3 by plotting $\ln(t_{0.5})$ vs T^{-1}. The activation energies obtained by these two methods are very close indeed. The activation energy for the non-steady-state process, E_τ, was obtained by applying Arrhenius-type of equation to the time required for the onset of crystallization. This is also shown in Fig. 3 by plotting $\ln(\tau)$ vs T^{-1}. It is seen that the activation energy for non-steady state process is much higher than that of overall crystallization. This indicates that two different mechanisms are involved during non-steady state and crystallization periods. A higher activation energy for non-steady state process suggests that it involves cooperative movement of all the atomic species which is probably necessary for the nucleation of the crystalline phases. Nevertheless, the activation energy for the non-steady state process is found to be similar in magnitude to that predicted by Chen [11] for stable glasses.

DISCUSSIONS

Analysis of Incubation or Transient Period

The presence of a well defined transient period, during which a steady state embryo distribu-

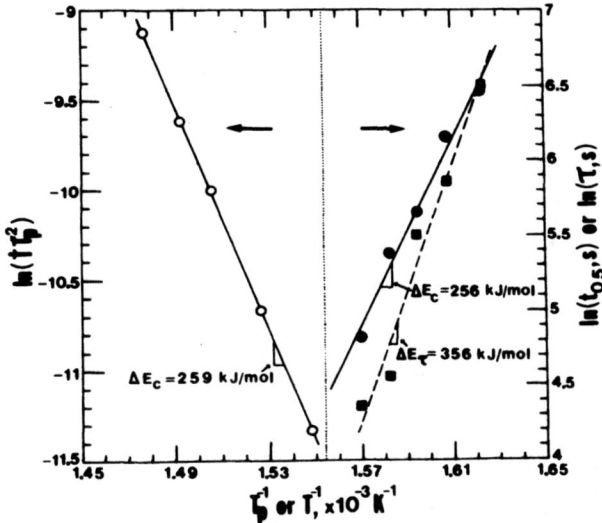

Fig. 3 Kissinger's and Arrhenius plots for determining the activation energies.

tion (characteristic of annealing temperature) is built up, was observed in the isothermal experiments. In the absence of quenched-in subcritical clusters, Kelton et al [12] showed that the analytical solution provided by Kaschiev [13] can be used as a good estimate of incubation period τ_k

$$\tau_k = \frac{4}{\pi^3 O_{n*} \dot{\gamma} Z^2}$$

where O_{n*} (= $4n^{*2/3}$) is the number of surface atoms in critical size nucleus and $\dot{\gamma}$ is the jump frequency of the atoms across the liquid/crystal interface. The Zeldovich factor Z is expressed as

$$Z = \left(-\frac{\Delta G_m}{6\pi R T n*}\right)^{1/2}$$

where n* is the number of atoms in a spherical shaped critical size nucleus, T is the absolute temperature and ΔG_m is the molar Gibbs energy difference between the liquid and the crystalline phase. The Gibbs energy parameters for all phases in the Ni-Zr system have been evaluated recently [14] and a procedure to evaluate O_{n*} and $\dot{\gamma}$ has been described elsewhere [6]. The calculated transient times are listed in Table 1 and are compared with the experimentally obtained incubation periods. The transient times were calculated for the nucleation of either $NiZr_2$ or (αZr) phase. In either case the calculated values are similar in magnitude and compare favorably with the experimentally observed incubation times. Another interesting point to be noticed in Table 1 that the experimentally observed incubation times lie between the calculated transient times for the nucleation of $NiZr_2$ and (αZr).

Table 1. Comparison between the observed incubation period and the calculated transient time as a function of temperature in $Ni_{24}Zr_{76}$ amorphous alloy.

Temp., K	Experimental incubation time (s)	Calculated transient times (s) for the nucleation of	
		$NiZr_2$	(αZr)
616.3	670	464	1104
622	351	222	595
627	245	130	353
632	95	77	212
637	78	46	128

Despite the reasonably good agreement between the calculated and experimental transient times in Table 1, however, strictly speaking the treatment of one component system may not be valid for the nucleation involving composition change in a binary system. Numerical simulation of transient nucleation kinetics in binary alloys has been proposed and such calculations are currently being underway for the Ni-Zr amorphous alloys.

Non-linearity of the KJMA plots

As shown in Fig. 2, KJMA plots exhibit significant departure from linear KJMA behavior which indicates possibility of presence of changing crystallization mechanisms, nucleation rate and growth morphology etc. In order to obtain further insight of the isothermal DSC results, recently Calka and Radlinski [15] have introduced the concept of local value of KJMA exponent, m_{loc}, defined as

$$m_{loc} = \frac{\partial \ln[-\ln(1 - X)]}{\partial \ln(t - \tau)}$$

Figure 4 shows the variation of m_{loc} with the progress of transformation. At the beginning of transformation m_{loc} is very close to 2, indicative of constant nucleation rate, interface controlled one-dimensional growth process, and it is also consistent with the observation of surface crystallization at the initial stages [6]. With the increasing extent of transformation, the contribution of bulk crystallization to the overall transformation increases leading to higher m_{loc} values. In order to investigate the microstructural evolution in the bulk and to understand the effect of microstructural aspects on the KJMA behavior, both conventional TEM and HRTEM were employed. Figure 5 is a typical TEM micrograph of partially crystallized sample, which demonstrates the non-spherical growth of the crystallites and the new nucleation of crystallites adjacent to the existing crystallites. A wide size distribution of the crystallites, consistent with the thermally activated nucleation and growth, may be noticed in Fig. 5.

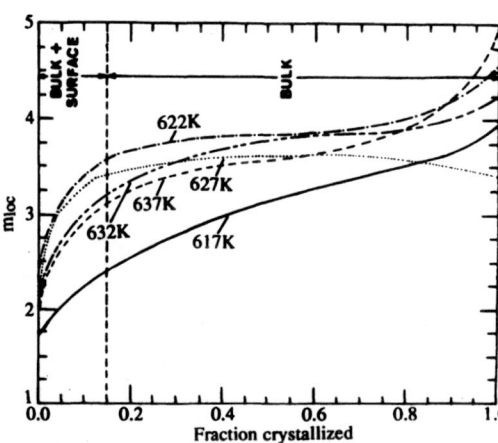

Fig. 4 Variation of local value of KJMA exponent, m_{loc}, with the progress of transformation.

Anisotropic growth could be due to anisotropic surface energy, preferred growth direction and anisotropy of the Young's modulus of the crystals [16]. Figure 6 is the HRTEM micrograph of $Ni_{24}Zr_{76}$, after annealing at 600 K for 1h, showing the brighter and darker area in the amorphous regions. Such changes in contrast are interpreted as due to the change in composition, and are consistent with the kinetic data that supports the idea that the rearrangement of the atomic species occurs during the incubation period. Figure 7 shows nano-scale $NiZr_2$ crystals and extensive short range ordering (SRO) and possibly medium range ordering (MRO) in the surrounding matrix after

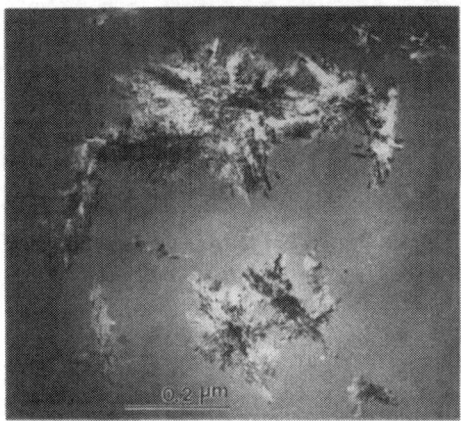 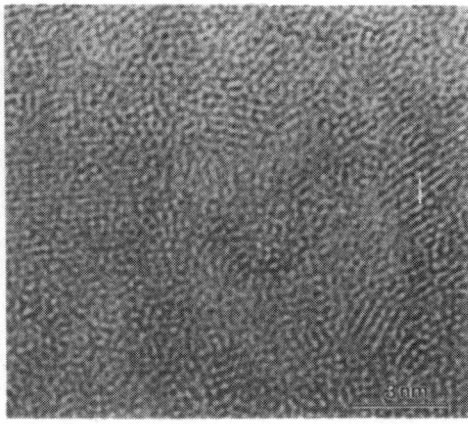

Fig. 5 Bright field TEM micrograph of $Ni_{24}Zr_{76}$ Fig. 6 HRTEM micrograph showing change
partially crystallized at 623 K for 12 mins.   in contrast, SRO and lattice fringes

annealing $Ni_{24}Zr_{76}$ at 600K for 1h. Such developments of SRO and MRO are believed to be a necessary step for the nucleation of the $NiZr_2$ phase. Figure 8 is the HRTEM micrograph of $Ni_{24}Zr_{76}$, after annealing at 600K for 2h, showing the eutectic growth of (α-Zr) and $NiZr_2$ phases. As shown, the presence of these two equilibrium phases in Fig. 8 is confirmed by image simulation.

For eutectic crystallization, which is the case for the present alloy, having constant growth rate and constant nucleation rate KJMA exponent is expected to be 4. But, as shown in Fig. 4, m_{loc} varies from 3.25 to 3.75 in the range of 10 to 80 pct crystallization. In other words, generally there is negative deviation from the predicted KJMA slope. Based on the HREM micrographs shown in Figs. 6 to 8, we arrive at the following conclusions : (a) nucleation and growth of (α-Zr) and $NiZr_2$ do not take place directly from a truly amorphous matrix, (b) crystalline phases do not nucleate and grow at constant state of disorder, (c) $NiZr_2$ crystals nucleate first and grow up to a few nanometers before nucleation of (αZr). The presence of SRO and MRO will possibly reduce the driving force for interface mobility, and hence the growth rate will be slower resulting in negative deviation from ideal KJMA behavior. Since the extent of SRO and MRO will vary with time, the relative effects of these on the nucleation and growth rates will be different, thus giving rise to different m_{loc} values with increasing crystallized fraction. Beyond about 80 pct crystallization, m_{loc} continues to increase upto 5.5 suggesting a trend of increasing nucleation rate, which could be due to the self-heating of the sample. Additionally, at the later stages of transformation the impingement effects become more and more severe, and can contribute significantly to the KJMA analysis. In fact, positive deviation from KJMA prediction has been confirmed by computer simulation and it has been suggested that KJMA extended-volume relation overcorrects for impingement [17].

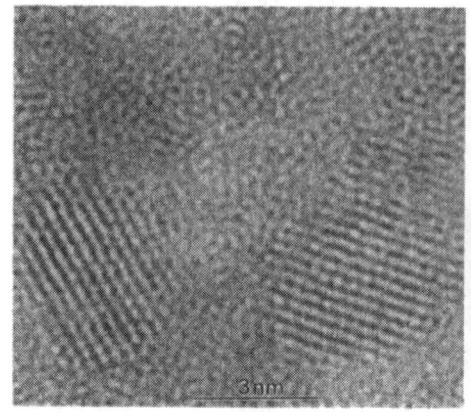

Fig. 7 HRTEM micrograph showing nano-scale NiZr$_2$ crystals.

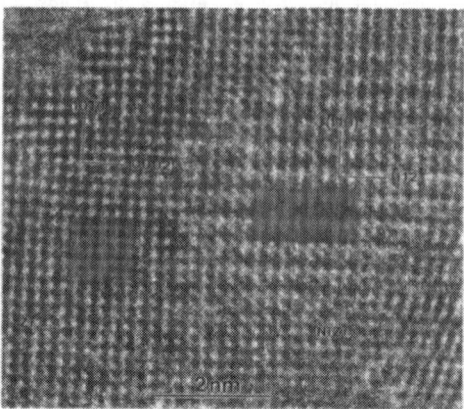

Fig. 8 HRTEM micrograph showing eutectic growth of (αZr) and NiZr$_2$ phases.

CONCLUSIONS

(i) Isothermal crystallization kinetics data of Ni$_{24}$Zr$_{76}$ amorphous alloy indicates that the activation energy for non-steady state process is much higher than that of overall crystallization. (ii) The experimentally observed incubation times are found to agree favorably with the calculated transient times. (iii) Non-linearity of the KJMA plots are thought to be due to the surface crystallization, non-random nucleation, anisotropic growth of the crystals, impingement effects, variation in the state of disorder of the matrix with time and variation of the nucleation rate. (iv) In an amorphous Ni$_{24}$Zr$_{76}$ alloy NiZr$_2$ nucleates first. As expected it corresponds to the phase with the maximum chemical driving force as well as minimum activation energy.

REFERENCES

[1] H. H. J. Buschow and N. M. Beekmans, Phys. Rev. B, 19, 3847 (1979).
[2] Y. D. Dong, G. Gregan and M. G. Scott, J. Non-Crstalline Solids, 43, 403 (1981).
[3] Z. Altounian, Tu Guo-Hua and J. O. Ström-Olsen, J. Appl. Phys., 54, 3111 (1983).
[4] G. K. Dey, E. G. Baburaj and S. Banerjee, Mat. Sci. Engg., 21, 117 (1986).
[5] A. Kolb-Telieps, Int. J. Rapid Solidification, 3, 109 (1987).
[6] G. Ghosh, M. Chandrasekaran and L. Delaey, Acta Metall. Mater., 39, 925 (1991).
[7] A. E. Kolgomorov, Akad. Nauk. SSSR. IZV. Ser. Mat, 1, 355 (1937).
[8] W. A. Johnson and R. F. Mehl, Trans. AIME, 135, 416 (1939).
[9] M. Avrami, J. Chem Phys., 7, 103 (1939).
[10] J. W. Christian, The Theory of Transformation in Metals and Alloys, Pergamon, Oxford, 2nd. Edn., 1975.
[11] H. S. Chen, Appl. Phys. Letters, 29, 12 (1976).
[12] K. F. Kelton, A. L. Greer and C. V. Thompson, J. Chem. Phys., 79, 6261 (1983).
[13] D. Kashchiev, Surface Science, 14, 209 (1969).
[14] G. Ghosh, to appear in J. Mater. Res. (1994).
[15] A. Calka and A. P. Radlinski, J. Mater. Res., 3, 59 (1987).
[16] U, Köster, in Proc.Phase Transformations, Cambridge, G. W. Lorimer, Ed., The Institute of Metals, London, pp.597 (1986).
[17] C. W. Price, Acta Metall., 35, 1377 (1987).

SIMULATION BY CELLULAR AUTOMATA OF THE (RE)CRYSTALLIZATION OF A MATRIX CONTAINING AN INERT SECOND PHASE

C.F. PEZZEE AND D.C. DUNAND
Department of Materials Science and Engineering, Massachusetts Institute of Technology, Cambridge, MA 02139

ABSTRACT

Two-dimensional cellular automata simulations were carried out to study the case of the crystallization (or recrystallization) of a matrix containing an inert, immobile second phase. A range of particle area fractions and aspect ratios were investigated under continuous grain nucleation conditions, assuming that the effect of particles is limited to geometric impingement upon contact with the growing grains. Systematic deviations from the classical Johnson, Mehl, Avrami, Kolmogorov equation for single-phase materials are observed with increasing particle aspect ratio and particle fraction. Inert particles also influence both mean size and mean aspect ratio of the final grains.

INTRODUCTION

The classical theory of Johnson, Mehl, Avrami and Kolmogorov (JMAK) (1) describes the kinetics of crystallization of liquids and amorphous solids and recrystallization of crystallin solids. Analytical JMAK-type solutions are available only for simple cases such as constant growth rate, continuous nucleation or site-saturated nucleation (2). More complex conditions (e.g., space- or time-dependent nucleation or growth rates) have been investigated by computer simulation techniques based on (i) the equation of motion of grain boundaries (3-5) (ii) binary tree construction on a grid (6-10) (iii) the Monte-Carlo method (11-14) and (iv) cellular automata (15-17).

A second phase can strongly influence the kinetics of a matrix undergoing recrystallization by increasing the rate of nucleation at the matrix-particle interface (2, 10) or by decreasing that rate through Zener pinning of the subgrains necessary for stable nuclei (4, 18). Rollett et al. (13, 14) have studied by two-dimensional Monte Carlo simulations the recrystallization kinetics of a matrix containing equiaxed particles much smaller than the final grain size. They concluded that, at high stored energies, the recrystallization growth is not affected by the particles. Grain coarsening during and after recrystallization was however found to be strongly inhibited by particles. Nes et al. (9) investigated the recrystallization behavior of a matrix containing stringers of densely spaced particles, and found that the growth rate was direction-dependent, leading to elongated grains in the direction of the stringers.

In the present study, we focus solely on the impingement exerted by particles on growing grains, in order to study this geometric effect independently of other effects found in experimental studies of crystallization and recrystallization in two-phase materials, e.g. particle-stimulated nucleation, grain-boundary pinning by particles or particle dragging or pushing by moving boundaries. Impingement by a second phase may become important when the specific area of matrix-particle interface is high, i.e., for high volume fractions, high aspect ratios and small sizes of particles, in such materials as those formed by phase separation through solidification or solid state precipitation, in metal- or ceramic matrix composites and in foamed materials.

COMPUTATIONAL PROCEDURES

Hesselbarth and Göbel (15) first used cellular automata to simulate recrystallization in a material consisting of a field of cells with two possible states - recrystallized and unrecrystallized - evolving with time according to local topological rules. Time is discretized in time-steps, which are divided in growth and nucleation events. During the nucleation event, single-cell nuclei are randomly dis-

tributed in the field; only those that are nucleated on an unrecrystallized cell are considered. During the growth event, the entire field is updated according to the following rules:
- a recrystallized cell remains recrystallized,
- an unrecrystallized cell becomes recrystallized if at least one of its neighbors is recrystallized. It becomes part of the same grain as the recrystallized neighbor.

We add to the original model by Hesselbarth and Göbel (15) two rules to take into account the presence of inert particles (17):

- at time $t=0$, second-phase particles are placed randomly on the field with a minimum spacing of two cells between particles; these particles do not grow.
- matrix grains neither nucleate nor grow within particles.

The following parameters were used in the present study:
  (i) a two-dimensional field of 262,144 (512^2) square cells oriented along orthogonal axes with periodic boundary conditions;
  (ii) homogenous nucleation conditions, with a matrix nucleation rate of $4 \cdot 10^{-4}$ for each nucleation step;
  (iii) a neighborhood of six alternating neighbors for the growth phase (15);
  (iv) an unrecrystallized cell with recrystallized neighbors belonging to more than one grain becomes part of any of the competing grains with the same probability.

Two parameters related to the geometry of the second phase were varied: particle area fraction f and particle aspect ratio r. A set of baseline parameters ($f=0.125$ and $r=4$) was chosen, and each parameter was varied separately according to Table I, keeping the size of the particles equal to 64 cells. The grain size A (measured in number of cells) and grain aspect ratio R - defined as the ratio (larger than 1) of the projections of the grain along the two orthogonal axis - were calculated for each grain at the end of the simulation and averaged over two simulations carried out with the same parameters but different spatial distributions of particles and nuclei (Table I).

Table I: Particle geometric characteristics, mean size $\overline{A}$ and mean aspect ratio $\overline{R}$ of recrystallized grains (* baseline condition).

Particle Parameter		$\overline{A}$ (cells)	$\overline{R}$
aspect ratio r	area fraction f		
4	0	225	1.28
	0.03125	224	1.29
	0.0625	222	1.34
	0.125*	215±1	1.36±0.005
	0.25	198	1.47
1	0.125	219	1.35
4*		215±1	1.36±0.005
16		201	1.70
64		172	3.16

RESULTS AND DISCUSSION

Grain Topology

In Figs. 1a-b, a third of the cellular automaton field is shown at different times in the simulation. The particles are filled, the grain boundaries are shown as lines and the origin of each grain is indicated as a filled cell within each grain. Figures 1a (i)-(iii) show a matrix without particles

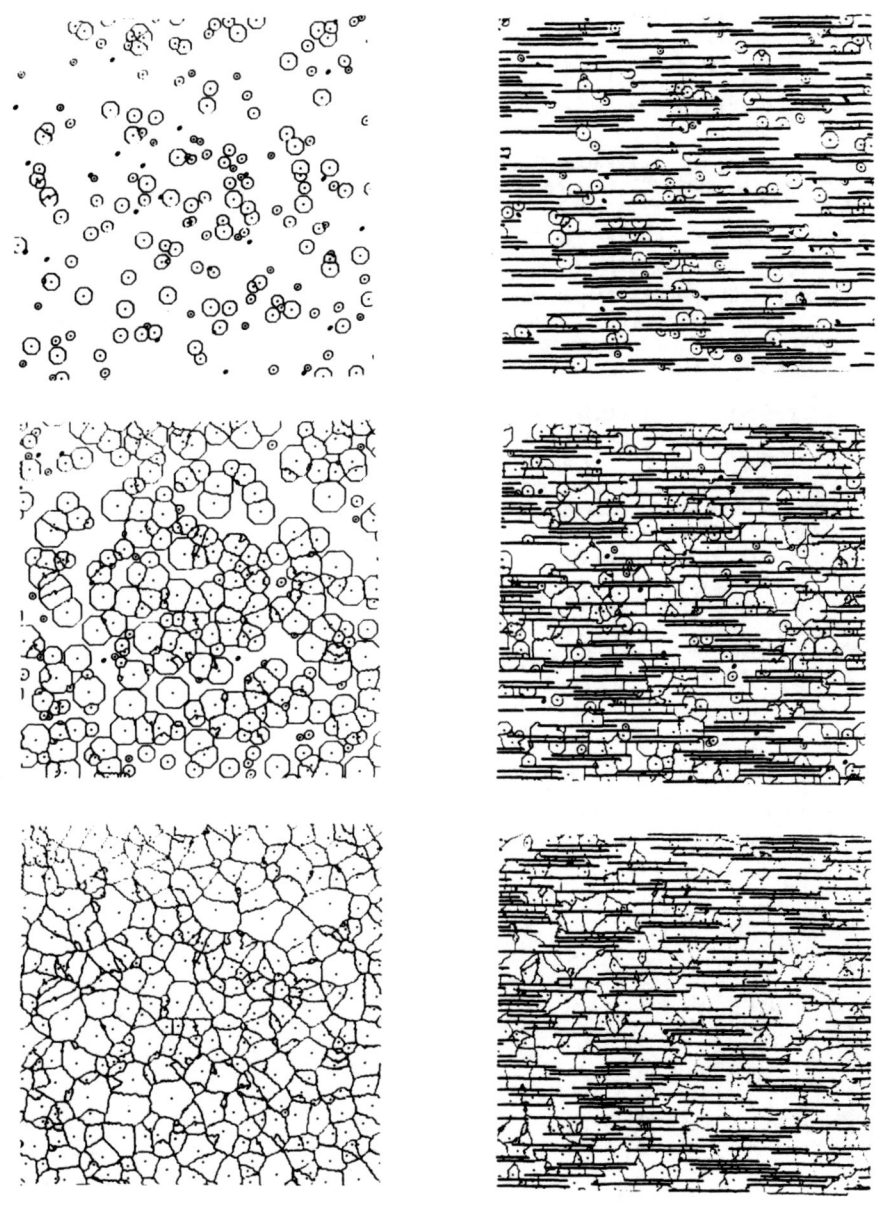

Figure 1: Section of the recrystallized field (a) without particles and (b) with particles, with parameters $a=64$ cells, $f=0.125$ and $r=64$ at different times:
(i) early stage, (ii) intermediate stage, (iii) complete recrystallization.

where impingement between grains occurs during growth, corresponding to the case explored by Hesselbarth and Göbel (15). Figures 1b (i)-(iii) correspond to a matrix containing high aspect ratio particles (r=64). Most grains contact at least one two-dimensional fiber early in their growth (Fig. 1b (i)) and become channeled by fibers (Fig. 1b (ii)), resulting in a final grain structure markedly elongated in the direction of the fibers (Fig. 1b (iii)).

The mean grain area $\overline{A}$ decreases and the mean grain aspect ratio $\overline{R}$ increases as the particle perimeter per unit area matrix increases, i.e. with increasing particle aspect ratio and area fraction (Table I). However, as shown in Ref. (17), the distribution of grain areas normalized by the mean area $\overline{A}$ is unchanged, while the distribution of normalized grain aspect ratio changes significantly with varying particle parameters. Increases in $\overline{A}$ and $\overline{R}$ are both due to the impingement by particles: (i) as growth is slowed by particles but the nucleation rate is constant, more grains nucleate during the course of recrystallization and the mean grain area is reduced, and (ii) as elongated particles have an anisotropic impingement effect, the shape of the grain becomes elongated in the direction of the particle long axis.

The latter effect can be illustrated by simulating the growth of a single grain between two isolated fibers (Fig. 2a). Five distinct stages are observed: (i) unimpeded, two-dimensional growth before contact with the fibers ($t\leq7$); (ii) channeled, one-dimensional growth between the fibers, with the mobile grain boundaries perpendicular to the fibers ($7<t\leq31$); (iii) two-dimensional growth from the recrystallized regions at both ends of the channel formed by the fibers ($31<t\leq63$); (iv) merging of these two distinct regions, with the formation of a groove on the grain boundary parallel to the fibers ($63<t\leq90$); (v) disappearance of the groove and further unimpeded growth of the grain ($t>90$). The resulting grain shows an flattened octagonal shape and an area smaller than if it had grown freely. For comparison, Fig. 2b shows the growth of the same grain placed between two arrays of small, equiaxed particles with the same total area as the fibers in Fig. 2a. The grain engulfs the particles as it grows, with the result that its shape and area are almost the same as those of a freely growing grain

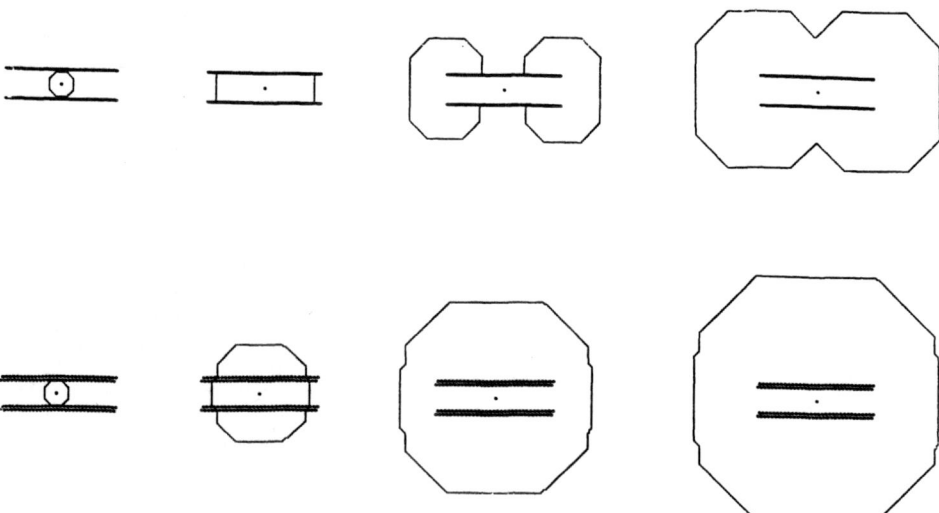

Figure 2: Different stages of growth of a single grain in a matrix containing (a) 2 large, elongated fibers of same total area and (b) 128 small, equiaxed particles.

Kinetics of Recrystallization

The fraction recrystallized matrix x is plotted in Figs. 3 as a function of time according to the JMAK equation:

$$x = 1 - \exp\left(-kt^n\right), \tag{1}$$

where k is a function of the nucleation and growth rates and n is the Avrami exponent. It is apparent from Figs. 3 that recrystallization is slowed with increasing particle area fraction and aspect ratio. As predicted by the JMAK theory, the theoretical value $n=3$ is found for the case where the matrix is particle-free (Fig. 3a, $f=0$), for matrix recrystallized fractions between $x=0.05$ and $x=0.95$ (deviations at low and high recrystallized fractions are due to the small number of grains growing). However, significant departures from the theoretical value are observed when particles are added to the field. As the particle area fraction f increases, the Avrami exponent plotted in Fig. 3a decreases, as expected from an increasing impingement between particles and grains. More important deviations from $n=3$ result from increasing particle aspect ratio r (Fig. 3b). For $r=16$, n takes values between 2 and 3, and for $r=64$, the minimum value of n is below 2. This is the result of the strong anisotropic impingement effect illustrated in Fig. 2b, whereby growth occurs in a regime intermediate between two-dimensional growth ($n=3$) and one-dimensional growth ($n=2$).

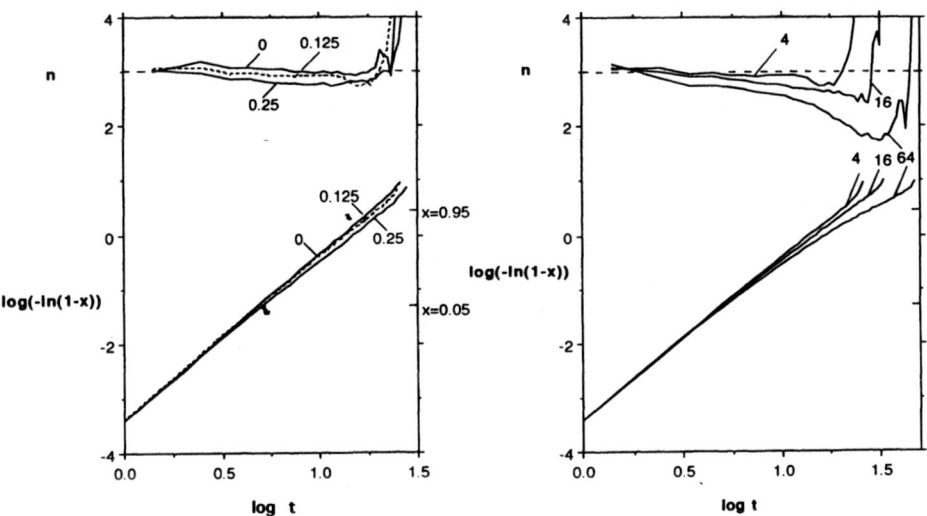

Figures 3: JMAK plots with the time evolution of the Avrami exponent n for various (a) area fraction f, (b) aspect ratio r.

CONCLUSIONS

The cellular automaton algorithm introduced by Hesselbarth and Göbel (15) for the two-dimensional recrystallization of single-phase materials is extended to two-phase materials. The effect of second-phase particles is assumed to be only by geometric impingement with growing grains.

Particles with various area fractions and aspect ratios are found to influence the kinetics of recrystallization by lowering the Avrami exponent and increasing the time necessary for full recrystallization: within the parameters explored, the effect is largest for particles with the highest aspect ratios which prevent growth in the direction perpendicular to their long axis. Increasing particle area fraction has a similar but smaller influence. Particles also influence both the mean size and aspect ratio of recrystallized grains, which decrease with increasing particle aspect ratio and area fraction.

Acknowledgments: C.F.P. gratefully acknowledges the support of the MIT Undergraduate Research Opportunity Program and D.C.D. acknowledges the support of AMAX in the form of an endowed chair.

REFERENCES

1. D. A. Porter and K. E. Easterling, Phase Transformations in Metals and Alloys, (Van Nostrand Reinhold, London, 1989), p. 288.
2. K. Marthinsen, T. Furu, E. Nes, and N. Ryum, in Simulation and Theory of Evolving Microstructures, edited by M.P. Anderson and A.D. Rollett (TMS, Warrendale, 1990), p. 87.
3. H. J. Frost and C. V. Thompson, Acta Metall. **35**, 529 (1987).
4. F. J. Humphreys, Mat. Sci. Technol. **8**, 135 (1992).
5. D. Juul Jensen, Scripta Metall. Mater. **27**, 1551 (1992).
6. K. W. Mahin, K. Hanson, and J. W. Morris, Acta Metall. **28**, 443 (1980).
7. K. Marthinsen, O. Lohne, and E. Nes, Acta Metall. **37**, 135 (1989).
8. T. O. Saetre, O. Hunderi, and E. Nes, Acta Metall. **34**, 981 (1986).
9. E. Nes, N. Ryum, and O. Hunderi, Acta Metall. **33**, 11 (1985).
10. T. Furu, K. Marthinsen, and E. Nes, Mat. Sci. Technol. **6**, 1093 (1990).
11. D. J. Srolovitz, G. S. Grest, and M. P. Anderson, Acta Metall. **34**, 1833 (1986).
12. D. J. Srolovitz, G. S. Grest, M. P. Anderson, and A. D. Rollett, Acta Metall. Mater. **36**, 2115 (1988).
13. A. D. Rollett, D. J. Srolovitz, M. P. Anderson, and R. D. Doherty, Acta Metall. Mater. **40**, 3475 (1992).
14. A. D. Rollett, D. J. Srolovitz, R. D. Doherty, M. P. Anderson, and G. S. Grest, in Simulation and Theory of Evolving Microstructures, edited by M.P. Anderson and A.D. Rollett (TMS, Warrendale, 1990), p. 103.
15. H. W. Hesselbarth and I. R. Göbel, Acta Metall. Mater. **39**, 2135 (1991).
16. M. Rappaz and C. A. Gandin, Acta Metall. Mater. **41**, 345 (1993).
17. C. F. Pezzee and D. C. Dunand, Acta Metall. Mater., accepted.
18. F. J. Humphreys, Scripta Metall. Mater. **27**, 1551 (1992).

TWO-STAGE CRYSTALLIZATION OF POLY(ETHER ETHER KETONE) AND ITS BLENDS WITH POLY(ETHER IMIDE)

HSIN-LUNG CHEN AND ROGER S. PORTER
Polymer Science and Engineering Department, University of Massachusetts, Amherst, MA 01003

ABSTRACT

Thermal mechanical analysis (TMA) has been used to study the crystallization behavior of poly(ether ether ketone) (PEEK) and its blends with poly(ether imide) (PEI). The two crystallization stages of PEEK are clearly distinguished by measuring the variation of film thickness with time during isothermal crystallization. Upon blending with PEI, the distinction of the two PEEK crystallization stages becomes obscure. This is attributed to the depressions in both nucleation density and spherulite growth rate upon blending with PEI.

An Avrami analysis, modified by considering both primary and secondary crystallization, is used to extract the respective kinetic behavior of these two crystallization stages. The results indicate that the secondary crystallization proceeded slower than the primary crystallization in the diffusion-controlled crystallization region. On the other hand, these two crystallization stages proceeded at comparable rate in the thermodynamically-controlled crystallization region. It is also found in the diffusion-controlled crystallization that blending with PEI induced a larger depression in the secondary crystallization rate than in the primary crystallization rate. Explanations for these observations are proposed and discussed.

INTRODUCTION

Poly(ether ether ketone) (PEEK) is a semicrystalline, aromatic polymer with good thermal and mechanical properties. The T_g of amorphous PEEK is 145 °C.[1] The normally observed crystallinity and melting point of PEEK are about 35% and 335 °C, respectively.[1] Morphological studies of crystalline PEEK have identified two distinct crystal populations of different thickness.[2-4] The thinner secondary crystals have been found to grow in the constraints imposed by the thicker primary crystals. The inhomogeneous morphology of PEEK has been considered to give rise to its well-known double melting endotherms observed by DSC.[5-7] Although these results have shown the existence of two lamellar morphologies in crystalline PEEK, other information, such as the respective crystallization kinetics and the origin of this two-stage crystallization behavior require further studies.

It is known that blending with a miscible polymer may have significant effects on the crystallization behavior of the crystalline polymer. These include changes in thermodynamic driving force and molecular mobility associated with crystallization. The blends of PEEK and poly(ether imide) (PEI) have been the subject of several investigations.[8-12] PEI is a linear, amorphous, aromatic polymer with a T_g of 215 °C. The blends of PEEK and PEI are miscible over the entire composition range.[8-12] Miscibility of PEEK and PEI in the amorphous phase decreases the crystallization rate of PEEK.[8,9,11] Blending with PEI may provide a method to systematically control the crystallization rate of PEEK. The investigation of the effect of PEI on the crystallization behavior of PEEK is thus of interest.

This paper reports the investigations of the two-stage crystallization behavior of PEEK and the effect of blending with PEI on this two-stage crystallization phenomenon. It will be shown that thermal mechanical analyzer (TMA) can be used to resolve the two crystallization stages for PEEK. This simple and convenient technique is a one-dimensional dilatometer, measuring the variation of film thickness with time during isothermal crystallization. The respective crystallization kinetics of these two stages will be revealed by a model considering both primary and secondary crystallization. The results will be discussed.

301

EXPERIMENTAL

PEEK powder was obtained from Imperial Chemical Industries (ICI), Wilton, U.K. The molecular weights are $M_n = 16,800$, $M_w = 39,800$. Poly(ether imide), PEI, was obtained from General Electric (GE, Ultem 1000). The molecular weights are $M_n = 12,000$, $M_w = 30,000$.

Blends of PEEK and PEI were prepared by solution-precipitation from dichloroacetic acid (boiling point = 194 °C). Fully amorphous films of PEEK and PEEK/PEI blends for TMA investigations were prepared by compression molding at 400 °C for 5 mins under vacuum followed by quenching in cold water.

A Perkin-Elmer TMS-2 Thermomechanical Analyzer was used to detect the thickness change of PEEK and PEEK/PEI blend films during isothermal crystallization at 162 to 225 °C. A PEEK or PEEK/PEI blend film of about 0.1 mm thick was placed on the platform of the sample tube and the probe placed in contact with the film. An oil bath was equilibrated at the desired crystallization temperature (T_c) with the control of ± 1°C. The oil bath was then quickly moved to immerse the sample, and the change of the specimen thickness with respect to time was recorded.

Crystallizations at high T_c (from 290 to 310 °C) of PEEK and PEEK/PEI blends were monitored by a Perkin-Elmer DSC 4. The sample was heated from 100 °C to 400 °C at ~ 200 °C/min, and was held at this temperature for 3 mins. The sample was then quickly cooled at ~ 200 °C/min to the desired T_c, and the isothermal crystallization exotherm was recorded.

RESULTS AND DISCUSSION

TMA Crystallization Curves

Fig. 1 displays the direct recorded TMA traces of amorphous PEEK and PEEK/PEI blend films during isothermal crystallization at 191 °C. It is seen that the two crystallization stages are clearly distinguished in the TMA trace of pure PEEK: an initial drop of specimen thickness corresponds to the first stage (primary) crystallization, a subsequent plateau indicates an induction period before the final drop of the specimen thickness, corresponding to the second stage (secondary) crystallization. When PEI is blended with PEEK, the TMA crystallization curves of PEEK are changed. It can be seen in Fig. 1 that the plateau between the first and the second stage crystallization becomes obscure in the blends. As the PEI concentration in the blends is increased to 50 wt%, the two crystallization stages become indistinguishable. Since two melting endotherms have also been observed for PEEK/PEI blends,[9,12] there is no doubt that double lamellar morphology still exists in the blends. The disappearance of the distinction of the two crystallization stages in PEEK/PEI blends is thus due to the significant overlap of these two stages in the crystallization curves.

The primary crystallization in polymers has normally been attributed to the formation of spherulites, and the secondary crystallization has been attributed to the crystallization taking place inside the spherulites (intraspherulitic crystallization).[13] The intraspherulitic crystallization in PEEK has also been observed by optical microscopy.[3] The overlap of the two PEEK crystallization stages in the blends may indicate that the intraspherulitic crystallization took place before the impingement of PEEK spherulites.

Fig. 2 shows the optical micrographs of PEEK spherulites grown from pure PEEK and PEEK/PEI 75/25 blend at $T_c = 250$ °C. The average size of the spherulites grown from the blends is larger than that grown from pure PEEK. This shows that blending with PEI decreases the nucleation density of PEEK crystallization. In addition, previous studies have shown the depression in spherulite growth rate of PEEK by blending with PEI.[11] These two observations indicate that the time required for the spherulites to impinge is shorter in pure PEEK than that in the blends. Therefore, for pure PEEK crystallized at 191 °C in Fig. 1, intraspherulitic crystallization actually took place after the impingement of the spherulites, so that the two crystallization stages can be distinguished. On the other hand, for the blends, intraspherulitic crystallization proceeded before the impingement of the spherulites because of lower nucleation density and slower

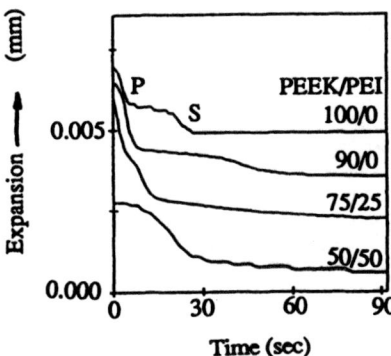

Figure. 1 Direct recorded TMA traces of PEEK/PEI blends isothermally crystallized at 191 °C. The primary crystallization is denoted by "P", and the secondary crystallization is denoted by "S".

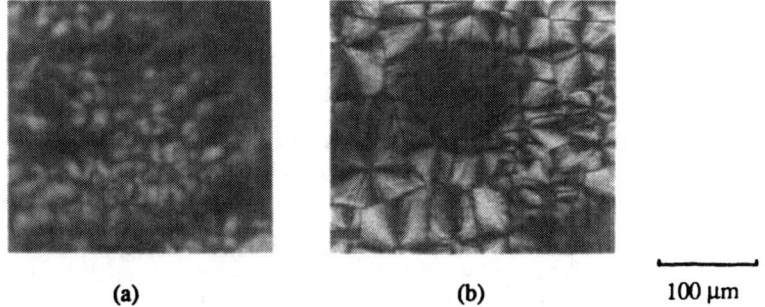

(a)   (b)   100 µm

Figure 2 Optical micrographs of PEEK spherulites formed at $T_c = 250$ °C : (a) grown from pure PEEK, and (b) grown from PEEK/PEI 75/25 blend.

spherulite growth. Therefore, an overlap of the primary and the secondary crystallization occurred making these two crystallization stages indistinguishable by TMA.

Two-Stage Crystallization Kinetics

TMA measures the thickness change of the specimen during isothermal crystallization, which can be viewed as a one-dimensional dilatometry . It is reasonable to assume that the crystallization proceeded isotropically in the film. The evolution of the relative degree of crystallinity can then be calculated from the change of specimen thickness with respect to time by

$$X(t) = \frac{1 - [l(t) / l_0]^3}{1 - (l_\infty / l_0)^3} \qquad (1)$$

where $l(t)$ is the specimen thickness at time t, l_0 is the initial specimen thickness, and l_∞ is the ultimate specimen thickness. Due to the limitation in obtaining high temperature by the oil bath, the T_c range studied by TMA was limited to 162 to 225 °C. The crystallization at high T_c (from 290 to 310 °C) was instead monitored by DSC. Although DSC is not sensitive enough to resolve the two PEEK crystallization stages at low T_c, the crystallization curves detected by DSC at high T_c should be close to the real crystallization curves due to the significant overlap of these two stages.[14]

A crystallization model proposed by Price[13] was adopted here for the data analysis. This model is an Avrami theory modified by considering the secondary crystallization. Spherulite formation is regarded as the primary crystallization, and the intraspherulitic crystallization is regarded as the secondary crystallization. When the crystallization time is shorter than the induction time for the intraspherulitic crystallization (t_i), only spherulite formation takes place and hence the crystallinity development is simply given by the Avrami equation:

$$X(t) = 1 - ce^{-k_s^n t^n} \qquad (2)$$

where c is the total relative crystallinity developed in the intraspherulitic crystallization, n and k_s are the exponent and rate constant of the spherulite formation, respectively. At time $t > t_i$, the total crystallinity is given by:

$$X(t) = 1 - ce^{-k_s^n t_i^n} + \int_{t_i}^{t} [1 - ce^{-k_i^m(t-\tau)^m}] nk_s^n \tau^{n-1} e^{-k_s^n \tau^n} d\tau \qquad (3)$$

where m and k_i are the exponent and rate constant of the intraspherulitic crystallization, respectively. The crystallization rate constants k_s and k_i are obtained by curve fitting. The curve fitting was performed by the least square method. The values of n chosen for the fit were estimated from the initial slopes of the Avrami plot. Thus, n =3 was chosen for this T_c range, which signifies a thermodynamically-controlled crystallization with heterogeneous nucleation and spherical growth geometry. For the crystallization at 162 to 225 °C, since the crystallization is diffusion controlled,[15] n = 1.5 was chosen to signify a diffusion-controlled crystallization with heterogeneous nucleation and spherical growth geometry. For the values of m, because of the spatial restrictions imposed by the primary crystals, the dimensionality of the growth is reduced.[16] Here m = 2 (thermodynamically-controlled with heterogeneous nucleation and disk growth geometry) was chosen for the T_c range of 290 to 310 °C, and m =1 (diffusion-controlled) was chosen for the T_c range of 160 to 225 °C. The initial guesses of the rate constants k_s and k_i have been recommended to be obtained from the two intercepts in the Avrami plot.[16]

Typical fits of the Price model to the data are shown in Fig. 3. It can be seen that the Price model does give a better fit to the data than the conventional Avrami analysis. Fig. 4 is an overlay plot of logarithmic rate constants vs. T_c for PEEK and 75/25 blend. It can be seen that in the thermodynamically controlled T_c range, k_i is close to k_s for both PEEK and the blend at a given T_c. However, in the diffusion controlled T_c range the difference between k_i and k_s becomes clear. The difference between k_i and k_s is larger for the blend than that for pure PEEK in this T_c range. A parameter β has been defined in the Price model to signify the relative rate of the intraspherulitic crystallization to the spherulite formation, $\beta = k_i / k_s$.[24] The average value of β of PEEK in the T_c range of 162 to 201 °C is about 0.44, whereas that of PEEK in the T_c range of 300 to 310 °C it is about 1.0. Comparing these values of β, it can be seen that the intraspherulitic crystallization proceeded about two times slower than the spherulite formation in the diffusion-controlled crystallization. On the other hand, these two crystallization stages proceeded nearly as fast as each other in the thermodynamically-controlled crystallization .

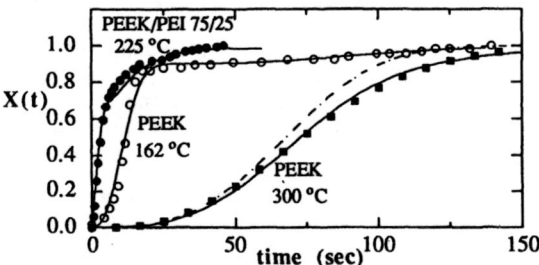

Figure 3 Relative degree of crystallinity vs. time plot of PEEK/PEI blends; the composition and the respective T_c are indicated in the figure. The solid lines are the results obtained by Price model fit. For PEEK T_c = 300 °C the dash line is the fit by the Avrami equation.

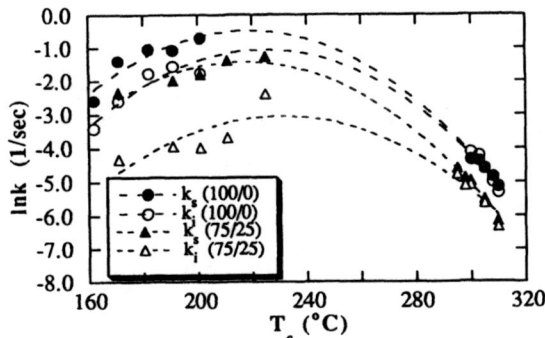

Figure 4 Overlay plot of logarithmic rate constants vs. T_c of PEEK/PEI blends.

The secondary crystallization of PEEK is restricted by the constraints imposed by the primary crystals. Therefore, the diffusion of the chain segments in the secondary crystallization is more difficult than those in the primary crystallization. In the diffusion-controlled crystallization, the chain mobility is the dominant role in determining the overall crystallization rate. The secondary crystallization is hence slower than the primary crystallization at low T_c. On the other hand, in the thermodynamically-controlled crystallization, the thermal energy imparted on the chains can easily overcome the diffusion barrier. Therefore, the rate of the secondary crystallization is comparable to that of the primary crystallization.

The average value of β of 75/25 blend in the T_c range of 171 to 225 °C is about 0.15, and that in the T_c range of 295 to 310 °C is about 0.85. Comparing the values of β of 75/25 blend with that of PEEK, it can be seen that β of the 75/25 blend is about 3 times smaller than that of pure PEEK in the diffusion-controlled crystallization. On the other hand, β of the blend is close to that of pure PEEK in the thermodynamically-controlled crystallization. This indicates that blending with PEI has induced a stronger depression in the secondary crystallization rate than in the primary crystallization rate.

The crystallization in the miscible blends requires a mutual diffusion to transport the crystallizable component to the growth front and reject the amorphous component away.[15] The PEEK primary crystals imposed a spatial restriction making this mutual diffusion in the interlamellar regions more difficult than in the fully amorphous phase. In addition, depending on the distance of segregation, the gradual rejection of PEI during PEEK crystallization could induce an accumulation of PEI in the crystal growth front. Due to this accumulation effect and the increased difficulty in mutual diffusion imposed by the primary crystals, the diffusion barrier in the secondary crystallization in the blend is increased by an even larger extent comparing to that in pure PEEK. Therefore, lower value of β is observed in the blend in the diffusion-controlled crystallization.

CONCLUSIONS

The two-stage crystallization kinetics of PEEK and PEEK/PEI blends have been studied. The depressions in both nucleation density and spherulite growth rate in the blends obscure the distinction of these two crystallization stages in the TMA crystallization curves. A crystallization kinetics study indicates that the PEEK chain segments participating in the secondary crystallization is more sensitive to mobility change. In the diffusion controlled T_c range, the secondary crystallization proceeded slower than the primary crystallization. This is due to the increased difficulty of diffusion as a result of the spatial constraint imposed by the primary crystals. Blending with PEI induces a larger depression in the secondary crystallization rate than in the primary crystallization rate. This is ascribed to the hindrance in mutual diffusion due to the spatial restriction imposed by the PEEK primary crystals and likely the accumulation of PEI in the amorphous phase during PEEK crystallization.

ACKNOWLEDGMENT

We are grateful for the financial support from the National Science Foundation Materials Research Laboratory at the University of Massachusetts at Amherst.

REFERENCES

1. D. J. Blundell and B. N.Osborn, Polymer **24**, 953 (1983).
2. D. C. Bassett, R. H. Olley, I. A. M. Al Raheil, Polymer **29**, 1745 (1985).
3. H. Marand and A. Prasad, Macromolecules **25**, 1731 (1992).
4. A. J. Lovinger, S. D. Hudson, and D. D. Davis, Macromolecules **25**, 1752 (1992).
5. S. Z. D. Cheng, M.-Y. Cao, and B. Wunderlich, Macromolecules **19**, 1868 (1986).
6. Y. Lee and R. S. Porter, Macromolecules **20**, 1336 (1987).
7. Y. Lee, R. S. Porter, and J. S. Lin, Macromolecules **22**, 1756 (1989).
8. J. E. Harris and L. M. Robeson, J. Appl. Polym. Sci. **35**, 1877 (1989).
9. G. Grevecoeur and G. Groeninckx, Macromolecules **24**, 1190 (1991).
10. S. D. Hudson, D. D. Davis, and A. J. Lovinger, Macromolecules **25**, 1759 (1992).
11. H.-L. Chen and R. S. Porter, Polym. Eng. Sci. **32**, 1870 (1992).
12. H.-L. Chen and R. S. Porter, Polym. Sci., Polym. Phys. Ed. **31**, 1845 (1993).
13. Price, F. P. J., Polym. Sci., Part A **3**, 3079 (1965).
14. H.-L. Chen and R. S. Porter, Polymer **34**, 4576 (1993).
15. M. Day, Y. Deslandes, J. Roovers, and T. Suprunchuk, Polymer **32**, 1258 (1991).
16. B. Hsiao, J. Polym. Sci., Polym. Phys. Ed. **31**, 237 (1993).

NUCLEATION AND GROWTH RATES IN ISOTHERMAL CRYSTALLIZATION OF AMORPHOUS Si$_{50}$Ge$_{50}$ FILMS

J. H. Song and James S. Im
Columbia University, Department of Chemical Engineering, Materials Science,
and Mining Engineering, New York, NY

ABSTRACT

Isothermal crystallization behavior of as-deposited thin amorphous Si$_{50}$Ge$_{50}$ films (~1000Å-thick) at 580°C has been investigated using transmission electron microscopy (TEM). The crystal counting method was employed in order to obtain directly the two-dimensional steady-state crystal nucleation rate of 3.9×10^3 #/cm^{2}sec (equivalent volumetric nucleation rate of 3.4×10^8 #/cm^{3}sec). The modified two-dimensional Johnson-Mehl-Avrami analysis, in which the growth rate of the crystals was the only adjustable parameter, and in which the time-dependent nucleation rate and the size effect associated with the onset of the observation are considered, was developed in order to extract the crystal growth rate of 16.5 Å/sec. When compared to the crystallization of a-Si films, these nucleation and growth rates confirm the observation that it is possible to achieve significantly faster crystallization at lower temperatures while producing substantially better microstructures (i.e., > 5 µm grain-sized poly-Si$_{50}$Ge$_{50}$ obtained within two hours at 580°C vs. 1-2µm grain-sized poly-Si obtained in about > 10 hours at 600°C).

INTRODUCTION

Poly-SiGe films are promising materials for the fabrication of thin-film transistor (TFT) devices, and are beginning to demonstrate potential as a legitimate contender against poly-Si films. However, because poly-SiGe TFT devices tested to date have been fabricated on grain-boundary-rich (i.e., small-grained) poly-SiGe films [1], the full potential of the material has not yet been tapped, and remains to be further explored. It is likely that improvement in the microstructure of poly-SiGe films would lead to a corresponding improvement in the performance of poly-SiGe TFT devices.

To its credit, SiGe is a processing-friendly material, which is compatible with existing Si processing technologies [2]. But most significantly, poly-crystal SiGe films may be an attractive material for TFT devices as they could hold property and process advantages over their Si counterparts. The materials process currently utilized to produce the best low temperature TFT devices — i.e., solid-phase crystallization of a-Si films — is not entirely satisfactory from a processing perspective as the method requires a long heat treatment period (several tens of hours) at a high processing temperature of 600°C [3].

Although there exists a small number of articles on the solid-phase crystallization (SPC) of a-SiGe films, they provide no information pertaining to the microstructure of the crystallized material [4, 5]. In particular, these investigations did not utilize TEM-based "crystal counting" analysis [3], and as a result, failed to provide any independent information about the incubation period, nucleation rate, and growth rate that characterize crystallization; such information is precisely what is needed in order for one to predict both (1) the processing time required for crystallization and (2) the microstructure of the crystallized poly-SiGe films.

Fundamentally, our investigation of crystallization of a-SiGe films may potentially provide scientifically long-awaited experimental information on the kinetics of nucleation in an ideal binary system (Si and Ge constitute an ideal binary solution [6]). Although theoretical development of this topic has continued since the 1950s [7-9], the lack of experimental data has become the bottleneck, which has rendered this subject matter incomplete.

Having been motivated by the above technological and fundamental needs for detailed characterization of the crystallization behavior, the specific goals of this investigation were to characterize explicitly the nucleation and growth rates of crystals using (1) experimental methods and (2) analysis based on the transformation formalism.

EXPERIMENTAL METHOD

Samples consisted of 1140Å-thick amorphous $Si_{50}Ge_{50}$ films, which were initially deposited on thermally oxidized Si wafers by co-evaporation of Ge and Si in UHV conditions. In order to ensure that deposited SiGe films were amorphous, deposition was carried out with no substrate heating — care was taken to ensure that the temperature of the sample during deposition never exceeded 100°C. The samples were then cut into small pieces and annealed in a UHV furnace ($\sim 10^{-8}$ torr) at 580°C for various durations. The analysis and quantification of the transformation kinetics were conducted via extensive and systematic use of planar view TEM analysis.

Very low magnification (3.3K) was employed during the analysis since the characteristic length scale associated with the resulting microstructures was very large. In addition, randomly scattered artifacts produced during the sample preparation stage were present. As a result, the average size of the crystals at which they were definitively identified (and thus counted) corresponded to a crystal radius of around 0.35 μm. This aspect of the experiment has important ramifications on the proper analysis of transformation kinetics, and further discussion will be provided later.

RESULTS

As Figures 1a to 1e illustrate, the amorphous-to-crystal transformation is characterized by the nucleation of crystals and the subsequent growth of the nucleated crystals. These figures correspond to bright-field TEM micrographs, where the dark grey area corresponds to the amorphous SiGe and the bright region with dark streaks corresponds to the crystal SiGe.

There are several observations, which can be readily made from these micrographs. (1) The shape of the crystals is significantly more isotropic than the dendritically-shaped crystals, which are observed in crystallization of Si films. (2) The size of the crystals is very large. (3) The nucleation of crystals in the untransformed matrix occurs throughout the duration of the transformation. (4) Crystallization is completed within a couple of hours. The last point is clearly seen in Figure 2, where the fraction of the film that is transformed is plotted as a function of time; the

Figure 1. Successive bright field TEM micrographs showing the isothermal transformation of initially amorphous 1000Å-thick $Si_{50}Ge_{50}$ to crystalline Si via nucleation and growth at T=580°C. (a), (b), (c), (d), and (e) correspond to annealing times of t=20, 40, 60, 80, and 100 min. The uniformly dark background area corresponds to the amorphous Si, and the bright and dark regions correspond to the crystal Si.

resulting sigmoidal curve clearly demonstrates that the transformation is indeed complete within two hours.

Inspection of the variations in the normalized density of the observable crystal grains over time confirms the general notion that nucleation is occurring in a non-site-saturated manner (Figure 3). Additionally, the plot further reveals what appears to be a time-dependent nucleation rate at the early stage of the transformation; after a short apparent incubation period, crystals appeared at a low nucleation rate, which then reached a much higher and what appears to be a constant and steady-state two-dimensional nucleation rate of 3.9×10^3 #/cm^{2}sec (or an equivalent volumetric nucleation rate of 3.4×10^8 #/cm^{3}sec).

DISCUSSION AND ANALYSIS

The unusual aspect of the above crystallization behavior pertains to the observation of a low nucleation rate period prior to the establishment of the steady-state nucleation regime. The origin of this behavior may possibly be attributed to the existence of small crystallites within the as-deposited films, which are supercritical (i.e., crystals are larger than the critical value) but whose size distribution is such that most of them are not large enough to grow at a constant rate [10]. This is to say that the grains, which appear during the initial non-steady state period, correspond to the grains, which grew from the pre-existing supercritical (but interfacial energy effected) crystals, while the grains, which appear during the subsequent steady-state period, correspond to the crystals, which have actually nucleated (i.e., originates from the steady-state embryo distribution). However, it is also possible, in principle, that such a behavior arises from participation of two distinct nucleation mechanisms — such as heterogeneous and homogeneous nucleation (or for that matter, two different types of heterogeneous nucleation). For that matter, it is also conceivable that we are actually resolving the consequences of the transient nucleation of crystals during which the

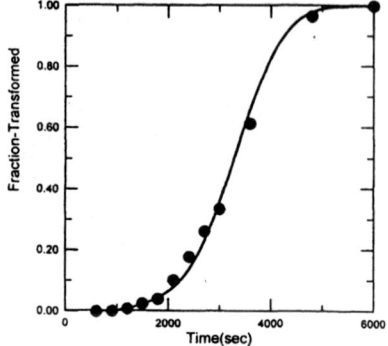

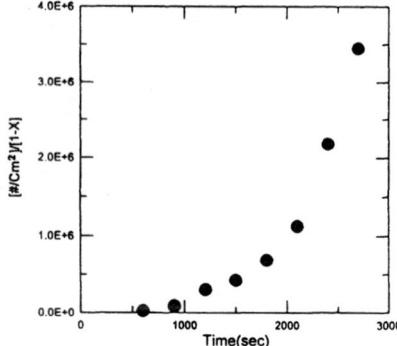

Figure 2. Fraction transformed vs. anneal time at temperatures of 580°C. The dark circles correspond to experimental data, and the solid curve corresponds to the theoretical fit that yields a growth rate of 16.5Å.

Figure 3. Variation of crystal density normalized by the fraction of untransformed amorphous $Si_{50}Ge_{50}$ region, as a function of time at temperatures of 580°C.

nucleation rate increases sigmoidally to the steady-state value [11]; further experimentation and analysis are needed in order to clarify unambiguously the origin of the observation.

We now turn to the formal theory of transformation kinetics, which is needed in order to determine the growth rate of crystals — the major remaining parameter of crystallization. The availability of information pertaining to the rate of appearance of crystals and the fraction crystallized (which is insensitive to negligence of the presence of small crystals) as well as the spherical and two-dimensional nature of the crystal grains enable us to attack rigorously the fraction transformation analysis (i.e., the so-called Johnson-Mehl-Avrami analysis) [12]. Here, regardless of the precise origin and mechanism, we must incorporate the consequence of the time-dependent crystal appearance rate in order to execute the analysis properly.

To this end, it is noted that the rate at which crystals appear during the early period varies slowly enough that we could approximate overall behavior as consisting of (1) an initial incubation period, θ_1, during which no significant nucleation occurs, (2) an early period (starting after θ_1) during which crystals appear at a rate of R_1 (estimated to be $\sim 4.7 \times 10^2$ #/cm²sec from Figure 3) and (3) a steady-state nucleation period (starting after θ_2) during which crystals appear at a rate of R_2 (estimated to be $\sim 3.9 \times 10^3$ #/cm²sec from Figure 3).

In the fraction transformed analysis in which the constant growth rate is used, the corresponding incubation period should refer to *the time at which nucleated crystals acquire a constant growth rate*. As such, the proper values of θ_1 and θ_2 (the values which are to be used in the fraction transformed analysis) must take into account the fact that the size at which crystals were counted (i.e., r_{exp}, which is 0.35 μm) is presumably significantly larger than the critical size above which crystals grow at a constant rate [10].

Additionally, since the sizes of the grains involved are significantly larger than the film thickness, we would utilize a two-dimensional treatment of the transformation analysis. Furthermore, we assume that the growth of crystals (when outside the size-dependent growth regime [10]) proceeds at a constant rate, and that R_2 corresponds to the steady-state nucleation rate, which is applicable until the end of the transformation. (We have verified the assumption of a constant growth rate via *in situ* TEM analysis and the observation of small crystals even when the fraction transformed is very high indicates but does not prove that constant nucleation proceeds throughout the transformation.)

When all the factors are taken into account, the following expression, which provides the fraction transformed, f, as a function of anneal time, t, is obtained.

$$ f(t) = 1 - \exp\left\{ -\frac{\pi}{3} v_g^2 \left(R_1 \tau_1^{\,3} + (R_2 - R_1)\tau_2^{\,3} \right) \right\} \text{,} $$

$$ \text{where} \qquad \tau_1 = \begin{cases} 0 & for \quad t < \theta_1 \\ t - \theta_1 & for \quad t \geq \theta_1 \end{cases} $$

$$ \text{where} \qquad \tau_2 = \begin{cases} 0 & for \quad t < \theta_2 \\ t - \theta_2 & for \quad t \geq \theta_2 \end{cases} $$

In the above equation, v_g is the growth rate and $\theta_1 = \theta_{1exp} - \delta\theta$ and $\theta_2 = \theta_{2exp} - \delta\theta$, where θ_{1exp} and θ_{2exp} refers to values which are directly extracted from Figure 3 (~600 sec. and 1990 sec., respectively) and $\delta\theta = (r_{exp} - r_g^*)/v_g$, where r_g^* refers to a size above which crystals grow at a constant rate. Since $r_{exp} >> r_g^*$, $\delta\theta \approx r_{exp}/v_g$. It is important to note that the only adjustable parameter in the above analysis is the growth rate, v_g. The result of the analysis, which was carried out in a self-consistent manner, yielded a growth rate of 16.5 Å/sec (and $\delta\theta$ of 212 sec.), and is plotted in Figure 2.

From a technological perspective, the observed transformation behavior is remarkable as it possesses all desired characteristics. Specifically, it is noteworthy that compared to the crystallization of pure a-Si films at 580°C [3], the volumetric crystal nucleation rate of ~3.4×10^8 #/cm^{3}sec corresponds roughly to an enhancement by a factor of ~40 and the growth rate of ~16.5 Å/sec corresponds to an enhancement by a factor of ~170.

The above equation clearly shows that a combination of higher values of nucleation and growth rates leads to rapid crystallization. (For example, it would have required a-Si films, with lower nucleation and growth rates, at 580°C greater than 100 hours to achieve the full transformation, as opposed to about two hours required for Si$_{50}$Ge$_{50}$.) An equally important point to note here is the fact that such *enhancement* in *crystallization kinetics* is observed while, at the same time, the *substantially better microstructure* (i.e., ultra large-grained poly-Si$_{50}$Ge$_{50}$ films) is obtained. The average grain diameter of the thin films obtained under constant nucleation rate, I, and growth rate is proportional to the ratio $(v_g/I)^{1/3}$ [3]; it predicts that grains, which are around 2 to 3 times larger, are expected in crystallized poly-Si$_{50}$Ge$_{50}$ at 580°C compared to crystallized poly-Si at 600°C. A comparison of our results (> 5 μm) to Si (1~2 μm, [3]) indicates that this is in fact the case.

CONCLUSION

We have found that ultra large-grained poly-Si$_{50}$Ge$_{50}$ films can be obtained via low temperature annealing of as-deposited amorphous Si$_{50}$Ge$_{50}$ films. TEM analysis (including the crystal counting method) was employed in order to characterize experimentally the evolution in crystal density and fraction transformed as a fraction of isothermal annealing time at 580°C. The modified two-dimensional Johnson-Mehl-Avrami analysis, in which the growth rate was the only adjustable parameter, was developed to yield a crystal growth rate of 16.5Å/sec. In summary, compared to the crystallization of a-Si films, the crystallization of a-Si$_{50}$Ge$_{50}$ can be carried out significantly faster at lower temperatures, while producing substantially better microstructures (i.e., > 5 μm grain-sized poly-Si$_{50}$Ge$_{50}$ obtained within two hours at 580°C vs. 1-2 μm grain-sized poly-Si obtained in about > 10 hours at 600°C).

It is shown that these characteristics — lower processing temperature, shorter processing time, and larger-grained material — are a fortuitous and straight-forward consequence of (1) higher crystal nucleation and growth rates of Si$_{50}$Ge$_{50}$ *and* (2) substantially lower nucleation-to-growth-rate ratios than that which is achievable with Si.

The authors gratefully acknowledge Dr. S. S. Iyer of IBM T. J. Watson Research Center for providing samples and Dr. Joanna L. Batstone of IBM T. J. Watson Research Center for helpful discussions.

REFERENCES

1. T.-J. King, K. C. Saraswat and J. R. Pfiester, *IEEE Elect. Dev. Lett.* **12** (1991) 584.
2. T.-J. King, James R. Pfiester, John D. Shott, James P. McVittie, and Krishna C. Saraswat, *IEEE DM*, (1990) 253.
3. R. B. Iverson and R. Reif, *J. Appl. Phys.* **62** (1987) 1675.
4. M. Maenpaa and S. S. Lau, *Thin Solid Films* **82** (1981) 343.
5. M. Maenpaa, L. S. Hung, M. G. Grimaldi, I. Suni, J. W. Mayer, M.-A. Nicolet, and S. S. Lau, *Thin Solid Films* **82** (1981) 347.
6. R. A. Swalin, *Thermodynamics of Solids*, (New York: Wiley, 2nd edition, 1972).
7. H. Reiss, *J. Chem. Phys.* **18**, 840 (1950).
8. K. C. Russell, *Acta Metall.* **16**, 761 (1968).
9. A. L. Greer, P. V. Evans, R. G. Hamerton, D. K. Shangguan, and K. F. Kelton, *J. Cryst.. Growth* **99**, 38 (1990).
10. K. F. Kelton and A. L. Greer , *J. Non-Cryst. Solids* **79**, 295-309 (1986).
11. D. Turnbull, *Metals Technology*, Technical Publication 2365 (1948).
12. J. W. Christian, *The Theory of Transformations in Metals and Alloys*, second edition, (Oxford: Pergamon Press, 1975).

AMORPHOUS/CRYSTALLINE STRUCTURE AND PHASE TRANSFORMATIONS IN METASTABLE SEMICONDUCTING $Ge_{1-x}Sn_x$

SUSANNE M. LEE* AND KATAYUN BARMAK**
*Lawrence Livermore National Laboratory, 7000 East Ave., Livermore, CA 94550 and Lawrence University, Dept. of Physics, P. O. Box 599, Appleton, WI 54912
**Department of Materials Science and Engineering, Lehigh University, 5 East Packer Avenue, Bethlehem, PA 18015.

ABSTRACT

The semiconducting crystalline alloys, $Ge_{1-x}Sn_x$, are of interest due to theoretical predictions about their electronic band structures which make them useful in infrared photodetectors. However the composition region where these alloys have the desired properties is greater than the equilibrium solid solubility limit of Sn in Ge ($x \leq 0.01$). We have circumvented the solubility limits and produced thin (2000Å) and thick (4-8μm) films of $Ge_{1-x}Sn_x$ ($x \leq 0.31$) by rf sputtering. Differential scanning calorimetry (DSC) measurements were performed to study grain growth and crystallization processes in these highly metastable semiconductors. X-ray and electron diffraction measurements indicated the materials were amorphous, but the fact that some of the films were fine grained polycrystalline samples only became apparent in their DSC spectra. We present models that describe quantitatively the transformation behavior in both sets of films.

INTRODUCTION

Metastable semiconducting crystalline $Ge_{1-x}Sn_x$ alloys have predicted direct electronic energy gaps in the mid to far infrared (IR) wavelength range, for Sn concentrations greater than 20at.%. Present IR photoconductive materials (tellurides) in this wavelength range are very expensive to mass produce due to uncontrollability of the constituent elements during deposition. Crystalline $Ge_{1-x}Sn_x$ is an alternative to these conventional materials, except that the solid solubility of Sn in Ge is less than 1at.% and for the desired IR sensitivity more than 20at.% is needed. We have successfully produced such high Sn concentration alloys in what we originally thought was an amorphous form. Isothermal annealing in a differential scanning calorimeter (DSC) produced a signal which decayed with time, contrary to the standard bell shaped signal expected from nucleation and growth of the crystalline phase. We will show here that this decaying signal is due to crystal growth and, together with x−ray diffractometry (XRD), indicates that the as-deposited material was already polycrystalline with an 18Å grain size.

EXPERIMENTAL

By rf sputtering in an Ar atmosphere, we successfully fabricated uniform alloys of $Ge_{1-x}Sn_x$ for Sn concentrations ranging from 0 to 31at.%. We report here on two specific concentrations and thickness regimes: 25at.%, 0.2μm thick, and 31at.%, 8μm thick. The deposition conditions for each film were identical except for the deposition time and the Sn concentration. It was therefore expected that the as-deposited structure of the films would be nearly identical. In addition the deposition conditions were chosen such that the initial structure of each set of films should have been amorphous. Electron diffraction performed on the thinner material produced diffuse rings; XRD on the thicker material showed broad diffraction peaks. Both sets of measurements indicated the films were "amorphous". For the desired IR photonic properties, the alloys need to be crystalline. We therefore examined the "crystallization" and subsequent Sn phase separation in a DSC.

We scanned each material at 10K/min as shown in Fig. 1. The thinner material clearly exhibits two well separated exothermic reactions which we identified with a series of electron diffraction measurements on samples annealed to various temperatures. The first exotherm corresponded to crystallization and the second to phase separation. Figure 1a also shows a second heating scan of the sample, used for baseline determination, which exhibits an endotherm corresponding to Sn melting resulting from the phase separation process. Figure 1b, on the other hand, shows only one broad exotherm with several "bumps" on it indicating the presence of more than one process occurring in the temperature range of the exotherm.

If such processes have slightly different thermal activation energies, *isothermal* anneals at various temperatures will permit time separation of the processes. With this as our motivation, we ramped the temperature of each sample at 10K/min to isothermal anneal temperatures 10–15K lower than the onset of the first transformation in the scanned data. The thin samples exhibited two, time-separated, bell-shaped exotherms (Fig. 2a), which were identified by high resolution transmission electron microscopy (HRTEM) to be crystallization and phase separation. The crystallization exotherm fits the standard Johnson-Mehl-Avrami nucleation and growth theory with a nucleation constant n=3.5.[1] Combining the isothermal DSC data with HRTEM, we found the number of nuclei decreased as a function of time and the crystal growth at all sites was three dimensional.

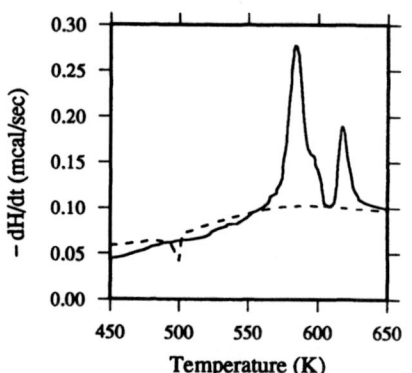 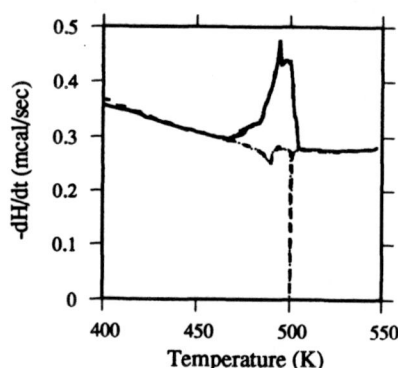

Fig. 1: (a) 10K/min scan of 0.2μm $Ge_{0.75}Sn_{0.25}$ and (b) 10K/min scan of 8μm $Ge_{0.69}Sn_{0.31}$, with the dotted line in each figure representing the second heating scan of the material.

The 450K isothermal anneal of the thick material is very different, see Fig. 2b. For approximately the first eight minutes, the rate of enthalpy release decays with time and then the material exhibits a "bumpy" but bell-shaped exotherm. XRD of a sample annealed partially through the decay (4 min.) indicates that no Sn phase separated from the sample. However, XRD of a sample annealed through the bell-shaped exotherm demonstrates that this exotherm is due to Sn phase separation, as was the case for the second exotherm in the thin sample. XRD of the sample annealed through four minutes of the decay also exhibited sharper diffraction peaks than in the as-deposited sample, indicating less disorder in the annealed sample. To determine the thermal response of the process occurring during this short decay, we lowered the isothermal anneal temperature to 410K, see Fig. 2c, which produced a much longer decay (~80 min. before the signal started to rise indicating the onset of phase separation). XRD of a sample annealed through 70 minutes of the 410K isotherm again displayed sharpened diffraction peaks with no Sn separation. The sharpening of these x-ray peaks was somewhat surprising since the thick and thin materials were deposited under identical deposition conditions and we expected the initial state of the thick sample to be amorphous like the thinner material.

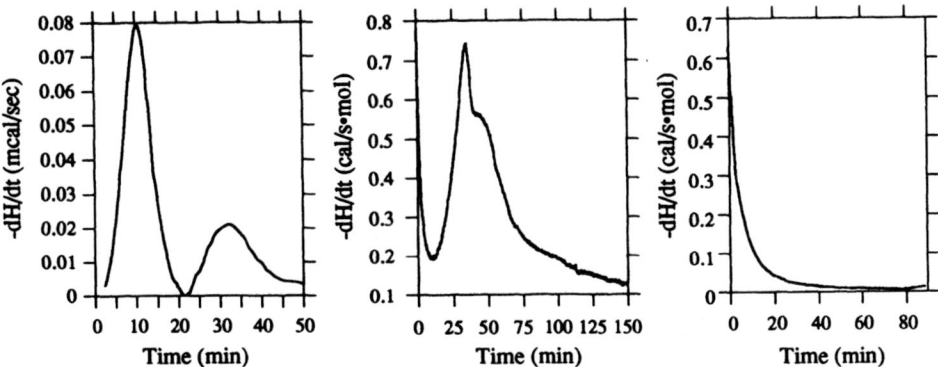

Fig. 2: (a) 540K isothermal DSC signal of 0.2μm thick $Ge_{0.75}Sn_{0.25}$; (b) 450K isothermal signal from 8μm thick $Ge_{0.69}Sn_{0.31}$ (c.) 410K isothermal signal from thick $Ge_{0.69}Sn_{0.31}$.

DISCUSSION

We postulated that this decay in rate of heat release in the thick samples was due to grain growth and will now show that theoretical simulations of such grain growth do indeed describe the experimental observations. Assuming grain growth, the sharpening of the x-ray peaks corresponded to an increase in the average crystallite size from 18Å to 45Å with no measurable simultaneous Sn separation. The subsequent analysis follows that of Chen and Spaepen.[2] Starting from the standard grain growth power law equation for the evolution of the average grain size, r, as a function of time, t, at temperature T gives

$$r^n(t) = r_0^n + t\, k(T) \tag{1}$$

where r_0 = the grain size in the as-deposited material and $k(T)$ = rate constant. The time evolution of the interfacial enthalpy, $H(t)$, as the grains grow at a given temperature T, is related to the enthalpic part of the interfacial tension, γ_H, the molar volume of the sample, V, and a geometrical factor $g = 1.3\pm0.2$ (for equiaxed grains)[2]

$$H_T(t) = g\gamma_H V / r_T(t) \tag{2}$$

$$H_T(t)\, r_T(t) = g\gamma_H V = \text{constant} = (H_o)_T (r_o)_T \tag{3}$$

where H_o = the initial enthalpy of the system. Combining these three equations, the rate at which heat is released from the sample during the grain growth process at temperature T is given by:

$$-\dot{H}_T(t) = \frac{(H_o)_T (r_o)_T}{n} \frac{k(T)}{(r_o^n + k(T)t)^{(n+1)/n}} \tag{4}$$

If a fitting parameter τ is defined such that

$$\tau \equiv \frac{(r_o)_T^n}{k(T)} \tag{5}$$

315

then the rate at which heat is evolved during grain growth at a given temperature is:

$$-\dot{H}_T(t) = \frac{(H_0)_T}{n\tau}\left(1 + \frac{t}{\tau}\right)^{-(n+1)/n}$$ (6)

Using the definition for the time evolution of the enthalpy during grain growth:

$$H_T(t) = (H_0)_T + \int_0^t \dot{H}_T(t)\, dt$$ (7)

$H_T(t)$ can be rewritten in terms of the fitting parameter τ

$$H_T(t) = (H_0)_T \left(\frac{t+\tau}{\tau}\right)^{-1/n}$$ (8)

A plot of ln H(t) versus ln(t+τ) should produce a straight line of slope (-n), if we do have grain growth and τ is chosen appropriately.

RESULTS

Since both isothermal anneal temperatures were well below the onset of the exotherm observed during the temperature scan of the sample, the total enthalpy of each isothermal decay was taken to be the same as that measured during the lower temperature isotherm (410K) and was 304cal/mol. Figure 3 shows the best fit to our 410K and 450K isothermal anneal data (Figs. 2b, c), with τ = 3.4-3.5min (correlation factor = 0.99976 over this range of τ) and 4.2-4.8min (correlation factor = 0.99994 over this range of τ) respectively.

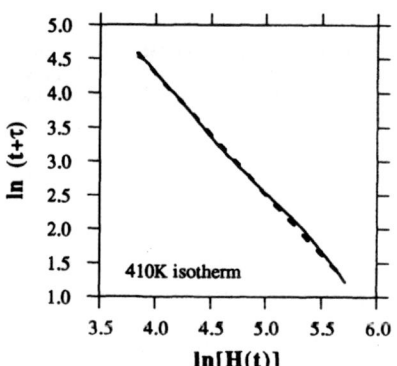 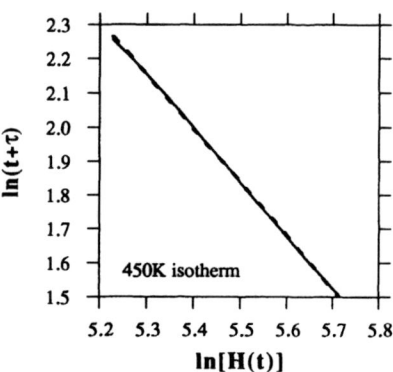

Fig. 3: (a.) dashed line shows fit to 410K isothermal DSC data (solid line) for τ = 3.4min; slope = −1.77 and (b.) dashed line shows fit to 450K isothermal DSC data (solid line) for τ = 4.5min, slope = −1.6.

The grain growth constant determined from these plots is n=1.76±0.01 for the 410K isothermal and n=1.6±0.1 for the 450K isothermal. These numbers are within the range of other

316

experimentally measured values for n (0.5 to 4).[3] Theoretically n should be 2. Using these values for n and τ, and the grain growth model discussed above, we simulated the DSC output using equation 6. The results are shown in Fig. 4.

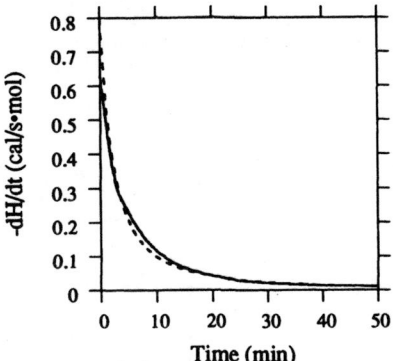

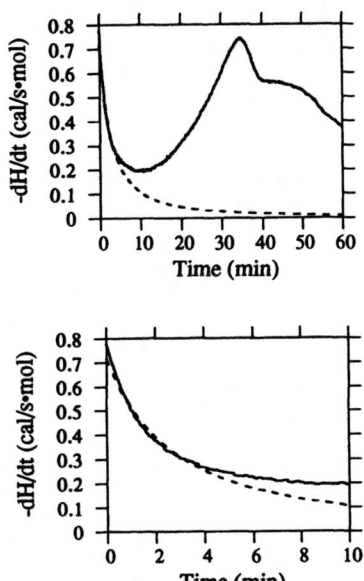

Fig. 4: (a.) (above) 410K simulation (dashed line) of the isothermal DSC data (solid line) using $\tau = 3.4$min and n = 1.76; (b.) (top right) shows the 450K simulation (dashed line, $\tau = 4.5$, n = 1.6) of the data (solid line); and (c.) (lower right) shows an expanded view of the decay portion and simulation of the 450K isotherm.

The remarkable agreement between the experimental data and the simulation, together with the sharpening of the x-ray peaks after annealing through each isothermal decay, shows that we are indeed justified in identifying the decay observed in the isothermal DSC data with grain growth of pre-existing crystalline grains.

With this value for the grain growth parameter, n, we estimate the activation energy for this process by assuming an Arrhenius temperature dependence for the rate constant

$$k(T) = k_o' \exp(-Q'/k_B T) \tag{9}$$

As Eyring[4] and Wert[5] have shown, it is more correct to assign an Arrhenius temperature dependence for the jump frequency instead of the rate constant, but the activation energies calculated either way are so close as to be within experimental error of each other and thus the difference can be ignored. If we let $t_{1/2}$ be the time at which the total enthalpy of the system has decreased to half its initial value and combining equations (1), (3), and (9), we find:

$$t_{1/2} H_o^n k_o' \exp(-Q'/k_B T) = \text{constant} \tag{10}$$

Thus a plot of $\ln(t_{1/2} H_o^n)$ versus $1/T$ should be a straight line with slope Q'/k_B. Such a plot for our data, which was inevitably a straight line since we performed only two different isothermal temperature anneals, produced an activation energy for grain growth of 0.5eV±0.2eV (the error in Q' is a result of the spread in τ values which gave the maximum correlation coefficient). A more accurate determination of this energy requires a minimum of three isothermal temperature anneals. Even so, our activation energy is reasonable in comparison with the activation energy for growth of $CoSi_2$ from an amorphous matrix (0.97-1.17eV) measured by other methods[6] and the 2.2eV that Chen and Spaepen determined for grain growth of the icosohedral phase of $Al_{0.83}Mn_{0.17}$ from an analysis similar to that presented here.[7]

The enthalpic part of the interfacial tension was also calculated using equation (3) with $g=1.3$, $r_o=18Å$, and $H_o=304cal/mol$. The molar volume was determined assuming 8 atoms per unit cell and a lattice constant of 5.918Å, found from the angular position of the (111) peak in the XRD pattern. The enthalpic part of the interfacial tension was $113mJ/m^2$, within the range of $140mJ/m^2$ found by Chen and Spaepen for their $Al_{0.83}Mn_{0.17}$ samples. It is interesting to note that the grain boundary energy for Sn is $164mJ/m^2$,[8] not too different than the $113mJ/m^2$ determined for our $Ge_{0.69}Sn_{0.31}$ alloys.

CONCLUSION

We have successfully produced alloys of $Ge_{0.75}Sn_{0.25}$ and $Ge_{0.69}Sn_{0.31}$ with an rf sputtering system. X-ray diffraction indicated that both samples were amorphous. However, isothermal DSC measurements permitted us to distinguish between these seemingly similar structures. The amorphous sample exhibited an isothermal bell-shaped crystallization exotherm which, when analyzed using the standard Johnson-Mehl-Avrami formalization, produced an Avrami $n=3.5$ and indicated that the nucleation rate of the crystalline state decreased with increasing time while the grains grew three dimensionally at all nucleation sites. The other $Ge_{0.69}Sn_{0.31}$ sample exhibited an isothermal decay in the rate of heat released during the beginning portion of the isothermal anneal and subsequently a bell-shaped curve, corresponding to phase separation of Sn from the alloy. Analysis of the decay, using the standard grain growth model of Burke, Turnbull, and others[9], gave a grain growth exponent of 1.7 ± 0.1, in reasonable agreement with $n=2$ of the models used; an estimated thermal activation energy for grain growth of 0.5eV; and an interfacial grain boundary energy of $113mJ/m^2$, within the range of the grain boundary energy for Sn, $164mJ/m^2$. These numbers together with a fitting parameter were used to simulate the rate at which heat would be released during an isothermal anneal assuming the only process occurring during the anneal was grain growth of pre-existing grains. Comparison of these simulations with the experimental isothermal DSC data showed a remarkable agreement between the two. This indicated that indeed only grain growth occurred during the decay portion of the isotherms, despite the indication from XRD that the materials in their as-deposited state were amorphous and therefore should have first undergone nucleation of the crystalline phase before grain growth occurred. XRD of the samples annealed through the isothermal decays showed a sharpening of the diffraction peaks corresponding to more than a doubling of the grain size from the as-deposited state with no attendant Sn separation from the metastable diamond cubic phase of the alloy.

REFERENCES

This work was performed under the auspices of the U.S. Department of Energy by Lawrence Livermore National Laboratory under contract number W-7405-ENG-48.

[1] S.M.Lee, J. Appl. Phys., publication date:1 Feb 1994.
[2] L. C. Chen and F. Spaepen, J. Appl. Phys. **69**, 679 (1991).
[3] H. V. Atkinson, Acta Metall. **36**, 469 (1988).
[4] H. Eyring, J. Chem. Phys. **3**, 107 (1935).
[5] C. A. Wert, Phys. Rev. **79**, 601 (1950).
[6] M. C. Ridgway, R. G. Elliman, R. P. Thornton, J. S. Williams, Appl. Phys. Lett. **56**, 1992 (1990); D. A. Smith, K. N. Tu, B. Z. Wiess, Ultramicroscopy **30**, 90 (1989); C. A. Hewett, I. Suni, S. S. Lau, L. S. Hung, and D. M. Scott, Mat. Res. Soc. Symp. Proc. **27**, 145 (1984).
[7] Chen and Spaepen, op. cit., 679.
[8] L. E. Murr, Interfacial Phenomena in Metals and Alloys, (Addison-Wesley, London, 1975).
[9] J. E. Burke and D. Turnbull, Prog. Metal Phys. **3**, 220 (1952); C. Wagner, Z. Elektrochem. **65**, 581 (1961); M. Hillert, Acta Metall. **13**, 469(1965); C. V. Thompson, H. J. Frost, and F. Spaepen, Acta Metall. **35**, 887 (1987).

THICKNESS EFFECT ON THE CRYSTALLIZATION KINETICS OF ELECTROLESSLY-DEPOSITED Ni(P) THIN FILMS

L.T. Shi and E.J.M. O'Sullivan
IBM Thomas J. Watson Research Center, Yorktown Heights, NY 10598

ABSTRACT

In order to understand thickness and interfacial effects on the crystallization kinetics of amorphous solids, Ni(P) thin films electrolessly deposited on Cu seed layers were annealed at constant heating rates or at constant temperatures in a DSC to obtain activation energies and Avrami exponents. It was found that the activation energy of crystallization in Ni(P) changes as a function of sample thickness when the sample thickness is less than 1.0 μm. Furthermore, the Avrami exponent was found to change not only as a function of thickness but also as a function of annealing temperature.

INTRODUCTION

Because of its relatively low deposition cost, technological importance, and simple amorphous phase, the crystallization of electrolessly-deposited (ELD) Ni(P) has been extensively studied using various analytical techniques[1-8]. It has been determined[3,7,8] that the crystallization kinetics of Ni(P) depend on the phosphorus content in the sample. However, some controversy still exists in the reported results measured from samples with almost identical composition; for example, the activation energy of crystallization for the ELD Ni(P) with 11 wt% P reported in ref. 2 (227 KJ/mole) is very different from that reported in ref. 8 (150-160 KJ/mole). In addition, the interfacial effect on phase transitions in thin films is an interesting subject. Its understanding is essential to many technological applications. In an attempt to account for the above discrepancy and to gain some understanding about the interfacial effect, we have conducted a kinetic study of crystallization in ELD Ni(P) samples with various thicknesses using differential scanning calorimetry (DSC).

EXPERIMENTAL

ELD Ni(P) films were deposited from an electrolessly plating bath with a pH value of 8.1 and maintained at 73°C. The bath solution contained $NiSO_4$, citric acid as complexant, boric acid as buffer, sodium hypophosphite as reducing agent and Pb as stabilizer. The plating rate was estimated to be 175 Å/min. Films were deposited on a 2 μm thick, sputtered Cu layer on a Si wafer, with 400Å Cr as an adhesion promoting layer (to Si wafer). Prior to plating, the Cu surface was catalyzed by immersion in an acidic, aqueous, Pd salt solution. The plated Ni(P) films contained about 11.5 wt% P and exhibited an amorphous structure. It is well known[4,9,10] that as-deposited Ni(P) films containing P greater than 7-8 wt% are generally amorphous.

Calorimetry analysis was conducted in a modified DuPont 910 DSC under a dry and pure nitrogen environment. Two different sample forms were used for annealing tests at constant heating rates: one was a free standing film detached from the Si substrate, and the other a film attached to the Si substrate (i.e. *in situ*). The latter was cut from the wafer in a size which could be fitted in the DSC pan, and was tested (without crimping the pan) with the Ni(P) film side facing down and contacting the pan. The purpose of using these two sample

319

forms was to examine the influence of the thick Si substrate ($\sim$390μm) on the DSC measurements. However, the *in situ* sample is not appropriate for isothermal annealing tests because it contains a relatively small amount of Ni(P) film (<0.8 mg) which can not generate a sufficient signal. Thus, only free standing samples were used in the isothermal annealing tests. It should be noted that in the free standing sample the ELD Ni(P) film was still adhered to the Cu seed layer so that the effect of interface between Ni(P) and Cu could be investigated. Samples with various thicknesses from 1000Å to 5 μm were used in this study.

RESULTS AND DISCUSSION

As expected, the heat flow peak corresponding to crystallization in the ELD Ni(P) films was found to change with the heating rate. The activation energy of crystallization in Ni(P) was measured by plotting $\ln[(1/T_p^2)(dT/dt)]$ vs. $1/kT_p$ based on the analysis by Kissinger[11] and Ozawa[12]. Figs. 1(a) and 1(b) show such plots obtained from free standing films and *in situ* samples with various ELD Ni(P) thicknesses, respectively. It is noted that the peak crystallization temperature decreases as the thickness of the Ni(P) film decreases, and the reproducibility of the tests is reasonably good as demonstrated in Fig. 1(b). The slopes of the lines, which were obtained by linear regression fitting, yielded an activation energy varying from about 1.7 eV for the thinnest Ni(P) film, to a limiting value of about 2.4 eV for thicker Ni(P) films, as shown in Fig. 2. The origin of the scattering for the samples with Ni(P) film thickness of 2000Å is unclear. However, the crystallization behavior of the thick Ni(P) films may be considered to be relatively independent of the Si substrate and Ni(P)/Cu interface effects. The activation energy measured in this work for thick Ni(P) films, i.e.

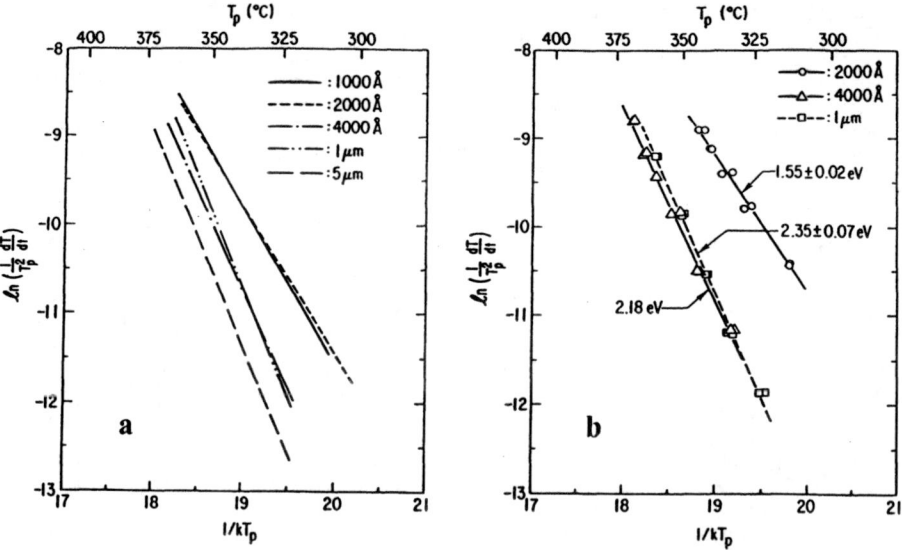

Fig. 1. Plots of $\ln[(1/T_p^2)(dT/dt)]$ vs. $1/kT_p$ for (a) free standing films and (b) films on Si substrates with various Ni(P) thicknesses, where T_p is the peak transition temperature of crystallization measured from DSC plots, and dT/dt the heating rate.

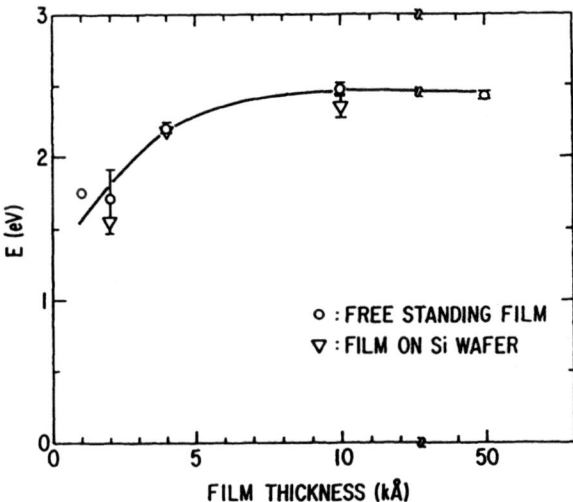

Fig. 2. Activation energy of crystallization in ELD Ni(P) as a function of deposited Ni(P) film thickness.

~2.4 eV, is close to the value of 2.32 eV obtained using DTA by Szasz et al.[5] for ELD Ni(P) of unspecified composition, and the value of 2.35 eV reported by Mahoney and Dynes[2] using DSC for a film with 11 wt% P. However, the value of 1.7±0.2 eV measured for the present thin samples is almost identical to the value (~1.6 eV) measured using DSC by Agarwala and Ray[8] for 11.5 wt% P films.

It has been pointed out[13,14] that the Avrami exponent (n) in an equation derived from the Johnson-Mehl-Avrami (JMA) analysis[15], i.e.

$$x = 1 - \exp(-kt^n), \tag{1}$$

can be used to explore the crystallization mechanism in amorphous solids; here x is the fraction of crystallization which has occurred in the amorphous solid, K is a temperature-dependent rate parameter, and t is the crystallization time. Thus, for a three dimensional phase formation process involving nucleation and growth, n should be 3 for a constant number of nuclei or 4 for a constant nucleation rate [13]. Here the results of heat flow as a function of annealing time obtained from the DSC isothermal annealing tests were used to derive the Avrami exponent. As is usually the case, the heat flow curves (obtained by subtracting the baseline curve obtained in the second run from the curve obtained in the first run) are bell shaped. However, it was noted that the heat flow curves measured from the same type of samples changed with the annealing temperatures, as an example shown in Fig. 3. The Avrami exponent n was measured by plotting ln (-ln(1-x)) vs. ln t, where x is measured by the integral of the bell-shaped heat flow curve with respect to the crystallization time t. Figure 4 shows such plots corresponding to the examples shown in Fig. 3. The slopes of the lines (also obtained by linear regression fitting) in Fig. 4 gave the corresponding Avrami exponent, and indicated that the Avrami exponent actually varies with the annealing temperatures. Fur-

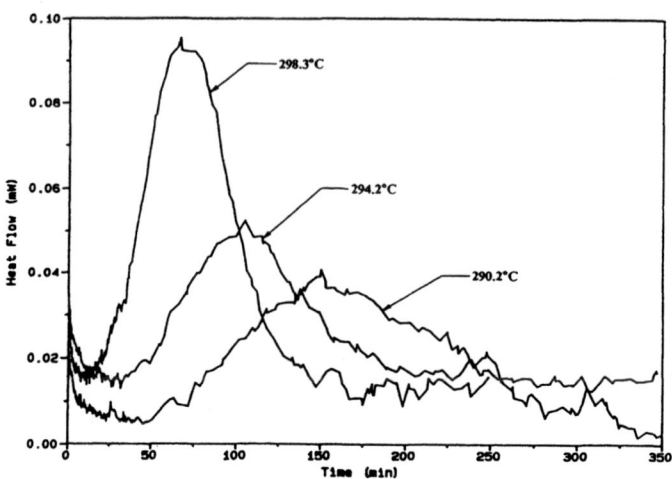

Fig. 3. Curves of heat flow vs. time measured from the free standing ELD Ni(P) samples with a thickness of 5 μm isothermally annealed at various temperatures.

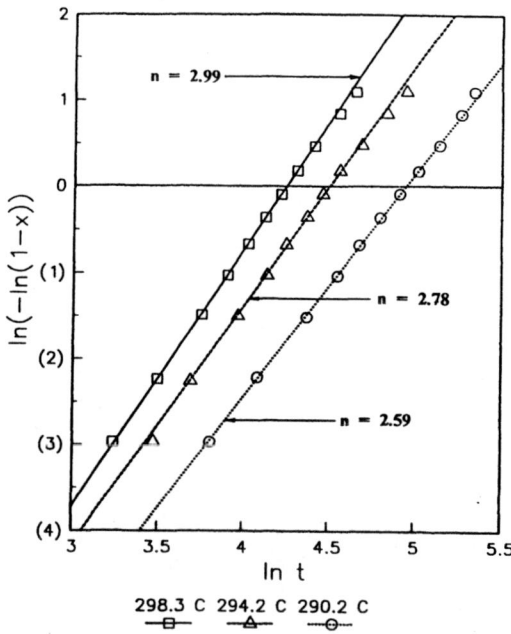

298.3 C 294.2 C 290.2 C
—□—   —△—   ⊙⊙

Fig. 4. Plots of ln(-ln(1-x)) vs. ln t corresponding to the heat flow curves shown in Fig. 3.

thermore, by collecting the data obtained from films with different thicknesses, it was revealed that the Avrami exponent plots also shift along the axis of the annealing temperature as the sample thickness changes, as shown in Fig. 5.

Due to the experimental limitations, some of the plots shown in Fig. 5 may not be extendable to higher temperatures. For example, the complete curve of the bell-shaped heat flow peak could not be obtained when the samples with a thickness of 2000Å were isothermally annealed at a temperature higher than 274 °C , since the crystallization was already partially completed in the sample before the desired temperature was reached. Nevertheless, a tendency of nearly parallel shift of the Avrami exponents towards lower annealing temperatures as the sample thickness decreases is quite obvious in Fig. 5. Thus, we may conclude that at a given temperature (say 280°C) the thinner sample has more degrees of freedom (higher value of Avrami exponent) than the thicker sample for crystallization to occur; this also reasonably explains why the smaller the sample thickness, the lower is the crystallization peak temperature measured in the constant-heating-rate annealing tests. However, it is not clear why and how the difference in the crystallization mode can quantitatively result in different activation energies, although the interface between Ni(P) and Cu must play a more important role in the thinner samples than in the thicker ones. Further investigation taking the interfacial stress into account is under way.

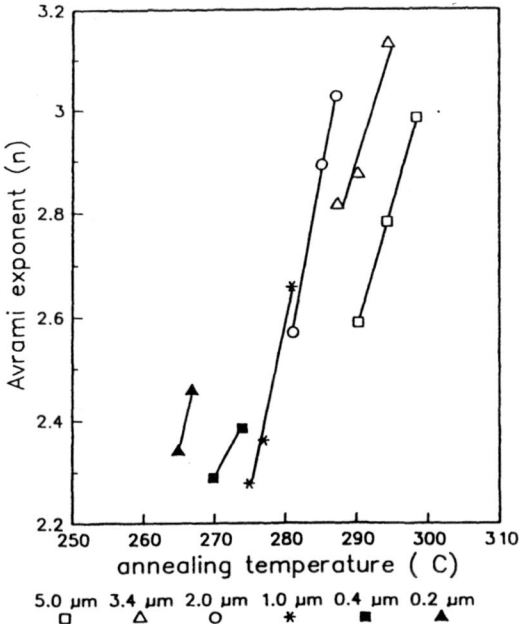

Fig. 5. Avrami exponent as a function of annealing temperature for samples with different thicknesses.

CONCLUSIONS

We have demonstrated that both the activation energy and the Avrami exponent of crystallization in ELD Ni(P) amorphous films are functions of sample thickness. In addition, the Avrami exponent measured on the same type samples changes as a function of isothermal annealing temperature.

ACKNOWLEDGMENTS

The authors would like to thank J.R. Marino for plating some of the Ni(P) films used, M. Plechaty for determining the composition of the Ni(P) material, K.N. Tu for inspiring and useful disussions, and L.T. Romankiw for a review of the manuscript.

REFERENCES

1. M.S. Grewal, S.A. Sastri, and B.H. Alexander, Thermochimica Acta, 14, 25 (1976).
2. M.W. Mahoney and P.J. Dynes, Scripta Metall., 19, 539 (1985).
3. I. Bakonyi, A. Cziraki, I. Nagy and M. Hosso, Z. Metallkd., 77, 425 (1986).
4. E. Ma, S. Lu, and P. Li, Thin Solid Films, 166, 273 (1988).
5. A. Szasz, X.D. Pan, J. Kojnok, and D.J. Fabian, J. Non-Cryst. Solids, 108, 304 (1989).
6. K.L. Lin and P.J. Lai, J. Electrochem. Soc., 136, 3803 (1989).
7. R.C. Agarwala and S. Ray, Z. Metallkd., 80, 556 (1989).
8. R.C. Agarwala and S. Ray, Z. Metallkd., 83, 203 (1992).
9. A.H. Graham, R.W. Lindsay, and H.J. Read, J. Electrochem. Soc., 112, 401 (1965).
10. J.P. Randin, P.A. Maire, E. Saurer, and H.E. Hintermann, J. Electrochem. Soc., 114, 442 (1967).
11. H.E. Kissinger, Anal. Chem., 29, 1702 (1957).
12. T. Ozawa, J. Thermal Analysis, 2, 301 (1970).
13. E. Ma and C.V. Thompson, J. Appl. Phys., 69, 2211 (1991).
14. K.N. Tu, Appl. Phys. A, 53, 32 (1991).
15. J.W. Christian, The Theory of Transformation in Metals and Alloys, 2nd ed. (Pergamon, Oxford, 1975).

CRYSTALLIZATION KINETICS OF $Fe_{80-y}V_yB_{12}Si_8$ AMORPHOUS RIBBONS

P. ALTÚZAR, C. VÁZQUEZ, L. BAÑOS AND R. VALENZUELA
Institute for Materials Research, National University of Mexico, Ap. Postal 70-360, Mexico D.F., 04510, Mexico.

ABSTRACT

The effects of vanadium in the crystallization kinetics amorphous ribbons $Fe_{80-y}V_yB_{12}Si_8$ ($0.5 \leq y \leq 15$) has been investigated by Differential Scanning Calorimetry, by using the Avrami, Kolmogorov-Johnson-Mehl-Avrami, and Calka and Radlinski equations. The addition of vanadium to Fe-B-Si alloys leads to an enhancement of stability against crystallization, as shown by an increase in the effective activation energy for crystallization ($E_{eff} = 3.6eV$) and an increase in the temperature for the first crystallization peak, as a function of vanadium content (from 780 to 855K). Results also show that the crystallization mechanism, nucleation rate and dimensionality of growth are constant throughout the crystallization process in the composition range investigated.

INTRODUCTION

Amorphous alloys in the system Fe-B-Si possess excellent soft ferromagnetic properties, such as an extremely high magnetic permeability and a good frequency stability, which make them useful in many technological applications[1,2]. Their thermal stability, however, has always been a source of concern, since the presence of crystals leads to considerable changes in magnetic properties[3]. In this paper, we present an investigation of the effects of vanadium on the crystallization kinetics of $Fe_{80-y}V_yB_{12}Si_8$ amorphous ribbons, by using the Avrami, Kolmogorov-Johnson-Mehl-Avrami (KJMA), and Calka and Radlinski equations. Vanadium results in an increased stability since the temperature of the first crystallization peak increases.

EXPERIMENTAL PROCEDURE

Amorphous ribbons in the system $Fe_{80-y}V_yB_{12}Si_8$ with $0.5 \leq y \leq 15$ were prepared by the melt-spinning technique. The ribbons width and thickness were approximately 2mm and 25μm, respectively. X-ray diffraction by means of a Siemens D500 diffractometer of both surfaces of samples exhibited the characteristic diffuse ring of the amorphous state. The elemental composition of ribbons was checked by electron microprobe analysis.

Samples of mass 3-4 mg were studied in a DSC 910 Dupont Differential Scanning Calorimeter. The activation energy for crystallization was determined from the Avrami equation[4] by the varying the heating rate from 5 K/min to 40 K/min. For the KJMA method, all the thermal treatments were performed *in-situ*, in the DSC apparatus. Samples were heated from room temperature to the annealing temperature at 40 K/min, and then subjected to an isothermal period. The annealing temperature was 40K below the temperature of the first crystallization peak for each composition. They were then heated again at 40 K/min up to the first crystallization peak. The crystallization fraction, x, was evaluated from the area under the crystallization peak.

Mat. Res. Soc. Symp. Proc. Vol. 321. ©1994 Materials Research Society

RESULTS

The DSC thermograms at 20 K/min, Fig. 1, exhibited two well-resolved crystallization peaks for $x \leq 6$. Crystallization temperatures showed an increase as a function of composition, Fig. 2.

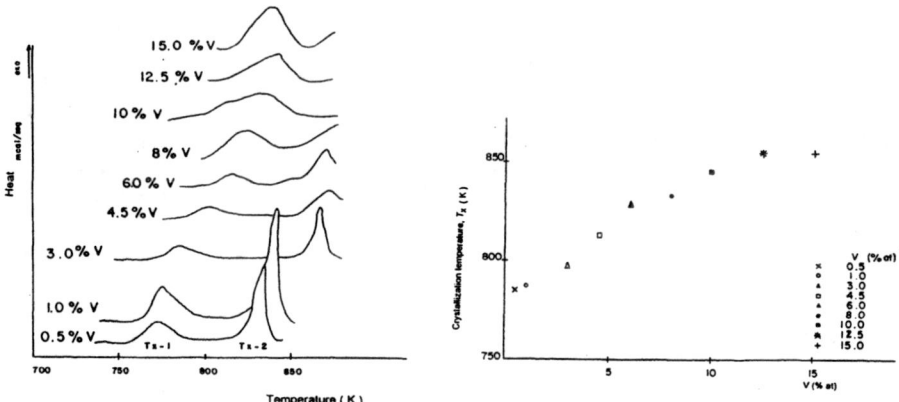

Fig. 1. DSC Thermograms at a heating rate of 20K/min for all the compositions.

Fig. 2. Temperature of the first crystallization peak as a function of composition.

The effective activation temperature for crystallization can be evaluated from the non-isothermal expression derived by Marseglia[4]:

$$d \ln (r/T) / d (1/T) = -E / nR = E_{eff} \tag{1}$$

where r is the heating rate, T the temperature, E the activation energy, n the Avrami exponent, R the gas constant and E_{eff} the effective activation energy. A different expression, the so-called Kissinger equation is widely used to evaluate the activation energy for crystallization, with very similar results. However, it has been shown[5] that this is due to the mathematical form of equations, but that the model physically correct is the one derived from the Avrami expression, Eq, (1). The latter is plotted in Fig. 3 as a function of composition. As shown in this figure, E_{eff} is relatively constant for the composition range investigated. The exponent n in the isothermal form of the Avrami expression:

$$1 - x = \exp (- K\ t^n) \tag{2}$$

where x is the crystallized fraction, K the Arrhenius rate constant and t the reaction time, can be evaluated from the KJMA expression[6]:

$$\ln [-\ln (1 - x)] = n \ln t + \ln K \tag{3}$$

by plotting the left-hand side term of Eq. (3) as a function of ln t, as shown in Fig. 4 for $y = 0.5$ and 6. The Avrami exponent is important since it contains information about nucleation and the growth mechanism, which can be obtained from the partition function[7]:

$$n = a + bp \tag{4}$$

where a refers to the nucleation rate, varying from 0 for quenched-in nuclei, to 1, for a constant nucleation rate with no pre-existing nuclei; b is associated with the dimensionality of crystal growth, having the values 1, 2, 3 for uni-, bi- and tridimensional growth; and p defines the mechanism controlling the growth: it has the value 0.5 for diffusion-controlled (parabolic) growth, and 1.0 for interfacial (linear in time) growth.

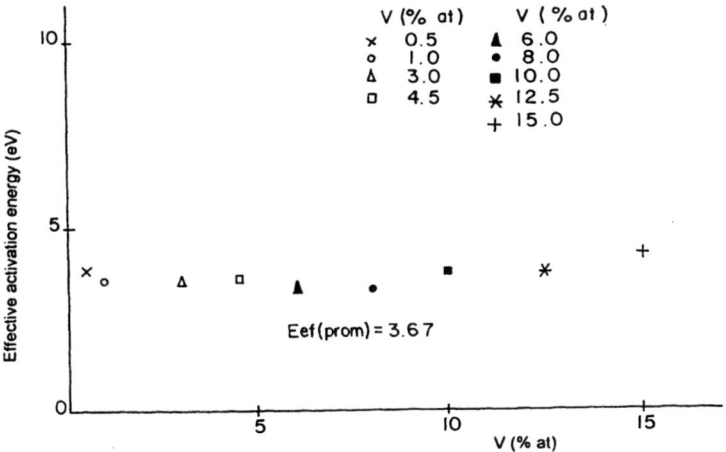

Fig. 3. Effective activation energy for the first crystallization peak.

Equation (3), however, is valid for a constant n, which is strictly true only for some limiting cases. The calculation of the partition function by means of an average value of n extending over a wide range of volume fraction transformed may be inappropriate if competing reactions or changes in growth mechanism or dimensionality occur during the progress of the transformation.

Calka and Radlinski[8] have devised a more sensitive approach by considering the first derivative of the Avrami expression as a function of the volume fraction transformed, to obtain a *local* Avrami exponent n_{loc}:

$$d \ln [-\ln (1-x)] / d \ln t = n_{loc}x \tag{5}$$

A "local" slope (for each fraction transformed) of the KJMA curve is therefore obtained by Eq. (5). If growth mechanism and morphology are the same throughout the process, n and n_{loc} are the same for any x. Equation (5) was used to obtain the n_{loc} value for several values of y, Fig. 5. It is immediately apparent that, except for both the low and high transformed fractions, where differences in dimensionality and/or growth mechanism can be expected, n_{loc} shows the same value than n, with considerably less dispersion.

X-ray diffraction experiments in samples annealed at temperatures just above the first crystallization peak confirmed that these crystals have the bcc structure of α-Fe, with a lattice parameter slightly above the value of pure α-Fe.

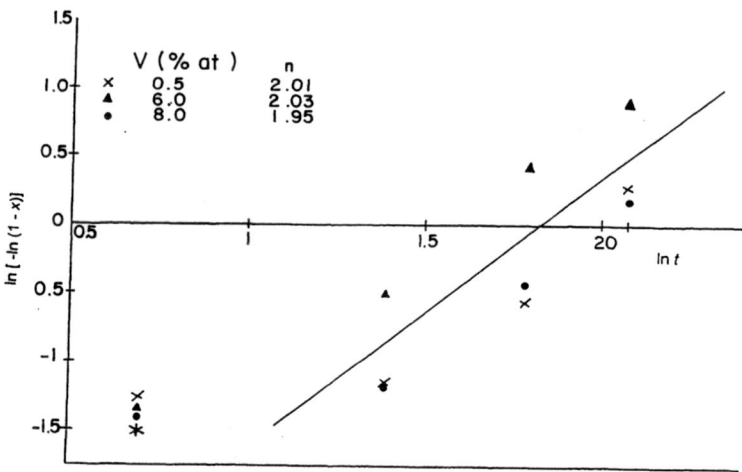

Fig. 4. Avrami exponent as evaluated from the KJMA expression.

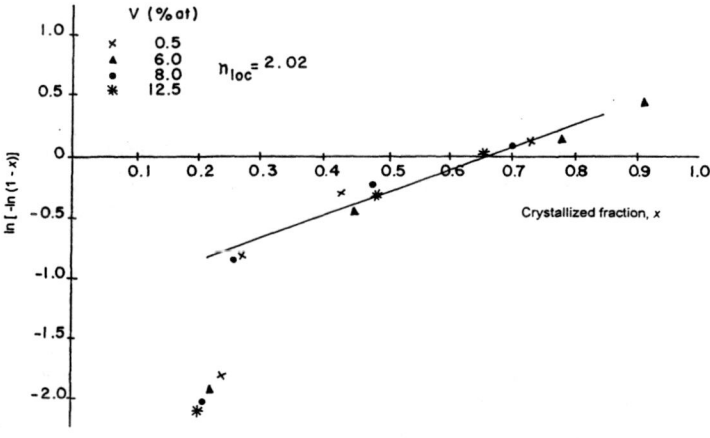

Fig. 5. Local Avrami exponent as a function of composition evaluated from the Calka and Radlinski expression (Eq. 5).

DISCUSSION

The crystallization temperature for the first peak increase as the vanadium content increases, and the values of effective activation energies for all the investigated samples (E_{eff} = 3.6eV) is larger than the reported activation energies for the Fe-B-Si system (E_{eff} = 2.1eV for $Fe_{78}B_{13}Si_9$). These two results point to an enhanced stability as a result of vanadium addition.

The first crystallization peak corresponds to the crystallization of solid solution of Si in α-Fe, which, as a primary crystallization, is characterized by a diffusion controlled mechanism with tridimensional growth. The observed n_{loc} value therefore leads to a decreased nucleation rate (0.5), also consistent with the enhanced stability of the amorphous phase as mentiones in the above paragraph. Finally, these dimensionality, controlling mechanism and nucleation rate are the same throughout most of the crystallization process.

Acknowledgments

Authors thank Dr. R. Krishnan (CNRS France) for providing the amorphous samples, M. Reyes-Salas and S. Jiménez-Cortés for technical help, and DGAPA-UNAM for partial funding.

References

[1] H. Warlimont and R. Boll. J. Appl. Phys. **27**, 97, (1978).

[2] G.E. Fish. J. Appl. Phys. **53**, 3569, (1982).

[3] R. Hasegawa, G.E. Fish and V.R.V. Ramanan. Proceedings of the 4th International Conference on Rapidly Quenched Metals, edited by T. Masumoto and H. Fujimori, Sendai, Vol. II, p. 929, (1981).

[4] E.A. Marseglia. J. Noncryst. Solids, **41**, 31, (1980).

[5] P. Altúzar and R. Valenzuela. Mat. Lett., **11**, 101, (1991).

[6] A.N. Kolmogorov. Bull. Acad. Sci. USSR, Phys. Ser., **3**, 555, (1937)

[7] J.W. Christian. The Theory of Transformations in Metals and Alloys, Pergamon Press, London, 489, (1965).

[8] A. Calka and A.P. Radlinski. Mater. Sci. and Engin. **97**, 242, (1988).

[9] V.R.V. Ramanan and G.E. Fish. J. Appl. Phys., **53**, 2273, (1982).

SOLID STATE REACTIONS IN BINARY MIXTURES OF NANOMETER-SIZED PARTICLES

W. DICKENSCHEID AND R. BIRRINGER
Universität des Saarlandes, Institut für Werkstoffphysik, Postfach 151150, D-66041 Saarbrücken, Germany

ABSTRACT

Solid state reactions in mixtures of nanometer-sized Cu and Zr as well as Ni and Zr crystallites - produced by inert-gas condensation followed by *in situ* compaction - have been investigated by x-ray diffraction and thermal analysis. The annealing behavior is compared to that of corresponding multilayer samples. The results are discussed with emphasis placed on the different parameters controlling solid state reactions.

INTRODUCTION

Solid state amorphization reactions (SSAR) of binary metallic systems are governed by thermodynamic as well as by kinetic factors [1,2]. Thermodynamics answers the question whether or not excess Gibbs free energy exists that can act as a driving force for the transformation of the system into the amorphous state. For this to occur, kinetic constraints must obtain that prevent the system from transforming into the energetically more favorable equilibrium state (intermetallic compound). Such kinetic boundary conditions may result from different diffusivities of both atomic species, i. e., only one type of atom has an appreciable mobility in a given temperature window, whereas the other atoms are virtually immobile in comparison to the fast-moving species.

The nucleation and growth of the amorphous phase by SSAR has previously been studied by investigating the time-temperature behavior of crystalline metallic multilayers (ML) [1,2]. The amorphous phase was found to be nucleated at defect sites in the lattice of the slow-moving species, and its growth is determined by the diffusivity of the fast-moving species through the already-grown amorphous interlayer.

Unlike ML's, nanocrystalline materials (NCM) are composed of 3-dimensional objects, i.e., nm-sized crystallites, separated by highly incoherent interfaces. The specific interfacial energy of NCM can be as high as twice as that of large-angle grain boundaries in conventional polycrystalline materials [3]. Also, the diffusivity in the grain boundaries of NCM has been found to be more than four orders of magnitude larger than that in equilibrated grain boundaries [4]; hence, we expect NCM to behave differently compared to multilayered materials with respect to SSAR.

The model systems chosen for studying SSAR in binary mixtures of nm-sized crystals were Cu-Zr and Ni-Zr - the former because its preparation is relatively simple, and the latter because numerous experimental data on SSAR in Ni_xZr_{1-x} ML's are available in the literature [1,2]. Both systems fulfill the necessary conditions for SSAR to occur; nonetheless, there are differences in the driving force as well as in the diffusivities (Table I).

Table I. Free energy of mixing for binary amorphous and intermetallic alloys of equiatomic composition. Calculation of the driving force ΔG was performed using the semiempirical Miedema model [5]. The diffusivities of Cu, Ni and Zr in the Zr-matrix are parameters for the factors determining the kinetics.

	driving force ΔG [kJ/mol] [5] amorphous intermetallic		Diffusivity in Zr-lattice [m²/s] [6]
Cu-Zr	-15	-34	$D_{Cu}(600K)=4.4 \times 10^{-18}$ $D_{Zr}(600K)=3 \times 10^{-21}$
Ni-Zr	-39	-72	$D_{Ni}(600K)=3.1 \times 10^{-16}$ $D_{Zr}(600K)=3 \times 10^{-21}$

EXPERIMENTAL

For the production of binary mixtures of nanocrystalline particles, we used a UHV vacuum chamber (base pressure less than 10^{-7} mbar) with two vapor deposition sources mounted face-to-face and a liquid-nitrogen-cooled cold finger situated between them. The elemental materials were simultaneously deposited in an inert-gas atmosphere of typically 0.1-10 mbar; Cu and Zr were dc-magnetron sputtered, while Ni was thermally evaporated.

During the inert-gas condensation process the atoms in the vapor phase condense to form small nm-sized crystallites that are collected on the rotating cold finger. Due to the rotation, the crystallites of the different species are randomly distributed in the resulting powder. Once a considerable amount of powder has been collected, it is stripped off the cold finger and transferred *in situ* to a UHV compaction device. Compaction is performed at a pressure of 2 GPa for 24 hours at 100°C, resulting in disk-shaped samples having a diameter of 5 mm and a weight between 30 and 50 mg. We shall refer to this state as the as-prepared state. A more-detailed description of the inert-gas condensation technique for preparing nanocrystalline materials may be found in Refs. [7,8].

Multilayered $Cu_{50}Zr_{50}$ (ml-$Cu_{50}Zr_{50}$) samples were prepared by dc-magnetron sputtering Zr and rf-sputtering Cu onto NaCl single crystals held at room temperature in an Ar atmosphere of 3×10^{-2} mbar. The base pressure of the chamber prior to sputtering was better than 10^{-8} mbar. A 150 nm Cu layer was deposited above and beneath the ML samples to prevent oxidation of the Zr during exposure to air. We prepared a ml-$Cu_{50}Zr_{50}$ sample with a composition wavelength of 50 nm and total thickness of 1μm. Consequently, the thickness of the individual layers was 17 nm for Cu and 33 nm for Zr.

X-ray diffraction scans were performed using Cu-Kα radiation on a Θ-2Θ diffractometer fitted with a position-sensitive detector. The chemical composition of the samples was determined by energy dispersive electron probe microanalysis (EPMA) in a scanning electron microscope (SEM) (Jeol SXA 480).

RESULTS

A. Cu_xZr_{1-x}

The reaction kinetics of a nanocrystalline $Cu_{25}Zr_{75}$ (n-$Cu_{25}Zr_{75}$) sample were investigated by a constant-heating-rate scan in a differential scanning calorimeter (Perkin Elmer DSC 7) (Fig. 1).

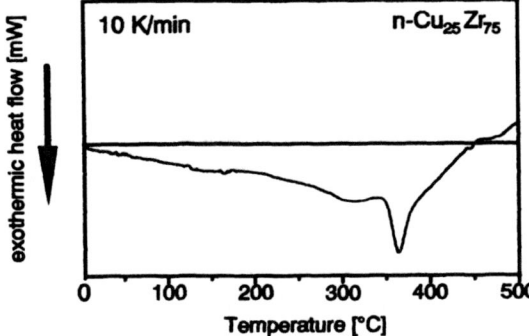

Fig 1:
DSC curve of n-Cu$_{25}$Zr$_{75}$, measured with a heating rate of 10 K/min, showing several exothermic reactions (see text).

The different reactions were subsequently identified with the help of x-ray diffraction. Above 100°C we observe an enthalpy release that is attributed to interfacial atoms moving to energetically more favorable positions, because no grain growth could be observed below 200°C. The maximum enthalpy release at 360°C is correlated to the formation of the intermetallic compound CuZr$_2$. At a slightly higher temperature an endothermic reaction becomes evident, which we interpret as a desorption of light-element impurities originating from the interfaces. Since such a behavior was also observed in n-Pt samples [3], it might be a typical feature for NCM prepared by inert-gas condensation.

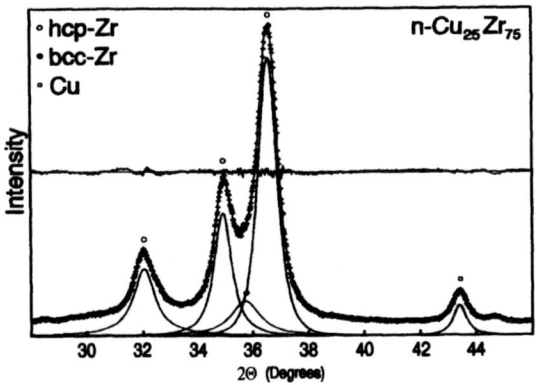

Fig 2:
X-ray scan (+) and profile fitting (solid lines) of n-Cu$_{25}$Zr$_{75}$. Hcp-Zr, bcc-Zr and Cu phases can be identified.

In the as-prepared state n-Cu$_{25}$Zr$_{75}$ shows only crystalline peaks of hexagonal Zr and face-centered-cubic Cu (Fig. 2). The grain size determined from the full width at half maximum (FWHM) of the hcp-Zr peaks using the Scherrer formula was 11 nm. There is also indication for the presence of the body-centered-cubic high-temperature Zr phase. It is well known that for small particles, a high-temperature phase like bcc-Zr may be stabilized at room temperature [9].

During subsequent annealing in a vacuum furnace (< 10^{-6} mbar) at appropriate temperatures (250°C, 17 h; 275°C, 40 h), we observed that the intermetallic compound CuZr$_2$ is nucleated directly without nucleation of an intermediate amorphous phase (Fig. 3a). For comparison we performed the same annealing experiments on a ml-Cu$_{50}$Zr$_{50}$ sample with a composition wavelength of 50 nm (Fig. 3b). In this case as well no SSAR occurred, the intermetallic compound having nucleated directly.

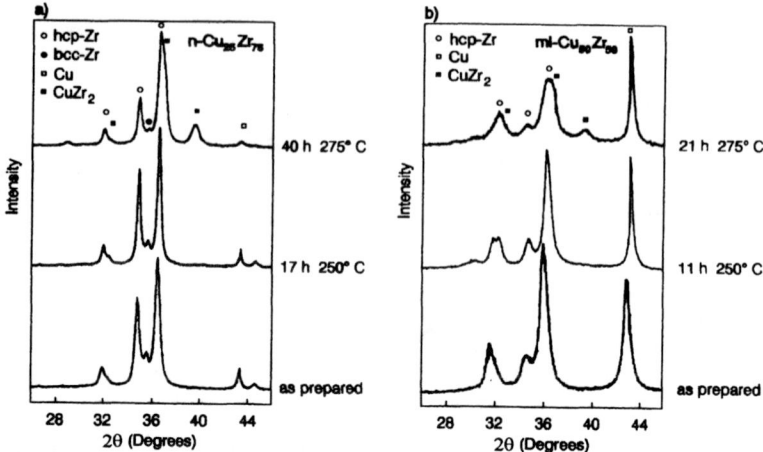

Fig 3:
a) Isothermal annealing of a n-$Cu_{25}Zr_{75}$ sample. Up to 250°C only grain growth can be observed. At 275°C, the intermetallic compound is nucleated.
b) Isothermal annealing of a ml-$Cu_{50}Zr_{50}$ sample showing the same behavior as the n-$Cu_{25}Zr_{75}$ sample. The high Cu peaks are due to the thick Cu layer on the top of the ML (oxidation barrier).

B. Ni_xZr_{1-x}

In the as-prepared state the n-Ni_xZr_{1-x} samples have a small amount of amorphous phase, as can be seen in Fig. 4a. Annealing of n-$Ni_{35}Zr_{65}$ in vacuum ($< 10^{-6}$ mbar) for 12 hours at 200°C caused the intensity of the amorphous phase to increase relative to that of hcp-Zr (Fig. 4b). The FWHM of the hcp-Zr peaks increases during annealing, which may be interpreted as shrinking of the Zr crystallites. The FWHM's of the peaks at $2\Theta=30°$, possibly a Zr-oxide, and at $2\Theta=44.5°$ (Ni-(111)) become smaller. This may result from grain growth made possible by compositional inhomogeneities, indication for which was observed by EPMA in the SEM. We assume that agglomerates of Ni crystallites a few tens of nanometers large exist in which grain growth occurs during annealing. Consistent with this explanation is the fact that, contrary to expectations, the sample did not transform completely to the amorphous state. Regardless of annealing time or temperature, some crystalline Ni and Zr could always be detected.

DISCUSSION

The nucleation rate for a heterogeneous nucleation process can be described by the following equation [2]:

$$\dot{N} = KD e^{\frac{-\Delta G_N}{k_b T}} \quad , \quad \text{with} \quad \Delta G_N \propto \frac{\sigma^3}{\Delta G^2} \tag{1}$$

a)

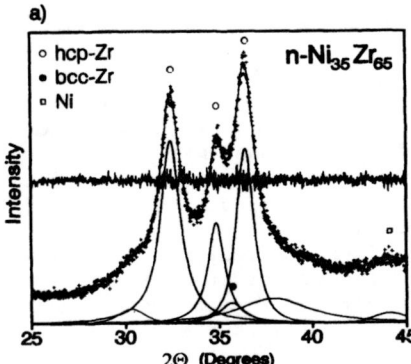

b)

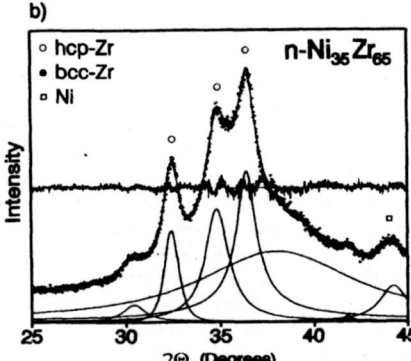

Fig 4:
a) n-Ni$_{35}$Zr$_{65}$ in the as-prepared state, already showing a small amount of amorphous phase.
b)After annealing (12 h at 200°C), the amorphous signal increased. The Ni peak (2Θ=44.5°) is sharper, indicating that grain growth has occurred.

where

K : proportionality constant,
D : diffusion coefficient,
ΔG_N: free energy barrier for nucleation,
σ : interfacial energy between the crystalline phase and the nucleating phases,
ΔG : thermodynamic driving force for the reaction.

Formation of the nucleus of the intermetallic compound requires both topological and chemical rearrangement of the atoms; therefore, both species have to move over distances at least as large as the size of the critical nucleus. Hence, D in Eq. (1) is determined by the diffusivity of the slower-moving species. Amorphous phase formation is topologically less restrictive, since mobility of only one species is required. In this case, D is controlled by the diffusivity of the fast-moving species. Thus, the larger the asymmetry of the diffusivities, the more that nucleation of the amorphous phase is favored.

The parameter ΔG_N depends on two variables: the driving force ΔG and the specific energy of the interface σ between the existing phase (at the nucleation site) and the phase to be nucleated. The larger the driving force ΔG and the smaller the interfacial energy σ, the smaller is the energy barrier ΔG_N for the nucleation process. Due to the higher power of σ, ΔG_N is dominated by σ. Hence, due to the lower interfacial energy for an amorphous/crystalline interface in comparison to that of a crystalline/crystalline interface, the nucleation rate for the amorphous phase should be larger. The formation of an intermetallic compound is only favored if low-energy interfaces, e.g. coherent interfaces between the starting crystalline phases and the intermetallic compound, exist.

We interpret our results in the following manner: For the CuZr system, neither in the nanocrystalline nor in the multilayered sample do we observe nucleation of an amorphous phase. We are not able to distinguish whether the reduced asymmetry of the diffusivities (see Table I) or the presence of low-energy interfaces inhibits the formation of the amorphous phase.

However, Atzmon et al.[10] reported SSAR in Cu_xZr_{1-x} ML's prepared by cold rolling of alternating elemental Cu and Zr sheets. This procedure resulted in the formation of an amorphous interlayer that grew during annealing, but the sample transformed only partially to the amorphous state. The preparation of samples by cold rolling or mechanical alloying differs from the above-described preparation techniques. Cold rolling and mechanical alloying have to be treated in the framework of externally driven systems. Especially for the cold-rolled samples one must consider the influence of dislocation-dislocation interactions and the interaction of dislocation cores with the different atomic species, since the kinetics of amorphous phase formation may be influenced by dislocation pipe diffusion and by dislocation networks acting as heterogeneous nucleation sites. Such influences are negligible if samples are prepared in the form of NCM or vapor-deposited ML's.

In the NiZr samples there is a certain amount of amorphous phase already in the as-prepared state. This may be related to the observation by Mazzone et al. [11] of the formation of amorphous phase in NiZr bilayers during plastic deformation. Free volume is created by plastic flow and acts as a region with enhanced diffusivity. The plastic flow occurring during compaction of our nanocrystalline samples could therefore lead to the formation of amorphous phase. During anealing at 200°C the amorphous phase grows at the expense of the elemental crystalline phases. This temperature is at least 100°C below the amorphization reaction temperature for multilayered Ni_xZr_{1-x} reported elsewhere [1,2]. This experimental finding may be explained by the enhanced mobility in grain boundaries of NCM.

ACKNOWLEDGMENTS

The authors wish to thank Prof. H. Gleiter, Dr. C. Krill and Dr. A. Tschöpe for many valuable discussions. This work was supported by the Deutsche Forschungsgemeinschaft (G.W. Leibniz-Programm) and the Fonds der Chemischen Industrie.

REFERENCES

1. W.L. Johnson, Progr. in Mat. Sci. 30, 81 (1986).
2. W.L. Johnson, in Materials Interfaces, Atomic-level Structure and Properties, edited by D. Wolf and S. Yip (Chapman & Hall, London, 1992), p. 516.
3. A. Tschöpe and R. Birringer, Acta metall. mater. 41, 2791 (1993).
4. W. Dickenscheid, R. Birringer, H. Gleiter, O. Kanert, B. Michel, B. Günther, Solid State Commun. 79, 683 (1991).
5. F.R. deBoer, R. Boom, W.C.M. Mattens, A.R. Miedema and A.K. Niessen, Cohesion in Metals, Transition Metal alloys, 1st ed.,(North Holland 1988), p. 375.
6. Landolt-Börnstein, New Series, Vol. 26, Diffusion in Solid Metals and Alloys, edited by H. Mehrer, (Springer 1990).
7. H. Gleiter, Progr. in Mat. Sci. 33, 223 (1989).
8. V. Haas and R. Birringer, Nanostr. Mat. 1, 491 (1992).
9. E.L. Nagaev, Phys. Rep. 4 & 5, 199 (1992).
10. M. Atzmon, J.D. Verhoeven, E.D. Gibson, and W.L. Johnson, in Rapidly Quenched Metals, edited by S.Steeb and H. Warlimont (Elsevier Science Publishers B.V. 1985), p. 1561.
11. G. Mazzone, A. Montone, and M. Vittori Antisari, Phys. Rev. Lett. 65, 2019 (1990).

STUDY OF NUCLEATION AND GROWTH IN AL-ZN
ALLOYS USING TEM

G.SUNDAR*, E.A.KENIK**,J.J.HOYT* AND S.SPOONER**
*Dept.of Mechanical and Materials Engg., Washington State Univ., Pullman,WA 99164
**Metals and Ceramics Division and **Solid State Division, Oak Ridge National Laboratory, Oak Ridge, TN 37831

ABSTRACT

Nucleation and growth studies were conducted on Al-Zn alloys at several temperatures using transmission electron microscopy (TEM) with an in-situ furnace. The value of the critical undercooling was established by noting the lowest temperature at which precipitates were no longer observed, following a quench into the two-phase metastable region. These results were compared with the Langer-Schwartz model of nucleation and growth in which it is predicted that the half-completion time (i.e, the time required for the supersaturation to reach half its initial value) diverges for initial supersaturations which are higher than those predicted by the classical nucleation theory.

INTRODUCTION

For most systems, the nucleation rate i.e the rate of formation of minority phase droplets, predicted by the classical theory of nucleation [1,2] varies from extremely low values to very large values through a narrow range of supersaturation, thus effectively defining the onset of nucleation, or cloud point. Although the classical theory is now over 60 years old, there have been very few studies done so far to compare its prediction with experimental data on solid-solid nucleation. Perhaps the first serious attempt made to compare theory and experiment in solid metals is that of Servi and Turnbull [3], reported in 1966. They investigated homogeneous nucleation kinetics in f.c.c Cu-rich Cu-Co alloys using resistivity measurements. Although their experiment was well designed, several factors such as the complicated steps involved in obtaining the particle number density and high Co concentrations leading to excessively rapid transformation kinetics, rendered their data imprecise. Another study of the comparison between homogeneous nucleation theory and experiment was conducted by Kirkwood and coworkers [4,5,6] not long afterwards, on ordered f.c.c Ni_3Al in disordered f.c.c α Ni-Al solid solutions, using transmission electron microscopy (TEM). Again, as in the previous case, over-rapid transformation kinetics were such a serious problem that nearly all data reported were obtained in the coarsening regime.

Considering the meager amount of success achieved in comparisons between theory and experiment in solid-solid transformations, one would therefore ask what has been achieved with parallel efforts on liquid-liquid and liquid-vapor transformations. From the outset, it would seem that these would follow nucleation theory better since they support no internal strains and the much higher diffusivities involved permit nucleation at very low supersaturations, where one would imagine the theory to be most accurate. Somewhat surprisingly, past experiments on liquid systems have also demonstrated serious discrepancies between theory and experiment [7-10].

It was Binder and Stauffer (BS) [11] who first recognized that the source of discrepancies lay in the experiments which consisted of measurements of the cloud points, that is, the temperature

337

at which nucleation first becomes profuse. Cloud point measurements essentially determine the product of nucleation rate and growth rate of droplets. In the light of this, BS argued that, since the nucleation rate strongly depends on the supersaturation (or undercooling), one should take into consideration the growth rate of nuclei too, besides the formation of stable nuclei, which deplete the surrounding supersaturated material. Their combined treatment of these two processes was further developed by Langer and Schwartz (LS) [12], the details of which are described below.

THE LANGER-SCHWARTZ MODEL

The LS model is a statistical theory of nucleation and growth which describes the kinetics of unmixing of slightly supersaturated off-critical fluids. The four relevant quantities considered in the kinetics of nucleation and growth are supersaturation, droplet size, droplet density and time. Using the concepts of scaling [13], LS introduced four reduced parameters. The first is the supersaturation

$$y(t) = x(t)/x_0 \qquad (1)$$

where

$$x(t) = \frac{2\delta C(t)}{\beta \Delta C} = \frac{2}{\beta}\frac{(C_m(t)-C_A)}{C_B-C_A}$$

The quantity $\beta \approx 1/3$ is the power law exponent of the miscibility gap, $C_m(t)$ is the solute concentration of the matrix at time t and C_A and C_B are the equilibrium solute concentrations of the matrix and the second phase respectively. The term x_0 is defined as

$$x_0 = 4\left(\frac{\sigma\zeta^2}{kT_c}\right)^{1/2} \qquad (2)$$

where ζ is the correlation length, σ the interfacial energy and T_c the critical temperature. The second scaled parameter defined by LS is the reduced time, which is given by

$$\tau = \left(\frac{Dx_0^3}{24\zeta^2}\right)t \qquad (3)$$

with D the diffusion coefficient. The third and the fourth scaled variables are the mean droplet radius ρ and the number density n which are given by

$$\rho = \frac{Rx_0}{2\zeta} ; \qquad n = 64\pi\left(\frac{\zeta}{x_0}\right)^3 N \qquad (4)$$

In all the expressions mentioned above, t, R and N are real time, real size and real number density respectively. The main results of their computation are presented in Fig.1. The quantity τ_c denotes the time required for the supersaturation to reach half its initial value. The solid line shows the LS prediction for half completion time vs. the initial supersaturation and the dash dotted curve depicts the classical theory curve.

338

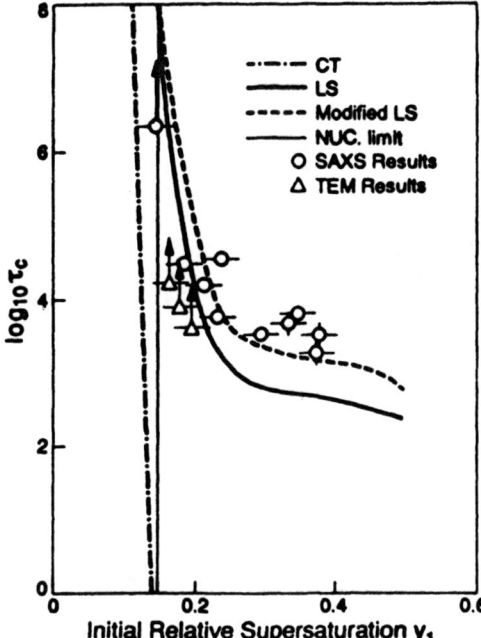

Fig.1. Log of the half completion time τ_c vs y_i as predicted by the classical theory (CT), the Langer-Scwhartz model (LS), and the LS model modified to include a time-dependent nucleation rate (dashed line). The open circles are the results from Al-Zn alloys using small angle x-ray scattering. The open triangles are the TEM results. The Vertical solid line illustrates the limit of initial supersaturation below which no nucleation is observed. The upward arrows on the TEM data points indicate the onset of nucleation.

The LS computation demonstrated that the numerous cloud point experiments do not show any drastic disagreement between the classical theory and experiment, provided the growth rate is taken into account.

In 1984, Legoues and Aaronson [14] studied the nucleation kinetics in Cu-rich Cu-Co alloys, following the lead of Servi and Turnbull [3], because these alloys provide a well established case of homogeneous f.c.c → f.c.c nucleation. These authors were careful to choose compositions and aging temperatures within a narrow "window" such that the effects of nucleation could be separated from particle growth. Their studies showed agreement within an order of magnitude of the classical prediction but did not attempt to measure changes in supersaturation as a function of time.

Shortly thereafter, Simon, Guyot and Ghilarducci de Salva [15] performed nucleation experiments on dilute Al-Zn alloys in terms of the LS model. Their results showed some discrepancies with the LS model, including one data point which seemed to confirm the classical theory without the inclusion of growth rate.

A more rigorous test of the LS model was carried out by the present authors [16] using SAXS (small-angle x-ray scattering). It was concluded that the model is accurate provided it is modified to include a time-dependent nucleation rate instead of a steady-state nucleation rate (see fig.1), since diffusivities in solid systems are not high enough to permit rapid unmixing.

In the present work, in-situ TEM was used as an additional tool to further investigate nucleation and growth kinetics in Al-Zn alloys, and thereby demonstrate the validity of the LS model.

EXPERIMENTAL PROCEDURE

Alloys of five average compositions (17,17.5,18 at% Zn) were rolled and strain annealed to produce an average grain size of ≈1mm, so that the bulk of the grains are devoid of any complicating effects due to the presence of grain boundaries. Disks of 3mm were punched from the rolled material and thinned by electropolishing to a thickness ideal for transmission electron microscopy.

The samples were heated and cooled by adjusting the current to the furnace in the Gatan heating holder. Temperatures were calibrated with the known boundary of the Al-Zn metastable miscibility gap [17]. Each sample was solutionized for approximately 0.5 h at 320° C and subsequently quenched to a temperature in the vicinity of 260° C (≈0.9 T_c).

The procedure employed in this study is different from the ones used in the past, which consisted of directly counting the number of precipitates per unit area, thereby obtaining the number density at any given time. In the present investigation, we simply noted the lowest temperature inside the miscibility gap at which precipitates were no longer observed and the time taken for the first precipitates to appear, that is, achieve experimentally detectable sizes.

The initial supersaturations were kept low to avoid over-rapid transformation kinetics. Also, this would ensure that the precipitates are far apart, thereby reducing the overlap of diffusional fields of adjacent growing precipitates.

The duration of the quench was about 30 sec, which is slow compared to other studies of phase separation in alloys [18] but is, in this instance, much faster than the kinetics being observed. Also, the relatively slow quench rate has the desirable effect of helping to eliminate the high temperature non-equilibrium concentration of vacancies. Excess vacancies are known to have a significant effect on the measured nucleation kinetics [14].

RESULTS AND CONCLUSIONS

Fig.2 shows a typical microstructure in the 17.5 % Zn alloy after quenching the sample to 261° C from 320° C. The precipitates visible in the microstructure appeared after approximately 5 min. of reaction time. Fig. 3 shows the microstructure in the same alloy after 10 min. of reaction time, following a quench to 263° C. Although the final quench temperature in this case is only 2° higher than the previous run, the nucleation rate has diminished so much that there is practically no transformation product. This illustrates the onset of nucleation alluded to in the Introduction. Fig.4 shows the microstructure in the same alloy, following a quench to 258°C. The precipitates appeared after merely 2 minutes of reaction time, and are much higher in density.

To convert the time required to form the first observable precipitates to the scaled time and supersaturation of Fig.1, one must obtain the various material parameters described in the LS model. Both the correlation length ζ and the critical temperature T_c have been measured in a study by Schwann and Schmatz [19]. The diffusion coefficient for Zn in Al is fairly well known [20], as is Δc at any temperature [17]. The surface energy was estimated by using a free-energy model [21], which yields a value of ≈1.75 for x_0. Since the error in the surface energy estimate represents the largest uncertainty in the data, x_0 was therefore taken to be in the range of 1.5-2.0.

The vertical solid line in fig.1 represents the limit of initial supersaturation below which no transformation is observed (see fig.3). The reaction times in fig.2 and fig.4 are plotted as open triangles in Fig.1, where the error bars indicate the uncertainty in σ discussed above. The upward arrow on these points indicate the onset of nucleation.

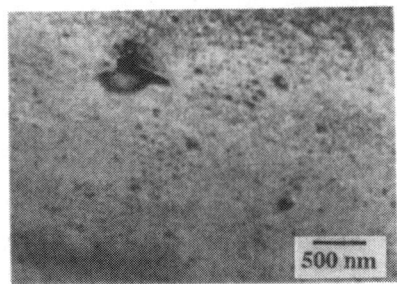

Fig.2. Precipitation in the 17.5% Zn alloy reacted 5 minutes at 261° C.

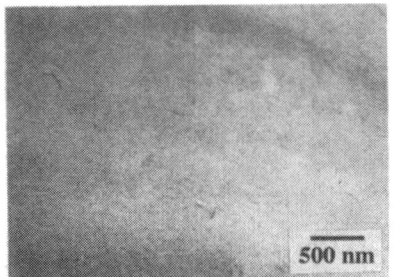

Fig.3. Microstructure indicating no precipitation in the 17.5% Zn alloy after 10min. at 263° C.

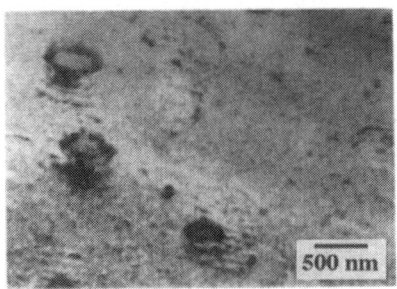

Fig.4. Precipitation in the 17.5% alloy after 2 minutes of reaction time at 258° C.

As expected, the TEM results do not quite match with the SAXS results since the former relates to the onset of nucleation and latter is a measure of the half-completion time.

The data confirms an important aspect of the LS model, namely that the onset of nucleation occurs at higher values of initial supersaturation than predicted by the classical theory.

This work was supported by the National Science foundation under Contract No. DMR 8919193, and by the Division of Materials Science, US Department of Energy through the SHaRE Program under contract DE-AC05-76OR00033 with the Oak Ridge Associated

Universities and under contract DE-AC05-84OR21400 with Martin Marietta Energy Systems for the U.S. Department of Energy.

REFERENCES

1. J.W. Gibbs, Collected works (Yale University Press, New Haven, CT,1948),p.1.
2. R. Becker and W. Doring, Ann. Phy.(N.Y) **24**, 719 (1935).
3. I.S. Servi and D. Turnbull, Acta Metall. **14**, 161 (1966).
4. D.H. Kirkwood, Acta Metall. **18**, 563 (1970).
5. D.H. Kirkwood and A.W. West, Scripta Metall. **10**, 687 (1976).
6. T. Hirata and D.H. Kirkwood, Acta Metall. **25**, 1425 (1977).
7. B.E. Sundquist and R.A. Oriani, J. Chem. Phys. **36**, 2604 (1962).
8. R.B. Heady and J.W. Cahn, J. Chem. Phys. **58**, 896 (1973).
9. J.S. Huang, S. Vernon and N.-C. Wong, Phys. Rev. Lett. **33**, 140 (1974).
10. J.S. Huang, W.I. Goldburg and M.R. Muldover, Phys. Rev. Lett. **34**, 639 (1975).
11. K. Binder and D. Stauffer, Adv. Phys. **25**, 343 (1976).
12. J.S. Langer and A.J. Schwartz, Phy. Rev.A 21, **948** (1980).
13. H.E. Stanley Introduction to Phase Transitions and Critical Phenomenon (New York: Oxford University Press, 1971).
14. F.K. Legoues and H.I. Aaronson, Acta Metall. **32**, 1855 (1984).
15. J.P. Simon, P. Guyot and A. Ghilarducci de Salva, Philo. Mag. A 49, **151** (1984).
16. G. Sundar, J.J. Hoyt and S. Spooner, Phys. Rev. B, **46**, 14266 (1992).
17. J. Lasek, Czech. J. Phys. **15**, 848 (1965).
18. G.B. Stephenson, K.F. Ludwig, J.L. Jordon-Sweet, S. Brauer, J. Mainville, Y.S. Yang, and M. Sutton, Rev. Sci.Instrum. **60**, 1537 (1989).
19. D. Schwann, and W. Schmatz, Acta Metall. **26**, 1571 (1978).
20. K.B. Rundman and J.E. Hiliard, Acta Metall. **15**, 1025 (1967).
21. J.J. Hoyt, M. Sluiter, M. Kraitchmann, and D. deFontaine, Acta Metall. **35**, 2315 (1987).

THE KINETICS OF THE ORDER-DISORDER TRANSFORMATION IN THE $Pt_1Co_{1-x}Ni_x$ ALLOY SYSTEM

Huan Tu, Eduardo O. Ruiz and J. Orehotsky
Wilkes University, Department of Engineering, Wilkes Barre, PA 18766

ABSTRACT

The kinetics of atomic ordering in a series of alloys in the $Pt_1Co_{1-x}Ni_x$ ternary system where $0<x<0.5$ were investigated by magnetic measurements. For all the alloys, the ordering kinetics as monitored by the saturation magnetization as a function of isothermal ordering time at a series of specified ordering temperatures, were found to be exponentially time dependent. The associated time constant was found to have an Arrhenius temperature dependence characterized by an activation energy that, with increasing nickel concentration, initially increased to maximum value at a nickel composition of $x \simeq .2$ and then systematically decreased. This result for the activation energy as a function of nickel content was found to correlate roughly with the behavior of the saturation magnetization of the ordered alloys, suggesting that the kinetics of the transformation in this alloy system is sensitive to the magnetic properties of the ordered state.

INTRODUCTION

Bilayered thin films where cobalt specifically is one of the constituent metallic layers have been receiving considerable attention recently for device applications (1-13) from magneto-optic and perpendicular recording to read heads based on the giant magnetoresistive effect. A considerable amount of activity has focused on periodic Pt/Co bilayer thin films (6-8) and on PtCo alloy thin films (9-13) for their potential promise as magneto-optic recording materials.

The mechanism responsible for the perpendicular magnetic anisotropy associated with these films and responsible for potential device use, is open to question. Interfacial considerations are frequently cited (6-7) as causing this anistropy in Pt/Co bilayer thin films. For the alloy thin films, the perpendicular anisotopy is either directly or indirectly (9-12) attributed to the presence or development of an ordered phase such as the face centered tetragonal $L1_0$ ordered phase that exists in the Pt_1Co_1 stoichiometric bulk alloy (14).

The platinum and cobalt lattice positions in this particular ordered phase consist of alternate planes of cobalt and platinum atoms parallel to the (001) basil plane of the tetragonal unit cell (14). This superlattice arrangement has suggestive implications when it is recognized that Pt/Co bilayer thin films most suitable for magneto-optic recording by virtue of their highly developed perpendicular magnetic anisotropy response features, consist of alternate layers of cobalt and platinum deposited in layer thicknesses that approach monolayer dimensions.

The $L1_o$ superlattice atomic arrangement of the ordered Pt_1Co_1 phase can be simulated in thin film form by depositing alternating monolayers of platinum and cobalt on an appropriate substrate, and annealing (11) to develop the superlattice simulation. This film would be characterized, to some extent, by the magnetic, magneto-crystalline and crystallographic features of the bulk ordered alloy. The troublesome feature of the bulk ordered alloy is that its Curie temperature is considerably above the range of values considered optimal for magneto-optical recording.

One approach to circumvent this difficulty in the Curie temperature is to shift research attention away from Pt/Co bilayer thin films reflective of the Pt_1Co_1 stoichiometric, ordered, tetragonally anisotropic crystal structure to PtCo alloy thin films of compositions nearer the Co_1Pt_3 stoichiometry which has a more favorable Curie temperature. Indeed, disordered CoPt alloy films around the 25 atomic percent cobalt composition appear to have favorable magneto-optic recording characteristics which were suspected to be somehow related to the ordered Co_1Pt_3 phase (10). The obvious unattractive aspects of these high platinum content alloys are that increasing the platinum increases the cost and decreases the magnetization response.

The other approach is to retain the bilayer concept of superlattice simulation by deposition techniques, but to partially substitute nickel for cobalt and create a Pt metal $/Co_{1-x}Ni_x$ alloy bilayer thin film structure reflective of a Ll_o ordered $Pt_1Co_{1-x}Ni_x$ superlattice that is known to exist also(15) in these bulk ternary alloys. One virtue of these alloys is that they are suspected to have a more favorable Curie temperature depending on the nickel content. The second virtue of these ordered $Pt_1Co_{1-x}Ni_x$ bulk alloys is that the tetragonality of the ordered crystal structure progressively increases with increasing nickel content (16) suggesting that the magnetocrystalline anisotropy will also increase with nickel content.

The object of this investigation is to examine the ordering kinetics of the order-disorder transformation in the bulk $Pt_1Co_{1-x}Ni_x$ alloy system to determine if the superlattice arrangement associated with the bulk system can be easily obtained by annealing the appropriately deposited periodic bilayer $Pt/Co_{1-x}Ni_x$ thin film structure.

Very little information is available in the literature on ordering kinetics in alloy systems containing platinum, nickel, or cobalt. A simple examination of the CoNi binary system (17) shows no ordered phases exist, suggesting that replacing cobalt with nickel in the Pt_1Co_1 stoichiometry may result in very sluggish ordering kinetics. No specific information is readily available in the literature on the ordering kinetics in the $Pt_1Co_{1-x}Ni_x$ alloy system. Ordering kinetics in the Pt_1Co_1 alloy system are available in the literature. When monitored by resistivity measurements, the ordering kinetics were found to be characterized by an activation energy of about 40 kcal/mole (18) which is surprisingly small relative to typical values of about 60 kcal/ mole for the activation energies characterizing platinum and cobalt self diffusion data. The size of this surprising difference was somewhat unexpected since both the ordering and self diffusion mechanisms are controlled by the same vacancy controlled atomic mobility mechanism. A systematic investigation of ordering kinetics not just for bilayer considerations, but in general for the PtNiCo alloy system would appear therefore to be most appropriate.

EXPERIMENTAL PROCEDURE

Five $Pt_1 Co_{1-x} Ni_x$ alloys of nominal stoichiometric compositions shown in Table I were made from their elemental constituents in a button arc melter. The melter's chamber was pumped to about 35 millitorr and then backfilled with an atmosphere of argon. This pumping and backfilling procedure was repeated about six times before the arc was struck. After striking the arc, a titanium button was melted to getter residual oxygen in the chamber. The elemental charge for each alloy was then melted. The resultant buttons were then cooled, turned over and remelted for homogeneity. The weight loss resulting from the melting operation ranged between 0 and 5 percent as shown in Table I. The samples typically weighed about 40 milligrams in order to be easily accommodated in the specimen holder of the vibrating sample magnetometer.

The alloys were obtained in their disordered state at room temperature by heating them in a small chamber, tightly wound, highly efficient furnace to 1000° C, holding for 30 minutes and then quenching into water. No attempt was made to verify by x-ray techniques the existence of the disordered crystal structure after this treatment. Verification of the disordered structure with this heat treatment for the $Pt_1 Co_1$ alloy was provided by measuring the saturation magnetization (Ms) which was found to have a value of 42 emu/gram which is reasonably consistent with the reported value (19) of 46 emu/gram at room temperature. The magnetization was measured as a function of field to 8 KOe in a vibrating sample magnetometer.

Table I

Sample stoichiometry, weight loss after arc melting, activation energy (Q) for the ordering kinetics, and saturation magnetization (Ms) of the ordered state in $Pt_1 Co_{1-x} Ni_x$ alloys.

Sample Stoichiometry	Weight Loss (%)	Q (kcal /mol)	Ms (emu/g)
Pt_1Co_1	4.5	58	21.5
$Pt_1Co_{.95}Ni_{.05}$	5.5	98	21.5
$Pt_1Co_{.9}Ni_{.1}$	0	---	32
$Pt_1Co_{.8}Ni_{.2}$	1.2	90	26.7
$Pt_1Co_{.5}Ni_{.5}$	1.0	73	20

The saturation magnetization of the ordered state in Pt_1Co_1 is reported to be 22 emu/gram so saturation magnetization measurements were employed to monitor the ordering kinetics. The monitoring by magnetic measurements was accomplished by rapidly heating the disordered alloy to a preselected ordering temperature from room temperature for an appropriate length of isothermal run time to achieve a partially ordered state, and then quenching into water. The saturation magnetization of the alloy sample in this partially ordered state was then measured at room temperature in the magnetometer. This isothermal heating, quenching and magnetic measuring procedure was repeated until the saturation magnetization as a function of accumulated run time did not continue to decrease and leveled off to a value representative of the completely ordered state. The alloy was then disordered again, and the procedure was repeated at another isothermal ordering temperature. These ordering temperatures were typically between 600° and 700° C for time convenience in monitoring the kinetics. All the heat treatments were done under a protective atmosphere of flowing hydrogen. The response time of the furnace was sufficiently rapid - approximately one or two minutes- to achieve isothermal conditions so that no appreciable ordering was expected on the heating approach to each particular ordering temperature employed in this investigation.

RESULTS AND DISCUSSION

The saturation magnetization values monitoring the ordering kinetics at various ordering temperatures for all the alloys were found to decrease exponentially with time at each ordering temperature:

$$Ms \sim \exp(-t/\tau) \tag{1}$$

The time constant (τ) of this exponential decay was found to obey an Arrhenius temperature dependency:

$$\tau = \tau_0 \exp(-Q/kT) \tag{2}$$

From these results for each alloy, the activation energy (Q) as a kinetic parameter characterizing the ordering process in each alloy, was determined. The results are presented as a function of the nickel content in Table I. The first feature apparent in the data is that the activation energies are generally consistent with activation energy values characterizing self diffusion results, suggesting that magnetization measurements may be more representative of the ordering process than resistance measurements which yielded an activation energy for the Pt_1Co_1 alloy (18) that was significantly smaller than typical self diffusion values. The second feature to notice is that partially replacing cobalt with nickel in the Pt_1Co_1 precursor alloy does not systematically make the ordering reaction more sluggish with increasing nickel content as was first expected based on the observation that the binary CoNi alloy system does not order (17). The activation energy is a parametric measure of the ordering kinetics, and it does not behave systematically with increasing nickel content as observed in Table I. It first increases as expected with increasing nickel content, suggestive of slower kinetics, but then decreases. This subsequent decreasing behavior could perhaps be explained by representing $Pt_1Co_{1-x}Ni_x$ as the psuedo-binary system Pt_1Co_1-Pt_1Ni_1, where it has been observed experimentally that both Pt_1Co_1 and Pt_1Ni_1 readily order (17).

The remaining result that naturally appears in this kinetic study is the behavior of the saturation magnetization in the ordered and disordered state of these alloys. The saturation magnetization of the ordered alloys is shown in Table I. Similar to the activation energy behavior with increasing nickel content, the saturation magnetization also initially increases and then decreases. This similar behavior may be purely coincidental or it may suggest that a correlation exists between magnetic properties and the ordering activation energy even though a direct correspondence does not exist as is clearly evident in Table I.

This observed rough correlation between the ordering activation energy and the saturation magnetization would imply that the activation energy for diffusion in solid state systems may be dependent on the magnetic state of the system. Conceptually the diffusion activation energies could then have different values above and below the Curie temperature in a ferromagnetic material. Surprisingly enough, there is some supportive experimental evidence. The activation energy for carbon diffusion in α-Fe well below the Curie temperature as determined by anelastic measurements, is significantly smaller than at higher temperatures (above $500^{\circ}C$) as determined by direct diffusion measurements (20). This observation for carbon diffusion in α–Fe above and below the Curie temperature is suggestively compelling in justifying the behavioral correlation observed in this investigation between magnetic and kinetic data. However, a simple conclusionary statement that a correlation does indeed exist or that carbon diffusion behavior in α-Fe is influenced by magnetic effects and not by other considerations, must be approached with extreme caution for a number of very good reasons.

SUMMARY

The ordering kinetics in the magnetic $Pt_1Co_{1-x}Ni_x$ (0<x<0.5) alloy system was found to be exponentially time dependent at isothermal ordering temperatures, with an associated time constant that was exponentially temperature dependent through an associated activation energy. The activation energy was found to increase initially and then decrease with increasing nickel content to x=0.5. This behavior of the experimentally determined activation energy did correlate roughly with the behavior of the saturation magnetization in the ordered alloys as a function of nickel concentration.

REFERENCES

(1)   M.N. Baibich, J.M. Broto, A. Fert, F. Nguyen Van Dau, F. Petroff, P. Eitenne, G. Creuzet, A. Friederich and J. Chazelas, Phys. Rev. Lett. 61, 2472, (1988).

(2)   S.S.P. Parkin, R. Bhadra and K.P. Roche, Phys. Rev. Lett. 66, 2152, (1991).

(3)   D.H. Mosca, F. Petroff, A. Fert, P.A. Schroeder, W.P. Pratt, Jr., and R. Laloee, J. Magn. Magn. Mater., 94, L1, (1991).

(4)   D. Greig, M. J. Hall, C. Hammond, B.J. Hickley, H.P. Ho, M.A. Howson, M.J. Walker, N. Wiser, and D.G. Wright, J. Magn. Magn. Mater., 111, L239, (1992).

(5)   M.E. Tomlinson, R.J. Pollard, D.G. Lord, and P.J. Grandy, J. Magn. Magn. Mater., 111, 79, (1992).

(6)   C.H. Lee, R.F.C. Farrow, C.J. Lin, and E.E. Marinero, Phys. Rev. B, 42, 11 384 (1990).

(7)   W.B. Zeper, F.J.A.M. Gredanas, P.F. Carcia, and C.R. Fincher, J. Appl. Phys., 65, 4971 (1989).

(8)   J.J. Greaves, A.K. Petford-Long, Y.-H. Kim, R.J. Pollard, P.J. Grandy, and J.P. Jakubovics, J. Magn. Magn. Mater. 113, 61, (1992).

(9)   D. Treves, J.T. Jacobs, and E. Swatzky, J. Appl. Phys., 46, 2760, (1975).

(10)   D. Weller, H. Brandle, G. Gorman, C.-J. Lin, and H. Notorys, Appl. Phys. Lett. 61, 2726, (1992).

(11)   B.M. Lairson, M.R. Visokay, E.E. Marinero, R. Sinclair, and B.M. Clemens, J. Appl. Phys. 74, 1992, (1993).

(12)   J.A. Aboaf, S.R. Herd, and E. Klokholm, IEEE Trans. Magn. MAG-19, 1514 (1983).

(13)   T.R. McGuire, J.A. Aboaf, and E. Klokholm, J. Appl. Phys., 55, 1951, (1984).

(14)   J.B. Newkirk, R. Smoluchowski, A.H. Geisler, and D.L. Martin, J. Appl. Phys., 22, 290, (1951).

(15)   J.C. Woolley and B. Bates, J. Less.-Common Metals, 2, 11, (1960).

(16)   W.B. Pearson, A Handbook of Lattice Spacings and Structures of Metals and Alloys. Pergamon Press.

(17)   M. Hansen. Constitution of Binary Alloys. McGraw-Hill, New York, 1958.

(18)   J.B. Newkirk, A.H. Geisler, D.L. Martin, and R. Smoluchowski, Trans. Am. Inst. Min. Metal. Pet. Eng., 188, 1249, (1950).

(19)   R.A. McCurrie and P. Gaunt, Phil. Mag., 13, 567 (1966).

(20)   A.E. Lord, Jr., and D.N. Beshers, Acta Met. 14, 1659, (1966).

HETEROGENEOUS NUCLEATION OF SUPERPARAMAGNETIC PHASES IN ^{57}FE-BEARING CORDIERITE-GLASSES

IRMGARD ABS-WURMBACH* AND CORNELIA BOBERSKI**
*TU-Berlin, Institute for Mineralogy and Crystallography, Ernst-Reuter-Platz 1, D-10587 Berlin, Germany
**Hoechst AG, PO Box 800320, D-65926 Frankfurt/Main, Germany

ABSTRACT

Glasses of stoichiometric cordierite composition $Mg_2(Al_{1-x}Fe_x)_4Si_5O_{18}$, containing very low iron contents ($x=0.015$ and 0.005) have been investigated by ^{57}Fe Mössbauer spectroscopy. At higher Fe concentrations spinel exsolution has been observed in X-ray powder patterns. To allow Mössbauer spectroscopy at very low Fe-concentrations (<0.8 weight% Fe_2O_3), starting materials were doped with 100% ^{57}Fe. Glasses prepared in air at 1560°C were also oxidized in water saturated O_2 stream. Glasses were crystallized to cordierite by heating in air at 1100 to 1400°C. ^{57}Fe-Mössbauer spectra of all samples are governed by a doublet typical for Fe^{2+} in octahedral coordination (IS: 1.09-1.15, QS: 2.00-2.31 in mm/s relative to metallic iron). X-ray powder patterns exhibit no additional phases. But, the 295 K and 5 K Mössbauer spectra of treated and untreated glasses and of cordierites exhibit broad lines, which have been fitted by applying internal magnetic hyperfine fields H of ca. 500 kG. These lines are attributed to heterogeneously nucleated, superparamagnetic $(MgFe)(AlFe)_2O_4$ spinels of complex compositions. From the calculated subspectra one may conclude that 35 to 65% of the total iron, depending on preparation conditions, is incorporated in the spinels.

INTRODUCTION

Diluted spinel particles in the matrix of cordierite based glass- and glass-ceramics improve their mechanical properties [1]. To control the formation processes of small particles in a glass matrix it is very important to understand the mechanism of nucleation. This includes knowledge about how and in which concentration the spinel forming elements are distributed in the glass matrix itself, and composition, homogeneity and grain sizes of the heterogeneous nucleated crystals. The introduction of transition elements in the investigated system allows the application of spectroscopic methods such as EPR-, optical- and Mössbauer-spectroscopy for analysing the reaction products. Thus one may detect even traces of heterogeneous phases. Our investigation concentrates on properties at very low iron contents which have not been studied until now.

In this context stoichiometric cordierite glasses on the composition

$$Mg_2(Al_{1-x}Fe^{3+}_x)_4Si_5O_{18} \tag{1}$$

have been investigated by ^{57}Fe-Mössbauer spectroscopy. Experimental results of Boberski [2] have demonstrated a very limited single phase range, $0.01<x<0.025$, in cordierites synthesized from glasses in air. At higher iron concentrations X-ray powder diffraction patterns exhibit complex spinels as additional phases beside cordierite. The compositional range of these minute spinels was determined by microprobe analysis as

$$(Mg_{0.77}Fe_{0.23})(Al_{0.52}Fe_{1.48})O_4 - (Mg_{0.77}Fe_{0.23})(Al_{1.08}Fe_{0.91})O_4. \tag{2}$$

Previous Mössbauer studies on melts in the system $MgO-Al_2O_3-SiO_2-Fe_2O_3$ have been performed by Doenitz et al. [3], and Mysen et al. [4] on composition of glasses different from that of cordierite and with iron contents varying from 1.5 to 10 weight% Fe_2O_3. In the present investigation properties at very low iron contents with less than 1 weight% Fe_2O_3 have been of main interest. Since the ratio of iron substitution is very small, starting materials were enriched by 100% of the Mössbauer isotope ^{57}Fe to increase the limit of detection by a factor of about 50. Cordierite glasses and their crystallization products have been investigated concerning valence state and coordination of incorporated iron. By means of this method it is also possible to detect traces of spinel phases having heterogeneously formed within the glasses. Therefore, Mössbauer spectra were also used to gather informations concerning quantity, homogeneity, grain sizes and reactivity of the exsolved spinels.

EXPRIMENTAL METHODS

Sample preparation
i) Glasses of the composition (1) with x=0.015 and 0.005 were prepared from oxide mixtures by melting in air at 1560°C ("untreated" glass). ii) It was attempted to increase the Fe^{3+}/Fe^{2+}-ratio by treating glasses at 750°C in a water saturated O_2 stream ("treated" glass). iii) Glasses were crystallized to cordierites by heating in air at 1100 to 1400°C and, iv) by hydrothermal runs (2 kbar, 700°C), under controlled oxygen fugacities (f_{O2}) using solid state buffers.

Analytical methods
Samples were examined by microscopic observations and X-ray powder diffraction. Additional methods are Mössbauer- and EPR-spectroscopy on powders. Gamma-ray source for Mössbauer experiments was ^{57}Co in Rh, which was always kept at room temperature, whereas the absorber temperature was varied in the range RT (295 K) and LHe (ca. 5 K). The velocity scale (mm/s) was calibrated against metallic alpha-iron. Spectra were fitted assuming Lorentzian lines.

RESULTS AND DISCUSSION

X-ray diffraction and optical examination indicate single phase Fe^{3+}-containing glasses and cordierites. But, all ^{57}Fe-Mössbauer spectra show: i) only a **doublet typical for Fe^{2+}** in octahedral coordination (Table I) instead of the expected doublet typical for Fe^{3+} and, ii) **additional broad lines** increasing in intensity with decreasing temperature (Figures 1, Table I) in spectra from glasses and from cordierites prepared in air. The observed spectra are different from those observed by Doenitz et al. [3] but have some similiarities with spectra described by Mysen et al. [4] concerning the Fe^{2+}-doublet.
i) Glasses exhibit rather undefined broad Fe^{2+}-lines, as is expected from structural relationships in a glass matrix [4]. This holds also at 5 K. Linear intensity I of these lines increases at 5 K by a factor of three in comparison to less than two in cordierite (Table I), thus indicating that the recoil-free fraction f is smaller in glasses than in crystallized substances. In contrast to [3 & 4] no Fe^{3+}-doublet could be refined in spectra obtained from "untreated" glasses. But, in "O_2-treated" glass (AC9, Fig.2), one underlay a second doublet which might be interpreted because of its small isomer shift of 0.70 mm/s (Table I) as a paramagnetic Fe^{3+}-doublet. The isomer shift of this

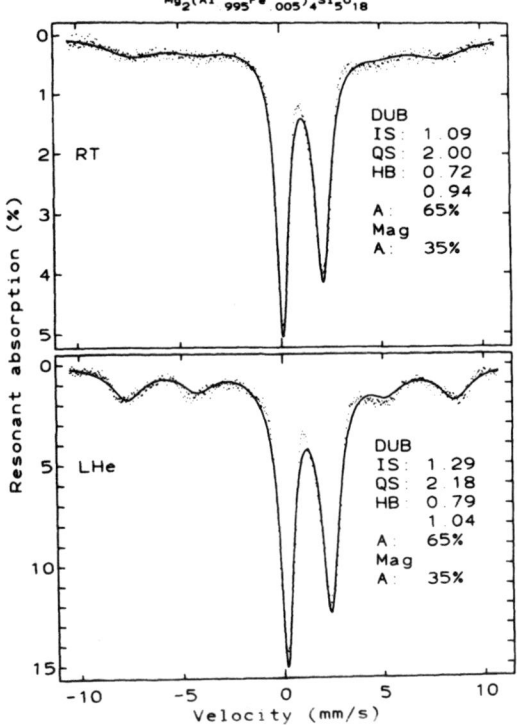

Ferri-Cordierite Glass 0.5

$Mg_2(Al_{.995}Fe_{.005})_4Si_5O_{18}$

RT

DUB
IS: 1.09
QS: 2.00
HB: 0.72
     0.94
A: 65%
Mag
A: 35%

LHe

DUB
IS: 1.29
QS: 2.18
HB: 0.79
     1.04
A: 65%
Mag
A: 35%

Resonant absorption (%)

Velocity (mm/s)

Fig.1: Mössbauer spectra of cordierite glass (x=0.005) at room- and liquid helium temperature. Notice the increase in the line intensities of the paramagnetic doublet and the magnetic sextet and furthermore that relative paramagnetic and magnetic intensities remain constant. The envelopes of the calculated subspectra (2 lines + 1 sextet) are shown. *Symbols:* A = Area; HB = Half width; IS = Isomer shift; QS = quadrupole splitting; Mag = sextets from magnetic hyerfine splitting.

spectrum with 0.70 mm/s (Fig.3) exceeds that of tetrahedrally coordinated Fe^{3+} (IS<0.3 mm/s or slightly more [4]), thus indicating either octahedral Fe^{3+} in the glass matrix [4] or a "superparamagnetic" doublet of a spinel-phase.

ii) The broad lines are attributed to the magnetic hyperfine splitting sextets of spinel phases, they are even present at iron contents as low as 0.5% of the hypothetical iron end member. Internal magnetic fields evaluated from the magnetic hyperfine splitting (Table I) agree well with known values for spinels rich in Fe^{3+} [e.g. 5]. The interpretation of these lines is a complex problem. a) They may only be caused by spinels of such minute crystals, smaller than a magnetic domain size, that because of thermal fluctuations they behave like a superparamagnetic phase. Thus, it might be concluded on particle sizes around 10 nm. b) Since spinels are diluted by diamagnetic ions (2), the variation of next neighbors of a paramagnetic ion may also lead to fluctuations of the internal magnetic field H, giving rise to sextets similar to superparamagnetic ones and, at elevated temperatures to an inward collapse of magnetic hyperfine splitting, resulting in a "paramagnetic" doublet. c) Some contribution to the broadness of the lines may result from variations in compositions of spinels inducing different isomer shifts, quadrupole splittings and magnetic internal field H. Presumably a combination of all these effects contributes to the observed phenomena. Since the electric field gradients depend on the different species of next neighbor cations, in complex spinels crystal field is no longer cubic and therefore, QS values unequal to zero are observed.

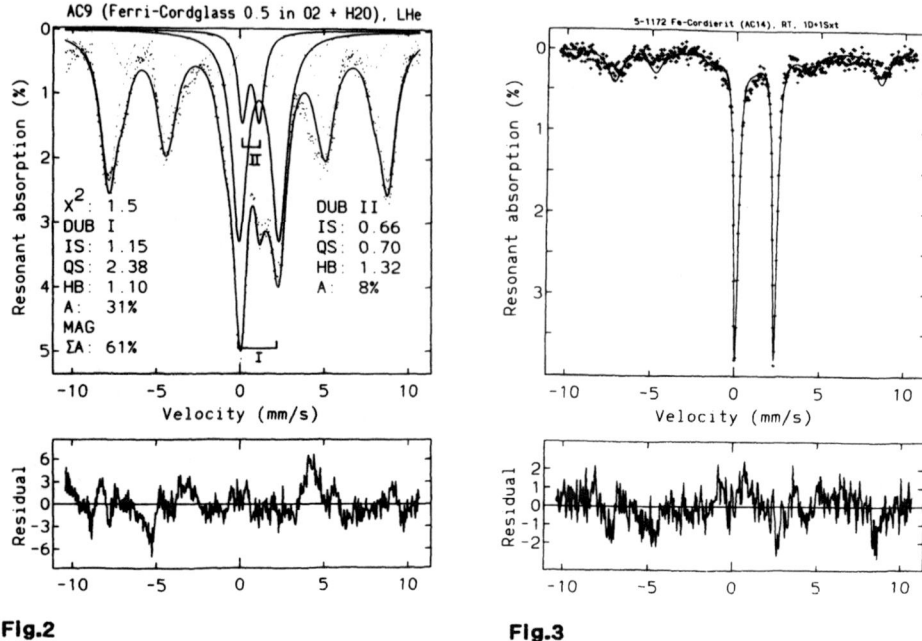

Fig.2

Fig.3

Fig.2: The Mössbauer spectrum of O_2-treated glass at 5 K (AC9). Fitting by two doublets should be taken as an approximation procedure. Contributions of paramagnetic lines of the magnetic phases to doublets I and II can't be excluded.

Fig.3: High-cordierite tempered from untreated glass in air exhibits a paramagnetic doublet with small line width indicating incorporation of Fe^{2+} in the octahedral M-position of cordierite.

Oxidation in O_2-stream (AC9, Table I) increases the amount of iron bound in magnetic phases from about 35% in untreated glass to about 60% in treated glass, as concluded from the area of the magnetic subspectra. Tempering of untreated glass to cordierite (AC 14, Table I) promotes growth of spinel crystallites (and homogenization?) as deduced from the greater area of the magnetic subspectrum, smaller lines and higher magnetic fields H in comparison to the starting material.

In cordierites synthesized under hydrothermal conditions, at the very high f_{O2}'s of the Mn_2O_3/MnO_2- and the Mn_3O_4/Mn_2O_3 (HC6, Table I)-buffers, spinel phases of the glasses used as starting material have disappeared. The expected Fe^{3+}-incorporation doesn't show up by a Fe^{3+}-doublet. First, these runs demonstrate that in presence of water the exsolved spinel phase is reactive enough to vanish under incorporation in the silicate. Second, Fe prefers to be incorporated as Fe^{2+} in the Mg-site instead of as Fe^{3+} in the T1-site.

To prove whether traces of Fe^{3+}, being below the detection limit of Gamma-ray spectroscopy, are present, EPR - spectra were measured. Indeed, some threevalent iron on tetrahedral position has been observed in glasses and in cordierites synthesized in air, but no Fe^{3+} can be determined in cordierites grown under hydrothermal conditions, indicating that at least in crystallized cordierite incorporation of Fe^{3+} might be a product of hampered reaction kinetics.

Table I: ^{57}Fe-Mössbauer parameters [mm/s relative to metallic iron] of selected samples (x = 0.05)

T	Sample	Chi²	Doublets					Sextets					
			IS	QS	HW$_{av}$ %	I$_{av}$ %	Area	IS	QS %	HW [kG]	I* %	H	Area
RT	glass	1.4	1.09	2.00	0.83	4.3	65	0.38	-0.06	3.2	0.3	475	35
5K		2.6	1.24	2.18	0.92	12.7	65	0.46	-0.01	1.9	1.6	504	35
5K	AC9	1.5	1.15	2.38	1.10	-	31	0.39	0.13	1.2	-	517	50
		+	0.66	0.70	1.32	-	8	0.4	0.2	1.0	-	465	11
RT	AC14	0.8	1.19	2.29	0.34	6.5	53	0.23	0.82	1.2	-	493	47
RT	HC6	0.7	1.19	2.31	0.29	6.5	100						
5K		0.7	1.34	2.58	0.28	10.5	100						

Symbols and abbreviations: * = glass treated in O₂; T = absorber temperature; Chi² = goodness of the fit; IS = isomer shift; QS = quadrupole splitting; HW$_{av}$ = average full width at half peak height; I$_{av}$ = average linear intensity; I* = intensity of the two outer lines; Area = area of a subspectrum relative to the total of calculated subspectra, *Remarks:* Linear intensities I are only given for samples measured at different temperatures. Errors are 1 or less in the last given figure.

IV: SUMMARY AND CONCLUSIONS

^{57}Fe-Mössbauer experiments on ^{57}Fe-enriched cordierite phases, $Mg_2(Al_{1-x}Fe^{3+}_x)_4Si_5O_{18}$, containing an extremely diluted amount of iron (theoretically a substitution of Al by 0.5 atom% Fe^{3+}) suggest that even this small amount may not be totally incorporated in glass and cordierites in dry runs performed in air. Instead, heterogeneously nucleated "superparamagnetic" spinel phases form and the incorporated iron is divalent. Still, spinel particles are very small, within the range of single domains and are correspondingly reactive in hydrothermal runs thus being completely absorbed by cordierite. But, even the iron of the spinels reacting with the silicate phases under very high oxygen-fugacities is incorporated only as Fe^{2+}. This indicates at low iron concentration the prefered thermodynamic stability of octahedrally coordinated divalent iron, at least in alkali free phases.

Since the observed heterogeneous spinels seem to be very complex, presumably in composition, degree of order and grain sizes, a separate study would be necessary to learn more about the nature of these phases. One might think about starting compounds, being richer in iron and on longer heating periods to increase grain sizes. Furthermore, one needs for comparison single phase spectra of well defined spinels to enable the gathering of the relevant magnetic and Mössbauer parameters.

The results obtained in this study help to characterize glasses with very low iron concentrations. Furthermore, they demonstrate how important it is to make visible phases, not detectable by conventional methods. Results obtained in this study make a new discussion of previous studies on iron richer compositions [2-4] necessary.

Acknowledgements: Mössbauer spectra were measured at the Institute for Mineralogy in Marburg. The authors wish to thank St. Hafner (Marburg) for valuable discussions. Sincere thanks are also due to S. Nagel, who developed the Lfit and Hfit programs, used for evaluating the Mössbauer spectra and H. Rager (Marburg) for measuring and interpreting the EPR spectra.

REFERENCES

[1] F.-D. Doenitz & W. Vogel, *Kristall und Technik* **15**, 891 (1980).
[2] C. Boberski, *Doctoral thesis, Ruhr-Universität-Bochum*, (1986).
[3] D. Doenitz, C. Russ, and W. Vogel, *Silikattechnik* **34**, 155 (1983).
[4] B.O. Mysen, D. Virgo, E.-R. Neumann, and F. Seifert, *Amer. Mineral.* **70**, 317 (1985)
[5] B.J. Evans and S.S. Hafner, *J. Appl. Phys.* **40**, 1411 (1967)

DSC AND MÖSSBAUER STUDIES OF $Fe_{80}B_{20}$ CRYSTALLIZATION

Federica Malizia and Franco Ronconi

University of Ferrara, Department of Physics, Via Paradiso 12, I-44100 Ferrara, Italy
Unità Consorzio Interuniversitario Nazionale Fisica della Materia and Gruppo Nazionale
Struttura della Materia (C.N.R.)

ABSTRACT

Differential Scanning Calorimetry has been used to investigate the mechanism of the isothermal crystallization kinetics in $Fe_{80}B_{20}$ metallic glass. It is shown that the whole crystallization analysis must include, not only a crystal nucleation-and-growth process, but also a grain-growth process and that these two processes are separated in time during isothermal annealing. These processes have been studied directly finding the parameters which characterize their mechanism. From the theoretical Johnson-Mehl-Avrami equation describing the nucleation-and-growth process, it was possible to calculate the evolution of the transformed fraction of the material as a function of the annealing time. To infer the meaning of the transformed fraction, samples subjected to different thermal treatments have been studied by Mössbauer Spectroscopy. Our results reveal that the transformed fraction is the sum of the crystalline component formed by all atoms located in the lattice of the grains and the interfacial component composed of atoms in the interfacial regions between grains.

INTRODUCTION

Metallic glasses are metastable materials which, if subjected to suitable temperature treatment, transform to a more stable crystalline state. Their crystallization processes have been the object of extended studies in the past few years because the accurate knowledge and control of the crystallization mechanism makes it possible to obtain stable nanostructured materials with desidered morphological characteristics which cannot be obtained by other preparation techniques [1]. The crystallization process can be followed using differential scanning calorimetry (DSC) because the isothermal signal is proportional to the enthalpy difference between the amorphous and crystalline phases [2]. This technique makes it possible to distinguish between nucleation-and-growth and grain-growth processes thanks to the qualitative differences in their isothermal calorimetric signals [3].

This paper expands our previous results on glass crystallization [4,5] and deals with the study of the evolution of transformed fraction as a function of the annealing time during the eutectic isothermal crystallization of $Fe_{80}B_{20}$ amorphous alloy at the temperature of $701\ K$.

EXPERIMENTAL

$Fe_{80}B_{20}$ amorphous ribbons were prepared by a melt spinning technique. The devitrification kinetics were studied using a Perkin-Elmer DSC 7 differential scanning calorimeter with computerized data acquisition under a pure nitrogen atmosphere. The DSC instrument was calibrated using the melting point of indium and zinc. Particulary attention was paid to determining the baseline with required accurancy. Partially crystallized samples were obtained by annealing the amorphous ribbons at the same temperature (701 K) for

Mat. Res. Soc. Symp. Proc. Vol. 321. ©1994 Materials Research Society

different times. This annealing temperature is between the glass transition ($T_g = 658K$ [6]) and the onset of crystallization temperature ($T_0 = 707 \pm 2K$). Mössbauer experiments have been performed at room temperature in a standard transmission geometry using a ^{57}Co in Rhodium matrix. Magnetic hyperfine field distributions have been evaluated from Mössbauer spectra using the Le Caer-Dubois method [7] with a smoothing parameter of 0.1 chosen according to the very low in the statistical noise in our Mössbauer spectra.

RESULTS AND DISCUSSION

Isothermal treatments at the temperature $T_{iso} = 701K$ were performed on amorphous samples for 35 minutes. The obtained isothermal signal (curve a) is shown in Fig.1. In the previous paper [4] we have shown that the early stages of $Fe_{80}B_{20}$ crystallization is governed by the nucleation-and-growth process of large regions, and the calorimetric signal can be completely reproduced considering only the expected signal from the Johnson-Mehl-Avrami (JMA) theory. The expected calorimetric signal $(\frac{dH}{dt})_{N.G.}$ from JMA theory can be written as:

$$(\frac{dH}{dt})_{N.G.} = \Delta H n b (t - t_0)^{n-1} exp(-b(t - t_0)^n) \qquad (1)$$

where ΔH is the total enthalpy difference between the transformed and untransformed state, n is the Avrami exponent which depends on both the nucleation mechanism and growth morphology [8], b is a parameter which depends on nucleation frequency and growth rate, t is the isothermal annealing time and t_0 is the incubation time. Performing a fit of isothermal data with eqn. (1), it was possible to find the parameters which characterize the nucleation-and-growth process and the range of time in which this process can alone describe the calorimetric signal. The optimization of the parameters of the fitting was performed using Simplex [9] and Fletcher's switching [10] methods. Fig. 1 shows the best fit curve (curve b) to the data. The estimated parameters characterizing this transformation were: $\Delta H = 107.9$ (Joule/g), $b = 17.5 \ 10^{-7} \ (s^{-1})$, $n = 2.64$ and $t_0 = 32.9$ (s). It must be noted that JMA model fits experimental data very well only in a limited range $(t_s - t_0)$ of the isothermal time, where t_s is the time in which experimental data begin to deviate from the model. To try to describe the whole isothermal signal, we have supposed that for time longer than t_s grain-growth starts to contribute to the calorimetric signal. We have tried to test our assumption by fitting experimental data for $t > t_s$ with the equation:

$$\frac{dH}{dt} = (\frac{dH}{dt})_{N.G.} + (\frac{H_0 r_0}{g}) \frac{g p \lambda^g}{[r_0^g + g p \lambda^g (t - t_s)]^{g+1/g}} \qquad (2)$$

where H_0 and r_0 are the initial enthalpy and grain radius respectively, g is the grain-growth exponent depending on the mechanism of crystal grain-growth [11], p is a parameter that depends on atom mobility and on interfacial tension, and λ is a parameter with length dimension. In Fig. 1 the best curve to the data for $t > t_s$ (curve c) is shown, the good agreement with the experimental data supports our assumption that grain-growth accompanies the nucleation and growth of the eutectic regions. The estimated parameters for the grain-growth process are: $r_0 = 0.62$ (nm), $g = 2.0$, $p = 0.009 \ (s^{-1})$, $\lambda = 0.28$ (nm) and $t_s = 289$ (s).

These calorimetric studies have shown that: i) the early stages of $Fe_{80}B_{20}$ crystallization is governed by nucleation and growth of large regions, and the calorimetric signal

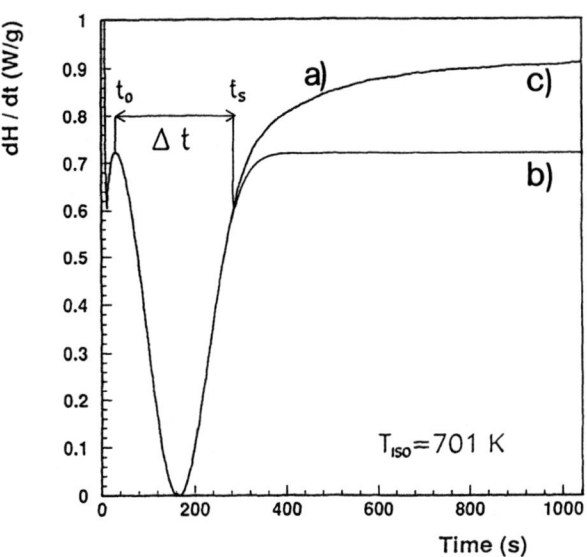

Fig. 1 . Rate of enthalpy change against time during isothermal annealing: experimental curve (a) obtained at 701 K, curve (b) and (c) are best fits to these data using eqns. (1) and (2), respectively.

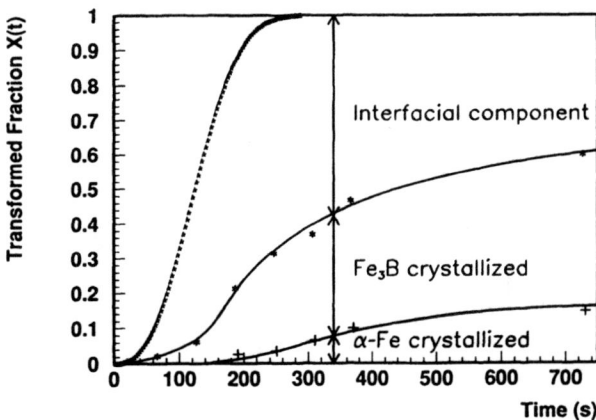

Fig. 2. Characteristic sigmoidal plots of the transformed fraction X(t), calculated by DSC, against elapsed time up to t_s at 701 K. The points (*) represent the volume fraction of the whole bulk crystalline component (V_{cry}); the points (+) show the only bcc-Fe volume fraction.

could be completely reproduced considering only the expected signal from the Johnson-Mehl-Avrami nucleation-and-growth theory; ii) in order to describe the whole isothermal signal it is necessary to consider also the contribution of the grain-growth process; iii) it is possible to find parameters characterizing these two processes by fitting the experimental isothermal signal with the ones expected from nucleation-and-growth and grain-growth theories.

Now, knowing the parameters which characterize the nucleation-and-growth process and the time range in which it works, it was possible to calculate the evolution of transformed fraction during isothermal annealing at $T_{iso} = 701K$ using the JMA equation:

$$X(t) = 1 - exp(-bt^n) \qquad (3)$$

where $X(t)$ is the fraction of transformed material. In Fig.2 it is shown a plot of the evolution of the transformed fraction in function of the annealing time.

In order to infer the meaning of the transformed fraction of the material we have prepared several samples subjected to isothermal treatments at $T_{iso} = 701K$ but for annealing times ranging from 60 and to 720 seconds. The obtained samples were studied by Mössbauer spectroscopy to gain information on the nature of the existing crystalline phases and their relative percentages.

Isothermal annealing of $Fe_{80}B_{20}$ amorphous alloy, at a temperature above the glass transition and below 773 K, is known to produce Fe_3B and α-Fe crystalline phases [12,13].

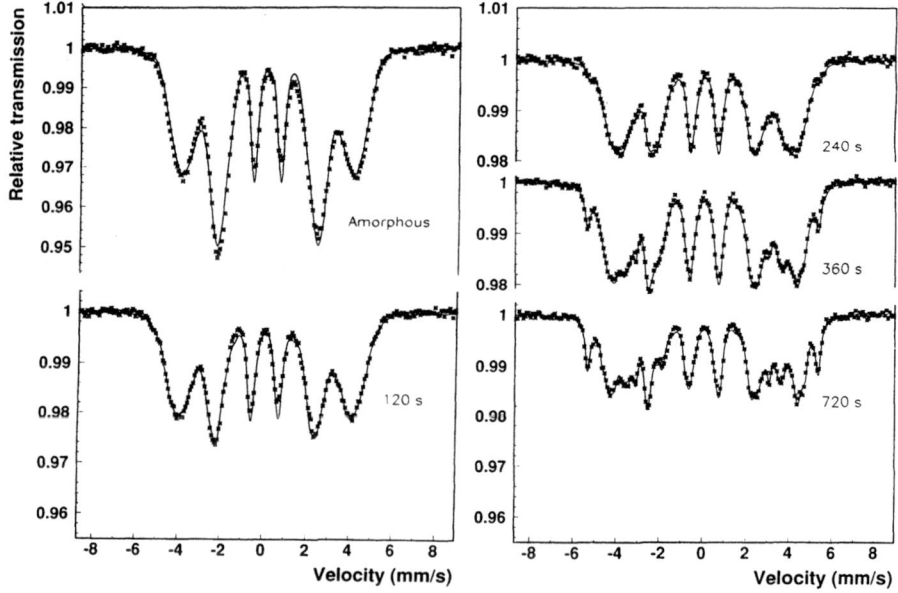

Fig. 3. Mössbauer spectra at room temperature after different isothermal annealing times at 701 K.

The hyperfine parameters of these two crystalline phases are well known [14], so it is possible to extract from Mössbauer spectra (see Fig. 3) their relative volume fractions (V_{cry}). Table I gives the values of V_{cry} for the crystalline phases and how they change as a function of the time during isothermal annealing. We can observe that there is a systematic difference between the value of $X(t)$, calculated by DSC, and the value of V_{cry} obtained by Mössbauer analysis. This fact is due to the contribution to $X(t)$ of all atoms which are not in the grain lattices but are in their interfacial regions. Mössbauer spectroscopy is a technique very sensitive to local atomic enviroment, so it can distinguish the bulk crystalline component from the interfacial one and separately to assign their contributions

Table I. Post-annealed structures as studied using Mössbauer spectroscopy. V_{cry} is the percentage of the total crystalline fraction due to bcc-Fe and to tetragonal boride Fe_3B which has three crystallographically different iron sides: Fe I, Fe II and Fe III with 2, 4, 3 metalloid nearest neighbours, respectively.

Anneal. Time (s)	Vcry (%)	(Fe3B) II (%)	(Fe3B) III (%)	(Fe3B) I (%)	bcc Fe (%)
60	25.1	21.6	2.2	1.3	0.0
120	28.9	24.6	0.6	3.7	0.0
180	26.2	18.6	3.6	2.7	1.3
240	34.6	23.6	5.8	3.1	2.1
300	35.6	18.5	8.5	3.5	5.1
360	45.3	22.9	11.7	2.1	8.6
720	58.6	22.9	16.1	6.2	13.4

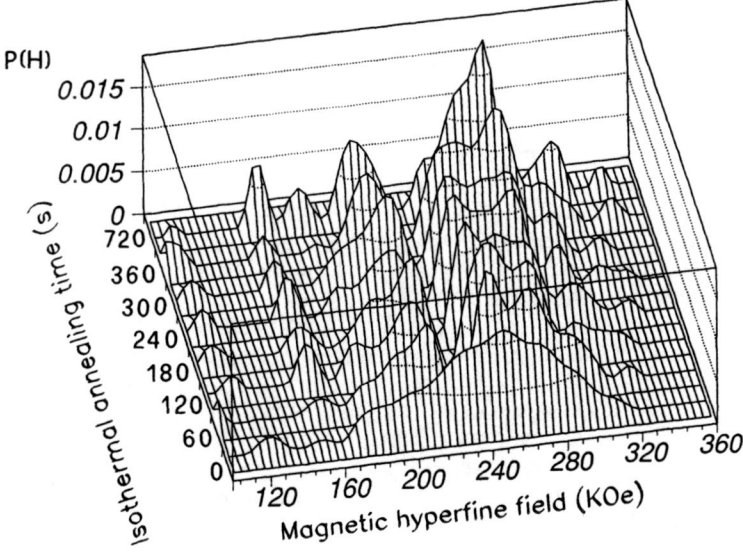

Fig. 4. The hyperfine magnetic field distribution of the extra-crystalline component after different annealing times at 701 K.

to the total transformed fraction $X(t)$. Fig. 2 showns the the transformed fraction of atoms in the crystalline state and in the boundary of the grains.

The structural characteristics of the interfacial component are deduced from magnetic hyperfine field distribution. The broad peaks of this distribution function, (see Fig. 4), support the idea of local fluctuations in the interatomic distances as a result of structural distortions of the interfacial component. The average magnetic hyperfine field of each peak is due to the contributions of resonating atoms in the same near-neighbour environments.The three principal peaks are ascribed to the atoms of same chemical composition of metastable tetragonal boride Fe_3B: Fe I, Fe II anf Fe III with 2, 4, 3 metalloid nearest neighbours respectively.

CONCLUSIONS

We have investigated the crystallization process of the $Fe_{80}B_{20}$ amorphous alloy using differential scanning calorimetry and Mössbauer spectroscopy. Our results are: i) during isothermal crystallization of $Fe_{80}B_{20}$ both nucleation-and-growth and grain-growth processes are active but they are separated in time; ii) the parameters characterizing these processes were found by fitting isothermal data with the signals expected from nucleation-and-growth and grain-growth theories; iii) using the obtained parameters it was possible to calculate the isothermal evolution of the transformed fraction of the material; iv) Mössbauer spectroscopy allowed us to show that the transformed fraction can be seen as the sum of the atoms which are in the crystalline phases and the atoms which are in the interfaces between the crystallites.

AKNOWLEDGEMENTS

This work has been supported by the 'Solid State Physics' C.N.R. project and by the 'Ministry of University and Scientific and Technological Research'.

REFERENCES

[1]   A.L.Greer, *Mater. Sci.* **97**, 285 (1988).
[2]   A.L.Greer, *Acta Metall.* **30**, 171 (1982).
[3]   L.C.Chen and F. Spaepen, *Nature* **336**, 366 (1988).
[4]   F.Malizia and F.Ronconi, *Phil. Mag. B* , in press (1993).
[5]   F.Malizia and F.Ronconi, *J. Appl. Phys.*, in press (1993).
[6]   T.Masumoto, *Sci. Rep. RITU* **29**, 265 (1981).
[7]   G.Le Caer and J.M.Dubois, *J. Phys.* **E12**, 1083 (1979).
[8]   J.W.Christian, "The theory of transformation in metals and alloys", (Pergamon press, Oxford, 1975).
[9]   J.A.Nelder and R.Mead, *Comput. J.* **7**, 308 (1965).
[10] R.Fletcher, *Comput. J.* **13**, 317 (1970).
[11] H.V.Atkinson, *Acta Metall.* **36**, 469 (1988).
[12] U.Köster, "Amorphous metals and non-equilibrium processing", Ed. M.Von Allmen (Les Editions de Physique, 1984), 175.
[13] E.E.Alp, K.M Simon, M.Saporoschenko and W.E.Brower, *J. Non-Crys. Sol.* **61-62**, 871 (1984).
[14] G.Le Caer and J.M.Dubois, *Phys. Stat. Sol.* **64**, 275 (1981).

PART V

Beam-Assisted Crystallization and Amorphization

ION BEAM SYNTHESIS OF LUMINESCENT SI AND GE
NANOCRYSTALS IN A SILICON DIOXIDE MATRIX

H.A. Atwater*, K.V. Shcheglov*, S.S. Wong*, K.J. Vahala*, R.C. Flagan*, M.L. Brongersma**
and A. Polman**
* Thomas J. Watson Laboratory of Applied Physics, California Institute of Technology,
Pasadena, CA 91125.
**FOM Institute for Atomic and Molecular Physics, Kruislaan 407, 1098 SJ Amsterdam,
Netherlands.

ABSTRACT

Ion beam synthesis of Si and Ge nanocrystals in an SiO_2 matrix is performed by precipitation from supersaturated solid solutions created by ion implantation. Films of SiO_2 on (100) Si substrates are implanted with Si and Ge at doses 1 x 10^{16}/cm^2 - 5 x 10^{16}/cm^2. Implanted samples are subsequently annealed to induce precipitation of Si and Ge nanocrystals. Raman spectroscopy and high-resolution transmission electron microscopy indicate a correlation between visible room-temperature photoluminescence and the formation of diamond cubic nanocrystals approximately 2-5 nm in diameter in annealed samples. As-implanted but unannealed samples do not exhibit luminescence. Rutherford backscattering spectra indicate a steepening of implanted Ge profiles upon annealing. Photoluminescence spectra are correlated with annealing temperatures, and compared with theoretical predictions for various possible luminescence mechanisms, such as radiative recombination of quantum-confined excitons, as well as possible localized state luminescence related to structural defects in SiO_2. Potential optoelectronic device applications are also discussed.

I. LUMINESCENT GROUP IV NANOCRYSTALS

Recently there has been considerable interest in luminescent group IV semiconductor nanocrystals inspired by a potential for fabricating integrated optoelectronic devices and optical amplifiers on silicon or other substrates. A considerable amount of this investigation has been devoted to investigation of characteristics and mechanisms of luminescence in porous silicon, for which high quantum efficiency photoluminescence has been reported[1], and this has invigorated efforts to make practical silicon-based light emitting devices. It has been suggested widely that this luminescence is due to 'quantum size effects' at nanometer-scale features in porous silicon. Theoretical work has suggested that sufficiently small crystals, of the order of a few nanometers, of Si and Ge should exhibit large oscillator strengths relative to bulk silicon, thus allowing efficient radiative recombination of electrons and holes at these nanocrystals[2,3].

A more promising approach for Si-based optoelectronic devices may be luminescence from Si nanocrystal structures embedded into SiO_2, since a dense, relatively chemically impervious SiO_2 film may have significant performance and reliability advantages over other nanocrystalline forms of Si, such as porous Si. This configuration may lend itself to practical optoelectronic devices on ultra-large-scale integrated circuits and in waveguide-based amplifiers. Indeed electroluminescence was first reported in such structures in 1984[6], well before the current renaissance of interest in optical properties of silicon[1]. In that work, electronically pumped Si nanocrystals inside the insulator of an MOS-type structure exhibited luminescence consisting of three peaks, the first of which, near 2 eV, is qualitatively similar to the

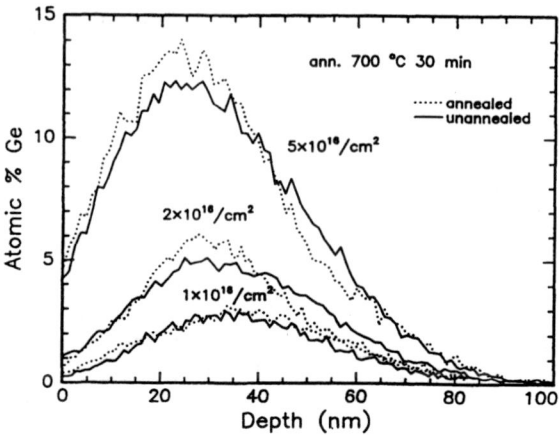

Figure 1: Concentration depth profiles for Ge-rich samples as-implanted and annealed at 700°C for 30 min, derived from RBS spectra. The respective peak concentrations for the three implant doses are 3, 6, and 13 at. % Ge. Upon annealing, a slight steepening of the concentration profile was observed.

luminescence reported in this work. Since then, several groups have reported the formation of nanocrystals in SiO$_2$ by annealing supersaturated solid solutions of group IV semiconductors prepared by rf co-sputtering of SiO$_2$ and different group IV semiconductors[7,9,10,11]. They observed visible photoluminescence from samples containing these nanocrystals, the spectra appearing quite similar to those of porous silicon. However, the specific relationship between nanocrystal formation and luminescence is not yet clear. Ion implantation of Si into SiO$_2$ also yielded visible photoluminescence, although detailed structural characterization was not carried out[13]. It has been suggested that this luminescence is also due to either transitions between discrete energy levels of a quantum dot, or recombination of molecular-like excitons trapped at these nanocrystals.

II. ION BEAM SYNTHESIS OF LUMINESCENT GROUP IV NANOCRYS-TALS

In this work, ion implantation was chosen to create supersaturated Si and Ge solid solutions because of the inherent high purity of the process, and the ease of controlling the concentration profiles of implanted material. In addition, vacuum annealing of ion implanted samples permits the structural and optical properties of nanocrystals to be characterized in the absence of hydrogen, thus enabling luminescence related to hydrogenated amorphous silicon and siloxene to be neglected. Silicon dioxide grown by thermal oxidation of Si is also an extremely pure and well-characterized material, and the Si-SiO$_2$ interface has a very low density of interface states. Films prepared in this manner also lend themselves to straightforward optoelectronic device fabrication via processes readily integrable with existing ultralarge-scale integrated circuit technology. Ion implanted structures may also enable planar waveguides, since a few atomic percent excess of Ge increases the refractive index of glass enough to produce a weak guide, as routinely practiced in optical fiber synthesis.

Thermally-grown silicon dioxide films 100 nm thick on P-doped (100) silicon wafers were ion implanted with ^{29}Si at 50 keV and with ^{74}Ge at 70 keV, with implant energies cho-

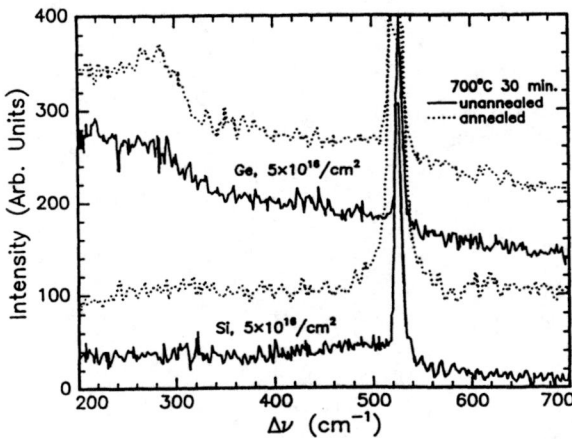

Figure 2: Raman spectra of Ge- and Si-implanted samples for doses of $5 \times 10^{16}/cm^2$, before and after annealing at 700°C for 30 min (Ge and Si spectra are offset relative to each other for clarity). The implanted, unannealed Ge spectrum shows a broad Ge-Ge shoulder below 300 cm^{-1}, believed to correspond to amorphous excess Ge. Upon annealing, this feature is narrowed, but is still broad, consistent with segregation and crystallization of implanted material. For Si-implanted unannealed samples, a broad asymmetric Si-Si shoulder is seen below 525 cm^{-1}, which evolves upon annealing into a broad peak centered at 525 cm^{-1}, superimposed onto the sharp Si-Si peak due to the Si substrate.

sen to place the concentration profile completely within the SiO_2 film thickness, at three doses, $1 \times 10^{16}/cm^2$, $2 \times 10^{16}/cm^2$, and $5 \times 10^{16}/cm^2$. The samples were subsequently annealed at temperatures 400 - 1000°C for 30 min or 40 min in an evacuated tube furnace with a base pressure of 10^{-7} Torr. Raman spectroscopy, high-resolution transmission electron microscopy (HRTEM) Rutherford backscattering spectrometry (RBS), and photoluminescence (PL) measurements were performed on as-implanted and annealed samples. The HRTEM was performed on a 300 keV electron microscope with 0.23 nm point-to-point spatial resolution. Raman spectra were measured using a 0.5 m double grating spectrometer, with data acquired in photon counting mode, and excitation was by the 441.6 nm line of a 50 mW He-Cd laser. Visible photoluminescence was excited by a 457.9 nm line of Ar ion laser, detected with a standard photomultiplier with a GaAs photocathode. Infrared photoluminescence was excited using the 514.5 nm Ar line and detected using a liquid nitrogen-cooled Ge detector. RBS was done with He^{++} at 2 MeV, with normal incidence and a 170° backscattered ion detector position.

Concentration depth profiles derived from RBS spectra are shown in Fig. 1 for Ge-rich samples, both as-implanted and annealed at 700°C for 30 min. The respective peak concentrations for the three implant doses are 3, 6, and 13 at. % Ge. Upon annealing, a slight steepening of the concentration profile was observed. This is probably due to the diffusion of Ge from regions with lower initial concentration and negligible nanocrystal nucleation to regions with high initial supersaturation and thus higher nucleation rate. The reduction of supersaturation following nucleation facilitates Ge diffusion against the initial concentration gradient. The slight shift of the peak with increasing dose is apparently due to partial sputtering of SiO_2 during high dose implantation. Monte Carlo simulations of collision

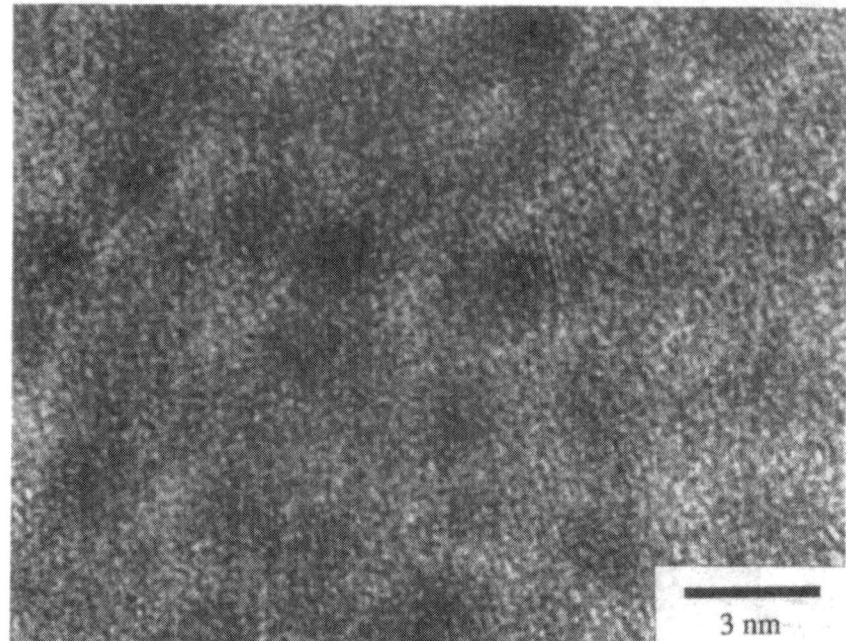

3 nm

Figure 3: High-resolution transmission electron micrograph of a sample implanted with $2x10^{16}/cm^2$ Ge and annealed at 1000°C. A dense array of crystalline Ge particles indexable to diamond-cubic structure are visible in the SiO_2 matrix, with average size of approximately 3 nm.

cascades performed using the TRIM[14] code indicated that as much as 15 nm of SiO_2 may be sputtered at the highest dose. The ratio of sputtering yields for oxygen and silicon exceeds 2, so the remaining oxide will be slightly Si-enriched at the surface. Thus, it is conceivable that for high Ge implantation doses, the excess Si may also precipitate, forming SiGe alloy nanocrystals.

III. STRUCTURAL AND OPTICAL CHARACTERIZATION

Raman spectra of Ge- and Si-implanted samples are shown in Fig. 2 for doses of $5x10^{16}/cm^2$, before and after annealing at 700°C for 30 min. The implanted, unannealed Ge spectrum shows a broad Ge-Ge shoulder below 300 cm^{-1}, believed to correspond to amorphous excess Ge. Upon annealing, there is a narrowed, but still broad, Ge-Ge peak at approximately 300 cm^{-1}, consistent with segregation and crystallization of implanted material. For Si-implanted unannealed samples, a broad asymmetric Si-Si shoulder is seen below 525 cm^{-1}, which turns into a broad peak centered at 525 cm^{-1} superimposed onto the sharp Si-Si peak due to the Si substrate. Raman spectra of samples implanted at lower doses show the same qualitative behavior although the features are less pronounced due to lesser amount of active material.

Bright field and dark field transmission electron micrographs for Ge implanted samples indicated nanocrystals ranging in size from approximately 1-5 nm, and transmission electron

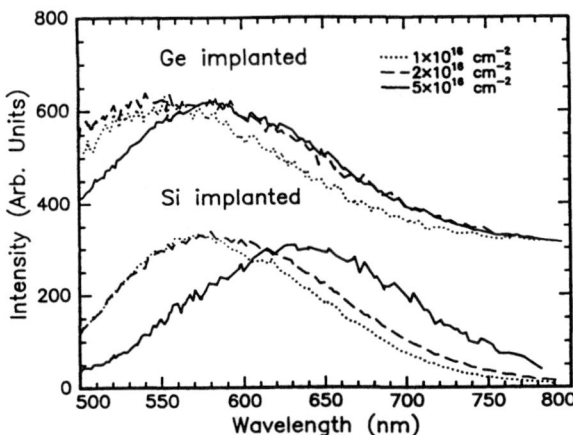

Figure 4: Visible photoluminescence spectra for Ge and Si samples implanted with doses from $1 \times 10^{16}/cm^2$ to $5 \times 10^{16}/cm^2$ annealed at 700 °C for 40 min (Ge and Si spectra are offset relative to each other for clarity). Similar annealing conditions produce a red shift of the luminescence peak energy with increasing dose, consistent with growth of larger nanocrystals at high doses.

diffraction indicates randomly oriented diamond cubic crystalline particles. A HRTEM image of a sample implanted with $2 \times 10^{16}/cm^2$ Ge and annealed at 1000°C is shown in Fig. 3. A dense array of crystalline Ge particles are visible in the SiO_2 matrix, with average size of 3 nm. We were unable to produce clear HRTEM images of Si nanocrystals, although transmission electron diffraction indicated diamond cubic structure, and nanocrystals were observable in dark field microscopy. We note that the imaging conditions here are considerably more difficult than for isolated free nanocrystals, due to the high noise background arising from the amorphous matrix. Unlike Ref. [8], no evidence was found in HRTEM for a systematically smaller Ge nanocrystal interplanar spacing at smaller nanocrystal size. No reflections were visible in transmission electron diffraction other than those allowed for the diamond cubic structure. It is also important to note that the resolution limits in HRTEM imposed by the scattering conditions specific to small nanocrystals in SiO_2 leave open the possibility that a significant population of nanocrystals exists at sizes below the minimum resolvable size.

A typical set of visible PL spectra are given in Fig. 4 for Ge and Si samples implanted with doses from $1 \times 10^{16}/cm^2$ to $5 \times 10^{16}/cm^2$ annealed at 700 °C for 40 min. It is apparent that similar annealing conditions produce a red shift of the luminescence peak energy with increasing dose, which is consistent with growth of larger particles at high doses. The luminescence power is linear with 457.9 nm Ar pump power in the range of 0.1-10 W/cm^2, as shown in Fig. 7, suggesting that luminescence does not arise from a second- or higher- order processes, such as stimulated Raman scattering. The relative luminescence intensity was found to be smallest for the highest dose implants, corresponding to the largest nanocrystals.

For a given implanted ion dose, the wavelength for peak luminescence intensity varies non-monotonically with increasing isochronal anneal temperature in the range of 400-1000°C. This is illustrated by the evolution of photoluminescence spectra with anneal temperature, represented as photoluminescence intensity surfaces in Fig. 5. Implanted samples were annealed at 400, 500, 600, 700 and 1000 °C. The photoluminescence intensities in Fig. 5(a),

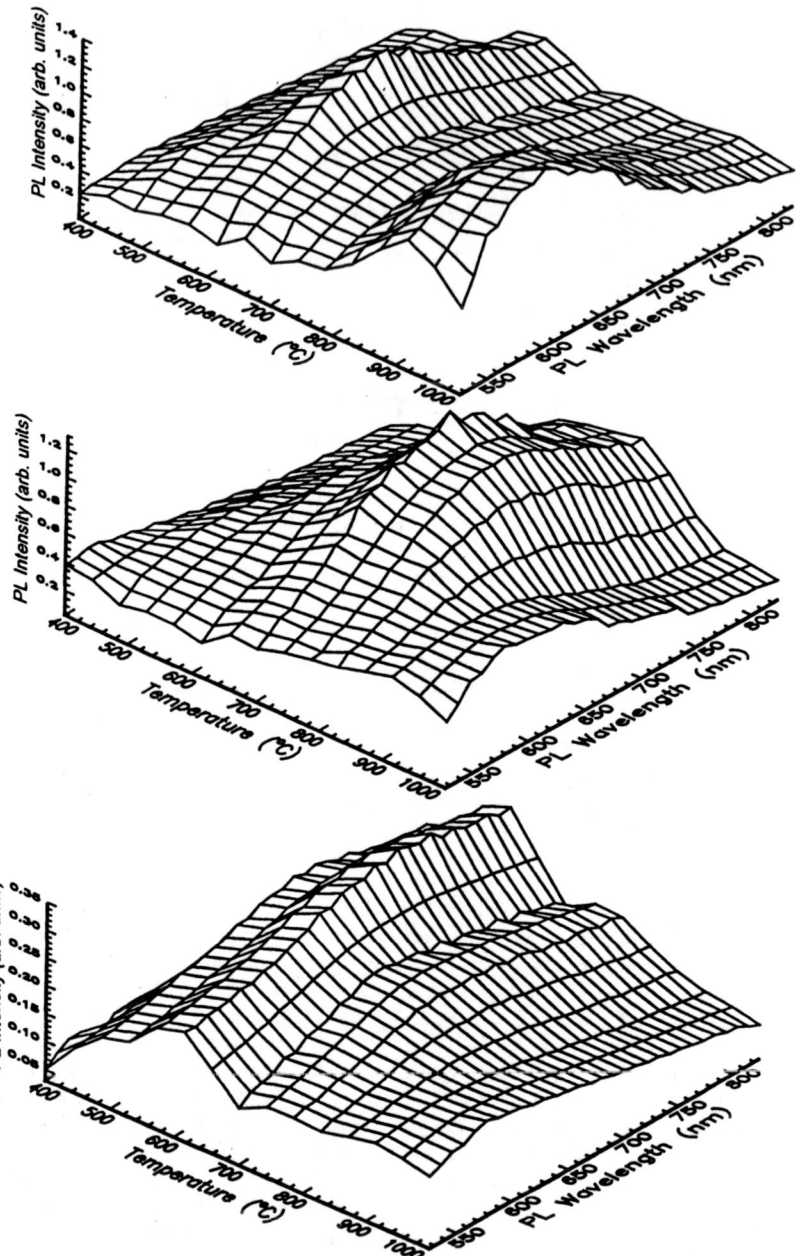

Figure 6: Variation of photoluminescence intensity with wavelength, as a function of annealing temperature for Ge-implanted samples with doses of (a) $1 \times 10^{16}/cm^2$, (b) $2 \times 10^{16}/cm^2$, and (c) $5 \times 10^{16}/cm^2$. All anneals were of 30 min duration. Note the difference in range of photoluminescence intensity for (a), (b) and (c).

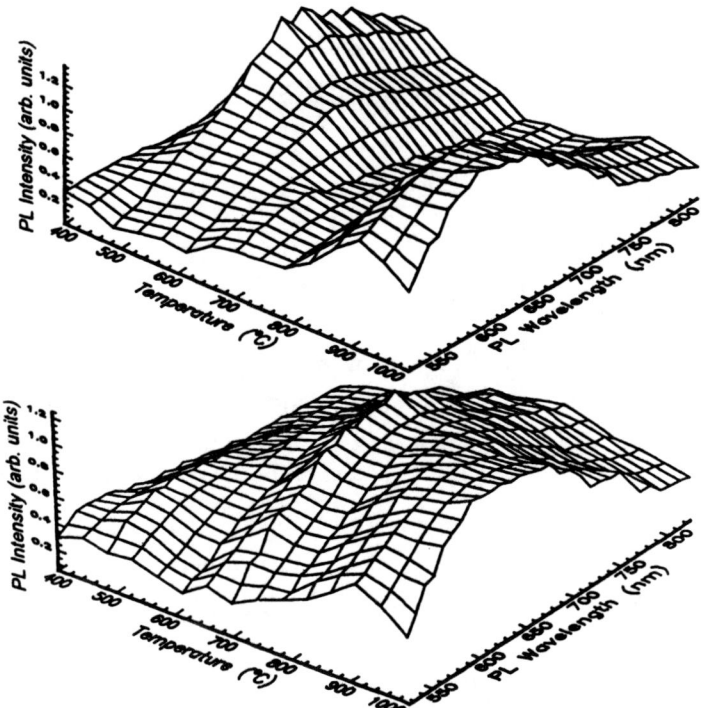

Figure 7: Variation of photoluminescence intensity with wavelength, as a function of annealing temperature for Si-implanted samples with doses of (a) $1 \times 10^{16}/cm^2$ and (b) $2 \times 10^{16}/cm^2$. All anneals were of 30 min duration. Note the difference in range of photoluminescence intensity for (a) and (b).

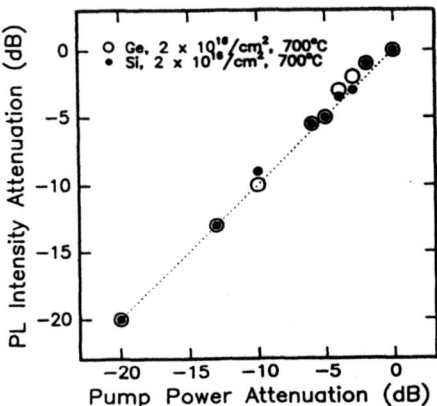

Figure 8: Variation in photoluminescence intensity with pump power for Ge-implanted and Si-implanted samples, both implanted with doses of $2 \times 10^{16}/cm^2$ and annealed at 700 °C for 30 min.

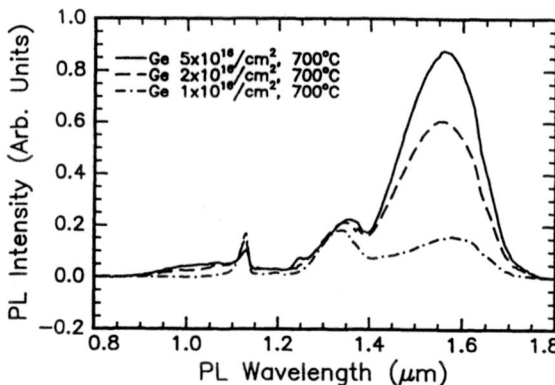

Figure 8: Infrared photoluminescence spectra measured at 77 K for Ge samples implanted with doses from $1 \times 10^{16}/cm^2$ to $5 \times 10^{16}/cm^2$ annealed at 700 °C for 40 min. Pump wavelength is 514.5 nm.

Fig. 5(b) and Fig. 5(c) are normalized relative to each other, the difference in ranges for photoluminescence intensity on the three figures should be noted. The photoluminescence spectra for a Ge dose of $1 \times 10^{16}/cm^2$, shown in Fig. 5(a), illustrate the non-monotonic behavior in the evolution of the wavelength for photoluminescence peak intensity. At an anneal temperature of 400 °C, the peak in photoluminescence intensity lies at approximately 575 nm, whereas for samples annealed at 600 °C, the peak in photoluminescence intensity lies at approximately 700 nm, and samples annealed at 1000 °C have a peak photoluminescence intensity at approximately 570 nm. The photoluminescence spectra for Ge doses of $2 \times 10^{16}/cm^2$ and $5 \times 10^{16}/cm^2$, shown in Fig. 5(b) and Fig. 5(c), respectively, indicate similar trends. Note that luminescence intensity for the $5 \times 10^{16}/cm^2$ dose is much smaller than that for doses of $1 \times 10^{16}/cm^2$ and $2 \times 10^{16}/cm^2$. For the dose of $1 \times 10^{16}/cm^2$, the variation of peak luminescence intensity with anneal temperature is also non-monotonic, with a drop in peak luminescence intensity occuring at 700 °C between the overall maximum at 600 °C and another relative maximum at 1000 °C. Also noteworthy is a monotonic red shift with increasing Ge dose of the peak in luminescence intensity for temperatures of 500-700 °C. Similar trends were observed for Si-implanted samples as well, as illustrated for doses of $1 \times 10^{16}/cm^2$ and $2 \times 10^{16}/cm^2$ in Fig. 6.

Infrared photoluminescence spectra measured at 77 K are shown in Fig. 8 for Ge samples implanted with doses from $1 \times 10^{16}/cm^2$ to $5 \times 10^{16}/cm^2$ annealed at 700 °C for 40 min. Three prominent features are observed at approximately 1.1 μm, 1.35 μm, and 1.55 μm. The first feature corresponds to luminescence from the band edge in bulk silicon, and most likely originates from the substrate. The second feature, occuring at approximately 1.35 μm, is interesting. Broad luminescence features are frequently observed in damaged bulk Si in this wavelength range. One possibility is that ion implantation of Ge at 70 keV into the 100 nm SiO_2 film produced damage in the underlying bulk Si substrate, although this is deemed unlikely since the damage profile has a peak at a depth of approximately 30 nm below the SiO_2 surface, and is confined within the SiO_2. Another possibility is that the diamond cubic Ge nanocrystals (or SiGe nanocrystals) in the SiO_2 matrix have optical active defects similar to those defects in bulk Si. Alternatively, the band around 1.35 μm may be associated with an irradiation-induced defect in the SiO$_2$. Also, the shoulder on the low-wavelength side of the band edge luminescence may also be attributed to such defects. The feature at approximately 1.55 μm may be the P(1) defect which has a peak in luminescence intensity at 1.57 μm, a phosphorus-related defect, which might arise as a result of ion irradiation of the SiO$_2$ which was thermally grown from a P-doped Si substrate.

IV. MECHANISMS FOR LUMINESCENCE

Two mechanisms have been proposed to account for the luminescence of the samples. One is the radiative recombination of excitons confined at the nanocrystals. More specifically, it is possible to characterize the nanocrystal as either (i) a blue-shifted, bulklike 'quantum dot' or (ii) a new, molecular-like state not related to the electronic structure of the bulk semiconductor material. Another possible mechanism is the luminescence of synthesis-induced (in this case, irradiation-induced) defects in SiO$_2$. The most widely studied defects in SiO$_2$ that exhibit luminescence at energies near those observed in the ion implanted samples appear to be those classified as E' defects[16]. This defect is formed when an oxygen atom is displaced leaving behind a silicon atom with a hole. This bound hole acts as a localized state that can trap electrons.

For the radiative excitonic recombination mechanism, emission energies in the range of 1-4 eV are predicted from both molecular orbital[2,4] and effective mass[3,5] calculations for diamond cubic Ge and Si, with emission energy depending on particle size. It should be noted that, at present, there is not good agreement among the various calculations as to exactly what the electronic structure should be(e.g., some effective mass[5] and molecular orbital[4] calculations indicate the energy gap is almost 1 ev higher than other effective mass[3] and molecular orbital[2] calculations). The predicted exciton binding energy of over 50 meV for particle size below 4 nm is consistent with radiative excitonic emission at room temperature[3]. At least qualitatively, the particle size of approximately 3 nm and observed luminescence energy of 1.6-1.9 eV are consistent with the radiative emission energy predicted for Ge and Si nanocrystals which exhibit excitonic emission. Molecular orbital calculations[2] predict oscillator strengths 1000 times large for 1 nm nanocrystals than for 3 nm nanocrystals. Thus if these predictions are interpreted literally, is difficult to understand why even a very small number of < 3 nm structures would not produce intense luminescence and thereby produce luminescence spectra which are not actually peaked in the range of 500-700 nm, but simply monotonically increasing towards decreasing wavelengths. It is important to point out, however, that these calculations assumed an isolated nanocrystal with a completely passivated surface. Electronic interaction with the host matrix, nearby nanocrystals and other defects may play a significant role in determining luminescence characteristics. Nonetheless, there is at least a qualitative agreement between a exciton recombination model and the present visible photoluminescence results, such as a red-shift and a reduced intensity for samples with higher concentration, presumably due to reduction of exciton energy and oscillator strength for particles of larger size. It is worth noting that if the visible luminescence in Fig. 4 is indeed due to quantum dot effects in the nanocrystals, the predicted relative blue shift at a given size for Ge is much larger than for Si[3,12]. This is consistent with the larger exciton Bohr radius in Ge (11.5 nm) than in Si (4.3); thus for a given size Ge shows a larger blue shift than Si[12]. Since inhomogeneous size broadening prevents the observation of narrow PL lineshapes, it is plausible that PL spectra from Si and Ge nanocrystals should appear essentially indistinguishable. Interestingly, Ge nanostructures formed by etch-

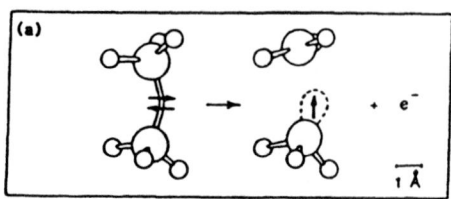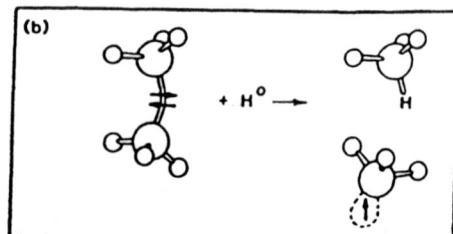

Figure 9: Schematic depictions of possible configurations for the E' center defect as (a) an electron bound to a hole resulting from a broken Si-Si bond, or (b) a hole resulting from a broken Si-Si bond, in which one of the Si atoms is coordinated by H(after Ref. [16]).

ing of bulk Ge also exhibited photoluminescence in the same wavelength regime seen in these experiments[17].

Ion implantation of SiO_2 produces a high density of displaced atoms and defects. Although the high temperature stability of radiation-induced visible luminescent centers has not be extensively studied, many of these defects –such as the E' center– are expected to anneal at temperatures above 600°C, although the high temperature stability of radiation-induced defects is not as well explored as their low-temperature behavior. Possible E' defect configurations are depicted in Fig. 9. Moreover, the E' center is known to luminesce in the blue region of the visible spectrum, not in the yellow-red range, where peaks of the luminescence were observed in this work. Previous ion implantation experiments reported emission at 2 eV in as-implanted SiO_2 samples[13]. No measurable luminescence was detected for as-implanted samples in the present experiments. The differences are possibly related to the difference in implantation conditions, as both the doses and energies were approximately an order of magnitude higher in the previous work, thus potentially yielding significantly different damage yields related to electronic energy loss.

We now consider the photoluminescence data of Fig. 5 in the context of these two mechanisms. First, we note that nanocrystals are formed via a nucleation and growth process, and that variation of the annealing temperature may alter the dynamics of nucleation and growth in a rather complex manner. Accepting for the moment the validity of an nanocrystal-related excitonic emission mechanism for luminescence, we might assume the following sequence of events would lead to the observed non-monotonic variation of the wavelength for peak photoluminescence intensity: (i) at temperatures of 400-500°C, nucleation occurs slowly, followed by incomplete nanocrystal growth during the anneal; (ii) at temperatures of 600-700°C, nucleation occurs relatively slowly, followed by more complete nanocrystal growth during the anneal, leading to a red shift of the wavelength for peak luminescence intensity; (iii) nucleation occurs very rapidly followed by complete nanocrystal growth (i.e., complete depletion of the atomic Ge supersaturation in the SiO_2 matrix), leading to a high density of small nanocrystals, causing a blue shift wavelength for peak luminescence intensity back to approximately 575 nm.

A second possible interpretation is that one luminescence mechanism is operative in the samples annealed from 400-600°C, and another luminescence mechanism is operative in the samples annealed at 1000 °C. Accepting for the moment the validity of a 'synthesis-induced defects in glass' mechanism, we might assume that the increase in annealing temperature from the range 500-700°C to 1000°C leads to the formation of a new defect complex. However, it is difficult to say at present what this defect might be, since it is stable at 1000°C. Also, if optically active defect complexes are being formed at high anneal temperatures from simple irradiation-induced defects, an increase in the implanted ion dose would be anticipated to give rise to an increase in the number of optically active defect complexes, and in turn the visible photoluminescence intensity. Instead, a decrease of the peak visible photoluminescence intensity at high anneal temperatures is seen with increasing implanted ion dose, as illustrated by Fig. 5. However, the infrared photoluminescence data indicate an increase in the intensity of the 1.3 μm and 1.6 μm features with increasing implantation dose, suggesting that these features may be related to irradiation-induced defects.

Complicated interactions may exist between nanocrystals and radiation-induced defects, and preferential sputtering of O during high dose implantation might create Si-rich regions, leading to Si-Ge nanocrystals for Ge implantation and Si nanocrystals for, e.g., noble gas implantation. Indeed non-cubic forms of Si such as planar and cubic Si clusters also give rise to luminescence in the vicinity of 2 eV, although to date these structures have only been prepared in such a way that the surface Si atoms are functionalized by bulky radicals which facilitate synthesis of such nonequilibrium structures[18]. The report of non-cubic Ge nanocrystals in SiO_2 prepared by sputtering is also intriguing[8], but more extensive work is required to corroborate this finding and understand its significance. Indeed, the distinction between extended defects and very small nanocrystals may not be a sharp one, and has not yet been made experimentally for Ge or Si in SiO_2.

V. SUMMARY AND FUTURE DIRECTIONS

In summary, we report synthesis and characterization of nanocrystals of Si and Ge in an SiO_2 host matrix by ion implantation to create a supersaturated solid solution, and subsequent precipitation through annealing. We have observed intense visible photoluminescence at room temperature from these samples with peaks between 550 and 700 nm, depending on implant dose and the annealing temperature. The luminescence data is qualitatively consistent with a mechanism of radiative excitonic recombination associated with nanocrystals, but some features suggest that synthesis-induced defects may play a role in luminescence, and that the detailed physical picture is much more complex. More extensive investigations of nucleation and growth kinetics are required, as are high spatial resolution probes of long range and local order.

ACKNOWLEDGEMENTS

One of us (K.V.S.) acknowledges the Fluor Foundation for fellowship support. This work was supported by the United States Department of Energy under grant DE-FG03-89ER45395. Work at FOM is part of the research program of FOM and was made possible by financial support from NWO, STW and IOP-Electro-Optics.

References

[1] L.T. Canham, Appl. Phys. Lett., **57** 1046 (1990).

[2] B.Delley and E.F.Steigmeier, Phys. Rev., **B47**, 1397 (1993).

[3] T.Takagahara and K.Takeda, Phys. Rev., **B46**, 15578 (1992).

[4] H. Kimura, S. Imanaga, Y. Hayafuji and H. Adachi, J. Phys. Soc. Jpn., **62**, 2663 (1993).

[5] N. Matsumoto, K. Takeda, H. Teramae and M. Fujino, Advances in Chemistry, **224**, ACS Books, (1989), p. 515.

[6] D.J. DiMaria, J.R. Kirtley, E.J. Pakulis, D.W. Dong, T.S. Kuan, F.L. Pesavento, T.N. Theis, J.A. Cutro, and S.D. Brorson, J. Appl. Phys. **56**, 401 (1984).

[7] Y. Maeda, N. Tsukamoto, Y. Yazawa, Y. Kanemitsu, and Y. Masumoto, Appl. Phys. Lett. **59**, 3168 (1991).

[8] Y. Kanemitsu, H. Uto, Y. Masumoto and Y. Maeda, Appl. Phys. Lett. **61** 2187 (1992).

[9] S.Hayashi, Y.Kanzawa, M.Kataoka, T.Nagareda and K. Yamamoto, Z. Phys. **D26**, 144 (1993).

[10] S.Hayashi, M.Fijii. K.Yamamoto, Jpn. J. Appl. Phys. **28**, 1464 (1989)

[11] S.Hayashi, T.Nagareda, Y. Kanzawa, and K. Yamamoto, Jpn. J. Appl. Phys. **32**, 3840 (1993).

[12] A.D. Yoffe, Advances in Physics, **42**, 173 (1993).

[13] T. Shimizu-Iwayama, M. Ohshima, T. Niimi, S. Nakao, K. Saitoh, T. Fujita, and N. Itoh, J. Phys. Cond. Matter **5**, L375 (1993).

[14] J.F. Ziegler, J.P. Biersack and U. Littmark, *The Stopping and Range of Ions in Solids*, Pergamon, New York, 1985.

[15] K.V. Shcheglov, S.S. Wong, K.J. Vahala, R.C. Flagan, to be published in Appl. Phys. Lett., 1994.

[16] K.S. Song and R.T. Williams, *Self-Trapped Excitons*, Springer Verlag, New York, 1993, Ch. 7, pp. 270-299.

[17] R. Venkatasubramanian, D.P. Malta, M.L. Timmons and J.A. Hutchby, Appl. Phys. Lett., **59**, 1603 (1993).

[18] Kanemitsu, K.Suzuki, H.Uto,Y.Masumoto,T.Matsumoto, S.Kyushin, K.Higuchi, and H.Matsumoto., Appl. Phys. Lett., **61** 2446 (1992); Y. Kanemitsu, K. Suzuki, H. Uto, Y. Masumoto, K. Higuchi, S. Kyushin and H. Matsumoto, Jpn. J. Appl. Phys. **32**, 408 (1993).

This article also appears in Mat. Res. Soc. Symp. Proc. Vol 316

SOLID-PHASE EPITAXIAL CRYSTALLISATION OF Ge_xSi_{1-x} ALLOY LAYERS

ROBERT G. ELLIMAN, WAH-CHUNG WONG and PER KRINGHØJ
Electronic Materials Engineering Department, Research School of Physical Sciences and
Engineering, Australian National University, Canberra, ACT 0200, AUSTRALIA.

ABSTRACT

Thermally-induced solid-phase epitaxial crystallisation (SPEC) and ion-beam-induced
epitaxial crystallisation (IBIEC) of amorphous Ge_xSi_{1-x} alloy layers is examined for three
different starting structures: a) strain-relaxed alloy layers of uniform composition, b) strained
alloy layers of uniform composition, and c) Ge implanted Si layers. Thermal annealing
experiments show that the activation energy for strain-relaxed alloys is higher than that expected
from a simple extrapolation between the activation energies of Si and Ge, and exceeds that of Si
for $x \leq 0.3$. Experiments on thin strained layers show that MBE grown strained layers which are
stable during annealing at 1100°C for 60 s are also fully strained after SPEC, whereas layers
which relax during annealing at 1100°C also relax during SPEC. Experiments on ion-implanted
Ge_xSi_{1-x} structures show that fully strained $Si/Ge_xSi_{1-x}/Si$ heterostructures can be fabricated
for ion fluences below a critical fluence, and as for uniform alloy layers that this critical fluence is
accurately predicted by equilibrium theory. Strain relaxation during SPEC of uniform alloys and
implanted structures is shown to be correlated with a sudden reduction in crystallisation velocity
which is believed to be caused by stress-induced roughening or faceting of the
crystalline/amorphous interface. IBIEC of thick (800 nm) implanted layers is shown to be limited
by competition from ion-beam induced random crystallisation, while thin (120 nm) uniform alloys
and implanted structures are shown to crystallise epitaxially and to exhibit similar behaviour to
thermally annealed samples under certain conditions.

INTRODUCTION

Solid-phase epitaxial crystallisation (SPEC) of amorphous semiconductor layers has been
extensively studied over the past few decades [1-5]. The crystallisation rate has been measured
over a wide temperature range and shown to be a thermally activated process characterised by an
activation energy of 2.68 eV [2]. The crystallisation rate has also been shown to depend on
crystallographic orientation, being fastest in the (100) orientation and slowest in the (111)
orientation, and on the presence of impurities [1]. Low concentrations (<< solid solubility limit)
of electrically active impurities have been shown to increase the crystallisation velocity whereas
electrically inactive impurities such as C and O decrease it [1]. At high concentrations ($\geq$ solid
solubility limit) most impurities are observed to retard SPEC, with some impurities being
segregated at the crystalline/amorphous interface during SPEC [1]. An impurity of particular
interest in this regard is H which has also been shown [3] to segregate at the crystalline
amorphous interface and to retard SPEC. In addition, recent experiments have also shown that
hydrostatic pressure and biaxial stress can increase the rate of SPEC [4,5]. The available data
suggests that SPEC is activated by a bond-breaking event at the crystalline/amorphous interface
[5] and that this event occurs at a specific 'defect' site which can exist in a charged state. Dopants
are believed to increase the SPEC rate by increasing the total concentration of these 'defect' sites
[5]. The effect of stress on SPEC is of particular interest since it provides information about the

shape and size of the activating defect [4,5]. GeSi alloy layers provide useful model systems in this regard, since the effect of a biaxial stress can be studied directly. SPEC of GeSi alloy layers also provides information about the influence of high impurity concentrations on SPEC, allowing the role of stress and strain-relief processes to be assessed without the complication of phase separation or electronic effects.

SPEC of intrinsic GeSi alloy layers has been studied previously for uniform alloy layers [6-11] and layers synthesised by ion-implantation [12-18]. It has been shown [6-11,13,16-18] that whilst GeSi strained layers grown by molecular beam epitaxy (MBE) or chemical vapour deposition (CVD) can exceed the theoretical critical thickness by as much as several orders of magnitude, layers grown by SPEC generally relax at, or near, the predicted critical thickness, despite the fact that the growth temperatures (~500-600°C) are similar in both cases. Strain relaxation during SPEC has been shown [8] to be preceded or accompanied by roughening or {111} faceting of the crystalline/amorphous interface. This faceting has been observed to retard SPEC during subsequent annealing. Ion-beam-induced epitaxial crystallisation (IBIEC) of amorphous GeSi alloy layers has also been studied [18-22] in an attempt to learn more about the atomistic processes responsible for SPEC.

This paper presents results of recent, and on-going, experiments on SPEC and IBIEC of GeSi alloy layers. SPEC and IBIEC are examined in strain-relaxed GeSi alloys; in thin, initially strained layers; and in ion-beam synthesised alloy layers.

EXPERIMENTAL

Three different Ge_xSi_{1-x}/Si structures were employed in the present study: a) 1.0 μm thick, strain relaxed alloy layers with x = 0.21 and 0.53, grown on a graded buffer layer of ~1.0 μm thickness, b) 120 nm thick, uniform alloy layers with x = 0.085, 0.17, 0.27, 0.37 and 0.67, each with a 5 nm Si capping layer, and c) ion-implanted structures produced by either 90 keV or 800 keV Ge implants into (100) Si. The thin, uniform layers and the fully relaxed alloy layers were grown by MBE at Aarhus University using a VG-80 MBE system and a growth temperature of ~550°C. Amorphous layers were produced by implanting samples at -196°C with Si ions. For fully relaxed layers, ~400 nm amorphous layers were produced by 200 keV Si ions implanted to a fluence of 1×10^{15} Si.cm^{-2}; whereas for the thin, uniform layers, 220 nm amorphous layers were produced by 90 keV Si ions implanted to a fluence of 5×10^{15} Si.cm^{-2}. In the latter case, the amorphous layer extended beyond the GeSi alloy layer into the Si substrate.

Thick implanted structures were fabricated by implanting 800 keV Ge ions into (100) Si at room temperature to fluences in the range 8.4×10^{16} Ge.cm^{-2} - 3.8×10^{17} Ge.cm^{-2}. Samples were subsequently implanted at -196°C with 800 keV Si ions to a fluence of 5×10^{15} Si.cm^{-2} in order to produce an abrupt amorphous/crystalline interface at a depth well beyond the implanted Ge distribution. To avoid contamination from recoil-implanted impurities, the native oxide was removed from samples prior to implantation by etching with a 10% HF solution. In addition, the vacuum pressure was maintained at $\leq 5 \times 10^{-7}$ Torr during implantation and all beam-defining apertures were made from Si to avoid contamination from aperture sputtering. Thin implanted structures were also fabricated by implanting 90 keV Ge ions into (100) Si.

SPEC was performed either in air during time-resolved reflectivity (TRR) analysis [2] or in a quartz tube furnace with a N_2 ambient at temperatures in the range from 300 to 600°C. For TRR analysis the sample was heated by contacting it to a preheated metal block and the sample reflectivity was monitored at two different wavelengths, 632.8nm and 1523nm. The low absorption infra-red wavelength used to monitor the growth of thicker layers and the visible wavelength used to monitor near surface (~300nm) crystallisation. The temperature of the TRR block was calibrated over the temperature range 500 - 650°C by comparison of Si SPEC rates with those of Olson and Roth [2]. Ion beam annealing was performed at temperatures in the range from 250°C to 350°C using 1.0 MeV Si ions. In-situ TRR analysis was also performed in this case but only at a wavelength of 632.8nm.

Rutherford backscattering and channelling analysis (RBS-C) was performed with 2.0 MeV He ions. Channelling was along the [100] zone axis normal to the sample surface. High-mass resolution spectra (scattering angle of 165°) and high depth resolution spectra (scattering angle 100°) were collected simultaneously. Electron microscopy was undertaken on plan-view and cross-section samples using a JEOL 2000EX operating at 200 kV.

RESULTS AND DISCUSSION

1) Solid Phase Epitaxial Growth

a) Strain Relaxed Layers

Before examining the influence of stress on SPEC it is instructive to examine the behaviour of unstrained (or strain-relaxed) $Ge_x Si_{1-x}$ alloys. Figure 1 shows Arrehnius plots of the SPEC velocity in strain-free Si, $Ge_{0.21} Si_{0.79}$, $Ge_{0.53} Si_{0.47}$ and Ge layers. The SPEC velocity, measured over the depth interval ~100 - 200 nm, is observed to increase with increasing Ge concentration, as expected, however, the activation energy does not vary monotonically between Si and Ge but appears to increase above that of Si for Ge concentrations ≤0.3. This is highlighted in Figure 2 which shows the activation energy for SPEC as a function of Ge concentration and in Table I which summarises v_0 and E_a for the different materials, where the SPEC velocity, v, is assumed to have the form $v = v_0 exp(E_a/kT)$. Similar effects have also recently been observed by others [23] in thin layers relaxed by high temperature annealing.

To confirm the strain-relaxed state of the alloy layers, the $Ge_{0.21} Si_{0.79}$ sample was thermally annealed at 1100°C for 60 seconds. (NB: Double crystal x-ray diffraction analysis confirmed that annealing under these conditions caused relaxation of a 120nm thick $Ge_{0.17} Si_{0.83}$ layer.). The annealed sample was then amorphised and SPEC measurements performed. The results are shown as the closed symbols in Figure 1. The pre-exponential factor and the activation energy were unchanged by this annealing procedure suggesting that little additional relaxation has occurred and hence that the layer was fully relaxed.

Since Si-Ge is close to an ideal solid solution, with an enthalpy of mixing of ~0.067 eV/atom or 0.034 eV/bond [24], the activation energy for solid phase epitaxy might reasonably be expected to scale between that of Si and Ge. The observed increase in activation energy, above that of Si, for x < 0.3 is particularly difficult to explain on the basis of existing bond-breaking models of SPEC. Particular since the Si-Ge bond strength is smaller than that of Si-Si.

b) Thin, Uniform Alloy Layers

Figure 3a shows a TRR spectrum of an x=0.085 alloy layer annealed at 588°C. The arrow indicates the time at which crystallisation reached the Si/GeSi interface. (Times less than this value correspond to SPEC in intrinsic Si whilst times greater that this value correspond to SPEC in the GeSi alloy.) Comparison of the SPEC rate before and after the interface shows a slight increase (~10%) in rate for the alloy layer. The TRR contrast is consistent with an abrupt

377

crystalline/amorphous interface and DCXRD analysis confirms that the alloy layer is fully strained after SPEC. The MBE-grown sample was also stable during thermal annealing at 1100°C for 60s. Figure 3b shows a TRR spectrum of an x=0.17 alloy annealed at 588°C. In this case the crystallisation rate decreases to 50% of the Si rate during the first ~60nm of the alloy layer before it increases to a rate 20% higher than that of Si during the final ~60nm of the layer. The poor TRR contrast in this case suggests that the crystalline/amorphous interface has roughened or faceted during crystallisation. DCXRD analysis after complete epitaxy showed that strain-relaxation had occurred. Note that the MBE-grown sample was also unstable during thermal annealing at 1100°C for 60s [25].

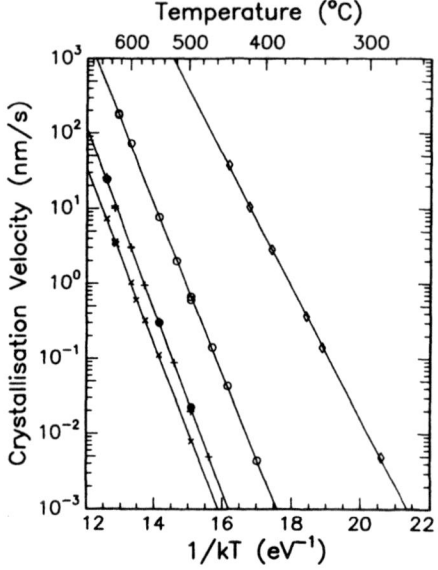 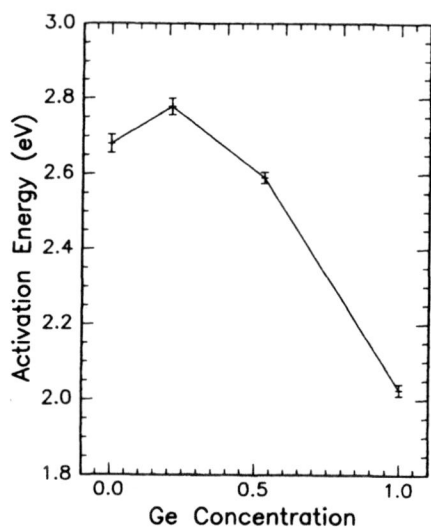

Figure 1: SPEC velocity as a function of reciprocal temperature for Si (x), Ge (◊) and fully relaxed alloy layers of composition x=0.21 (+), and x=0.53 (O). Data is also shown for x=0.21 samples preannealed at 1100°C for 60 s (•) prior to amorphisation.

Figure 2: Activation energy for solid phase epitaxy as a function of Ge concentration. The error bars represent the standard deviation of the statistical fit to the data.

Table I: Pre exponential factor, v_0, and activation energy, E_a, for SPEC of Si, Ge and strain-relaxed alloys of composition $Ge_{0.21}Si_{0.79}$ and $Ge_{0.53}Si_{0.47}$.

Composition	v_0 (cm/s)	E_a (eV)
Si	$3.1(\pm1.0)\times10^8$	2.68 ± 0.02
$Ge_{0.21}Si_{0.79}$	$3.5(\pm1.1)\times10^9$	2.78 ± 0.02
$Ge_{0.53}Si_{0.47}$	$6.4(\pm1.4)\times10^9$	2.59 ± 0.02
Ge	$6.1(\pm1.6)\times10^8$	2.02 ± 0.02

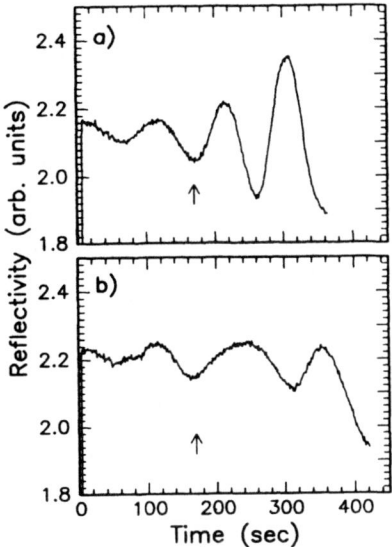

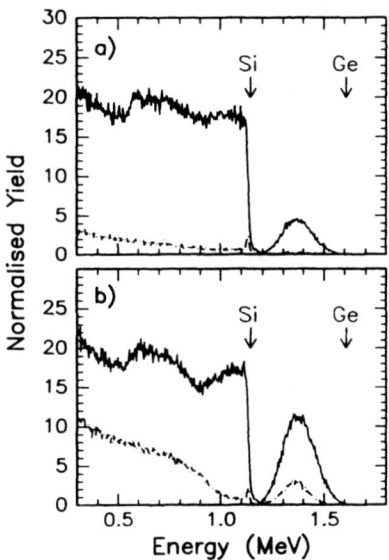

Figure 3: TRR (λ=632.8nm) spectra during SPEC of a) intrinsic Si, b) thin alloy layer with x=0.085, and c) thin alloy layer with x=0.17. The arrows indicate the position of the Si substrate/GeSi alloy interface.

Figure 4: RBS-C spectra (θ_s=165°) of samples as-implanted with 800 keV Ge ions to fluences of a) 8.4x10^{16} Ge.cm^{-2}, and b) 2.3x10^{17} Ge.cm^{-2} (solid) and after complete crystallisation at 600°C (dashed).

The velocity reduction observed during SPEC of the x=0.17 sample is consistent with stress-related retardation of SPEC, whilst the subsequent increase in velocity is believed to be a consequence of strain-relaxation. Several other studies [7-11] have demonstrated velocity reductions during SPEC of strained layers and reported increases in the activation energy for SPEC in such cases. However, Paine et al [7,8] have shown that such retarded epitaxy is generally associated with {111} faceting of the crystalline/amorphous interface. Hence the mode of crystallisation changes from (100) planar growth to {111} faceted growth in the region where the velocity reduction is observed. This raises the question as to whether or not the reported increases in activation energy can be attributed to fundamental thermodynamical effects, as reported in Si and Ge [4,5], or whether it is a consequence of the change in the mode of crystallisation. Comparison of the various reports suggests that the latter interpretation must be seriously considered.

c) Ion-Implanted Samples

It has previously been shown [13-18] that fully strained GeSi alloy layers can be produced by ion-implantation and SPEC provided the Ge fluence is kept below a critical fluence. This critical fluence can be calculated for a given implant energy from a model proposed by Paine et al [13]. The model is based on the minimum strain energy considerations but allows for the fact that there is no well defined alloy/substrate interface for structures synthesised by ion-implantation. For implant fluences above the critical fluence the model can also be used to calculate the depth at which relaxation should occur. For the 800 keV Ge implants used in the present study the model predicts a critical fluence of 1.1×10^{17} Ge.cm^{-2}, or a peak Ge concentration of 6.6%. For Ge fluences below this value SPEC should result in fully strained layers whilst for fluences above this value it should lead to partial relaxation.

Figure 4 compares RBS-C spectra from samples implanted with fluences below (8.4×10^{16} Ge.cm^{-2}) and above (2.3×10^{17} Ge.cm^{-2}) the critical fluence after thermal annealing at 600°C. Clear differences between the samples are evident. In particular, the ratio of the channeled to random yields for Ge and Si (at the Ge projected range) are 4.9% for the low fluence sample and 27% for the high fluence sample. In addition, the dechanneling rate in the high fluence case suggests the presence of extended defects extending to a depth of ≥470 nm, consistent with the presence of strain relieving defects. TEM (not shown) of sample cross-sections confirm this interpretation, showing defects extending to a depth of 555 nm, consistent with a depth of 561 nm predicted by the Paine model [13].

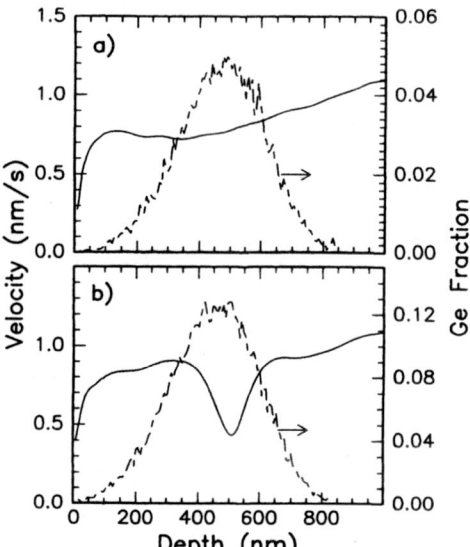

Figure 5: Crystallisation velocity and Ge concentration as a function of depth for sample implanted with a) 8.4×10^{16} Ge.cm^{-2} and b) 2.3×10^{17} Ge.cm^{-2} and annealed at 600°C.

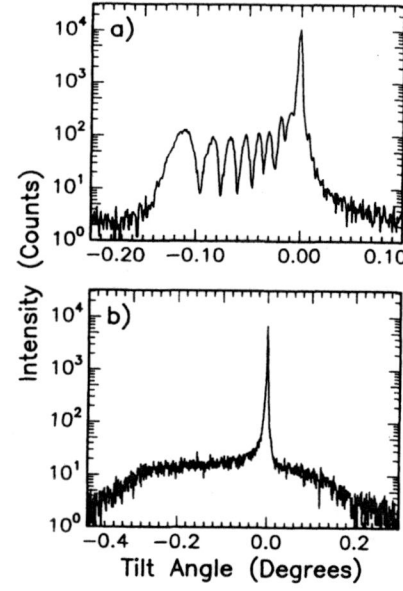

Figure 6: Double crystal x-ray diffraction spectra of sample implanted with a) 8.4×10^{16} Ge.cm^{-2} and b) 2.3×10^{17} Ge.cm^{-2} after complete annealing at 600°C.

Figure 5 shows the SPEC velocity as a function of depth, and the implanted Ge distribution for the samples described above. For the low fluence sample the SPEC rate is seen to decrease monotonically as epitaxy proceeds towards the sample surface. (Similar behaviour is observed for SPEC of thick amorphous Si layers due to H segregation at the crystalline/amorphous interface [3]). For the high fluence sample the SPEC rate initially decreases in a manner similar to that for the low fluence sample, but then exhibits a sudden decrease, beginning at a depth of 600 nm and reaching a minimum at a depth of 520 nm, before increasing to a near-constant value again, at a depth of 375 nm. The TRR spectrum from which this velocity distribution was determined also exhibits a reduction in peak-to-valley contrast in the region of the velocity reduction, with the contrast increasing again as the velocity recovers. This suggests that the crystalline/amorphous interface is rougher in the region of the velocity reduction, consistent with {111} faceting during strain relief [7,8].

a) 1.2×10^{17} Ge.cm^{-2}

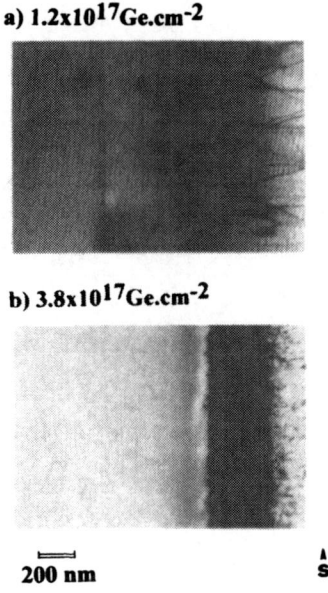

b) 3.8×10^{17} Ge.cm^{-2}

200 nm

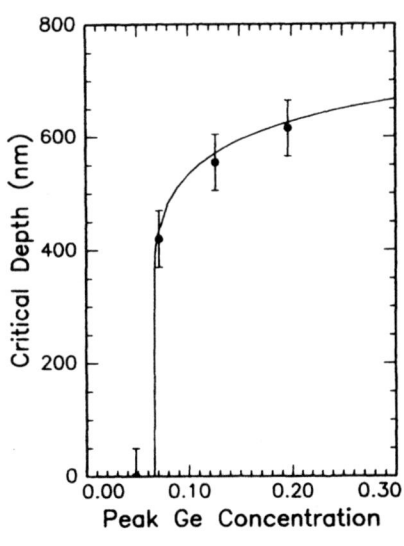

Figure 7: Bright field TEM images of a sample implanted with 800 keV Ge to a fluence of 3.8×10^{17} Ge.cm^{-2} and annealed at 600°C. Cross-sectional view a), and plan view b).

Figure 8: Critical thickness calculations for 800 keV Ge ions implanted into Si. Calculations are for stacking faults and 90° partial dislocations using the theory of Paine et al [8]. Experimental points extracted from cross-sectional TEM are also shown (•).

The double crystal x-ray diffraction spectra in Figure 6 show that the low fluence sample is fully strained after SPEC whereas the high fluence sample has relaxed. This is consistent with the RBS-C and TRR data presented above and with the critical fluence predicted by the equilibrium theory of Paine et al [8].

TEM analysis of samples after SPEC shows that the velocity reduction observed by TRR analysis is indeed associated with strain relaxation. In all cases examined, strain relieving defects were found to extend from the surface to the depth where the velocity reduction was observed. Typical TEM images are shown in Figure 7 for samples implanted to fluences of 1.2×10^{17} Ge.cm^{-2} and 3.8×10^{17} Ge.cm^{-2}. The depth to which the defects extend is in excellent agreement with the predictions of equilibrium theory, as shown in Figure 8 and Table II which summarise results for a range of implant fluences.

In summary, fully strained GeSi alloy layers can be fabricated by ion-implantation and subsequent SPEC for Ge fluences below a critical value. For fluences higher than the critical value, strain relaxation occurs during crystallisation and is correlated with a sudden reduction in the crystallisation velocity caused by roughening or faceting of the crystalline/amorphous interface.

Table II: Summary of experimental results and theoretical predictions as a function of Ge fluence. R_p and ΔR_p are the projected range and standard deviation of the Ge, respectively; y_c(TEM) is the depth to which dislocations extend, from TEM micrographs; y_c(Theory) [8] is the depth at which strain relief is predicted to occur; v_{onset} is the depth at which the velocity begins to decrease; and v_{min} is the depth at which the velocity reaches a local minimum.

Fluence (cm^{-2})	Peak Conc. (at. %)	Rp (nm)	ΔRp (nm)	y_c (TEM) (nm)	y_c (Theory) (nm)	v_{onset} (nm)	v_{min} (nm)
8.4x10^{16}	4.8	475	141	n/a	n/a	n/a	n/a
1.2x10^{17}	7.1	473	141	420	417	400	310
2.3x10^{17}	12.6	466	141	555	561	618	508
3.8x10^{17}	19.6	458	144	616	627	663	569

2) Ion-Beam Induced Epitaxial Crystallisation.

The crystallisation environment during IBIEC is very different to that experienced during thermal annealing; the temperature is considerably lower, ~300°C instead of ~600°C, and epitaxy proceeds in a material laden with point defects. The sensitivity of the crystallisation rate to stress and the resulting defect structures might reasonably be expected to reflect these differences.

a) Thin, Uniform Alloy Layers

Figures 9 shows TRR spectra ($\lambda = 632.8$nm) from alloy layers with x = 0.085 and 0.17 during IBIEC at 250°C. Surprisingly, these spectra show the same basic behaviour as samples annealed thermally, (see figure 3). For the x = 0.085 sample, IBIEC proceeds through the Si/GeSi interface without interruption, increasing in velocity (~14%) in the alloy. This increase is consistent with the expected increase due to the higher Ge concentration and with the additional ~4% increase in nuclear stopping power for 1.0 MeV Si ions in the alloy. For the x = 0.17 sample, IBIEC is initially retarded in the alloy layer but increases again after ~60 nm of growth. This is very similar to the thermal behaviour reported in Figure 3b, however, the TRR contrast is greater for the IBIEC sample than for the thermally annealed sample, suggesting that the crystalline amorphous interface is smoother during IBIEC. Since the measured roughness is believed to be associated with {111} faceting of the crystalline/amorphous interface this can be explained in terms of the orientation dependence of crystallisation velocity. SPEC is 25 times faster on {100} oriented material than on {111} material [1], whereas IBIEC is 2 times faster on {100} materials than on {111} material [26]. Roughening will therefore occur more quickly during SPEC.

The effects of stress during higher temperature IBIEC are less apparent than for the samples annealed at 250°C. As an example, Figure 10 shows TRR spectra for samples ion-beam-annealed at 350°C. The spectra show near-constant IBIEC velocities for x=0.085 and x=0.17 samples, with little evidence for stress related retardation. Two possible mechanisms are proposed to account for this temperature dependence: a) the nature and concentration of the irradiation induced defects is sufficiently different at the two temperatures to fully relax the strain during IBIEC at 350°C and not at 250°C, or b) the temperature dependence of the roughening/faceting transformation results in faceted growth during IBIEC at 250°C and not at 350°C. This temperature dependent behaviour is the subject of an on-going study.

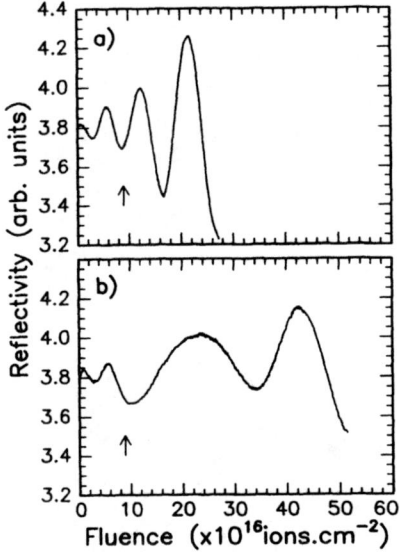

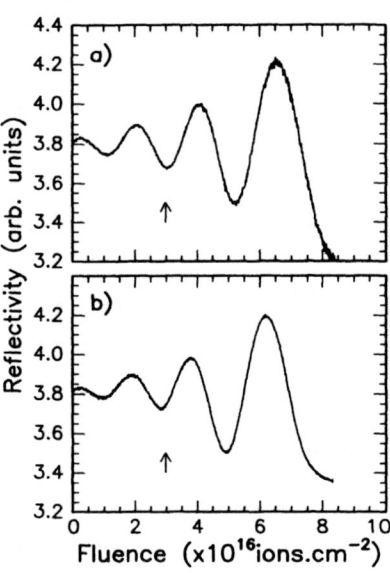

Figure 9: TRR spectra (λ = 632.8nm) of uniform alloy layers with a) x = 0.085, and b) x = 0.17, during IBIEC induced by 1.0 MeV Si ions at 250°C

Figure 10: TRR spectra (λ= 632.8nm) of uniform alloy layers with a) x = 0.085, and b) x = 0.17, during IBIEC induced by 1.0 MeV Si ions at 350°C

b) Ion-Implanted Samples

    IBIEC of ion-implanted structures has been explored for comparison with SPEC studies and to ascertain whether or not the equilibrium theory of Paine et al [8] is applicable in this case. IBIEC of samples implanted with 800 keV Ge ions to fluences of 1.2×10^{17}Ge.cm^{-2} (Peak concentration of 7.1%) were explored, however, in this case, IBIEC at 300°C using 2.0 MeV Si ions resulted in only partial crystallisation, with in-situ TRR analysis showing that epitaxy ceased after a Si fluence of 2×10^{17}Si.cm^{-2}. TEM analysis showed that beam induced random crystallisation competes with IBIEC for these thick layers. Note also that increasing the Ge concentration increased the rate of random crystallisation.
    That ion-implanted structures can be epitaxially crystallised by ion-irradiation is illustrated in Figure 11 which shows RBS-C spectra of samples implanted with 90 keV Ge to a fluence of 1×10^{17}Ge.cm^{-2} (Peak concentration ~30%) after thermal annealing at 600°C and IBIEC at 300°C. The spectra show that crystallisation proceeds to completion in both cases, consistent

with in-situ TRR analysis (not shown). A higher direct scattering peak and a higher rate of dechanneling are evident for the thermally annealed sample, consistent with the presence of a higher concentration of strain-relieving defects. TEM analysis, Figure 12, shows the presence of defects in both samples, although the concentration of defects appears greater, and the size distribution smaller, for the thermally annealed sample. TEM images using (400) two-beam diffraction conditions shows Moire fringes for the thermally annealed samples and no such fringes for the ion-beam annealed samples, implying more extensive relaxation in the thermal case. Unfortunately, the layers produced by 90 keV Ge implants are too thin to accurately compare the depth of relaxation with theoretical predictions.

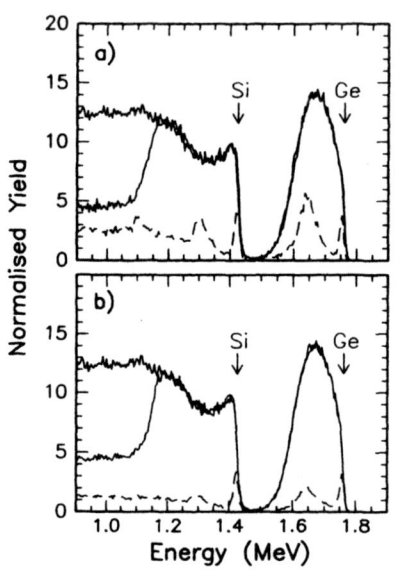

a) Thermally annealed at 600°C.

b) Ion-Beam annealed at 300°C.

g=(220)   200 nm

Figure 11: RBS-C spectra of sample implanted with 90 keV Ge to a fluence of 1×10^{17} Ge.cm^{-2} after: a) SPEC at 600°C and b) IBIEC with 1.0 MeV Si at 300°C.

Figure 12: Bright field TEM images of samples implanted with 90 keV Ge to a fluence of 1×10^{17} Ge.cm^{-2} after: a) SPEC at 600°C and b) IBIEC with 1.0 MeV Si at 300°C.

CONCLUSIONS

A survey of recent and on-going experimental studies has been presented. Specific conclusions at this stage include: 1) The activation energy for solid phase epitaxial crystallisation of strain-free GeSi alloy layers exceeds that calculated from a linear interpolation between the activation energies of Si and Ge and increases above that of Si for alloy compositions of $x<0.3$; 2) MBE-grown alloy layers which are thermodynamically stable at 1100°C for 60 s can be recrystallised in a fully strained configuration by SPEC, whereas layers which relax during annealing at this temperature also relax during SPEC; 3) Fully strained layers can be fabricated by ion-implantation and thermal annealing provided the Ge fluence is below a certain critical fluence, where the critical fluence is accurately predicted by equilibrium theory [8]; 4) Strain relaxation during SPEC is correlated with a sudden reduction in crystallisation velocity, and it is speculated that this velocity reduction is caused by stress-induced roughening or faceting of the crystalline/amorphous interface; 5) IBIEC of GeSi alloys is possible and exhibits very similar behaviour to thermally annealed samples under certain conditions; and 6) Ion-beam induced random crystallisation competes with IBIEC for thick layers and for high Ge concentrations.

ACKNOWLEDGMENTS

The authors should like to thank Mr. John Glasko for assistance with the ion-beam-annealing experiments and the Electron Microscope Unit at ANU for photographic services. The authors also gratefully acknowledge Dr. Arne Nylandsted Larsen at Aarhus University for providing the MBE grown samples.

REFERENCES

1. J.S.Williams, in *Surface Modification and Alloying by Laser, Ion, and Electron Beams*, (Plenum Press, NY, 1983) Chapter 5 (and references therein).
2. G.L.Olson and J.A.Roth. Mat. Res. Rep. **3**, 1 (1988).
3. J.A.Roth, G.L.Olson, D.C.Jacobson and J.M.Poate. Appl. Phys. lett., **57**, 1340 (1990).
4. M.J.Aziz, P.C.Sabin and G-Q,Lu. Phys. Rev. **B44**, 9812 (1991).
5. G-Q,Lu, E.Nygren and M.J.Aziz. J. Appl. Phys. **70**, 5323 (1991).
6. B.T.Chilton, B.J.Robinson, D.A.Thompson, T.E.Jackman and J.-M.Baribeau. Appl. Phys. Lett. **54**, 42 (1989).
7. D.C.Paine, N.D.Evans and N.G.Stoffel. J. Appl. Phys. **70**, 4278 (1991).
8. D.C.Paine, D.J.Howard, N.G.Stoffel and J.A.Horton. J. Mater. Res. **5**, 1023 (1990).
9. Q.Z.Hong, J.G.Zhu, J.W.Mayer, W.Xia and S.S.Lau. J. Appl. Phys. **71**, 1768 (1992).
10. C.Lee, T.E.Haynes and K.S.Jones. Appl. Phys. Lett. **62**, 501 (1993).
11. D.C.Paine, JOM, February, p55 (1993)
12. G.Mezey, S.M.Matteson and J.Gyulai. Nucl. Instr. Meth. **182/183**, 587 (1981)
13. D.C.Paine, D.J.Howard and N.G.Stoffel. J. Electronic Materials **20**, 735 (1991).
14. K.M.Yu, I.G.Brown and S.Im. Mat. Res. Soc. Symp. Proc. **235**, 293 (1992)
15. F.Corni, S.Fabboni, G.Ottaviani, G.Queirolo, D.Bisero, C.Bresolin, R.Fabbri and M.Seridori, J. Appl. Phys. **71**, 2644 (1992).
16. R.G.Elliman and W.C.Wong. Nucl. Instr. Meth. **B80/81**, 768 (1993).
17. R.G.Elliman and W.C.Wong, Presented at the International Conference on Defects in Semiconductors, Austria (1993), Materials Science Forum (In Press).
18. R.G.Elliman and W.C.Wong, Proceedings of the International Conference on Ion Beam Analysis, Hungary, (1993). Nucl. Instr. Meth. (In Press).
19. R.G.Elliman, M.C.Ridgway, J.S.Williams and J.C.Bean. Appl. Phys. Lett. **55**, 843 (1989).
20. A.Y.Yu, J.W.Mayer, D.J.Eaglesham and J.M.Poate. Appl. Phys. Lett., **54**, 2342 (1989).
21. A.J.Yu, PhD Thesis, Cornell University (1989).
22. R.G.Elliman, M.C.Ridgway and J.S.Williams. Mat. Res. Soc. Symp. Proc. **157**, 105 (1990).
23. E.Nygren, and T.E.Haynes et al. (private communication).
24. *Binary Alloy Phase Diagrams*, edited by T.B.Massalski, J.L.Murray, L.H.Bennett and H.Baker, American Physical Society for Metals, (1986), p1249.
25. P.Kringhoj, R.G.Elliman and J.L.Hansen. Mat. Res. Soc. Symp. Proc. (This symposium).
26. D.M.Maher, R.G.Elliman, J.Linnros, J.S.Williams, R.V.Knoell and W.L.Brown. Mat. Res. Soc. Symp. Proc., **93**, 87 (1987).

This article also appears in Mat. Res. Soc. Symp. Proc. Vol 316

ION BEAM INDUCED EPITAXIAL CRYSTALLIZATION OF SINGLE CRYSTALLINE 6H-SiC

V. HEERA*, R. KÖGLER*, W. SKORUPA* AND E. GLASER**
* Research Center Rossendorf Inc., P.O.B. 510119, D-01314 Dresden, Germany
** FSU Jena, Institute of Solid State Physics, Max-Wien-Platz 1, D-07743 Jena, Germany

ABSTRACT

For the first time, ion beam induced epitaxial crystallization (IBIEC) has been found in SiC. The effect of 300 keV Si^+ irradiation through an amorphous surface layer in single crystalline 6H-SiC at $477 \pm 5^\circ C$ has been investigated by RBS/C and XTEM. A shrinkage of the amorphous layer was found after ion irradiation at this temperature which is caused by both an ion dose independent thermal regrowth of about 20 nm and an additional ion beam induced epitaxial crystallization with a rate of about 1.5 nm/ 10^{16} cm^{-2}.

INTRODUCTION

Silicon carbide (SiC) exists in many polytypes [1] and is a promising wide band gap semiconductor for high temperature, high power and high speed electronic application [2,3]. However, single crystalline bulk material with high quality sufficient for industrial purposes has been available only recently as 4H- and 6H-polytype 1" wafers [4].

Compared to silicon, relatively little is known about the formation and annealing of implantation damage in SiC. The critical energy density for amorphization by ion implantation at room temperature was found to be about $2 \cdot 10^{21}$ keV/cm^3 [5]. Post implantation annealing at $300^\circ C$ led to an appreciable recovery of the irradiation damage only if the amount of deposited damage energy was less than the critical value. It has been shown by several authors [6,7,8,9] that the complete regrowth of implantation produced amorphous SiC layers by solid phase epitaxy (SPE) needs temperatures higher than $1400^\circ C$. Moreover, a perfect SPE regrowth, without visible remaining damage, occurred only at $1800^\circ C$ [7]. However, at such temperatures SiC begins to sublimate and dopant atoms are redistributed.

It is known that ion irradiation may be used for low temperature recrystallization of amorphous surface layers in some semiconducters. The ion beam induced epitaxial crystallization (IBIEC) has been extensively investigated in silicon material [10]. It was shown that IBIEC can take place in Si already at a temperature of $150^\circ C$, whereas a temperature of $550^\circ C$ is necessary for thermally induced SPE in this material. The IBIEC effect has been found to occur also in the compounds $Ge_{1-x}Si_x$ [11], GaAs [12], InP, $NiSi_2$, $CoSi_2$ [13], Al_2O_3 [14] and BP [15] at temperatures much lower than necessary for thermal SPE. The recent results on IBIEC of BP at $400^\circ C$ are of particular interest, because, similar to SiC, BP is a wide band gap semiconductor which usually needs processing temperatures far beyond $1000^\circ C$.

Encouraged by the sucess in ion beam induced recrystallization of several semiconductors, we have studied the effect of 300 keV Si^+ ion irradiation through an amorphous surface layer in single crystalline 6H-SiC at a temperature of $477^\circ C$ by Rutherford backscattering/ channeling (RBS/C) and cross-section transmission electron microscopy (XTEM).

EXPERIMENTAL

Previously, it was difficult to obtain reliable and comparable results about SiC because of the lack of single crystals with defined quality. Therefore, we used in our investigations 1" wafers of single crystalline 6H-SiC ((0001) orientation, one side polished, n-type, Si surface, off-axis) from Cree Research Inc. [4].

Ge$^+$ ions were chosen for the amorphization of 6H-SiC . On the one hand, their mass is sufficiently high to produce completely amorphous layers and, on the other hand, Ge is a group IV element which should not dope SiC. The critical amorphization dose for 200 keV Ge$^+$ implantation into 6H-SiC at room temperature was experimentally determined to be $3 \cdot 10^{14}$cm^{-2}. Therefore, amorphous surface layers were produced by 200 keV Ge$^+$ implantation with fluences of $5 \cdot 10^{14}$cm^{-2} and $1 \cdot 10^{15}$cm^{-2}, respectively. The amorphicity of the layers was tested by RBS/C and XTEM.

The wafer was then cut into pieces of about 1 cm^2 and half-capped by Si wafer pieces to preserve unirradiated reference regions for the RBS/C and XTEM analysis after the IBIEC irradiation. After a surface cleaning procedure the SiC specimens were mounted on a heating stage in the implantation chamber and pre-heated to a temperature of about 400°C. Subsequently the specimens were subjected to 300 keV Si$^+$ irradiation in the dose range between $1 \cdot 10^{16}$cm^{-2} and $3 \cdot 10^{17}$ cm^{-2}. The ion flux was about $3 \cdot 10^{12}$cm^{-2} s^{-1} and led to an additional temperature increase. During the ion irradiation the total temperature was adjusted to 477$\pm$5°C.

The amorphous surface layers were analyzed with 1.2 MeV He$^+$ RBS/C. In the case of the low amorphization dose additional XTEM studies were performed. In order to obtain information on the density changes of the irradiated SiC step height measurements with a surface profilometer (DEKTAK 8000) were carried out.

RESULTS AND DISCUSSION

Fig. 1 shows an overview of the RBS/C spectra taken from the low amorphization dose series which involves the aligned and random spectrum of the as-amorphized state as well as from the covered (thermal) and uncovered part (ibiec) of the specimen which was irradiated with the highest Si$^+$ dose. It can be seen from the random and aligned spectra of the as-amorphized specimen that an amorphous surface layer of about 140 nm thickness was produced by the 200 keV Ge$^+$ implantation into SiC. In the case of the higher amorphization dose the same amorphous layer thickness was found. However, the edge in the RBS/C spectrum that characterizes the a/c interface becomes somewhat steeper.

In the high energy part of the RBS/C spectra in Fig. 1 a Gaussian-like distribution of Ge atoms with a peak at a depth of about 95 nm can be seen which corresponds very well with the result of a TRIM calculation. The maximum Ge concentration can be estimated to be smaller than 0.1%. However, there is a discrepancy between the amorphous layer thickness expected from the TRIM calculation and the critical energy density for amorphization [5], which should be about 100 nm, and the 140 nm found in the RBS/C analysis. Such discrepancies between TRIM calculations and RBS/C results in SiC have already been found by Edmond et al. [6]. For high dose implantation these deviations could originate from a density reduction during amorphization of SiC from 3.2 g/cm^3 to 2.7 g/cm^3, which was proved by step height measurements [9]. We have also carried out step height measurements and found a surface step of about 20 nm between crystalline and amorphous regions in SiC.

This corresponds just to the density change mentioned above when assuming a linear swelling perpendicular to the surface.

The 300 keV Si$^+$ irradiation at 477°C leads to an obvious shrinkage of the initial amorphous layer thickness. In the case of $3 \cdot 10^{17}$ Si$^+$/cm^2 nearly half of the initial amorphous layer is recrystallized, as can be seen from Fig. 1 (RBS spectrum denoted with "ibiec"). It can be excluded that the amorphous layer shrinkage is due to sputtering. According to our TRIM calculation the sputtering rate should be only about 1 nm/ 10^{17}cm^{-2}. As can be seen in Fig. 1, the dechanneling rate increases substantially in the region behind the amorphous layer. This is caused by the Si irradiation damage. The RBS/C spectrum touches the random level at a depth of about 320 nm, which is in good agreement with the calculated mean projected range of 300 keV Si$^+$ in SiC. The concentration of silicon excess atoms in SiC estimated from the TRIM results for $3 \cdot 10^{17}$ Si$^+$/cm^2 is less than 2% in the region where IBIEC occurs and about 20% in the maximum of the range distribution. Indeed, this is a large deviation from the SiC stoichiometry in the layer around the mean projected range of 300 keV Si$^+$ and, therefore, should lead to a distinct density change. Actually, we found a step height of about 100 nm between the as-amorphized region and the region implanted with $3 \cdot 10^{17}$ Si$^+$/cm^2. The reason for the strong swelling during ion implantation into SiC seems to be the extremely high atomic density of $9.66 \cdot 10^{22}$ atoms/cm^3, which has to be compared e.g. with the Si density of only $5.02 \cdot 10^{22}$ atoms/cm^3. There is no free space for additional atoms introduced in the very close-packed SiC lattice.

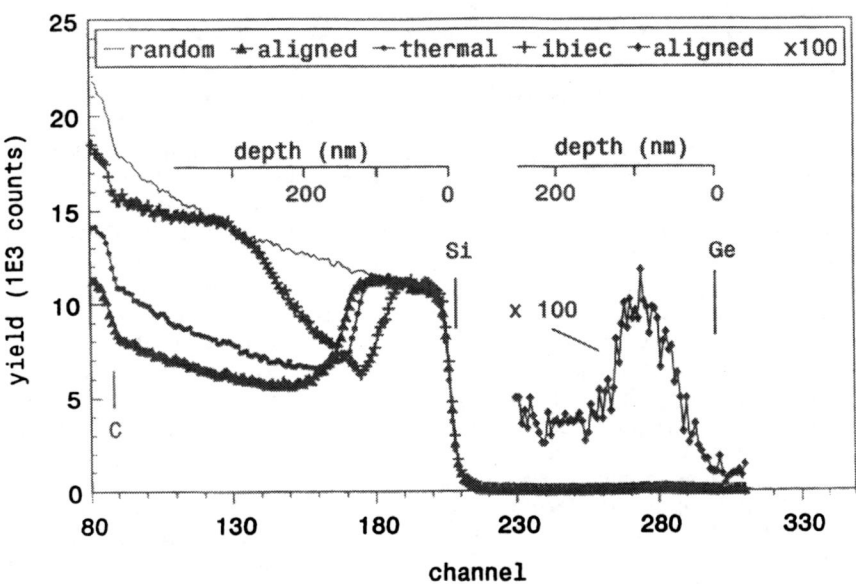

Fig. 1: Overview of the RBS/C spectra obtained in our experiments. The random and aligned spectrum of the the as-amorphized state are shown together with the spectra taken from the sample irradiated with $3 \cdot 10^{17}$ Si$^+$/cm^2 ("ibiec") and from the corresponding reference region ("thermal"). Note, that the Ge part of the spectrum is multiplied by 100.

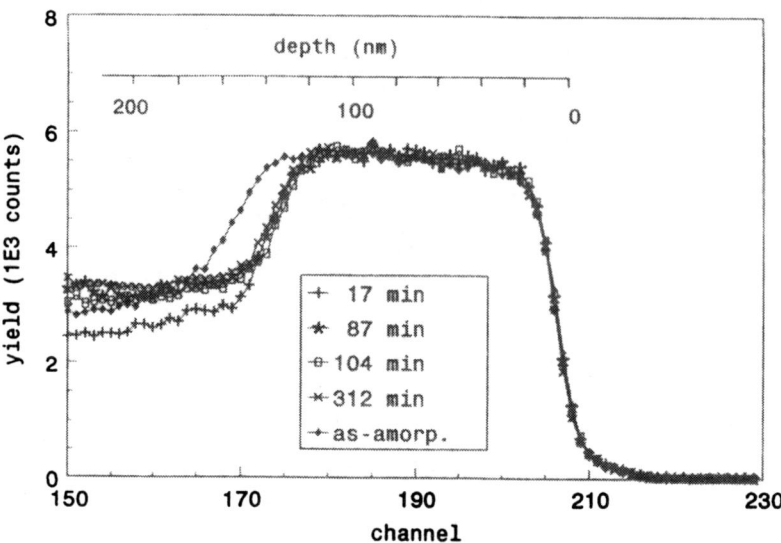

Fig. 2: RBS/C spectra measured in the unirradiated reference regions of the samples, which were irradiated with different Si$^+$ doses. The irradiation times are given in the inset.

Surprisingly, we observed a small recrystallized layer also in the region of the specimen that was covered to remain a reference sample without Si ion irradiation (see RBS/C spectrum denoted as "thermal" in Fig. 1). The only plausible explanation for this finding is that thermally stimulated SPE takes place already at the temperature of 477°C. In Fig. 2 the RBS/C results of the thermal recrystallization found in the reference regions of the samples irradiated with different ion doses are summarized. It can be seen that the thermally regrown layer thickness is about 20 nm and does not depend on the annealing time. Similar results for thermal annealing of amorphous layers in 3C-SiC were found by Bohn et al. [8] in the temperature range between 300° and 1400°C. The amorphous layer thickness was found to decrease during 10 minutes of annealing with a rate between 0.015 to 0.035 nm/°C. However, the regrowth process stops after a certain thickness and could not be continued below 1450°C even for much longer annealing times [9]. The limited low temperature SPE might be explained either by a not completely amorphized transition region at the a/c interface or by the formation of stable interface damage during the regrowth process, which cannot be removed at temperatures below 1450°C.

The RBS/C results of the IBIEC regrowth in 6H-SiC obtained by 300 keV Si$^+$ irradiation with doses of $1 \cdot 10^{16}$cm^{-2}, $5 \cdot 10^{16}$cm^{-2}, $1 \cdot 10^{17}$cm^{-2} and $3 \cdot 10^{17}$cm^{-2} are presented in Fig. 3. For comparison, the spectrum for the thermal regrowth is shown, too. There is no difference in the regrown thickness for the two lowest ion doses. However, the slope of the edge that characterizes the a/c interface decreases and the dechanneling rate behind the a/c interface increases. This may be attributed to a broadening of the interface region, which becomes more pronounced with increasing ion dose. For the sample irradiated with a dose of $5 \cdot 10^{16}$cm^{-2}, the XTEM investigations showed a rather rough a/c interface region of about 15 nm thickness. The electron scattering pattern revealed that the recrystallized material is

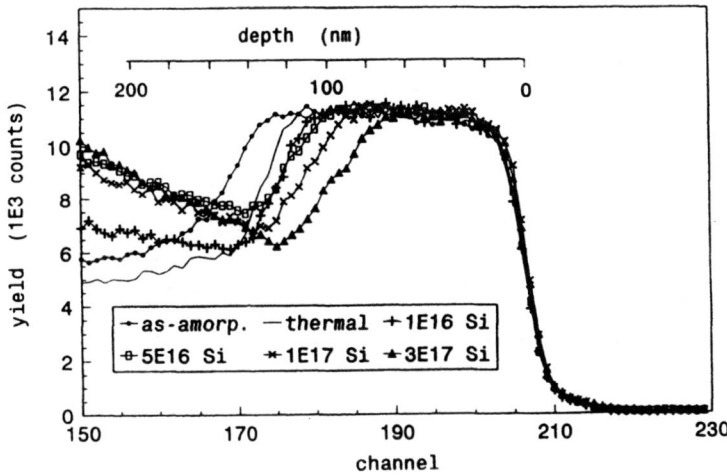

Fig. 3: RBS/C results for the IBIEC regrowth of the amorphous surface layer. The corresponding irradiation doses are given in the inset. For comparison, the thermal part of the regrowth is also shown in the figure

again of 6H polytype structure. The IBIEC process proceeds with increasing ion dose. A total recrystallized layer thickness (thermal and IBIEC part) of 57 nm was obtained for the dose of $3 \cdot 10^{17} cm^{-2}$. In Fig. 4 the IBIEC part of the regrown layer thicknesses (total layer thickness minus thermal regrown one) is given in dependence of the ion dose. There is an initial IBIEC rate of about 8 nm/ $10^{16} cm^{-2}$. Only a very small additional regrowth is obtained for $5 \cdot 10^{16} cm^{-2}$. For the ion doses higher than $1 \cdot 10^{16} cm^{-2}$ the mean IBIEC rate amounts to about 1.5 nm/ $10^{16} cm^{-2}$. This is, compared to IBIEC in Si, a very low recrystallization rate. It should be noted, however, that the temperature at which we have investigated the IBIEC process in SiC is about 1000° below the threshold temperature for complete thermal SPE. We believe, that the IBIEC process in SiC is stimulated by point defect migration and, therefore, at higher temperatures higher recrystallization rates can be achieved.

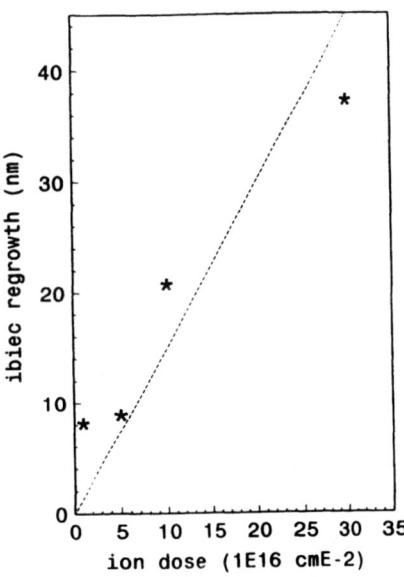

Fig. 4: Layer thickness regrown by IBIEC (without thermal part) in dependence of dose. The broken line corresponds to an IBIEC-rate of 1.5 nm/10^{16} cm^{-2}.

CONCLUSIONS

For the first time, we have shown that IBIEC occurs in 6H SiC at a temperature as low as 477°C. The recrystallization process of SiC involves three stages: (i) A thermal recrystallization, which is independent on the annealing time and ceases after a regrowth of a layer thickness of 20 nm, (ii) an initial IBIEC regrowth with a rate of 8 nm/ 10^{16}cm^{-2} and (iii) a subsequent IBIEC regrowth with a lowered rate of about 1.5 nm/ 10^{16}cm^{-2}, which seems to remain nearly constant for a wide dose range. The reasons for these distinct regrowth stages are not yet clear. However, it can be speculated that the structure of the a/c interface region is changed by the regrowth process. Indeed, the XTEM results show a growing roughness of the a/c interface with increasing irradiation dose.

The IBIEC rate found in 6H SiC is only about 1/100 of the the typical values found in Si. It should be emphasized, however, that we have found the IBIEC effect 1000° below the threshold temperature for complete SPE in SiC, whereas the typical temperatures for IBIEC in Si are only about 300° below the threshold temperature for SPE. Therefore, we believe, that complete IBIEC regrowth of amorphous layers in SiC could be obtained with lower irradiation doses when applying higher temperatures. However, further investigations are necessary to clarify this point.

ACKNOWLEDGEMENT Part of this work was supported by the Bundesministerium für Forschung und Technologie (Contract 03-SK3ROS-5)

REFERENCES

1. R. F. Davies and J. T. Glass, Advances in Solid State Chemistry **2**, 1 (1991).
2. G. Pensl and R. Helbig, in Festkörperprobleme/ Advances in Solid State Physics, Vol. **30**, ed. by U. Rössler (Vieweg, Braunschweig 1990), pp. 133-156.
3. J. A. Powell, P.G. Neudeck, L.G. Matus and J.B. Petit, Mat. Res. Soc. Symp. Proc. **242**, 495 (1992).
4. Cree Research, Inc., Durham, NC 27713, USA
5. J.A. Spitznagel, S. Wood, W. J. Choyke, N.J. Doyle, J. Bradshaw and S. G. Fishman, Nucl. Instr. Meth. **B16**, 237 (1986)
6. J. A. Edmond, S. P. Withrow, H. S. Kong and R. F. Davies, Mat. Res. Soc. Symp. Proc. **51**, 395 (1986).
7. J. Ryu, H. J. Kim and R. F. Davies, Mat. Res. Soc. Symp. Proc. **52**, 165 (1986).
8. H. G. Bohn, J. M. Williams, C. J. McHargue and G.M. Begun, J. Mater. Res. 2, **107** (1987).
9. C. J. McHargue and J. M. Williams, Nucl. Instr. Meth. **B80/81**, 889 (1993).
10. F. Priolo and E. Rimini, Mater. Sci. Rep. **5**, 319 (1990).
11. R.G. Elliman, M.C. Ridgway, J. S. Williams and J.C. Bean, Appl. Phys. Lett. **55**, 843 (1989).
12. S. T. Johnson, J.S. Williams, E. Nygren and R.G. Elliman, J. Appl. Phys. **64**, 6567 (1988).
13. J. S. Williams, M.C. Ridgway, R. G. Elliman, J.A. Davies, S.T. Johnson and G. R. Palmer, Nucl. Instr. Meth. **B55**, 602 (1991).
14. W. Zhou, D. K. Sood, R. G. Elliman and M. C. Ridgway, Nucl. Instr. Meth. **B80/81**, 1104 (1993).
15. N.Kobayashi, H.Koboyashi and Y. Kumashiro, Nucl. Instr. Meth. **B40/41**, 550 (1989).

This article also appears in Mat. Res. Soc. Symp. Proc. Vol 316

INDUCED CRYSTALLIZATION IN CW LASER-IRRADIATED SOL-GEL DEPOSITED TITANIA FILMS

GREGORY J. EXARHOS AND NANCY J. HESS
Materials Sciences Department, Pacific Northwest Laboratory, PO BOX 999,
MS K2-44, Richland, WA 99352

Abstract

Isothermal annealing of amorphous TiO_2 films deposited from acidic sol-gel precursor solutions results in film densification and concomitant increase in refractive index. Subsequent heating above 300°C leads to irreversible transformation to an anatase crystalline phase. Similar phenomena occur when such amorphous films are subjected to focused cw laser irradiation. Controlled variations in laser fluence are used to densify or crystallize selected regions of the film. Low fluence conditioning leads to the evolution of a subtle nanograin-size morphology, evident in AFM images, which appears to retard subsequent film crystallization when such regions are subjected to higher laser fluence. Time-resolved Raman spectroscopy has been used to characterize irradiated regions in order to follow the crystallization kinetics, assess phase homogeneity, and evaluate accompanying changes in residual film stress.

Introduction

A rapid, inexpensive procedure for the deposition of metal oxide coatings is based on the sol-gel processing technique. In this method, a precursor reagent, usually the alkoxide of the target metal oxide, is dissolved in a suitable reagent where hydrolysis and condensation reactions are controlled in order to develop the oxide phase and accompanying microstructure. [1-4] Amorphous films are formed readily by means of spin-casting or dip-coating cleaned substrates. Such films usually exhibit refractive indices much lower than amorphous coatings deposited using the more energy intensive PVD methods and film adhesion to the substrate appears to be somewhat poorer. However, adhesion can be improved markedly through solution processing at relatively low pH. [5] Deposition of low index coatings using the sol-gel process offers the possibility for the production of spatially-resolved dense coating regions through subsequent thermal processing routes.

Isothermal heating of amorphous sol-gel titania films above 300°C results in an irreversible transformation to the crystalline anatase phase characterized by a spheroidal grain morphology with grain sizes less than 100 nm. [6] Film densification accompanies this transformation and intact films are produced provided that the overall thickness is kept to less than 500 nm. The growth rate of evolved crystallites is retarded owing to diminished transport of the amorphous matrix to the densified nucleation center and is well-represented by a modified Avrami equation with a critical exponent on the order of two. [6] The magnitude of residual stress in these films and how it varies during crystallization has been followed in real time by means of Raman scattering measurements. [7] This previous work demonstrated that simultaneous

Mat. Res. Soc. Symp. Proc. Vol. 321. ©1994 Materials Research Society

measurement of two Raman mode frequencies is sufficient to uniquely determine both temperature and residual stress as the amorphous film crystallizes. Since time-resolved Raman measurements can be carried out on very short time scales, this method is appropriate for following crystallization phenomena in amorphous films subjected to rapid step increases in temperature.

Focused laser radiation offers the capability for producing rapid heating of films as well as the possibility for imprinting surface regions of variable refractive index. In work reported here, irradiation of amorphous sol-gel deposited films with a visible cw laser results in localized densification and crystallization at the irradiation sites. A critical laser fluence must be exceeded before crystallization can occur. Time-resolved Raman measurements acquired during sample irradiation are used to characterize surface temperature, the residual stress state of the film and how it evolves in time. Ongoing work involves applications of this technique for writing micrometer size images into refractory films deposited on silica or silicon substrates.

Experimental

Sol-gel titania films were prepared from highly acidic ethanol solutions of the titanium ethoxide precursor. [7] Homogeneous films were formed on silica or silicon(100) cleaned substrates at room temperature by spin casting at rates between 1200 and 1700 rev·min^{-1} for 60 s. Film thicknesses ranged from 200 to 1200 nm as determined from uv-vis-ir transmission measurements[8] and ellipsometry (632.8 nm and 70° incidence angle). Refractive indices averaged about 1.65 immediately after casting, and were found to increase somewhat with time at room temperature or during gentle heating.

Isothermal heating of films to 120°C was achieved by means of a heat lamp located 25 cm above the coated substrates. Heating to higher temperatures involved the use of a muffle furnace or a resistively wound optical furnace for the *in situ* Raman measurements. [6] Surface morphology after heating was characterized by means of Transmission Electron Microscopy and Atomic Force Microscopy. [6,9]

Laser heating involved imaging cw all-line emission from an Ar-ion laser onto the sol-gel coated substrate by means of an aberration-corrected 10x microscope objective having a small numerical aperture. Beam diameters were estimated at 10 μm. Under these conditions, laser fluences approaching several megawatts per cm^2 could be achieved. Film irradiations were performed at a number of fluences ranging from 0.1 to 5 MW·cm^{-2} and over times ranging from a few seconds to minutes.

Spatially resolved Raman measurements of the laser irradiated regions were collected using an optical microscope interfaced to the entrance port of a Spex triple spectrometer. Spectra were excited using 100 mW of 514.5 nm argon ion laser radiation focused onto the sample by means of a 40x objective. Raman scattering was collected in a backscattering geometry. These conditions did not lead to densification or crystallization of the films during measurement. At significantly higher laser fluences through the microscope, changes to the film became evident.

Results

The refractive index of amorphous sol-gel deposited titania films increases from 1.52 to 1.76 over a period of 60 hours upon drying at room temperature. An increase in index to about 2.00 is observed after films have been heated to temperatures near 100°C. Extinction coefficients nearly double as the films densify and contract to about half of their original as-deposited thickness. These results are in agreement with those of Thomas, et al, who observed similar effects in sol-gel films upon mild heat treatment or exposure to uv light. [10] The nature of the substrate has little effect on these changes which are seen in films deposited on silica as well as Si(100) surfaces.

Isothermal heating to temperatures in excess of 300°C leads to irreversible crystallization of the amorphous film to an anatase crystal phase. The kinetics of the transformation are controlled by the length of solution equilibration time prior to casting, drying time and temperature, and the substrate temperature during deposition. [6,11] The spheroidal microstructure which evolves (Fig. 1) is characterized by uniform close packed crystal grains having diameters less than 100nm. In general, the rate constant for the fraction of transformed material, F, is described by a modified Avrami equation (eqn. 1) where the temperature dependent rate constant for crystallization is k and the critical exponent, c, describes the crystallite habitat and includes mass transport effects. [6] Critical exponents are on the order of 2. The activation energy for crystallization is 34 kcal/mole.

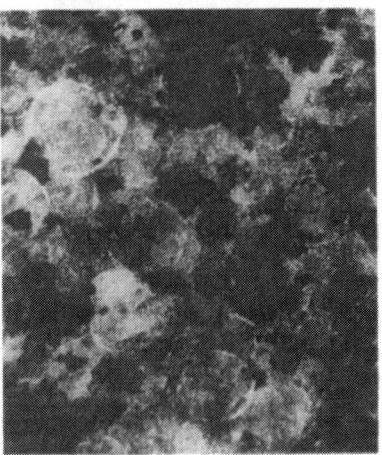

Fig. 1 TEM image of an 80nm thick anatase film removed from the silica substrate. Spheroidal grains average 50nm in diameter.

$$F = 1 - \exp(-k \cdot t^c) \qquad (1)$$

The two types of phenomena, conditioning and crystallization, illustrated above also are seen when amorphous as-deposited titania films are subjected to cw laser irradiation. For fluences below 1 MW·cm^{-2}, films were observed to densify accompanied by increases in both refractive index and extinction coefficient. At fluences exceeding this threshold, irreversible transformation to the anatase crystalline phase was observed. An increase in index to 2.40 was determined from ellipsometry measurements. Points A and B in Fig. 2 refer to regions of the film which have been conditioned or which have been irreversibly transformed to the anatase phase. Irradiation at fluences above 3MW·cm^{-2} led to catastrophic damage to the film which included material ablation, and recrystallization of the metastable amorphous and anatase phases to the rutile crystalline phase. The presence of multiple phases is readily discerned from Raman microprobe measurements of the damaged area. [12]

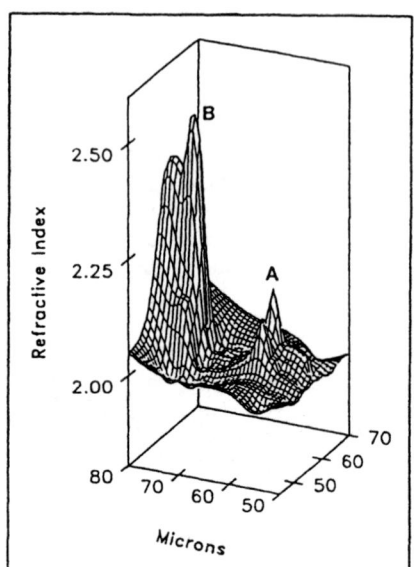
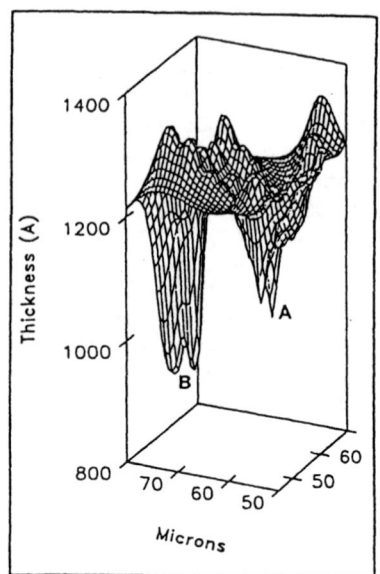

Fig. 2 Spatially resolved ellipsometric measurements of cw laser irradiated titania sol-gel coatings. (A) Conditioned region. (B) Transformed region.

Under irradiation conditions where the film does not suffer catastrophic damage, time-resolved Raman Spectroscopy has been used to follow induced crystallization phenomena. Fig. 3 illustrates the kind of information such measurements provide.

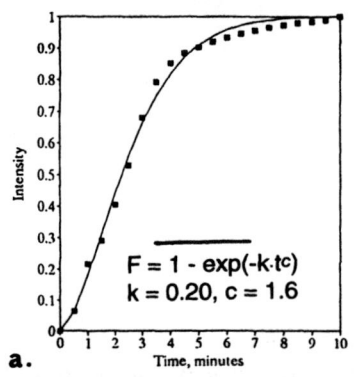

$$F = 1 - exp(-k \cdot t^c)$$
$$k = 0.20, c = 1.6$$

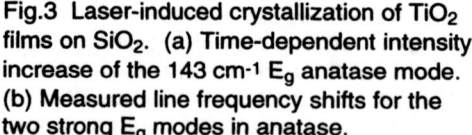

a.

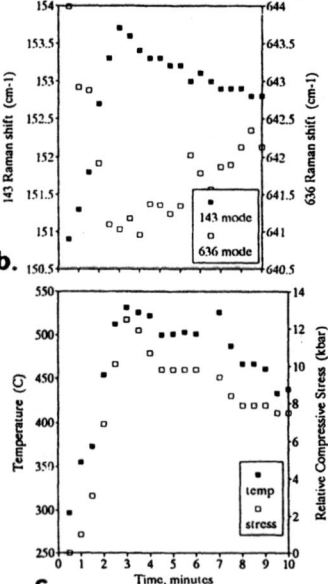

Fig.3 Laser-induced crystallization of TiO_2 films on SiO_2. (a) Time-dependent intensity increase of the 143 cm^{-1} E_g anatase mode. (b) Measured line frequency shifts for the two strong E_g modes in anatase. (c) Associated film temperature and residual film stress after reference 7.

At a fluence exceeding 1 MW·cm^{-2}, but less than 2 MW·cm^{-2}, the irradiated region completely transforms to the anatase phase in about ten minutes and the crystallite ingrowth rates are described by eqn. 1 with a critical exponent of 2. Frequencies for the two strongest anatase modes associated with each intensity measurement vary similarly in time and are used to determine film temperature and residual stress during crystallization by means of the contour analysis method developed previously. [7]

The film temperature in the irradiated region increased to over 500°C and then gradually decreased to a steady state value of about 425°C. Relaxation of residual film compressive stress by about 4 kbar also was seen. Similar results were found for films subjected to fluences between 2 and 3 MW·cm^{-2}. However, complete crystallization was completed within a matter of seconds. Fig. 4 illustrates the observed time dependence of the 143 cm^{-1} mode intensity and frequency. The film reached a maximum temperature of 550°C and then cooled to a steady state value of about 475°C. Residual stress also was found to relax by about 5 kbar.

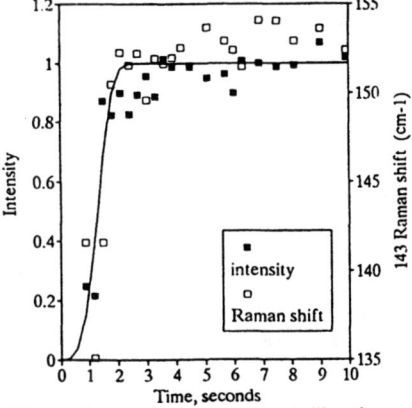

Fig. 4 Laser-induced crystallization of TiO$_2$ films on SiO$_2$. Changes in 143 cm^{-1} mode frequency and intensity.

Discussion

Modification to film optical properties and thickness during drying at moderate temperatures likely results from solvent and water loss. This effectively raises the glass transition temperature of the amorphous film and thereby acts to retard crystallization when films are heated to higher temperatures. During this *annealing* process, the extended network structure of the amorphous film will relax to a lower energy state, forming an extended network structure more resistant to crystallization. Under laser irradiation at relatively low fluences, the irradiated region densifies in analogy to thermal conditioning, and this treatment appears to retard crystallization of these regions at higher fluences when compared with unconditioned areas of the film.

At fluences beyond a critical threshold, irreversible crystallization of amorphous titania films ensues. Absorption of the energy leads to marked increases in film temperatures within the footprint of the focused laser beam. In these regions, large increases in refractive index are observed and are characteristic of the anatase phase. Transformation rates depend on laser fluence. Beyond a critical threshold, catastrophic damage to the film results and transformation to a rutile crystalline phase is indicated from Raman microprobe measurements within the damage region.

Conclusions

Irradiation of sol-gel deposited titania films at increasing laser fluence has been shown to lead to (i) conditioning with associated increased index, (ii) crystallization with associated index increase, and (iii), catastrophic damage depicted by material ablation and the presence of multiple crystalline phases. Time-resolved Raman measurements acquired during irradiation verify localized temperature increases and relaxation in residual compressive film stress. Results suggest that image inscription into amorphous sol-gel deposited films is possible by means of controlled laser irradiation. The inscribed images will show marked variation in refractive index when compared with unirradiated regions of the film.

Acknowledgement

This work has been supported by the Materials Sciences Division of the Office of Basic Energy Sciences, U.S. Department of Energy. Pacific Northwest Laboratory is operated by Battelle Memorial Institute for the U.S. Department of Energy under Contract DE-AC06-76-RLO 1830.

References

1 I.M. Thomas, *Opt. News* **7**:18 (1986).
2 S. Sakka, *Ceram. Bull.* **64(11)**:1463 (1985).
3 I.M. Thomas, *Proc. SPIE* **1438**:484 (1990).
4 T. Assih, A. Ayral, M. Abenoza, and J. Phalippou, *J. Mater. Sci.* **23**:3326 (1988).
5 K.F. Ferris, G.J. Exarhos, and C. Nguyen, *Influence of solution chemistry on the microstructure of sol-gel derived films*, in NIST Sp. Pub. **756**, eds. H.E. Bennett, A.H. Guenther, D. Milam, and B.E. Nemnam, US Dept. Commerce, pp. 272-278 (1987).
6 G.J. Exarhos, and M.J. Aloi, *Thin Solid Films* **193-194**:42 (1990).
7 G.J. Exarhos, and N.J. Hess, *Thin Solid Films* **220**:254 (1992).
8 J.C. Manifacier, J. Gasiot, and J.P. Fillard, *J. Phys. E., Sci. Instrumen.* **9**:4002 (1976).
9 N.J. Hess, G.J. Exarhos, and M.J. Iedema, *Proc. SPIE* **1848**:243 (1992).
10 I. M. Thomas, J. Wilder, R. Gonzales, and D. George, *1064 nm and 350 nm damage thresholds of high index oxide films deposited from organic solutions and sols*, in NIST Sp. Pub. **752**, eds. H.E. Bennett, A.H. Guenther, D. Milam, and B.E. Newnam, US Dept. Commerce, pp. 361-364 (1987).
11 G.J. Exarhos, N.J. Hess, and S.M. Wood, *Proc. SPIE* **1624**:444 (1992).
12 P.L. White, G.J. Exarhos, M. Bowden, N.M. Dixon, and D.J. Gardiner, *J. Mater. Res.* **6(1)**:126 (1991).

This article also appears in Mat. Res. Soc. Symp. Proc. Vol 316

DENSITY REDUCTION:
A MECHANISM FOR AMORPHIZATION AT HIGH ION DOSES

E.D. SPECHT, D.A. WALKO*, AND S.J. ZINKLE
Oak Ridge National Laboratory, P.O. Box 2008, Oak Ridge, TN 37831-6118
*Present address: Department of Physics, University of Illinois, Urbana, IL 61801-3080

ABSTRACT

At cryogenic temperatures, the accumulation of vacancy-interstitial pairs in Al_2O_3 from atomic displacements associated with ion implantation produces amorphization. At room temperature, these pairs recombine, and amorphization occurs only at high doses. X-ray reflectivity measurements show that amorphization of the surface of Al_2O_3 implanted at room temperature with 160 keV Cr^+ ions is preceded by a progressive reduction in near-surface density. Monte Carlo simulations show that this density reduction can be accounted for by high-energy-transfer collisions which knock atoms deep into the target, leaving widely separated vacancies and interstitials, which do not recombine. Electron microscopy shows that at least some of these vacancies condense into voids. We propose that this reduction in near-surface density can lead to amorphization at high doses. We present simple approximations for the density reduction expected for different ions and targets.

INTRODUCTION

The mechanisms by which ion implantation causes amorphization can be broadly categorized as structural and chemical. Chemical effects occur in the end-of-range (eor), where implanted ions come to rest, changing the chemical composition of the target. Structural effects are dominant in the midrange: as the ions pass through they displace target atoms from lattice sites. At relatively low temperatures, midrange amorphization typically requires a dose producing on the order of one displacement per target atom (dpa), while at higher temperatures structural damage is annealed out, and amorphization does not occur.[1-4]

Room temperature amorphization occurs at much higher ion doses for Al_2O_3, ZrO_2, MgO, and WC. While a dose ~1 dpa produces amorphization at cryogenic temperatures, ~100 dpa is required at room temperature.[1,5-8] The mechanism for this high-dose amorphization has been unclear. We address the origin of this anomalously high-dose amorphization by examining the density and microstructure of Al_2O_3 films implanted with ion fluences below the amorphization threshold. Implantation with Cr^+ was chosen to minimize chemical effects: it has been shown[9] that elements such as Cr which occupy substitutional sites induce amorphization at a higher dose. Room temperature amorphization of Al_2O_3 occurs at a dose between 1×10^{21} and 6×10^{21} Cr^+/m^2 [Ref. 6]; we examine samples implanted with 4×10^{20} and 1×10^{21} Cr^+/cm^2.

We summarize x-ray reflectivity, electron microscopy, and Monte Carlo simulation results[10] which show that amorphization is preceded by a progressive reduction in near-surface density. We derive approximate formulae for the expected density change as a function of the choice of target and incident ion.

X-RAY REFLECTIVITY

X rays with wavelength λ are totally reflected by a material with density ρ, average atomic number Z, and average atomic weight M below a critical incident angle $\theta_c = (A\rho Z\lambda^2/M)^{1/2}$, where $A = 2.70 \times 10^{10}$ m.[11] Because x rays penetrate only ~5 nm under conditions of total reflection, the critical angle for total external reflection provides a sensitive measure of near-surface density. As shown in Fig. 1, implantation increases this critical angle. The calculated reflectivity profiles (Fig. 1, lines) are derived from simple models[10] including surface roughness, the thickness of the implanted layer, and the degree to which the midrange density is decreased and the eor density is increased. Here we defer discussion of complete models of the density profile and emphasize the simple, model-independent correspondence between the decrease in reflectivity at the critical angle (triangles, Fig. 1) and the near-surface density. The critical angles observed in Fig. 1 are used to compute the near-surface densities in Fig. 2.

Note that the density decrease is *not* caused by a lattice expansion, in contrast to observations at low temperature.[12] It has been shown using glancing-incidence x-ray diffraction[13] that lattice expansion accounts for just 0.3% of the 12% reduction in density produced by 10^{21} Cr$^+$/cm^2. At low temperature, vacancy-interstitial pairs produced by implantation lead to a large lattice expansion, while at room temperature these pairs recombine. Rather, the density reduction results from the accumulation of excess vacancies in the midrange.

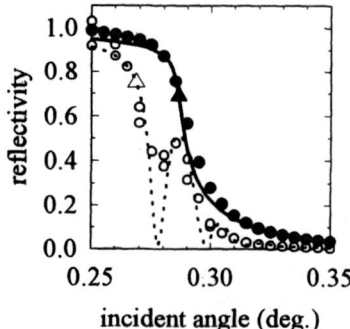

Fig. 1 X-ray reflectivity from Al$_2$O$_3$. ($\bullet$) unimplanted, ($\circ$) implanted with 10^{21} Cr$^+$/m^2, ($\blacktriangle$,$\triangle$) critical angles, lines are models from Ref. [10].

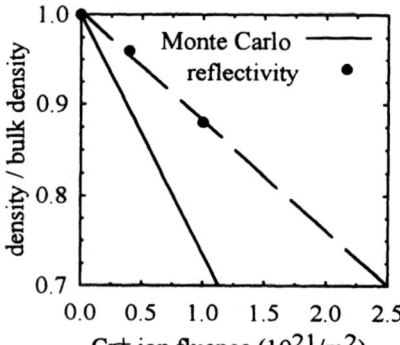

Fig. 2 Near-surface density as measured by x-ray reflectivity and computed by Monte Carlo simulations. Dashed line is a guide to the eye.

MONTE CARLO SIMULATIONS

Monte Carlo simulations of atomic displacements provide a quantitative model for the density reduction observed with x-ray reflectivity. We have applied the TRIM89 code[10,14] to calculate the effects of implantation. Because each ion is assumed to pass through structureless Al_2O_3, unaffected by implantation, the results are linear in incident ion fluence. The effects of substrate crystal structure are neglected, but because the density reduction is predominantly caused by high-energy-transfer collisions which do not depend on crystal structure, we expect the results to be accurate. More seriously, the calculation does not include recovery mechanisms. The simulated density profile is shown in Fig. 3, and the near-surface density is plotted in Fig. 2. The fluctuations in the Monte Carlo result in Fig. 3 is statistical, because we simulated only 30,000 ions.

ELECTRON MICROSCOPY

Monte Carlo simulations demonstrate that an excess of vacancies over interstitials is produced as atoms are pushed deeper into the target. Transmission electron micrographs (TEM) of cross-sections of implanted samples are used to determine how these excess vacancies are accommodated. Figure 4 shows a TEM of Al_2O_3 implanted with 4×10^{20} Cr^+/m^2. While a dense array of dislocations is apparent in the eor (60 - 230 nm), the midrange (0 - 60 nm) is free of visible dislocations and contains a high density (~5 x $10^{23}/m^3$) of small (~2 nm) features which exhibit contrast consistent with that of voids. Electron microdiffraction demonstrated that the midrange remains crystalline for both doses.

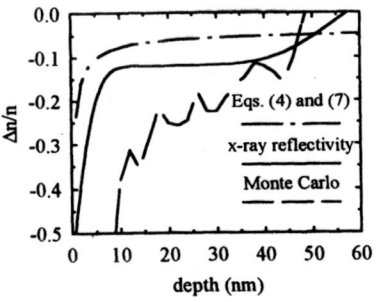

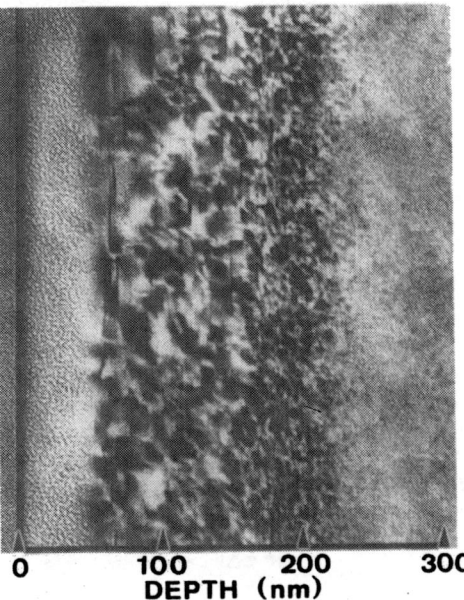

Fig. 3 Density profiles calculated based on Monte Carlo simulations, x-ray reflectivity data, and approximate Eqs. (4) and (7).

Fig. 4 Cross-section TEM showing the microstructure observed in Al_2O_3 following implantation with 4×10^{20} Cr^+/m^2.

ANALYTIC APPROXIMATION

While density changes can be calculated most accurately using Monte Carlo simulations, a simple order-of-magnitude approximation may prove useful. We consider a model where ions of mass M_1, atomic number Z_1, fluence f/a_B^2, and energy $E_0 E_R$ impinge on an elemental target with atomic mass M_2, atomic number Z_2, and number density n/a_B^3. Factors of the Rydberg energy $E_R = 13.6$ eV and the Bohr radius $a_B = 5.3 \times 10^{-11}$ m are included to make parameters dimensionless. We neglect straggling and use the approximate ranges given by Sigmund.[15] For the incident ions the range is

$$R_i(E_0)/a_B \approx 0.12 E_0 \, \bar{Z}^{1/3}(1 + M_2/M_1)/(n \, Z_1 \, Z_2) \,, \tag{1}$$

where $\bar{Z} = (Z_1^{2/3} + Z_2^{2/3})^{3/2}$. When an incident ion transfers energy TE_R to a target atom, the atom penetrates to depth

$$R_s(E_0,T)/a_B \approx 0.34 T^{3/2}/(n E_0^{1/2} Z_2^{5/3} \mu^{1/2}) \,, \tag{2}$$

which includes a factor of $2\sqrt{T/T_m}$ to approximately account for the effect of scattering angle on penetration depth; $\mu = M_1 M_2/(M_1 + M_2)^2$, and $T_m = 4\mu E_0$ is the maximum energy transfer to target atoms. Because we will consider high-energy-transfer collisions, we use the unscreened Rutherford cross-section,

$$d\sigma/d\tau \approx 4\pi Z_1^2 Z_2^2 a_B^2 \mu/(E_0 T^2) \,. \tag{3}$$

These approximations are valid for $0.001 < E_0 \bar{Z}^{1/3}/Z_1 Z_2 < 10$.

Two effects contribute to a midrange density reduction. The first depends on the proximity of the target surface:

$$\frac{\Delta n_1(x)}{n} \approx f \int_0^{T(E_0, x)} \frac{d\sigma}{dT} dT - f \int_0^{T_m(E_0)} \frac{d\sigma}{d\tau} d\tau \approx \frac{-6.2 f n^{1/3} Z_1^2 Z_2^{8/9} \mu^{2/3}}{E_0^{4/3}(x/a_B)^{2/3}} \,, \tag{4}$$

where $T(E_0,x)$ is the energy required for a scattered ion to penetrate to depth x. The first integral is the density gain to shallower target atoms being scattered *to* depth x; the second integral is the density loss due to target atoms being scattered *from* depth x deeper into the target. Due to the surface, no high-energy atoms are scattered to shallow depths, and density is reduced. Near the surface, Δn_1 becomes large in magnitude; its average over the ion penetration depth is

$$<\Delta n_1>/n \approx -75.3 f M_1^{1/3} \mu Z_1^{8/3} Z_2^{14/9}/(E_0^2 M_2^{1/3} \bar{Z}^{2/9}) \,.$$

A further reduction in midrange density arises because the scattering cross-section increases with decreasing ion energy. Because the incident ion energy is reduced as the ion penetrates the target, again more atoms are scattered *from* any depth, deeper into the target, than are scattered from shallower depths *to* that depth. At depth x, target atoms are scattered deeper by an average distance

$$M(x) \approx f \int_0^{T_m(E(x))} R_s(E,T)\frac{d\sigma}{dT}dT \tag{6}$$

where $E(x)$ is the incident ion energy at depth x, and density will be reduced by

$$\Delta n_2/n \approx (dE/dx)(dM/dE) \approx -142fZ_1^3Z_2^{4/3}M_1^{1/2}/(E_0^2\bar{Z}^{1/3}M_2^{1/2}\mu^{3/2}) . \tag{7}$$

Noting that $M \sim 2Z$, we see that the surface effect Δn_1 dominates for $Z_1 << Z_2$ and $Z_1 >> Z_2$, and is of similar magnitude to the energy-loss effect Δn_2 for $Z_1 \approx Z_2$.

In Fig. 3, we compare measured density changes with the more accurate Monte Carlo result with the approximations of Eqs. (4) and (7). While the Monte Carlo approximation overestimates the density reduction due to neglect of recovery processes, the analytic approximation used here underestimates the density reduction due to the neglect of multiple scattering processes.

DISCUSSION

The pairs of vacancies and interstitials created when ion bombardment displaces target atoms may be accommodated in three ways. Firstly, pairs may simply recombine. While thermal diffusion in Al_2O_3 is negligible at room temperature,[16,17] and the low amorphization threshold of Al_2O_3 at cryogenic temperature demonstrates that ion bombardment alone does not bring about vacancy-interstitial recombination, the resistance of Al_2O_3 to amorphization at room temperature shows that thermal effects and ion bombardment synergistically cause recombination.

Secondly, vacancies or interstitials may precipitate into dislocation loops. Electron microscopy (Fig. 4) shows that dislocations form in the eor but not in the midrange. Because implantation leads to an excess of vacancies in the midrange and of interstitials in the eor, this is consistent with observations that dislocation loops in Al_2O_3 are formed from interstitials but not from vacancies.[18,19]

Finally, isolated vacancies or interstitials may condense into other defects such as microvoids, vacancy clusters, or regions of amorphous Al_2O_3. While recombination or dislocation loop formation will restore the density to its unimplanted value, formation of these other defects will preserve the reduced midrange density. Because Monte Carlo simulations predict twice the observed density reduction, we conclude that half the excess midrange vacancies recombine, either with interstitials deeper in the target or at the surface, while half are stabilized as low-density point defects. Since observable voids account for only ~0.2% density reduction, whatever defect accounts for most of the density reduction is not apparent in TEM. For targets which are resistant to low-dose amorphization, such as Al_2O_3 at room temperature, the accumulation of these low-density defects may eventually destabilize the lattice and amorphize the target.

Self-implantation (e.g. stoichiometric implantation of Al_2O_3 with Al and O) minimizes chemical effects; this might provide clearer evidence that high-dose amorphization is due to displacement effects. Self-implantation, however, requires higher doses to produce large density changes than does implantation with heavier ions. Effective sample cooling would be required to achieve such high doses in a reasonable time, to avoid beam heating associated with high-flux irradiation..

ACKNOWLEDGEMENTS

This work was sponsored by the Division of Materials Sciences and by the Office of Fusion Energy, U.S. Department of Energy, under contract DE-AC05-84OR21400 with Martin Marietta Energy Systems, Inc.

REFERENCES

1. P.J. Burnett and T.F. Page, Radiat. Eff. 97, 283 (1986).
2. E. Glaser, G. Götz, N. Sobolev, and W. Wesch, Phys. Status Solidi A 69, 603 (1982).
3. C.W. White, C.J. McHargue, P.S. Sklad, L.A. Boatner, and G.C. Farlow, Mater. Sci. Rep. 4, 41 (1989).
4. J.L. Brimhall and E.P. Simonen, Nucl. Instrum. and Meth. B16, 187 (1986).
5. P.J. Burnett and T.F. Page, J. Mater. Sci. 20, 4624 (1985).
6. M.E. O'Hern, C.J. McHargue, C.W. White, and G.C. Farlow, Nucl. Instrum. and Meth. B46, 171 (1990).
7. S. Noda, H. Doi, and O. Kamigaito, J. Mater. Res. 4, 671 (1989).
8. S.J. Bull, J. Mater. Sci. 26, 3086 (1991).
9. C.J. McHargue, et al., Nucl. Instrum. and Meth. B16, 212 (1986).
10. E.D. Specht, D.A. Walko, and S.J. Zinkle, Nucl. Instrum. and Meth. B (in press).
11. L.G. Parratt, Phys. Rev. 95, 359 (1954).
12. T. Hioki, A. Itoh, M. Ohkubo, S. Noda, H. Doi, J. Kawamoto, and O. Kaigaito, J. Mater. Sci. 21, 1321 (1986).
13. E.D. Specht, C.J. Sparks, and C.J. McHargue, Appl. Phys. Lett. 60, 2216 (1992).
14. See J.P. Biersack and L.G. Haggmark, Nucl. Instrum. and Meth. 174, 257 (1980) for the original TRIM program. TRIM89 has been modified by T.P. Sjoreen (unpublished) to include different displacement thresholds for elements in compound materials.
15. P. Sigmund, Rev. Roum. Phys. 17, 823 (1972).
16. Y. Oishi and W.D. Kingery, J. Chem. Phys. 33, 480 (1960).
17. A.E. Paladino and W.D. Kingery, J. Chem. Phys. 37, 957 (1962).
18. T.D. Gulden, Philos. Mag. 15, 453 (1966).
19. W.E. Lee, M.L. Jenkins, and G.P. Pells, Phil. Mag. A51, 639 (1985).

This article also appears in Mat. Res. Soc. Symp. Proc. Vol 316

AMORPHIZATION AND DYNAMIC RECOVERY OF A2BO4 STRUCTURE TYPES DURING 1.5 MeV KRYPTON ION-BEAM IRRADIATION

L.M. Wang, W.L. Gong and R.C. Ewing
Department of Earth and Planetary Sciences, University of New Mexico,
Albuquerque, NM 87131

ABSTRACT

The temperature dependence of the critical amorphization dose, D_c, of four A_2BO_4 compositions, forsterite (Mg_2SiO_4), fayalite (Fe_2SiO_4), synthetic Mg_2GeO_4, and phenakite (Be_2SiO_4) was investigated by in situ TEM during 1.5 MeV Kr^+ ion beam irradiation at temperatures between 15 to 700 K. For the Mg- and Fe-compositions, the A-site is in octahedral coordination, and the structure is a derivative hcp (Pbnm); for the Be-composition, the A- and B-sites are in tetrahedral coordination, forming corner-sharing hexagonal rings (R3). Although the D_c's were quite close at 15 K for all the four compositions (0.2-0.5 dpa), D_c increased with increasing irradiation temperature at different rates. The D_c-temperature curve is the result of competition between amorphization and dynamic recovery processes. The D_c rate of increase (highest to lowest) is: Be_2SiO_4, Mg_2SiO_4, Mg_2GeO_4, Fe_2SiO_4. At room temperature, Be_2SiO_4 amorphized at 1.55 dpa; Fe_2SiO_4, at only 0.22 dpa. Based on the D_c-temperature curves, the activation energy, E_a, of the dynamic recovery process and the critical temperature, T_c, above which complete amorphization does not occur are: 0.029, 0.047, 0.055 and 0.079 eV and 390, 550, 650 and 995 K for Be_2SiO_4, Mg_2SiO_4, Mg_2GeO_4 and Fe_2SiO_4, respectively. These results are explained in terms of the materials properties (e.g., bonding and thermodynamic stability) and cascade size which is a function of the density of the phases. Finally, we note the importance of increased amorphization cross-section, as a function of temperature (e.g., the low rate of increase of D_c with temperature for Fe_2SiO_4).

INTRODUCTION

Recent ion beam studies on multi-cation, complex ceramics have shown that these materials are susceptible to irradiation-induced amorphization, either directly within the displacement cascade or due to attaining a critical defect density [1-4]. During ion beam irradiation, two competing processes, amorphization and relaxation/annealing occur dynamically. Although the diffusion driven recovery is increased by irradiation-enhanced diffusion, this process is suppressed at low temperatures. Thus, the analysis of the temperature dependence of the critical amorphization dose, D_c, allows the two processes to be examined separately. In general, D_c increases with the increasing temperature, but the slope of the D_c-temperature curve is indicative of a material's ability to recover. Although there has been detailed work on the temperature dependence of irradiation induced amorphization in Si [5] and intermetallics [6], the studies of complex ceramics have only recently been initiated [4,7]. Further, there are fundamental differences between defect production and recovery in simple materials, such as Si or intermetallic compounds, and structurally/compositionally complex ceramics [1]. In simple materials, an interstitial may recombine with vacancies at any site without causing large distortions of the crystalline structure (although chemical disordering, as in the intermetallics, can lead to amorphization), as is the case with "replacement events". However, in ceramics, the directional bonding of covalent-ionic compounds and the topologic complexity of multi-cation structures restrict recombination events, and interstitial defects can lead to relatively large distortion of the structure.

The temperature dependence of D_c for four A_2BO_4 compositions was investigated by in situ TEM during ion beam irradiation in order to investigate structural and bonding controls on amorphization and recovery processes. These preliminary results are part of a comprehensive study of the A_2BO_4 system which is characterized by a rich variety of compositions (e.g., A = Mg, Fe, Mn, Ca, Ni, Co, Be, Li, and REE; B = Si, Ge, Be and P), extensive solid-solutions between pure, end-member compositions, and a variety of structure types (e.g., the hcp derivative olivines, the ccp spinels, and the hexagonal phenakite structure). The phases studied were: forsterite (Mg_2SiO_4), fayalite (Fe_2SiO_4), phenakite (Be_2SiO_4) and synthetic Mg_2GeO_4. The $(Mg,Fe)_2SiO_4$ compositions are the end-members of a complete solid-solution series with the olivine structure. This structure (*Pbnm*) is a derivative of *hcp* of the oxygens with Mg and Fe in octahedral coordination and Si in tetrahedral coordination. The SiO_4 tetrahedral monomers are isolated from one another (nesosilicates), but are joined along three edges and one apex to the edge-sharing chains (parallel to *c*) of A-site octahedra [8]. Mg_2GeO_4 has the olivine structure (Ge substitutes for Si in tetrahedral sites), but the increased covalency of the Ge-O bond leads to lengthening of the shared edges. Recent studies of olivine have shown that nuclear interactions between energetic particles and target atoms (displacement damage), rather than ionization effects, are mainly responsible for ion beam-induced amorphization [2], and amorphization occurs directly within the displacement cascade under Kr^+ ion irradiations [9]. Phenakite (Be_2SiO_4) has a structure ($R\bar{3}$) in which the A- and B-site cations are all in tetrahedral coordination. The tetrahedra form corner-sharing 6- and 4-membered rings (perpendicular to *c*) in which the Be and Si atoms alternate. The rings are stacked (parallel to *c*) to form a corner-sharing three dimensional network [10].

EXPERIMENTAL PROCEDURES

The fayalite (Fe_2SiO_4), phenakite (Be_2SiO_4) and Mg_2GeO_4, are synthetic end-member compositions. The forsterite (Mg_2SiO_4) is natural and has an actual composition of $(Mg_{0.88}Fe_{0.12})_2SiO_4$ as determined by analytical electron microscopy.

TEM samples prepared by Ar ion milling were irradiated with 1.5 MeV Kr^+ ions in the HVEM-Tandem Facility at Argonne National Laboratory [11] at a dose rate of ~3.4×10^{11} ions/cm^{2}s. The facility consists of a high-voltage electron microscope (HVEM) connected to a tandem ion accelerator; thus, the selected area electron diffraction (SAD) pattern can be monitored *in situ* during ion irradiations to determine the D_c. The HVEM was used at an accelerating voltage of 300 keV. The SAD patterns were observed at the maximum observable sample thickness, as thinner regions become amorphous at lower doses due to surface effects. The irradiations were performed between 15 to 700 K, using a liquid helium-cooled cold stage or a hot stage. The maximum temperature increase due to beam heating was 60 °C. At 1.5 MeV, most of the Kr^+ ions completely penetrate the electron transparent thickness (~ 250 nm) of the samples, and the Kr concentration introduced into the sample is negligible. Because of differences in the density and displacement energy of the different targets, the per ion energy loss and damage production in the observable sample thickness are different. TRIM [12] calculations were performed to convert the critical ion dose to a damage dose in displacements per atom (dpa). However, this conversion may not be accurate, as the displacement energy (E_d) of 15 eV was assumed for all the materials. Nevertheless, the variations in the densities of the materials have been accounted for by this calculation.

Before and after ion irradiation, the samples were also examined by high resolution TEM (HRTEM) with a JEOL 2000FX electron microscope at the University of New Mexico, which was operated at 200 kV. Amorphization from the ion milling process was not detected.

RESULTS AND DISCUSSION

Amorphization was apparent from changes in the SAD pattern during the course of Kr^+ irradiation. Initially, a diffuse halo appeared, and its intensity increased with increasing ion dose. Concurrently, the diffraction maxima from the crystalline volume became progressively less intense and completely vanished at the critical amorphization dose, D_c [2]. The progressive amorphization process is also apparent in the HRTEM images taken before and after various ion doses (Fig. 1). Amorphization reached completion at much lower doses at the thin edges where the HRTEM images were taken than in the thicker regions where the SAD patterns were obtained, the measured D_c's are from the SAD observations.

The temperature dependence of the critical amorphization dose (converted to dpa from TRIM calculations) for the four A_2BO_4 materials are shown in Fig. 2. The doses required for amorphization of the four materials were quite close to each other at lower temperatures, and they increase at different rates with the increasing temperature.

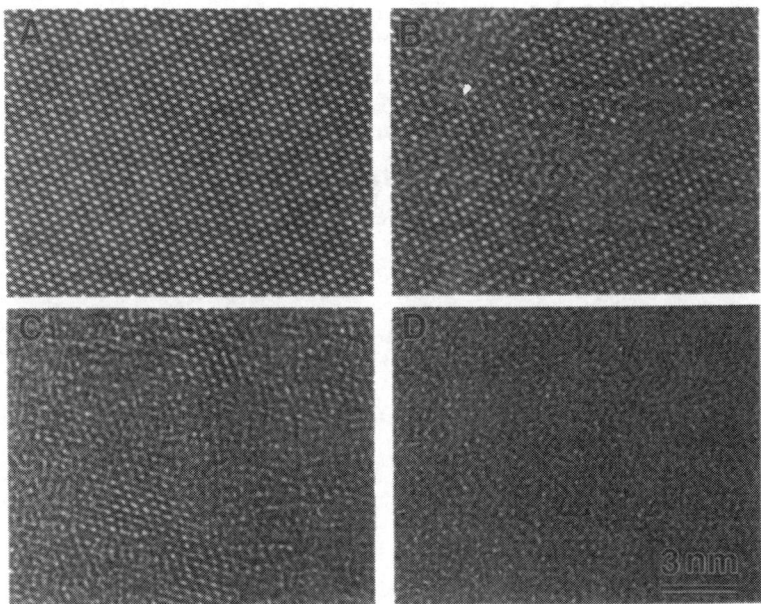

Fig. 1. HRTEM images of the Mg_2SiO_4 samples which are taken before (A) and after 1.5 MeV Kr^+ ion beam irradiations at room temperature to (B) 2.6×10^{12}, (C) 1.7×10^{13} and (D) 1×10^{14} ions/cm^2.

According to a model developed by Weber [4], the rate of direct-impact amorphization, with consideration of thermally activated and irradiation enhanced recovery processes, is

$$df_a/dt = (\phi\sigma - \tau^{-1} \exp[-E_a/kT])(1-f_a) \qquad (1)$$

where ϕ is the ion flux, σ is the cross-section for direct amorphization, $\tau^{-1} \exp[-E_a/kT]$ is the recovery rate in which E_a is the activation energy of the recovery process. Using this relationship, E_a can be determined from the temperature-D_c curve. The relationship between temperature and dose required for complete amorphization is approximately

$$\ln(1 - D_o/D_c) = C - E_a/kT \qquad (2)$$

where C is a constant dependent on ion flux and amorphization cross-section and D_o is the amorphization dose at T = 0 K. An Arrhenius plot of $\ln(1 - D_o/D_c)$ versus $1/kT$ yields $-E_a$ as the slope (Fig. 3). In addition to the activation energy, the critical temperature, T_c, above which amorphization will not be completed (D = ∞) is

$$T_c = E_a/kC. \qquad (3)$$

The critical amorphization doses at room temperature, as well as E_a and T_c, are given with key physical and thermodynamic parameters of the four studied phases in Table 1.

The amorphization doses at the room temperature are discussed first, as most of the physical and thermodynamic parameters available are only for this temperature. Be_2SiO_4 had the highest D_c (1.55 dpa), followed by Mg_2SiO_4 (0.79), Mg_2GeO_4 (0.76), and Fe_2SiO_4 (0.22). Thus, D_c increased with the decreasing Gibbs free energy (ΔG_f), average M-O bond distance, bond angle variance from ideal polyhedra, average bond compressibility (β), and with increasing repulsive energy (E_{rep}), bulk modulus (κ) and the difference between the unit cell volume at room temperature and the critical unit cell volume at melting (319 Å^3).

The relationship between D_c and the tabulated thermodynamic and physical parameters can be understood qualitatively. From a thermodynamic view, in order to make a crystalline material amorphous, the ΔG_f has to be raised to a level at least equal to that of the amorphous phase. Thus, the ΔG_f may be used to evaluate the stability of crystalline phases relative to that of the amorphous phase. The lower the Gibbs free energy the more stable the phase for given conditions (P-T-irradiation conditions). Similarly, the lower the bond angle variance from ideal polyhedral configurations and the higher bulk modulus, the more stable the phase. Thus, the greater the stability of a phase, the higher the required D_c. The M-O bond distance, the repulsive energy (short range lattice energy), and compressibility are all qualitative measures of bond strength which must be directly related to the displacement energy (E_d). If the amount of collisional energy deposited per ion is fixed (a function of the mass and energy of the ion), a material with a lower E_d will experience a higher number of displacements per ion. Thus, decreases in E_d cause a higher cross-section for direct amorphization (note, for the TRIM calculation E_d was assumed to be 15 eV for all the four phases).

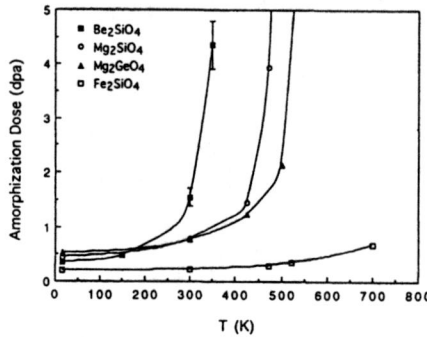

Fig. 2. Temperature dependence of amorphization dose, D_c, for 1.5 MeV Kr$^+$ ion beam-irradiated A_2BO_4 materials.

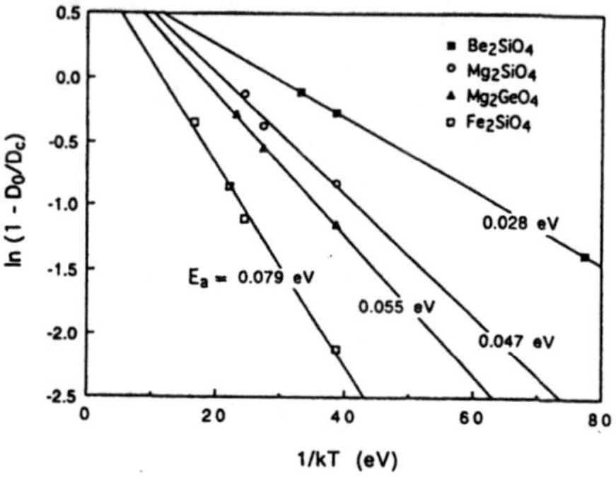

Fig. 3. Arrhenius plots of ln (1 - D_0/D_c) versus 1/kT calculated from the D_c-temperature curve (Fig. 2). The slopes equal - E_a.

Table 1. Amorphization parameters (vacancies/ion, D_c, E_a and T_c) compared to the physical, structural and thermodynamic parameters for the four phases of this study.

Material (A_2BO_4)	Density (g/cm^3)	Melting Point (K)	Average Bond Ionicity	ΔGf, 298K. (Kcal/Mol)	Avg. M-O Distance (Å)	Bond Angle Variance*	Average Bond $\bar{\beta}$ (Mbar^{-1})†
Be_2SiO_4	2.960		0.553	-485.8	1.644	6.6	0.067
Mg_2SiO_4	3.271	2183	0.668	-491.2	2.118	93	0.124
Mg_2GeO_4	4.018		0.658				0.128
Fe_2SiO_4	4.392	1490	0.504	-329.6	2.168	125	0.134

Material (A_2BO_4)	E_{rep}§ (Kcal/Mol)	K (Mbar)	V_{298}-V_m (Å^3)	Vac./ion in 200 nm foil	D_c @ 298 K (dpa)	E_a (eV)	T_c (K)
Be_2SiO_4		2.086		2200	1.55	0.029	390
Mg_2SiO_4	32.10	1.289	29.1	2800	0.79	0.047	550
Mg_2GeO_4		1.226		3300	0.76	0.055	650
Fe_2SiO_4	26.69	1.143	11.4	3500	0.22	0.079	995

* Bond angle variance represents the polyhedral distortion [13].
† Average compression coefficient for cation-anion bond [14].
§ Repulsive energy for M-O interactions.

Previous work on the Mg-Fe olivine solid-solution series has shown that D_c increases with the increased average bond ionicity [2]. As shown in Table 1, this is true for the three isostructural olivine phases, but not for phenakite. Phenakite has a much lower ionicity than Mg_2SiO_4 and Mg_2GeO_4 (closer to that of Fe_2SiO_4), and yet has the highest D_c at room temperature. In phenakite, all cations are in tetrahedral coordination, while in olivine the A-site cations are in octahedral coordination. These are two very different structures. Perhaps, the topologic simplicity of the phenakite structure as compared to that of olivine [1] offsets the effect of the lower bond ionicity in regards to its higher D_c. Such observations emphasize the need for systematic studies of different (but related) structure types over a range of compositions

The unit cell volumes of the olivines are quite different at room temperature, and they increase with the increasing temperature. Hazen [14] has extrapolated the cell parameters and M-O distances of various ferromagnesian olivines to their melting points and has shown that they have similar cell parameters, cell volumes and M-O distances at their respective melting temperatures. Thus, olivines have the same critical volumes at melting. Meng et al. [15] have shown that the volume expansion associated with hydrogen-induced amorphization is comparable in magnitude to that which occurs during the irradiation-induced amorphization of intermetallic compounds. Previous studies have also indicated a strong correlation between the onset of solid-state amorphization and critical volume expansion [16,17]. The thermodynamic parallels between solid-state amorphization and melting have been discussed, and volume expansion to the point of a shear instability has been proposed as an underlying cause for disordering and amorphization [17,18]. From this perspective, the difference between the unit cell volume at temperature, T, and at melting (ΔV_{T-m}) can be used as an indication of susceptibility to amorphization. The olivine composition with the greater ΔV_{298-m} (Mg_2SiO_4 vs. Fe_2SiO_4) is more resistant to amorphization at room temperature because more defects are required to expand the cell volume to that characteristic of the crystalline-to-amorphous transformation (e.g., similar to that at the temperature of melting).

The critical amorphization doses, D_c, for these phases converge as the temperature decreases. Considering the experimental error, the critical doses are virtually identical at 15 K, except for Fe_2SiO_4 which is lower by a factor of two. This suggests that (1) the damage recovery existing at room temperature is almost completely suppressed at 15 K; and (2) the physical and thermodynamic properties of these phases are also very similar at low temperature.

The slope of the D_c-temperature curves is also a function of the density of the material. The lower the target density, the faster the curve rises (resulting in a lower activation energy for recovery and lower critical amorphization temperature). The effect of the lower target density is very similar to the observations of Koike et al. [6] on the effect of smaller projectile-mass on the amorphization dose for CuTi. This dependence and similarity can be qualitatively explained in terms of the collision cascade size. For a fixed target, the cross-section for nuclear collision is larger when a greater projectile mass is used, thus, creating a larger displacement cascade. The same relation is true for a fixed projectile mass in a target of greater density. As shown in table 1, the per ion vacancy production in a 200 nm thick foil varied from 2200 for Be_2SiO_4 (2.96 g/cm^3) to 3500 for Fe_2SiO_4 (4.392 g/cm^3). The greater the per ion vacancy production, the larger the displacement cascade. Because the activation energies for the recovery process calculated using Weber's model are too low for crystal nucleation (usually several eV), it is reasonable to assume that the recovery in the amorphous cascade region is mainly through the irradiation-enhanced epitaxial regrowth of the surrounding crystalline volume. The recovery rate for a smaller cascade (smaller amorphous domain) can be faster than that of a larger one because the surface area to volume ratio is larger for smaller cascades, thus allowing more efficient epitaxial recrystallization (i.e., shrinkage of the amorphous volume).

When one considers thermally-activated and irradiation-enhanced annealing, one commonly considers diffusion coefficients for atomic species and defect migration energies. Materials with higher diffusion coefficients and lower defect migration energies are supposed to anneal more easily or to have a smaller activation energy for recovery. However, such a model is not consistent with the data obtained for fayalite. Morioka [19] has determined that the the metal cation diffusion coefficient in fayalite is more than two orders of magnitude higher than that of forsterite at all temperatures. Also, the vacancy migration energy is usually proportional to the melting temperature (T_m) [20, 21], and fayalite has a much lower T_m than forsterite. However, E_a for fayalite is much larger (0.079 eV) than that of forsterite (0.047 eV). Such discrepancies cannot be fully attributed to the larger cascade size in fayalite. One plausible explanation for this discrepancy lies in the consideration of the effect of temperature on damage production. With increasing temperature, the unit cell volume increases and with the concomittant increase in M-O bond lengths, the displacement energy, E_d, decreases. This has the effect of increasing the cross-section for direct amorphization (i.e., σ in eq. (1)). Thus, in the absence of recovery, the amorphization rate will increase at higher temperatures. The competition between this increased amorphization cross-section and the increased annealing rate at elevated temperatures, rather than the annealing rate alone, determines the slope of the D_c-temperature curve.

CONCLUSIONS

The temperature dependence of amorphization dose, D_c, for four A_2BO_4 compositions, forsterite (Mg_2SiO_4), fayalite (Fe_2SiO_4), phenakite (Be_2SiO_4) and Mg_2GeO_4, were investigated by *in situ* TEM during 1.5 MeV Kr^+ ion beam irradiation between 15 to 700 K. At low temperature (near 15 K), the D_c's were nearly identical, but D_c increased with increasing temperature at different rates. Through the analysis of the D_c-temperature curves, we conclude:

(1) The Gibbs free energy of formation, bulk modulus, bond angle strain, and volume difference between that of the phase at room temperature and at its melting point correlate with the material's susceptibility to amorphization.

(2) In addition to diffusivities of cations, the size of displacement cascades, which vary with target density under the same ion beam, affects the recovery rate of the amorphous domain. The larger the cascade, the slower the recovery rate.

(3) Increased temperature may not only increase the annealing rate, but may also increase the amorphization cross-section by reducing the displacement energy.

ACKNOWLEDGMENTS

The authors are very grateful to H.R. Westrich of Sandia National Laboratories and H.W. Green, II, of the UC-Riverside for providing the synthetic samples of the silicates and germanate, respectively, and to the HVEM-Tandem Facility Staff at Argonne National Laboratory for assistance during ion irradiations. S.X. Wang prepared most of the TEM samples. The electron microscopy was completed, in part, at the Electron Microbeam Analysis Facility of the Department of Earth and Planetary Sciences of the University of New Mexico supported by NSF, NASA, DOE-BES, and the State of New Mexico. This work was supported by the Office of Basic Energy Sciences, US Department of Energy under grant DE-FG03-93ER45498.

REFERENCES

1. L.M. Wang, R.K. Eby, J. Janeczek and R.C. Ewing, Nucl. Instr. and Meth. **B59/60**, 395 (1991).
2. L.M. Wang and R.C. Ewing, Mat. Res. Soc. Symp. Proc. **235**, 333 (1992).
3. L.M. Wang and R.C. Ewing, MRS Bulletin **XVII** (5), 38 (1992).
4. W.J. Weber, R.C. Ewing and L.M. Wang, J. Mater. Res., in press.
5. F.F. Morehead, Jr. and B.L. Crowder, Radiat. Eff. **6**, 27 (1970).
6. J. Koike, P.R. Okamoto and L.E. Rehn, J. Mater. Res. **4**, 1143 (1989).
7. W.J. Weber and L.M. Wang, Mat. Res. Soc. Symp. Proc. **279**, 523 (1993).
8. J.J. Papike, Reviews of Geophysics, **25**, 1483 (1987).
9. L.M. Wang, M.L. Miller and R.C. Ewing, Ultramicroscopy **51**, 339 (1993).
10. R.M. Hazen and A. Y. Au, Phys. Chem. Minerals **13**, 69 (1986) .
11. C.W. Allen, L.L. Funk, E.A. Ryan and S.T. Ockers, Nucl. Instr. and Meth. **B40/41**, 553 (1989).
12. J.F. Ziegler, J.P. Biersack and U. Littmark, The Stopping and Range of Ions in Solids (Pergamon, New York, 1985).
13. K. Robinson, G.V. Gibbs and P.H. Ribbe, Science **122**, 567 (1971).
14. R.M. Hazen, Am. Miner. **62**, 286 (1977); R.M. Hazen and C.T. Prewitt, ibid., **62**, 309 (1977).
15. W.J. Meng, P.R. Okamoto, L.J. Thompson, B.J. Kestel and L.E. Rehn, Appl. Phys. Lett. **53**, 1820 (1988).
16. A. Seidel, G. Linker and O Meyer, J. less. Comm. Met. **145**, 358 (1988).
17. P.R. Okamoto and M. Meshii, Proc. ASM Symp. on "Science of Advanced Materials", edited by H. Wiedersich and M. Meshii (Chicago, IL, 1989).
18. D. Wolf, P.R. Okamoto, S. Yip, J.F. Lutske and M. Kluge, J. Mater. Res. **5**, 286 (1990).
19. M. Morioka, Geochim. Cosmochim. Acta. **45**, 1573 (1981).
20. G.M. Hood, J. Nucl. Mater. **139**, 179 (1986).
21. R.I.M.A. Rashid and N.H. March, Phys. Chem. Lig. **19**, 41 (1989).

This article also appears in Mat. Res. Soc. Symp. Proc. Vol 316

LAYER BY LAYER AMORPHIZATION IN Si:
TEMPERATURE, ION MASS AND FLUX EFFECTS

A. BATTAGLIA, G. ROMANO, S.U. CAMPISANO
Dipartimento di Fisica, Università di Catania, Corso Italia 57 I95129, Catania, Italy

ABSTRACT

The layer-by-layer amorphization process is explored in a temperature range in which the kinetics of crystallization can be neglected. It has been found that the pure amorphization rate increases exponentially as the substrate temperature is decreased with an apparent activation energy of 0.48 eV. Moreover the rate increases with both the ion flux and the energy deposited into elastic collisions. A phenomenological model is proposed to explain the experimental results.

INTRODUCTION

Ion stimulated phase transitions from crystalline to amorphous silicon and vice versa represent a quite remarkable example of reversible phase transition occuring in conditions far from the thermodynamical equilibrium. The mechanism originating the phase transition and the kinetics of formation of the amorphous phase have been investigated in many details and several models have been proposed [1-4]. In particular it has been demonstrated that this transition may be described by a classical mechanism of nucleation and growth. The nucleation has been ascribed to the prompt part of the collision cascade while the growth of the amorphous layer has been ascribed to long living defects [5]. By measuring the onset of amorphization it has been found that, at a given temperature and for a fixed total fluence, there is a critical flux at which a continuous amorphous layer is obtained [6-8]. This critical flux depends exponentially on temperature with an activation energy ranging from 1.2 to 0.7 eV, depending on the ion species and/or on the ion fluence applied for inducing the process [7,8]. More recently the process of nucleation and growth has been studied by changing the ion flux and, assuming a constant value for the nucleation rate, the amorphization velocity has been found flux dependent [9].

In order to have a better understanding on the flux and temperature dependences of the pure amorphization process, it is necessary to measure the amorphization velocities independently from the pure nucleation process.

In the case of a pre-existing crystal-amorphous (c-a) interface ion flux noticeably influences the kinetics of the phase transformation [10-12]. At a fixed ion flux, crystallization or layer by layer amorphization occurs depending on the substrate temperature. The critical temperature is therefore defined as the temperature at which no interface movement occurs. The dependence of the critical temperature upon ion flux was found to be controlled by an activation energy of 1.2 eV. This value was entirely ascribed to those defects giving rise to the crystallization process [11,12].

With the aim to clarify the role of temperature and ion flux on the amorphization process we have performed accurate measurements of the kinetics of layer by layer amor-

411

Mat. Res. Soc. Symp. Proc. Vol. 321. ©1994 Materials Research Society

phization in a temperature range for which the pure crystallization kinetics is negligible. The effects of ion flux, temperature, and energy deposited into elastic collisions are investigated and the experimental results are explained on the basis of a phenomenological model.

EXPERIMENTAL

Si samples (100) in orientation, were amorphized by multiple Ge or Xe implantations in order to obtain samples with pre-existing a–layers both at the surface ($\sim$ 17 nm thick) and buried ($\sim$ 170 nm below the surface). A crystalline layer, $\sim$ 100 nm thick, was thus embedded between the two amorphous layers. Ion-assisted layer-by-layer amorphization was induced by irradiation with 600 keV Kr^{++} ions at ion fluxes ranging from $3 \times 10^{11}/cm^2 sec$ to $5 \times 10^{12}/cm^2 sec$. During irradiations the samples were mounted onto a resistively heated copper block whose temperature, controlled by a thermocouple, was varied in the range 50-140 °C. The kinetics of amorphization was determined by *in situ* reflectance measurements described elsewhere [13]. Some selected samples were analyzed by 2.0 MeV He^+ channeling in the backscattering geometry in order to further check the accuracy of the method. The amorphization velocities, determined by the two methods, differ by less than 10 % and thus all the measured values will be quoted without referring to a particular measurement method.

In another set of experiments, layer-by-layer amorphization was induced by implanting different ions at a fixed ion flux ($1 \times 10^{12} ions/cm^2$) through a c-a interface lying at $\sim$ 600nm from the surface. The process of amorphization was characterized by 2.0 MeV He^+ Rutherford Backscattering Spectroscopy (RBS) in combination with the channeling technique.

ION FLUX AND TEMPERATURE EFFECTS

At a given the substrate temperature we have investigated the flux dependence of the layer by layer amorphization process. In Fig.1 values for the interface velocity as a function of the ion flux ($\dot{\phi}$) are reported for two different irradiation temperatures. Data can be fitted by assuming a $\dot{\phi}^{2/3}$ law; the rate of amorphization therefore depends less than linearly upon ion flux at temperatures at which the process of crystallization is fully inhibited.

The temperature dependence has been investigated in the range 80-140 °C and the data are summarized in Fig.2. The absolute values of the interface velocity are reported as a function of $1/T$ for different fluxes. Data taken by 600 keV Kr^{++} irradiation, refer to both crystallization [13] and amorphization. At each ion flux an exponential increase in the amorphization velocity by decreasing temperature is observed. In the investigated temperature range an activation energy of $\sim$ 0.48 eV is determined.

The most widely accepted model [12] for ion beam induced crystal-amorphous transition in silicon assumes the rate R as given by the balance between crystallization (R_c) and amorphization (R_a) rates:

$$R = R_c - R_a \tag{1}$$

The model assumes that the rate of crystallization is governed by recombination of pairs of

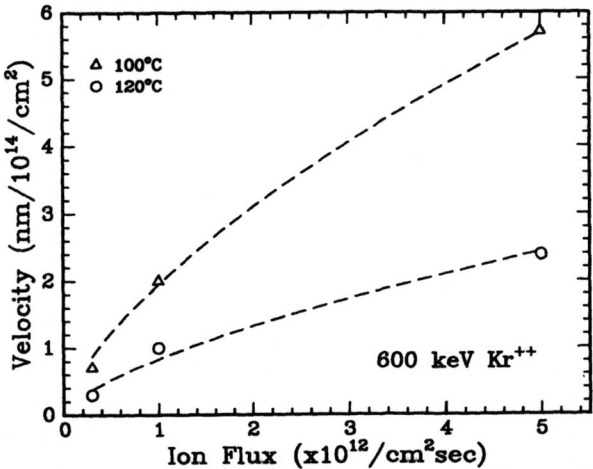

Fig.1 The interface velocities are reported as a function of the ion flux for two different irradiation temperatures.

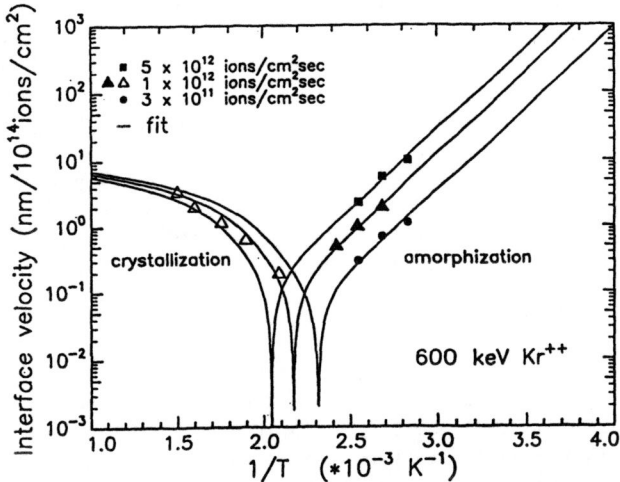

Fig.2 Absolute values of the interface velocity in the crystallization and in the amorphization regime as a function of 1/T. Symbols refer to the experimental data obtained under different irradiation conditions, lines represent fits to the experimental data obtained by applying the presented model.

defects and the rate of amorphization is constant . Temperature and ion flux dependeces are correlated only with the crystallization rate determining the defect decay rate and the defect generation rate.

On the basis of our experimental results, however, also the amorphization rate is flux and temperature dependent. Therefore we have modified the model by introducing the experimental dependence for the amorphization velocity upon flux and temperature:

$$R_a = V_o \cdot \dot{\phi}^n \cdot exp\left(\frac{E_a}{kT}\right) \qquad (2)$$

By considering the Jackson's expression for the crystallization rate [12] and equ.2 for the amorphization rate we can fit our experimental data reported in Fig.2. Lines represent fits to the data obtained with literature values [10,12] for the parameters of the crystallization term and with E_a=0.48 eV and n=2/3. The only fitting parameters were the activation energy for the defect decay rate in the crystallization term (E_c) and the pre-exponential V_o in the amorphization rate. By assuming E_c =1.28 eV and V_o= 0.56×10^{-6} ($nm/10^{14}ions/cm^2 \cdot (10^{12}ions/cm^2 sec)^{-2/3}$), a good agreement is obtained and both the temperature and flux dependence of the amorphization rate and the data relative to the crystallization process can be taken into account by the present description.

Few things can be pointed out from our data. First the interface movement is governed by ion generated defects which can be produced either in the crystalline side and/or in the amorphous side of the interface promoting amorphization and/or crystallization. In particular, at low temperatures, defects generated in the crystalline side of the interface will accumulate with a rate determined by the 0.48 eV activation energy, giving rise to the measured amorphization kinetics. At higher temperatures, defects generated in the amorphous side of the interface will recombine with a rate determined by the 1.28 eV activation energy determining the cristallization kinetics. At any temperature and flux for which $R_c \simeq R_a$ the overall kinetics is quite complex as shown by the full lines drawn in Fig.2. Far from these conditions the rate is determined by either one of the two classes of defects generated in the amorphous or in the crystalline side of the interface.

The accumulation rate depends also upon the ion flux suggesting that defects should live at least for a time ($\sim$ 1 sec for the considered ion fluxes) as long as the arrival of two successive ions in the same cascade area ($\sim 10^{-12} cm^2$ for 600 keV Kr^{++} ions).

ION MASS EFFECT

In order to have a better understanding on the mechanisms of defects generation and accumulation for low temperature ion irradiation we have measured the layer by layer amorphization as a function of the energy deposited into elastic collisions.

The amorphization velocities have been determined by performing accurate Rutherford Backscattering measurements on samples having a crystal-amorphous interface lying at $\sim$ 600 nm from the surface. The process was induced by irradiating the samples at a fixed substrate temperature (50°C) and for a fixed ion flux ($1 \times 10^{12}/cm^2 sec$) with 1.0 MeV Si, 2.5 MeV Co and 2.0 MeV Ge ions. These irradiations produce a different number of displacements per unit lenght (N_d) at the c-a interface. The calculated values (TRIM 89) of displacements are reported on Fig.3 as a function of depth.

At $\sim$ 600 nm from the surface, Ge ions produce about a factor of 6 more vacancies with respect to the Si ions. Moreover by changing the starting interface depth we can

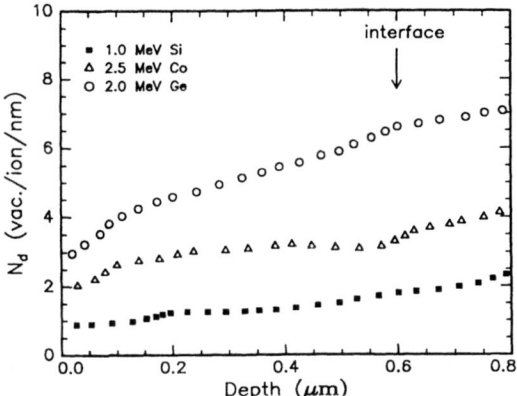

Fig.3 Calculated values (TRIM89) of displacements per unit lenght produced by different ion irradiations in Si

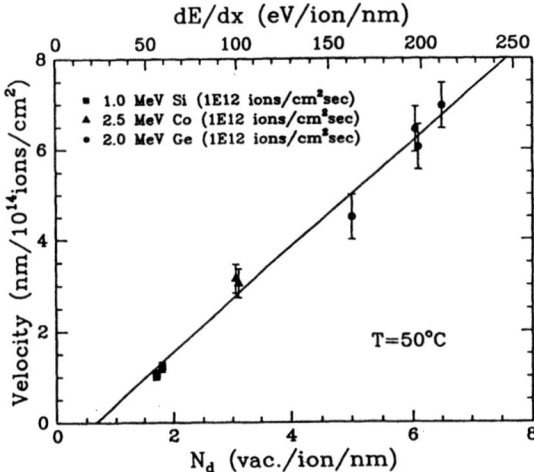

Fig.4 The amorphization velocity is reported as a function of the number of displacements generated, at the c-a interface, by the impinging ions per unit lenght.

follow the amorphization kinetics as a function of N_d with a fixed ion. The experimental results are summaraized in Fig.4. The interface velocity is reported as a function of the number of displacements generated by the impinging ions per unit lenght. A linear trend is found suggesting that defect accumulation is directly related with defects production at the c-a interface. Moreover, below a critical number of displacements (0.7 vac/ion/nm) no interface movement can be detected. This indicates that for low values of energy deposited into elastic collisions the concentration of defects is not enough to promote amorphization.

This lower limit may be related to the effects produced by light ion implantation. It is known that, even at room temperature, boron implantation in silicon does not produce continuous amorphous layers. As a matter of fact, for 100 keV B implantation the number of displaced Si atoms is about 0.8 vac/ion/nm, very close to the measured threshold value. Therefore, the concentration of defects, generated by light ions, may not be enough to promote amorphization.

CONCLUSIONS

The amorphization kinetics, at temperatures at which the crystallization is fully inhibited, depends on flux, temperature, and energy deposited into elastic collisions. These experimentally determined dependeces are incorporated in a semiempirical model describing crystallization and amorphization induced by ions passing through a c-a interface. All the data will provide the basis for a complete description of ion induced amorphization.

REFERENCES

[1] J.R. Dennis and E.B. Hale, J. Appl. Phys.,**49** 1119 (1978)
[2] J.F. Gibbons, Proc. IEEE **60**,1062 (1972)
[3] F.F. Morehead and B.L. Crowder, Rad. Eff. **25**,49 (1980)
[4] M.L. Swanson, J.R. Parsons and C.W. Hoelke Rad. Eff. **9** , 249, (1971)
[5] S.U. Campisano, S. Coffa, V. Raineri, F. Priolo and E. Rimini Nucl. Instrum. and Meth. **B 80/81**, 514 (1993)
[6] O.W. Holland. D. Fathy, J. Narayan and O.S. Oen, Rad. Eff. **90**, 127 (1985).
[7] P.J. Schultz, C. Jagavish, M.C. Ridgway, R.G. Elliman and J,S. Williams Phys. Rev. B, **44**, 9118 (1991)
[8] J.S. Williams, R.G. Elliman, M.C. Ridgway, C. Jagadish, S. Ellingboe, R. Goldberg, M. Petrovich, W.C. Wong, Z. Dezhang, Nucl. Instrum. Meth. **B 80/81**, 507 (1993)
[9] A. Battaglia, S.U. Campisano, J. Appl. Phys. in press
[10] F. Priolo, E. Rimini, Mater. Sci. Rep. **5**, 319, (1990)
[11] J. Linnros, R.G. Elliman and W.L. Brown J. Mater. Res. **3**, 1208 (1980)
[12] K.A. Jackson, J. Mater. Res. **3**, 1218 (1988)
[13] A. Battaglia, F. Priolo, E. Rimini, Appl. Surf. Sci. **56-58**, 577 (1992)

This article also appears in Mat. Res. Soc. Symp. Proc. Vol 316

A Critical Regime for Amorphization of Ion Implanted Silicon

R. D. Goldberg*, J. S. Williams and R. G. Elliman

Department of Electronic Materials Engineering, Australian National
University, Canberra, 0200, Australia.
*Present Address: Department of Physics, The University of Western
Ontario, London, Ontario, Canada, N6A 3K7.

ABSTRACT

A critical regime has been identified for ion implanted silicon where only slight changes in temperature can dramatically affect the levels of residual damage. In this regime decreases of only 5° C are sufficient to induce a crystalline-to-amorphous transformation in material which only exhibited the build-up of extended defects at higher temperatures. Traditional models of damage accumulation and amorphization have proven inapplicable to this regime which exists whenever dynamic defect annealing and damage production are closely balanced. Irradiating ion flux, mass and fluence have all been shown to influence the temperature—which varies over a range of 300° C for ion species ranging from C to Xe—at which the anomalous behaviour occurs. The influence of ion fluence suggests that complex defect accumulation plays an important role in amorphization. Results are presented which further suggest that the process is nucleation limited in this critical regime.

INTRODUCTION

Traditionally, the amorphization of silicon bombarded with medium to heavy ions has been viewed in two ways: i) in terms of increasing free energy with damage accumulation, which results in the lattice collapsing into the amorphous phase when energetically favourable [1], and ii) the accumulation and overlap of discrete zones of damage produced along the track of each incoming ion until the whole layer becomes amorphous [2]. Both of these models have proven successful in understanding the onset of ion induced amorphization, especially at low temperatures (LN$_2$) where the generated defects are relatively immobile. However, at higher temperatures the models become inadequate. For example, implantation flux, a parameter not considered in earlier models, is shown to be an important factor controlling amorphization at both a pre-existing interface [3] and at the irradiating ions' end-of-range (eor) [4, 5]. Indeed, it has been shown that the amorphous phase can be difficult to nucleate under such conditions and that the presence of pre-existing interfaces, surfaces and crystal damage can provide appropriate nucleation sites [4]. Furthermore, a critical set of irradiation conditions exists such that, a slight decrease in temperature, or increase in flux, will result in amorphization which otherwise would not occur. This interdependence between substrate temperature and irradiation flux has been interpreted in terms of an interplay between defect production, controlled by the ion flux, and dynamic defect annealing, which is moderated by the changing defect types and mobilities that accompany shifts in temperature.

417

The present work examines damage accumulation in silicon and the effects that ion flux, fluence and species have upon it, in the regime where dynamic defect annealing plays a significant role. The effects of pre-existing damage on the amorphization process will be used to illustrate that amorphization is nucleation limited in this regime.

EXPERIMENTAL PROCEDURE

Implantations involving rare gas ions were undertaken using a 200 keV Whickham ion implanter, located at the Royal Melbourne Institute of Technology, while all other implantations were performed on the 1.7 MeV NEC 5SDH tandem accelerator located at the Australian National University. In all cases the bombarded wafer was thermally coupled to a copper block with conducting paint and clips. The block was either heated or cooled to maintain the silicon samples at the required temperature which ranged from 0° to 340° C. A constant ion flux was maintained throughout each irradiation.

Rutherford backscattering channeling (RBS-C) and cross-section transmission electron microscopy (TEM) were used to analyze the resultant damage. The number of displaced atoms (N_d) was calculated using the computer program Nd. This program [6], based on work by Ziegler [7], fits a dechanneling component from RBS-C spectra before extracting N_d over a given range. The depth of material used to determine N_d was constant for each ion type but varied between ion species, due to the difference in their implantation ranges. TRIM [8] was used to calculate the energy deposition and implanted ion profiles.

RESULTS AND DISCUSSION

Figure 1 displays N_d resultant from an 80 keV Si bombardment at a series of temperatures and ion fluences. The data shown is for irradiations ranging from 2 to 10×10^{15} ions/cm^2 which were all delivered at a rate of 2.7×10^{13} ions/cm^2/s. The data for each fluence have been fitted with the same function as obtained for the 1×10^{16} ions/cm^2 series. Examining the data for each ion fluence it becomes clear that three distinct regions exist. These are as indicated on the 1×10^{16} ions/cm^2 data set. In region I the wafer is totally amorphized over the irradiated region (confirmed by TEM) and no temperature dependence is observed in the residual damage structure. Hence N_d values in this region appear similar for all implant doses. This region reflects a dominance of defect production and amorphization over annealing. At higher temperatures another region (III) exists, for a specific fluence, where temperature also has little effect on the final value of N_d. In contrast to region I, dynamic defect annealing is dominant and the sample remains totally crystalline, although, TEM results (figure 2a) have revealed a band of interstitially-based extended-defects. Despite dynamic defect annealing dominating in this regime the fact that N_d is non-zero implies the annealing is not complete. In region II (the critical regime) the extreme sensitivity of N_d to even slight changes in temperature suggests that defect production and dynamic defect annealing are closely balanced.

The delivered ion fluence has a marked influence on determining the damage regime. For example, at 150° C, as the dose is increases from 2×10^{15} to 10×10^{15} ions/cm^2, the resultant damage moves from region III through region II to region I. This suggests that the changing damage structures in the sample with increasing dose have a strong effect on the defect kinetics and hence amorphization occurring during irradiation. Figure 3 shows the damage created by the 170° C implants in figure 1. Damage accumulation is evident both at the surface (most probably at the Si/naturally occurring SiO$_2$ interface) and the ions' eor

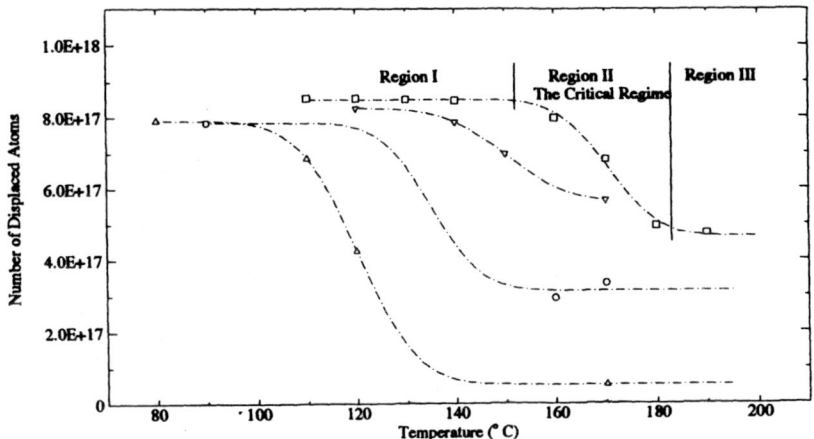

Figure 1: The number of displaced atoms, plotted as a function of temperature for various doses ($2(\triangle)$, $4(\circ)$, $8(\triangledown)$ and $10(\square)$) $\times 10^{15}$ ions/cm^2) of 80 keV Si ions. A constant flux of 2.7×10^{13} ions/cm^2/s was used for all the irradiations. Marked on the 10×10^{15} data are the three 'damage regimes' discussed in the text.

(0.12 μm, as calculated by TRIM). Intriguingly, when a buried amorphous layer forms (as confirmed by TEM at depths between 0.8 μm and 0.14 μm at $>4.4 \times 10^{15}$ ions/cm^2), it does so not at the point of maximum energy deposition ($\sim$0.6 μm) but at the site within the defective band where energy deposition is greatest (see figure 3). Contrastingly, in the deeper regions of the defect band where the energy deposition is less (0.14μm to 0.19 μm), the crystal does not amorphize. (Such behaviour is illustrated in figure 2b.) This indicates that energy deposition is an important parameter with respect to amorphization, while the initial formation of the amorphous layer within the defect band implies that the process is nucleation limited.

It is evident from figure 1 that increasing the total fluence has two effects: to move the critical regime to higher temperatures and to increase the saturation damage level of N_d in region III. Both of these effects can be understood if the damage, as it accumulates within the crystal, does so in the form of extended defects (as figure 2a shows) and not as amorphous zones. These defects increase in complexity and density as the fluence is increased and, thereby, eventually alter the balance between dynamic defect annealing and production within the sample. Table 1, lists the temperature at which the critical region is centered (defined as the temperature at which N_d lies half way between its value in region I and III) for a series of implant species. From the two Ar data sets it is clear that the position of the critical regime depends on the ion flux. Examining the data for C and Xe, which were delivered at very similar fluxes, reveals that the critical regime is shifted by $\sim$300° C between the two. Implicit in this ion species dependence is that, defect accumulation/annihilation processes are strongly dependent on the energy deposited within individual ion cascades. The denser cascades of heavier ions clearly alter the type of defect structures which form and agglomerate and hence, the nature of the defect band and nucleation of the amorphous layer.

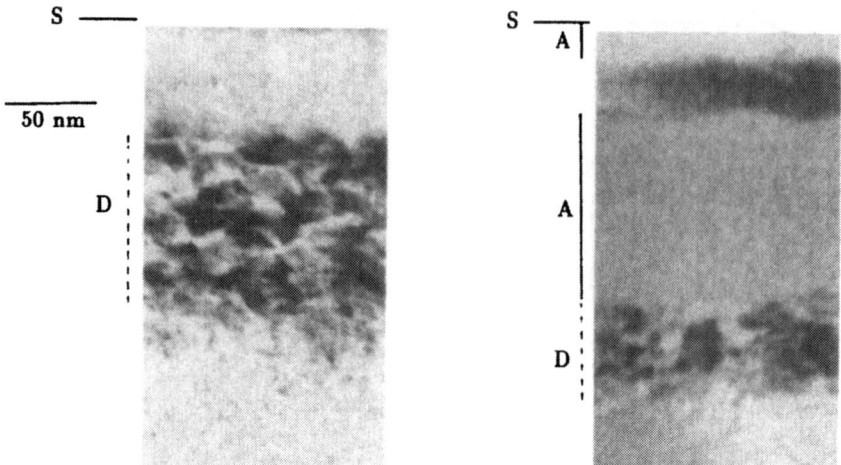

Figure 2: Cross-sectional TEM showing the damage structures found in regions III (a) and II (b). The region III structure reveals a defect band at the implanted ions end-of-range, while the region II micrograph shows the formation of amorphous material at both the surface and the defect band. 'S'–signifies the surface, 'A'–amorphous material and 'D'–extended defects. Both specimens were produced by bombardment with 80 keV Si ions at a rate of 2.7×10^{13} ions/cm^2/s. The samples were irradiated to a fluence of 4.4×10^{15} (a) and 10×10^{15} ions/cm^2 (b) at 170° and 160° C respectively. RBS-C spectra corresponding to (a) is shown in figure 3, while the 1×10^{16} ions/cm^2 spectra of the same figure is similar in nature to that of (b).

The abruptness of the transition between regions II and III and its dependence on energy deposition is intriguing. Once an irradiation is commenced, the dynamic annealing rates will increase with defect concentration until a 'steady state' is reached [10]. During this period, the free energy within the crystal will undergo an initial sharp increase, due to the presence of the defects and then climb only slowly during the 'steady state' phase, as the concentrations of complex defects increase. The magnitude of the initial increase in the free energy will be controlled by both the ion flux and mass [11] as well as the substrate temperature. As only slight changes in structure occur in the 'steady state' regime, if the irradiation is halted at this stage, the material will appear similar for a range of temperature, fluence and flux conditions. This is exactly what is observed in region III. If on further irradiation the free energy increases above a certain level (E_α), amorphization may occur. At lower temperatures, the reduced defect annealing within the crystal will mean that the 'steady state' free energy will lie closer to E_α. As a result, it will take less irradiation time (fluence) in region III before E_α is reached. This behaviour is manifested in figure 1 by the increases in N_d in region III and the increase of critical regime temperatures for the higher fluence irradiations. If the temperature is lowered enough, E_α will be reached before a steady state sub-amorphous defective band is achieved and damage accumulation will appear almost constant up to the point of amorphization, a feature indicative of low temperature irradiations. For irradiations where dynamic defect annealing and damage

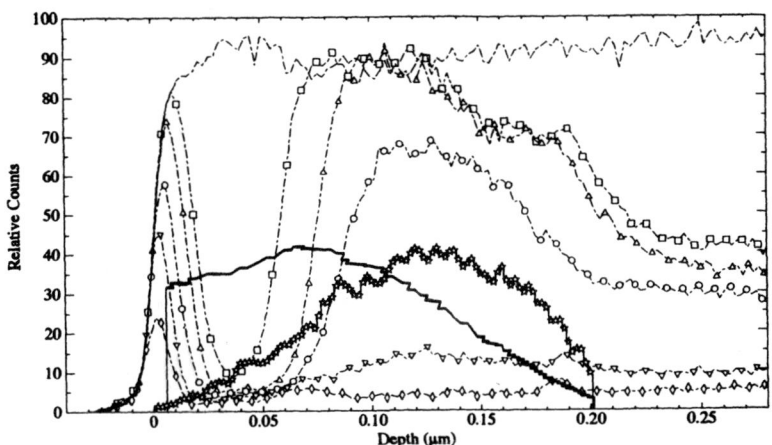

Figure 3: RBS-C spectra showing the accumulation with fluence of damage in a Si crystal resulting from bombardment with an 80 keV Si beam. Data for irradiations of 1×10^{16} ($\square$), 6.9×10^{15} ($\triangle$), 4.4×10^{15} (o) and 2.3×10^{15} (∇) ions/cm^2 are shown overlayed with the implanted ion profile ($\star$) and the energy deposited into nuclear collisions (—), as calculated by TRIM[8]. A virgin ($\diamond$) and random (- - -) spectra are shown for comparison. In all cases shown the substrate temperature was maintained at 170° C and the irradiating flux was 2.7×10^{13} ions/cm^2/s.

production are balanced, the free energy will approach E_α only slowly. It is proposed that, under these conditions, the free energy within the crystal may rise to the level required for amorphization but will not undergo the transformation without an appropriate defective band to nucleate the process. Indeed, it has been observed that amorphous material can be nucleated on pre-existing defects (eg: dislocations) which are well removed from the irradiating ions' end-of-range [10, 12]. The increase of the critical regime temperature with ion mass can, therefore, be understood in terms of the higher numbers of defect complexes that form in the damage cascades of these ions.

CONCLUSION

Under the irradiation conditions where dynamic defect production and annealing are closely balanced, a critical region has been identified where only slight changes in implant conditions have a marked effect on the amount of residual damage within the sample. A model has been proposed to explain the influence of the irradiating ion flux, fluence and species on the temperature of this regime. Results have shown that within this regime, complex defect accumulation is necessary to nucleate the amorphous phase. Under conditions where dynamic defect annealing dominates, the free energy of the defective crystal rises only slowly and the amorphous phase can be difficult to form in the absence of specific defect structures which nucleate the transition.

Ion Species	Ion Flux (ions/cm^2/s)	Critical Regime Centre ($^\circ$ C)
C	6.2×10^{12}	25
Si	3.1×10^{12}	65
Ar	1.2×10^{12}	110
Ar	4.8×10^{12}	125
Kr	8.2×10^{12}	235
Xe	7.3×10^{12}	325

Table 1: The shift observed in the central temperature of the critical regime, resultant from variations in ion species and flux. The beam energy was 80 keV in all cases and the irradiations were to a fluence of 1.0×10^{15} ions/cm^2 for all of the implants except carbon, which was to 2×10^{15} ions/cm^2.

References

[1] J. R. Dennis and E. B. Hale, J. Appl. Phys. 49(3), 1119 (1978).

[2] J. F. Gibbons, Proc. IEEE 60, 1062 (1972).

[3] J. Linnros, R. G. Elliman and W. L. Brown, J. Mater. Res. 3(6), 1208 (1988).

[4] J. S. Williams, K. T. Short, R. G. Elliman, M. C. Ridgway and R. Goldberg, Nucl. Instr. and Meth. B48, 431 (1990).

[5] Peter J. Schultz, C. Jagadish, M. C. Ridgway, R. G. Elliman and J. S. Williams, Phys. Rev. B 44(16), 9118 (1991).

[6] Written by R. Brown, Ph.D. student, The University of Melbourne, Melbourne, Australia.

[7] James F. Ziegler, J. Appl. Phys. 43(7), 2973 (1972).

[8] J. P. Biersack and L. G. Haggmark, Nucl. Inst. and Meth. 174, 257 (1980).

[9] R. G. Elliman, J. Linnros and W. L. Brown, Mater. Res. Soc. Proc.100, 363 (1988).

[10] R. D. Goldberg, J. S. Williams and R. G. Elliman, to be published.

[11] K. A. Jackson, J. Mater. Res. 3(6), 1218 (1988).

[12] R. D. Goldberg, Ph.D thesis, The University of Melbourne.

This article also appears in Mat. Res. Soc. Symp. Proc. Vol 316

DEFECT INDUCED AMORPHIZATION IN SILICON:
A TIGHT BINDING MOLECULAR DYNAMICS SIMULATION

D. MARIC
Swiss Scientific Computing Center, CH-6928 Manno (Switzerland)
L. COLOMBO
Dipartimento di Fisica, Universita' di Milano, via Celoria 16, I-20133 Milano (Italy)

ABSTRACT

We present an investigation on the amorphization process of crystalline silicon induced by ion beam bombardment by simulating the insertion of self-interstitials at different temperatures. The simulation is carried out by tight-binding molecular dynamics which allows for a detailed characterization of the chemical bonding and electronic properties of the irradiated samples. The irradiation process consists of two steps: (i) insertion of defects at a constant rate; (ii) annealing of the sample and observation of its structural properties. Thanks to the large size of the simulation cell (up to 276 atoms) we can characterize the amorphous network both on the short-range and medium-range length scale. Electronic properties are investigated as well and their evolution is monitored during the insertion process. Finally, we present a thorough comparison of the structural properties of the irradiated sample with amorphous silicon as obtained by rapid quench from the melt.

INTRODUCTION

Electron-beam and ion-beam bombardment of solids is well know to induce a crystal-to-amorphous transition in irradiated samples.[1,2] Such a phenomenon has attracted a huge number of experimental and theoretical investigations because of its fundamental interest from the point of view of the formation, nature and stability of the amorphous state of matter.

The key role in such a solid-state amophization process is played by point defects (interstitials or vacancies) and Frenkel pairs that are introduced in the host lattice. In this paper we present a molecular dynamics (MD) simulation of the response of a crystalline silicon lattice to random insertion of self-interstitials. Our main goals are to find the treshold concentration of defect above which the amorphization is observed and to study the dependence of the crystal-to-amorphous transition upon the temperature of the substrate and the rate of insertion of self-interstitials. Moreover we present a detailed characterization of both structural and electronic properties of the irradiated samples and show a comparison to amorphous silicon (a-Si) as obtained by direct quenching from the melt.[3,4]

MD simulations are particularly suitable to describe at the atomic level defect-induced structural relaxations and defect-defect interactions.[3,5,6] In this paper we make use of the tight-binding molecular dynamics (TBMD) scheme that we have successfully applied to study a-Si obtained by overcooling liquid silicon (l-Si).[7] The reliability of our method has been extensively discussed elsewhere.[3] In particular, we remark that the present TB Hamiltonian has been successfully applied to study intrinsic point defects in crystalline Si, like vacancies, interstitials, divancancies and Frenkel pairs.[5]

Mat. Res. Soc. Symp. Proc. Vol. 321. ©1994 Materials Research Society

In the TBMD simulations the atomic trajectories are calculated by means of an hamiltonian consisting in three contributions: (i) ionic kinetic energy; (ii) electronic band-structure energy that is, in turn, calculated by a semi-empirical sp^3 tight-binding Hamiltonian taken from ref.(8); (iii) short-ranged repulsive potential. The electronic degrees of freedom do not directly enter in the dynamics and, consistently, time-steps as large $3\ 10^{-15}$ s can be used. On the other hand, the calculation of the interatomic forces governing the atomic dynamics is derived from the underlying electronic structure of the system through the term (ii) of our Hamiltonian. This gives to TBMD the accuracy needed to describe covalent materials with directional bonding and places this scheme in the quantum-mechanical regime.

Taking full advantage from the reduced computational workload of TBMD, we made use of large simulation cells of 216 atoms where up to 60 self-interstitials have been inserted. Such unprecedented large simulation boxes guarantee that our results are only slightly affected by size effects.

SAMPLE PREPARATION

The full simulation of the irradiation process consists in three steps. First of all a number of self-interstitials (ranging from 5 to 60) was randomly inserted into a crystalline sample that was previously equilibrated at the desired temperature. In the present calculation we have performed our computer experiments at two different temperatures: 300 K and 600 K. The defects have been inserted into the host lattice with zero kinetic energy at distance from lattice sites larger that 0.5 A. This latter feature guarantees that no excess kinetic energy is pumped into the systems because of strong repulsive interactions between very close particles. During the insertion step the volume of the simulation box has been scaled linearly with the number of defect in order to keep the density of the sample constant. The insertion rate was $5\ 10^{13}$ atoms/s. Even if our insertion rate was considerably higher than in current ion beam bombardments, the resulting computational effort was formidable. As a matter of fact, the full simulation required several tens of CPU hours on a NEC SX-3 supercomputer.

After irradiation, we have equilibrated the sample for 6.6 ps. The temperature was maintained constant by rescaling the atomic velocities at each time-step of the TBMD run. Finally, we have observed the system for 2.4 ps and calculated the relevant structural and electronic properties.

STRUCTURAL PROPERTIES

As for the overall structural properties, we found that those systems where up to 15-18 defects were inserted (i.e. concentration of self-interstitials close to 7-8%) behave like distorted diamond lattices. In fact, the pair correlation function g(r) (here not shown) is very similar to that one of crystalline silicon (c-Si) up to the shell of the fourth nearest neighbours. Only the bond angle distribution function shows the presence of short-range non-tetrahedral structures related to the presence of interstitials.

By increasing the absolute number of defects, we have observed that the crystal-to-amorphous transition occurs at defect concentration above 20%. This is clearly shown in Fig.1 where we report the g(r) for three different samples where 20, 30, 60 defects have been inserted (samples a,b,c respectively). The simulation has been repeated at T=300 K (left) and T=600 K (right). In Fig.1 we report also the pair correlation function of a-Si as obtained by l-Si (dashed line). We can clearly see that both samples c are pretty nicely disordered by the huge number of interstitials showing no structural correlations beyond the shell of the second nearest neighbours, falling at ~4 A. Moreover, it is worth noticing that the position of both the onset and the first minimum of g(r) are the same as in the case of the overcooled liquid (dashed line). This feature indicates that the fraction and distribution of the empty space in a-Si is independent of the sample preparation.

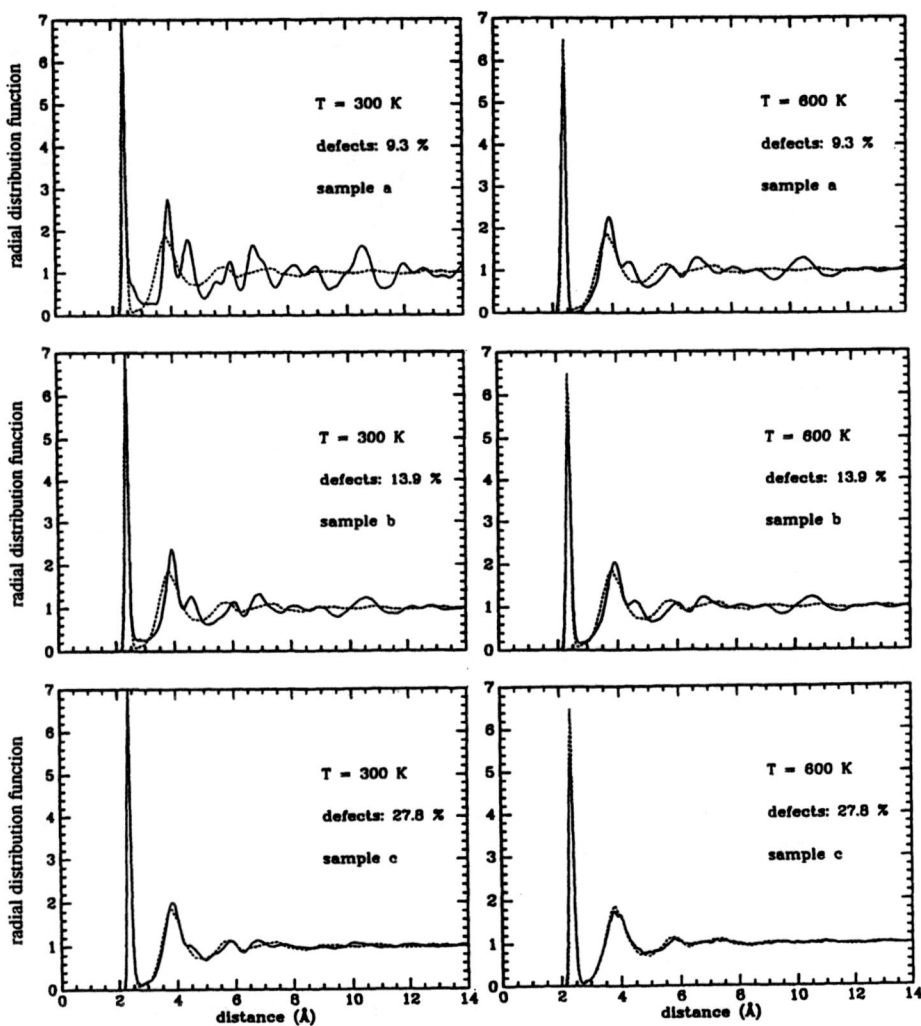

Fig.1 - Pair correlation function of three irradiated samples at different temparatures (full line). Dashed line: g(r) of overcooled l-Si (ref.[3]).

The effects due to the different temperature of the substrate are more evident for lower and intermediate defect concentration (sample a and b, respectively). We observe that sample a is much more distorted when the temperature is T=600 K. However, the structural quality of the system is less affected by further insertion of defects if compared to the room remperature experiment.

More detailed information about the defect distribution and the topology of the amorphous network can be obtained from Table I where we report the relative occurrence of under-coordinated (n<4), fourfold-coordinated (n=4) and over-coordinated (n>4) atoms in the two systems with defect concentration of 27.8% (sample c). The cutoff distance has been fixed at 2.8 A, corresponding to the first minimum of $g(r)$ (see Fig.1). The average coordination number n_{ave} is reported as well.

	n<4 (%)	n=4 (%)	n>4 (%)	n_{ave}
sample c T=300 K	5	85.2	9.8	4.05
sample c T=600 K	6.5	89.1	4.4	4.01
overcooled liquid (ref.[3]) T=300 K	4.9	90.4	4.7	3.99
CP (ref.[4]) T=300 K	0.2	96.6	3.2	4.03
expt. (ref.[9]) T=300 K	–	–	–	3.97

Table I - Relative occurrence of under-, fourfold- and over-coordinated atoms. The value of the average coordination number n_{ave} is reported as well. CP data refers to a first-principle simulation based on the Car-Parrinello method.

There is an overall agreement among the different sets of data confirming that defect-induced amorphization produces and a-Si sample with structural properties similar to those ones of overcooled liquid. However, Table I shows that the detailed distribution of dangling bonds and over-coordinated atoms differs from sample to sample and depends upon temperature.

ELECTRONIC PROPERTIES

So far we have discussed the structural properties of the irradiated samples. Another important question addressed in the present paper concerns the modification of the electronic structure of silicon caused by the insertion of interstitials. To this aim we have calculated the total electronic density of states (DOS) by averaging over all the different configurations explored by the samples during the observation run. The results are shown in Fig.2 where the same labelling of Fig.1 has been adopted and the DOS of overcooled liquid[3] (dashed line) is reported as well. The Fermi level is indicated by a vertical dash-dotted line.

At room temperature the sample irradiated with 20 interstitials (defect concentration: 9.3%) still present a close similarity to the crystalline case: in fact, the three main structure at ~-12 eV ~-6 eV and ~-3 eV are in rather good agreement to experimental valence x-ray photoelectron spectroscopy data. However, we note that there is a sizeable amount of gap states, particularly at the Fermi level. These states correspond to the dangling bonds associated to under-coordinated atoms. By increasing the number of defects, the crystal-like structures in the DOS become less and less evident and, finally, in the case of sample c the amorphous situation is recovered.

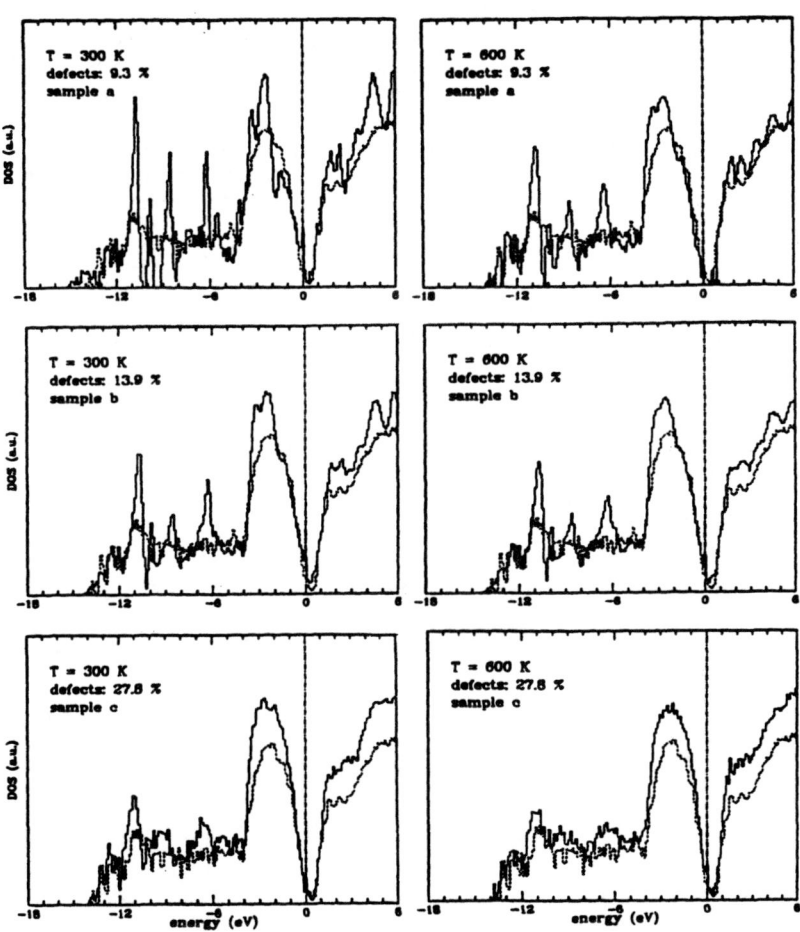

Fig.2 - Total electronic density of states of three irradiated samples at different temperatures (full line). Dash-dotted line: Fermi level.

Finally, at T=600 K the situation is slightly different: the crystal-like features as much less pronounced even at low irradiation dose. However, the small peak at the Fermi level is again evident and results to be not affected by varying the number of defects and the temperature of the substrate. This is the main difference with respect to a-Si obtained by quench from the melt where no gap states where found. As far as the gap is concerned, we note that it is narrower that in crytalline case in rather good agreement to experimental findings[10] and first-principles simulations.[4]

CONCLUSIONS

We have simulated the dynamical response of a silicon crystal lattice to the insertion of random self-interstitials. We have found that the defect concentration threshold for the crystal-to-amorphous transition is above the 20%. For a concentration of 27.8% the lattice periodicity is completely destroyed and the resulting amorphous network has structural properties very similar to overcooled liquid silicon. The electronic properties are more sensitive to the irradiation dose and substrate temperature. Evidence of persistent gap states has been found.

Finally, we are considering further work including vacancies and Frenkel pairs in order to draw physical conclusions on the amorphization process closer to the experimental situation.

REFERENCES

1. Phase tansformations during irradiation, edited by F.V. Nalfi (Applied Sciences, Englewood, NJ, 1983)
2. W. Eckstein, Computer simulation of ion-solid interactions (Springer Series in Materials Science vol.10, Springer-Verlag, Heidelberg, 1991)
3. G. Servalli and L. Colombo, Europhys. Lett. **22**, 107 (1993)
4. I. Stich, R. Car and M. Parrinello, Phys. Rev. B**44**, 11092 (1991)
5. E.G. Song, E. Kim, Y.H. Lee, Y.G. Hwang, Phys. Rev. B**48**, 1486 (1993)
6. C.Z. Wang, C.T. Chang and K.M. Ho, Phys. Rev. Lett. **66**, 189 (1991)
7. L. Goodwin, L. Skinner and A.J. Pettifor, Europhys. Lett. **9**, 701 (1989)
8. D. Chadi, Phys. Rev. Lett. 41, 1062 (1978); Phys. Rev. B**29**, 785 (1984)
9. Y. Waseda and K. Suzuki, Z. Phys. B**20**, 339 (1975)
10. Tetrahedrally bonded amorphous semiconductors, edited by D. Adler and Fritzsche (Plenum, New York, 1985)

This article also appears in Mat. Res. Soc. Symp. Proc. Vol 316

428

PHOTOINDUCED AMORPHOUS ↔ CRYSTALLINE TRANSITIONS IN $Sb_X Se_{1-X}$ FILMS

PAUL STRADINS*+, OJARS BALCERS*‡, AND VYATCHESLAV GERBREDER**
*Physics Institute, Salaspils, Latvia ;
**Department of Physics, Daugavpils Pedagogical Institute, Daugavpils, Latvia

ABSTRACT

We present a study of laser pulse induced crystallization and amorphization in Sb_xSe_{1-x} films. The time required to reach stable amorphous state after the pulse increases with exciting pulse length and becomes constant when stationary temperature field is approached by the end of the pulse. The time dependence of the excited spot's local temperature is deduced directly from amorphization threshold intensity dependence on the pulse length and is used to calculate the cooling times after the amorphizing pulse. Two photocrystallization regimes are distinguished depending on whether the melting starts before or after the end of crystallization, the condition depending on crystallization tendency for given composition x. The occurrence of melting limits the maximum optical contrast during photocrystallization. The results show that the crystallization tendency of Sb_xSe_{1-x} rises with x and has a local maximum between $x = 0.5$ and $x = 0.7$.

INTRODUCTION

It is now understood that the fastest laser-induced crystallization and amorphization processes suitable for optical memory applications are the diffusionless transitions with almost the same short and medium range order in amorphous and crystalline states [1]. These transitions can involve the formation of metastable crystalline phases [2].

On the other hand, the micron-size laser beam induced heating and cooling technique itself provides a powerful tool for investigating the amorphization tendency of various substances because of very high (up to 10^{10} K/s) and controllable heating and cooling rates.

In this work we have studied the laser-induced crystallization and amorphization in Sb-Se system in which fast crystallization can take place near $x = 0.67$ [3] and is most likely due to the diffusionless formation of metastable crystalline phase Sb_2Se in two possible modifications [3,4]. We investigated the amorphization kinetics after the amorphizing pulse, the amorphization threshold intensity dependence on the pulse length and the photocrystallization time as well as the maximal optical contrast dependence on light intensity inducing photocrystallization. Similar time - resolved techniques have been applied by other authors to measure the crystallization and amorphization kinetics in Ge-Te [5] and In-Sb-Ge [6] systems.

We also tried to describe in simple terms the heating and cooling processes during and after the laser pulse to which the sample is subjected. The calculations of the excited spot's temperature are often complicated by the lack of knowledge about the thermal properties of evaporated films [7]. We deduce the temperature time dependence directly from laser-induced amorphization data and make a distinction between the effect of external conditions (heating and cooling rates etc.) and the intrinsic phase-change properties of Sb_xSe_{1-x}.

Present addresses:
+ James Franck Institute, University of Chicago, Chicago, IL 60637, USA
‡ Dept. of Materials Science and Mechanics, Michigan State University, East Lansing, MI 48824,USA

Mat. Res. Soc. Symp. Proc. Vol. 321. ©1994 Materials Research Society

EXPERIMENTAL DETAILS

Sb_xSe_{1-x} thin film samples were prepared by thermal evaporation in 10^{-7} Torr vacuum from Sb and Se sources. Rotating sample was exposed alternately to each source, the average increase in thickness due to one exposure being about 20 Å. Two types of sample structures were used. The first consisted of a single 35 nm thick Sb_xSe_{1-x} layer on glass substrate. On these samples the photocrystallization kinetics, optical contrast and amorphization threshold measurements were performed. Other samples were multilayered. A 100 nm thick Sb_xSe_{1-x} layer was sandwiched between 200 nm ZnS layers on glass. These samples were used for time-resolved amorphization measurements. Thermal crystallization of amorphous films was performed in the oven at a constant heating rate 3.5 K/min. The crystalline structure was analyzed by powder X-ray diffraction.

Photoinduced changes studied were induced by a focused pulsed Kr+ laser beam (λ=647.1nm). Rectangular light pulses were formed by an acoustooptical modulator giving tunable pulse duration from 50 ns and rise and fall times of 25 ns. The focusing of the beam on the sample was visually controlled by a microscope in the transmission mode. The excited spot's diameter was about 2.4 μm if measured as the size of amorphized spot in the crystalline film.

By tuning the microscope, we projected the excited spot's real image onto a small circular diaphragm. Thus, the light power mainly from the central part of the spot was registered by a photodiode placed after the diaphragm. In this way, the optical contrast became more pronounced as it was less affected by the periphery of the spot unable to undergo structural changes during the pulse.

T_a and T_c being the optical transmissivities of amorphous and photocrystallized (up to saturation) spot, the optical contrast is defined as $C = (T_a - T_c) / T_a$. We defined the crystallization time t_C as a time required to reach the optical contrast $0.15 C_0$, where C_0 was the absolute maximum obtainable contrast due to photocrystallization over the whole intensity range investigated for a given composition x (see Results and Discussion and Fig.4).

Amorphization occurs by rapid cooling of the excited spot which has been melted by the laser pulse. The amorphization time was determined as a time in which the optical transmission after the amorphizing laser pulse reached $(T_a + T_{ml}) / 2$, where T_{ml} is the optical transmissivity of the melted excited area at the end of the pulse. The amorphization kinetics was measured by means of the second (test) pulse formed by the same acoustooptical modulator at variable delay with respect to the end of the first (amorphizing) pulse [8].

RESULTS AND DISCUSSION

X-ray diffraction analyses show that all as-deposited samples studied are amorphous. After thermocrystallization in samples with x = 0.64, 0.67, 0.76 only the strongest line corresponding to interplane distance 3.07 Å is seen. This is close to the strongest line of metastable phase Sb_2Se in both modifications [3] and [4]. x = 0.55 and x = 0.50 give the most apparent lines of Sb_2Se[4]. x = 0.40 contains the stoichiometric crystalline Sb_2Se_3 phase. Thermocrystallized x = 0.52 is amorphous by X-ray diffraction probably due to the small grain size (<100 Å). In all samples, when thermocrystallized, a distinct drop of optical transmission near the crystallization temperature (180-220°C) is observed.

Short and intense laser pulse amorphizes the excited area in a previously crystallized film. For a given pulse length τ the absorbed light intensity threshold I_t exists at which a small uniform circle with higher optical transmission occurs in the center of the excited spot after the pulse. This circle disappears by heating the sample above the thermocrystallization temperature or by applying a longer, low-intensity laser pulse which both restore the initial polycrystalline structure.

The dependence of I_t on the amorphizing pulse length τ is shown on Fig.1 for x = 0.67 and 0.64. Amorphization is possible with the pulses in the range from 5×10^{-7} s (the upper limit is set

430

by the laser power output) to 2×10^{-5} s. For longer pulses, a small ablative hole is formed instead of the amorphous circle at the center of the spot.

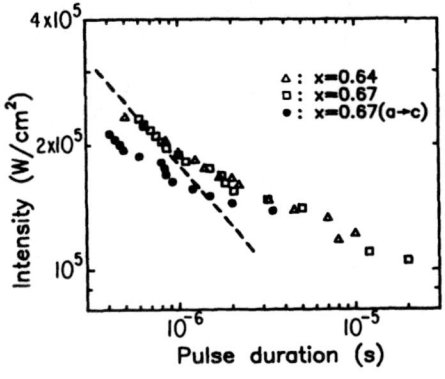

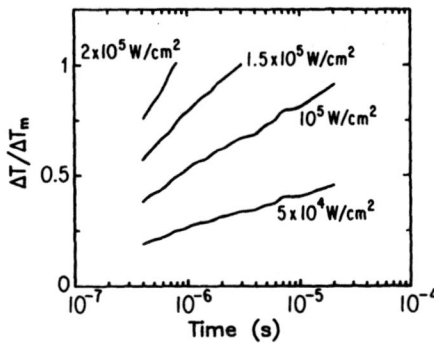

Fig.1. Amorphizing threshold intensity versus laser pulse duration (triangles, squares) for x= 0.64 and 0.67.

Fig.2. Time dependences of the excited spot's temperature calculated from data of Fig.1 and Eq.(4).

The $I_t(\tau)$ corresponds to the condition that at the given absorbed light intensity I and pulse length τ the melting point is just reached. The amorphization tendency of material will be reflected as a termination of the curve $I_t(\tau)$ at a certain pulse length at which the cooling will be no longer fast enough to amorphize the melted spot. In other words, the cooling time to glass transition temperature should be close to or smaller than the minimum crystallization time for the given composition in order to amorphize the spot. From Fig.1 we see that amorphization is still possible for τ as long as several 10^{-5} s whereas the same composition x = 0.67 exhibits minimal crystallization times on the order of 10^{-6} s (see Fig. 5).

The reason for this behavior is that the spot's cooling time after the pulse quickly saturates with increasing the pulse length τ. For long pulses, an almost stationary temperature field is established in the excited spot and the cooling time no longer depends on the heating history. The spot's cooling time is then approximately equal to the time at which the stationary temperature field is established in the excited spot when heated by the laser beam. It can be estimated as $r^2 / a \sim$ 3 μs (r is the spot's radius and a is the thermal diffusivity of the substrate).

In general, the cooling kinetics in the absence of phase changes can be derived using the linearity of the heat conduction equation [9] by representing a rectangular heat pulse with temporal length τ as a superposition of step function heat source H and equal strength sink S switched on at t = 0 and t = τ, respectively. Both H and S are distributed in the excited area of the film according to the absorbed intensity profile. The solution of the heat conduction equation after the pulse is (see e.g. [4])

$$\Delta T_a(t) = \Delta T_H(t) + \Delta T_S(t - \tau) = \Delta T_H(t) - \Delta T_H(t - \tau) \qquad (1)$$

where $\Delta T_H(t)$, $\Delta T_S(t)$ represent the excess temperature fields due to the heat source H and sink S, respectively, and $\Delta T_a(t)$ is the resulting excess temperature field. For long pulses such that $\Delta T_H(\tau) \approx$ const the cooling time after the pulse is close to the time at which $\Delta T_H(t)$ saturates. For shorter pulses, the "heat" term in (1) changes near t = τ and contributes to the cooling rate.

With decreasing pulse length τ below r^2 / a the heat flow in the substrate tends to become 1 - dimensional. At the condition that most of the supplied heat is transferred to the substrate by the end of the pulse ($\tau \geq 10$ ns) the excited spot's excess temperature approximately follows the time dependence $\Delta T(\tau) \sim I \cdot \tau^{0.5}$ [9]. It can be represented as the average temperature of the heated part of the substrate with the thickness $(a\tau)^{0.5}$ by the heat supplied to the unit area of the film $I \cdot \tau$. For amorphization threshold we have $\Delta T(\tau) = \Delta T_m = T_m - T_0 = const$ (where T_m and T_0 are the melting and ambient temperatures), therefore $I_t (\tau) \sim \tau^{-0.5}$. The measured $I_t (\tau)$ of Fig.1 indeed tends to this dependence (dashed line) at $\tau \leq 1$ μs.

From the amorphization threshold intensity dependence on pulse length (Fig.1) the time dependence of the excited spot's temperature during the laser pulse can be obtained. Namely, we assume that the local heating by a focused laser beam giving the absorbed intensity I can be expressed as:

$$\Delta T (t) = I \cdot f (t) \tag{2}$$

(this condition means that the heat transport properties of the system have a weak temperature dependence and that distributed absorption of the beam in the film is the only heat source). Amorphization threshold dependence on pulse length (Fig.1) provides the function $f (t)$:

$$f (\tau) = \Delta T_m / I_t (\tau) \tag{3}$$

Therefore, the time dependence of the spot's temperature is

$$\Delta T (t) = \Delta T_m \cdot I / I_t (t = \tau) \tag{4}$$

In Fig.2 the calculated dependencies $\Delta T(t) / \Delta T_m$ are shown for several absorbed light intensities. All the plotted $\Delta T(t)/\Delta T_m$ increase considerably over the first microsecond or less and tend to saturate at longer t (note the logarithmic time scale).

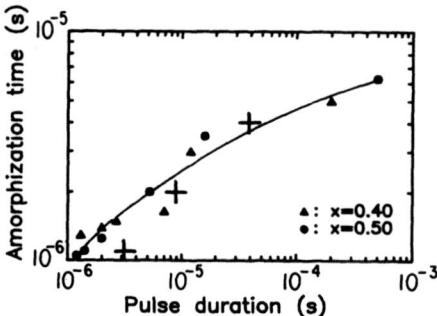

The dependence of the amorphization time on the amorphizing pulse length is shown on Fig.3 for $x = 0.40$ and $x = 0.50$. These measurements were performed on ZnS sandwiched samples at $I \approx (1.3 - 1.5) \cdot I_t$ in order to fully amorphize the spot. For microsecond pulses the amorphization after the pulse proceeds also in microseconds. For longer pulses the saturation of the cooling time with increasing the pulse length is clearly seen: the amorphization time remains ≤ 7 μs for pulses up to 100 μs long.

In order to validate our thermal approach we calculated the cooling time necessary to reach the glass transition temperature T_g after the amorphizing pulse τ at the intensity close to the amorphization threshold $I_t (\tau)$.

Fig.3. Amorphization time (triangles, circles) and calculated cooling time after the pulse (crosses) versus the amorphizing pulse length.

Assuming that for all samples used $T_g \approx 450K$ and $T_m \approx 840K$ we have $\Delta T_g / \Delta T_m \approx 0.3$. The cooling times were calculated for 3 intensities ($I = 10^5$, 1.25×10^5, 1.5×10^5 W/cm^2) using the Equations (1) - (4) and plotted on Fig.3 (crosses). The calculated thermal cooling times agree reasonably well with the experimentally measured amorphization times.

From the compositions studied, samples with $x = 0.40, 0.50, 0.64$ and 0.67 could be amorphized over the almost entire range of available cooling times (1 μs - 7 μs) and have similar I_t and amorphization time values. Samples with $x = 0.55$ and 0.76 could not be amorphized under our experimental conditions.

One can see that for more complete amorphization tendency analysis a wider range of available cooling times is needed. In our study this range is limited by the laser power output from below (≈ 1 μs) and by the spot's geometry from above (<10 μs). More promising would be the controllable change of spot's diameter, substrate thermal properties, switch-off time of the pulse or spot's spatial movement .

The photocrystallization studies were performed on amorphous samples with x = 0.52, 0.55, 0.67, and 0.76. For relatively low light intensities, the optical transmission after the beginning of the laser pulse drops due to crystallization of the spot and reaches the saturation. However, the saturation value of optical contrast depends on the light intensity (see Fig.4).

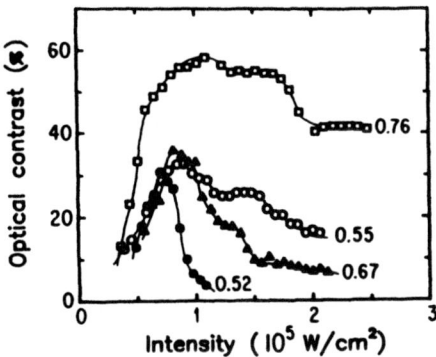

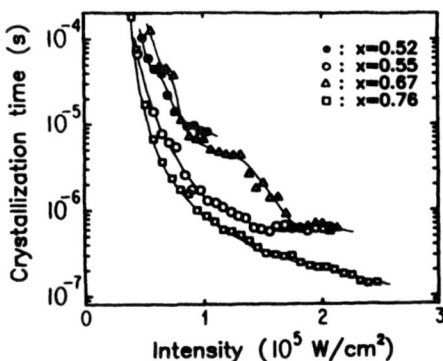

Fig.4. The change in the maximum optical contrast with absorbed light intensity for photocrystallization (x values indicated).

Fig.5. Photocrystallization time dependence on the absorbed light intensity.

For all samples the intensity dependence of the maximum obtainable optical contrast $C(I)$ passes through the maximum near the absorbed intensity value $I_m \approx 8 \times 10^4$ W/cm^2; for x = 0.76 and x = 0.55 the maxima are spread. These maxima divide two types of behavior in crystallization kinetics. For $I < I_m$ the optical transmission $T_{opt}(t)$ decreases monotonically down to saturation. When $I > I_m$ the optical transmission drops, goes through the minimum and rises again before saturation has been reached. Microscopic examination in the latter case shows that before reaching the minimum in $T_{opt}(t)$ the treated spot is partly crystalline. If the light pulse is switched off after this minimum, a small ablative hole is formed in the center of the spot. We suppose that in this case temperatures close to melting point (or interval) T_m are reached in the central part of the excited area before the crystallization has been completed.

The dependence of the absorbed intensity I_c (τ) necessary to reach the minimum in T_{opt} at the end of the pulse τ is plotted on Fig.1 (filled circles). The closeness of I_c (τ) and amorphization threshold intensity $I_t(τ)$ demonstrates that for both curves almost the same temperature near T_m is reached . The fact that $I_c(τ)$ is less than $I_t(τ)$ can be explained by the fact that the ablative hole can be formed in amorphous film at $T < T_m$ due to the high atomic mobility, and by the effect of the released latent heat during the crystallization (more significant for shorter pulses).

The maxima in $C(I)$ can be explained by the two regimes of crystallization. The first occurs for relatively slow heating when the laser heated spot becomes crystallized before its temperature has reached T_m. The second takes place for rapid heating when the spot has not crystallized fully before T_m is reached in the center.

A maximum of $C(I)$ at $I = I_m$ corresponds to the most fully crystallized spot. Subsequent decrease of contrast at intensities higher than I_m occurs because the time necessary for reaching T_m reduces and the degree of crystallization reduces, too. The decrease of $C(I)$ for intensities lower than I_m can be due to the radial distribution of intensity and temperature inside the spot. For these intensities, the periphery of the spot is not able to crystallize completely during the pulse.

In Fig.5 the crystallization time dependence on absorbed intensity is shown. The crystallization time at a given light intensity increases as the composition x is changed in the sequence $0.76 \rightarrow 0.55 \rightarrow 0.67; 0.52$. High crystallization tendency for x = 0.76 and 0.55 is consistent with the widespread maxima in the optical contrast C(I) for these compositions (see Fig.4) which means that in this case the crystallization is so fast that the melting temperature is hardly reached before the end of the crystallization.

Considering that even at the highest cooling rates available (5×10^8 K/s) we could not amorphize samples with x = 0.76 and 0.55 we conclude that the crystallization tendency is the highest for x = 0.76, lower for x = 0.55 and the lowest for x = 0.67 and 0.52. One can notice, however, that the shortest crystallization times for x = 0.67 (near $I = 2 \times 10^5$ W/cm^2) are the same as for x = 0.55 and therefore x = 0.67 sample also could not be amorphized at our cooling rates on the contrary to the observations (Fig.1). The reason why this does not take place is that the crystallization times for x = 0.67 are short only at the highest light intensities, i.e. near the melting temperature. When the temperature is lowered, the crystallization time for x = 0.67 rises considerably faster than for x = 0.55 (see Fig.5). In the cooling of melted spot, the temperature interval close to T_m is passed in a short time (according to Eq.(1) and behavior of $\Delta T(t)$), not allowing to take full use of the shortest crystallization times; therefore, the amorphization tendency for studied x = 0.67 is higher than for x = 0.55.

To summarize, the observed laser-induced crystallization and amorphization properties of $Sb_x Se_{1-x}$ films can be divided into two groups. The first originates from the heating and cooling processes during and after the pulse and depends on the sample and excitation geometry, thermal properties etc. These are responsible for amorphization threshold intensity values, amorphization kinetics (if amorphization is possible at available cooling rates) and the increase in optical contrast due to photocrystallization with light intensity for low intensities. The second is associated with the crystallization tendency itself. For given "extrinsic" conditions it defines the crystallization kinetics, the possibility to amorphize and the onset of the second photocrystallization regime when the melting starts before the end of crystallization. The crystallization tendency of $Sb_x Se_{1-x}$ in the x range studied generally rises with x and has a maximum between x = 0.5 and 0.7 most likely due to the possibility to form a single metastable crystalline phase from amorphous state.

ACKNOWLEDGMENTS

The authors wish to thank Prof. G. Ozolins and Dr. A. Krumina for X - ray diffraction analyses and Prof. K. Shvarts for general support.

REFERENCES

1. S. R. Ovshinsky, J. of Non-Cryst. Sol. **141**, 200 (1992)
2. G. Fuxi, X. Songsheng, Ann. Phys. **1**, 391 (1992)
3. R. Barton, Ch.R. Davis, K. Rubin, G .Lim, Appl. Phys. Lett. **48**, 1255 (1986)
4. K.K. Shvarts, P.J. Stradins, Radiation Effects and Defects in Solids **119** (P2), 881 (1991)
5. Y. Nakayoshi, Y. Kanemitsu, Y. Masumoto, Y. Maeda,
   Jpn.J.Appl.Phys.**31** (P1,2B), 471 (1992)
6. J. Solis, C. Ortiz, C.N. Alfonso, F.Catalina, Appl. Phys. **A54**, 279 (1992)
7. K.Rubin, D.P. Birnie III, M. Chen, J.Appl. Phys. **71** (8), 3680 (1992)
8. O. Balcers, P .Stradins, V. Gebreder, Latvian J. of Phys. Sci. **1**, 11 (1992)
9. H. S. Carslaw, J. C. Jaeger, <u>Conduction of Heat in Solids,</u> 2nd ed.
   (Oxford University Press 1959) ,Ch.X.

ION-BEAM-DRIVEN AMORPHIZATION OF $Ca_2La_8(SiO_4)_6O_2$ SINGLE CRYSTALS

W. J. WEBER[*], N. J. HESS[*], and L. M. WANG[**]
[*]Pacific Northwest Laboratory, P.O. Box 999, Richland, WA 99352, USA
[**]Department of Earth and Planetary Sciences, University of New Mexico, Albuquerque, NM 87131, USA

ABSTRACT

Single crystals of $Ca_2La_8(SiO_4)_6O_2$, with 1% Nd substituted for La, were irradiated with 0.8 MeV Ne^+ and 1.5 MeV Kr^+ ions over the temperature range from 15 K to 773 K. The irradiations were carried out using the HVEM-Tandem Facility at Argonne National Laboratory. The structural changes and the ion fluence for complete amorphization were determined by *in situ* transmission electron microscopy. The ion fluence for complete amorphization increased with temperature in two stages associated with defect annealing processes. The critical temperature for amorphization increased from ~360 K for 0.8 MeV Ne^+ to ~710 K for 1.5 MeV Kr^+. During *in situ* annealing studies, irradiation-enhanced recrystallization was observed at 923 K. Spatially-resolved fluorescence spectra of the Nd ion excited with 488.0 nm laser excitation showed marked line-broadening toward the center of the amorphous regions. Initial measurements indicate the subtle shifts of the $^9I_{9/2}$ groundstate energy levels can be measured by pumping directly into the excited state $^4F_{3/2}$ manifold suggesting that the line broadening observed originates from a distribution of geometrically distorted Nd sites.

INTRODUCTION

Understanding irradiation-induced amorphization and recrystallization in ceramics is important to the use and reliability of advanced ceramics in nuclear power application (fission and fusion), to the safe storage of nuclear waste materials, and to the modification and tailoring of ceramic properties by energetic ion beams. A recent study of polycrystalline $Ca_2Nd_8(SiO_4)_6O_2$ doped with ^{244}Cm [1] showed that amorphization in this material occurs directly within the collision cascade of the Pu recoil nucleus emitted during alpha decay of Cm and that the fully amorphous state is reached at a dose equivalent to 0.4 dpa. Recrystallization of the fully amorphous state in this material occurs in a single recovery stage with an activation energy of 3.1 ± 0.2 eV [2]. In order to better understand the effect of temperature on the amorphization process, the effect of irradiation on recrystallization processes, and changes in the local structure, a systematic study of ion-beam-induced amorphization in single crystal $Ca_2La_8(SiO_4)_6O_2$ has been undertaken as part of a larger collaborative effort that includes other orthosilicates [3,4]. This paper expands on previous results [5,6], addresses the effects of temperature on irradiation-induced amorphization and recrystallization in $Ca_2La_8(SiO_4)_6O_2$, and reports on the use of spatially-resolved fluorescence techniques to investigate local distortions.

EXPERIMENTAL PROCEDURES

The single crystals of $Ca_2La_8(SiO_4)_6O_2$ were grown from a stoichiometric melt by the Airtron Corp. and have the hexagonal apatite ($P6_3/m$) structure with lattice parameters: $a_o =$

0.9651(3) nm and c_o = 0.7151(2) nm. Specimens with [001] orientations were cut as thin sections from the crystals and prepared as TEM specimens by Ar^+ ion milling. The irradiations were carried out at the HVEM-Tandem Facility at Argonne National Laboratory, which consists of a modified Kratos/AEI EM7 high voltage electron microscope (HVEM) and a 2 MV tandem ion accelerator [7]. Irradiations with 0.8 MeV Ne^+ and 1.5 MeV Kr^+ were performed 5 to 10° off the zone axis of the oriented specimens over the temperature range from 15 K to 773 K to investigate the temperature dependence of amorphization. The displacement dose (dpa) for complete amorphization in the electron transparent thickness (~200 nm) was calculated using TRIM-90 (full cascades) [8] and an assumed displacement energy of 25 eV. The progression of the amorphization process during irradiation of the single crystal $Ca_2La_8(SiO_4)_6O_2$ specimens was followed in the electron transparent thickness of the specimens by *in situ* transmission electron microscopy and selected area electron diffraction. Post-irradiation characterization of the structural changes at intermediate dose levels was carried out by HRTEM and spatially-resolved fluorescence spectroscopy.

The spatially-resolved fluorescence spectra were collected at room temperature with a Raman microprobe fitted with 40X objective, which was used to image the laser excitation on to the post-irradiation TEM disks. The fluorescence from the specimens was collected in backscattering geometry and imaged onto the slits of a 0.5 meter triple spectrometer equipped with 300 and 150 groves/mm gratings in the dispersive and filter stages, respectively. This choice of gratings allowed collection of a spectral window from approximately 855 to 930 nm. The fluorescence, which was detected with a liquid nitrogen cooled Charged Coupled Device (CCD), was excited with approximately 50 mW of both 488.0 radiation from an argon ion laser and tunable 800 to 818 nm radiation from a solid state titanium-doped sapphire laser. All spectra were collected with 50 μm exit slits.

RESULTS AND DISCUSSION

At low doses, a diffuse halo associated with the presence of amorphous material appears in the electron diffraction patterns. As shown previously [6], the intensity of this diffuse halo increases with increasing dose, while the intensity of the diffraction maxima from the remaining crystalline material decreases and eventually disappears at the dose for complete amorphization. Post-irradiation HRTEM revealed amorphous regions 2 to 3 nm in size at low doses. At higher doses, the specimens are completely amorphous in the thin regions; however, a few randomly oriented nanocrystallites are occasionally observed, suggesting recrystallization may occur within the core of some displacement cascades.

The dose for complete amorphization in $Ca_2La_8(SiO_4)_6O_2$ increases with temperature in two stages under both 0.8 MeV Ne^+ and 1.5 MeV Kr^+ irradiations, as shown in Fig. 1. At 15 K, the dose for complete amorphization is 0.36 and 0.32 dpa for the 0.8 MeV Ne^+ and 1.5 MeV Kr^+ irradiations, respectively, indicating that within experiment error (10%) the amorphization dose is independent of recoil-energy spectra at this temperature. Stage I occurs below ~200 K and has been speculated [6] to be due to intracascade close-pair recombination. In Stage II (above ~200 K), the amorphization dose continues to increase with temperature due to the mobility of other intracascade defects. Irradiation with 1.5 MeV Kr^+ to 4.4 dpa at 773 K produced no observable amorphization. As discussed previously [6], the more rapid increase in amorphization dose with temperature for the 0.8 MeV Ne^+ irradiation is due to the larger fraction of recoverable defects and smaller cascades produced by the low energy recoils.

The effects of temperature on the dose for complete amorphization can be described by

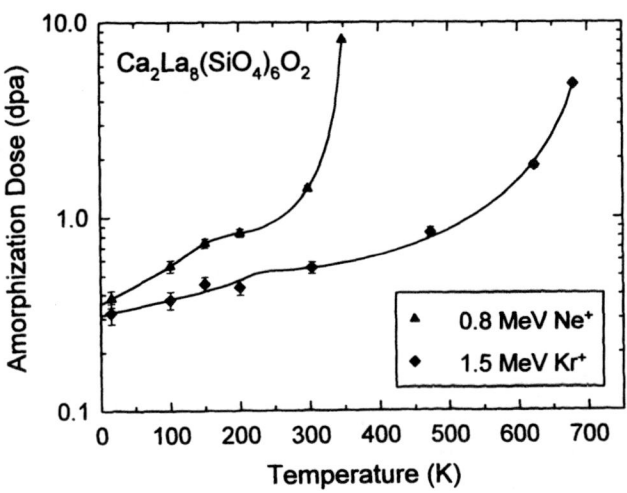

Fig. 1 Temperature dependence of the dose for complete amorphization.

kinetic processes with activation energies, E_a, associated with each annealing stage. As described in more detail elsewhere [5,6], the relationship between temperature, T, and dose, D, for complete amorphization is given by:

$$\ln(1 - D_o/D) = \ln(1/\phi\sigma\tau) - E_a/kT \qquad (1)$$

where D_o is the dose for complete amorphization at T = 0 K (defined for each annealing stage), ϕ is the ion flux, σ is the cross section for direct amorphization, τ is a time constant, and k is the Boltzmann constant.

By applying the above model (Eq. 1) to the data in Fig. 1, the activation energies, E_a, can be determined from an Arrhenius-type plot of $\ln(1 - D_o/D)$ versus $1/kT$. The amorphization dose, D_o, at T = 0 K associated with Stage I annealing is estimated to be 0.36 and 0.32 dpa for the 0.8 MeV Ne$^+$ and 1.5 MeV Kr$^+$ ions, respectively. For Stage II annealing, D_o is estimated to be 0.80 and 0.52 dpa for the Ne$^+$ and Kr$^+$ irradiations, respectively. The results of this analysis, along with a linear fit, are shown in Fig. 2. The activation energies determined for each stage are nearly identical for both the Ne$^+$ and Kr$^+$ irradiations, consistent with the presence of similar irradiation-produced defects. The activation energy determined by linear regression for Stage I annealing is 0.01 ± 0.003 eV, and the activation energy determined for Stage II annealing is 0.13 ± 0.02 eV. These values are considerably less than the activation energy for thermal recrystallization (3.1 eV) of fully amorphous $Ca_2Nd_8(SiO_4)_6O_2$ [2], but are similar in magnitude to the activation energies determined previously for zircon [4]. The Stage II activation energy is also similar in magnitude to activation energies previously reported for irradiation-enhanced epitaxial recrystallization in other materials [9,10].

In addition to the activation energies, the critical temperature, T_c, above which complete amorphization does not occur (D = ∞), can be determined from the linear fit to the Stage II data

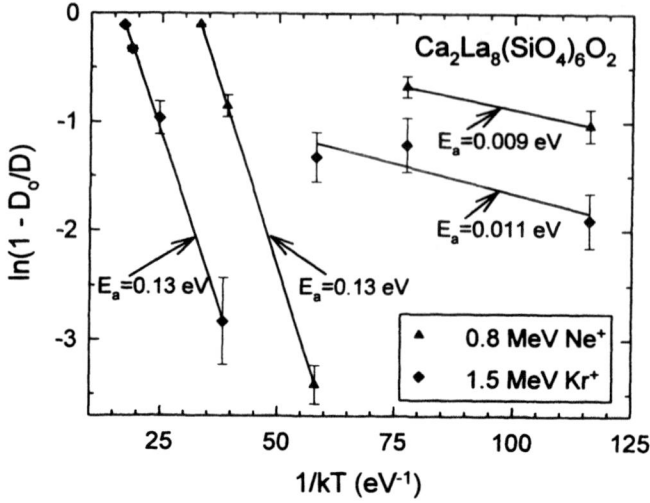

Fig. 2 Arrhenius-type plot of $\ln(1 - D_o/D)$ versus $1/kT$.

in Fig. 2. Since $\ln(1 - D_o/D) \to 0$ as $D \to \infty$, Eq. 1 yields the following expression for T_c:

$$T_c = E_a / [k \ln(1/\phi\sigma\tau)]. \tag{2}$$

Based on the activation energies, E_a, and intercepts, $\ln(1/\phi\sigma\tau)$, determined for Stage II annealing, T_c is ~360 K for the 0.8 MeV Ne$^+$ irradiations and ~710 K for the 1.5 MeV Kr$^+$ irradiations. This increase in T_c with projectile mass is similar to behavior previously observed in Si [11] and CuTi [12].

In situ recrystallization of a fully amorphous specimen was studied in the HVEM/Tandem facility as a function of temperature, with and without a 1.5 MeV Kr$^+$ beam. At 823 and 873 K, annealing for 30 minutes with and without the ion beam produced no observable recrystallization. Annealing for 30 minutes at 923 K without the ion beam on the specimen also produced no observable recrystallization; however, at 923 K with the ion beam on the specimen, recrystallization commenced almost immediately. After 2 minutes at 923 K with the ion beam on, fine crystallites of ~50 nm diameter were observed in the bright-field image and crystalline diffraction spots were observed in the electron diffraction pattern. This irradiation-enhanced recrystallization process continued over a 38 minute period as a randomly-oriented polycrystalline structure developed with crystallites of ~300 nm diameter. The activation energy for this recrystallization process can be estimated, as shown previously [2], following the method of analysis of Primak [13], where the most probable activation energy, E_o, for an annealing process occurring at a particular temperature, T, and time, t, is given by the expression:

$$E_o = kT \ln(Bt) \tag{3}$$

438

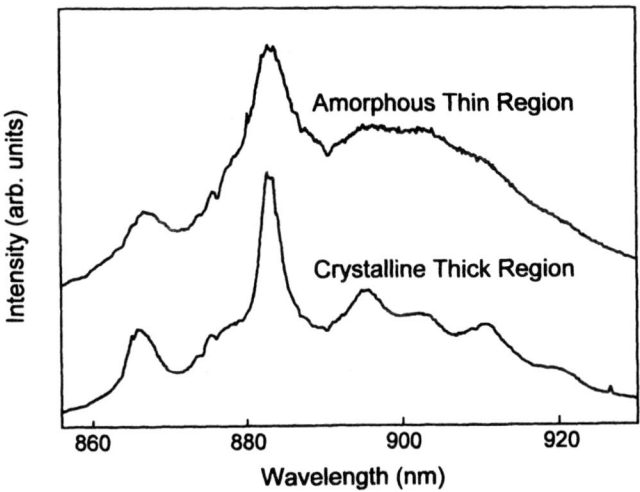

Fig. 3 Fluorescence spectra in $Ca_2La_8(SiO_4)_6O_2$ irradiated with 1.5 MeV Kr^+ to 0.37 dpa at 100 K.

where B is a frequency factor. The frequency factor for recrystallization of the radiation-induced amorphous state in $Ca_2Nd_8(SiO_4)_6O_2$ has been previously determined to be 5×10^{14} s^{-1} [2]. Using this value and the above results, the activation energy for recrystallization in $Ca_2La_8(SiO_4)_6O_2$, based on Eq. (3), is estimated to be 3.2 eV, which is within the experimental uncertainty of the value (3.1 ± 0.2 eV) rigorously determined for Cm-doped $Ca_2Nd_8(SiO_4)_6O_2$ [2].

The absorption spectra of a thin polished section of unirradiated $Ca_2La_8(SiO_4)_6O_2$ single crystal were collected at room temperature on a NIR/UV-VIS double-beam spectrophotometer from 300 to 1000 nm. High resolution absorption spectra were collected from 650 to 950 nm to identify the location and structure of the electronic excited state multiplets, namely the $^4F_{7/2}$ and $^4S_{3/2}$ manifold spanning 740 to 752 nm, the $^4F_{5/2}$ and $^2H_{9/2}$ manifold from 798 to 806 nm, and the $^4F_{3/2}$ multiplet at 868 to 894 nm. In the unirradiated single crystal, the fluorescence emission centered at approximately 900 nm, which indicates that it originates from the $^4F_{3/2}$ multiplet at 868 to 894 nm and represents the transition to the $^4I_{9/2}$ groundstate multiplet. Eight easily resolvable peaks are observed in this region indicating that the J-degeneracy is completely removed by crystal field effects. Based on the absorption and the fluorescence spectra, the $^4F_{3/2}$ level is estimated to be split by 130 cm^{-1} and the five levels of the $^4I_{9/2}$ multiplet to be split by approximately 220 cm^{-1} each. However, low temperature measurements are needed for an accurate determination of the crystal field splitting.

Using the Raman microprobe, spatially-resolved fluorescence measurements excited with 448.0 nm radiation were made in the thin regions of a TEM specimen irradiated with 1.5 MeV Kr^+ to 0.37 dpa at 100 K to a fully amorphous state in the electron transparent thickness. Spectra were taken at 10 μm intervals from the edge of the electron transparent (amorphous) region to thicker regions. In the thin regions, where the material is fully amorphous, the fluorescence spectra exhibited marked line-broadening, as shown in Fig. 3. The spectra collected further away from the thin region displayed progressively narrower line shapes as the specimen thickness increased and more crystalline material is probed by the laser. The narrower lineshapes in the thicker regions reflect the larger contribution of the crystalline material, beneath the amorphous surface damage, to the spectra. Using the tunable red excitation, fluorescence spectra were collected at a location within the ion beam irradiated region that contained a large amorphous component. As the excitation wavelength was tuned within the $^4F_{5/2}$ and $^2H_{9/2}$ manifold, subtle frequency and linewidth changes were observed in the emission spectra suggesting that the fluorescence spectrum originates from a distribution of distorted sites. Similar experiments conducted at liquid He temperature are planned.

ACKNOWLEDGEMENTS

The authors would like to thank the HVEM-Tandem Facility staff at Argonne National Laboratory for assistance during the ion irradiations. This research was supported by the Division of Materials Sciences, Office of Basic Energy Sciences, U. S. Department of Energy under Contract DE-AC06-76RLO 1830.

REFERENCES

[1]   W.J. Weber, J. Am. Ceram. Soc. **76** [7], 1729 (1993).

[2]   W.J. Weber, Radiat. Eff. **77**, 295 (1983).

[3]   L.M. Wang and R.C. Ewing, MRS Bulletin, **XVII** [5], 38 (May, 1992).

[4]   L.M. Wang, R.C. Ewing, W.J. Weber, and R.K. Eby, in: *Beam-Solid Interactions: Fundamentals and Applications,* edited by M.A. Nastasi, L.R. Harriott, N. Herbots, and R.S. Averback (Mater. Res. Soc. Proc. **279**, Pittsburgh, PA, 1993) pp. 451-456.

[5]   W.J. Weber and L.M. Wang, in: *Beam-Solid Interactions: Fundamentals and Applications,* edited by M.A. Nastasi, L.R. Harriott, N. Herbots, and R.S. Averback (Mater. Res. Soc. Proc. **279**, Pittsburgh, PA, 1993) pp. 523-528.

[6]   W. J. Weber and L. M. Wang, Nucl. Instr. and Meth. **B** (1994) in press.

[7]   C.W. Allen, L.L. Funk, E.A. Ryan, and A. Taylor, Nucl. Instr. and Meth. **B40/41**, 553 (1989).

[8]   J.F. Ziegler, J.P. Biersack, and U. Littmark, *The Stopping and Range of Ions in Solids* (Pergamon Press, New York, 1985).

[9]   G. Carter and M.J. Nobes, J. Mater. Res. **6** [10], 2103 (1991).

[10]   J. Delage, O. Popoola, J.P. Villain, and P. Moine, Mater. Sci. and Engr. **A115**, 133 (1989).

[11]   F.F. Morehead and B.L. Crowder, Radiat. Eff. **6**, 27 (1970).

[12]   J. Koike, P.R. Okamoto, and L.E. Rehn, J. Mater. Res. **4** [5], 1143 (1989).

[13]   W. Primak, *The Compacted States of Vitreous Silica* (Gordon and Breach, New York, 1975) p. 131.

ION BEAM INDUCED GRAPHITIZATION OF PHENOLFORMALDEHYDE

D. ILA, A. L. EVELYN AND G. M. JENKINS
Center for Irradiation of Materials, Alabama A&M University, P.O. Box 1447, Normal, AL 35762-1447, U.S.A.

ABSTRACT

We have studied MeV ion beam induced phase transformation in phenolformaldehyde cured at 170°C by both Raman microprobe spectroscopy and *in situ* resistance measurement of the irradiated area. Samples were irradiated using various doses of protons, alphas and nitrogen beams. Irradiated volumes in each sample were tested *in situ* for enhanced electrical conductivity and later on by Raman microprobe spectroscopy. The results have shown changes in the resistance as much as seven orders of magnitude by alpha particles, six orders by nitrogen bombardment and three orders by hydrogen. Raman microprobe spectroscopy of the darkened phase shows development of the D- and G-lines which are characteristic of the production of a carbonized resin. These spectra indicate that maximum carbonization was caused by the nitrogen beam.

INTRODUCTION

Ion bombardment and implantation have been widely used to alter the electrical properties of polymers and the results have been reviewed elsewhere [1-3]. Ion implantation produces accurate location of dopant depth and doping with ions that cannot be achieved chemically. Most workers report enhanced conductivity which saturates after high fluence. This enhancement is associated with darkening. Ion implantation causes physical damage which includes breaking covalent bonds, scissoring polymer chains, cross-linking and release of volatile species. It is also reported that conductivity enhancement by ion bombardment is a result of energy deposition and densification rather than ion doping [4]. Generally, the darkening has been attributed to beam induced carbonization, without specifying the form of carbon produced or studying the mechanism of carbonization. It has also been reported that such darkened polymers are resistant to oxidation and retain enhanced conductivity over many months [5]. This enhanced conductivity induced by ion bombardment decreases with time [6]. We have been concentrating on the effect of MeV ion bombardment which allows us to alter the polymer structure locally well below the surface; the higher the energy the deeper the damaged volume.

A recent study [7] has shown that the electrical conductivity of a polyimide increases several orders of magnitude after ion beam bombardment. Surface conductivity change is accompanied by a shift to lower binding energy of all core level spectra for both normal and grazing emission. The authors ascribe these changes to the formation of a conductive phase of interconnected aromatic carbon characterized by C1s spectra which approach those of graphite. Formation of this conducting phase seems to be a common effect of inert ion beam treatment of resins and depends on ion fluence.

We have studied this effect in phenolic resin using various bombarding ions because we had already studied the thermal conversion of this resin into a polymeric glassy carbon [8] and

441

wished to compare this with irradiation induced conversion. We are also considering this resin for encapsulation of nuclear waste. Precursor resol $C_7H_8O_2$, a liquid, converts to fully cured phenolic resin C_7H_6O of specific gravity 1.25 on heating at 170°C. This resin further transforms without disruption and with no change in shape to glassy carbon of specific gravity 1.45 on heating to 1000°C. Heat treatment above 550°C produces an intermediate conducting hydrocarbon medium which improves progressively in conductivity as the temperature is raised further and hydrogen is released. The final product consists of long ribbon-like molecules of sp^2 carbon atoms aggregated locally to form sub-crystalline domains which are arranged randomly in space [8]. Our main objective was to obtain qualitative information on changes induced in the resin precursor to this structure on bombardment by MeV protons, alphas and nitrogen ions.

EXPERIMENTAL PROCEDURES

Resin film about 30μm thick was spin-coated on glass slides, and heated from ambient to 170°C in argon at 10°C per hour. Conducting aluminum electrode strips were evaporated onto these slides. Using TRIM [9] code, we predetermined the thickness of these electrodes for each ion species to ensure that none of the selected incident ions could pass through them. The TRIM prediction for damage caused by 0.5, 1.0, 2.0 and 3.5 MeV alphas in a C_7H_6O resin showed that maximum damage occurs at depths of 2.4, 4.2, 8.7 and 17.3 μm and very few alphas penetrate further than the maximum damage depth. Most helium atoms will thus be lodged in the damage volume, from which they will probably migrate to the surface. Protons will leave hydrogen atoms in a similar volume which will, in all probability, combine with surrounding carbon.

We also used TRIM to choose the energy of protons (1 MeV) and alphas (3.9 MeV) to have the same shallow depth penetration of 17 μm. To compare nitrogen irradiated with alpha irradiated sample we chose 1.3 MeV alphas and 6.5 MeV nitrogen ions to give 4.5 μm penetration depth. Resistivities of the irradiated resin films were measured in situ using an I-V curve tracer connected to the aluminum strips.

We compared the radiation effects of protons and alphas, using a 200 nA proton beam current and a 20 nA alpha beam current to avoid cracking the films by rapid carbonization and to allow thermal diffusion due to bombardment in the radiated volume. We then compared irradiation induced carbonization by 1.3 MeV alphas and 6.5 MeV nitrogen ions, using 50 nA of alphas and 500 pA of nitrogen ions, again to avoid cracking. Using TRIM we predicted the ratio of ballistic damage by alphas to that by protons and the ratio of ballistic damage by nitrogen to that by alphas for the same fluence. This information was used to select the fluence needed for each irradiation for like damage by these three ions at the above energies.

Structural changes were studied by Raman microprobe spectroscopy using 514 nm laser excitation, concentrating on spectral features between 1000 cm^{-1} and 2000 cm^{-1} related to the electrical properties measured in situ.

RESULTS

The integrated charge on each irradiated resin film and the change in its resistance was measured in situ over several days with the above beam parameters. The behavior of the measured I-V for each sheet was monitored to register cracking in the irradiated film which manifested as a noisy and irregular I-V curve. As soon as these features were seen the experiment was halted to make direct visual confirmation of damage.

In all cases the radiation induced darkening of the resin film; in many cases no film disruption developed even after high doses. This encourages us to proceed further with the induction of finely defined conducting paths in the resin insulator. Figure 1 shows the plot of resistance versus ion fluence for samples bombarded with 1 MeV protons, 3.9 MeV alphas and 6.5 MeV nitrogen ions. As indicated in this figure, the form of the resistance/dose curve was similar for all bombarding particles. There was little effect until a critical fluence was experienced; thereafter resistance reduced rapidly. The critical fluence for rapid decrease in resistance was lowest for nitrogen ions. The alpha and proton critical doses were three orders of magnitude grater than that of nitrogen. So far the greatest increase in conduction without cracking the carbonized film has been recorded by alphas.

The Raman spectrum of thermally carbonized phenolic resin contains two peaks: at 1595 cm^{-1}, representing the G/D' line which is attributed to in-plane vibrations of aromatic layers, and at 1350cm^{-1}, representing the D line which is attributed to "disorder" in such layers [10], but could also be related to lower stiffness in bonds at the edges of aromatic layers. Figure 2 shows the growth of both peaks upon heat treatment to 525°C. We emphasize that these pyrolyzed resins are not pure carbons and still contain much hydrogen.

Figure 3 shows the Raman spectra for proton irradiated (5.7 x 10^{17} ions/cm^2), alphas irradiated (3.2 x 10^{17} ion/cm^2) and nitrogen irradiated (1.4 x 10^{14} ions/cm^2) samples. There is a stronger G/D' signal for the nitrogen irradiated sample, perhaps due to high damage. A D-peak also appears in both alpha and nitrogen spectra. The G/D' signal from nitrogen bombarded resin was stronger than from samples bombarded by alphas and also stronger than that from samples bombarded by protons. The D-line could not be discriminated in the sample bombarded by protons, whereas there was a strong indication of a D-line in nitrogen irradiated samples, somewhat less in alpha irradiated samples.

DISCUSSION

The low critical fluence for rapid reduction in resistance observed using nitrogen ions suggests that conduction aromatic layers are quickly established for incident ions with high stopping power. The high rise in conduction found for the sample irradiated with a high fluence of alphas contrasts with the small decrease observed by others on irradiating fully carbonized resin [11,12]. This observation, in addition to the Raman spectrum from this sample, suggests that localized carbonization has been effected which may eventually produce polymeric carbon. However, we can distinguish only between graphitizable and non-graphitizable carbon by completing the carbonization and annealing at high temperatures (2500°C).

It is appropriate to discuss what possible form of carbon may be produced by irradiation damage. If this causes local heating the resulting carbon will be polymeric by an analogy to previous results on pyrolysis of this resin. However, bombardment does produce C-C disruption and polymer debris may recombine to form lower energy graphitic carbon. The change induced by proton bombardment is much less than by alphas, perhaps because stopped protons tend to react with carbonized resin, although their concentration is small.

CONCLUSIONS

We have demonstrated that phenolic resin films can be bombarded with MeV ions without causing disruption. The irradiated volume deep below the surface is proved to contain carbonized material by measurement of a large local increase in electrical conductivity and by

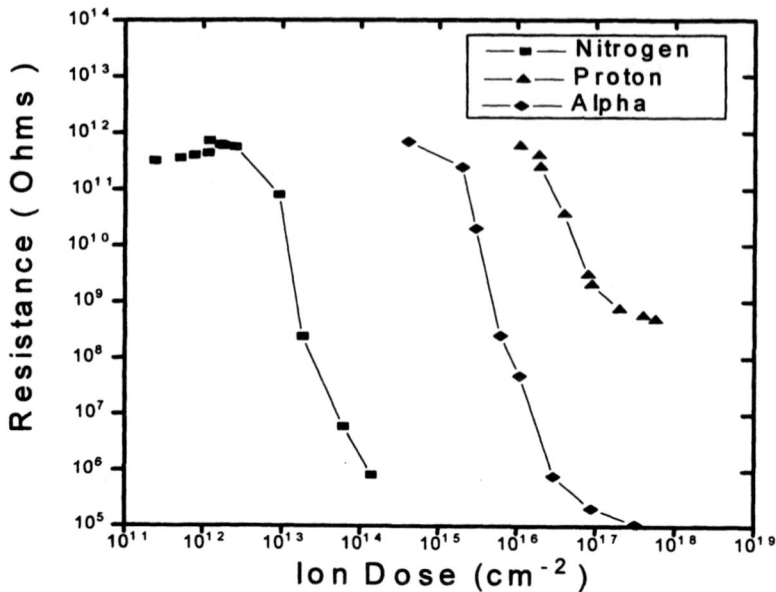

Figure 1. Plot of resistance versus ion fluence for samples bombarded with 1 MeV protons, 3.9 MeV alphas and 6.5 MeV nitrogen ions.

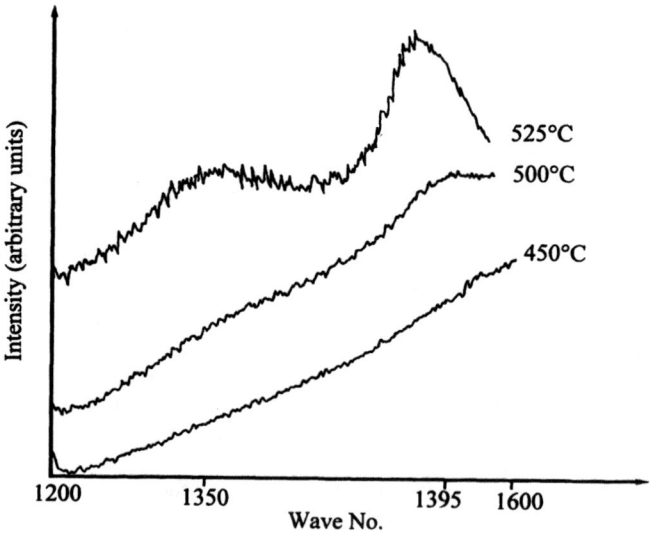

Figure 2. Raman spectra of thermally carbonized phenolic resin showing peak growth at 1350 cm^{-1} and at 1595 cm^{-1}.

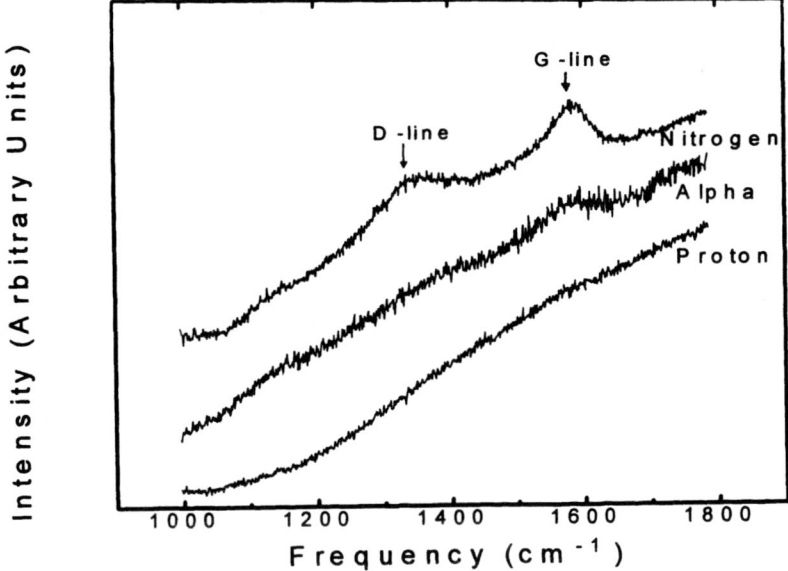

Figure 3. Raman spectra for 1 MeV proton irradiated (5.7×10^{17} ion/cm^2), 1.3 MeV alpha irradiated (3.2×10^{17} ion/cm^2) and 6.5 MeV nitrogen irradiated (1.4×10^{14} ion/cm^2) samples.

Raman structural analysis. We hope that in the future we will be able to carbonize the resin locally, with fine control of translation beam movement and depth (by varying beam energy) using a microbeam focusing assembly, to develop nano-circuitry and nano-fabrication of carbon devices.

ACKNOWLEDGMENTS

The authors acknowledge the help of E. K. Williams, A. Crastes and Y. Qian in sample preparation and discussion of the results. This project is supported by National Science Foundation EPSCoR-II (Alabama) Grant No. EHR-9108761 and the Howard J. Foster Center for Irradiation of Materials at Alabama A&M University.

REFERENCES

1. M. S. Dresselhaus, B. Wasserman and G. E. Wnek (Mater. Res. Soc. Symp. Proc. **27**, Elsevier Science Pub., New York) pp. 413.
2. L. Calcagno and G. Foti, Nucl. Instr. Methods **B59/60**, 1153 (1991).
3. Y. Wang, S. S. Nohite, L. B. Bridewell, R. E. Giedd and C. J. Schofield, J. Mater. Res. **8**, 388 (1993).
4. C. J. Sofield, C. J. Bedell, P. R. Graves and L. B. Bridwell, Nucl. Instr. and Meth. **B67**, 432 (1992).

5. T. Venkatesan, R. P. Levi, T. C. Banwell, T. A. Tomberello, M-A. Nicolet, R. Hamm and A. E. Meixner (Mater. Res. Soc. Proc. **45**, Pittsburgh, PA, 1985) pp. 189.
6. M. Döbeli, T. J. Jones, A. Lee, R. P. Levi and T. A. Tombrello, Rad. Eff. Defects Solids **118**, 325 (1991).
7. H-S. Jeoung and R. C. White, in *Phase Formation and Modification by Beam-Solid Interactions*, edited by G. S. Was, L.E. Rehn and D. M. Follstaedt (Mater. Res. Soc. Proc. **235**, Pittsburgh, PA, 1991) pp. 787-792.
8. G. M. Jenkins and K. Kawamura, *Polymeric Carbons–Carbon Fibre, Glass and Char* (Cambridge University Press, 1976).
9. J. F. Ziegler, J. P. Biersack and U. Littmark, *The Stopping and Range of Ions in Solids* (Pergamon, New York, 1985).
10. R. Vidan and D. B. Fischbach, J. Am. Chem. Soc. **61**, 13 (1978).
11. D. McCulloch and S. Prawer in *Beam-Solid Interactions: Physical Phenomena,*edited by J. A. Knapp, P. Børgesen and R. A. Zuhr (Mater. Res. Soc. Proc. **257**, Pittsburgh, PA, 1990) pp. 825-830.
12. G. M. Jenkins, D. Ila and E. K. Williams in *Polymer/Inorganic Interfaces*, edited by R. L. Opila, F. J. Boerio and A. W. Czanderna (Mater. Res. Soc. Proc. **304**, Pittsburgh, PA, 1993) pp. 173-177.

PART VI

Interface Phenomena

QUESTIONS AND ANSWERS ON THE ACTIVATION STRAIN

MICHAEL J. AZIZ
Division of Applied Sciences, Harvard University, 29 Oxford St., Cambridge MA 02138

ABSTRACT

The activation strain tensor describes the effect of nonhydrostatic stresses on atomic or interfacial mobilities. It has been measured for solid phase epitaxial growth of crystalline Si (001) into amorphous Si. The activation strain concept is explained and some subtle points are discussed. Implications for proposed mechanisms of solid phase epitaxy are reviewed, and new implications for combined bulk and interfacial control are presented. Questions raised during the oral presentation are answered.

INTRODUCTION

Nonhydrostatic stresses are common during epitaxial growth as well as in other crystal growth situations. An understanding of their effects on diffusion and crystallization is therefore an important part of the study of the stability of epitaxially grown materials. There has been a lot of work on the effect of stress on the thermodynamics of growing solids, but the effects of stress on the atomic or interfacial mobilities themselves has not been addressed.

Today all silicon integrated circuits are doped by ion implantation. Some regions get high enough implant doses to create an amorphous surface layer. The implantation damage can be removed by Solid Phase Epitaxial Growth (SPEG), in which the crystal grows back into the amorphous phase as the crystal/amorphous interface moves in a plane-front manner back to the free surface. There is also interest in SPEG for lateral overgrowth in the production of SOI structures. SPEG is conceptually and experimentally among the simplest cases of crystal growth to study, because you can do it in a pure monatomic system — by implanting silicon instead of a dopant, for example — so that long range transport of impurity atoms to or from the interface doesn't complicate the picture; also because the thermodynamics of the two phases are well understood, and because the interface remains planar during growth. Solid Phase Epitaxy can therefore be thought of as one of your prototypical crystal growth processes.

We have been able to rule out a number of proposed atomistic mechanisms for SPEG. We drew these conclusions from the results of our experiments on the dependence of the SPEG rate on hydrostatic pressure, doping, and state of relaxation of the amorphous phase [1]; a nonhydrostatic-stress dependence measurement [2]; and an extension of transition-state theory to nonhydrostatic stress states to develop a formalism for the interpretation of nonhydrostatic stress effects on kinetic processes [2]. The initial stages of the work are published in [1,2]. What makes this development of particular significance is that we have been able to use these measurements and the activation strain interpretation to successfully make predictions for the behavior under conditions where it had not yet been measured. The formalism is completely general, and our procedures can be readily generalized to other situations involving thermally activated atomic transport processes such as diffusion and creep, in addition to crystal growth. In this paper I shall focus on the unique aspects of the nonhydrostatic stress work, which is newer and generally less familiar than the other work.

EXPERIMENTS

The interface velocity v during SPEG of Si and Ge (001) is *enhanced exponentially by hydrostatic pressure* in the 0-5 GPa (0-50 kbar) regime [1]. The magnitude of the enhancement in Si is approximately a factor of ten per 5 GPa of pressure. The results are characterized by an activation volume ΔV^* that is large and negative, being -28% of the atomic volume in Si and -45% of the atomic volume in Ge.

449

In contrast, the growth rate *decreases with uniaxial compression* in the plane of the interface, and is enhanced by uniaxial tension, as shown in Fig. 1 [2]. The bar-shaped samples in Fig. 1 (a) were deformed elastically at temperatures high enough for solid phase epitaxy to proceed but low enough to prevent plastic deformation of the crystal. The stress state so induced is uniaxially compressive in the x direction (left-right on the page) on the top side of the wafer and uniaxially tensile on the bottom side; the magnitude of the stress varies linearly with position from the center out to the left and right edges. These apparently contradictory effects of hydrostatic vs. nonhydrostatic compression have been explained by an extension of the theory of thermally activated growth to nonhydrostatic stress states.

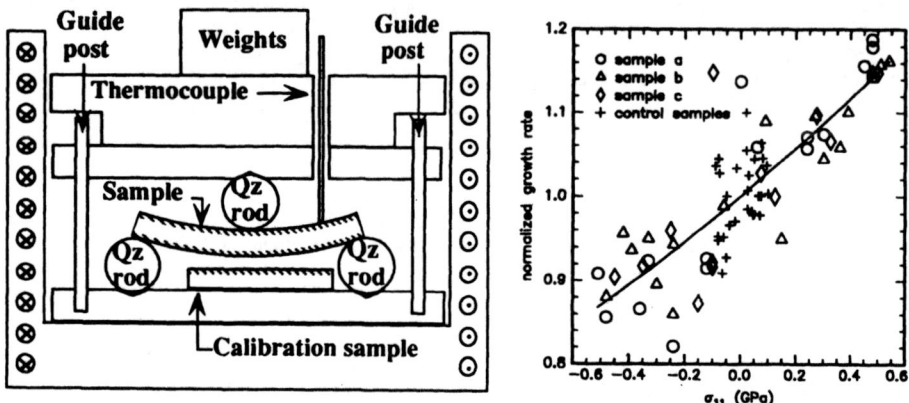

Fig. 1. (a) Schematic cross section of three-point bending apparatus for annealing wafers under nonhydrostatic stress. Cross-hatched areas correspond to amorphous Si. (b) Variation of average solid phase epitaxy rate with applied stress. Positive stresses are tensile.

THEORY OF THERMALLY ACTIVATED GROWTH

The rate of a kinetic process is generally written as the product of a mobility for that process with some measure of the driving force for the process. In this work we are addressing the dependence of the mobility, rather than of the driving force, on nonhydrostatic stress. In the theory of thermally activated growth, applied to any defect-mediated, unimolecular growth process occurring under hydrostatic pressure, the rate is given by

$$\text{Rate} = (\text{constant}) \cdot C \cdot m \cdot [1 - \exp(\Delta G / kT)] \tag{1}$$

where C is the defect concentration, m the defect mobility, kT has the usual meaning, and ΔG in the thermodynamic factor is the Gibbs free energy of the growing phase minus that of the parent phase ($\Delta G < 0$). The defect mobility is proportional to a Boltzmann factor in the barrier height, $\exp(-\Delta G_m / kT)$, where ΔG_m is the standard Gibbs free energy of the system in the transition state (saddle point in configuration space) minus that in the initial state [3]. If the defect concentration equilibrates rapidly, C is proportional to a Boltzmann factor in the standard Gibbs free energy of defect formation, $\exp(-\Delta G_f^0 / kT)$. It is also possible that the time scale for equilibration of the defect concentration is much longer than the duration of the experiment, in which case C is a constant. The formation and migration terms are often combined into a free energy of activation, resulting in

$$\frac{\text{Rate}}{\text{thermodynamic factor}} \equiv M = (\text{constant}) \exp(-\Delta G^* / kT), \tag{2}$$

where M is the mobility for the reaction and $\Delta G^* = \Delta G_m + \Delta G_f^0$ or $\Delta G^* = \Delta G_m$ as the case may be. In SPEG, the very slow variation of the thermodynamic factor (compared to that of the Boltzmann factor) with pressure and temperature permits us to treat the term $[1-\exp(\Delta G/kT)]$ as a constant, as illustrated in the Questions section below.

The familiar activation enthalpy ΔH^* comes from the dependence of ΔG^* on reciprocal temperature; ΔH^* becomes the activation energy ΔE^* at zero pressure. The activation volume ΔV^* comes from the dependence of ΔG^* on pressure using the thermodynamic identity $V = \partial G/\partial p|_T$:

$$-kT \ \frac{\partial \ell n \, \upsilon}{\partial p} = \Delta V^*. \tag{3}$$

When ΔG^* is the sum of a formation and migration term, so also will be ΔH^*, ΔE^* and ΔV^*.

The formation energy is the change in the energy of the system upon the formation of a defect in its standard state. The formation volume ΔV_f^0 is the corresponding volume change, which is close to $+1 \, \Omega$, where Ω is the atomic volume, for vacancy formation and close to $-1 \, \Omega$ for self-interstitial formation, as illustrated in Fig. 2 (a) and (c). The application of pressure will reduce the equilibrium vacancy concentration and increase the equilibrium self-interstitial concentration.

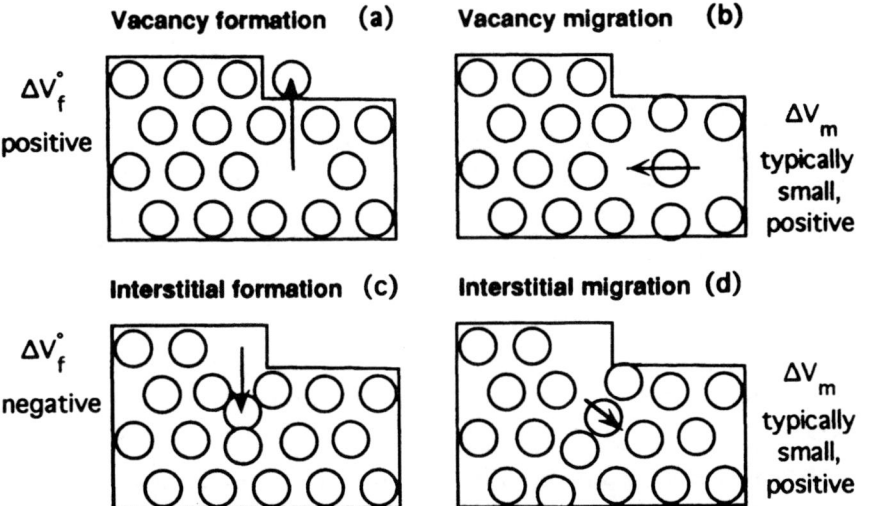

Fig. 2. *Schematic of formation and migration volumes for self diffusion by vacancy and interstitial mechanism in a system with a hard-sphere potential .*

The migration energy is the energy of the system in the transition state minus that of the system in the initial state. The migration volume ΔV_m is the volume of the system in the transition state minus that in the initial state. If the transition state involves a local dilation of the structure, as illustrated in Fig. 2 (b) and (d), then pressure will decrease the rate of fluctuation to such a high-volume transition state, resulting in a decreased defect mobility. Point defect migration volumes tend to be somewhat smaller in magnitude than formation volumes, so the sign of the formation volume usually determines the sign of the activation volume.

THE ACTIVATION STRAIN

When nonhydrostatic stresses are applied to the system, the *shape* of the volume changes discussed above makes a difference; it's not just the hydrostatic component that matters. That is, if the sample undergoes a transition involving a unidirectional shape change, then tension or compression in that direction will alter the transition rates more than tension or compression in orthogonal directions, as shown in Fig. 3. When transition state theory is generalized to nonhydrostatic stress states, the result is that we replace the activation volume by the activation strain tensor, ΔV_{ij}^*. It is worth noting that the trace of the activation strain tensor is the conventional activation volume characterizing the response to hydrostatic pressure. The resulting growth equation for the interface velocity is

$$ \upsilon(\sigma_{ij}) = \upsilon(0) \exp \frac{\sigma_{ij} \Delta V_{ij}^*}{kT}, \tag{4} $$

where σ_{ij} is the applied stress, and summation over all i and j is implicit.

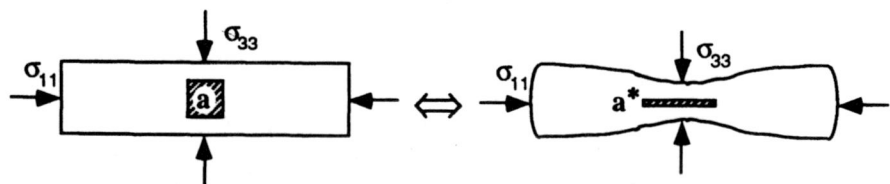

Fig. 3. *Illustration of activation strain for case in which strain of migration is anisotropic. State "a" is starting state; subsystem (cross-hatched region) undergoes structural transition to transition state "a*" which is short and fat. Induced shape change in surrounding elastic solid interacts with externally applied stresses. In case shown, rate of fluctuation from a up to a* will be enhanced by compressive σ_{33}, reduced by compressive σ_{11}, and, if the volume of a* is smaller than that of a, enhanced by hydrostatic pressure.*

The curve in Fig. 1(b) is a fit of eq. (4) to the $\upsilon(\sigma_{11})$ data, yielding a best-fit value of $\Delta V_{11}^* = +0.15 \, \Omega$ for SPEG of Si (001). Because the [100] and [010] directions in the (001) interface are equivalent, $\Delta V_{22}^* = \Delta V_{11}^*$. The trace of the matrix is obtained from the hydrostatic pressure measurements reported above, permitting a determination of ΔV_{33}^*. Off-diagonal elements represent shear strains of state a* relative to state a in Fig. 3; even if each individual transition state is sheared on an atomic scale, on average for the (001) interface between a cubic crystal and an amorphous phase there must be as many positively as negatively sheared states. All off-diagonal elements must then be zero, resulting in a complete determination of the activation strain tensor:

$$ \Delta V_{ij}^* = \begin{bmatrix} +0.15 & 0 & 0 \\ 0 & +0.15 & 0 \\ 0 & 0 & -0.58 \end{bmatrix} \Omega \tag{5} $$

From this result, we predicted that (i) biaxial stress of the type normally encountered in heteroepitaxy will have the square of the effect of uniaxial stress; (ii) uniaxial compression in the direction normal to the surface will enhance the growth rate even more than does hydrostatic pressure. The first prediction is consistent with some experiments in Si_xGe_{1-x} alloys [4-6]. Preliminary results from a test of the second prediction [7] are in quantitative agreement with the predicted value of ΔV_{33}^*.

IMPLICATIONS FOR SOLID PHASE EPITAXY

We have published detailed analyses [1,2] of the implications of these and other measurements for proposed SPEG mechanisms, assuming that there is a single rate-limiting step involving the generation and transport of these defects. The conclusions are summarized in Table I; the reasons listed in the third column are elaborated on in ref. [1]. Note that there are many ways to rule out bulk crystal point defect mechanisms, but that the only way to definitively rule out all bulk amorphous point defect mechanisms is from the activation strain measurements. This occurs because of the unlimited variety of point defects that may be postulated to occur in an amorphous phase, and because of how little we know about them. Hence I want to concentrate here on the implications of the nonhydrostatic stress measurements for bulk amorphous point defect mechanisms.

Table I. Tenability of point defect mechanisms for SPEG. Conclusions and reasons from ref. [1]

Defect Governing SPEG	Conclusions	Reasons
Vacancies in **crystal**	Si: Highly Implausible Ge: Impossible	ΔV^*, P_{max}, v/D_{bulk}, $sin(\theta)$ ΔV^*
Interstitials in **crystal**	Highly Implausible	v/D_{bulk}, $sin(\theta)$
Any point defect in **amorphous**	Impossible	tetragonal ΔV_{ij}^*
Any point defect in **amorphous**	Unlikely	relaxation, $sin(\theta)$
Dangling bonds at c/a **interface**	Plausible	ΔV^*, ΔE^*, v_0, $sin(\theta)$
Kink sites at c/a **interface**	Possible	special case of dangling bond mechanism

The activation strain has tetragonal symmetry, with two positive elements and one negative element. The formation volume (or "formation strain") of a point defect in either the amorphous or the diamond cubic crystal phases should be isotropic. Even inherently anisotropic point defects such as divacancies should be distributed equally among their possible orientations, resulting in the same concentration change per unit of uniaxial stress in any direction. As shown in Fig. 4 (a), the migration volume (or "migration strain") in the crystal or in the amorphous phase, if the stress has not relaxed, is not necessarily isotropic. In-plane tension can enhance the mobility of some species toward the interface, e.g. by opening channels for their migration. This explanation would be in accord with our experimental observations of an enhanced growth rate under these conditions. If this were the case, however, hydrostatic compression would be expected to *reduce* the mobility toward the interface, as shown in Fig. 4 (b), contrary to our experimental observations. Furthermore, according to the work of Witvrouw and Spaepen, the amorphous phase is fully stress-relaxed under the conditions in our experiments [8]. Some of our observations discussed in the Questions section below support this conclusion. In this case, neither the mobility nor the concentration should be affected by uniaxial stress.

453

(a) **(b)**

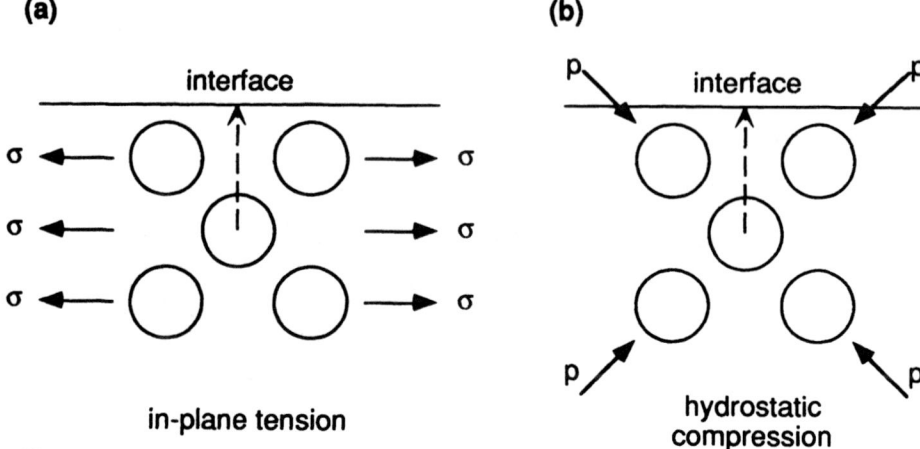

Fig. 4. (a) In-plane tension can enhance mobility toward interface, but then (b) hydrostatic compression should reduce mobility toward interface.

COMBINED BULK AND INTERFACE CONTROL

We have ruled out all mechanisms in which there is a *single* rate-limiting step that occurs in the bulk of either phase. The rate-limiting step cannot be, for example, transport to the interface of dangling bonds from the bulk of the amorphous phase. However, it is possible to have two processes occurring in series in which both processes are partially rate-limiting, e.g., transport of dangling bonds to the interface in series with an interfacial reaction. If d represents the rate of bulk generation and transport to the interface of defects, and K represents the interface reaction rate, then under steady-state conditions the rate will be given by

$$v = \frac{K \cdot d}{K + d}.$$ (6)

If both d and K have Arrhenius form ($K = K_0 \exp(-Q_K/kT)$, $d = d_0 \exp(-Q_d/kT)$), then in general v will not have Arrhenius form, which is contrary to the experimental results of Olson and Roth [9], who measured Arrhenius behavior over ten decades in v. Arrhenius form will be recovered from (6) if K « d or if d « K, making either K or d, respectively, the single rate-limiting reaction, or if K and d have the same activation energy. Considering the large range of v over which Arrhenius behavior is observed, we anticipated a negligibly small permissible difference in activation energies, as is implicit in the arguments presented in our earlier publication [1]. Closer examination [7], however, reveals a significant range of permissible difference in activation energies. The data of Olson and Roth are reproduced in Fig. 5 along with the predictions of eq. (6) for $Q_d - Q_K = 0$, 0.5, and 1.0 eV. Although the data are best fit by a strictly Arrhenius form ($Q_d - Q_K = 0$), the curve for 1.0 eV has only barely too much curvature to be consistent with the data, and the curve for 0.5 eV has little enough curvature to fit the data. In addition, the temperature uncertainty in the high temperature data may be great enough to permit a difference of $Q_d - Q_K = 1$ eV.

The activation energy for dilute hydrogen diffusion in a-Si has recently been shown [10] to be 2.7 eV, which is the same as that for SPEG. It is possible that both H diffusion and SPEG require the migration of unpassivated dangling bonds in the bulk and at the interface, respectively [10]. The migration energies in these different locations might well be within 0.1 eV of each other. Our analysis could not rule out this or other similar mechanisms.

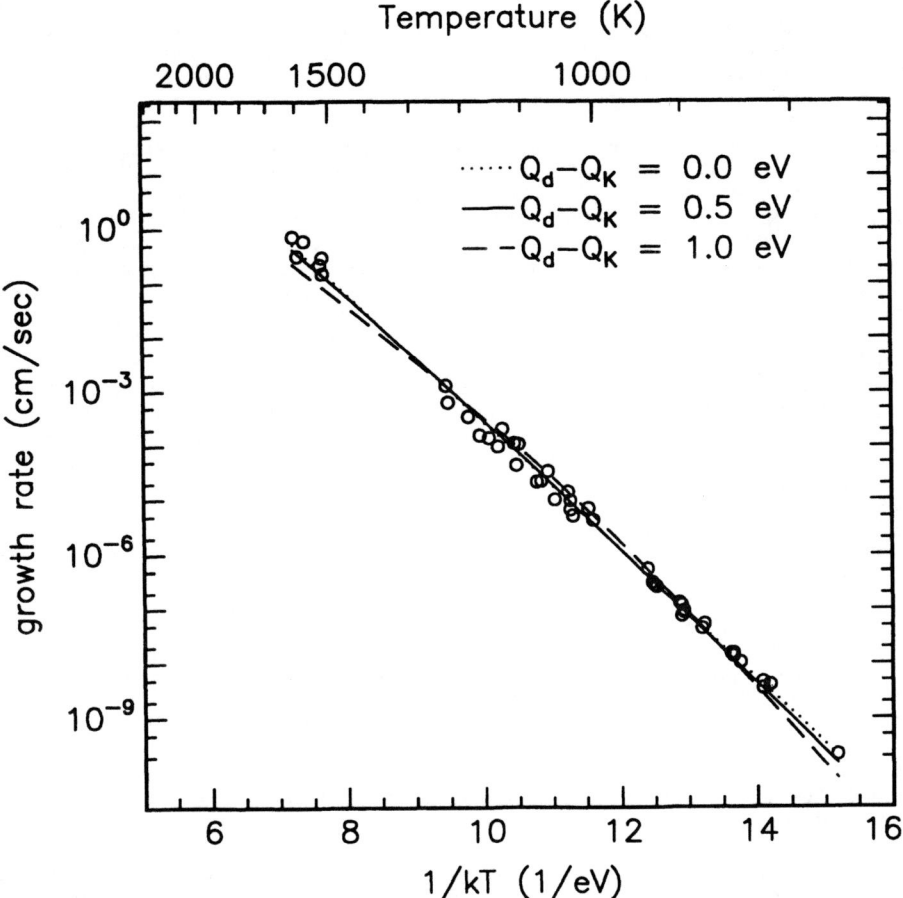

Fig. 5. Arrhenius dependence of SPEG is consistent with combined control of bulk and interface processes so long as their activation energies are close and their prefactors assume corresponding values. Data from Olson and Roth[9]. Curves from eq. (6). Prefactors are adjusted in this case so that $K = d$ at $1/kT = 10$ eV^{-1}.

SUMMARY

Point defect mechanisms are reflected in the dependence of the rate on temperature, hydrostatic pressure and nonhydrostatic stress. Measurements of the latter two, while more difficult than the former, tell us more. Such measurements have been made for SPEG of Si (001). The results are useful for model-testing and have been used to rule out a number of proposed mechanisms. The formalism is completely general, and our procedures can be readily generalized to other situations involving thermally activated atomic transport processes such as diffusion and creep, in addition to crystal growth.

GENERALIZATIONS

This model-testing process is quite general, and it has tremendous potential. A model-builder proposes a particular process: this atom goes here, that atom goes there. We can now ask: how are temperature, pressure, and nonhydrostatic stresses predicted to affect the rate of that proposed process, and how does that compare with the observed behavior? The advent of total-energy calculations should make it much easier to answer such questions very accurately for any proposed atomistic mechanism for any kinetic process. A single test may be sufficient to rule out a model, but to establish a model a battery of different tests must be performed, and some will be more discerning than others. The nonhydrostatic stress test might be performed in the following manner: the atomic displacements in the reaction (such as the migration of an interfacial dangling bond) are prescribed, and the relaxation of the first couple of shells of neighbors is determined theoretically. This process would have to be extended out far enough so that the forces are Hookean farther away (in practice, this should not be very far). Continuum elasticity could then be used to determine the displacements of the sample surface, which would be compared directly with the measured activation strain tensor. In many cases, the situation could be greatly simplified by the judicious usage of periodic boundary conditions. It is this latter method that we plan to use in further examining proposed mechanisms for SPEG. Note that we used a more general procedure in our earlier work to rule out all single-rate-limiting-step mechanisms involving point defects in the bulk of the amorphous phase. Ruling out *all* possible point defects in the amorphous phase by the defect-by-defect calculations described here would have been impossible, given the variety of postulated point defects in the amorphous phase and the additional possibility that a new defect might be postulated at any time. We were able to make more general conclusions by exploiting the shape of the measured activation strain. Only the mechanisms that cannot be ruled out by such general arguments need be tested by an actual predictive calculation of the activation strain.

QUESTIONS AND, ANSWERS

To close this paper I would like to answer some commonly asked questions, several of which were asked during the MRS meeting. I hope that the answers to these questions will clarify some of the more subtle points associated with the activation strain concept and its application.

Q: What is wrong with a model whereby floating bonds must migrate to the interface in order for SPEG to occur but where the rate-limiting step is the interface reaction, not the generation and migration in the bulk?

We have ruled out all mechanisms in which there is a *single* rate-limiting step that occurs in the bulk of either phase. In fact, any mechanism in which there is a single rate-limiting step that occurs *at the interface* is still in accord with all the experimental results. However, if the single rate-limiting step occurs at the interface (such as dangling or floating bond migration), then isn't it completely superfluous to model the bulk process unless you can create experimentally, or predict theoretically, a situation in which the bulk process becomes significantly rate-limiting? Otherwise, aren't we arguing about something that cannot be measured? For example, nobody advances a model for how, or how rapidly, a defect gets *down* from the saddle point in its migration process, because the rate-limiting step is getting *up* to the saddle point. Nobody advances a 'thermal energy model' for how the system acquires enough thermal energy for the defect to surmount a migration barrier, because acquiring thermal energy, although undoubtedly necessary, is not the rate-limiting step. Likewise for a SPEG model involving the acquisition of defects from the bulk *if* it is not the rate-limiting step. So for such a mechanism to be considered seriously, a prediction must be made of a situation in which the bulk process becomes measurably rate-limiting.

Note, however, that bulk generation and transport can be partially rate-limiting, and the interface reaction can be partially rate-limiting, *if* the prefactors and activation energies fall within the range of values demarcated earlier. Considering that Arrhenius behavior is observed over 10 decades in growth velocity, the permissible difference in activation energies between bulk and interfacial steps is surprisingly large.

Note also that although nobody has proposed an interfacial floating bond mechanism, in which the rate limiting step is the creation and migration of floating bonds at the interface (much as in the interfacial dangling bond model) it makes sense to consider such a mechanism because nothing that has yet been done can rule it out.

Q: Why do you believe that you're studying effects on mobilities rather than on driving forces?

SPEG is an ideal system for studying interface mobilities because under all experimental conditions the driving force is very large and doesn't vary much. It will be more difficult to separate driving force effects and mobility effects on an interface near equilibrium. How little does the driving force vary?

(1) If we consider temperature changes, $\Delta G/kT$ is diminished by about a factor of 3.2 [11]and [$1-\exp(\Delta G/kT)$] is diminished by only about 53% as the temperature increases from 800 to 1600 K, whereas the growth rate increases by about ten orders of magnitude. So long as the mobility M is defined as the ratio of velocity to free energy difference, or as the ratio of velocity to any reasonable thermodynamic function of free energy difference (such as [$1-\exp(\Delta G/kT)$], as we have done in eqs. (1-2)), then the lion's share of the effect must be attributed to changes in the mobility and not to changes in the thermodynamic factor.

(2) Because a-Si and c-Si have very similar densities the driving force is almost independent of pressure, yet the growth rate goes up markedly with pressure. Amorphous Si is actually slightly less dense than c-Si, by about 1.8% [12], so the driving force for crystallization increases very slightly with increasing pressure. If we assume identical compressibilities for the two phases, then over the entire pressure range covered experimentally (0-5 GPa),the free energy difference at 800 K goes from 94 only to 100 meV per atom. With a greater compressibility for the amorphous phase than for the crystal, the change with pressure will be smaller still. The thermodynamic factor $1-\exp(\Delta G/kT)$ changes from 0.58 to 0.61, or by about 4%, over this pressure range. Because the velocity increases by more than a factor of ten over this same range, we attribute these changes in growth rate to a change in mobility with pressure.

(3) The highest level of uniaxial stress reached in our experimental work is 0.6 GPa, which is about an order of magnitude lower than in the hydrostatic pressure work. No matter how you choose to define chemical potentials in the nonhydrostatically stressed system, the magnitude of the change with stress will be much smaller than the minuscule change with pressure illustrated above.

Q: When you rule out bulk defect mechanisms in the amorphous phase you say the a-Si is stress-relaxed, yet you invoke nonzero stress in the material surrounding the transition state in the development of the theory. Is there an inconsistency here?

Not really. The bulk of the a-Si film is stress-relaxed. However, the first few atomic layers of a-Si adjacent to the c/a interface are not likely to be stress-relaxed because these layers are intimately bonded to the crystal. This stress will decay away with distance from the interface. That the stress is relaxed in the *bulk* of the amorphous film rules out a bulk point defect mechanism in the amorphous phase. Admittedly, you could postulate a point defect mechanism whereby the rate-limiting step is the generation and transport of defects from only the stressed boundary layer in the amorphous film; however, you've still got to come up with an explanation for why uniaxial compression in the plane of the film retards the SPEG rate but hydrostatic pressure enhances the rate (Fig. 4). Finally, the strain resulting from an atomic fluctuation from the starting state to the transition state is unlikely to relax significantly, even in the bulk of the amorphous phase, on the time scale of the life of such a fluctuation.

Q: In determining the second diagonal element of the activation strain tensor, you invoke fourfold rotational symmetry for the (001) interface between a-Si and c-Si, but on a microscopic level there's only twofold symmetry.

That's absolutely correct. At this point, our experiments can only resolve macroscopic growth rates, and as a result we have measured a macroscopic activation strain tensor with tetragonal symmetry. If an experiment were to be performed with monolayer resolution during layer-by-layer growth, alternate monolayers would be characterized by their own activation strain tensor lacking tetragonal symmetry.

Q: According to your current description, each orientation of interface is characterized by its own activation strain tensor. Is it possible that the activation strain tensor for every interface can be expressed as n · T, *where* n *is the unit vector normal to the interface, and* T *is a higher-rank tensor that is only materials-dependent and not orientation-dependent?*

That would be nice, but it might not necessarily work out that way. Such behavior might follow from a terrace-ledge-kink model of the interface, in which all the action happened at kinks and only the number of kinks varied with orientation. However, if the interface structure is fundamentally different over some range of orientations than over another (e.g., ledges of differing orientations or heights), I don't see how a single T could ever account for the behavior over both ranges. Perhaps each range will have its own T. We are currently examining such a TLK model theoretically and are also attempting to measure the orientation-dependence of the activation strain.

Q: In your talk you showed us a micrograph in which the crystal contains some dislocations, yet you claim that the strain was completely elastic in the wafer bending experiment.

The micrograph that I showed you is not from the bending experiment; it's from preliminary work of W.B. Carter [7] on loading the interface with uniaxial compression in the z direction (normal to the interface). I have not included it in this paper because it is, after all, preliminary. It was created by squeezing a Si wafer between the flattened tips of two diamond anvils in the absence of any pressure-transmitting medium; the stress in this case was not very uniform. Work is under way on making the stress more uniform. The interesting aspect of this micrograph is that it shows a (001) interface between a-Si and c-Si that is faceted in some places, which invites speculation about the origin of the faceting. The wavelength of the faceting is the same order of magnitude as the break-even wavelength for balancing elastic strain relief against increased interfacial area. Because, however, $\Delta G/kT$ is of order unity under these conditions, kinetic considerations may dominate.

In the three-point bending experiment, on the other hand, no dislocations were observed in TEM. An additional observation in this experiment also indicates that the crystal did not deform plastically: A sample cooled down in the furnace while maintained under stress is found to retain some curvature after the load is removed. Upon further annealing in the absence of this load, the sample straightens itself out. This observation indicates that the amorphous phase undergoes stress relaxation between room temperature and SPEG temperatures, while the crystal undergoes only elastic deformation.

ACKNOWLEDGMENTS

I thank John Roth for a stimulating discussion about H diffusion in a-Si and its possible relation to SPEG. The experimental work reviewed here has been done in collaboration with William B. Carter, Guo-Quan Lu, Eric Nygren, and Paul C. Sabin. This research has been supported by NSF-DMR-89-13268.

REFERENCES

[1] G.-Q. Lu, E. Nygren and M.J. Aziz, J. Appl. Phys. **70**, 5323 (1991). References apparently missing from this list can be found in this paper, which contains 98 references.
[2] M.J. Aziz, P.C. Sabin and G.-Q. Lu, Phys. Rev. B **44**, 9812 (1991).
[3] G.H. Vineyard, J. Phys. Chem. Solids 3, 121 (1957).
[4] D.C. Paine, D.J. Howard, N.D. Evans, D.W. Greve, M. Racanelli and N.G. Stoffel, Mat. Res.

Soc. Symp. Proc. **202**, 561 (1991).

[5]F. Corni, S. Frabboni, G. Ottaviani, G. Queirolo, D. Bisero, C. Bresolin, R. Fabbri and M. Servidori, J. Appl. Phys. **71**, 2644 (1992).

[6]Q.Z. Hong, J.G. Zhu, J.W. Mayer, W. Xia and S.S. Lau, J. Appl. Phys. **71**, 1768 (1992).

[7]W.B. Carter and M.J. Aziz, (unpub).

[8]A. Witvrouw and F. Spaepen, J. Appl. Phys. **74**, 7154 (1993).

[9]G.L. Olson and J.A. Roth, Mater. Sci. Rep. **3**, 1 (1988).

[10]J.A. Roth, G.L. Olson, D.C. Jacobson and J.M. Poate, Mat. Res. Soc. Symp. Proc. **297**, 291 (1993).

[11]E.P. Donovan, F. Spaepen, D. Turnbull, J.M. Poate and D.C. Jacobson, J. Appl. Phys. **57**, 1795 (1985).

[12]J.S. Custer, M.O. Thompson, D.C. Jacobson, J.M. Poate, S. Roorda, W.C. Sinke and F. Spaepen, Appl. Phys. Lett. **64**, (1994).

THE EFFECT OF STRAIN AND STRAIN-GRADIENTS ON THE CRYSTALLISATION KINETICS OF $Si_{1-x}Ge_x$ ALLOY LAYERS

PER KRINGHØJ*, ROBERT G. ELLIMAN* AND JOHN L. HANSEN**
*Electronic Materials Engineering Department, Research School of Physical Sciences and Engineering, Australian National University, Canberra, ACT 0200, Australia.
**Institute of Physics and Astronomy, Aarhus University, DK-8000 Aarhus C, Denmark

ABSTRACT

A comparison of strain relief in SiGe alloy layers during high temperature annealing and solid phase epitaxial crystallisation (SPEC) has shown that layers which are thermodynamically stable are also fully strained following SPEC, whereas metastable layers that relax at high temperatures also relax during SPEC. This is illustrated by the fact that a uniform SiGe layer with x=0.085 is stable during annealing at 1100°C for 60 sec and is fully strained following SPEC. In contrast, a uniform layer with x=0.17, which was fully strained as-grown by molecular beam epitaxy (MBE), is shown to relax during high temperature annealing and during SPEC.

A depth dependent SPEC velocity is observed for metastable layers, with a decrease in velocity as the alloy layer begins to crystallise and an increase in velocity as strain relaxation proceeds.

INTRODUCTION

Solid phase epitaxial crystallisation (SPEC) of Si and SiGe alloys has been widely studied [1-7]. The influence of strain on SPEC has also been characterised and it has been established that strained SiGe layers regrow slower and with an activation energy higher than both Si, and the corresponding relaxed alloy[6-9].

Growth of strained SiGe layers by MBE or chemical vapour deposition can be achieved for Ge compositions well above the theoretical limit for pseudomorphic growth. These layers are not thermodynamically stable, however, and high temperature annealing results in the formation of strain relieving defects. These layers are referred to as metastable SiGe layers.

In the present study SPEC of stable and metastable SiGe layers, with both uniform and depth dependent Ge composition has been examined and compared with high temperature stability of the layers.

EXPERIMENTAL PROCEDURE

Two types of $Si_{1-x}Ge_x$ epitaxial layers were investigated: i) uniform layers with a thickness of 120 nm and an x-value of either 0.085, 0.17, 0.27, 0.37, or 0.67, and ii) linearly-graded layers with a triangular Ge distribution, each with a peak-concentration of 10% and alloy layer width of 190, 370, and 940 nm. All layers were MBE-grown at Aarhus University in a VG-80 system. The substrates used were <100> oriented p-type Si and the growth-temperature was ~550°C. The uniform layers were grown with a Si capping layer of 5 nm and the triangles with a Si capping layer of 100 nm.

Amorphous layers were created by irradiation with Si-ions with the samples maintained at LN_2-temperature during implantation. The ion energy and dose was chosen in order to ensure complete amorphization: Either 30 keV; $8 \cdot 10^{14}$ cm^{-2} or 90 keV; $5 \cdot 10^{15}$ cm^{-2} for the uniform layers, and 900 keV; $5 \cdot 10^{15}$ cm^{-2} for the triangles. The two different energies employed for amorphisation of the uniform layers, 30 and 90 keV, result in partial amorphization of the epitaxial layer and complete amorphization, extending into the substrate, respectively. SPEC of the amorphous layers was measured over the temperature range 500-650°C using in-situ time-resolved reflectivity (TRR), at two different wave-lengths (λ=0.6238 μm and λ=1.523 μm). Crystal-quality, and the composition of SiGe layers were measured with Rutherford backscattering spectrometry and channeling (RBS-C) using 2 MeV-He ions and the strain was determined with double crystal x-ray diffraction (DCXRD) measurements. In order to investigate the temperature stability of the above mentioned layers, MBE-grown layers were rapid thermal annealed in the temperature range from 700 to 1100°C. The temperature was maintained for one minute and the annealing ambient was Ar.

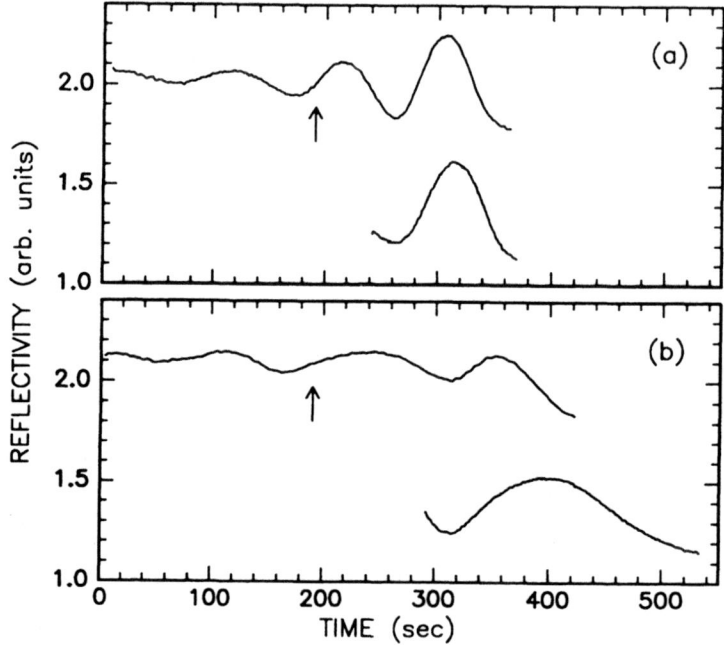

Fig.1 TRR spectra recorded at 588°C for samples amorphized with 90 keV ($5 \cdot 10^{15}$ Si/cm^2) and 30 keV ($8 \cdot 10^{14}$ Si/cm^2) irradiation, a) is for an 8.5% Ge layer, and b) is for a 17% Ge layer.

RESULT AND DISCUSSION

Fig.1(a) shows TRR spectra from a uniform, x=0.085, layer amorphized with 90 and 30 keV Si ions. In the 90 keV case the amorphous layer thickness is 220 nm resulting in amorphization past the interface (the position of the interface is indicated on the figure), whereas in the 30 keV case the amorphous layer thickness is 80 nm leaving ~40 nm of crystalline SiGe material. The time-scale for the 30 keV-spectrum has been shifted in order to overlay the starting depth (the distance between constructive and destructive interference in the spectra is $\lambda/4n_i$=33 nm, where λ is the wavelength and n_i is the refractive index of the amorphous layer; n_i is slightly composition dependent being 20% higher for Ge, relatively to Si). Similar SPEC behaviour is observed in both cases, and similar behaviour was found over the complete temperature range investigated.

TRR spectra for a layer with a Ge composition of 17% are shown in Fig.1(b). In the 90 keV case a significant reduction in SPEC velocity is observed as crystallisation proceeds into the alloy layer. The average regrowth velocity for the first 30 nm is one half of the corresponding velocity in the Si substrate, and a lower peak-to-valley intensity is observed in the TRR spectrum, indicating roughening of the crystalline/amorphous interface. An increase in the SPEC velocity in the near surface (last 60 nm) is also observed, with the average regrowth velocity for the last 60 nm being 20% higher than the velocity in Si. In the 30 keV case a significant reduction in the regrowth velocity, relatively to both the 90 keV and pure Si, is seen in the near surface. The average velocity for the last 60 nm is reduced by a factor of two, relative to the velocity at the corresponding depth in the 90 keV case. Although the TRR-spectra of Fig.1(b) were recorded at 588°C, the overall picture is again similar for all temperatures investigated.

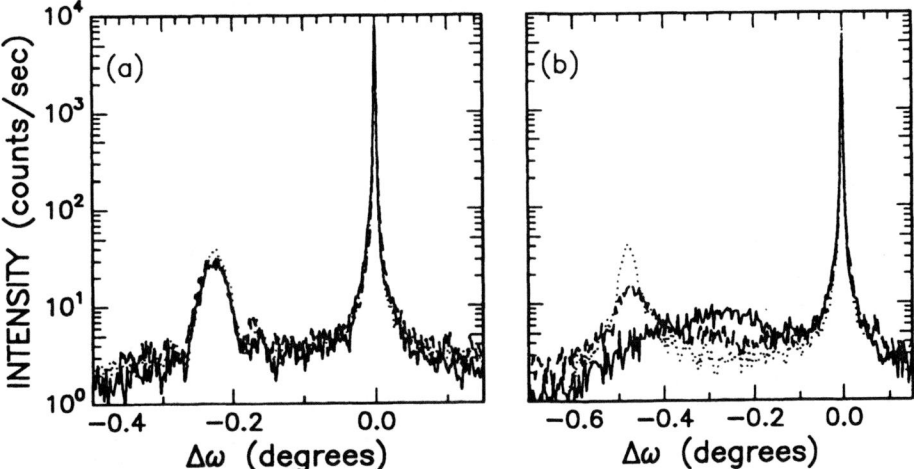

Fig.2 DCXRD spectra for samples with a uniform Ge content. The three curves represent MBE-grown (dotted curve), SPEC after amorphization with 30 keV (dashed curve) and 90 keV (full curve). Ge content with x=0.085 (a), and Ge content with x=0.17 (b).

DCXRD measurements on both MBE- and SPE-grown samples are shown in Fig.2. No significant difference between MBE- and SPE-grown samples is observed for the x=0.085 case, with similar strain observed in both cases, independent of the amorphization conditions. In addition no significant relaxation was observed after RTA at temperatures up to 1100°C for 60 sec, in good agreement with ref [10], where a similar alloy layer was examined. For the x=0.17 case a significant relaxation is observed in the SPE-grown sample amorphized with 90 keV. The partial amorphized sample contains residual strain after SPEC, although significant relaxation has occurred. Significant relaxation is also observed in this case after RTA at 1100°C. Relaxation after SPEC was also observed by Chilton et al. [3] for thinner SiGe alloy layers with a higher Ge concentration.

The TRR spectra for the low Ge content sample, x=0.085, show constant SPEC velocity with depth, in agreement with the regrowth behaviour of a fully strained SiGe layer. The higher

Ge content sample, x=0.17, shows retarded growth, relative to Si, for the first ~30 to 50 nm, again in agreement with the regrowth behaviour of strained SiGe, however, after a certain thickness the crystalline/amorphous interface becomes rough and the strain is partly relieved, whereafter the regrowth velocity increases. This is consistent with the result of Hong. et al ref.6, where strain relief during SPEC of completely amorphized SiGe layers, with a Ge content of 10% and thickness of 350 nm, was seen and where increased SPEC velocity in thermally stress-relaxed SiGe layers were observed. In the case where the layer is only partly amorphous (30 keV), retarded SPEC is seen all the way to the surface due to the presence of strain in the underlying crystalline SiGe layer and due to the low amorphous layer thickness, which either prevents the roughening and relaxation or coercing it to occur at the surface.

The temperature stability of the above mentioned layers is shown in Fig.3, where RBS-C result from layers RTA-processed at 700 and 1100°C are shown together with SPE-grown samples (the peak at channel 237 is due to oxygen on the surface). For the low Ge content sample, x=0.085, no difference between the MBE-, SPE-grown, and RTA-processed samples, is observed, whereas the spectra from the high Ge content sample, x=0.17, clearly indicate the presence of defects at the Si/SiGe interface in the SPE-grown and RTA processed samples. The spectra also suggest that relaxation in the SPE-case is greater than for the MBE-grown sample annealed at 1100°C, even though the growth temperatures for SPEC and MBE are similar.

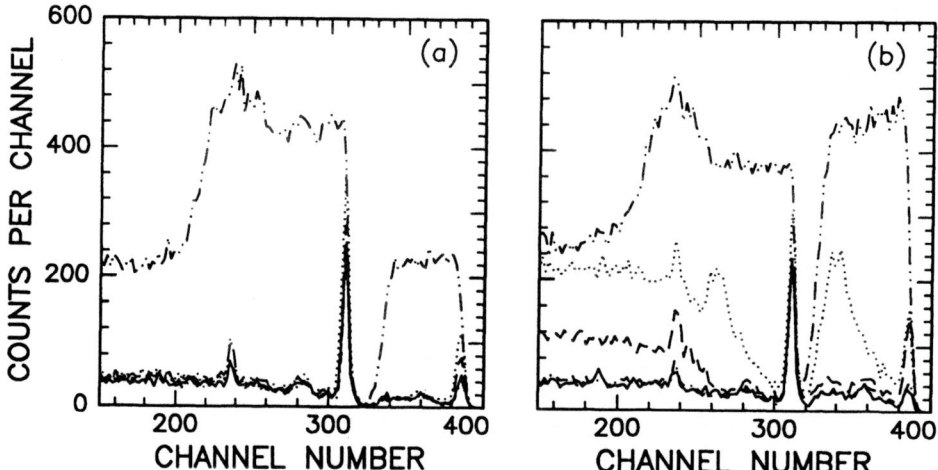

Fig.3 RBS-C results for samples with uniform Ge content, as-implanted after 90 keV (double dot-dashed), SPE-grown at 588°C (dotted), MBE-grown and RTA at 700°C (dot-dashed), and RTA at 1100°C (dashed). The scattering angle was 100 degree and the as-implanted spectrum has been reduced a factor of two. Ge content with x=0.085 (a), and Ge content with x=0.17 (b).

For higher Ge content samples, x≥0.27, strain relaxation occurs during MBE-growth. TRR spectra for x=0.27 were, however, similar to the sample with x=0.17, with respect to the difference in SPEC velocity between amorphization with 30 and 90 keV. This is believed to be caused by the presence of residual strain in the layer. For x≥0.37 no difference in SPEC velocity, between the partially and completely amorphized samples, was observed. Probably, due to lower levels of residual strain preventing strain retardation in the 30 keV case and due to a low critical thickness allowing the strain to be released in the first 10 to 20 nm in the 90 keV case, resulting in a SPEC velocity, for the partly amorphized and completely amorphized sample, close to fully relaxed SiGe [11].

The triangular alloys, also fall into the categories of stable and metastable, with the thinner being stable with respect to annealing temperature, and the two others being unstable, resulting in strain relief at 1100°C.

In Fig.4 the crystallisation velocity as a function of depth is shown for a SiGe alloy with a triangular Ge distribution (total thickness of the SiGe layer is 940 nm, and the amorphous layer thickness is 1300 nm). A significant decrease in regrowth velocity is observed at a depth of 400-600 nm, whereafter the velocity increases again. As for the uniform layers, the decrease in SPEC velocity was observed after a roughening transition of the crystalline/amorphous interface. DCXRD measurements show significant relaxation of the SiGe layer after SPEC and a defect band is observed at a depth of ~450 nm with RBS-C. The SiGe alloy with a thickness of 190 nm has a constant velocity versus depth curve and no difference between MBE-grown and SPE-grown can be determined with either DCXRD or RBS-C.

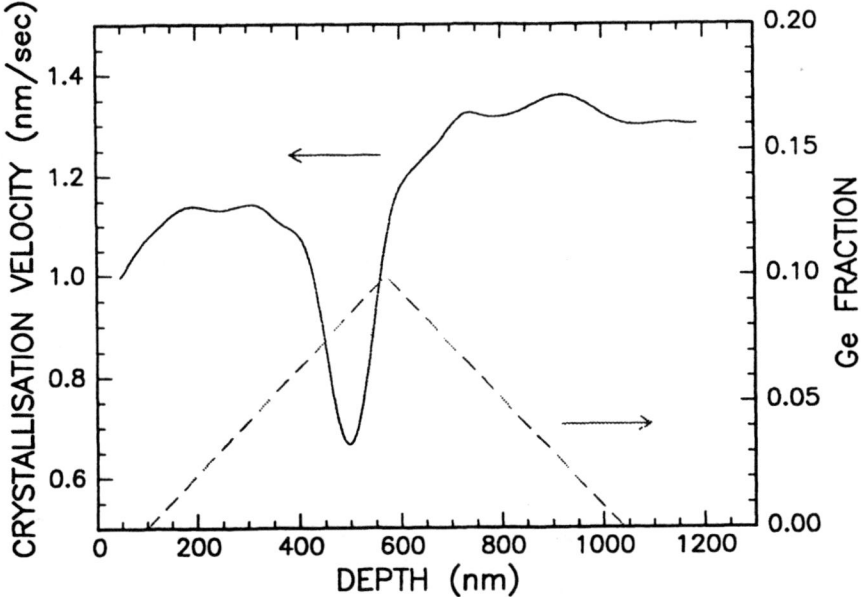

Fig.4 *SPE-velocity versus depth, at 598° C, for a sample with a triangular Ge composition, together with the Ge distribution as deduced from RBS-C.*

The measurements above on the uniform and triangular shaped SiGe alloy layers suggest that a strain induced roughening of the crystalline/amorphous interface retards the SPEC process, probably via {111} faceting of the interface, which is believed to occur during strain relief [4,5]. Note that the regrowth velocity is reduced by {111} faceting simply by the significantly lower regrowth velocity in the <111> direction [1]. After strain relief the driving force for faceting is reduced and as a consequence the SPEC velocity increases and the crystalline/amorphous interface becomes smooth, due to the higher growth velocity in the <100> direction. In agreement with the observation that the same activation energy is found before and after strain relief [11].

CONCLUSION

A uniform SiGe layer with a Ge composition of 8.5% was found to be thermodynamically stable and exhibited a constant regrowth velocity with depth, independent of amorphization conditions. The layer was also found to be fully strained. A SiGe layer with a composition of 17% was found to relax during thermal annealing and exhibited depth dependent regrowth velocity. The velocity was reduced, relative to Si, in the first 30 to 50 nm of growth, and increased as strain relief progressed. The TRR contrast decreased prior to strain relief, consistent with concomitant roughening of the crystalline/amorphous interface.

SiGe layers with a Ge content of 27% exhibited similar behaviour to the sample with x=0.17; a reduction in the near surface regrowth velocity was observed for the partially amorphized sample, relative to the completely amorphized sample. This was ascribed to the presence of residual strain in this sample. No strain related effects were observed for higher Ge concentration samples, where residual strain is expected to be lower.

Similar results were observed for SiGe strained layers with triangular Ge distributions; the thinnest layer was found to be stable during both thermal annealing and SPEC whereas the thicker layers relaxed during both treatments (RTA and SPEC). Comparison of the SPEC velocities for the different samples enabled the velocity to be measured at the same Ge concentration for each of the three concentration gradients. However, no difference was observed.

The above results show that despite the similar growth temperatures employed during MBE and SPEC of SiGe strained layers the extent of metastability which can be achieved varies significantly. MBE-grown layers can exceed the equilibrium critical thickness by more than an order of magnitude whereas little or no metastability is possible during SPEC. The most obvious difference between the two crystallisation processes is that during MBE the alloy layer is growing at an atomically sharp crystalline/vacuum interface whereas for SPEC, crystallisation occurs at a crystalline/amorphous interface. TRR measurements during SPEC show that relaxation is accompanied, or preceded by, roughening of the crystalline/amorphous interface. It therefore appears probable that this roughening or faceting transition holds the key to the difference between growth by MBE and SPEC, enabling the formation of strain relieving defects in the SPEC case.

REFERENCES

1. J.S. Williams, in *Surface Modification and Alloying by Laser, Ion, and Electron Beams*, (Plenum Press, NY, 1983). Chapter 5 (and references therein).
2. G.L. Olson and J.A. Roth. Mat. Res. Rep. 3, 1 (1988).
3 B.T. Chilton, B.J. Robinson, D.A. Thompson, T.E. Jackman, and J.-M. Baribeau, Appl. Phys. Lett. 54, 42 (1989).
4. D.C. Paine, N.D. Evans, and N.G. Stoffel, J. Appl. Phys. 70, 4278 (1991).
5. D.C. Paine, D.J.Howard, N.G. Stoffel and J.A. Horton. J. Mater. Res. 5, 1023 (1990).
6. Q.Z. Hong, J.G. Zhu, J.W. Mayer, W. Xia, and S.S. Lau, J. Appl. Phys. 71, 1773 (1992).
7. C. Lee, T.E. Haynes, and K.S. Jones, Appl. Phys. Lett. 62, 501 (1993).
8. Z. Atzmon, M. Eizenberg, P. Revesz, J.W. Mayer, S.Q. Hong, and F. Schaffler, Appl. Phys. Lett. 60 2243 (1992).
9. C.S. Pai, S.S. Lau, I. Suni, and L. Csepregi, Appl. Phys. Lett. 47, 1214 (1985).
10. E.V. Thomsen, A. Nylandsted Larsen, J.L. Hansen, P. Kringhøj, S.Y. Shiryaev, and G. Weyer, *17th International Conference on Defects in Semiconductors*, (Gmunden, Austria, July 1993).
11. P. Kringhøj and R.G. Elliman, to be published.

STABILITY AND PRECIPITATION KINETICS IN Si$_{1-y}$C$_y$/Si AND Si$_{1-x-y}$Ge$_x$C$_y$/Si HETEROSTRUCTURES PREPARED BY SOLID PHASE EPITAXY

J.W. STRANE*, S. T. PICRAUX**, H. J. STEIN**, S. R. LEE**, J. CANDELARIA***, D. THEODORE***, AND J. W. MAYER****
*Cornell University, Dept. MS&E, Ithaca NY 14853.
**Sandia National Laboratories, Albuquerque NM 87185.
***Motorola Inc., MRT, Mesa, AZ, 85202.
****Arizona State University, Dept. of Chem. Bio. and Matls. Eng., Tempe AZ 85287

ABSTRACT

This study investigates the stability of metastable Si$_{1-y}$C$_y$/Si heterostructures during rapid thermal annealing (RTA) over a temperature range of 1000 - 1150° C. Heterostructures of Si$_{1-y}$C$_y$/Si and Si$_{1-x-y}$Ge$_x$C$_y$/Si (x=0.077, y ≤ .0014) were formed by solid phase epitaxy from C implanted, preamorphized substrates using a 30 minute 700° C anneal in N$_2$.The occupancy of C in substitution lattice sites was monitored by Fourier Transform Infrared Absorption spectroscopy. The layer strain was monitored by rocking curve x-ray diffraction and the structural changes in the layers were determined using plan-view and X-sectional transmission electron microscopy (TEM). For anneals of 1150° C or above, all the substitutional C was lost from the Si lattice after 30 seconds. TEM verified that the strain relaxation was the result of C precipitating into highly aligned βSiC particles rather than by the formation of extended defects. No nucleation barrier was observed for the loss of substitutional C. Preliminary results will also be discussed for Si$_{1-x-y}$Ge$_x$C$_y$/Si heterostructures where there is the additional factor of the competition between strain energy and the chemical driving forces.

INTRODUCTION

Tailoring of electronic band structures and strain in Si heteroepitaxial structures can be accomplished by creating IV-IV alloys[1]. We have recently shown that C can be alloyed into Si and Si$_{1-x}$Ge$_x$ in concentrations several orders of magnitude above the solid solubility limit using solid phase epitaxy (SPE) of amorphous layers[2,3]. In these alloys C resides primarily on substitutional lattice sites and strains Si and SiGe as predicted by Vegard's law ($\delta a_o/\delta c \approx 2.0 \times 10^{-3}$ nm/at. % C where $\delta a_o/\delta c$ is the change is lattice parameter per addition of substitutional C). In order to introduce C into device structures a good understanding of the stability limits of C in Si and Si$_{1-x}$Ge$_x$ is needed. Our initial studies on the SPE produced Si$_{1-y}$C$_y$/Si heterostructures investigated the stability over the temperature range of 700° C-950° C[5]. It was found that heat treatment induced the loss of substitutional C by the precipitation of 2-4 nm diameter particles of cubic βSiC. The βSiC precipitates were oriented in the same directions as the Si matrix. It was proposed that these precipitates form through the loss of coherency of Guinier-Preston zones. The loss of substitutional C showed no incubation period which would be indicative of a transformation barrier. As the substitutional C was lost from the alloy, the layer strain correspondingly was reduced. No extended defects were found in the microstructure indicating that the clustering of C and precipitation was resposible for the strain relaxation. It was concluded that chemical rather than mechanical driving forces induced layer relaxation. In the

467

present work we extend the study of the thermal stability of $Si_{1-y}C_y$/Si heterostructures to higher temperature though the use of rapid thermal annealing (RTA). Finally preliminary results on the stability of $Si_{1-x-y}Ge_xC_y$/Si heterostructures are reported. In the ternary system the effects of strain and chemical instability can be separately evaluated.

EXPERIMENTAL PROCEDURE

Metastable $Si_{1-y}C_y$ alloy layers were prepared by ion implantation of C into preamorphized layers of <100> float zone Si (ρ = 100-150 Ω-cm). Carbon was implanted at 25 keV and a dose of 5.25×10^{15} /cm^3. This implant procedure was expected to yield a peak C concentration of 1.0 at. %. Surface contamination from the amorphization and implantation processes was removed with a 4 min 1000 watt rf generated O_2 plasma. Regrowth of the implanted layers was obtained using a two anneal sequence in an N_2 ambient. First, a 30 minute 450° C anneal was used to clean-up end of range damage from the implantation and smooth the amorphous/crystalline interface. Next, a 30 minute 700° C anneal was used for SPE regrowth. The regrowth of the layer and C incorporation was confirmed by ion channeling and Fourier Transform Infrared Absorption (FTIR) prior to annealing. To study the high temperature stability, rapid thermal anneals (RTA) were performed over the temperature range 1000° C - 1150° C for times of 30 to 150 seconds.

The $Si_{1-x-y}Ge_xC_y$ alloy layers were formed by C implantation into 100 nm of $Si_{0.923}Ge_{0.077}$ deposited by chemical vapor deposition on <100> Si. Silicon implantation was used to first amorphize a 150 nm thick layer. The C was implanted with multiple energies to make a flat profile where the C resides primarily in the SiGe epilayer. Layers with C concentrations of 0.46, 0.93 and 1.4 at. % were created. The cleaning and annealing procedures were the same as for the $Si_{1-y}C_y$ layers.

The concentration of substitutional C in the heterostructures was measured using a Nicolet 60SX Fourier Transform Infrared Spectrometer. The localized vibrational mode (LVM) for substitutional C lies directly on top of the Si-Si two phonon absorption mode, therefore, in order to characterize the C content subtraction of an identical reference without C implantation was performed. Rocking curve double crystal X-ray diffraction was used to characterize the strain in the heterostructures. The evaluation of the scans was performed using kinematical theory and has been previously reported[5]. The final microstructure was examined using transmission electron microscopy (TEM) on a JEOL 200CX. Both plan view and cross-section specimens were prepared. Final thinning for these specimens was performed using ion milling.

RESULTS

I. RTA of $Si_{1-y}C_y$/Si Heterostructures

For the 1000° C RTA, the reduction of the LVM absorbance signal for C substitutional in Si was measured by taking the full width half maximum multiplied by the height. Triangles (Δ) in Figure 1 mark the evolution of this signal with increasing anneal time. This signal has been shown to be directly related to the substitutional C content in the epitaxial layer.[5] The loss of substitutional C initiates immediately after annealing begins with the most rapid loss of substitutional C occurring in the first stages of annealing. After 150 seconds of annealing the substitutional C signal has been reduced by over 50%. The reduction of the LVM absorbance peak is accompanied by the appearance of a broad

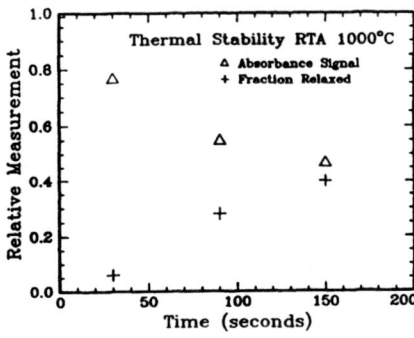

Figure 1 Thermal stability of $Si_{0.99}C_{0.01}/Si$; the (Δ) corresponds to the remaining substitutional fraction of C; the (+) corresponds to the amount of strain relaxation.

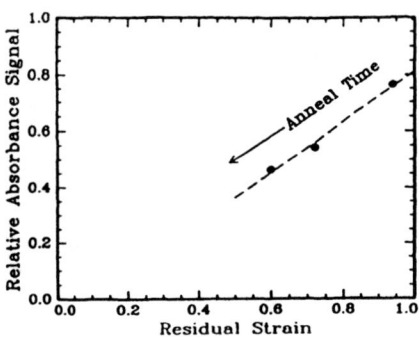

Figure 2 The relative absorbance corresponding to the substitutional C LVM plotted versus the residual layer strain in the $Si_{0.99}C_{0.01}/Si$ layer.

absorbance band located at 12.2 μm. During furnace anneals of supersaturated C in Si, this band has been previously correlated with the βSiC phonon absorption mode[6].

X-ray rocking curve scan for {004} and {224} reflections were used to show that the layer remained commensurate with the substrate and did not develop lattice tilt during the anneals. The relaxation of the layer strain in the (004) planes is shown in Figure 1 with the crosses(+). After 150 seconds, approximately 40% of the layer strain was relaxed. The relaxation of the strain did not exhibit the rapid reduction in the initial stages that was observed for the loss of substitutional C. Plan-view and cross-section TEM of the annealed layers indicated that extended defects such as misfit dislocations were not formed and therefore could not account for the strain relaxation. This relaxation occurs exclusively via the loss of substitutional C and eventual clustering and precipitation of βSiC.

Figure 2 plots the loss in absorbance signal versus the residual strain for the annealed layers. After the initial rapid loss in absorbance signal, the strain and absorbance signal tail off in an almost linear fashion. The large loss in absorbance signal relative to the strain relaxation during short anneals, indicates that initial C lost from the lattice does not relax the layer. Formation of coherent precipitates (i.e. Guinier-Preston zones) would explain this behavior.

The 1000° C RTA results from the absorbance measurements are compared to those for a similar layer annealed in a conventional furnace in Figure 3. The plot shows the time for 50% reduction

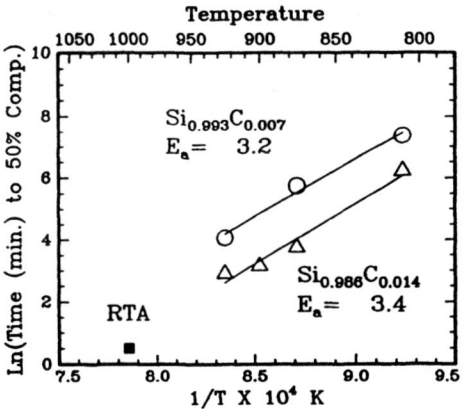

Figure 3 Arrhenius plot of the time to 50% loss in substitutional C versus reciprical Temperature.

of the substitutional C versus reciprocal temperature in Arrhenius fashion. The furnace results extrapolate well to the RTA result, indicating an activation energy of 3.3 ± 0.5 eV. The activation energy of C diffusion in C is 3.1 eV[4]. The similarities in activation suggest that the loss of substitutional C from Si is diffusion limited.

The absorbance spectra for alloys annealed above 1000°C (i.e. 1075 and 1150°C) showed a complete loss of the LVM absorbance peak for substitutional C. The complete loss of this signal took place in less than 30 seconds. The loss in the peak is accompanied by high absorbance changes over the full infrared range, indicating that the heat treatments are altering the absorbance in the substrate. These changes in the absorbance spectra were accompanied by the loss of a side peak due to the epilayer in the X-ray rocking curve spectra. This verifies that the higher temperatures anneals relax the film and substitutional C is completely lost for RTA anneals as short as 30 seconds.

II. $Si_{1-x-y}Ge_xC_y/Si$ Heterostructures

The LVM for substitutional C in Si can also be used in Si-Ge-C ternary systems to quantify C content. Significant broadening of the absorption peak relative to $Si_{1-y}C_y$ was observed, therefore, quantification requires calibration to the implant doses. It was found that the full width half maximum multiplied by the peak height yields a signal which is linear with C content up to nearly 1.5 at. % C. This absorbance signal was used to monitor the substitutional C content in SPE formed $Si_{1-x-y}Ge_xC_y/Si$ heterostructures. Figure 4 shows the loss of the substitutional C during a 925° C anneal for $Si_{0.986}C_{0.014}/Si$ and $Si_{0.91}Ge_{0.076}C_{0.014}/Si$ heterostructures. The shape of the depletion curves are similar for both the ternary and binary structures, both having sharp losses in C content during the initial anneal times. After 150 minutes both the heterostructures have lost all substitutional C from the Si lattice. During the annealing, a broad absorption band forms at 12.2 μm. This band is identical to the one which formed in the $Si_{1-y}C_y/Si$ heterostructures and is due to the formation of small βSiC precipitates[5]. There is no shift in shape or position in the SiC phonon band for the Ge containing alloys, suggesting the precipitates are similar. Note that despite the reduction of lattice strain in the heterostructure, the ternary alloy loses substitutional C at least as fast as the binary. This indicates that the strain energy plays a very small role in the transformation process.

TEM was used to investigate the microstructure after annealing unfer the conditions of Figure 4 to the point where absorption measurements indicated all substitutional C is lost. Figure 5a shows the cross-section of a $Si_{0.915}Ge_{0.076}C_{0.09}$ layer

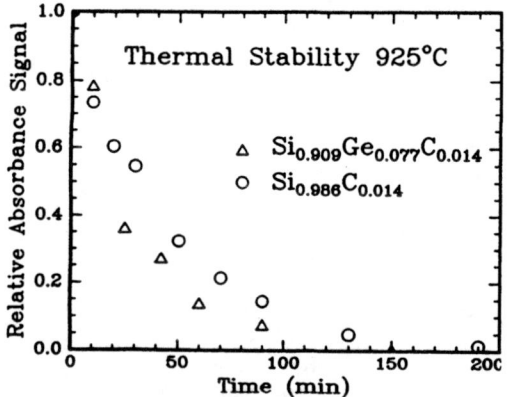

Figure 4 Comparison of the depletion of the absorbance signal for $Si_{0.91}Ge_{0.076}C_{0.014}$ and $Si_{0.986}C_{0.014}$ versus annealing time at 925° C.

470

50nm

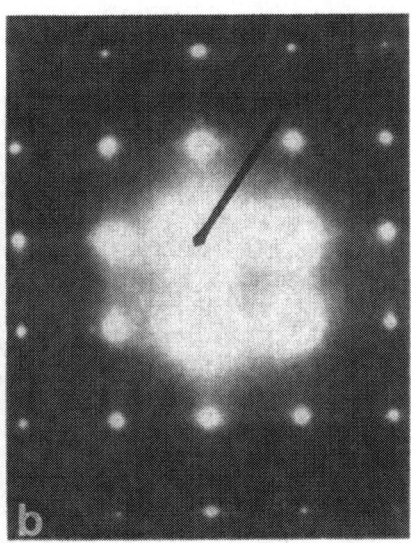

Figure 5 a) TEM X-section of $Si_{0.915}Ge_{0.076}C_{0.009}$/Si heterostructure annealed at 925° C for 130 min. b) Diffraction pattern from plan-view of same layer. The small satellite spots around the Si spots are from βSiC.

after annealing 925° C for 130 minutes. The surface region of the film which was implanted with C is speckled with small 2-4 nm diameter precipitates. The diffraction pattern from the plan-view of the same specimen (Figure 5b) show that these particles are cubic βSiC. βSiC has the diamond structure and the precipitates are observed to be aligned with the Si matrix lattice. There is no evidence of strain around the precipitates indicating the interface between the Si lattice and the precipitates must be incoherent. The precipitates reside primarily in the top 100 nm thus there is no significant diffusion of C into the bulk. There is a thin line of precipitates along the initial amorphous/crystalline interface. This layer of precipitates results from the gettering of C by the Si interstitials in the end of range region of the implant. This C gettering at the end of range region has been also observed in the $Si_{1-x}C_x$/Si heterostructures grown by SPE[5]. In short, the TEM micrographs show that the loss in substitutional C follows the same

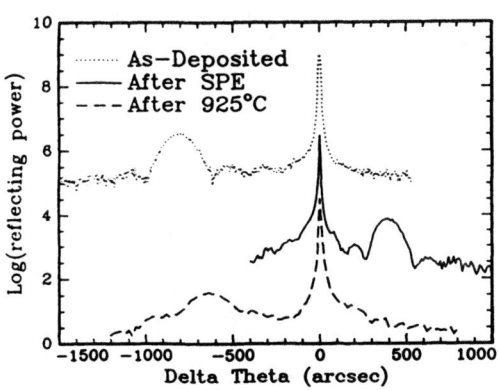

Figure 6 (004) Rocking curve X-ray diffraction scans for $Si_{0.91}Ge_{0.076}C_{0.014}$/Si before and after 925° C for 130 min. The as grown $Si_{0.923}Ge_{0.077}$ is also shown.

471

path in $Si_{1-x-y}Ge_xC_y$ as previously observed in $Si_{1-y}C_y$.

Figure 6 shows the changes in the rocking curve diffraction scan along the (004) reflection with the depletion of substitutional C by annealing. The uppermost scan is from the as deposited $Si_{0.923}Ge_{0.077}$ layer and is included as a reference. The middle and bottom scans are from $Si_{0.91}Ge_{0.076}C_{0.014}$/Si heterostructures just after SPE growth and after 925°C for 130 min respectively. Evaluations of all three of these heterostructures using rotated (004) and (224) reflections revealed that all the layers are nearly commensurate and have no lattice tilt. The as deposited $Si_{0.923}Ge_{0.077}$ layer has compressive strain due to the incorporation of Ge. The SPE ternary heterostructure has an average X-ray strain of (0.28 % tensile) due to the overcompensation of the Ge with the smaller C atom. After annealing at 925° C for 130 min., the strain has become compressive (0.4 %). Thus, the layer strain nearly returns to the strain expected by the $Si_{0.923}Ge_{0.077}$ binary heterostructure on Si. Thus, Ge appears to act essentially as a spectator to the precipitation and strain evoution in the layers heavily supersaturated with C.

CONCLUSIONS

We have extended the study of $Si_{1-y}C_y$/Si heterostructure stability to rapid thermal processing temperature regime. These results show that extrapolation of the lower temperature furnace results adequately describes the stability up to 1000° C. At temperatures above 1000°C, the heterostructures lose substitutional C by C clustering into G. P. zones and precipitation of βSiC. The $Si_{1-x-y}Ge_xC_y$/Si heterostructures were found to lose substitutional C identically to $Si_{1-y}C_y$/Si. The strain compensation of Ge has little effect on the kinetics involved in the loss of substitutional C.

ACKNOWLEDGEMENTS

This research was supported by the Cornell Semiconductor Research Corporation. Work at ASU supported in part by the Advanced Research Project Agency and monitored by the Air Force Office of Scientific Research under contract No. F49620-939-C-0018 (G. Witt). Work at Sandia National Laboratories supported by DOE Basic Energy Sciences/ Materials Science under contract DE-AC04-76DP0078. The wafer processing was performed by the ACT line at Motorola Inc. Mesa Arizona.

REFERENCES

1. R. F. Soref, J. Appl. Phys. **70** (4), 2470, (1991).
2. J. W. Strane, S. R. Lee, H. J. Stein, B. L. Doyle, S. T. Picraux, and J. W. Mayer, Appl. Phys. Lett. **63** (20), 2789, (1993).
3. J. W. Strane, W. Edwards, H. J. Stein, S. R. Lee, S. T. Picraux and J. W. Mayer, in Evolution of Surface and Thin Film Microstructure, eds. H. A. Atwater, E. Chason, M. Grabow, M. Lagally (Mater. Res. Soc. Proc. 280, Pittsburgh, PA, 1992) pp. 609-613.
4. R. C. Newman, and J. Wakefield, in Metallurgy of Semiconductor Materials, ed. J. B. Shroeder, **15**, 201, (1962).
5. J. W. Strane, H. J. Stein, S. R. Lee, S. T. Picraux, J. Watanabe, and J. W. Mayer to be submitted J. Appl. Phys.
6. A. R. Bean, and R. C. Newman, J. Phys. Chem. **32**, 1211, (1971).

THE APPLICATION OF SOLID PHASE EPITAXY FOR THE INCORPORATION OF SUBSTITUTIONAL CARBON IN SILICON

JON J. CANDELARIA*, J. K. WATANABE *, N. DAVID THEODORE*, RICHARD B. GREGORY*, DIETER K. SCHRODER**, LAWRENCE M. STOUT,** AND NIGEL G. CAVE*.

*Materials Research and Strategic Technologies, Motorola Inc., 2200 W. Broadway, Mesa, AZ.
**Department of Electrical Engineering, Arizona State University, Tempe, AZ.

ABSTRACT

Carbon was substitutionally incorporated into silicon using ion implantation and solid phase epitaxy (SPE) to regenerate a high quality crystalline substrate. Carbon was implanted into Si(100) substrates using a single implant of 25 keV at doses ranging from 1.75×10^{15} to $1.05 \times 10^{16}/cm^2$. After carbon implantation half of the substrates were amorphized using a silicon implant. All of the wafers were subjected to a 700°C anneal in N_2 ambient for 30 minutes to induce SPE regrowth of the implanted regions. FTIR, SIMS, RBS, and TEM were used to characterize the samples. Results indicate that carbon was substitutionally incorporated into the silicon lattice, but that some carbon did precipitate to form silicon carbide. Post-amorphization improved regrowth of implanted regions in lower dose implanted wafers. Electrical measurements on diode structures indicate that the band gap was reduced for carbon incorporation at these concentrations.

INTRODUCTION

Carbon in silicon alloys ($Si_{1-y}C_y$) are being investigated as band gap engineered materials with possible applications in silicon based heterojunction bipolar transistors (HBTs) [1-6]. It would be expected that increasing the carbon concentration in unstrained silicon would raise the band gap from the value for silicon (1.1 eV) to that of SiC (2.5 eV) and finally to that of diamond (5.5 eV) [7]. However, in strained $Si_{1-y}C_y$ alloys the band gap initially decreases with carbon incorporation [5,8]. This interesting phenomenon could be exploited for fabrication of Si HBT devices using $Si_{1-y}C_y$ alloys in the base.

In order to achieve a band gap reduction, carbon must be incorporated into the silicon lattice substitutionally. The difference in the covalent radii of carbon and silicon strains the lattice around the carbon atoms. The presence of strain eliminates the degeneracy of energy levels associated with unstrained silicon, and thus reduces the band gap. Deposition of $Si_{1-y}C_y$ alloys has been achieved by using remote plasma enhanced chemical vapor deposition [1], molecular beam epitaxy (MBE) [2-5] and by carbon implantation followed by an anneal to induce solid phase epitaxial (SPE) regrowth [6].

In this study, carbon was implanted from 1.75×10^{15} to $1.05 \times 10^{16}/cm^2$ through a 400Å screen oxide into 16-20 Ω-cm Si(100) substrates at 0º tilt. After removing the screen oxide, one wafer with each carbon dose was then implanted with 25 keV, $4.0 \times 10^{15}/cm^2$ $^{29}Si^+$ to amorphize the wafer at 0°. This process, which will be referred to as "post-amorphization", is thought to improve recrystallization. The screen oxide was removed to ensure that the silicon implant was deeper than the carbon implant. All of the wafers received a 700ºC, 30 minute anneal in N_2 ambient to induce SPE regrowth of the implanted region. A rapid thermal anneal (RTA) was then performed at 1000°C for 30 seconds.

Information on the microstructure of the wafers was obtained from TEM analysis and RBS channeling measurements. SIMS measurements provided the concentration profile of carbon within the silicon substrate. FTIR analysis determined the relative levels of carbon incorporated into substitutional lattice sites. Electrical measurements on n+/p+ diodes were used to estimate the band gaps of the silicon/carbon alloys.

RESULTS AND DISCUSSION

The RBS/ion channeling spectra of the non-amorphized wafers are displayed in figure 1. The large surface peaks of the 7.0 and $8.75 \times 10^{15}/cm^2$ C-implanted samples relative to the starting material indicates that these samples remained heavily damaged after annealing. Thus at doses of $7.0 \times 10^{15}/cm^2$ or greater, the carbon inhibited effective SPE regrowth. The surface peaks of the three samples implanted with ≤5.25 x $10^{15}/cm^2$ appeared nearly identical, but overall were larger than the surface peak for the starting material. This suggested a small amount of damage remained in the carbon-implanted regions of these wafers.

TEM images (not shown) of the non-amorphized 3.5, 5.25 and $7.0 \times 10^{15}/cm^2$ C-implanted wafers showed the damage detected in the RBS/ion channeling analysis. For implant doses of 1.75 to $5.25 \times 10^{15}/cm^2$, an increasing amount of end-of-range (EOR) and peak-of-implant (POI) damage became visible. The defect density was much higher in the ≥5.25 x $10^{15}/cm^2$ C-implanted samples than for the lower doses. EOR damage as well as a high density of damage above the EOR band was present with dense thickets of microtwins and dislocations extending down from the surface of the wafer.

TEM images of the post-amorphized carbon-implanted samples showed only a single band of EOR damage for carbon doses of up to $5.25 \times 10^{15}/cm^2$, whereas the non-amorphized sample had damage in the region above the EOR band. Post-amorphization improved the crystalline quality of the implanted region for carbon doses up to $5.25 \times 10^{15}/cm^2$ by providing a surface amorphous region through which epitaxial regrowth could occur more effectively. At doses above $5.25 \times 10^{15}/cm^2$, the carbon atoms were more efficient at disordering the lattice during implantation, and so there was no visible difference in the defect density between the "non-amorphized" and post-amorphized wafers at the higher doses.

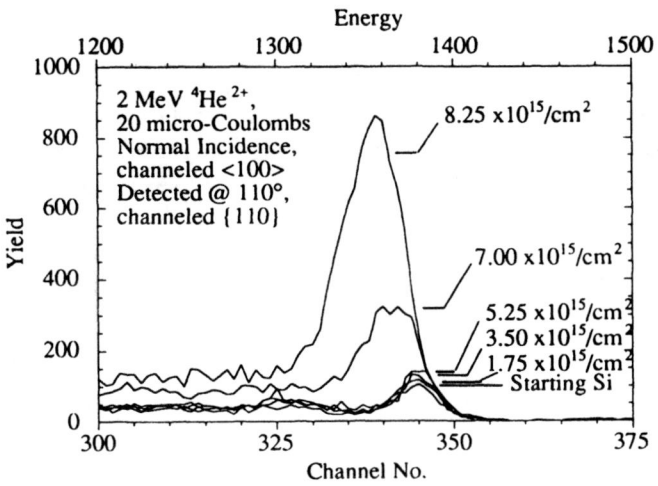

<u>Figure 1</u>   RBS/ion channeling spectra of the non-amorphized wafers.

The SIMS profiles showed that the peak of the implant was located approximately 400 Å below the surface with the end of range damage near 700 Å. This agreed well with the TEM images for the depths of the POI and EOR damage regions. The lowest carbon dose had a smooth profile, but at higher doses discontinuities became apparent, indicating that some carbon was gettered to the POI defects at the lower doses and towards the EOR band at higher doses (figure 2). The discontinuities in the carbon profile increased in depth as the carbon dose was increased. This observation is contrary to what would be expected as the implant depth would generally decrease with dose due to increased substrate amorphization. This phenomenon is currently being investigated to determine the reason for the observed shifts. Peak concentrations for doses of 1.75, 3.5, 5.25, 7.0, and 8.75 x 10^{15}/cm^2 were calculated as 2.5e20, 5.5e20, 7.2e20, 9.1e x 10^{20}, and 1.1 x 10^{21}/cm^3 respectively. This result indicated a carbon concentration threshold for poor SPE regrowth of 7 x 10^{20}/cm^3 (1.4%) which is similar that obtained by Strane et $al.$ [6]. Upon post-amorphization, the carbon for all doses gettered to the EOR damage band (profiles not shown). For the lower-dosed wafers, the POI damage was no longer present and thus could not act as a gettering site. These observations suggest that the silicon implant created an EOR band of damage, but aided in the elimination of damage due to the carbon implant.

The infra-red spectra obtained from the 1.75, 5.25, and 7.0 x 10^{15}/cm^2 dosed, non-amorphized samples are presented in figure 3. The peak due to the local vibration mode (LVM) of substitutional carbon in silicon at 607 cm^{-1} reached a maximum intensity for the 5.25 x 10^{15}/cm^2

dose. The intensity then progressively decreased as the carbon dose was increased. An additional broad peak at approximately 750 cm⁻¹ was assigned to the Si-C (stretching) vibration indicating that carbon precipitated to form silicon carbide. The peak position was lower than expected for crystalline silicon carbide (790-800 cm⁻¹) which suggested that the silicon carbide was amorphous. The presence of the silicon carbide peak indicated that the substitutional carbon was not thermally stable as previously shown by Powell *et al.* [5]. This was not surprising since the solid-solid solubility of carbon in silicon was exceeded by over four orders of magnitude at the implant doses used. The intensity decrease of the carbon LVM peaks at the higher doses was accompanied by an increase in the intensity of the Si-C peak, indicating increased precipitation. Amorphizing the substrate after carbon implantation increased the amount of carbon incorporated substitutionally into the wafer. This result again indicated that post-amorphization improved the SPE regrowth of this material.

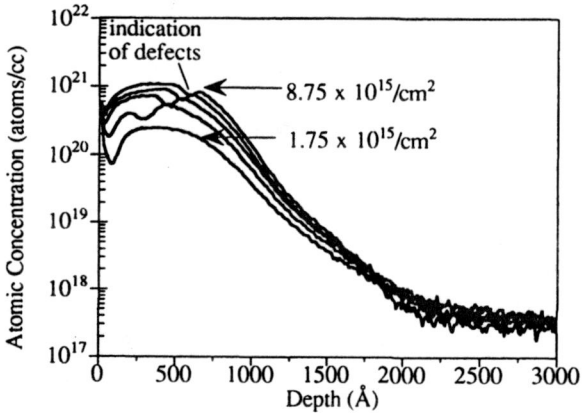

Figure 2   SIMS carbon concentration depth profiles of the non-amorphized wafers.

Electrical measurements were made on n+/p+ vertical diodes. Both reverse and forward biasing was applied to the diodes to look for any evidence of recombination current and band gap shift respectively. No evidence was found of any significant recombination, or reverse-bias leakage current. An estimate of the electrical band gap (E_g) was extracted from the forward current characteristics due to the exponential dependency of the diode's saturation current (I_o) on the band gap of the material. The band gap is obtained by plotting $\ln(I_o/T^4)$ versus $1/T$. The slope (m) of this linear plot then yields the electrical band gap ($E_g = -m_* k_B$ where k_B is the Boltzmann constant) provided the diode exhibits ideal characteristics. A temperature range of 36-128°C was used in the current work. A general reduction in the band gap with carbon implant dose was observed as shown in figure 4. Surprisingly no significant differences were detected in the band gaps obtained from the non-amorphized and amorphized wafers. This result is currently under investigation as the increased substitutional carbon content in the amorphized wafers would have been expected to cause a greater reduction in the band gap [5] due to the increased strain. One point to note is that this technique yields the band gap at the junction and so any variation in the doping profiles from sample to sample would result in a different bandgap.

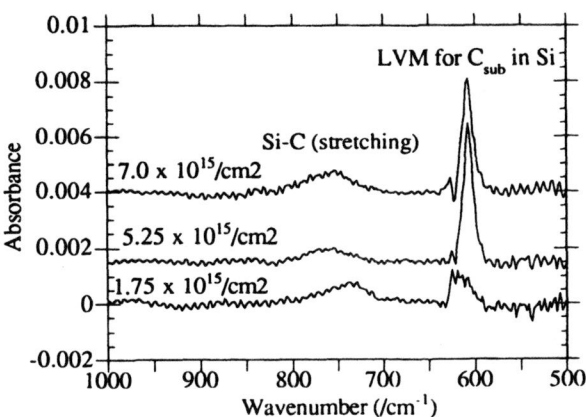

Figure 3   FTIR spectra of the 1.75, 5.25, and 7.0 x 10^{15}/cm^2 C-implanted non-amorphized wafers.

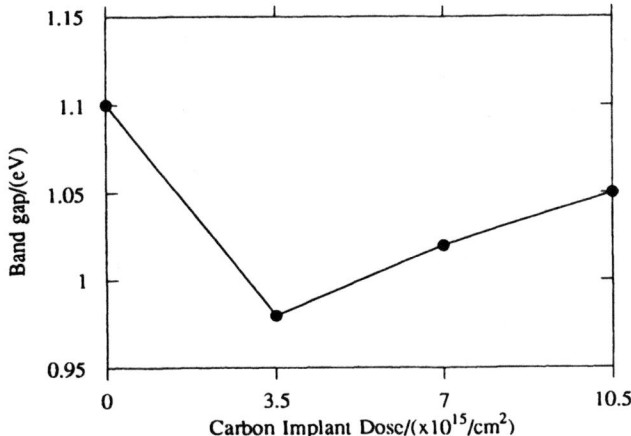

Figure 4   Typical variation of the band gap as a function of carbon implant dose (±0.05 eV).

CONCLUSIONS

Incorporation of substitutional carbon into the silicon lattice was achieved using ion implantation and solid phase epitaxy. The best microstructural quality was obtained for carbon implant doses of $5.25 \times 10^{15}/cm^2$ or less and with the use of a post-amorphization silicon implant. Post-amorphization also increased the amount of carbon incorporated into substitutional sites of the lattice. The greatest amount of substitutional carbon was measured in the post-amorphized $5.25e15/cm^2$ dosed wafer. At higher doses carbon precipitated to form silicon carbide which possibly inhibited the SPE process. Electrical measurements indicate that the band gap was reduced for the carbon implant doses used in this study. The reduction in the band gap indicated that this material could have applications as the base region of a heterojunction bipolar transistor.

ACKNOWLEDGMENTS

The authors would like to thank Ray Doyle for preparation of XTEM specimens, and Denise Cantu and Lorraine Johnston for printing of TEM micrographs.

REFERENCES

1.   J. B. Posthill, R. A. Rudder, S. V. Hattengady, G. G. Fountain, and R. J. Markunas, Appl. Phys. Lett. **56** (8), 734 (1990).

2.   S. S. Iyer, K. Eberl, M. S. Goorsky, F. K. LeGoues, J. C. Tsang, and F. Cardone, Appl. Phys. Lett. **60**(3), 356 (1992).

3.   K. Eberl, S. S. Iyer, J. C. Tsang, M. S. Goorsky, and F. K. Legoues, J. Vac. Sci. Tech. **B10**(2), 932 (1992).

4.   J. C. Tsang, K. Eberl, S. Zollner, and S. S. Iyer, Appl. Phys. Lett. **61**, 961 (1992).

5.   A. R. Powell, K. Eberl, F. E. LeGoues, B. A. Ek, and S. S. Iyer, J. Vac. Sci. Tech. **B11**(3), 1064 (1993).

6.   J. W. Strane, W. J. Edwards, H. S. Stein, S. R. Lee, B. L. Doyle, S. T. Picraux, and J. W. Mayer in Evolution of Surface and Thin Film Microstructure edited by . H. A. Atwater, E. Chason, M. Grabow, M. Lagally (Mater. Res. Soc. Proc. **280**, Pittsburgh, PA, 1992) pp. 609.

7.   R. A. Soref, J. Appl. Phys. **70**, 2470 (1991).

8.   A. A. Demkov and O. F. Sankey, Phys. Rev. B, **48**, 2207 (1993).

TRENDS IN SOLUTE SEGREGATION BEHAVIOR
DURING SILICON SOLIDIFICATION

RICCARDO REITANO†, PATRICK M. SMITH*, AND MICHAEL J. AZIZ
Division of Applied Sciences, Harvard University, Cambridge MA 02138.
† Present address: Dipartimento di Fisica, Universita di Catania,
  Corso Italia 57, I-95129 Catania, Italy.
* Present address: Lawrence Livermore National Laboratory, Livermore CA 94550.

ABSTRACT

At the high growth rates accessible during pulsed-laser induced melting and solidification and explosive crystallization, crystal growth kinetics are dominated not by equilibrium thermodynamics, but by the atomistic mechanisms by which crystallization proceeds. These mechanisms can be probed by testing the predictions of solute trapping models based on various crystal/melt interface structures against measurements. We have measured the dependence of solute trapping of several group III, IV, and V elements in silicon on both interface orientation and crystallization speed. The Aperiodic Stepwise Growth Model of Goldman and Aziz accurately fits both the velocity and orientation dependence of the solute trapping observed in these systems. The success of the model implies a ledge structure for the crystal/melt interface and a step-flow mechanism for crystal growth. In addition, we have observed an empirical inverse correlation between the two free parameters ("diffusive speeds") in this model and the equilibrium solute partition coefficient of a system. This correlation may be used to estimate values of the diffusive speeds for other systems in which solute trapping has not been or cannot be measured.

INTRODUCTION

In recent years a fairly thorough understanding of the redistribution of ion-implanted dopants in silicon by pulsed-laser melting (PLM) has been developed. The initial discovery that implantation damage could be repaired by pulsed-laser processing [1] was followed by several years of experimental measurements and the development of models to explain the phenomenon [2,3]. However, a complete description of the redistribution of solute atoms during rapid solidification of silicon remains elusive. It has become clear that the enhanced dopant supersaturations observed in pulsed-laser melted ion-implanted silicon result from an increase in the partition coefficient during rapid solidification of silicon. The partition coefficient k is defined as the ratio of the solute concentration in the solid (x_S) to that in the liquid (x_L) at the solid/liquid interface:

$$k = x_S/x_L \tag{1}$$

At the extremely high solidification speeds (up to 5 m/s) obtained in pulsed-laser melting, the partition coefficient increases from its equilibrium value k_e toward unity. This trapping of solute during rapid solidification, observed several years earlier in Zn-Cd alloys [4], results in the observed enhanced concentrations of dopant elements.

Several models have been advanced which attempt to explain the increase in k with solidification rate [5,6,7,8,9,10,11]. Of these, the Aperiodic Stepwise Growth Model (ASGM) has been the most successful at matching experimental solute trapping data in silicon [10]. While the other models all predict that k should rise with increasing

479

solidification velocity, only the ASGM accounts for the observed dependence of solute trapping on crystallographic orientation. This model assumes that silicon solidification proceeds by the aperiodic lateral passage of (111) steps, as shown in figure 1.

Figure 1: Solidification at a speed v *is accomplished by the lateral passage of (111) steps along the interface. Solute atoms are incorporated into the steps, and can only escape back to the liquid (at the ledge or on the terrace) until another step passes over the solute atom, permanently trapping it.*

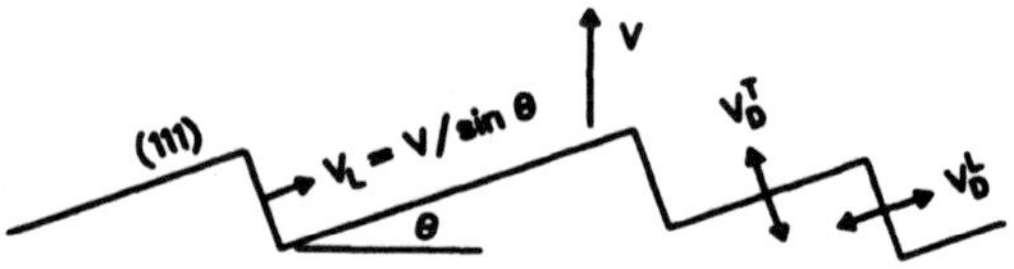

The ASGM predicts that the partition coefficient should exhibit the following dependence on solidification speed and growth direction:

$$k(v,\theta) = \frac{k_e + v/(v_D^T \cos\theta)\left(\dfrac{k_e + v/(v_D^L \sin\theta)}{1 + v/(v_D^L \sin\theta)}\right)}{1 + v/(v_D^T \cos\theta)} \qquad (2)$$

Here v is the solidification speed, θ is the angle between the growth direction and the (111) axis of the crystal, and v_D^T and v_D^L are "diffusive speeds," the speeds at which a solute atom can diffuse from the solid into the liquid across the terrace and at the ledge, respectively. The orientation dependence predictions of the ASGM have been verified for the Si-Bi system, and in this work we test its validity for a much wider range of dilute Si alloys: As, Ga, Ge, In, Sb and Sn. In addition to testing the applicability of the ASGM, we hope to gain some insight into what determines the values of the terrace and ledge diffusive speeds.

EXPERIMENTAL TECHNIQUES

Silicon samples were cut from a single boule at misorientations in 5 degree increments from (110) through (111) to (100). The samples were then ion-implanted with the desired solute (As, Ga, Ge, In, Sb, or Sn) to a depth of about 600-1000Å, depending on the ion type. Ion doses ranged from 1×10^{15} atoms/cm^2 for Ga and In to 1×10^{16} atoms/cm^2 for As, with the doses for Ge, Sb and Sn in the range 2.5-3.0$\times10^{15}$ atoms/cm^2. Individual samples were pulsed-laser melted with a 30ns XeCl (308nm) pulse of 1.0-1.5 J/cm^2 using the laboratory depicted in figure 2.

The surface reflectance was measured with the 488nm line of an argon ion laser and recorded by a 750MHz digitizing oscilloscope. Since the reflectivity of silicon increases substantially upon melting, the reflectance measurement allows us to determine the "melt duration," the amount of time the surface remains molten. Since the thermal and optical properties of both solid and liquid silicon are quite well known, the melt depth as a function of time can be calculated using one-dimensional heat-flow simulations [12]. The simulated melt durations are then matched to the observed melt duration, allowing the melt depth (and solidification velocity) to be determined. The solidification velocity could be decreased by over an order of magnitude by pre-heating the samples from behind for up to several seconds with a 50-watt continuous-wave CO$_2$ laser; this reduces the thermal gradients and thermal conductivities in the samples, thereby impeding heat flow away from the surface and decreasing the solidification velocity.

Solute atom depth profiles were measured before and after irradiation by grazing-exit Rutherford Backscattering Spectrometry (RBS). Partition coefficients were determined by comparing measured solute depth profiles to simulated profiles. One-dimensional diffusion and segregation simulations were used to diffuse the initial solute profile while the sample was molten, and segregation of solute in the simulation at the solid/liquid interface was controlled by specifying a value for the partition coefficient k. The melt history (melt depth as a function of time) for each sample was taken from the heat flow simulations discussed above. Several diffusion/segregation simulations, each using a different value of k, were run for each sample, and the best value of k was determined by selecting the simulated profile which best matched the solute profile measured after irradiation.

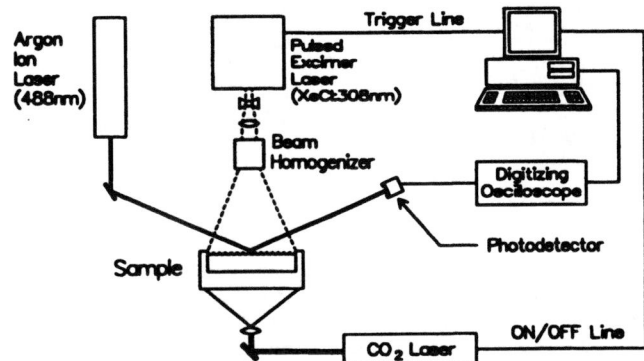

Figure 2: Schematic drawing of pulsed-laser laboratory used in these experiments. After the sample is pre-heated from behind by the CO_2 laser, it is melted by the excimer laser pulse. Surface melting and solidification are detected by measuring the reflectance with the 488nm line of an argon ion laser.

While it is possible to select a value for the partition coefficient simply by fitting simulated solute depth profiles to the measured profile, this method is quite sensitive to errors in the estimate of the RBS depth resolution. The depth resolution affects the shape of the surface peak in the profile, and since this near-surface region is the section of the profile that contains all the segregated solute, it is the most important portion of the profile for calculating k. We use another method, less susceptible to such errors, to calculate k. Rather than simply fitting simulated profiles to measured profiles, we compare the amount of solute swept to the surface during solidification to the amount left behind in the bulk of the sample. For each of the simulated profiles, this ratio R_A is calculated by integrating the fractional solute concentration $x(z)$ over a "surface" region S and a "bulk" region B (shown in figure 3):

$$R_A = \int_S x(z)dz \bigg/ \int_B x(z)dz \qquad (3)$$

The ratios R_A for each simulated profile are then compared to the R_A for the measured profile, and the value of k for which the simulated profile's R_A matches that of the measured profile is selected. In addition to not being affected by errors in the estimate of the RBS depth resolution, this method is less sensitive to other systematic errors such as inaccurate melt-depth measurements at large depths.

RESULTS

The irradiation conditions for each sample type were determined by the amount of solute segregation observed in that system. Since there must be a measurable surface accumulation of solute after irradiation to accurately measure k, the range of solidification velocities was varied to produce a suitably large surface accumulation. For three of the systems (Si-Sn, Si-Ga and Si-In) the surface peak in the solute depth profile (depicted in figure 3) for samples solidified at 4.5 m/s was large enough to calculate values of the partition coefficient. However, it was necessary to decrease the solidification rate to less than 0.6 m/s to obtain measurable surface solute peaks for the other three systems (Si-As, Si-Sb and Si-Ge).

Figure 3: A sample As profile after irradiation, with a simulated profile. Note the surface and bulk regions (S and B, respectively) used in the area-ratio calculation of the partition coefficient (equation 3).

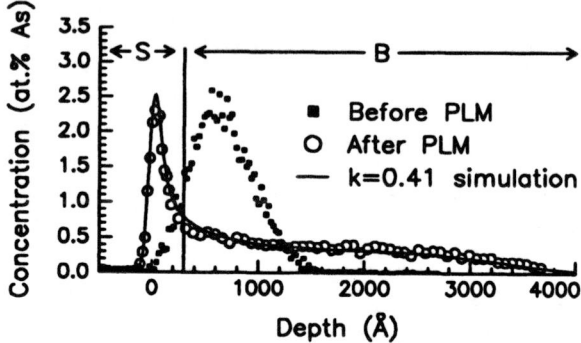

The partition coefficients measured for the Si-Sn system at several orientations are plotted in figure 4; the Si-Sn system is representative of the three systems irradiated without pre-heating of the samples. The data are sharply peaked for solidification in the (111) direction, and are fit quite well by the ASGM (equation 2). For growth in the (111) direction the solid/liquid interface contains almost no ledges, so the terrace diffusive speed v_D^T is determined by the value of k at $\theta=0$. The ledge diffusive speed v_D^L is then varied to obtain the best fit over all the orientations. The $k(\theta)$ data measured for the Si-In and Si-Ga systems are qualitatively quite similar, and are also fit well by the ASGM.

Figure 4: Partition coefficients measured for Sn in Si at various orientations are shown in circles, with the ASGM fit. All the samples were solidified at the same velocity (4.2 m/s). The terrace and ledge diffusive speeds for this fit are 5 m/s and 7 m/s, respectively.

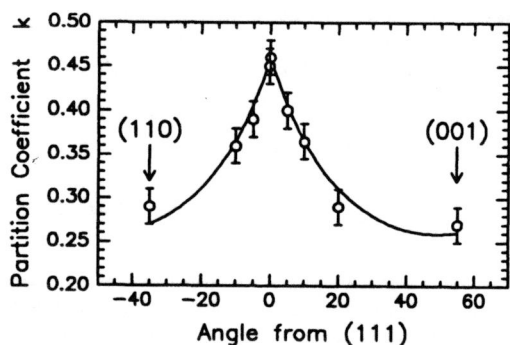

Samples from the other three systems (Si-As, Si-Sb and Si-Ge) irradiated at room temperature showed no discernable surface solute peaks, so these samples were pre-heated to temperatures of 1270-1500K to decrease the solidification velocity into the 0.1-0.6 m/s range. Solute depth profiles measured before and after pre-heating for samples that were not irradiated with an excimer laser pulse verified that the CO_2 laser pre-heating altered the solute profiles only negligibly. No direct measurements of the pre-heat temperatures were made, but the substrate temperatures could be extracted from the transient reflectance data. Several heat-flow simulations were run for each sample to be analyzed, and the experimental conditions input to each simulation were identical, with the exception of the initial sample temperature. The best-fit initial temperature was then determined by selecting the simulation for which the simulated melt duration matched that observed in the experiment. These pre-heat temperatures, and the resulting solidification velocities, are listed in Table I.

The ASGM was able to fit the $k(\theta)$ data for the Si-As and Si-Sb systems irradiated at the higher temperatures, but we did not observe any dependence of k on orientation for the Si-Ge system. For the Si-Ge system, partition coefficients of 0.62-0.76 were measured over the entire range of orientations. The partition coefficients, solidification velocities, liquid diffusivities and ASGM diffusive speeds for all the systems studied are listed in Table I.

DISCUSSION

The orientation dependence of the partition coefficient is fit quite well by the ASGM in all the systems we studied (Si-As, Si-Ga, Si-Ge, Si-In, Si-Sb and Si-Sn) with the exception of Si-Ge, for which no orientation dependence was observed. The velocity dependence $k(v)$ is also accurately fit by the ASGM in the systems in which it has been measured (Si-Bi [10], Si-As [13] and Si-Sn [14]). However, the two free parameters (the ledge and terrace diffusive speeds) of the ASGM limit its use in practical applications. We have therefore searched for correlations between these diffusive speeds and known alloy properties, to enable the model to be used for systems in which measurements of the diffusive speeds have not been made. We see no apparent correlation between either of the diffusive speeds and the liquid or solid-phase solute diffusivities. However, there appears to be an inverse correlation between the diffusive speeds and the equilibrium partition coefficient k_e, as shown in figure 5. Both $\log(v_D^T)$ and $\log(v_D^L)$ appear to be approximately linearly related to $\log(k_e)$; fits to the data yield $\log(v_D^T) = -1.347 - 0.783 \log(k_e)$ and $\log(v_D^L) = -0.556 - 0.624 \log(k_e)$.

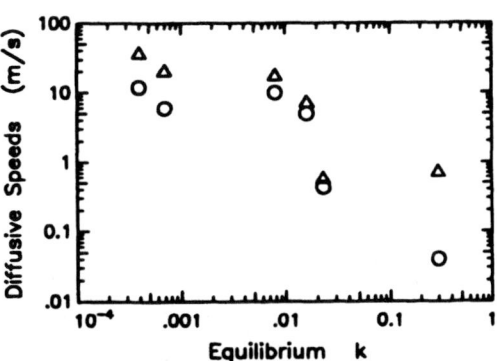

Figure 5: Ledge diffusive speeds v_D^L (triangles) and terrace diffusive speeds v_D^T (circles) plotted as functions of the equilibrium partition coefficient k_e.

This correlation should not be that surprising, as it indicates that the stronger a solute atom's preference for the liquid, the faster one must solidify to trap the solute. In terms of the transition-state basis for the ASGM, the barrier to solute escape into the liquid (which is related to the diffusive speed) is lowered as the driving force for solute redistribution increases (at smaller values of k_e). A similar correlation between the diffusive speed and the equilibrium partition coefficient has been observed in solute trapping in aluminum alloys, and possible origins of this correlation are described in ref. [15].

CONCLUSION

The Aperiodic Stepwise Growth Model has been shown to accurately model the solute trapping behavior of several solutes (As, Ga, In, Sb and Sn) in dilute concentrations in silicon. This model, which assumes that silicon growth occurs by the lateral growth of (111) terraces across the solid-liquid interface, fits the dependencies of the partition coefficient on both interface orientation and solidification velocity quite well, and is the only model which can account for the observed orientation dependence. We have observed an inverse correlation between the equilibrium partition coefficient and the two free parameters in this model, the ledge and terrace diffusive speeds. This research was supported by the NSF under grant NFS-DMR-92-08931.

Table I: Summary of solute trapping data for dilute silicon systems.
Data for Si-Bi are taken from [10].

Solute	k_e	Substrate Temp. (K)	Velocity (m/s)	D_{LIQ} (cm^2/s)	v_D^T (m/s)	v_D^L (m/s)
As	0.3	1300 - 1500	0.1 - 0.4	3.0×10^{-4}	0.04	0.7
Ga	0.008	300	4.3	4.8×10^{-4}	10	17
Ge	0.33	1270 - 1450	0.13 - 0.61	3.0×10^{-4}	---	---
In	0.0004	300	4.5	4.0×10^{-4}	12	36
Sb	0.023	1350 - 1420	0.17 - 0.28	2.0×10^{-4}	0.43	0.56
Sn	0.016	300	4.2	2.5×10^{-4}	5	7
Bi	0.0007	300	1.7 - 5.0	2.0×10^{-4}	6	20

REFERENCES

[1] I. B. Khaibullin, E. I. Shtyrkov, M. M. Zaripov, M. F. Galyautdinov and G. G. Zakirov, Sov. Phys. Semicond. 11, 190 (1977).
[2] J. M. Poate and James W. Mayer, eds., Laser Annealing of Semiconductors (Academic Press, New York, 1982).
[3] C. W. White and P. S. Peercy, eds., Laser and Electron Beam Processing of Materials (Academic Press, New York, 1980).
[4] J.C. Baker and J.W. Cahn, Acta metall. 17, 565 (1969).
[5] J. W. Cahn, S. R. Coriell and W. J. Boettinger, "Rapid Solidification" in Laser and Electron Beam Processing of Materials, eds. C. W. White and P. S. Peercy (Academic Press, New York, 1980).
[6] K. A. Jackson, "Crystal Growth and Phase Formation" in Surface Modification and Alloying by Laser, Ion, and Electron Beams, eds. J. M. Poate, G. Foti and D. C. Jacobson (Plenum Press, New York, 1983).
[7] R. F. Wood, Appl. Phys. Lett. 37, 302 (1980).
[8] M. J. Aziz, J. Appl. Phys. 53, 1158 (1982).
[9] M. J. Aziz and T. Kaplan, Acta. metall. 36, 2335 (1987).
[10] L. M. Goldman and M. J. Aziz, J. Mater. Res. 2, 524 (1987).
[11] A. A. Wheeler, W. J. Boettinger and G. B. McFadden, Phys. Rev. E 47, 1893 (1993).
[12] M. J. Aziz, C. W. White, J. Narayan and B. Stritzker, in Energy Beam-Solid Interactions and Transient Thermal Processing, editted by V. T. Nguyen and A. G. Cullis (Les Éditions de Physique, Paris, 1985), p. 231.
[13] J. A. Kittl, M. J. Aziz, D. A. Brunco and M. O. Thompson, submitted to J. Mater. Res.
[14] D. E. Hoglund, M. J. Aziz, S. R. Stiffler, M. O. Thompson, J. Y. Tsao and P. S. Peercy, J. Cryst. Growth. 109, 107 (1991).
[15] P. M. Smith and M. J. Aziz, submitted to Acta metall. mater.

DOPANT ACTIVATION AND EPITAXIAL REGROWTH IN P-IMPLANTED PSEUDOMORPHIC $Ge_{0.12}Si_{0.88}$ LAYERS ON Si(100)

D.Y.C. Lie*, T. K. Carns**, N. D. Theodore***, F. Eisen*, M.-A. Nicolet* and K. L. Wang**
*California Institute of Technology, M/S 116-81, Pasadena, CA 91125
**University of California, Los Angeles, CA 90024
***D. N. Theodore, Motorola Inc., Mesa, Arizona, AZ 85202

Abstract

A pseudomorphic $Ge_{0.12}Si_{0.88}$ film 265 nm thick grown on a Si(100) substrate by molecular beam epitaxy was implanted at room temperature with a dose of 1.5×10^{15} cm² of 100 keV P ions. The projected range of the ions is about 125 nm, which is well within the film thickness. Only the top portion of the $Ge_{0.12}Si_{0.88}$ layer was amorphized by the implantation. Both implanted and non-implanted samples were subsequently annealed in vacuum for 30 minutes from 400 °C to 800 °C. Values of electron Hall sheet mobility and concentration in the implanted $Ge_{0.12}Si_{0.88}$ epilayer were measured after annealing. The solid phase epitaxial regrowth is complete at 550 °C, where the implanted phosphorus reaches ~ 100 % activation. The regrown $Ge_{0.12}Si_{0.88}$ layer exhibits inferior crystalline quality to that of the virgin sample and is relaxed, but the non-implanted portion of the film remains pseudomorphic at 550 °C . When annealed at 800 °C, the strain in the whole epilayer relaxes. The sheet electron mobility values measured at room temperature in the regrown samples ($T_{ann} \geq 550$ °C) are about 20% less than those of pure Si.

Introduction

Si/Ge_xSi_{1-x} heterostructures are attractive for high performance electronic and optical devices and circuits, since it is a silicon-based technology that can be utilized to adjust bandgaps and enhance carrier mobility for the improvement of designs. Numerous Si/Ge_xSi_{1-x} devices have been made, and the fabrication of many of them requires the use of ion implantation. Recent studies revealed that the implantation-induced damage and the subsequent annealing behavior of Si/Ge_xSi_{1-x} heterostructures are different from those of Si. It is therefore important to study how to properly apply ion implantation to Si/Ge_xSi_{1-x}. For example, to achieve best dopant activation and minimize the residual damage generated by energetic ions, in Si IC processing one usually amorphizes the doped region first and subsequently anneals it to let epitaxial regrowth take place. However, it was reported that for strained Ge_xSi_{1-x} layers irradiated by Si ions, the regrowth results in poor crystalline quality [1-4].

Even though there are several studies on the solid-phase epitaxial regrowth for Ge_xSi_{1-x} layers [5-8], studies of dopant incorporation and their activation which take advantage of this regrowth in strained Ge_xSi_{1-x} layers have only been reported by Atzmon et al [9-10] and Hong et al [11] where Sb ions where used. In this paper, we study the implantation of the practically useful P ions. We use 100 keV ^{31}P ions to amorphize a pseudomorphic $Ge_{0.12}Si_{0.88}$ layer without damaging the Si-GeSi interface. This layer is later annealed in vacuum at various

485

temperatures. The focus of this study is on the electrical and material characterization and the stability of the heterostructure after ion implantation and annealing. We measure dopant activation efficiency and sheet electron mobility values of the regrown GeSi layers and compare our results with the existing data in the literature for Si. The material properties of the layers were monitored with MeV ^{4}He ion channeling/backscattering spectrometry, double crystal x-ray diffractometry, and cross-sectional transmission electron microscopy.

Experimental Procedure

Non-intentionally doped pseudomorphic $Ge_{0.12}Si_{0.88}$ layers on Si(100) substrates were grown by ultra-high vacuum molecular beam epitaxy at UCLA. The thickness of the film is 265 nm, as determined by both MeV ^{4}He backscattering spectrometry and cross-sectional transmission electron microscopy. These samples are of excellent crystalline quality, with a minimum channeling yield of ~ 3 % for both Si and Ge signals. Double-crystal x-ray rocking curves exhibit the pseudomorphic nature of the heterostructure, with zero parallel strain within the experimental sensitivity (~ 10^{-4}), and a perpendicular strain value of 0.87 %. The measured strain values are in excellent agreement with the value of 0.88% calculated from linear elasticity theory. The dislocation density in this pseudomorphic film is below the detection limit of the x-ray double-crystal diffractometry, i.e., 10^6/cm². This Si/GeSi heterostructure was implanted at room temperature in high vacuum (~ 10^{-7} Torr) with 100 keV ^{31}P ions to a dose of 1.5 x 10^{15} /cm². The samples were chemically cleaned before loading into the implanter. During implantation, the sample normal was tilted by 7° with respect to the direction of the incident beam to minimize channeling. The ion doses reported here are within ± 5% accuracy. The projected range and the straggling of the P ions, according to the TRIM-92 simulation code, are about 125 nm and 43 nm, respectively [12]. Both virgin and implanted samples were annealed in vacuum (~ 3 x 10^{-7} Torr) for 30 min from 400 °C to 800 °C. X-ray double crystal rocking curves of both (400) symmetrical and (311) asymmetrical diffraction were taken at room temperature in air as little as one hour after implantation, as well as several months later, and the spectra remain the same over this time period. There is no detectable increase in the channeling yield as the result of the ^{4}He irradiation. The electrical properties of the layers were characterized by Hall effect measurements and sheet resistance measurements using the van der Pauw patterns.

Results and Discussions

Figure 1 shows the cross-sectional transmission electron micrographs of a Si/Ge$_{0.12}$Si$_{0.88}$ sample (a) as-implanted and (b) annealed at 550 °C for 30 min in vacuum. Fig. 1 (a) clearly indicates that the as-implanted GeSi film is not amorphous all the way down to the Si-GeSi interface, and the thickness of the amorphous layer is about 195 nm. Point-defect clusters are present in the transition region between the amorphous and the crystalline GeSi, or immediately below this transition region. Fig. 1 (b) indicates that the amorphous GeSi layer has recrystallized entirely. The recrystallized layer contains a high density of threading dislocations (~10^{10} cm^{-2}). No threading dislocations are found in the bottom GeSi layer. There is also no

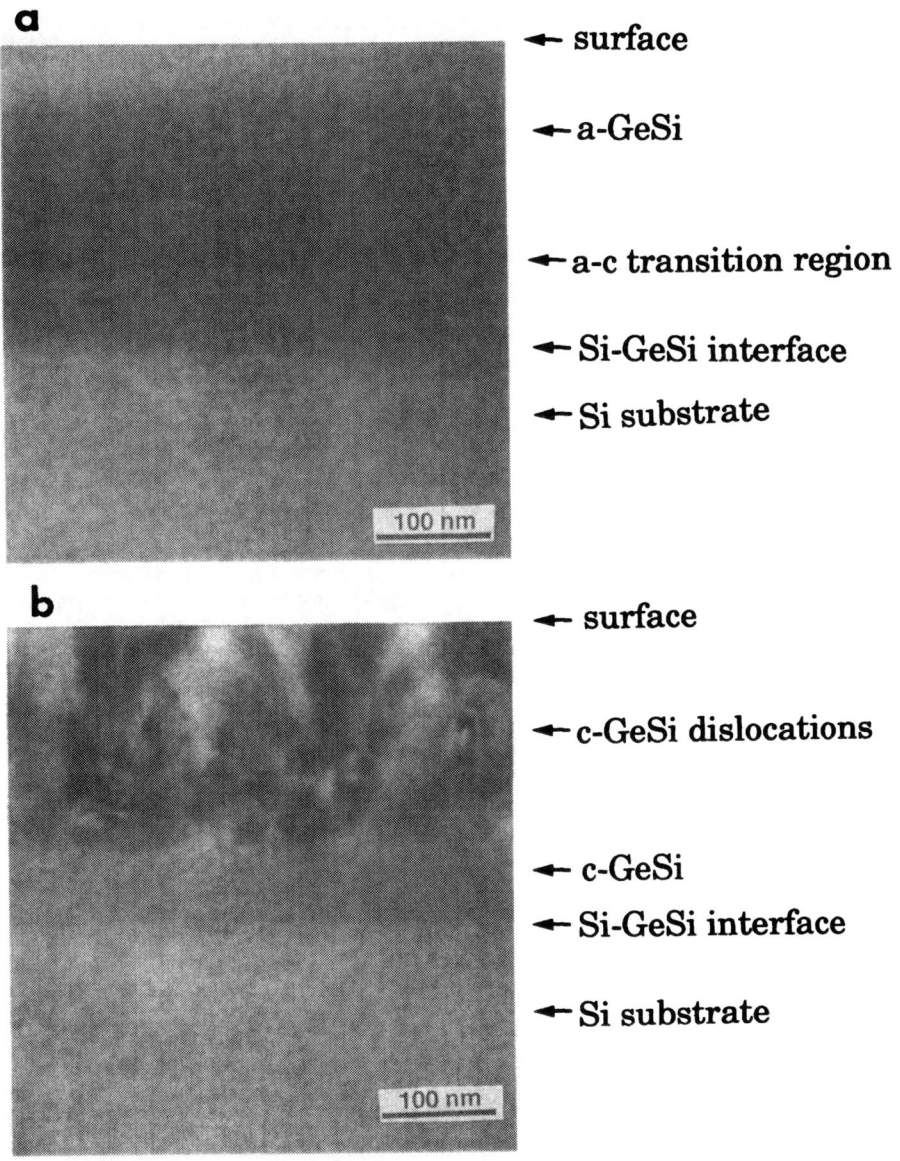

a

← surface

← a-GeSi

← a-c transition region

← Si-GeSi interface
← Si substrate

100 nm

b

← surface

← c-GeSi dislocations

← c-GeSi
← Si-GeSi interface

← Si substrate

100 nm

Fig. 1. Bright-field cross-sectional transmission electron micrographs of a Si(100)/Ge$_{0.12}$Si$_{0.88}$ sample (a) as-implanted with 100 keV 1.5 x 10^{15} P ions/cm^2 at room temperature, and (b) after annealing in vacuum at 550 °C for 30 minutes.

evidence of interfacial misfit dislocations at the Si-GeSi interface, within the resolution limit of electron microscopy.

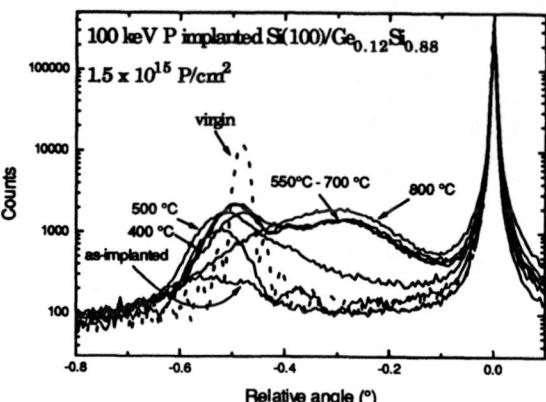

Fig. 2. Double-crystal x-ray rocking curves of (400) symmetrical diffraction for the samples in Fig. 1. The non-implanted $Si(100)/Ge_{0.12}Si_{0.88}$ (dotted line) shows excellent crystallinity and pseudomorphicity. This sample is amorphized by 100 keV 1.5 x 10^{15} P ions/cm^2 at room temperature, and subsequently annealed in vacuum for 30 minutes from 400 - 800 °C.

Figure 2 shows the change of x-ray rocking curves for (400) diffraction with implantation and annealing. The rocking curve for a non-implanted virgin sample is also plotted for reference. We can interpret the x-ray results in light of cross-sectional transmission electron micrographs. The rocking curve of a virgin GeSi film has a sharp peak at -0.49 °, which implies that the perpendicular strain in the film is quite uniform. The rocking curve of an as-implanted sample shows a weak and broad peak, which suggests that some GeSi of imperfect crystallinity remain after the implantation, consistent with the observation of Fig. 1(b). After annealing at 400 °C, the intensity of this peak increases by a factor of four and the peak sharpens at ~ -0.52°, which indicates that some of the implantation-induced defects in the layer are removed by annealing. The residual damage increases the perpendicular strain in the layer, as is reflected by the wider separation between the peak of that layer and the substrate peak [13-14]. At 500 °C, the intensity of this peak further doubles, and the diffraction intensity in the range ~ - 0.17° to - 0.52° also increased smoothly. This spectrum indicates that additional defects have disappeared at 500 °C and the strain in the crystalline region becomes non-uniform in depth. A sample annealed at 550 °C exhibits two peaks ; that located at ~ - 0.28 ° is very broad. This angle is the position of the Bragg angle of a fully relaxed $Ge_{0.12}Si_{0.88}$ film. By the position and the intensity of this broad peak, the regrown layer is relaxed and has a crystalline quality inferior to that of the as-grown sample. This finding is consistent with results obtained from transmission electron micrographs (Fig. 1 (b)). A second peak is located at ~ - 0.49 °, where the Bragg angle of a pseudomorphic $Ge_{0.12}Si_{0.88}$ film should be. Correlating this peak with Fig. 1 (b) indicates that it must originate from the lower crystalline portion of the GeSi film. At 800 °C, x-ray rocking curves as well as cross-sectional transmission electron micrographs (not shown here) indicate that the whole epilayer is relaxed.

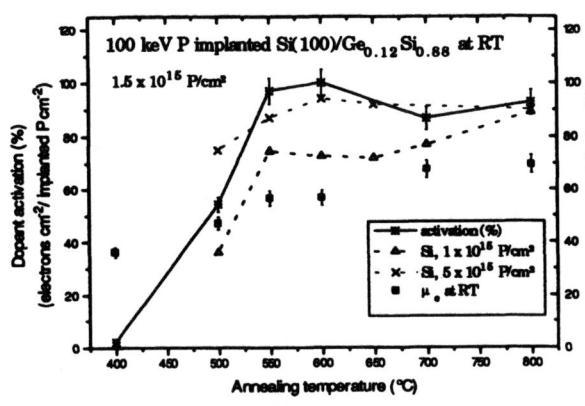

100 keV P implanted Si(100)/Ge$_{0.12}$Si$_{0.88}$ at RT

1.5 x 10^{15} P/cm^2

Dopant activation (%)
(electrons cm^2/ implanted P cm^2)

Sheet electron mobility (cm^2/V.s)

Annealing temperature (°C)

- activation (%)
- Si, 1 x 10^{15} P/cm^2
- Si, 5 x 10^{15} P/cm^2
- μ_e at RT

Fig. 3 The ratio of electrons/cm^2 (obtained from Hall measurements) to implanted atoms/cm^2 for ^{31}P implantation into Si as a function of the isochronal annealing temperature. The implantation was done at room temperature and the ion energy is 100 keV. The duration of annealing at each temperature was 30 min. The data for 280 keV P implanted Si are plotted for comparison (dotted line, from ref. [Crowder]). The values of measured sheet electron mobility in Ge$_{0.12}$Si$_{0.88}$ layers (filled squares) are also plotted here.

Figure 3 plots the fraction of activated phosphorus atoms versus the annealing temperatures. The dotted lines are the data taken from reference [15] for 280 keV P implanted Si annealed for 30 min in vacuum. The measured sheet electron mobility values for GeSi layers are also plotted in Fig. 3. About 50% of the dopants are activated at 500 °C, and almost all of them are activated at 550 °C. The dopant activation curve for this pseudomorphic GeSi layer implanted with 1.5 x 10^{15} P ions/cm^2 is very similar to that of Si implanted with 5 x 10^{15} P ions/cm^2. Maszara et al [16] reported that for 300 keV Si self-implantation at room temperature, a dose of 1 x 10^{15} Si ions/cm^2 is not sufficient to form a continuous amorphous layer in Si, but a higher dose of 5 x 10^{15} Si ions/cm^2 can result in an ~ 500 nm amorphous layer extending from the surface of the sample. As reported in reference 13, the amorphization dose in Ge$_{0.10}$Si$_{0.90}$ for room temperature 100 keV Si implantation is about three times smaller than that in Si. The extent of amorphization introduced by 5 x 10^{15} P ions/cm^2 in Si may thus be about the same as that created by 1.5 x 10^{15} P ions/cm^2 in Ge$_{0.12}$Si$_{0.88}$. It is known that solid-phase epitaxial regrowth of amorphized Si or Ge facilitates the activation of dopants [17, 18]; we suggest that the same fact holds for Ge$_x$Si$_{1-x}$.

Fig. 3 also shows the sheet electron mobility values measured at room temperature in implanted GeSi samples after annealing. After regrowth (T$_{ann}$ ≥ 550 °C), the mobility values are about 20% less than those obtained in Si, which is around 90 cm^2/Vs. Even though we believe that ionized impurity scattering is the dominant scattering process for both heavily doped Si and Ge$_x$Si$_{1-x}$, non-negligible alloy scattering at room temperature and scattering by defects in the layers should lower the electron mobility values in regrown GeSi [19-21]. For T$_{ann}$ < 550 °C, the values of electron mobility decrease and the dopant-activation is poor, which is likely due to the residual damage in the layer.

We have thus demonstrated that if a continuous amorphous layer is formed in a pseudomorphic GeSi layer as a result of P implantation, the electron mobility in the regrown layer is not enhanced, and the layer is full of dislocations. Since high dopant concentrations such as 10^{19}/cm^{-3} to 10^{21}/cm^{-3} are typically needed to form reliable contacts to semiconductors, contacts for strained Ge$_x$Si$_{1-x}$ layers made with implantation at room temperature may inevitably induce amorphization. The highly damaged contact areas after annealing and regrowth may thus pose a potential reliability issue for devices made with strained-layer Ge_xSi_{1-x}.

ACKNOWLEDGMENTS

This work was supported by the Semiconductor Research Corporation under a coordinated research program at Caltech and at UCLA, contract no. 93-SJ-100. The authors would like to thank Dr. C. J. Tsai for providing the x-ray simulation program and R. Gorris, M. Easterbrook for repairing the equipment.

REFERENCES :

1. Q. Z. Hong, J. G. Zhu, J. W. Mayer, W. Xia and S. S. Lau, J. Appl. Phys. **71**, 1768 (1992)
2. D. C. Paine, D. J. Howard, N. G. Stoffel, and J. H. Horton, J. Mater. Res. **5**, 1023 (1990)
3. G. Bai and M.-A. Nicolet, J. Appl. Phys. **71**, 4227 (1992)
4. S. Mantl, B. Holländer, W. Jäger, B. Kabius, H. J. Jorke, and E. Kasper, Nucl. Instr. Meth. B **39**, 405 (1989)
5. C. Lee, T. E. Haynes, K. S. Jones, Appl. Phys. Lett. **62**, 501 (1993)
6. B. T. Chilton, B. J. Robinson, D. A. Thompson, T. E. Jackman, and J.-M. Baribeau, Appl. Phys. Lett. **54**, 2 (1989)
7. C. S. Pai, S.S. Lau, I. Suni, and L. Csepregi, Appl. Phys. Lett. **47**, 1214 (1985)
8. Stella Q. Hong, Q. Z. Hong, and J. W. Mayer, Appl. Phys. Lett. **63**, 2053 (1993)
9. Z. Atzmon, M. Eisenberg, P. Revesz, J. W. Mayer, S. Q. Hong, and F. Schäffer, Appl. Phys. Lett. **60**, 2243 (1992)
10. Z. Atzmon, M. Eisenberg, Y. Shacham-Diamand, J. W. Mayer and F. Schäffer, Appl. Phys. Lett. **61**, 2902 (1992)
11. Stella Q. Hong, Q. Z. Hong, and J. W. Mayer, J. Appl. Phys. **72**, 3821 (1993)
12. J. F. Ziegler, J. P. Biersack and U. Littmark, *The Stopping and Range of ions in Matter*, (Pergamon Press, London,1985)
13. D. Y. C. Lie, A. Vantomme, F. Eisen, M.-A. Nicolet, T. K. Carns and K. L. Wang, J. of Appl. Phys. **74**, 6039 (1993)
14. D. Y. C. Lie, A. Vantomme, F. Eisen, M.-A. Nicolet, V. Arbet-Engels, and K. L. Wang, Mater. Res. Soc. Symp. Proc. **262**, 1079 (1993)
15. B. L. Crowder and F. F. Morehead, Jr. Appl. Phys. Lett, **14**, 313 (1969)
16. W. P. Maszara and G. A. Rozgonyi, J. Appl. Phys. **60**, 2310 (1986)
17. J. Gyulai in *Ion Implantation - Science and Technology*, edited by J. F. Ziegler, (Academic Press, Orlando, 1984), pp139-210.
18. J. W. Mayer, L. Eriksson and J. A. Davis, *Ion Implantation in Semiconductors - Silicon and Germanium*, (Academic Press, New York), pp200-208.
19. G. Masetti, M. Severi and S. Solmi, IEEE, Trans. Electron Devices, **30**, 764 (1983)
20. Fisful, *Heavily Doped Semicondcutors*, (Plenum Press, New York, 1969), p133.
21. T. Manku and A. Nathan, IEEE Trans. Electron Dev., **39**, 2082 (1992)

STRAIN RELAXATION DURING SOLID-PHASE EPITAXIAL CRYSTALLISATION OF Ge$_x$Si$_{1-x}$ ALLOY LAYERS WITH DEPTH DEPENDENT Ge COMPOSITIONS

WAH-CHUNG WONG AND ROBERT G. ELLIMAN
Electronic Materials Engineering Department, Australian National University, Canberra, ACT 0200, Australia

ABSTRACT

Solid-phase epitaxial growth (SPEG) of amorphous GeSi alloy layers has been examined. It is shown that fully strained depth dependent GeSi alloy layers can be produced by multiple ion-implantation and SPEG for implant doses below critical values. For doses above these critical values strain relaxation is shown to occur during SPEG at a well defined depth, and to be correlated with a reduction in the SPEG velocity caused by roughening or faceting of the crystalline/amorphous interface. The velocity reduction is shown to be a reliable indicator of strain relaxation. Both the critical dose and the depth at which strain relaxation occurs are shown to be in excellent agreement with equilibrium critical thickness theory.

INTRODUCTION

Si/Ge$_x$Si$_{1-x}$ strained-layer heterostructures have many potential electronic and optoelectronic applications [1,2], however, the growth of fully strained GeSi alloy layers is limited to a certain range of compositions and thicknesses [3,4]. Above these limits strain relaxation can occur. Understanding the properties of strained layers and the mechanisms by which strain relief occurs is essential for full exploitation of these materials.

GeSi heterostructures are usually grown either by molecular beam epitaxy (MBE) or chemical vapour deposition (CVD) techniques. It is clear that such techniques offer the necessary flexibility for the growth of complex multilayered structures, however, for certain simple structures ion-implantation could offer an attractive alternative. The big advantage being that it is simple, and is compatible with existing processing facilities and procedures. In addition, the Gaussian distribution of implanted Ge could be used to advantage in certain device structures. For example, in a graded-base heterojunction bipolar transistor (HBT), where the graded GeSi base region is employed to reduce base transit times [5]. Various test devices employing Ge implantation have been fabricated and the initial results look encouraging [6,7].

In addition to the potential application of implant synthesised GeSi alloy layers, such structures also offer useful model systems for studying the effect of stress on solid phase epitaxial growth (SPEG) and for examining strain relief mechanisms during SPEG of amorphous alloy layers. For example, several studies [8-21] have shown velocity reductions during SPEG of strained layers. Increases in the activation energy for SPEG have also been reported in such cases [12,13,16]. However, it has been shown [10-12,16] that such retarded epitaxy is generally associated with roughening or {111} faceting of the crystalline/amorphous interface. It is therefore possible that the reported increases in activation energy are not attributable to fundamental thermodynamical effects but are a consequence of the changing mode of crystallisation. Further work is required to confirm this interpretation.

491

EXPERIMENTAL

Dual Gaussian Ge profiles were produced by a combination of 800 keV and 300 keV ^{74}Ge implants into Si (100) wafers at room temperature. In all cases, the 800 keV implant was performed first, followed by the shallower 300 keV implant. After all Ge implants were completed, the whole thickness of GeSi was amorphised at -196°C by implanting with 800 keV Si ions to a dose of 5×10^{15} cm^{-2}. The resulting amorphous layer was around 1.2 μm thick, extending at least 250 nm beyond the Ge distribution. This ensured that SPEG was initiated in intrinsic Si, and provided a region where the crystallisation velocity of intrinsic Si could be determined.

All samples were annealed in air at 600°C while the crystallisation velocity was monitored in-situ using time resolved reflectivity (TRR) [22]. TRR measurements were performed at two wavelengths, 632.8 nm and 1523 nm, with the visible wavelength being used to monitor the near-surface (~300 nm) and the longer wavelength being used to monitor greater depths.

Rutherford backscattering and channeling (RBS-C) spectrometry was performed with 2.0 MeV He ions employing a scattering angle of 165°. Channeling was along the [100] zone axis. Double crystal x-ray diffraction (DCXRD) was used to measure the strain distribution in fully crystallised GeSi layers. X-ray rocking curves were measured in the vicinity of the Si (400) Bragg reflection using a CuKα_1 source. Transmission electron microscopy (TEM) was performed on a JEOL 2000EX microscope operating at 200 kV.

Calculation of equilibrium critical depths was performed using the model of Paine et al [11] assumed that the dominant strain relieving defects consisted of stacking faults and associated 90° partial dislocations.

RESULTS AND DISCUSSION

Equilibrium theory [11] predicts a critical dose of 1.1×10^{17} Ge.cm^{-2} (peak concentration of 6.6%) for 800 keV Ge implants and a critical dose of 6.0×10^{16} Ge.cm^{-2} (peak concentration of 7.4%) for 300 keV Ge implants. Three dual implant configurations are considered in the present report: a) both implants below the critical dose, b) both implants above the critical dose, and c) the 300 keV implant above and the 800 keV implant below the critical dose.

I. Both implants below the critical dose.

Figure 1 shows RBS-C spectra (Fig. 1a), a DCXRD spectrum (Fig. 1b) and SPEG velocity data (Fig. 1c) for alloy layers prepared by a double implant of 800 keV Ge to a dose of 6.4×10^{16} Ge·cm^{-2} (peak concentration of 3.5%) and 300 keV Ge to a dose of 2.6×10^{16} Ge·cm^{-2} (peak Ge concentration of 3.1%). The RBS-C spectra of post-annealed samples show near perfect crystal after SPEG with a minimum yield of ~3.6% in the Si portion of the spectrum compared to ~3% for an unimplanted sample. Dechanneling within the GeSi region is also low.

That the layer has crystallised in a fully strained configuration is illustrated by the DCXRD spectrum in Fig. 1b. The x-ray rocking curve shows clear strain related oscillations at angles less than the (400) substrate peak, indicating a depth dependent compressive strain in the GeSi layer. The actual strain distribution can be obtained by comparing the measured spectrum with simulated rocking curves calculated using dynamical diffraction theory [24-26]. For a fully strained layer the strain at a particular depth is given by 0.043 times the local Ge concentration.

In this case, a good match (Fig. 1b) is achieved between the experimental and simulated rocking curves. The SPEG rate for this sample, as extracted from TRR measurements, is shown in Fig. 1c. Also included is a plot of the SPEG rate for an intrinsic amorphous Si reference sample grown under identical condition. Both samples show a slight reduction in growth rate due to hydrogen [23] but no significant difference is observed between the Ge implanted sample and the Si reference sample. This shows that a Ge concentration of ~3.5% (or a peak strain of ~0.15%) has no detectable effect on SPEG rate. The velocity reduction observed in the near-surface region of the Ge implanted sample is a consequence of recoil-implanted impurities and depends on the vacuum pressure during implantation.

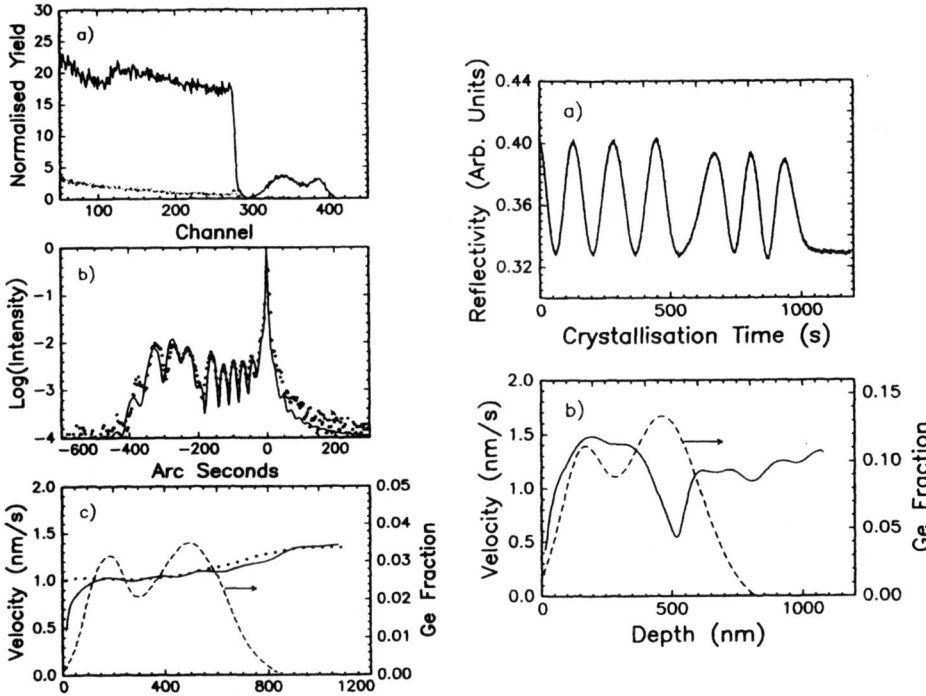

Figure 1: a) RBS-C spectra before and after annealing, b) DCXRD spectrum after annealing, and c) SPEG velocity for a sample implanted with 800 keV Ge ions to a dose of 6.4×10^{16} Ge.cm^{-2} and 300 keV Ge to a dose of 2.6×10^{16} Ge.cm^{-2}.

Figure 2: a) TRR spectrum and b) extracted SPEG velocity for a sample implanted with 800 keV Ge ions to a dose of 2.6×10^{17} Ge.cm^{-2} and 300 keV Ge to a dose of 8×10^{16} Ge.cm^{-2}. The Ge distribution is also shown for reference.

II Both implants above the critical dose.

Figure 2 shows a TRR spectrum and the extracted SPEG rate for a sample implanted with 800 keV Ge to a dose of 2.6×10^{17} Ge·cm^{-2} and 300 keV Ge to a dose of 8×10^{16} Ge·cm^{-2}. The

800 keV implant is approximately 2.5 times the estimated critical dose of $1.1\text{x}10^{17}$ Ge·cm^{-2} [11]. The SPEG velocity distribution, shown in Fig. 2b, shows a sudden decrease in velocity, commencing at a depth of ~600 nm. It reaches a minimum at a depth of ~520 nm and then increases again. This has previously been shown [10-12,16-21] to be a consequence of roughening or faceted of the crystalline/amorphous interface during strain relaxation. As such this velocity reduction represents a distinct signature for strain relief. (Note that the TRR spectrum shows a reduction in contrast in the region of the velocity reduction, consistent with roughening or faceting of the crystalline/amorphous interface.) From previous studies [17-21], strain relieving dislocations are expected to extend to a depth corresponding to the midpoint between the onset of the velocity reduction and the minimum velocity. In this case this corresponds to a depth of ~560 nm. Using the measured Ge distribution the equilibrium theory developed by Paine et al [11] predicts a critical depth of 577 nm, in excellent agreement with experiment. Such agreement has previously been reported for single 800 keV Ge implants. DCXRD measurements confirm that the sample is strain-relaxed after SPEG.

Figure 2b also shows that the crystallisation rate increases after strain relaxation has commenced and that it reaches a value higher than that of intrinsic Si in the depth range 100 - 400 nm. This is consistent with the fact that the SPEG rate of strain-relaxed alloys increases with increasing Ge concentration [20]. In the present case the maximum rate is ~1.5 times that of intrinsic Si for a Ge concentration of ~11%, in excellent qualitative agreement with the factor of 3 increase reported for a fully relaxed alloy with a Ge concentration of 21% [20].

III. 800 keV implant below and 300 keV implant above the critical dose.

Figure 3 compares the SPEG velocity extracted from in-situ TRR measurements with the Ge distribution measured by RBS for a sample implanted with 800 keV Ge to a dose of $6.4\text{x}10^{16}$ Ge.cm^{-2} (peak concentration of 3.5%) and 300 keV Ge to a dose of $8\text{x}10^{16}$ Ge.cm^{-2} (peak concentration of 9%). In this case there is a sudden decrease in velocity at a depth of ~200nm. Although occurring much closer to the sample surface, this reduction is similar to that observed at a depth of ~600 nm for the high-dose 800 keV implant, (see Fig. 2b). The subsequent increase in velocity observed in the latter case is absent in this case as the velocity minimum lies within 100 nm of the surface. The velocity measurement is also complicated by recoil implanted impurities in this region. TEM analysis of sample cross-sections confirmed the presence of strain-relieving defects in the near-surface region. DCXRD analysis also provides evidence for strain-relaxation, as shown in Fig. 4c.

In order to understand the x-ray rocking curve for the dual-implant sample, Fig. 4c, the rocking curves from samples implanted with a single implant of 800 keV Ge (Fig. 4a) or 300 keV Ge (Fig. 4b), are plotted for comparison. Figure 4a shows experimental and the simulated rocking curves for the 800 keV implant only. The simulated spectrum was constructed by assuming that the layer was fully strained and using the Ge profile extracted from RBS measurements. The excellent agreement between the measured and simulated spectra confirms that the regrown layer is fully strained. This is expected as the peak Ge content of just 3.5% is well below the critical value of 6.6% for 800keV implants. Figure 4b shows measured and simulated rocking curves for the 300 keV implant only. In this case, the model of Paine et al [11] predicts a critical peak Ge concentration ~7.4%. Since the peak concentration in the implanted sample is 9% some relaxation is expected in this case. This was confirmed by TEM analysis. However, despite some relaxation, the DCXRD spectrum in Fig. 4b shows that the sample contains a high degree of residual strain. Comparison with the simulated spectrum shows that the experimental data represents only the envelope of the simulated curve, with a broad plateau extending to angles of -900 arc seconds. Since the total spread of satellite peaks is an indication of peak strain in the structure, the spread of ~900 arc seconds reflects a much higher peak strain than for the sample implanted with 800 keV Ge. It should be noted, however, that the strain relieving defects generated during SPEG are observed to extend only to the depth at which they form [18-20]; they do not glide deeper to relieve additional strain. Therefore high levels of residual strain are expected to result from the crystallised portion of the layer. Since strain relief appears to occur in the region between the surface and the peak of the Ge distribution in the case of the 300 keV implant, the maximum measured strain should correspond to that predicted, as observed. The loss of structure in the curve presumably results from strain relaxation in the near-surface region. Figure 4c shows rocking curves for the dual-implant sample. Strain contributions from the separate implants can readily be discerned, with the peaks extending to -400 arc seconds being dominated by the low strain 800 keV implant and the broader envelope, extending to -900 arc seconds, coming from the 300 keV, high dose implant.

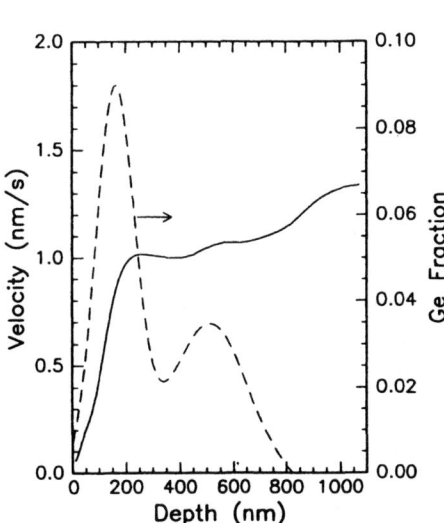

Figure 3: Ge distribution and SPEG velocity for sample implanted with 800 keV Ge ions to a dose of 6.4×10^{16} Ge.cm^{-2} and 300 keV Ge to a dose of 8×10^{16} Ge.cm^{-2}.

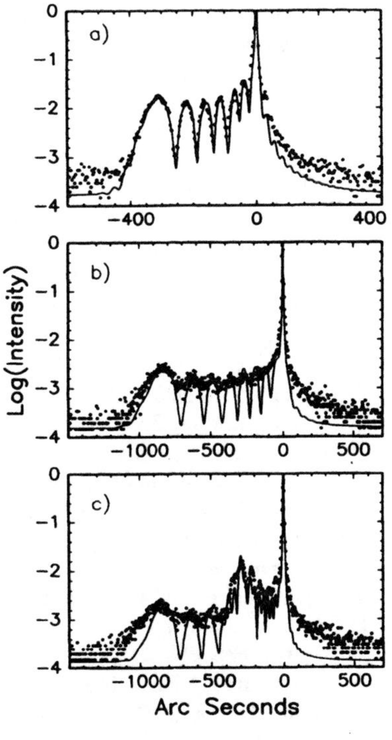

Figure 4: DCXRD spectra for samples implanted with a) 800 keV Ge ions to a dose of 6.4×10^{16} Ge.cm^{-2}, b) 300 keV Ge ions to a dose of 3×10^{16} Ge.cm^{-2}, and c) both the 800 keV and 300 keV implants listed above.

CONCLUSIONS

It has been shown that fully strained depth dependent GeSi alloy layers can be produced by multiple ion-implantation and SPEG for implant doses below critical values. For doses above the critical values, it was shown that strain-relaxation occurs during SPEG at a well defined depth, and is associated with a reduction in the crystallisation velocity. It is speculated that this velocity reduction is a consequence of stress induced roughening or faceting of the crystalline/amorphous interface. Both the critical dose and the depth at which strain relaxation occurs were shown to be accurately predicted by an equilibrium model developed by Paine et al [11].

REFERENCES

1. J.C.Bean, Proceedings of the IEEE, **80**, 571 (1992).
2. E.A.Fitzgerald, Y.-H.Xie, D.Monroe, P.J.Silverman, J.M.Kuo, A.R.Kortan, F.A.Thiel and B.E.Weir, J.Vac.Sci.Technol., **B10**, 1807 (1992).
3. S.C.Jain, J.R.Willis and R.Bullough. Advances in Physics **39**, 127 (1990).
4. E.A.Fitzgerald, Mat. Sci. Rep., 7, 87 (1991).
5. E.F.Crabbe, J.D.Cressler, G.L.Patton, J.M.C.Stork, J.H.Comfort and J.Y.-C.Sun. IEEE Electron Device Letters, **14**, 193 (1993).
6. A.Gupta, C.Cook, L.Toyoshiba, J.Quiao, C.Y.Yang, K.Shoji, A.Fukami, T.Nagano and T.Tokuyama. J. Electr. Mat., **22**, 125 (1992).
7. C.R.Selvakumar and B.Hecht, IEEE Electron Device Letters, **12**, 444 (1991).
8. G.Mezey, S.M.Matteson and J.Gyulai. Nucl. Instr. Meth. **182/183**, 587 (1981)
9. B.T.Chilton, B.J.Robinson, D.A.Thompson, T.E.Jackman and J.-M.Baribeau. Appl. Phys. Lett. **54**, 42 (1989).
10. D.C.Paine, D.J.Howard, N.G.Stoffel and J.A.Horton. J. Mater. Res. **5**, 1023 (1990).
11. D.C.Paine, D.J.Howard and N.G.Stoffel. J. Electronic Materials **20**, 735 (1991).
12. D.C.Paine, N.D.Evans and N.G.Stoffel. J. Appl. Phys. **70**, 4278 (1991).
13. Q.Z.Hong, J.G.Zhu, J.W.Mayer, W.Xia and S.S.Lau. J. Appl. Phys. **71**, 1768 (1992).
14. F.Corni, S.Fabboni, G.Ottaviani, G.Queirolo, D.Bisero, C.Bresolin, R.Fabbri and M.Seridori, J. Appl. Phys. **71**, 2644 (1992).
15. K.M.Yu, I.G.Brown and S.Im. Mat. Res. Soc. Symp. Proc. **235**, 293 (1992)
16. C.Lee, T.E.Haynes and K.S.Jones. Appl. Phys. Lett. **62**, 501 (1993).
17. R.G.Elliman and W.C.Wong. Nucl. Instr. Meth. **B80/81**, 768 (1993).
18. R.G.Elliman and W.C.Wong. Nucl. Instr. Meth. (In Press).
19. R.G.Elliman and W.C.Wong. Materials Science Forum. (In Press)
20. R.G.Elliman, W.C.Wong and P.Kringhoj. Mat. Res. Soc. Symp. Proc. (This proceedings).
21. P.Kringhoj, R.G.Elliman and J.L.Hansen. Mat. Res. Soc. Symp. Proc. (This proceedings).
22. G.L.Olson and J.A.Roth. Mat. Res. Rep. **3**, 1 (1988).
23. J.A.Roth, G.L.Olson, D.C.Jacobson and J.M.Poate. Appl. Phys. lett., **57**, 1340 (1990).
24. S.Takagi, Acta Cryst. **15**, 1311 (1962)
25. S.Takagi, J. Phys. Soc. Japan, **26**, 1239 (1969)
26. D.Taupin, Bull. Soc. Fr. Mineral. Crystallogr., **87**, 469 (1964).

SOLID PHASE EPITAXY PROCESS OF Ar-ION BOMBARDED SILICON SURFACES AND RECOVERY OF CRYSTALLINITY BY THERMAL ANNEALING OBSERVED WITH SCANNING TUNNELING MICROSCOPY

Katsuhiro UESUGI, Masamichi YOSHIMURA and Takafumi YAO
Department of Electrical Engineering, Hiroshima University, Higashi-Hiroshima 724, Japan

ABSTRACT

The solid-phase epitaxy (SPE) process of Ar^+-ion bombarded Si(001) surfaces and recovery of crystallinity by thermal annealing are studied " in situ" by using a scanning tunneling microscope (STM). As-bombarded surfaces consist of grains of 0.63–1.6 nm in diameter. The grains gradually coalesce and form clusters of 2–3.6 nm in diameter at annealing temperature of 245 °C. (2x1) and (1x2) reconstructed regions surrounded by amorphous regions are partially observed on the surface by prolonged annealing, which suggests the onset of SPE. Successive observation reveals that the smoothing of the surface occurs layer by layer. As annealing temperature is raised up to 445 °C, the amorphous layer epitaxially crystallizes up to the topmost surface, and (2x1) reconstructed surface with monatomic-height steps is observed. The smoothing of the surface structures and the formation of nucleation of Si islands are observed during annealing at 500 °C.

Introduction

Solid-phase epitaxy (SPE) technique is widely used to produce new large scale integrated circuits and fabrication of crystalline Si films on insulator structures. The kinetics of crystallization in amorphous Si layer have been studied by many techniques such as Rutherford back scattering (RBS),[1-3] low-energy electron diffraction (LEED),[4-6] scanning electron microscopy (SEM)[2] and transmission electron microscopy (TEM).[1,3] The microscopic mechanisms of SPE processes are still unclear. In a previous study, we showed " in situ" microscopic investigation of Ar^+-ion bombarded Si surfaces for a high dose case and its regrowth processes by using a scanning tunneling microscope (STM).[7] Although STM gives the information on surface morphology not at the interface on an atomic scale, the STM investigation at elevated temperature can reveal the essential feature of SPE processes from the very beginning of epitaxy. We found that the Si surface bombarded by Ar^+ ion at 3 keV to a dose of $5x10^{15}$ cm^{-2} was covered with grains of 1–2 nm in diameter. The surface morphology became continuous and smoother by thermal annealing at ~500 °C. The onset of SPE occurred at around 590 °C, at which temperature a (2x2) reconstructed region surrounded by amorphous regions was partially observed. During annealing at 590–620 °C, we observed the development of the areas of the (2x2) and c(4x4) reconstruction surrounded by amorphous regions.

In this study, we have carried out detailed studies of the surface for the case of lower Ar^+-ion dose using an STM operated at elevated temperatures. The evolution of the surface morphology from an amorphous to a crystalline layer during annealing below 500 ° C and recovery of crystallinity by annealing are investigated.

Experimental

The instrument used in the present experiment is an ultra-high vacuum (UHV) STM system (JEOL Ltd., JSTM-4500VT) which consists of an STM chamber, a preparation chamber equipped with an Ar^+-ion sputtering system, and a load-lock chamber. The base pressure of the STM and preparation chambers was of the order of $1.8x10^{-8}$ Pa. A specimen was an n-type Si(001) of 0.02–0.04 $\Omega \cdot cm$. Before being loaded into the load-lock chamber, the specimen was cleaned in pure acetone solution. The clean surface was obtained by repeated flash heating to ~1200 °C after prebaking at ~500 °C for 12 h. After the clean (2x1) surface was confirmed from an STM image, it was bombarded in the preparation chamber by Ar^+ ions at 3 keV to a dose ranging from 1.9–3.8x10^{14} cm^{-2}. The impact angle of Ar^+-ion beams measured from the surface is about 30°. The thermal annealing of the specimen was performed at 245–500 °C in the STM chamber under UHV. The specimen was heated by direct heating and its temperature was measured using an optical pyrometer. Dynamic STM observations were performed in a constant tunneling current mode at elevated temperatures. A tip used was a tungsten wire sharpened by electrochemical etching.

Results and Discussion

As-bombarded Si(001) surfaces are covered with granular grains. Figure 1 shows a typical topographic STM image of the surface at a dose of 1.9×10^{14} cm^{-2}. As bombarded surfaces show essentially the same features within the range of dose examined. The size of the grain is 0.63–1.6 nm in diameter and its density is about 2×10^{15} grains/cm^2. Most of the grains showed bright contrast independent of the sample bias polarity. However, a few grains changed contrast as the polarity of the bias is reversed: some of them were bright at a negative sample bias, while the others were bright at a positive bias. These grains are possibly generated by adsorbed residual gas molecules (contamination) or residual Ar ions in the amorphous layers, since Si particles should show the same contrast for both polarity. Although it is difficult to elucidate the origin of the grains from the present experiments, we note that the density of these grains was of the order of 6.2×10^{13} cm^{-2} for the Ar dose of 1.9×10^{14} cm^{-2}. We note that in the case of lower Ar$^+$-ion dose, step-like structures are observed on as-bombarded surfaces. Two distinct types of step-like structures along the [110] direction are observed (A and B), where the A step is straight, while the B step is jagged. Both types of steps are alternately aligned along the [110] direction and the width of terraces is 15–22 nm. After annealing at 500 °C, we observed terraces with 18–23 nm width separated by monatomic height steps. We note that the step-like structures on the Ar$^+$-ion bombarded surface indicate the presence of monatomic-height steps at the amorphous/crystal interface.

When the annealing temperature increases up to 245 °C, the clusters of 2–3.6 nm in diameter are observed on the surface as shown in Fig. 2. The size of the cluster is about three times lager than that of the grain and the density of the cluster is about half the density of the grain before annealing (1.1×10^{15} cm^{-2}). These facts suggest the coalescence of amorphous grains by thermal annealing. Prolonged annealing at 245 °C promotes crystallization of the amorphous layers. (2×1) reconstructed regions surrounded by amorphous regions are partially observed on the surface which suggests the onset of SPE. Figure 3 shows a typical topographic STM image of the surface

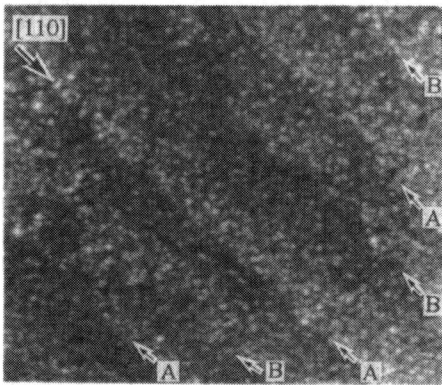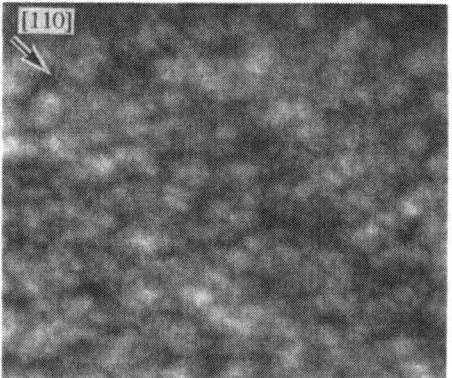

Fig. 1. STM image of 93x80 nm^2 area of an Ar$^+$-ion bombarded Si surface at 3 keV to a dose of 1.9×10^{14} cm^{-2}. The sample bias is –2 V and the tunneling current is 0.1 nA. The surface is covered with grains of 0.63–1.6 nm in diameter.

Fig. 2. STM image of 43x37 nm^2 area of the surface during annealing at 245 °C. The sample bias is – 0.9 V and the tunneling current is 0.3 nA. The grains coalesce by thermal annealing. The size of clusters is 2–3. 6 nm in diameter.

during annealing at 245 °C. Figure 3 (b) shows the surface morphology of the same area of Fig. 3 (a) taken after 35 sec. It is observed that the surface becomes smoother with time. The surface reconstruction of the small regions (A) as indicated by the square changes from (2x1) to (1x2). Judging from contrast, the depth of the recessed regions (B) becomes shallow. The dissociation of a cluster (C) occurs to yield (1x2) reconstruction. The dissociated Si has nucleated narrow island such as a single dimer-wide string (D). The direction is perpendicular to the dimer rows of the second layer. The surface becomes smoother layer by layer through dissociation of clusters, which suggests that the amorphous/crystal interface moves towards the topmost surface.

As annealing temperature increases, the reconstructed regions develop and a mixture of (2x1) and (1x2) reconstruction is observed, where the amorphous layers epitaxially crystallize up to the topmost surface. Figures 4 show a typical topographic STM images of the surface during annealing at 445 °C. (2x1) and (1x2) reconstructions are separated by monatomic-height steps. We note that the surface is defective and contains many monatomic-height islands of (2x1) reconstruction. The terrace and step structures does not develop during annealing at this temperature.

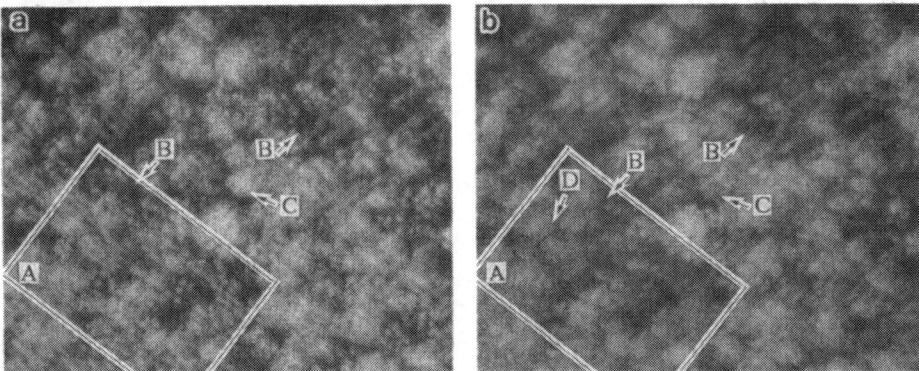

Fig. 3. STM images of the surface (41x34 nm^2) during annealing at 245 °C. The sample bias is 1 V and the tunneling current is 0.3 nA. (2x1) reconstructions surrounded by amorphous regions are partially observed.

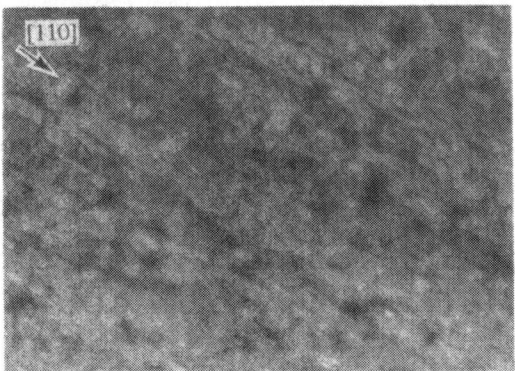

Fig. 4. STM image of 82x60 nm^2 area of the surface during annealing at 445 °C. The sample bias is – 1.6 V and the tunneling current is 0.3 nA. The amorphous layer epitaxially crystallizes up to the topmost surface.

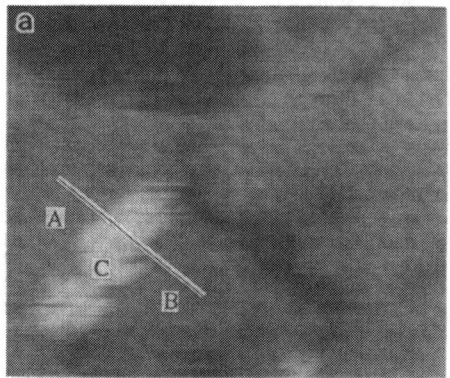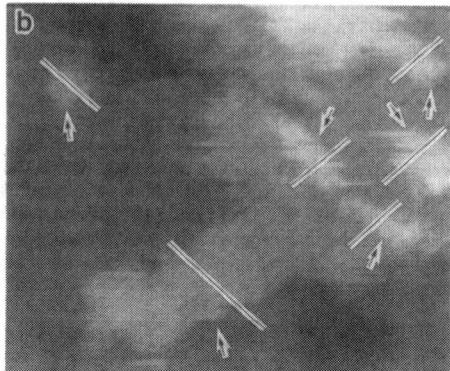

Fig. 5. STM images of the surface during annealing at 500 °C. The sample bias is −0.74 V and the tunneling current is 0.5 nA. (a) 20x17 nm^2, (b) 29x25 nm^2.

In many of the STM images, the monatomic-height island is often accompanied by the formation of antiphase boundaries. Figures 5 show typical topographic STM images of the surface during annealing at 500 °C. The dimer rows of the islands are clearly observed and are running perpendicular to the underlying dimer rows, which indicate the monatomic-height island. In Fig. 5 (a), the (2x1) domain in region A has an antiphase of that of B. The solid line shows the dimer row of the (2x1) B domain. It is noted that the dimer row of the A domain is antiphase of B. Hence the antiphase boundary exists in this layer before growth of island C. Most of the islands on the surface nucleates at the antiphase boundaries in the underlying layer as indicated by arrows in Fig. 5 (b).

We observe " in situ" smoothing of the surface structures as the annealing temperature increases up to 500 °C. Figures 6 show successive images during annealing at 500 °C taken every 35 sec. The smoothing of the surface, such as decreasing of a hollow region with monatomic height (A) and defects (B), are observed. The nucleation of Si island is also observed at the antiphase boundaries (C and D). These observation clearly indicate that the antiphase boundaries act as effective nucleation sites for epitaxial growth on Si(001)-(2x1) surfaces, which is consistent with the observation of epitaxial growth of Si on Si(001) surfaces at low temperatures.[8]

Conclusions

Ar$^+$-ion bombarded Si(001) surface and its SPE process by thermal annealing in UHV have been investigated " in situ" using an STM. The SPE process of as-bombarded surface is summarized as follow: Ar$^+$-ion bombarded Si surface consists of grains of 0.63– 1.6 nm in diameter. As annealing temperature is raised up to 245 °C, coalescence of fine grains occurs, and the top surface is partially occupied by crystallized regions of (2x1) reconstruction, which suggests the onset of SPE. The surface becomes smoother layer by layer through dissociation of clusters. The amorphous layer epitaxially crystallizes up to the topmost surface by thermal annealing at 445 °C. It is observed that the nucleation of islands initiates at the antiphase boundary. At annealing temperature of 500 °C, the smoothing of the surface structures and the nucleation of Si islands are observed real time.

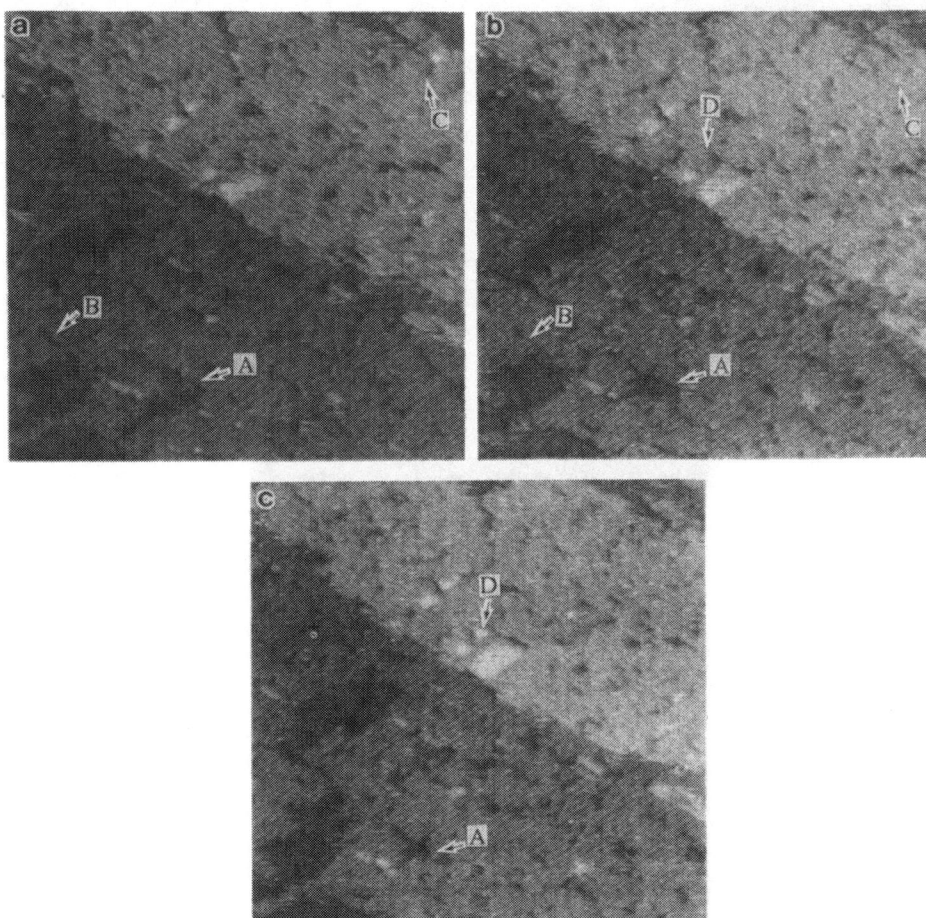

Fig. 6. Successive STM images of 60x60 nm² area of the surface during thermal annealing at 500 °C taken every 35 sec. The sample bias is –0.45 V and the tunneling current is 0.3 nA.

References

[1] J.C. Bean, G.E. Becker, P.M. Petroff and T.E. Seidel, J. Appl. Phys. **48** (1977) 907.
[2] L. Csepregi, E.F. Kennedy, J.W. Mayer and T.W. Sigmon, J. Appl. Phys. **49** (1978) 3906.
[3] S.S. Lau, S. Matteson, J.W. Mayer, P. Revesz, J. Gyulai, J. Roth, T.W. Sigmon and T. Cass, Appl. Phys. Lett. **34** (1979) 76.
[4] K. Sumitomo, K. Tanaka, I. Katayama, F. Shoji and K. Oura, Surf. Sci. **242** (1991) 90.
[5] W. Bock, H. Gnaser and H. Oechsner, Surf. Sci. **282** (1993) 333.
[6] T. Shigeta, Appl. Phys. Lett. **52** (1988) 619.
[7] K. Uesugi, T. Yao, T. Sato, T. Sueyoshi and M. Iwatsuki, Appl. Phys. Lett. **62** (1993) 1600.
[8] R.J. Hamers, U.K. Köhler and J.E. Demuth, J. Vac. Sci. Technol. A **8** (1990) 195.

INTERFACE STABILITY DURING SOLID PHASE EPITAXY
OF STRAINED Ge_xSi_{1-x} FILMS ON Si(100)

XIAOBIAO ZENG, TAN-CHEN LEE, JOHN SILCOX AND MICHAEL O. THOMPSON
Department of Materials Science and Engineering, Cornell University, Ithaca, NY 14853

ABSTRACT

Strained solid phase epitaxial (SPE) regrowth of amorphous Ge_xSi_{1-x} on Si(100) substrates was studied using time-resolved reflectivity (TRR). Films of CVD-grown $Ge_{0.13}Si_{0.87}$ on Si were amorphized by Si ion implantation, and subsequently regrown at temperatures between 550 °C and 610 °C. Information on regrowth dynamics and interface roughness evolution was obtained by accurately modeling the complicated TRR data for Ge_xSi_{1-x} regrowth using a moving, statistically roughening interface. The SPE regrowth rate slowed as the interface crossed into the Ge_xSi_{1-x} layer and the originally planar interface roughened, as confirmed by transmission electron microscopy. A minimum in the regrowth velocity was observed after regrowing approximately 60 nm into the Ge_xSi_{1-x} layer; the SPE rate subsequently increased to a final, thickness-dependent velocity that was still below that for pure Si. Upon entering the Ge_xSi_{1-x} layer, the interface roughened quickly to a 15–20 nm amplitude, increasing only slightly more during the remainder of regrowth. The degree of roughening and velocity reduction was found to be dependent on the anneal temperature. In contrast, samples with low Ge concentrations (< 3 at.%) prepared by ion implantation exhibited minimal interface roughening and essentially identical SPE velocities as pure Si.

INTRODUCTION

The growth of strained Ge_xSi_{1-x} heterostructures has advanced rapidly in the past five years. Such structures have already been demonstrated in high electron mobility bipolar transistors[1] and, potentially, may be useful in novel optical and electronic device applications [2]. While most of these device demonstrations have used as-grown structures, it is likely that practical application of the material will require more traditional semiconductor processing including ion implantation and the accompanying requisite solid phase epitaxial (SPE) growth to activate the dopants and restore crystallinity. Of particular importance will be the use of low temperature SPE (500–700 °C) to avoid thermal relaxation of the strained interfaces. This manuscript reports on velocity and interface measurements of Ge_xSi_{1-x} during regrowth in this regime.

In addition, there is as yet no detailed microscopic understanding of SPE in even pure Si [3]. Studies of SPE in strained and unstrained Ge_xSi_{1-x} alloys may thus not only provide understanding of the limits to heteroepitaxy in Ge_xSi_{1-x} and other group IV alloys, but may also elucidate the mechanisms of SPE and interface breakdown during SPE.

Previous studies have been carried out on SPE of strained Ge_xSi_{1-x} alloys in both compositionally abrupt alloy films [4, 5], and compositionally graded layers formed by Ge-implantation into Si(100) [6, 7]. In general, it has been found that the interface velocity, compared to pure Si, is retarded while growing in the strained layers and that the interface roughens by faceting on [111] planes. While early studies were based on furnace regrowth, more recently the time-resolved reflectivity (TRR) technique developed by Olson et al. [8] has been used to follow the velocity changes in-situ. With such time-resolved measurements,

Lee *et al.* [9] recently showed that the velocity retards upon entering the Ge_xSi_{1-x} layer, but recovers substantially during continued regrowth.

In this paper, we report the time-resolved evolution of the velocity and average amorphous/crystal (a/c) interface roughness during regrowth of abrupt $Ge_{0.13}Si_{0.87}$ films and Ge-implanted films with peak concentrations of ≈ 3 at.% Ge. TRR data obtained during SPE regrowth were analyzed using a statistically roughening interface model. Parameters for the interface position and roughness were determined by matching simulations with experimental data. These measurements indicate dramatic changes both in the velocity and roughness as the interface enters a compositionally sharp strained Ge_xSi_{1-x} layer, with subsequent recovery of the velocity but continued interface roughness. In contrast, regrowth in the low Ge concentration films remains essentially unchanged from Si regrowth.

EXPERIMENTAL

Compositionally abrupt epitaxial CVD films of $Ge_{0.13}Si_{0.87}$ (157 nm) and ion implantation produced compositionally varying films were studied. The non-abrupt Ge samples were prepared by implantation of 7.2×10^{15} Ge/cm^2 into Si(100) wafers at 80 keV; this produced samples with a peak Ge concentration of ~ 3 at.% located ~ 54 nm below the surface. $Ge_{0.13}Si_{0.87}$/Si and reference Si wafers were amorphized to a depth of ~ 380 nm by implantation of 5×10^{15} Si/cm^2 at 200 keV with the wafers held at nominally liquid nitrogen temperature. The Ge implanted wafers (and additional reference Si wafers) were implanted under identical conditions with 150 keV Si ions, resulting in 300 nm amorphous layers. These implantation energies were chosen so the amorphous zone extended well into the underlying Si substrates. In all cases, SPE regrowth began from pure Si crystal substrates with steady state regrowth established prior to entering Ge containing regions.

Simultaneous SPE regrowth velocity measurements were made on a Ge containing sample and an identically implanted Si sample. The two samples were attached with carbon paint to a larger Si wafer to ensure a common temperature during regrowth. Annealing was carried out in a vacuum furnace (10^{-7} Torr) and heater temperatures of 550–610 °C. Samples, sitting directly on the heater, were further enclosed in a molybdenum black-body shroud with only small apertures for access by HeNe TRR probe beams.

Two HeNe (632.8 nm) laser beams were reflected at near-normal incidence from the surface of the two samples. Incident and reflected intensities were monitored with photodiodes and digitized for computer analysis. The absolute furnace temperature was determined in each run by measuring the pure Si(100) regrowth rate and calculating the temperature from the known SPE activation energy (2.68 eV) [8]. Temperature differences between the two samples (typically less than 3 K) were estimated by measuring the initial velocity in pure Si during regrowth of Ge_xSi_{1-x} samples. These simultaneous measurement of a Si reference and Ge_xSi_{1-x} sample, coupled with the initial regrowth in bulk Si for Ge_xSi_{1-x}, permit relative velocities to be determined to better than 5% accuracy [10]. In addition to the TRR measurements, the SPE regrowth of several samples was halted at various stages to permit cross-sectional transmission electron microscopy (XTEM) of the evolving a/c interfaces.

RESULTS

Fig. 1 shows typical TRR curves for $Ge_{0.13}Si_{0.87}$/Si and pure Si annealed at nominally 562 °C and a Ge-implanted sample regrown at 595 °C. Constructive and destructive interference oscillations, arising from reflections at the air/amorphous and a/c boundaries, are readily apparent in the TRR as the a/c interface advances toward the surface. Because of attenuation in the amorphous layer, the interference contrast in pure Si increases as the interface approaches the surface; the absence of this increase in Ge_xSi_{1-x} indicates interface roughening during regrowth. In the case of pure Si SPE, the a/c interface is planar and its TRR trace can be well fitted with standard thin-film reflection formulas [8, 11]. For 632.8 nm light, each successive maxima in the interference pattern corresponds to regrowth of approximately 65.2 nm. At 562 °C, the fringes are evenly spaced and increase in amplitude consistent with a planar interface moving at a constant 0.2 nm/s.

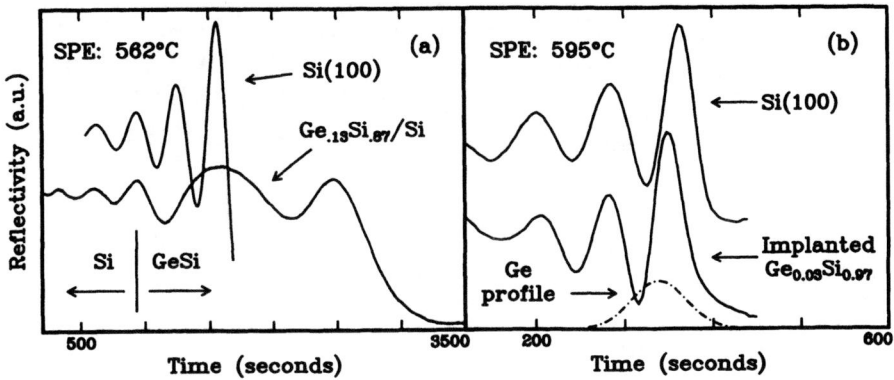

Figure 1: TRR curves for (a) $Ge_{0.13}Si_{0.87}$ and Si(100) annealed at 562 °C and (b) implanted Ge_xSi_{1-x} and Si(100) annealed at 595 °C.

For the $Ge_{0.13}Si_{0.87}$ films (Fig. 1a), SPE began in the Si substrate and the regrowth behavior was essentially identical to the Si control sample. However, once SPE proceeded into Ge_xSi_{1-x} layer, the interference oscillations increased in period, corresponding to interface retardation, and the amplitude failed to increase, corresponding to interface roughening. The final regrowth fringe, however, shows the regrowth velocity increasing again as the interface approaches the surface.

In contrast to the markedly different TRR traces in Fig. 1a, the regrowth of Ge-implanted films is nearly identically to the growth of pure Si as shown in Fig. 1b. The amplitude of the interference contrast for the Ge-implanted sample is again consistent with a planar interface. The raw data indicate a small velocity difference (10%) between the Ge containing sample and the reference; this difference, however, is the result of small differences in the sample temperature (3 K) as determined by simultaneous measurements of two reference samples. Within our absolute accuracy (5%), there is no difference between the regrowth velocity of the Ge-implanted samples and pure Si.

Cross-sectional TEM further confirmed the basic features of the regrowth described above. A series of partially regrown Ge_xSi_{1-x} samples were prepared by halting SPE, while monitoring the TRR signal, at various stages of regrowth. While SPE proceeded inside the Si(100) substrate (Fig. 2a), the a/c interface remained absolutely planar. However, as it regrew into the Ge_xSi_{1-x} layer, the interface remained planar only for the first 10–20 nm of regrowth. As regrowth proceeded, the interface roughened substantially after 100 nm of regrowth (Fig. 2b). In the fully regrown sample, defects were observed which originated after ∼30 nm of "ideal" growth (Fig. 2c).

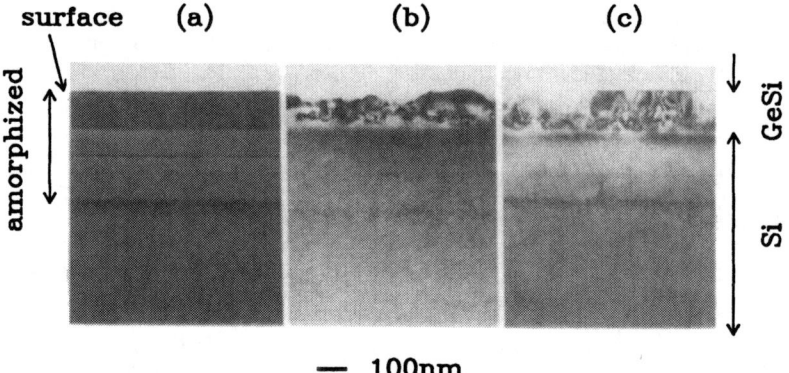

Figure 2: XTEM images of partially regrown $Ge_{0.13}Si_{0.87}/Si$, with a/c interface (a) in Si substrate, (b) in GeSi layer and (c) fully regrown.

SIMULATION AND DISCUSSION

The XTEM studies clearly showed the evolution of the interface roughness, but quantitative measurements by XTEM were not feasible over a wide range of conditions. The TRR interference contrast, however, contains information on both the velocity and the interface roughness. Consequently, we developed a model to fit the complicated behavior of the roughening interface and extract quantitative roughness and position values. The XTEM studies suggested that the rough interface could be approximated as a normal distribution of planar interfaces, characterized by a mean position x_0 and a normal deviation σ. If the lateral scale for the thickness variations is greater than the wavelength of the light (approximately true), the average reflectivity R_{ave} from the interface can be approximated by a convolution of the reflection $R(x)$ from a planar interface with the probability distribution $P(x)$ of interface positions (here given for a Gaussian):

$$R_{ave} = \int_0^\infty R(x)P(x)\,dx \quad , \quad P(x) = A\exp\left(-\frac{(x-x_0)^2}{2\sigma^2}\right),$$

where A normalizes the probability to unity. (To match the final peaks, care must be taken to allow $P(x)$ to approach $\delta(x = 0)$ as the interface approaches the surface.) For calculations, the temperature dependent complex index for a-Si and c-Si were taken to be $n_{a-Si} = (4.39 + 5 \times 10^{-4}\,T) - 0.61\,i$ and $n_{c-Si} = (3.72 + 4.88 \times 10^{-4}\,T) - 0.018\,i$ [8, 12].

Initial estimates for the interface position and roughness were determined from the extrema from the TRR curves. At each order extrema, there is a unique pair of x_0 and σ values (x_0 in particular shifts considerably with roughness for extrema near the surface). These values were initial guesses for knots in a spline representation of $x_0(t)$ and $\sigma(t)$. The knots were subsequently varied until a minimum deviation between simulated TRR curve and the experimental data was achieved. Results of fitting a Ge_xSi_{1-x} sample regrowing at

609 °C by this method are shown in Fig. 3. To some degree, the excellent fit justifies the assumptions used in this model.

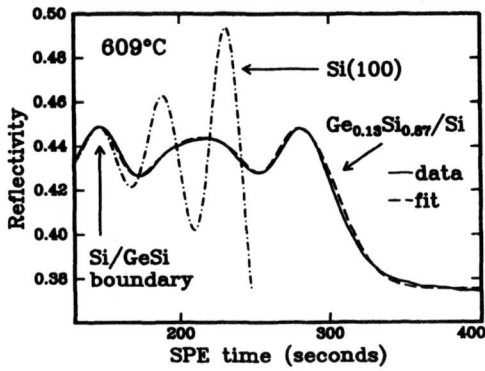

Figure 3: An example of TRR fitting: solid line is experimental data and dashed line is calculated curve.

The interface depth and roughness as a function of anneal time and regrowth velocity as a function of interface depth were extracted from TRR curve fitting and are shown in Fig. 4 for $Ge_{0.13}Si_{0.87}/Si$ annealed at various temperatures. From Fig. 4b, the SPE velocity is seen to decrease, relative to pure Si, after only a few tens of nm regrowth in Ge_xSi_{1-x}. The regrowth velocity reaches a minimum after regrowing ~60 nm in GeSi, followed by a velocity recovery as growth continues to the surface. For low temperature (~560 °C) anneals, the velocity decrease is substantial, becoming less pronounced at high temperatures (> 600 °C). Interface roughness begins to develop in the first 10–20 nm in GeSi and quickly builds to an almost constant level, 15–20 nm, for all temperatures. Continued regrowth results in only slight roughness increases. The apparent jump in roughness near the surface for some samples is an artifact of the fitting algorithm. Near the surface, the final peak is often too low to fit and the fitting algorithm compensates by over-estimating the roughness without substantial improvement in χ^2. The origin of this problems continues to be examined.

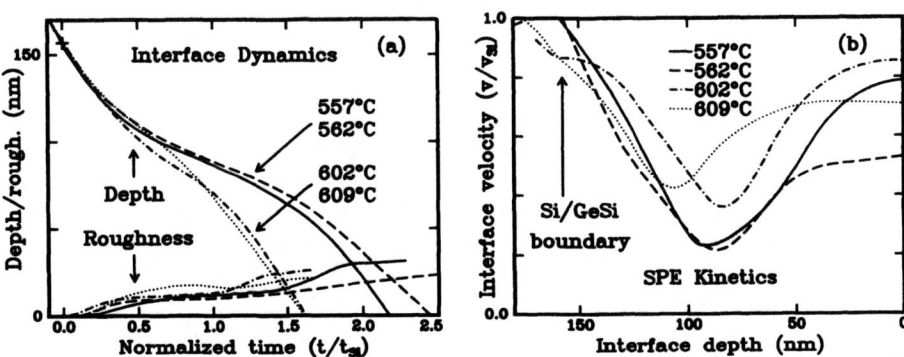

Figure 4: (a) Interface depth and roughness vs. anneal time (normalized to Si regrowth time) and (b) SPE rate (normalized to that of Si) vs. depth for $Ge_{0.13}Si_{0.87}/Si$.

During regrowth in this temperature regime, the SPE rate drops to 20–40% of the bulk Si velocity during the first 60 nm of growth. Continued regrowth allows the velocity to recover to 60–80% of the bulk values. Furthermore, there appears to be a strong correlation between roughness and the interface velocity retardation. Interface roughness develops during the initial stages of regrowth in Ge_xSi_{1-x}. For high temperature anneals, substantial velocity retardation begins and defects are observed by XTEM after ~30 nm of regrowth, approximately the same time that the interface roughness reaches its plateau. Whether defects occur when the roughness reaches a certain level, or if the formation of defect prevents the roughness from increasing further, is presently unknown.

SUMMARY

The SPE regrowth behavior of compositionally sharp and graded Ge_xSi_{1-x} interfaces have been studied using time-resolved reflectivity. A model to quantitatively extract the interface position and roughness from the non-trivial Ge_xSi_{1-x} TRR traces was described. In $Ge_{0.13}Si_{0.87}$ samples, regrowth was observed to proceed initially into Ge_xSi_{1-x} almost unretarded for ~30 nm. Interface roughness developed rapidly after this "ideal" growth, jumping to a nearly constant value of 15–20 nm. After the roughness fully developed, the velocity decreased to 20–40% of pure-Si rates, with the degree of reduction depending on the annealing temperature. Continued regrowth allowed the interface velocity to recover to 60–80% after an additional growth of ~50 nm. XTEM micrographs corroborate the amplitude and development of the roughness obtained by TRR analysis. For non-abrupt Ge_xSi_{1-x} films formed by ion implantation with a peak concentration of ~3 at.% Ge, the SPE velocity is virtually indistinguishable from that of pure Si.

We would like to acknowledge Dr. C. Tracy at Motorola for the CVD Ge_xSi_{1-x}/Si samples. Some work was performed at the Cornell National Nanofabrication Facility supported by the NSF (ECS-8619049), Cornell University and industrial affiliates. This work was supported by the SRC through the Cornell Microscience & Technology Program (93-SC-069).

REFERENCES

[1] C. A. King, J. L. Hoyt, C. M. Gronet, J. F. Gibbons, M. P. Scott, and J. A. Turner, IEEE Electron. Dev. Lett. **10**, 52 (1989); S. S. Iyer, G. L. Patton, J. M. C. Stork, B. S. Meyerson, and D. L. Harame, IEEE Trans. Electron. Devices **ED-36**, 2043 (1989).

[2] K. L. Wang, J. Park, S. S. Rhee, R. P. Karunasiri, and C. H. Chern, Superlattices and Microstructures **5** 201 (1989); T. L. Lin, R. W. Fathauer, and P. J. Grunthaner, App. Phys. Lett. **55**, 795 (1989).

[3] See, for example, papers by M. J. Aziz and J. S. Custer in these proceedings.

[4] D. C. Paine, N. D. Evans, and N. G. Stoffel, J. Appl. Phys. **70**, 4278 (1991).

[5] Q. Z. Hong, J. G. Zhu, J. W. Mayer, W. Xia, and S. S. Lau, J. Appl. Phys. **71**, 1768 (1992).

[6] D. C. Paine, D. J. Howard, N. G. Stoffel, and J. A. Horton, J. Mater. Res. **5**, 1023 (1990).

[7] F. Corni, S. Frabboni, G. Ottaviani, G. Queirolo, D. Bisero, C. Bresolin, R. Fabbri, and M. Servidori, J. Appl. Phys. **71**, 2644 (1992).

[8] G. L. Olson, and J. A. Roth, Mater. Sci. Reports **3**, 1 (1988).

[9] C. Lee, T. E. Haynes, and K. S. Jones, Appl. Phys. Lett. **62**, 501 (1993).

[10] J. A. Yater, Ph.D. Thesis, Cornell University (1992).

[11] M. Born, and E. Wolf, *Principles of Optics*, 6th ed. (Pergamon Press, London, 1980) ch. 1.

[12] G. E. Jellison Jr. and F. A. Modine, Phys. Rev. **B27**, 7466 (1983).

PART VII

Microstructural Development During Crystallization

BULK SEMICRYSTALLINE POLYMERS: A STATUS REPORT

H.D. KEITH
Institute of Materials Science, University of Connecticut, Storrs, CT 06269-3136, USA

ABSTRACT

Discovery several decades ago that long-chain molecules crystallize by folding prompted vigorous basic research on structure-property relations in bulk semicrystalline polymers with expectation of insights that, paralleling experience in metal physics, would lead to better materials for technology. However, complexity of the crystallization process and difficulty in exploring morphology in adequate detail were both underestimated. Important aspects of behavior were recognized but improved polymeric materials have resulted more from exploratory engineering with composites and blends than from basic research on homopolymers. A summary of this research is presented with emphasis upon developments key to present status; also upon the truly formidable challenges that remain, most readily accessible questions for research being of limited technological significance.

INTRODUCTION

As an opening to this symposium on crystalline microstructure in polymers Professor Cebe has asked me to present something largely "tutorial" in nature. My involvement in this subject, in which I have been interested for almost four decades, has centered upon crystallization from quiescent (unsheared) melt when many basic principles and molecular processes are manifested in their simpler, though far from simple, forms. I have chosen to attempt a reflective overview of basic research in this area where we still have problems to solve, new phenomena to ponder and new avenues to explore. It is instructive for us to re-evaluate from time to time our limited understanding and the focus of our aims in ways that best serve ongoing technological development; that, after all, is what materials science is all about. In that regard, incidentally, I have a few concerns about some recent trends in research and interpretation which, put in historical perspective, seem rather unhelpful.

EARLY ASPIRATIONS

In present context we may date the beginning of morphological research on synthetic semicrystalline polymers to the mid-1950's, and especially to the impetus then given it by Keller. He drew attention to general occurrence of spherulitic aggregates when crystallization takes place in melts and also to their radiating fibrillar structures [1]. Soon thereafter, he pioneered the growth of polymer single crystals from solution [2] and confirmation that unexpectedly regular habits which these crystals often exhibit (lamellae of polyhedral shape about 10 nm thick) are attributable to chain folding. It was soon established in many laboratories that lamellar habits, also involving some measure of chain folding, are commonly produced during crystallization within viscous melts. It was recognized that fibrillar structures as, for example, seen earlier by electron microscopy in

511

cast films of polyesters and polyamides, are in fact clusters of highly elongated but very narrow lamellae and that, in general, extended radial units in polymer spherulites mostly consist of ribbon-like lamellae in various states of aggregation. In one move, chain folding offered a rational explanation, not only for compact molecular packing within lamellae of regular habit despite participation of chains of widely varying lengths, but also for a previously puzzling prevalence in polymer spherulites of tangential chain orientation.

The degree of order apparently signalled by these various structures contrasted sharply with earlier perceptions, and in the scientific climate of the time occasioned a keen sense of expectancy and excitement. During the preceding decade there had been striking progress in condensed matter physics, notably in connection with structure-property relations in metals and alloys, and against this background the new revelations about chain molecules crystallizing by folding were bound to excite curiosity about potential for parallel achievements in relation to polymers. Much of the ensuing research was spurred, I recall, not only by seeking understanding of this "new" state of matter (as we tended to think of it) but also by conscious desire to achieve better control over properties. These were fine aspirations, pragmatic enough to elicit approval even today; nevertheless, they were overly optimistic, not least because first deductions about structural order based upon electron microscopy and electron diffraction were at times quite misleading.

REALITIES

Initial reaction to the concept of chain folding was "mixed"; in fact, it extended over the broadest conceivable spectrum. At one extreme it was held for some years that crystallizing chains fold sharply and regularly on growth faces. In time, however, measurements of density and analysis of small-angle x-ray scattering showed that, as a general matter, such an inference is untenable even in the case of single crystals, especially in relation to supposed sharpness of folding. Departures from regularity of folding when aggregates of crystals grow in viscous melts must inevitably be more pronounced. At the other end of the spectrum, skeptics of chain folding expressed concern about feasibility of chain dynamics required during rapid crystallization in melt; their objections have been addressed theoretically although, in my view, they may not have been met completely — a point to which I shall return.

Insofar as differences of viewpoint still remain about folding they apparently relate to questions such as: if folding is irregular, how irregular is that and in which specific context? Experimental input is still indirect and inferential. However, few would now argue with the following: crystals of lamellar habit in polymer spherulites involve an appreciable measure of chain folding, but such matters as fluctuations in stem length, adjacency of re-entry, conformations at folds, and molecular connectivity between adjacent crystals have for the moment to be discussed statistically. It is significant that permissible slip processes by dislocation glide in these crystals are severely limited, and appreciable shear must of necessity involve a disruption of structure [3].

By the mid-1960's it was evident that, far from introducing simplifying order, chain folding had added new diversity to characterization of morphology in semicrystalline polymers and to interpretation of their properties. Further, there was inevitably a close coupling with other variables associated with molecular structure, polydispersity and the possibility of crystallization over extremely wide ranges of supercooling. Basic research was facing formidable challenges even in simpler situations and, although not of immediate consequence, even more so in relation to a burgeoning technology concerned with polymers which have been "formed" mechanically or incorporated in blends or composites.

SPECIFIC TOPICS

My principal topics are morphology and its variability, concerns that are basic to an understanding of properties. Crystallization kinetics, inevitably, are also involved in important ways. I believe I may assume my audience to have more than slight familiarity with these subjects and, as materials scientists, likely to be receptive to some emphasis upon parallels between polymeric and monomeric materials. Proceeding to specific points, I shall take chain folding as already outlined to be a given and add further detail where appropriate.

Lessons from Kinetics of Crystal Growth in Monomeric Solids

Growth rates of polymer crystals (whether precipitating from solution or extending radially in spherulites) obey nucleation-controlled kinetics accurately over uniquely large ranges of supercooling. For many purposes it makes little difference whether primary nucleation is homogeneous or (as commonly the case) heterogeneous; however, rate of surface nucleation on lateral (side) surfaces of thin chain-folded lamellae is of prime importance. As is well known, maximization of this rate, and hence of overall growth rate, is held to be a major factor in causing chains to fold in the first place. The relevant theory, originated by Lauritzen and Hoffman [4], is an adaptation of a classical theory devised for surface nucleation on monomeric crystals. In the latter context one considers monomolecular layers spreading in two dimensions from nuclei which, though small when of critical size, nevertheless comprise many crystallizing units. In a polymer on the other hand, one addresses the advance of a strip initially comprising one molecule folding back and forth across the entire width of a narrow substrate; the unit of repetitive addition is taken to be one stem together with one fold, but the critical activation barrier is crossed merely by attachment of the first stem. Otherwise, there is common ground in that, as soon as a surface nucleus exceeds critical dimension, additional molecules can add readily to it so that a new monolayer is considered to spread across the full extent of the growth face, this face again becoming molecularly smooth and awaiting another nucleation event. That, at least, is what is supposed to happen, but in crystals of perceptible size it almost never does, in monomerics or in polymers. The models ignore some pertinent factors.

Under constant thermodynamic driving force observable growth rates are independent of crystal size whereas the models predict accelerated growth as crystal faces expand and probabilities of nucleation should increase accordingly. In monomerics we know that sporadic nucleation is commonly outpaced by steady generation of new layers at growth-promoting centers, often associated with screw dislocations, and this results in significant departures from predicted kinetics both as regards crystal size and temperature dependence. In a polymer lamella, however, it is apparent that dislocations cannot influence growth rate. We recall that giant screw dislocations (in a topological sense only) initiate development into multilayer structures, but growth rates of individual layers are not affected. Screw dislocations with Burgers vectors in the plane of a lamellae, even if able to overcome seemingly irresistible image forces preventing their formation, would only provide short steps on growth faces and be ineffective as nuclei. Moreover, no other irreducible growth-promoting mechanism is evident. No wonder then that predicted nucleation kinetics are followed so tenaciously, at least in terms of temperature dependence. However, at crystallization rates fast enough to be of practical interest, multiple nucleation occurs to a degree that growth faces become roughened (stepped) by layers spreading simultaneously from many points of origin and temperature dependence of growth rate is modified (in modest quantitative terms but not in functional form) because of redundancy among competing nuclei. This is called regime II and further increase in growth rate leads eventually to regime III where something akin to a nucleative frenzy sets in [5]. As is evident, growth in either of these regimes is independent of crystal dimension at all readily observable sizes. Only with very slow growth at small supercooling (regime I), to which theory is supposed to apply without modification, is there concern about constant rate regardless of size. The matter has received much attention but is not completely resolved; limited coherence lengths on substrates are invoked and for present purposes uncertainty as to their origin is of no consequence.

Existence of the three distinguishable regimes, although unrecognized before about 1972, is now well founded and should no longer occasion surprise. In my view, purely kinetic studies to establish regime transitions in more and more polymers tend to be overdone. Morphological consequences of these transitions and potential influence on properties are at least equally deserving of close attention, as we shall see later. It is highly desirable that morphological changes associated with regime transitions be reported so that useful information does not needlessly pass unnoticed.

There can be no question from the foregoing that, in the technical sense used in crystal growth, polymer crystals are "facetted". This implies that interface morphology is governed by spreading of coherent layers and not, as in unfacetted metal crystals in contact with melt, by a relatively isotropic surface energy associated with disordered surface layers. Protuberant apices of facetted crystals can therefore act as growth-promoting centers if sufficiently favored by gradients of concentration to generate surface nuclei significantly faster than elsewhere on growth faces. Further, there are interesting aspects to dynamics of layer spreading in these circumstances. The theory (for monomerics) was outlined by Frank in 1958 [6] and, in a form particularly relevant to this discussion, later elaborated in an analysis of variability of habit in snow crystals [7]. New layers may originate fairly regularly in time at an apex but in spreading they tend to bunch up to form relatively thick multi-layer steps which in time slow to a virtual halt producing, in Frank's word, "lacunae". The faster the growth the more pronounced the bunching, and the steeper the concentration gradient along the surface the shorter the distance between source and lacuna. In present context, this has a lot to do with dendritic growth in polymer lamellae precipitated rapidly from solution, and is also important in another way to be addressed presently. (I note in passing that the presence of growth-promoting centers in monomerics was revealed by spreading of bunched multilayer steps readily visible by optical microscopy at modest power.)

It will be evident that in referring to fast and slow crystal growth I did not mean comparison between polymers of different chemical type at corresponding supercoolings but, rather, variation within some given but unspecified polymer as temperature or composition (e.g. molecular weight distribution, branch content) are changed. In terms of molecular processes these terms are always to be considered in relation to molecular mobility. The fastest observed growth in isotactic polystyrene (deep in regime II, perhaps in III), about 5 nm s^{-1}, would be quite slow for polyethylene (and certainly in regime I). From our discussion, and remembering that nucleation often involves cilia associated with chains already partly incorporated in a crystal (as we know from concentration dependence of growth rates in dilute solution) we see, then, that "faster" crystallization (below regime I) will inevitably involve superfolding (chains folding back on themselves), nonadjacent re-entry and extended loops in a major way, even though chain statistics and/or density considerations at fold surfaces seem to place a lower limit to probability of adjacent re-entry at about ⅔ [8]. To this admittedly clouded picture must be added another consideration, unobtrusive perhaps, but nonetheless of vital importance. The ductility of bulk polymers, particularly those exhibiting high degrees of crystallinity, indicates that they are more than assemblies of lamellar single crystals. They may be brittle if of low molecular weight or crystallized much more slowly than in engineering practice, but this merely emphasizes that in the general case there must be a considerable measure of direct molecular coupling between crystals. How this comes about is not yet evident from present deliberations.

From Crystal to Spherulite

Pioneers of crystallography like Bravais and Groth examined objects of regular geometrical form easily measurable with great accuracy and developed an exact science which they expounded in precise terminology; they provided a sound basis for all that followed with the advent of x-ray diffraction. It is more difficult to be comparably precise with descriptive morphology of crystals or crystalline aggregates that are neither homogeneous in structure nor rigidly regular in form especially when, as with polymers, we have inadequate means for observation or measurement. Even so, our collective attempts at communication have on occasion been less than distinguished.

The most favorable objects for study, naturally, are regularly polyhedral lamellae precipitated from dilute solution, and there was much preoccupation with these in early days. This meant a concentration of attention upon polymers whose crystal structures have high axial symmetry, although polyethylene, partly for historical reasons and partly because its molecule is the prototypical organic chain, received more attention than any other, despite having a lower symmetry (orthorhombic, pseudohexagonal). A prominent concern was the contrast between habits of single crystals in these polymers and occurrence in spherulites of the same polymers, when crystallized from melt, of crystals seemingly of highly elongated "fibrillar" form; what did the difference signify about unresolved details of structure in spherulitic bulk polymer, about how they arise and what influence they might have? There were to be no easy answers.

Generally speaking, direct observation of crystallization as a dynamic process is restricted to optical microscopy (OM) which has limited resolution, and the higher resolution afforded by transmission electron microscopy (TEM) is available only for very thin specimens already crystallized to the full extent or for surface replicas taken from bulk specimens in similar condition. In either case, interpretation is likely to be easiest when crystallization has been isothermal. Even then, with crystallization from the melt, secondary crystallization occurring during and following primary crystallization (radial expansion) at temperature, and also during subsequent cooling, is a complicating factor. Scanning electron microscopy has not had major impact in these contexts. On the other hand, relatively recent development of combined fixatives and electron stains [9], and especially of a permanganic acid etch for polyolefins and other chemically inert polymers [10], have been important innovations extending applicability of TEM to examination of internal structure. However, there are two circumstances surrounding these later developments which have not been without consequence. First, we note that two decades elapsed between early studies and introduction of the newer methods. Secondly, disparity of order 10^3 in resolution between OM and TEM tends to concentrate observation on two scales so different that it is often difficult to relate corresponding perceptions of the same structure one to the other. There is now a tendency to neglect information about structure on the coarser scale and to overemphasize different, though still ambiguous, indications of structure seen on a fine scale. This can easily lead to distorted interpretation, and I shall presently illustrate how subtly misleading it can be (Figure 1).

In his earliest papers on polymer spherulites Keller noted close similarities with spherulites in monomeric compounds, even to common occurrence of banding (extinction behavior in polarized light microscopy attributable to twisting orientation about radii). There was already an extensive literature and, as far as geometry is concerned, no mystery as to how these radiating aggregates develop by repeated branching and fanning of elongated crystals. What was lacking was understanding of why this strange polycrystalline habit should be adopted quite commonly, particularly when it arises consistently in some circumstances in materials which are perfectly capable under other conditions of producing equiaxed polyhedral single crystals. Concern about relating lamellar single crystals in polymers to corresponding spherulites therefore amounted to little more than reappearance of an old problem in new guise.

It was in this vein that Padden and I spent much time during the years 1957-60 surveying spherulitic crystallization under diverse conditions in a large variety of systems and, having discovered what we thought were key common elements and how they relate to one another, we devoted more time to testing our ideas before publishing our often cited papers of 1963-4 [11,12]. Commonality of behavior was attributed (as it had to be) to factors independent of specific crystallography of candidate materials and, while discussion of monomerics and polymers was closely intermingled, it naturally reflected a knowledge of classical patterns of crystal growth that was clearer than our insight into behavior of the newer materials. What was quite clear in general, however, was the prevalence of segregation processes at growth fronts and, in consequence of appreciable melt viscosity, production of highly localized, and hence unusually steep, concentration gradients at these fronts. Accompanying this there was a redistribution of species between radial crystals and intervening melt in a manner we described as cellular and, in this way, the role of diffusion in producing elongated habits regardless of crystallographic symmetry could be reconciled with constancy of radial growth rate under isothermal conditions. As a major factor contributing to spherulitic development our phenomenological theory stressed morphological instability — now called pattern formation — but, most emphatically, this differed fundamentally from what later became known as Mullins-Sekerka instability in unfacetted crystals, even though there are many similar consequences. Frank's discussion of interrupted layer spreading on facetted surfaces now seems strong reinforcement of a cruder argument upon which we based selection of diffusion length as an adequate and appropriate scaling parameter for onset of instability.

We are here interested only in implications of relevance to polymers. I shall not dwell on predictions (a) that segregation accompanies crystallization from melt even in stereoregular homopolymers in the form of fractionation, or (b) that it should be possible to grow relatively large (~10μm) single crystals of characteristic polyhedral habit even from melt by crystallizing very slowly at small supercooling so that diffusion length is greatly increased, concentration gradients being correspondingly decreased and, as a consequence, onset of morphological instability being delayed. These now appear in some literature under the "it is well known that" category although I point out that justification for (a), originally deduced by us from morphological and kinetic indications and soon confirmed directly by Wunderlich and others, is still a stubborn challenge to theorists. Experimental confirmation of (b) satisfied our initial curiosity in what, after the event, seemed an almost trivial way. It is noted, however, that protracted growth, fanning and branching of lamellae in regime I, as also in crystallization from solution, produces what are unquestionably spherulites, albeit of exceptionally coarse texture.

Lamellae having elongated habits occur in polymer spherulites, if for no other reason, because ongoing branching at giant screw dislocations necessarily implies repeated interruption within each layer of lateral (tangential) growth, growth whose rate is also sensitive to local build-up of any species segregated during radial growth. An important consideration in our work was interpretation of variable coarseness of texture in terms of clustering of such lamellae on various scales to form birefringent entities analogous to radial fibers in monomeric spherulites, fibers which become wider when cellular segregation takes place on a larger scale in response to increasing diffusion length. Observed texture responded to changes in crystallization temperature and in molecular weight (viscosity) as would then be expected (assuming no transition in crystal structure or preferred growth direction). With the advent of permanganic etching it became clear that scaling commonly involves correspondence between diffusion length and thickness of lamellar stacks [13].

Spherulitic Texture

One of my objects is to attempt a constructive bridging of earlier work and current research in the area of description and causative analysis of spherulitic texture where, I believe, there is terminological confusion and needlessly conflicting interpretation. Several contributory causes can easily be identified. No technique presently available gives anything like adequate information to illuminate questions of practical import in clear and unambiguous form. All have shortcomings and have tended to concentrate attention on those limited aspects for which they are best suited. It is also unfortunate, though understandable, that so much basic research on crystallization has centered upon polymers of higher crystallographic symmetry and, to an inordinate degree, upon linear polyethylene whose slim unbranched molecule gives it many exceptional properties. Among these are an abnormally large temperature coefficient of growth rate and an ability of chain-folded crystals to undergo very rapid isothermal thickening. Being only pseudo-hexagonal it has non-equivalent growth faces and a crystal habit that changes rapidly with crystallization temperature, even producing curved faces in a manner that is a striking object lesson in importance of layer spreading in facetted crystals.

Difficulty in relating evidence provided by OM and by TEM is perhaps well illustrated in Figure 1 showing micrographs taken from a current study; the optical micrographs (a) - (e) are all at magnification 450x and the electron micrograph (f) is at 21,000x. The polymer is nylon 6 crystallized in all cases at 190°C; in frames (d) and (e) it has been blended with 2.5% w/w with poly(vinylpyrrolidone) (PVP) serving here as an "active" diluent (see later). (a) and (b) show in polarized light (crossed polars oriented 45° apart) the full complement of birefringent entities near the surface of a film at least 6μm thick (a minimal estimate from measured path differences and known birefringence of highly oriented nylon 6 fibers). (c) shows the same field in phase contrast and its appearance, we note, is unchanged if a

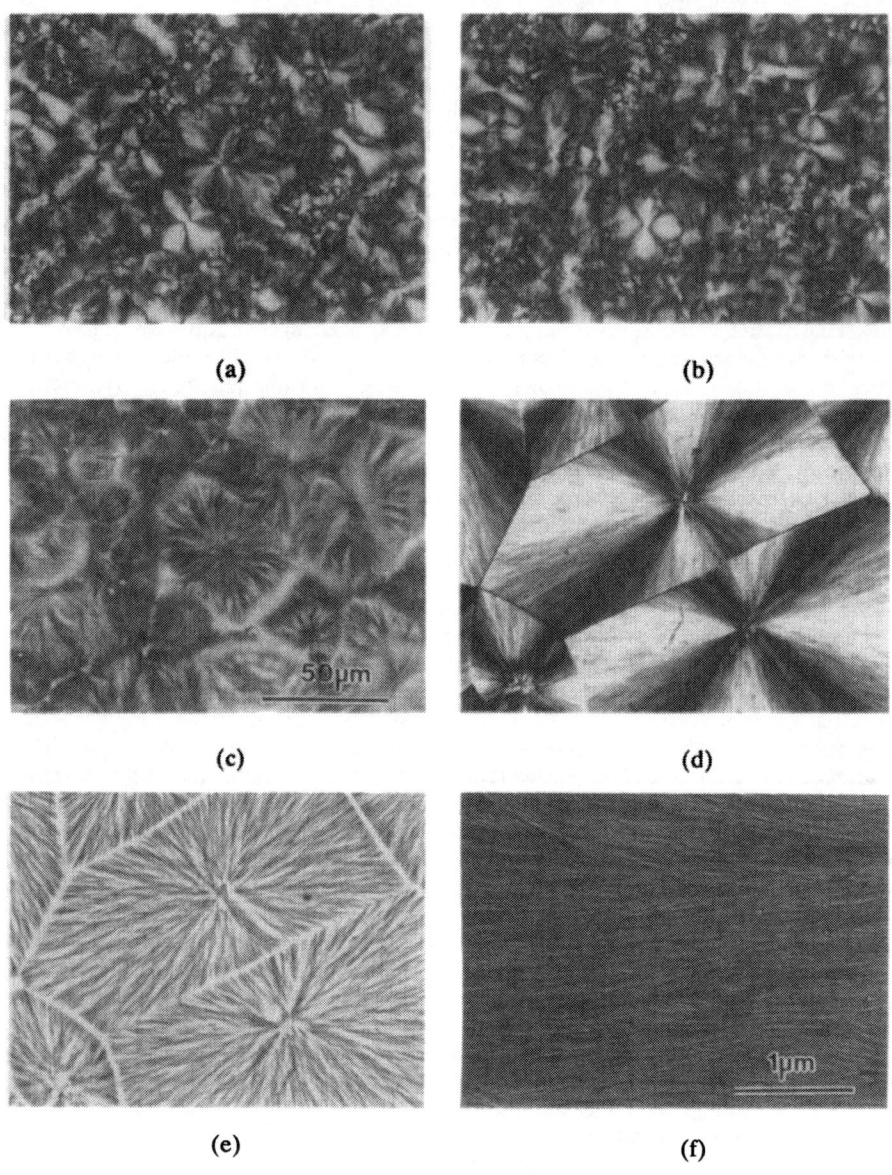

Figure 1: Optical (all same magnification) and Electron Micrographs of Nylon 6
(see text)

polarizer is inserted or is then rotated. Texture is drastically changed by the diluent and for a comparably thick film is shown in (d) and (e) in polarized light and in positive phase contrast respectively. To our great surprise TEM of surface replicas show textures of spherulites as seen in (c) and (e) to be all but indistinguishable (minute surface pits resulting from removal of diluent from the surface by rinsing in methanol alone distinguish the blend unambiguously). In both cases there are lamellar edges in profusion but no clear indication of difference between coarseness of fibrillar texture as evident in comparing (c) and (e); indeed, (f) shows part of such a spherulite in the unblended polymer. Contrast in (c) apparently results from raised features that reflect local clustering in surface layers of lamellae oriented more edge-on and showing more intense birefringence; the larger spherulite centered in (a) - (c) shows this clearly. The facts that in TEM one cannot easily relate the inclinations of thin lamellae to appearance of their traces at a surface and that depth of field is much larger than resolution have in (f) combined to obscure useful information.

The first application of permanganic acid etch was to cut surfaces in linear polyethylene where within banded spherulites it reveals abundant lamellae closely layered locally. Twisting orientation helps identify growth directions and lamellar inclinations to the exposed surface though less precisely than might be desired; nevertheless, important new information about chain tilts and non-planarity in lamellae has been gained. Degrees of crystallinity are high in these structures making it difficult to discern which lamellae grew first and which filled in later in the course of solidification. There is much less ambiguity with specimens crystallized very slowly in regime I (often for days or weeks) and then quenched. There then are striking views of multilayer precursors of spherulites often fanning quite openly and clearly distinguished from the finer texture produced by the quench. A similar pattern is followed, except that twisting is absent, in isotactic polystyrene where splaying of lamellae upon branching can be very pronounced. In these special circumstances (melt composition being homogeneous because of large diffusion length) the basic pre-requisite for spherulite formation in a polymer—lamellar growth with branching at screw dislocations—is met in its simplest form; it has been described as outward growth of dominant and divergent (splaying) subsidiary lamellae [14]. Indeed, such a simple scheme has been promoted as sufficient explanation of spherulitic growth in general circumstances, but clearer definition and a fuller description are both needed. At lower crystallization temperatures growth fronts comprise densely packed clusters of lamellae with little splaying— which are then dominant and in what sense? When two lamellae diverge from a branching site one presumably is subsidiary only if clearly outstripped for some reason by the other, seldom the case with crystallization in regime I. The term subsidiary lamella would seem better reserved for later, possibly secondary (in the sense of Avrami kinetics) additions to existing structure; "infilling subsidiary" is then an appropriate description, sometimes used but not consistently. Whether intended or not, description by the simplest of verbal cartoons creates an impression of development of compact texture at lower crystallization temperature being merely a scaling down of what happens in regime I crystallization. Padden and I regard any such view as seriously misleading; in Sclair polyethylene, for example, we have found a pronounced change to occur between crystallization at 126°C, where texture is still open up to 5 μm behind leading growth fronts, and 124°C where growth rate is six times faster yet compact stacks of lamellae extend to within less than 1 μm of such fronts [15]. Our contrasting views and supporting evidence have been discussed elsewhere [13,15]. Unfortunately, the cartoon description, which still features in some current literature, has at times been presented in terms dismissive of evidence from optical microscopy; resolution may be limited but, quite definitely, segregation processes do occur and close clustering of elongated lamellae in fibrous bundles is not an artefact associated with crystallization in thin films or in blends with uncrystallizable diluents.

Much of the early work featured crystallizations likely in regime II, conditions that produce coarser texture for convenient observation but can nevertheless be encountered in molded plastics of moderate dimension. As significant factors in considering structure-sensitive properties it highlighted processes that may impact upon redistribution of shorter or stereoirregular chains and upon conditions obtaining when intercrystalline connectivity is established; in other words, it addressed the important question as to why adequate connectivity fails to develop in regions of weakness. However, it must be acknowledged that, despite its importance, we are still uncertain as to exactly how giant screw dislocations are generated repetitively in growing crystals. To my knowledge, interpretation in terms of shear at reentrant corners on roughened growth faces hinges solely upon evidence from polyethylene and poly(isoprene) in which incidence of branching increases dramatically as crystallization conditions change from regime I to regime II; hence my appeal for more evidence about morphological changes near newly established regime transitions.

There appears to be a fairly smooth gradation in passing to crystallization in regime III, which generally produces tougher material. Lamellae then grow rapidly in compact arrays, possibly with considerable lattice disorder accommodating species which might segregate if growth were slower and, almost certainly, with abundant formation of intercrystalline tie molecules. How such connectivity comes about is only one, but vitally important, aspect that is little understood about crystallization under conditions generally of technological interest. A close stacking of lamellae notwithstanding, one may well wonder how chains being drawn in to growth faces by enforced reptation [16] might commonly bridge crystals unless these faces are generally aligned one above the other in a proximity very close even on a molecular scale, a situation not usually seen in micrographs. One might appeal to the special case of polyethylene in which rapid isothermal thickening and mutual inter-penetration by chains is known at higher temperatures, but what of lower temperatures, and of other polymers? Again, if the point of attachment of a folding molecule moves quickly up and down a growth face as it "reels in a dragging tail", how does it manage to bypass other chains in its neighborhood? Other questions about current theory also come to mind, and closer analysis of molecular dynamics at moving interfaces, though difficult, is greatly needed.

Mention has already been made of banding (twisting orientation) which is not uncommon in polymer spherulites. Associated effects in polarized light usually vary with conditions of crystallization in parallel manner with apparent coarseness of texture and have been most useful as sensitive indicators of morphological change; a striking example from recent work will be mentioned presently. The twisting has been attributed by Keith and Padden [17] to surface stresses occasioned by congested molecular packing at fold surfaces of lamellae and especially to asymmetry in stress at opposite fold surfaces when there is appreciable chain tilt (obliquity of chain axis to lamellar normal). This explanation was prompted originally by Bassett's observations of nonplanar habits in polyethylene [18] and for this polymer there is now supporting evidence for structural asymmetry in opposing fold surfaces and for mechanical response to predicted bending moments [19]. Measured against twisting in polymer spherulites in general we find many observations conforming, and none as yet clearly in conflict, with the interpretation. Nevertheless, there have been attempts to advance alternative analyses using arguments specific to polyethylene and seemingly not transferable to other polymers which show vigorous twisting, e.g. to aliphatic polyesters [20,21].

It was from pronounced and quite unprecedented regularity of banding that a new aspect of spherulitic crystallization came to light in aliphatic polyesters and polyamides [22]. Small concentrations ($\sim$1-5%) of certain ("active") polymeric diluents induce growth with strikingly increased compactness of lamellar organization even at temperatures where coarse texture would normally be produced; concomitantly, spherulite size is increased as initial nucleation is somewhat suppressed. Contrast between (a) and (d) or (c) and (e) in Figure 1 illustrates these effects in nylon 6 which, lacking chain tilt, never does exhibit banding. Such diluents operate by active participation in molecular processes at growth fronts, but mechanisms are still unclear. A study is in hand to explore influence on mechanical properties, as yet preliminary and without result. A possible improvement in toughness would be pleasing not only for practical reasons but, in principle, as a polymer analogue to, say, alloying aluminum with a little copper or iron with a little carbon. It would also reinforce a general inference that close proximity of overlapping lamellae at growth fronts should enhance molecular connectivity.

SUMMARY

Plastics engineers may be disappointed, but surely not seriously inconvenienced, by limited progress overall in our basic studies. Wherever I look, and I trust I may be forgiven for concentrating on topics close to my own work and interest, I sense there is a long way to go. What is now needed, in my view, both in theory and in the planning and interpretation of experiments, is conscious effort to analyze structure as it develops cooperatively in closely associated clusters of lamellae. This will be difficult but it does address the real situation upon which properties characteristic of bulk polymer crucially depend; from a materials science standpoint, models based upon lamellae considered as essentially independent entities are becoming sterile. To those who will carry on the work I would emphasize the merits of maintaining broad perspective, of paying careful attention to terminology and, above all, of adopting "robust" methods in evaluation and interpretation of evidence. Here I borrow a term from statisticians to whom it signifies approaches which work well and are broadly of substantial value even if founded on premises that may not be rigidly valid in all aspects; it implies proof against invalidation evoked merely by looking a little further afield.

REFERENCES

1. A. Keller, J. Polymer Sci. **17**, 293, 447 (1955).

2. A. Keller, Phil. Mag. **2**, 1171 (1957).

3. H.D. Keith and E. Passaglia, J. Res. Natl. Bur. Stds. **68A**, 513 (1964).

4. J.I. Lauritzen, Jr. and J.D. Hoffman, J. Chem. Phys. **31**, 1680 (1959).

5. J.D. Hoffman, Polymer **24**, 3 (1983).

6. F.C. Frank, in <u>Growth and Perfection of Crystals</u>, edited by R.H. Doremus, B.W. Roberts and D. Turnbull (Wiley, New York 1958) pp. 411-17.

7. F.C. Frank, Contemp. Phys. **23**, 3 (1985).

8. C.M. Guttman, E.A. DiMarzio and J.D. Hoffman, Polymer **22**, 1466 (1981).

9. G. Kanig, Kolloid Z. **251**, 782 (1973).

10. R.H. Olley and D.C. Bassett, Polymer **23**, 1707 (1982).

11. H.D. Keith and F.J. Padden, Jr., J. Appl. Phys. **34**, 2409 (1963).

12. H.D. Keith and F.J. Padden, Jr., J. Appl. Phys. **35**, 1270, 1286 (1964).

13. H.D. Keith and F.J. Padden, Jr., Polymer **27**, 1463 (1986).

14. D.C. Bassett, CRC Crit. Rev. Solid State Phys. Mater. Sci. **12**, 97 (1984).

15. H.D. Keith and F.J. Padden, Jr., J. Polym. Sci. Phys. Ed. **26**, 703 (1988); ibid **25**, 2371 (1987).

16. J.D. Hoffman, Polymer **23** 656 (1982).

17. H.D. Keith and F.J. Padden, Jr., Polymer **25**, 28 (1984).

18. D.C. Bassett, A.M. Hodge and R.H. Olley, Discussions Faraday Soc. **68**, 218 (1979).

19. H.D. Keith, F.J. Padden, Jr., B. Lotz and J.C. Wittmann, Macromolecules **22**, 2230 (1989).

20. D.C. Bassett, R.H. Olley and I.A.M. Al Raheil, Polymer **29**, 1539 (1988).

21. A. Toda and A. Keller, Colloid Polym. Sci. **271**, 328 (1993).

22. H.D. Keith, F.J. Padden, Jr. and T.P. Russell, Macromolecules **22**, 666 (1989).

SHEAR INDUCED MORPHOLOGY OF
SEMICRYSTALLINE BLOCK COPOLYMERS

PETER KOFINAS[†] AND ROBERT E. COHEN [*]
† Department of Materials Science and Engineering
* Department of Chemical Engineering
Massachusetts Institute of Technology, 77 Massachusetts Avenue, Cambridge,
Massachusetts 02139

ABSTRACT

A series of semicrystalline diblock copolymers of poly(ethylene) / poly(ethylene-propylene) (E/EP) were subjected to high levels of plane strain compression using a channel die. Deformations were imposed both below and above the melting point of the E block. The crystallographic and morphological textures were examined using wide-angle x-ray diffraction pole figure analysis and two dimensional small-angle x-ray scattering. The lattice unit cell orientation of the crystallized E chains with respect to the lamellar superstructure was determined, as well as the lamellar orientation relative to the specimen boundaries. When the diblocks are textured above the E block melting point at various compression ratios, the lamellae orient perpendicular to the plane of shear, while texturing below T_m causes the lamellae to orient parallel to the plane of shear. The orientation of the crystallized E chains was perpendicular to the lamellar normal, irrespective of the texturing temperature. The various shear-induced lamellar morphologies have potential applications in gas transport control to develop membranes for use in gas separations or as barrier materials.

INTRODUCTION

In previous investigations in this laboratory [1, 2], the lattice unit cell orientation with respect to the lamellar microstructure was determined for semicrystalline diblock copolymers containing an ethylene crystallizable block. The orientation of the crystallized ethylene chains was found to be perpendicular to the lamellar normals. This unusual chain alignment was attributed to the influence of interface - dominated nucleation and topological constraints on growth when the ethylene block chains crystallize within the amorphous lamellar microdomains present in the heterogeneous melt phase of the block copolymers. Bates and co-workers [3, 4, 5] have studied the lamellar orientation of nearly symmetric amorphous poly(ethylethylene) / poly(ethylene-propylene) (EE/EP) diblock copolymer samples, which were textured using large strain dynamic shear. Near the order-disorder transition (ODT) temperature, and at low shear frequencies, the lamellae arrange parallel to the plane of shear, while higher frequency processing leads to lamellae perpendicular to the plane of shear. At temperatures further below the ODT the parallel lamellar orientation is obtained at all shearing frequencies. These interesting and unexpected results

Mat. Res. Soc. Symp. Proc. Vol. 321. ©1994 Materials Research Society

prompted us to enquire into the possibility that semicrystalline block copolymer systems might also exhibit the perpendicular lamellar morphology under shear.

We have determined the lamellar orientation and chain organization upon crystallization for various processing histories in a series of diblock copolymers having crystalline quasi-poly(ethylene) (E) blocks and amorphous poly(ethylene-propylene) (EP) blocks. Mechanical properties of E/EP diblocks and triblocks have been reported [6, 7], and some work has been done to characterize the morphology [7] [8]. We have previously examined in considerable detail the gas transport properties of the E/EP polymers used in the present study [9]. Using the results presented below we demonstrate that changes in the temperature of plane strain compression processing can be used to force the lamellae to orient either perpendicular or parallel to the plane of shear; the orientation of the crystallized E chains, however, always remains parallel to the plane of the lamellar superstructure irrespective of the processing temperature.

EXPERIMENTAL

The E/EP block copolymers were synthesized by hydrogenation of 1,4-poly (butadiene) / 1,4-poly (isoprene) block copolymers. The butadiene block consists of 10% 1,2 , 35% trans 1,4 and 55% cis 1,4 PB, while the isoprene block contains 93% cis 1,4 and 7% 3,4 PI. The catalytic hydrogenation procedure is described in detail elsewhere [10, 11]. Hydrogenated PB thus resembles low-density polyethylene (E) and hydrogenated PI is essentially perfectly alternating ethylene propylene rubber (EP). All the E/EP diblocks have a total molecular weight of 100,000 g/mol and the block compositions range from 30 to 70 % E. The molecular weight values were determined from Gel Permeation Chromatography measurements on the polydiene precursors (first block and diblock), from knowledge of reactor stoichiometry and conversion, and from a previous demonstration [10] that little or no degradation occurs during the hydrogenation reactions. The melting points of the crystallizable E blocks of the series of E/EP diblocks, were all between 99 and 103 °C, as determined by Differential Scanning Calorimetry.

A channel die, the description of which is given in detail elsewhere [12, 13, 14, 15], was used to subject the polymers to plane strain compression up to compression ratios of 11. Figure 1 shows a sketch of the channel die and defines the three principal directions of deformation, i.e., the lateral constraint direction (CD), the free (or flow) direction (FD), and the loading direction (LD). The channel die was maintained at a selected constant temperature during the compression flow, and the load was applied continuously until the desired compression ratios were achieved. The compressed specimens were quenched under load to room temperature, followed by load release. The final compression ratio was determined from the reduction of the thickness of the samples.

The change in lamellar orientation due to deformation was studied by means of small-angle x-ray scattering (SAXS). The SAXS measurements were performed on a computer-controlled system consisting of a Nicolet two-dimensional position-sensitive detector associated with a Rigaku rotating-anode generator operating at 40 kV and 30 mA and providing Cu Kα radiation. The primary beam was collimated by two Ni mirrors. In this way the x-ray beam could be effectively focused onto a beam stop with a very fine size without losing much intensity. The specimen to detector distance

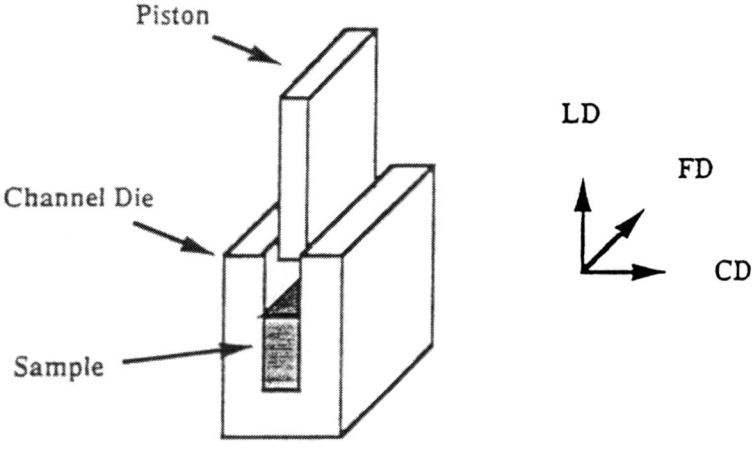

Figure 1: Channel die apparatus

Figure 2: Optical micrograph E/EP 60/40 crystallized from the melt

was 2.7 m, and the scattered beam path between the specimen and the detector was enclosed by an Al tube filled with helium gas in order to minimize the background scattering.

A separate Rigaku wide-angle x-ray diffractometer with a rotating anode source was employed. The Cu Kα radiation generated at 50 kV and 60 mA was filtered using a thin-film Ni filter to remove the Kβ signal. A Rigaku pole figure attachment was controlled on-line, and x-ray diffraction data were collected by means of a Micro VAX computer running under DMAXB Rigaku-USA software. The slit system that was used allowed for collection of the diffracted beam with a divergence angle of less than 0.3^o. Complete pole figures [14] were obtained for the projection of Euler angles of sample orientation: β from 0° to 360° with steps of 5°, and α in the range 0° to 90° also in 5° steps. x-ray data from the transmission and reflection modes were connected at the angle $\alpha = 50^o$.

RESULTS AND DISCUSSION

Crystallization of the diblock copolymers occurs almost instantaneously as the polymers are cooled below their melting point; varying the thermal history has little observable effect on the degree of crystallinity. Using polarized light microscopy we observed that all diblocks exhibit spherulitic morphology when crystallized from the melt, even in the sample containing as little as 30% polyethylene. These observations suggest that a lamellar morphology predominates over the entire composition range examined here. A representative micrograph of the spherulitic morphology of the E/EP polymers is shown in Figure 2.

The channel die experiments were conducted at compression ratios of $\lambda = 4$ to $\lambda = 12$ with no observed change in the SAXS pattern. There is no significant scattering when the x-ray beam is parallel to the constraint direction; spread-out spots on the 2D detector are observed when the sample is irradiated in the flow direction in contrast to the sharp dots obtained when the x-ray beam is along the loading direction. Pole figure analysis revealed that the (200) and (020) poles are concentrated along the constraint and the flow direction respectively, indicating that the chains are oriented in the loading direction.

The E/EP 50/50 sample when textured at 80 ^{o}C, shows a set of SAXS patterns which are completely different from the results obtained from the 150 ^{o}C channel die compression. The view from the loading direction reveals no scattering, arcs are observed along the loading direction upon irradiation parallel to the constraint direction, and broad spots are obtained when the x-ray beam is parallel to the flow direction. The SAXS patterns thus reveal that the morphology changes from lamellae oriented perpendicular to the plane of shear when the specimen is deformed above the E block melting point, to lamellae oriented parallel to the shear plane for samples textured below the melt. The pole figures for the E/EP 50/50 specimen textured below the E block melting point, have the (200) poles in the loading direction and the (020) poles in the constraint direction, which implies that the c-axis of the polyethylene unit cell is oriented in the flow direction. Plane strain compression below the E melting point thus also results in a crystallized E chain orientation which is parallel to the plane of the lamellar superstructure.

The results for the lamellar orientation and the unit cell orientation within the crystallized lamellae in the textured E/EP 100 K diblocks, as deduced from the pole figure and SAXS analysis, are summarized schematically in Figure 3.

We believe that all of our block copolymers are being processed in the 150 °C channel die experiments exactly in the regime near the ODT in which fluctuations [5] dominate the lamellar structure of the melt. In this regime of the morphological phase diagram, the diblock copolymers have already been shown to respond to shear deformation by organizing their lamellar structure perpendicular to the plane of shear [3]. This peculiar orientation was attributed [5] to the disordering ('melting') of the lamellae with immediate regrowth. The thermodynamic barrier to disordering was overcome only near the ODT, thus production of perpendicular lamellae only occurs near the ODT [5]. We find the same behavior here for the case of plane strain compression of our 100 K specimens at 150 °C. We therefore conclude that our E/EP diblocks form ordered heterogeneous melts under the shear field imposed from the channel die, when deformed at 150 °C above the E block melting point. As the sample is cooled under load the perpendicular lamellar phase is subsequently 'frozen -in' at the onset of crystallization. The crystallization therefore is required to occur in the presence of this pre-existing lamellar morphology, a situation which has already been shown [1, 2] clearly to lead to the type of unit cell orientation where the chains are perpendicular to the lamellar normals.

The deformation mechanisms involved in producing the channel die samples below the E block melting point, which result in a 'parallel lamellar' morphology, are mostly crystallographic in nature [14], similar to those found in the sub T_m plastic deformation of many polycrystalline materials. The crystallized E chains are once again always perpendicular to the normals of the lamellar superstructure, but in this case the crystallographic texture is deformation-induced unlike the interface dominated crystallization of confined chains mentioned above.

SUMMARY

We have shown by SAXS that two distinct lamellar orientations can be produced when a series of semicrystalline E/EP diblock copolymers of varying E block content and 100,000 g/mole total molecular weight are subjected to high levels of plane strain compression. When the diblocks are deformed above the E block melting point at various compression ratios, the lamellae orient perpendicular to the plane of shear, while texturing below the melt causes the lamellae to orient parallel to the plane of shear. The morphology produced above the melting point was attributed to proximity of the order-disorder transition to the processing temperature. Although several examples of the unexpected perpendicular orientation have been reported for amorphous diblock copolymers, to the best of our knowledge this is the first experimental documentation of perpendicular orientation in sheared semicrystalline block copolymer lamellar phases. The semicrystalline diblocks exhibit the perpendicular lamellar orientation for a much broader composition range than any amorphous diblock system reported to date. Furthermore, the semicrystalline systems offer the advantage that the crystallographic texture, which is eventually locked into the materials when cooled below T_m, provides an independent set of clues regarding the orientation of the lamellae at the point when crystallization takes place. Conversely, the deformation mechanisms at $T < T_m$, which lead to the series lamellar morphology, are crystal-

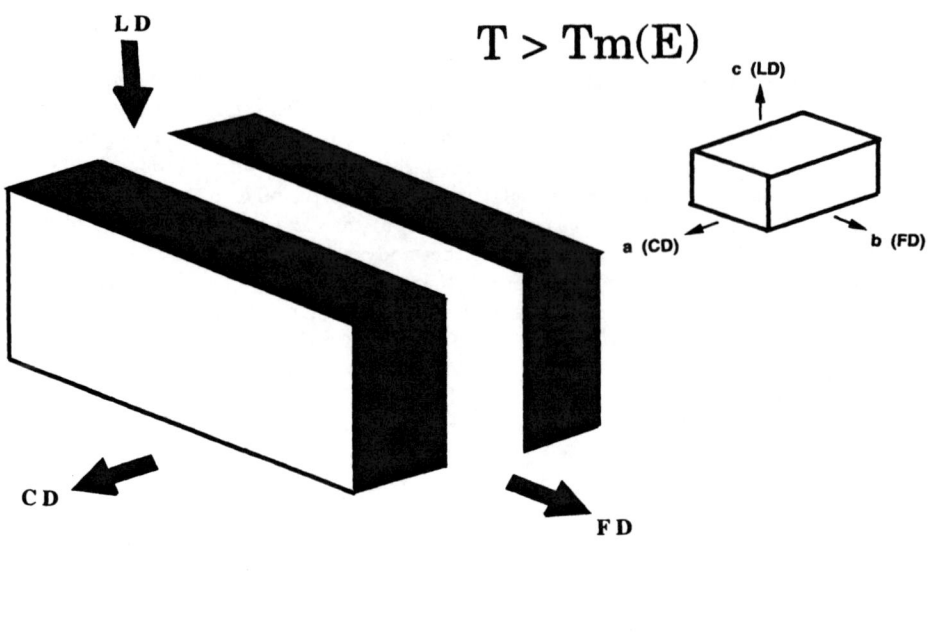

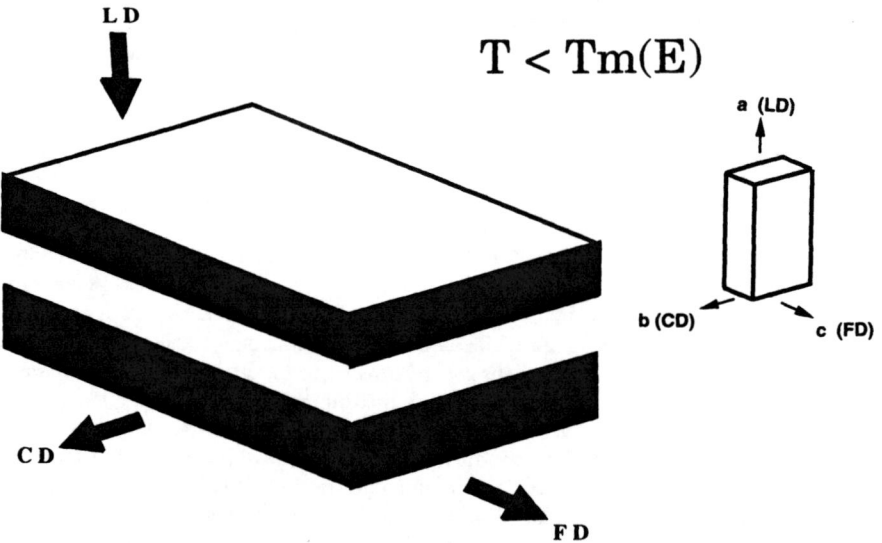

Figure 3: Sketch of lamellar and unit cell orientation in E/EP specimens processed above (a) and below (b) the E block melting point

lographic in nature. WAXS pole figure analysis has revealed that the orientation of the crystallized E chains is perpendicular to the lamellar normal, irrespective of the deformation temperature. When the processing is carried out above T_m, the heterogeneous melt orients perpendicular to the shear planes and then upon cooling the E block chains crystallize within the amorphous lamellar microdomains, a process which has been shown to generate a crystallographic texture with chains perpendicular to the lamellar normals [1, 2].

The unusual shear-induced perpendicular lamellar morphology may have some potential advantages; for example, it provides a 'parallel' material from the perspective of transport through a film. We have made use of this structure in an earlier study [9] of gas transport in these same semicrystalline block copolymers. The structure in which the alternating amorphous and semicrystalline lamellae are oriented normal to the film surfaces enables the membrane designer to enjoy the structural and thermal stability offered by the semicrystalline regions without having them interfere with the gas flux through the amorphous lamellae.

In our work to date we have examined only diblock copolymer structures. Ongoing research includes studies of tri- and tetra- block copolymers of E/EP covering a wide range of composition and molecular weight.

ACKNOWLEDGEMENT

This research has been supported by the Office of Naval Research and the Goodyear Tire and Rubber Company.

References

[1] K. C. Douzinas, R. E. Cohen, *Macromolecules*, *25*, 5030 (1992).

[2] R. E. Cohen, A. Bellare, M. A. Drzewinski, *Macromolecules*, submitted.

[3] K. Almdal, K. A. Koppi, F. S. Bates, *Macromolecules*, *25*, 1743 (1992).

[4] K. A. Koppi, M. Tirrell, F. S. Bates, *Phys. Rev. Lett.*, *70(10)*, 1449 (1993).

[5] K. A. Koppi, M. Tirrell, F. S. Bates, K. Almdal, R. H. Colby,
Journal De Physique II, *2(11)*, 1941 (1943).

[6] Y. Mohajer, G. L. Wilkes, J. E. McGrath, *Polymer*, *23*, 1523 (1982).

[7] R. Seguela, J. Prud'homme, *Polymer*, *30* 1446 (1989).

[8] P. Rangarajan, R. A. Register, L. J. Fetters, *Blends of Amorphous and Crystalline Polymers Symposium*, ACS National Meeting, August 1992, Washington, DC, Division of Polymer Chemistry, *Polymer Preprints*, *33(2)* (1992), *Macromolecules*, in press.

[9] P. Kofinas, R. E. Cohen, A. F. Halasa, *Polymer*, in press.

[10] A. F. Halasa, U.S. Patent 3 872 072.

[11] R. E. Cohen, P. -L. Cheng, K. Douzinas, P. Kofinas, C. V. Berney,
Macromolecules, *23*, 324 (1990).

[12] L. Lin, A. S. Argon, *Macromolecules*, *25*, 4011 (1992).

[13] H. H. Song, A. S. Argon, R. E. Cohen, *Macromolecules*, *23*, 870 (1990).

[14] A. Galeski, A. S. Argon, R. E. Cohen, *Macromolecules*, *25*, 5705 (1992).

[15] R. W. Gray, R. J. Young, *J. Mater. Sci*, *9*, 521 (1974).

CRYSTALLIZATION FROM POLYMER BLENDS

RICHARD S. STEIN
University of Massachusetts, Amherst, MA 01003

ABSTRACT

Binary polymer blends in which one or both components crystallize have been considered. These may be miscible or immiscible in the amorphous state, and phase separation may occur along with crystallization. The rates of these two processes vary differently with temperature. so the consequent morphology and properties depend upon the temperature history. For miscible blends, the glass transition temperature depends upon composition. This affects diffusion so that crystallization kinetics varies as composition changes. If only one component crystallizes and the other is rejected, this rejected component diffuses away from the growing crystalline regions in a manner dependent upon the interaction between components.

These phenomena are studied using a variety of techniques such as calorimetry, x-ray diffraction and scattering, and small-angle light and neutron scattering for a number of polymer systems.

INTRODUCTION

Binary polymer blends may or may not be miscible in the amorphous state depending on the value of the Flory interaction parameter, χ. Phase separation may occur upon increasing or decreasing the temperature, depending upon the temperature dependence of χ, so that a lower (LCST) or upper (UCST) critical solution temperature may be observed[1]. If one or both components crystallize, such crystallization may occur in the miscible or immiscible region, and will usually result in the rejection of the non-crystallizing component which must then diffuse into the surrounding amorphous phase[2]. This results in a change in the glass transition temperature, T_g, of this phase, with its consequent effect on the rate of crystallization[3].

Thus, the crystallization kinetics and the resulting morphology will depend in a complex way upon these phase relationships, and will depend upon the thermal history of the sample during the phase separation and crystallization processes.

CRYSTALLIZATION FROM THE MISCIBLE STATE

Let us consider the case of a miscible binary blend in which one component crystallizes. Some examples are poly(vinylidene fluoride)/poly(methyl methacrylate) (PVDF/PMMA), poly-ε-caprolactone/poly(vinyl chloride) (PCL/PVC), and isotactic/atactic polystyrene (i-PS/a-PS). The crystallization rate depends upon the crystal melting point, T_m, and the T_g of the amorphous phase, both of which depend upon blend composition. As shown by Scott[4], T_m is depressed by blending according to[5]

$$[1 - T_d^\circ/T_m^\circ] = -(B\ V_2/\Delta H_{2u})\phi_1^2 \qquad (1)$$

where T_m° is the melting point of the unblended crystallizable component, T_d° is T_m° for the blend. The constant B, the interaction energy density, is given by[5] $\chi_{12}RT/V_{1u}$. ΔH_{2u} is the enthalpy of melting of the crystallizable component. The specific volume is V_2, the

Mat. Res. Soc. Symp. Proc. Vol. 321. ©1994 Materials Research Society

volume fraction is ϕ_1. The amount of melting point depression depends upon χ as can be seen from the above relation, as was shown by Morra and Stein[5] in their study of PVDF/PMMA blends. It is important in such studies to achieve equilibrium melting conditions using procedures such as the Hoffman-Weeks extrapolation[6].

The variation of T_g with composition mat be described by equations such as that of Flory and Fox[7],

$$[\ 1/T_g\ ] = [\ \phi_1/(T_g)_1]\ + [\ \phi_2/(T_g)_2] \tag{2}$$

where ϕ_1 and ϕ_2 are the volume fractions of the two components and $(T_g)_1$ and $(T_g)_2$ are their respective T_g's.

It has been proposed that crystallization kinetics may be described in terms of a reduced temperature[8],

$$\Theta = (T - T_g)/(T_m - T_g) \tag{3}$$

to account for the effect of composition upon both T_m and T_g. A generalized description of how crystallization rate depends upon Θ is shown in Fig. 1. It goes to zero at $T = T_g$ (where $\Theta = 0$) and below where the polymer is glassy and diffusion is negligible. It also becomes zero at $T = T_m$ (where $\Theta = 1$) and above when the crystals are melted. The rate is a maximum at some temperature, T_c, between T_g and T_m.

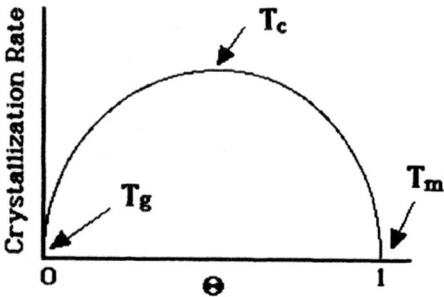

Fig. 1. The variation of crystallization rate with the reduced
temperature, Θ.

The effect of blending on changing T_g is usually greater than that for changing T_m. For both PVDF/PMMA and PCL/PVC, introducing the non-crystallizing component slightly lowers T_m but appreciably raises T_g, usually decreasing the crystallization rate[3]. In both these cases, there is no possibility of inclusion of this component in the crystal, so it must be rejected to the surrounding amorphous phase. Thus, the concentration of this high T_g phase at the growing crystal front will slow down the transport of crystallizable polymer to the growing crystal. It will tend to diffuse away from this region, and its effect will depend on the relative rates of this diffusion and of crystallization. At low supercooling, where crystallization is slow, diffusion may avoid formation of this high T_g region. However, at high supercooling where crystallization is fast and diffusion is slow, the rejected component may accumulate slowing down the growth. Thus, the temperature dependence of crystallization rate for such a blend may differ from that of the pure crystallizable polymer in that the rate will be lowered more at lower temperature. Of course, if the rejected component has a lower T_g than the crystallizable one, the effect may be opposite and the crystallization rate may be speeded up.

For iPS/aPS blends, χ is small so there is little depression of T_m. The T_g's are similar for the two components so T_g is not very dependent upon composition. Thus, in this case, the principal effect of blending on the crystallization kinetics will be that of dilution of the iPS, and its rate of crystallization will not be affected as much as in the previous cases.

The usual morphology of a semi-crystalline polymer is that of lamellae[9], exhibiting alternate layers of crystalline and amorphous polymer. The average separation is called the *identity period* , d, which gives rise to intensity maxima in the small-angle x-ray scattering (SAXS) given approximately by the Bragg relationship[10],

$$n \ \lambda = 2 \ d \ \sin \ \theta \tag{4}$$

where n is an integer, the "order" of the diffraction, λ the wavelength of the x-rays, and θ, the scattering angle [Fig. 2]. The integrated intensity or the invarient is defined as

$$Q = \int \ I(q) \ q^2 \ dq \tag{5}$$

where $q = (4\pi / \lambda) \sin (\theta/2)$. The integral should be over all values of θ from 0 to ∞, but in practice, the maximum obtainable value (when $\theta = 180°$) is $(4\pi / \lambda)$. For a two phase system with sharp boundaries, Q is given by

$$Q = K \ \phi_1 \ \phi_2 (\rho_1 \ - \rho_2)^2 \tag{6}$$

where ϕ_1 and ϕ_2 are the volume fractions of the two phases and ρ_1 and ρ_2 are their electron densities.

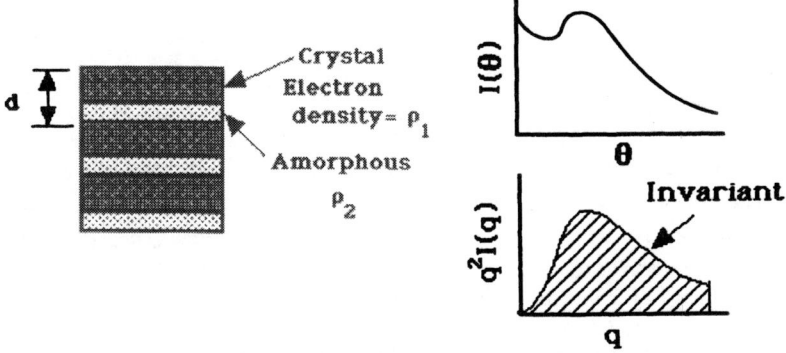

Fig. 2. The identity period, d, for a stack of lamellae, the resulting variation of SAXS intensity with scattering angle, θ, the variation of $q^2I(q)$ with q, and the invariant, Q

Observations of the way in which d and Q vary with composition serve to characterize the morphology of the crystallized blends. The rejected component initially appears in the interlamellar region [Fig. 3]. The separation of the lamellae is usually of the order of 100Å or less, comparable with the molecular radius of gyration. Thus, the rejected molecules reside in a confining space, so their configurational entropy will be reduced. Diffusing out of such spaces represents an entropy increase which will be a driving force. The free energy change accompanying this process, ΔG_d, depends upon the enthalpy, ΔH_d, which is related to the χ parameter for the blend components. For both PVDF/PMMA and PCL/PVC, χ is negative so that ΔH_d is positive, leading to a positive ΔG_d. Consequently, the rejected component stays between the lamellae[2,12].

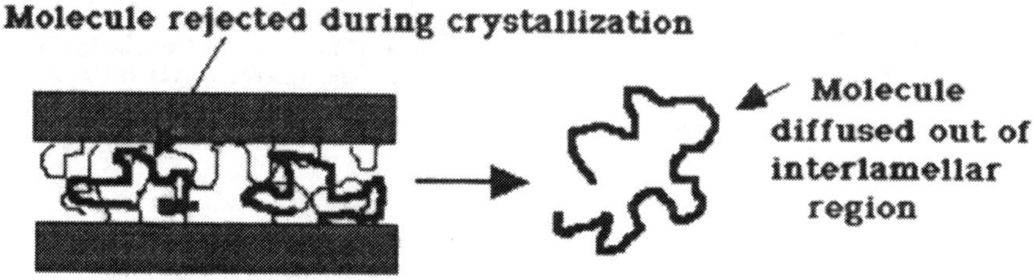

Molecule rejected during crystallization

Molecule diffused out of interlamellar region

Fig. 3. Rejection of component of blend during crystallization and its diffusion out of interlamellar region.

This conclusion is supported by two observations: (1) d increases with the concentration on the non-crystallizable species in a manner such as to accomodate it in the interlamellar region, and (2) Q varies with composition in a different manner for the two blends. For PVDF/PMMA, the electron density of the PMMA is less than that of the PVDF. Thus, if ρ_1 is the electron density of the PVDF crystal, then ρ_2 is that of the amorphous phase consisting of a mixture of the two components, so

$$\rho_2 = \phi_{PVDF,\,am}\ \rho_{PVDF} + \phi_{PMMA,am}\ \rho_{PMMA} \tag{7}$$

Thus, as $\phi_{PMMA,am}$, the volume fraction of PMMA in the amorphous phase, increases, ρ_2 decreases, so $(\rho_1 - \rho_2)$ increases as does Q in agreement with observation. However, in the case of PCL/PVC, the electron density of the PVC is *greater* than that of the PCL, so its rejection into the interlamellar region results in an *increase* in ρ_2 and a *decrease* in $(\rho_1 - \rho_2)$. With sufficient concentration of PVC in the blend, $(\rho_1 - \rho_2)$ becomes zero after which ρ_2 becomes *greater* than ρ_1 leading to a negative $(\rho_1 - \rho_2)$. Since Q depends upon $(\rho_1 - \rho_2)^2$, this accounts for the observation that Q decreases to zero and then increases again with increasing PVC concentration in the blend[2].

In the case of iPS/aPS, there is negligible change in d with composition, suggesting that in this case, the rejected aPS does not reside between lamellae[13]. This is understandable since the χ between these almost identical species is very small so ΔH_{mix} and hence, ΔH_d will be close to zero. Thus, the negative ΔS_d will dominate leading to a negative ΔG_d. This provides a driving force for the aPS to diffuse out of the interlamellar regions. Here, the Q determined from SAXS is not helpful, since the electron densities of the aPS and iPS are essentially identical so ρ_2 for the amorphous phase does not vary with composition. Small-angle neutron scattering (SANS) using deuterium-substituted aPS has been employed in which case[13],

$$Q_{SANS} = \Phi_1\Phi_2 (a_1 - a_2)^2 \qquad (8)$$

Here, the a_i's are neutron scattering lengths which depend upon deuterium concentration. For the interlamellar amorphous phase,

$$a_1 = \Phi_{d\text{-aPS,am}}\, a_{d\text{-aps}} + \Phi_{h\text{-iPS,am}}\, a_{h\text{-iPS}} \qquad (9)$$

where, for example, $\Phi_{d\text{-aPS,am}}$ is the volume fraction of deuterous a-PS in the amorphous interlamellar phase. Since for neutrons, $a_{d\text{-aPS}}$ and $a_{h\text{-iPS}}$ are different, a_1, and thus Q_{SANS} should vary with $\Phi_{d\text{-aPS,am}}$. Since it is observed not to change, one concludes that $\Phi_{d\text{-aPS,am}}$ does not vary, supporting the conclusion that the aPS diffuses out of the interlamellar regions.

While the aPS diffuses out of the interlamellar regions, it cannot diffuse over dimensions comparable with spherulite sizes. In fact, there is little driving force for it to do so. Thus, the aPS remains within the spherulites of the iPS, residing in irregular domains other than interlamellar. This leads to disorder in the internal structure of the spherulite. A consequence in the light scattering from the spherulites is an increase in the diffuseness of the scattering pattern with the four-leaf clover type H_v pattern becoming less distinct and the ratio of V_v to H_v intensities increases, suggesting increasing density fluctuations in internal structure. The texture arising from this exclusion and diffusion process at the growing crystal front occurs by a "fingering" process as described by the Keith-Padden δ parameter and its modifications.

Of course, with the PVDF/PMMA and the PCL/PVC blends, the excluded PMMA and PVC also remains within the PVDF and PCL spherulites, respectively. These high T_g amorphous polymers residing between the lamellae of the crystals within the spherulites serve to rigidify the spherulites. They form a kind of structured crystal/amorphous composite, affecting mechanical properties.

CRYSTALLIZATION FROM THE IMMISCIBLE STATE

If a blend of a crystallizable and non-crystallizable polymer is phase separated prior to crystallization, one phase will be richer in the crystallizable component and the other poorer in a manner determined by the nature of the phase diagram (assuming equilibrium). The sizes of the phases may be determined by the mixing conditions or by the kinetics of the phase separation process leading to their formation. Usually, only the phase richer in the crystallizable component crystallizes with a rate dependent upon the concentration and T_g of the other component.

The usual process of crystallization is "nucleation and growth", with nucleation being most often heterogeneous. Heterogeneous nuclei are often impurities, catalyst fragments, etc. residing in the polymer. If a crystallizable polymer is subdivided into small regions, there is some probability, as shown by Cormia, Price and Turnbull[14], that some of these may be sufficiently small so as not to contain a heterogeneous nucleus. It must then crystallize homogeneously at a slower rate. It seems likely that this situation may arise in a phase separated blend.

Another phenomenon occurring with polymer crystallization is surface nucleation. A nucleus may form with higher probability at a surface or interface than in the bulk of the polymer. This leads to the frequently observed "trans crystallized" layers at sample surfaces[9]. The interfaces of phase separated polymers may serve as such nucleation sites with an effectiveness dependent upon the nature of the second phase. In this case, the crystallization rate may increase with decreasing domain size because of the increased surface/volume ratio.

As a consequence of viscosity differences, local orientation may occur within regions of a phase separated polymer. These could also influence crystallization rates because of "stress-induced crystallization" effects.

It follows that the size of a crystalline region will be limited by the size of the crystallizing domain. Its crystal morphology will depend upon its size and shape, the number of nuclei that it contains, and on the nucleation activity of its surfaces. If the phase is disperse, each domain will crystallize independently, and a nucleus in one domain will not influence crystallization in another. On the other hand, if the crystallizable domains are interconnected, the crystal growth following nucleation may extend throughout the sample. Thus, a discontinuity in crystallization kinetics might be expected at the percolation threshold when phases become interconnected.

CRYSTALLIZATION ACCOMPANYING PHASE SEPARATION

It is possible that a miscible polymer pair may both crystallize and phase separate. An example is iPS/PVME where iPS may crystallize and where phase separation may occur between the amorphous pair. A typical phase diagram that might be encountered is shown in Fig. 4a. This exhibits the variation with temperature and composition of the most usual LCST-type binodal and spinodal, the crystal melting point, and the glass transition temperature of the blend. The relative positions of these curves depend upon the particular blend system.

Consider the case for starting with the miscible amorphous blend at point A. If the temperature is increased past the binodal and spinodal to point B where phase separation occurs by spinodal decomposition, leading to two phases with initial compositions indicated by C and D. This leads to the development of a spatial periodicity in concentration, initially constant, with an amplitude which grows with time and which depends upon the temperature at B. The rate of the process increases with increasing temperature because of the greater thermodynamic driving force and the greater diffusion rate at the higher temperature. At later stages, the size of the phase separated regions grow as hydrodynamic and surface energy forces become effective. At any stage of this phase separation process, the temperature is rapidly dropped through the crystallization region to a point like E below T_g, the morphology will be "frozen in".

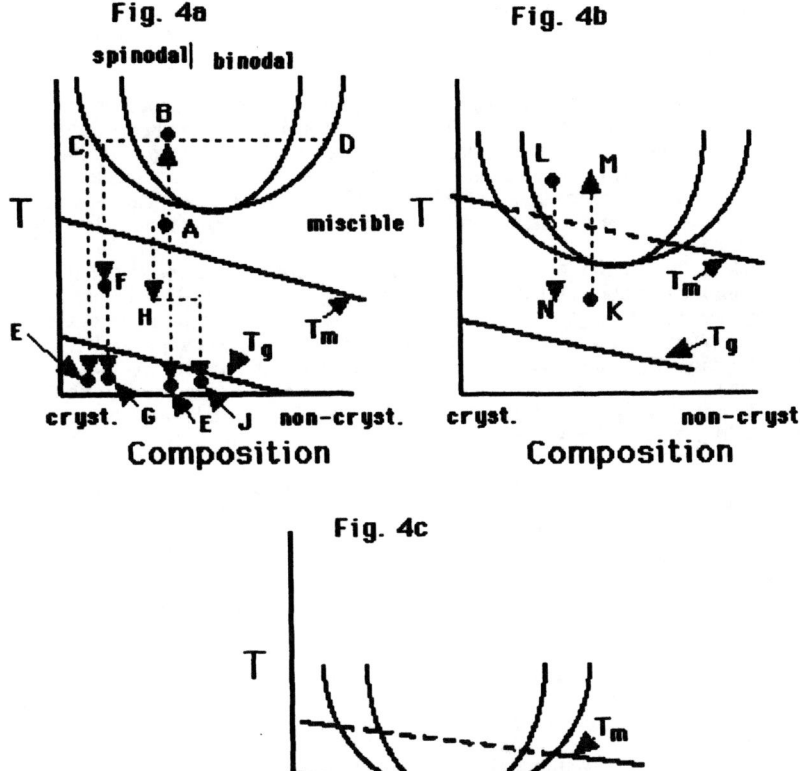

Fig. 4. Various paths for changing temperature for blends which phase separate as well as crystallize.

If instead of lowering the temperature to E, it is lowered to F, below T_m but above T_g, crystallization occurs, probably within the phase with the higher concentration of the crystallizable component. The considerations previously discussed for the crystallization of immiscibile systems apply. As discussed previously, the crystallization rate passes through a maximum at some temperature, T_c, between T_g and T_m. If, after crystallization has proceeded to some extent, the temperature is lowered to that at G, below, T_g, the morphology is again "frozen in" leading to a phase separated system of controlled dimensions with one of the phases crystallized to a desired extent.

An alternate path, starting at A, is to lower (rather than raise) the temperature to that at H where the one phase sample crystallizes. As this happens, the remaining amorphous phase becomes richer in the non-crystallizable component, so that the melting point of the crystals decreases until I is reached at which time crystallization stops. One achieves a morphology consisting of crystals of one component immersed in a miscible amorphous blend. This structure could be "frozen in" by dropping the temperature to J below T_g.

Another possibility for the phase diagram is that of Fig. 4b, where the crystal melting point may be above the binodal. In this case, there is no stable one phase region. The blend may either be crystallized, with a crystalline and a miscible amorphous phase, as at K, or amorphous but phase separated as at L. If the crystallized sample at K is heated to M above the binodal, phase separation occurs in the amorphous region. On the other hand, if the phase separated polymer at L is cooled to below T_m, to N, crystallization occurs, probably in one of the separated phases. In both cases, if the samples are quenched below T_g, the resulting structure may be frozen in.

A third possibility, illustrated in Fig. 4c, is where the binodal extends below T_g. In this case, it is not possible to achieve a miscible state, so the above discussion of crystallization of immiscibile systems applies.

It is evident that for a given composition blend, it is possible to achieve different morphologies and hence, different properties by subjecting samples to different thermal histories.

CASE WHERE BOTH COMPONENTS CRYSTALLIZE

An example of a case where both components crystallize is the blend of polycarbonate (PC) with poly-ε-caprolactone (PCL)[15]. In this case the PC crystallizes slowly, but addition of miscible PCL lowers the T_g and increases its crystallization rate. Of course, with sufficient addition of PCL, the PC becomes diluted and its T_m is lowered, and its crystallization rate will eventually be reduced.

If PC is added to PCL rich blends, the crystallization rate of the PCL will be lowered as a consequence of the raising of the T_g. Thus in such systems, one finds decreasing amounts of PCL crystallinity with increasing PC concentration. However, while the growth rate of the PCL crystals is reduced by adding PC, their nucleation rate is increased, presumably because of the PC crystals serving to nucleate the crystallization of the PCL.

In the range of 10% to 50% PC, both species crystallize. Synchrotron SAXS studies reveal that the PCL crystals grow between lamellae of the PC, so that a single broad SAXS intensity peak is obtained.

A case of interest is that of polyethylene blends such as the mixture of linear high density polyethylene (HDPE) and branched low density polyethylene (LDPE)[16,17]. These have been shown by SANS, using deuterium-substituted HDPE, to normally be miscible in the melt[16]. When the temperature is lowered, the HDPE, having the higher T_m, crystallizes more rapidly. At higher crystallization temperatures, segregation of the crystalline species of the two species occur[17]. However, upon quenching the miscible melt to lower temperature, some co-crystallization occurs, as demonstrated by differential scanning calorimetry and by crystal field splitting of infrared bands obtained when using deuterous HDPE in the blend[18].

It should be pointed out that other workers, primariily those at the University of Bristol, believe that there is less miscibility between linear and branched polyethylene than indicated by our work. However, more recent work, shortly to be published, is consistant with our studies. We believe differences arise from an inadequate interpretation of morphology changes occurring during the course of crystallization.

When the HDPE crystallizes, the lower melting LDPE is initially excluded into the interlamellar regions. Because of the positive DS_d and the negligible DH_d, this LDPE diffuses out from between the lamellae. Whether it crystallizes between the lamellae forming a mixture of layers of HDPE and LDPE crystals or whether it diffuses out from between the HDPE crystal forming segregated layers of LDPE crystals depends upon the competition between diffusion and crystallization rates. At high temperatures, diffusion is rapid and crystallization is slow, so segregation occurs. However, at low temperatures, diffusion is slower and crystallization more rapid, so interlamellar crystallization occurs.

This is confirmed by SAXS experiments[16,19]. Segregation leading to separate crystalline stacks of HDPE and LDPE crystals leads to two SAXS intensity peaks, whereas formation of mixed layers gives a single broad intensity peak. Experiments in real time using synchrotron radiation permit following the development of these structures.

While diffusion can take place over distances comparable with lamellar dimensions during the time of crystallization, it is unlikely that it can occur over the larger distances associated with superstructures such as spherulites. These grow from primary nuclei, often heterogeneities, and consist of radial branched crystalline lamellae with amorphous polymer in between them. With polyethylene blends, both species of crystals reside within the same spherulite. Segregation at the spherulite level is seldom, if ever, observed. The development of this spherulitic structure may be readily followed using the small-angle light scattering (SALS) technique[20].

The SALS from spherulitic polymers has been studied in this laboratory for over 30 years[21]. The light scattering patterns obtained with crossed (H_v) polarization show a characteristic four-leaf clover shaped pattern with an intensity maximum occurring at a scattering angle, q_{max}, given by

$$(4\pi R_s/\lambda) \sin (\theta_{max}/2) = 4.08 \qquad (10)$$

where R_s is the radius of the spherulite. A measure of the variation of θ_{max} with time, conveniently done using a computer interfaced CCD detector, serves to characterize the radial growth rate of the spherulite. The intensity at the maximum of this component is given by[22, 23]

$$I_{max}(H_v) = K \ N_s R_s^6 (n_r - n_t)^2 \qquad (11)$$

where N_s is the number of spherulites/cm^3 and $(n_r - n_t)$ is the anisotropy of the spherulite, the difference between its radial, n_r, and tangential, n_t, refractive index. This, in turn, assuming that the anisotropy arises from oriented crystals within the spherulite, is given by

$$(n_r - n_t) = \phi_{cs} \, \phi_{cs} \, (n_1 - n_2)_{cr} \tag{12}$$

where ϕ_{cs} is the volume fraction of crystals in the spherulite, ϕ_{cs} is the orientation function of the optic axis of the crystals with respect to the spherulite radius, and $(n_1 - n_2)_{cr}$ is the difference between the principal refractive indices of the crystal.

Additional information is obtained using parallel (V_v) polarizers. In this case, there are two contributions to the scattering. One, which is circularly symmetrical about the incident beam, arises from the difference between the average refractive index of the spherulite and that of the surrounding amorphous phase. Its intensity is a maximum when the volume fraction of spherulites is about 0.5. The other component which as two-fold symmetry arises from the anisotropy of the spherulite.

Thus, from observations of both the H_v and V_v components of scattering, it is possible to obtain information about the radius of the spherulites, their volume fraction, and their degree of crystallinity. From these, one may resolve the crystallization process into its components of nucleation, radial growth, and secondary crystallization.

A consequence of such a study on a blend of HDPE and LDPE reveals that if the degree of supercooling is not too great[23]:

1. The radial growth of the spherulites reach their limit and their volume fraction approaches one before appreciable crystallization of the LDPE occurs,

2. the crystallization of the LDPE mostly occurs through an increase in the anisotropy of the spherulites without change in their size, and

3. the ultimate spherulite radius achieved for HDPE is about twice that for LDPE. With blends of the two, the radius is close to that for HDPE over most of the range of compositions.

On the basis if these observations, the following mechanism for the crystallization of the blends has been proposed[23]:

1. With isothermal crystallization at a temperature below the melting points of both species, the HDPE, having the higher melting point, and consequently, the greater degree of supercooling, crystallizes first.

2. The spherulites which grow consist of crystalline lamellae of HDPE with molten LDPE residing between the lamellae.

3. If the degree of supercooling is not too great, these spherulites become volume filling before appreciable crystallization of the LDPE occurs. Thus, their size is characteristic of the HDPE.

4. The subsequent crystallization of the LDPE occurs within the already grown spherulites, consequently, there is no change in their size, just in their crystalline content and anisotropy.

One can understand on the basis of this proposed mechanism why the spherulite size is characteristic of that of the HDPE. It is of intererest to note that this is so even with comparatively low amounts (5-10%) of HDPE in the blend. In these cases, the volume filling spherulites which first form are of very low (3-8%) crystallinity. Even so, they serve as the "skeleton" of the morphology and control the size of the strucures formed by the crystallization of the LDPE which is the major component.

Thus, it is evident that the two species in the blend serve different morphological roles. The HDPE forms the framework and the LDPE the filler of this crystalline composite. It is apparent that their contributions to mechanical properties will differ.

CONCLUSIONS

The crystallization of polymer blends may lead to a variety of morphologies, dependent upon both the thermodynamics and the kinetics of the phase separation and crystallization processes. By understanding these and suitably adjusting compositions and thermal histories, it is possible to affect the morphologies in a manner so as to achieve desirable properties.

ACKNOWLEDGEMENT

The support of the National Science Foundation, Division of Materials Research, the Materials Research Laboratory of the University of Massachusetts, the Exxon Chemical Company and the Novacor Chemical Company is greatfully acknowledged. The author is indebted to the many graduate sudents, postdoctoral fellows and research associates who participated in the studies reported here.

REFERENCES

1. I. C. Sanchez in, Polymer Blends, Ed., D. R. Paul and S. Newman, Vol.1, Academic Press, New York (1978)

2. R. S. Stein, F. B. Khambatta, F. Warner and T. P. Russell, J. Polym. Sci., Phys. Ed., **14,** 1391 (1976)

3. B. S. Morra, Ph. D. Dissertation, University of Massachusetts, Amherst, MA (1980)

4. R. L. Scott, J. Chem. Phys., **17,** 279 (1949)

5. R. S. Stein and B. S. Morra, J. Polym. Sci., Polym. Phys. Ed., **20,** 2243 (1982)

6. J. D. Hoffman and J. J. Weeks, J. Res. Nat. Bur. Stand. U.S., **66,** 13 (1962)

7. T. G. Fox and P. J. Flory, J. Appl. Phys., **21,** 581 (1950)

8. R. S. Stein and A. Escala, in Multiphase Polymers S. L. Cooper and G. M. Estes Eds., Advances in Chemistry Series, American Chemical Society, Washington D. C., Volume 176, 455 (1979)

9. P. H. Geil, Polymer Single Crystals, (Polymer Reviews, Vol. 5) Wiley-Interscience, 1963

10. A. Guinier and G. Founet, Small Angle Scattering of X-Rays, John Wiley and Sons, New York (1955)

11. R. S. Stein, C. T. Murray and J. W. Gilmer, Macromolecules, **18**, 996 (1985)

12. R. S. Stein and B. S. Morra, J. Polym. Sci., Polym. Phys. Ed., **20**, 2261 (1982)

13. W. Herman, Ph. D. Dissertation, University of Massachusetts, Amherst, MA, (1988)

14. R. L. Cormia, F. P. Price and D. Turnbull, J. Chem. Phys., **37**, 1333 (1962)

15. Y. W. Cheung, R. S. Stein, G. D. Wignall and H. E. Yang, Macromolecules, **26**, 5365 (1993)

16. M. Ree, Ph. D Dissertation, University of Massachusetts, Amherst, MA (1987)

17. R. S. Stein, S. R. Hu and T. Kyu, J. Polym. Sci., Polym. Phys. Ed., **25**, 71 (1987). R. S. Stein, T. Kyu and S. R. Hu, ibid, p. 89

18. K. Tashiro, R. S. Stein and S. L. Hsu, Macromolecules, **25**, 1801 (1992)

19. K. Tashiro, M. M. Satkowski, R. S. Stein, Y. Lie, B. Chu and S. L. Hsu, Macromolecules, **25**, 1809 (1992)

20. R. S. Stein and M. B. Rhodes, J. Appl. Phys., **31**, 1873 (1960)

21. R. S. Stein and M. Srinivasarao, J. Polym. Sci., Polym. Phys. Ed., (in press, Dec. 1993)

22. R. S. Stein, R. J. Tabar, and P. Leitte-James, J. Polym. Sci., Polym. Phys. Ed., **23**, 2085 (1985)

23. R. S. Stein, M. Ree and T. Kyu, J. Polym. Sci., Polym. Phys. Ed., **25**, 105 (1987)

CRYSTALLIZATION PROCESSES IN POLY(ETHYLENE TEREPHTHALATE) / POLYCARBONATE BLENDS

VERONIKA E. REINSCH AND LUDWIG REBENFELD
TRI/Princeton, and Department of Chemical Engineering, Princeton University
P.O. Box 625, Princeton, NJ 08542.

ABSTRACT

Blends of poly(ethylene terephthalate), or PET, and polycarbonate (PC) over a range of compositions were studied in isothermal crystallizations from the melt using differential scanning calorimetry (DSC). Both crystallization rate and degree of crystallinity of PET depend on blend composition. The glass transition temperature, T_g, of PET and PC in blends and pure polymer were also measured by DSC. Elevation of the T_g of PET and depression of the T_g of PC are observed upon blending. In cooling scans, dynamic crystallization from the melt was observed. In PET/PC blends with high PC content, a novel dual-peak crystallization of PET was observed. The effects of thermal history on crystallization kinetics and degree of crystallinity were also determined in isothermal crystallization studies. For melt processing times between 1 and 30 min and for processing temperatures between 280 and 300 °C, melt processing temperature was seen to have a stronger effect than processing time.

INTRODUCTION

Blends of poly(ethylene terephthalate) (PET) and polycarbonate (PC) have been the subject of numerous studies, many of which focus on component miscibility and transesterification [1-4]. In the absence of substantial transesterification, PET and PC are immiscible [1]. The melting behavior of PET in PET/PC blends has been examined, and blending with PC is found to depress melting temperature and heat of fusion [5]. To date, we have been unable to find any studies concerning the crystallization kinetics of PET in PET/PC blends. In this study, we attempt to quantify the effect of blending and blend composition on the crystallization kinetics and degree of crystallinity of PET. We also examine the effect of thermal processing, in terms of both melt hold time and melt temperature, on the crystallization of a PET/PC blend.

EXPERIMENTAL

Samples of PET, PC, and PET/PC blends of composition 80/20, 60/40, 40/60, and 20/80 (wt% PET/wt% PC) were supplied in pellet form by G.E. Plastics. All samples contained a small amount of transesterification inhibitor. Samples were dried for 16 hr in a vacuum oven at 100 °C, and then molded into film in a two stage process for a total of 9 min at 275 °C and 200 psi pressure. Sheets of Teflon®-coated glass fabric were used as a release agent. Specimens of each blend film were prepared for study by differential scanning calorimetry (DSC) using a Perkin-Elmer DSC-4. Total specimen weights varied from 4 to 20 mg, in order that the PET weight in each specimen was roughly 4 mg.

Samples were first scanned in the DSC from 25 °C to 280 °C at 10 °C/min measuring heat flow as a function of temperature to determine the glass transition temperature, T_g, of each component. In cases where the recrystallization exotherm of PET overlapped with the T_g of PC, the sample was heated to 150 °C, cooled at 320 °C/min to 90 °C, and scanned again at 10 °C/min to 280 °C. In isothermal crystallization experiments, samples were scanned to 280 °C, held for 5 min, and then quenched at 320 °C/min to the crystallization temperature of interest. Data of heat flow as a function of time were collected. Crystallization temperatures in the range of 210 to 235 °C were used. When heat flow no longer changed with time, the sample was cooled to 15 °C below the

543

crystallization temperature, and then scanned at 10 °C/min to 280 °C to collect melting data. To avoid cumulative thermal history effects and degradation, a new sample was used for each crystallization experiment. All crystallizations were repeated three times. Dynamic crystallization from the melt was also studied: samples were scanned to 280 °C, held for 5 min, and then cooled at 5 °C/min to 25 °C measuring heat flow as a function of temperature.

The effects of melt processing on crystallization behavior were examined for the 60/40 PET/PC blend. The samples were scanned to a chosen melt processing temperature, held for a given melt time, then quenched to 220 °C. Isothermal crystallization at 220 °C was observed, collecting data of heat flow as a function of temperature. Melt processing temperatures were 280 and 300 °C, and melt times were 1, 5, 10, 20, and 30 min. When crystallization was complete, fusion data for the crystallized sample were collected by scanning from 205 °C to 280 °C, measuring heat flow as a function of time. A new sample was used for each treatment/crystallization, and each set of conditions was repeated three times.

RESULTS

Glass Transition Temperature

Scans of heat flow as a function of temperature show either two glass transition temperatures for PET and PC, or a single glass transition for PET and recrystallization of PET which masks the glass transition of PC. These distinct T_g's, summarized in Table I, indicate that PET and PC are immiscible over the range of compositions studied. In the blends, it is interesting that the T_g of PET is slightly higher than that of 100% PET, while the T_g of PC is lower than that of 100% PC. The same shifts in component T_g's have previously been observed in PET/PC blends [1]. Since the glass transition of PET takes place in the presence of *glassy* PC, the rigid PC "matrix" could contribute toward increasing the T_g of PET. Conversely, the glass transition of PC in the blend takes place in the presence of *rubbery* PET, which may effectively plasticize the PC. Also, it is possible that some transesterification had occured in the blend samples during processing. This would cause some convergence in the T_g values through homogenization introduced by the newly created copolymer.

Isothermal Crystallization Kinetics of PET

The crystallization half-time is determined from a crystallization exotherm as the time at which half the crystallization peak area has evolved. The crystallization rate can be defined as the inverse of the crystallization half-time. Figure 1 shows the crystallization rate of PET as a function of crystallization temperature in the systems studied. Crystallization rate in all systems decreases with increasing crystallization temperature, since the driving force for crystallization in this regime is the degree of undercooling from the melt. For blends containing less than 80 wt % PC, the crystallization rate of PET is enhanced by blending with PC. We would suggest that the surfaces

Table I: Glass transition temperature of PET and PC in blend samples determined by DSC scans in heating mode with scan rate of 10 °C/min.

Sample	$T_{g,PET}$ (°C)	$T_{g,PC}$ (°C)
PET	75	–
80/20 PET/PC	77	140
60/40 PET/PC	76	141
40/60 PET/PC	76	140
20/80 PET/PC	76	143
PC	–	145

Average standard deviation for these measurements = 1 °C

of PC "islands" in PET act as heterogeneous nucleation sites for PET crystallization. Nucleation effects would be expected to strongly enhance overall crystallization rate in the early stages of crystallization which are predominant in half-time measurements. It is also possible that the presence of an amorphous phase can affect the crystal growth rate of the crystallizable polymer phase. Enhancement of spherulite growth rate has been reported for PET/poly(methyl methacrylate) blends [6], while a depression of growth rate was observed in polypropylene/rubber and polypropylene/polyisobutylene blends [7].

Isothermal crystallization exotherms for 20/80 PET/PC were not observed, as crystallization was very weak and could not be separated from the DSC machine response in quenching to the crystallization temperature. This strong suppression of crystallization rate at high PC content may reflect excessive dilution of the PET in the blend. It is possible that nucleation at the interface between PC and PET is overwhelmed by the interference of PC with PET crystallization at sufficiently high PC content. Previous studies using infrared spectrophotometry have also concluded that blends with high PC content display a suppression or alteration of PET crystallinity [8,9].

Degree of Crystallinity

The degree of crystallinity of PET achieved during isothermal crystallization is determined from a fusion scan following crystallization. The degree of crystallinity is taken to be the heat of fusion of the crystallites per gram of PET in the sample divided by the heat of crystallization of PET per gram if the sample were 100% crystalline. We use a value of 33 cal/g for the heat of fusion of perfectly crystalline PET [10]. The degree of crystallinity is found not to depend on crystallization temperature in a systematic way, so results are averaged for a given blend over the crystallization temperature range studied. The results are given in Table II. The value of degree of crystallinity for the 20/80 PET/PC blend was determined using the heat of fusion from the initial scan, since no crystallization exotherms could be observed. The degree of crystallinity of PET in the blends is depressed as compared with 100% PET. This behavior has previously been reported for these blends [1,5]. Degree of crystallinity begins to drop off sharply as PC content in the blends approaches 80%. These results, like the crystallization rate results, indicate that in high PC blends the amorphous phase interferes with the development of crystallinity in the PET phase. Alternatively, it is possible that substantial transesterification occured, and this, too, would impair the ability of the PET phase to crystallize. Any transesterification removes PET available for crystallization, depressing ultimate crystallinity.

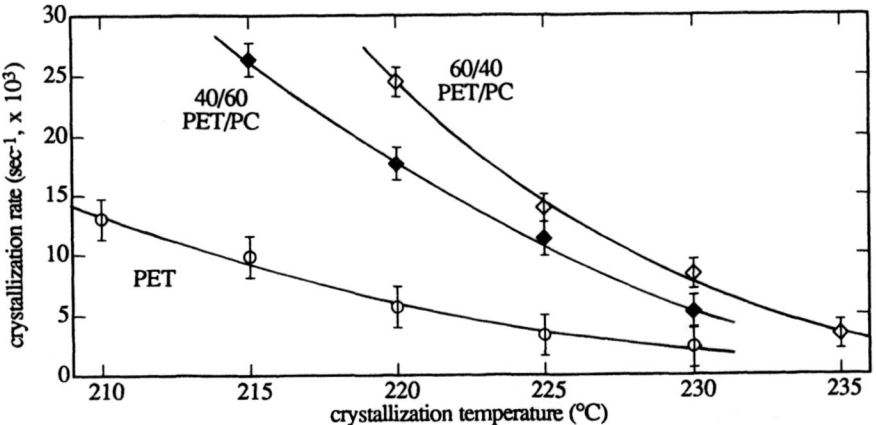

Figure 1: Crystallization rate (inverse of crystallization half-time) as a function of crystallization temperature for PET, 60/40 PET/PC and 40/60 PET/PC.

Table II: Degree of crystallinity, X, of PET in blend samples as determined from DSC fusion endotherms following crystallization.

Sample	X (%)
PET	36
80/20 PET/PC	32
60/40 PET/PC	33
40/60 PET/PC	30
20/80 PET/PC	12*
PC	–

Average standard deviation for these measurements = 1%
* From heat of fusion of quenched sample after scan from 25 °C.

Effect of Thermal Treatment on Crystallization

The effect of thermal treatment on the crystallization rate of 60/40 PET/PC is shown in Figure 2. The melt processing temperature is seen to have a much stronger effect on crystallization rate than melt hold time. All samples held at 300 °C before crystallization crystallize more slowly than those held at 280 °C, regardless of the time they were held in the melt. However, there is some effect of melt hold time. For both melt temperatures, the longer the melt hold time, the slower the crystallization. There are two factors which may contribute to this depression in crystallization rate. Higher melt processing temperatures and longer processing times result in the destruction of more residual crystallites. This reduces the density of nuclei, thus slowing subsequent crystallization. Also, longer melt times and higher temperatures may contribute to transesterification and other degradation reactions, thus slowing crystallization.

Figure 3 shows the effect of thermal processing on degree of crystallinity. Again, melt temperature has a stronger effect on crystallinity than melt time. The marked depression in final degree of crystallinity in the samples held at 300 °C seems to be strong evidence that transesterification or other degradation has occured. If the effect of melt treatment were simply the destruction of residual crystallites, it is not likely that the ultimate crystalline perfection of the material in subsequent crystallizations would be affected. The observed depression is more likely

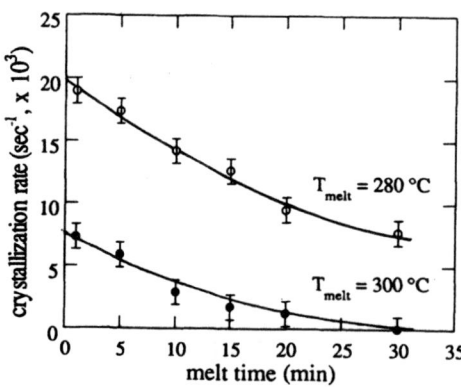

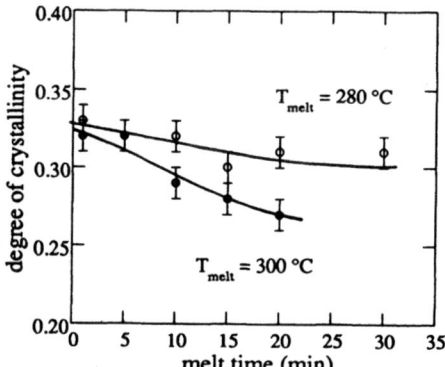

Figure 2: Crystallization rate $(1/t_{1/2})$ of 60/40 PET/PC as a function of time held in the melt prior to crystallization. Crystallization temperature = 220 °C.

Figure 3: Degree of crystallinity of 60/40 PET/PC as a function of time held in the melt prior to crystallization. $T_c = 220$ °C.

546

the result of reactions occuring in the melt which prevent the material from crystallization. In a previous study, complete suppression of PET crystallization following extended (160 min) heat treatment was observed and attributed to transesterification [2].

Dynamic Crystallization in Cooling Scans

Dynamic crystallization from the melt was studied in cooling scans. Fig. 4 shows typical cooling scans for these blends. The step change in baseline seen at roughly 140 °C is related to Tg of PC. Tg of PET is not observed, presumably because of masking by PET crystallinity developed earlier in the scan. In PET and in high PET content blends, we observe a crystallization exotherm at approximately 220 °C. In the 20/80 PET/PC scan, the PET crystallization exotherm is much smaller (lower degree of crystallinity), and shifted to lower temperatures (slower crystallization). This agrees with the results of the isothermal crystallizations discussed above. A very interesting and unexpected result for the 20/80 blend is the appearance of a second exotherm at 155 °C immediately above the Tg of PC. To better understand possible causes of this second exotherm, two further blend samples were produced. A sample with approximate composition 10/90 PET/PC was produced by combining films of 100% PC and 20/80 PET/PC during the molding process. Similarly, a sample with approximate composition 30/70 PET/PC was prepared by combining a 20/80 PET/PC film with a 40/60 PET/PC film. It was assumed that the molding process would provide adequate mixing.

The dynamic scans for these materials are shown in Fig. 5. It is interesting that the size of the low temperature exotherm appears to increase with increasing PC content of the sample. On a per gram PET basis, the heat evolved during the low temperature process is 1.0 cal for the 70% PC blend, 1.8 cal for the 80% PC blend, and 2.5 cal for the 90% PC blend. In contrast, the high temperature exotherm is larger in 30/70 PET/PC than in 20/80 PET/PC, and the peak maximum moves closer to the maximum of the 40/60 PET/PC sample exotherm. The high temperature exotherm is no longer distinguishable in the 10/90 PET/PC scan. These results confirm that in the blends with PET contents of 30% and less, there are two exotherms observable in dynamic crystallization from the melt. The first exotherm at roughly 210 °C appears to be PET crystallization, while the second at 155 °C may be related to PC or a co-crystalline form of PET and PC. We hope to learn more about this dual crystallization behavior through future WAXS studies.

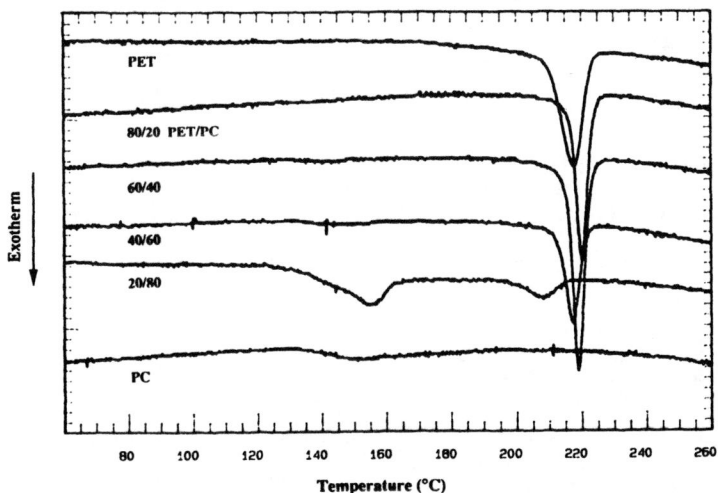

Figure 4: DSC cooling scans (rate = 5 °C/min) showing dynamic melt crystallization in PET and PET/PC blends.

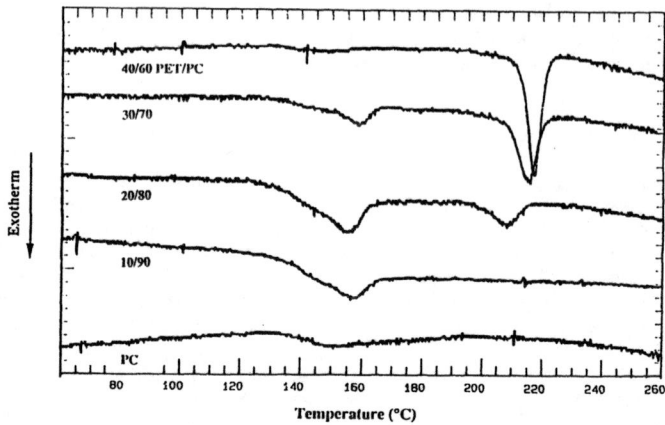

Figure 5:   DSC cooling scans (rate = 5 °C/min) showing dual peak behavior in
the dynamic melt crystallization of PET/PC blends with low PET content.

CONCLUSIONS

An increase in T_g of PET and decrease in T_g of PC in blended samples over pure polymer have been observed. These shifts may be the result of glass-rubber interactions or of transesterification. For blends containing more than 20 wt% PET, blending with PC results in an enhancement of isothermal PET crystallization rate from the melt. PET crystallization is strongly suppressed in blends containing 20 wt% PET. A depression in degree of crystallinity of PET is observed in blended PET as compared with pure PET, particularly for high PC content blends. These results reflect a less perfect final crystalline form of PET in the blends, which may be the result of transesterification or simply of an interference of the amorphous PC phase with the development of crystallinity. Strong effects of thermal history on blend crystallization kinetics were observed, particularly for higher melt temperatures. In dynamic crystallization experiments from the melt, for PET compositions between roughly 10% and 30% in the blend, two crystallization exotherms at approximately 210 °C and 155 °C are observed. The high temperature exotherm appears to be related to normal PET crystallization. However, the lower temperature exotherm (just above T_g of PC) increases in magnitude with PC content and may reflect a co-crystalline form of PET and PC.

The authors acknowledge the General Electric Company for providing the blend samples used in this study.

ADDED IN FINAL PROOF

The authors have recently discovered that the blend samples examined in this study contain an antioxidant and transesterification inhibitor not present in the 100% PET sample. It is possible that some of the effects observed in the blends are the result of these additives. Studies are now underway of PET containing these additives and will be reported soon.

REFERENCES

1. B.D. Hanrahan, S.R. Angeli, and J. Runt, Polym. Bull., **15**, 455 (1986).
2. Z.H. Huang and L.H. Wang, Makromol. Chem., Rapid Commun., **7**, 255 (1986).
3. R.S. Porter and L.-H. Wang, Polymer, 33, 2019 (1992).
4. T. Suzuki, H. Tanaka, T. Nishi, Polymer, 30, 1287 (1989).
5. S.R. Murff, J.W. Barlow, and D.R. Paul, J. Appl. Polym. Sci., **29**, 3231 (1984).
6. V.M. Nadkarni, V.L. Shingankuli and J.P. Jog, J. Appl. Polym. Sci., **46**, 339 (1992).
7. E. Martuscelli, Polym. Eng. Sci., **24**, 563 (1984).
8. X.-Y. Chen and A.W. Birley, British Polym.J., **17**, 347 (1985).
9. A.W. Birley and X.-Y. Chen, British Polym.J., **16**, 77 (1984).
10. A. Mehta, U. Gaur, B. Wunderlich, J. Polym. Sci., Polym. Phys. Ed., **16**, 289 (1978).

SMALL-ANGLE SCATTERING INVESTIGATIONS OF CRYSTALLINE BLEND MORPHOLOGIES OF POLY(ε-CAPROLACTONE) (PCL)/POLYCARBONATE (PC) BLENDS

Y. WILSON CHEUNG*, R.S. STEIN*, G.D. WIGNALL** AND J.S. LIN**
*University of Massachusetts, Department of Polymer Sci. and Eng., Amherst, Mass. 01003
**Oak Ridge National Laboratory, Solid State Division, Oak Ridge, TN. 37831

ABSTRACT

Crystalline morphologies of poly(ε-caprolactone) (PCL) and deuterated polycarbonate (PC) blends in both the semicrystalline/amorphous state and semicrystalline/semicrystalline state were probed by small-angle neutron and X-ray scattering (SANS and SAXS). Due to the different contrast between the phases for neutrons and X-rays, SANS exhibited a monotonic drop in intensity with increasing scattering angle while SAXS showed lamellar (peak) scattering.

Crystal- and amorphous-phase thickness were determined from the correlation function calculated from SAXS. This correlation function analysis suggested a transition from interlamellar exclusion to interlamellar incorporation of amorphous PCL in the PC lamellae. A two-correlation length model provided an excellent fit for the SANS data over the entire composition range. This model not only reproduced the shape but also the absolute magnitude of the scattering profiles. The long range correlation length ($\sim 10^2$ Å) and the short range correlation length (~ 10 Å) derived from this model were inferred to be associated with the crystalline PC domains and the local clusters found in the amorphous phase, respectively.

INTRODUCTION

The phase behavior of the PCL/PC blend system is extremely complicated and enormously rich.[1-3] The PCL crystallization kinetics in the PCL-rich blends have been investigated with both DSC[4] and synchrotron small-angle X-ray scattering (SAXS).[5] The focus of this paper is to elucidate the structure of the amorphous phase and the crystalline morphologies of the deuterated (d-PC)/PCL blends in the semicrystalline/amorphous (above T_m of pure PCL) state. Detailed discussion of the semicrystalline/semicrystalline blend morphologies as revealed by SAXS and small-angle light scattering (SALS) is described in a separate paper.[6]

Small-angle neutron scattering (SANS) is a powerful technique for probing polymer blend morphology. Unique to SANS is the specificity with which the labeled component can be examined. Application of SANS and SAXS for studying polymer blend morphology has been demonstrated to be highly effective in yielding unambiguous structural information.[7-8] This synergism comes about because the contrast mechanism in SAXS is very different from that in SANS. SAXS arises from electron density differences (e.g. amorphous and crystalline region), and whereas SANS arises from scattering length density (SLD) differences. The latter are dominated by the substitution of deuterium for hydrogen, due to the change of sign in the scattering length (b_D = 0.667X10^{-12} cm; b_H = -0.374X10^{-12} cm). SANS probes the spatial arrangement between deuterium and hydrogen or d-PC and PCL since PC is completely deuterated and whereas SAXS probes the crystalline/amorphous region.

THEORY

SAXS: Traditionally, the SAXS profiles are analyzed by: 1) computing the long period or the interlamellar spacing directly from the scattering curve, and 2) evaluating the electron density correlation function[9] from which the amorphous- and crystal-phase thickness may be obtained. The long period is calculated from the equation $L = (2\pi/q^*)$, where q^* is the peak value found in the Lorentz-corrected $q^2I(q)$ versus q plot. The normalized one-dimensional correlation can be calculated from the scattered intensity $I(q)$ by the following equation[9]

$$\gamma(z) = \frac{1}{\gamma(0)} \int_0^\infty q^2 I(q) \cos(qz) dq$$

(1)

where z is along the direction from which the electron density distribution is measured.

SANS: In the case of a highly complex heterogeneous system, the two-correlation length model[10] based on an extension of the Debye-Bueche random two phase model[11] is often used and is described below:

$$\gamma(r) = f \exp\left(-\frac{r}{a_1}\right) + (1-f) \exp\left(-\frac{r^2}{a_2^2}\right)$$

(2)

where a_1 is termed the short range correlation length and a_2 is the long range correlation length. The parameter f is defined as the fractional contribution of the exponential correlation function. The intensity distribution based on the above correlation function yields a sum of a Lorentzian and a Gaussain function as shown below:

$$I(q) = \frac{A_1}{(1+q^2 a_1^2)^2} + A_2 \exp\left(-\frac{q^2 a_2^2}{4}\right)$$

(3)

where A_1 and A_2 can be computed independently from the scattering length density difference.[10] In order to confirm the applicability of the above model, the zero angle scattering intensity derived from the calculation is compared to that extrapolated from the SANS profile.

EXPERIMENTAL

Sample Preparation: Both PCL and PC were obtained from Scientific Polymer Products. The deuterated PC was synthesized by solution polymerization of deuterated bisphenol with phosgene at 0 °C in methylene chloride.[12] The molecular weights of the homopolymers are summarized in Table I. Samples were prepared by dissolving the two homopolymers in methylene chloride. The blend solution was then cast on an aluminum surface. The samples were first dried under ambient conditions overnight and then vacuum dried at 90 °C for two days to ensure complete removal of residual solvent. Scattering samples were typically compression molded at T_g+50 °C under vacuum. Samples were then rapidly transferred to a metal surface and quenched to room temperature. Sample disc dimension was about 1 mm in thickness and 15 mm in diameter.

Thermal Analysis: Thermal transitions of the blends were measured with a Dupont-10 differential scanning calorimeter at a heating rate of 20 °C/min. The glass transition temperatures were obtained from the second heating scan and determined from the midpoint of the change in heat capacity. The melting endotherms of both components were recorded from the first heating scan and the melting point was obtained from the peak temperature of the endotherm.

Small-angle Scattering: Both SANS and SAXS measurements were performed at the Oak Ridge National Laboratory (ORNL) in Tennessee.[12] SAXS experiments were performed on the ORNL 10m spectrometer[13] operating at an accelerating voltage of 40 KeV and a current of 100 mA. The instrument was operated with a sample-detector distance of 5.17 m using $Cu_{k\alpha}$ radiation (λ = 1.54 Å) and a 20X20 cm^2 area detector with cell (element) size ~ 3 mm. SANS experiments were conducted on the W.C. Koehler 30m spectrometer[14] operating at a sample-detector distance of 10 m. The neutron wavelength was 4.75 Å ($\Delta\lambda/\lambda$ = 5%) and the source (3.5 cm dia.) and sample (1.0 cm dia.) slits were separated by a distance of 7.5 m.

Table I

Molecular weights for poly(ε-caprolactone) (PCL), hydrogeneous polycarbonate (h-PC), and deuterated polycarbonate (d-PC).

Polymer	M_W	M_n	M_W/M_n
PCL	23,700	16,500	1.44
h-PC	23,100	36,400	1.57
d-PC	134,000	15,600	8.59

RESULTS AND DISCUSSION

The phase transitions for the d-PC/PCL blends are summarized in Figure 1. In accordance with previous results,[1-3] this blend system is miscible in the amorphous phase, as demonstrated by the presence of a single glass transition temperature T_g, over the entire composition range. The composition dependence of T_g and the phase behavior of the blends will be critically analyzed in a forthcoming publication.[15] Due to solution-induced crystallization and annealing effects, the d-PC was semicrystalline in all the blends. In the case of the PCL-rich blends, both components were semicrystalline.

Correlation function analysis on the SAXS profiles provides very localized morphological information regarding the organization of the lamellae. The SAXS profiles and the corresponding correlation function results are described in reference 6. The interlamellar spacing, crystal- and amorphous-phase thickness for the solution cast (d-PC/PCL) blends are shown in Figure 2. For the PCL-rich blends, the long period and the associated morphological parameters were almost independent of composition. The invariance of the amorphous-phase thickness with composition indicated that PCL was rejected from the interlamellar region. In contrast, the amorphous-phase thickness for the PC-rich blends increased with increasing PC whereas the crystal-phase thickness remained fairly constant. This phenomenon suggested that PCL was incorporated within the inter-lamellar region resulting in an increase in the amorphous-phase thickness.

Interlamellar exclusion is always entropically favorable provided the amorphous component has sufficient mobility to diffuse away from the crystalline lamellae. However, interlamellar inclus-ion occurs when either the amorphous component is highly immobile or the specific interactions are so favorable that the enthalpic term dominates the entropy contribution. On the basis of this reasoning, the mode of interlamellar arrangement was postulated to be governed by the competition between entropy and mobility.

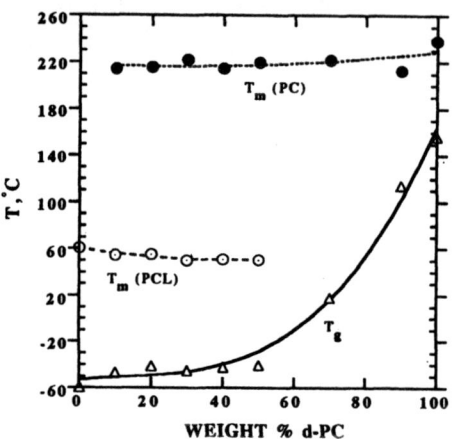

Figure 1. Thermal Transitions for d-PC/PCL Blends

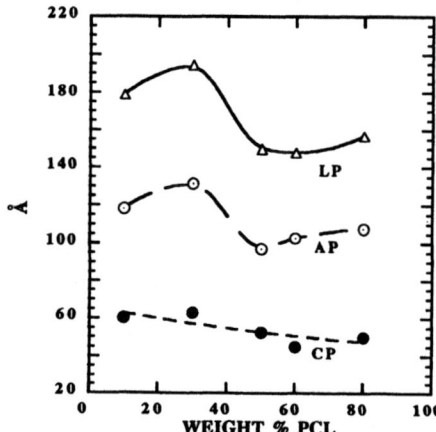

Figure 2. Correlation function results for d-PC/PCL blends in the semi-crystalline/amorphous state. LP = PC long period, CP = PC crystalline-phase thickness, AP = PC amorphous-phase thickness.

The two-correlation length model yielded a very good fit to the SANS data. A typical fit of Eqn. 3 to the SANS data is illustrated in Figure 3. More detailed discussion of the two-correlation length analysis may be found in reference 12. The short range and long range correlation length derived from the two-correlation length analysis are shown in Figure 4. Based on the magnitude and the composition dependence of the correlation length, one may associate the long range correlation length to be the PC crystalline domain and the short range correlation length as the correlation distance between d-PC and PCL in the amorphous phase. The long period of PC measured from SAXS is also plotted in this figure. Both the long range correlation length and the long period exhibit identical composition dependence and thereby further supports the applicability of this two-correlation length model and our interpretation.

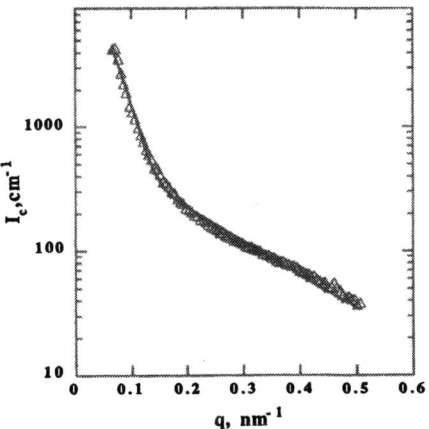

Figure 3. A two-correlation model fit for a 50%d-PC/50%PCL blend. The solid line indicates the fit.

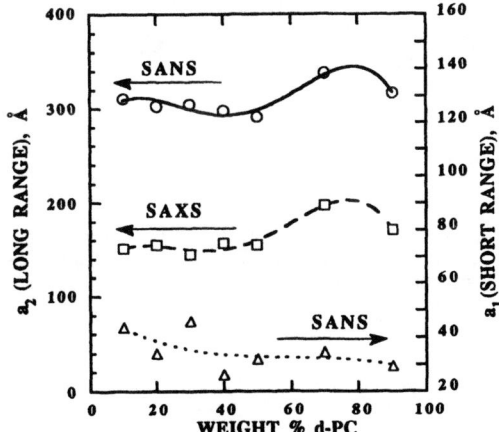

Figure 4. Long range and short range correlation length derived from SANS and the PC long period measured from SAXS for dPC/PC blends.

553

In the theoretical framework of the Debye-Bueche model, the system is assumed to be phase separated where the size of the heterogeneity is measured by the correlation length. On the basis of the SAXS results, the long range correlation length was attributed to the d-PC crystalline domain. A cursory examination of Figure 4 reveals that the long range correlation length is about two times larger than the long period. This discrepancy may be related to crystallization-induced phase separation where there exists a d-PC rich region, near the crystalline d-PC, within which the PCL is excluded. By a similar argument, the short range correlation length could be related to the local cluster found in the amorphous phase. This short range clustering could also result from localized phase separation. It is well know that crystallization is an ordering process and could be classified as a form of phase separation. As the blend undergoes crystallization, preferential enrichment of the crystallizing component over the other component may occur in the vicinity of the incipient crystal nuclei. These heterogeneities may develop into local clusters rich in the crystallizing component. Since thermal analytical techniques such as DSC and DMA have spatial resolution of only about few hundred Angstroms, these local clusters, roughly 30Å in size, will not be detectable and the amorphous phase may still appear as homogeneous based on the thermal measurements.

REFERENCES

1. C.A. Cruz, J.A. Barlow, D.R. Paul, Macromolecules **12**, 726 (1979).
2. M.M. Coleman, P.C. Painter, J. Appl. Spect. Reviews **20(3&4)**, 255 (1984).
3. J.M. Jonza, R.S. Porter, Macromolecules **19**, 1946 (1986).
4. Y.W. Cheung, R.S. Stein, in preparation.
5. Y.W. Cheung, R.S. Stein, B. Chu, G.W. Wu, Macromolecules, submitted.
6. Y.W. Cheung, R.S. Stein, J.S. Lin, G.D. Wignall, Macromolecules, in press.
7. T.P. Russell, H. Ito, G.D. Wignall, Macromolecules **21**, 1703 (1988).
8. G.D. Wignall, M. Farrar, S. Morris, J. Mater. Sci. **25**, 69 (1990).
9. G.R. Strobl, M. Schneider, J. Polym., Sci. Polym. Phys. Ed.**18**, 1343 (1980).
10. M. Moritani, T. Inoue, H. Kawai, Macromolecules **3(4)**, 433 (1970).
11. P. Debye, H.R. Anderson, H. Brumberger, J. Appl. Phys. **28**, 6 (1957).
12. Y.W. Cheung, R.S. Stein, G.D. Wignall, H.E. Yang, Macromolecules **26**, 5365 (1993).
13. G.D. Wignall, J.S. Lin, S. Spooner, J. Appl. Cryst. **23**, 241 (1990).
14. W.S. Dubner, J.M. Schultz, G.D. Wignall, J. Appl. Cryst. **23**, 469 (1990).
15. Y.W. Cheung, R.S. Stein, in preparation.

ACKNOWLEDGMENTS

Research is supported in part by Novacor Inc. and also by the division of Materials Sciences, U.S. Department of Energy under Contract No. DE-AC05-84OR21400 with Martin Energy Systems Inc. The authors thank H.E. Yang and P. Yazobucci of Eastman Kodak Co. for the deuterated polycarbonate synthesis.

SMALL ANGLE X-RAY SCATTERING
OF CRYSTALLINE POLYMER BLENDS

PETER P. HUO[#,*] PEGGY CEBE[#,‡] AND MALCOLM CAPEL[##]
Department of Materials Science and Engineering, Room 13-5082, Massachusetts Institute of Technology, Cambridge, MA 02139
Biology Department, Brookhaven National Laboratory, Upton, NY 11973

ABSTRACT

Binary blends of a semicrystalline polymer, poly(butylene terephthalate), PBT, with an amorphous polymer, polyarylate, PAr, were studied for a wide range of compositions. Small angle x-ray scattering, thermal and mechanical analyses were conducted to determine the location of the uncrystallizable PAr relative to the crystal lamellae of PBT, for blends crystallized by heating from the quenched state. Comparison of experimental and calculated long periods indicates that interlamellar PAr placement becomes less probable as the PAr composition increases. When PBT crystallizes in these blends, most PAr is rejected from the regions near the crystal surfaces when PAr fraction exceeds about 0.40.

INTRODUCTION

Polymer blends have been the focus of research for several decades, due to the possibility of obtaining property enhancement over that of the constituent homopolymers. Several important commercial binary blends comprise a crystallizable homopolymer and a non-crystallizable homopolymer [1-9]. The morphology of the crystals and the nature of the amorphous regions will both contribute to the ultimate properties of these complex materials. Our research group has been studying blends of poly(butylene terephthalate), PBT, with polyarylate, PAr, using x-ray scattering [10], solid state ^{13}C NMR[11], and thermal analysis [12]. PBT crystallizes very rapidly and imparts solvent resistance to the blend. PAr improves the mechanical properties by increasing the glass transition temperature. The chemical structures of the homopolymers are shown below:

Poly(butylene terephthalate), PBT   Polyarylate, PAr

PBT structure contains no bulky side groups, and the PBT chains can pack together to form crystals of the triclinic system [13-17]. PBT exists in two crystal modifications. Alpha phase, which is the phase that grows from melt crystallization, has the methylene sequence conformation GGTTGG. Beta phase, which is usually seen under conditions of applied stress[18,19], has extended, all trans methylene conformation. The PAr polymer contains two bulky methyl side groups that hinder interchain packing. The chain regularity is further disrupted by the 1:1 ratio of isophthalic (meta-phenyl) and terephthalic (para-phenyl) linkages of the monomer, resulting in a random copolymer that is unable to crystallize.

PBT/PAr blends have been shown to be miscible in the amorphous state at all composition ratios [20-23]. From the quenched state, as temperature increases above the glass transition, PBT

---

‡ To whom correspondence should be addressed
* Present address: W. R. Grace Corp., Analytical Division, Columbia, MD

Mat. Res. Soc. Symp. Proc. Vol. 321. ©1994 Materials Research Society

crystallizes and a phase separated system results[21,22]. A sketch of the intermediate stages of PBT crystal growth is shown in Figure 1. The location of the non-crystallizable PAr component will be determined by the kinetics of crystallization which is dominated by the rate of formation of secondary nuclei [24]. We would expect the crystal growth to be affected by the local molecular environment in the melt state. If a PBT primary nucleus is surrounded by a sufficiency of mobile PBT chains, we suggest that crystallization will be rapid, and there will be too little time for the PAr to diffuse away from the crystals. In this case, we expect to find the PAr located in interlamellar regions, which are designated as (1) in Figure 1. However, if a primary PBT nucleus is surrounded mostly by PAr chains, crystal nucleation will be hindered, growth rate will be reduced, and PAr may become phase separated into interfibrillar (2), or interspherulitic (3) regions [25].

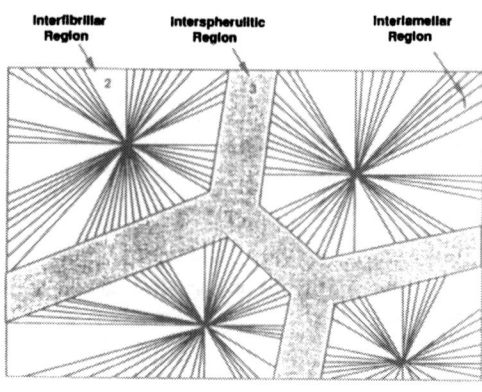

Figure 1: Intermediate stage of PBT crystal growth showing possible locations of PAr: (1) interlamellar, (2) interfibrillar, and (3) interspherulitic.

EXPERIMENTAL SECTION

Poly(butylene terephthalate) was obtained from Polysciences. PAr was obtained from Amoco, and was a random copolymer with a 1:1 ratio of isophthalic (meta-phenyl) and terephthalic (para-phenyl) moieties. Blends were prepared in a wide composition range, by coprecipitation from phenol/tetrachloroethane into methanol, according to the method of Kimura, et al.[20]. Blend powder was recovered by filtration, dried in a vacuum oven, and compression molded into sheets. Molding under pressure was done at 250°C for 60-90 sec. after which time the blends were quenched into ice water. Small areas were cold crystallized by heating from room temperature to 180°C and holding for one-half hour. As described previously [11,12], we used abusive DSC and NMR studies to determine that no transesterification occurred in blends prepared this way.

Thermal analysis was performed using a Perkin Elmer DSC-4 with scan rate of 20°C/min. Calibration of temperature and heat flow was made using Indium. The degree of crystallinity of the blends was determined from the total endothermic area using 34.0 cal/g [26] as the heat of fusion of the perfect crystal. Wide angle x-ray scattering (WAXS) studies were performed in reflection geometry using a Rigaku RU300 generator with Cu Kα radiation, and a diffracted beam graphite monochromator. The scan rate was 1°/min with a step scan interval of 0.02°.

Small angle x-ray scattering studies were performed at the Brookhaven National Synchrotron Light Source. The system was equipped with a two-dimensional position sensitive histogramming detector. Sample to detector distance was about 120 cm, calibrated using collagen fiber. X-ray wavelength was 1.28Å. Intensity patterns were isotropic, and circular integration was used to increase the signal to noise ratio.

RESULTS AND DISCUSSION

Figure 2a shows the endothermic heat flow from 150°C to 250°C for the semicrystalline blend films. No crystallization exotherm was seen at lower temperature during DSC scanning, indicating the blends were fully crystallized at 180°C. Nearly all the blends exhibit multiple melting endotherms. For 100/0, 80/20, and 60/40 there is a small endotherm just above the cold crystallization temperature, followed by a dominant higher temperature endotherm. Blend 40/60 has a double endotherm with equal peak heights, and 20/80 has a single endotherm. The location

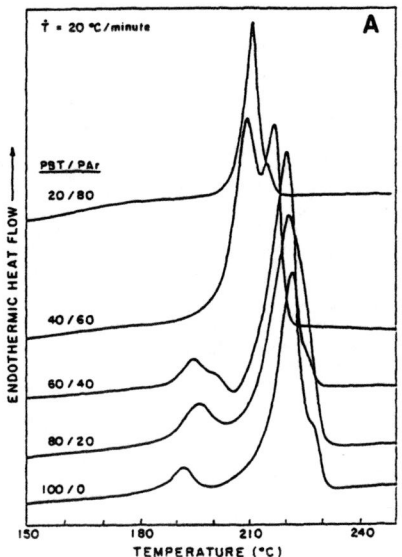
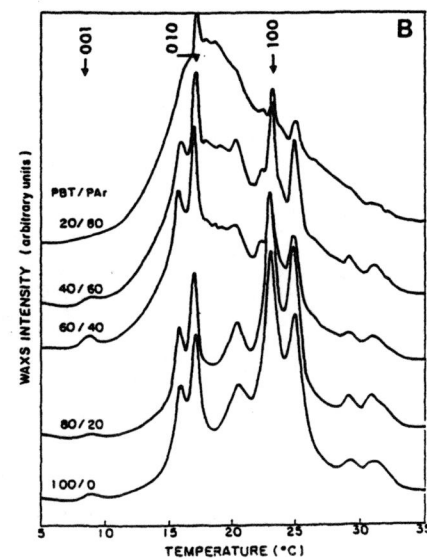

Figure 2: a.) Endothermic heat flow vs. temperature, and b.) WAXS intensity vs. scattering angle for cold crystallized PBT/PAr blends.

of the uppermost endotherm shifts to lower temperature as the PAr fraction increases. Based on the similar DSC thermograms, we may group the 100/0, 80/20, and 60/40 blends together. Results are consistent with a reorganization/recrystallization model for cold crystallized PBT when this material is heated above its prior treatment temperature [23,27-34].

In Figure 2b the WAXS intensity vs. scattering angle, 2θ, is shown for the cold crystallized materials. Six sharp crystal reflections are seen in PBT/PAr blends, and all are indexed to the triclinic alpha phase unit cell [13-17]. As the PAr content increases, the degree of crystallinity decreases based on comparison of the crystal peak area to the total area. In 20/80, the large amorphous halo is dominated by the PAr amorphous scattering. In spite of the large PAr content, the PBT can still form crystals whose full width at half maximum is not significantly different from PBT homopolymer.

The dynamic mechanical loss tangent has been reported previously for these blends[11]. All semicrystalline films showed two relaxation peaks, which are associated with two glass transitions. The lower peak shifted to higher temperature as the PAr fraction increased, while the uppermost peak was close to the glass transition temperature of PAr homopolymer. Two glass transitions suggest that there is amorphous phase heterogeneity in the cold crystallized blends. One amorphous region (lower Tg) contains both amorphous PBT and PAr, and this Tg shifts with increasing PAr fraction. Another amorphous region (higher Tg) contains a large proportion of PAr (probably also some amorphous PBT) and is relatively insensitive to PAr fraction. Though the observation of two glass transitions suggests an amorphous phase heterogeneity, the dynamic mechanical analysis alone can not be used to determine the location of the PAr. From Figure 1, any of the three placements of PAr could give a double glass transition. Tg relaxations may come from: PBT interphase near the crystal surfaces, interlamellar PBT and PAr farther from the crystal surfaces, and larger scale (interfibrillar or interspherulitic) phase separated regions. Even in PBT homopolymer, the mobile amorphous and the rigid amorphous phases may give different Tgs [27].

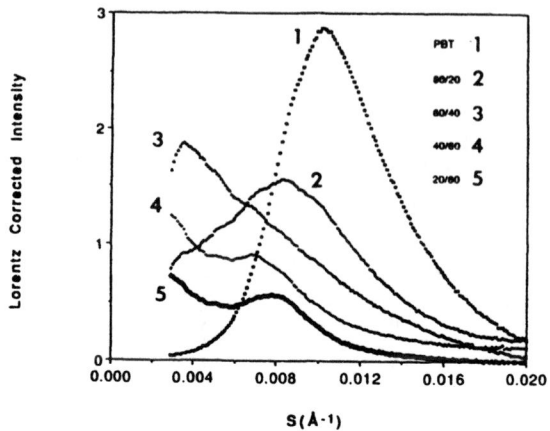

Figure 3: Lorentz corrected intensity vs. scattering vector,
s (s = 2sinθ / λ) for cold crystallized PBT/PAr.

We used SAXS to elucidate the structure of these blends by comparing the measured long period to that calculated on the assumption of an interlamellar placement of PAr. Figure 3 shows the Lorentz corrected intensity vs. scattering vector, s (s=2sinθ/λ). PBT homopolymer and blend 80/20 have intensities which decrease on both sides of the maximum. For blends 60/40, 40/60, and 20/80 the peak position does not change (only a weak shoulder is seen in 60/40) but the intensity increases as s--->0. The long period, L, obtained from the intensity maximum as a function of blend composition is shown in Figure 4 by the open triangles. The long period for PBT homopolymer is 100Å, while for the blends L ranges from 120-150Å.

In Table 1, we list the measured L from application of Bragg's Law. Next, $w_{c,p}$, the partial mass fraction crystallinity, is found from the total area under the DSC endotherm. l_c is determined from Scherrer analysis [35] of the width of the (001) reflection after corrections for instrumental broadening (20/80 was not analyzed this way because of difficulty in the peak separation). The last entry is the long period calculated on the assumption of an interlamellar placement of PAr. If all PAr were arranged in region (1) (see Figure 1) relative to PBT crystals, the long period L could be calculated directly from $L=l_c/v_c$, where l_c is the thickness of the crystal and v_c is the total volume

Table 1: Long period, L, partial mass fraction crystallinity, $w_{c,p}$, and lamellar thickness, l_c, for PBT/PAr blends.

Sample	L# (Å)	$w_{c,p}$	l_c‡ (Å)	L* (Å)
100/0	100	0.45	40	100
80/20	120	0.49	40	120
60/40	150	0.43	40	180
40/60	140	0.40	40	300
20/80	130	0.26	- - -	- - -

#	Measured by application of Bragg's Law to SAXS intensity maximum
‡	Determined from width of WAXS (001) reflection using Scherrer equation [35]
*	Calculated assuming an interlamellar placement of PAr
- - -	Not measured

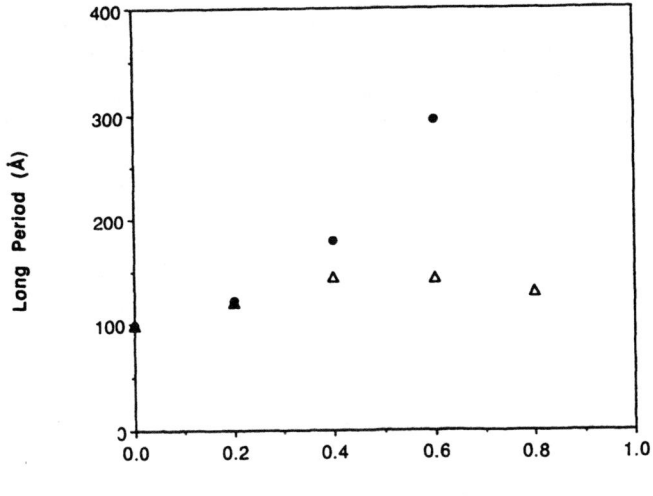

Figure 4: Long period, L, as a function of PAr composition
for cold crystallized PBT/PAr blends.

fraction crystallinity. If $L < l_c/v_{c,t}$, then at least some PAr is located in interfibrillar or interspherulitic regions. Here the total volume fraction of crystallinity is obtained by the following:

$$\frac{1}{v_{c,t}} = 1 + (\frac{1-w_{c,p}}{w_{c,p}})(\frac{\rho_1}{\rho_2}) + (\frac{x}{1-x})(\frac{\rho_1}{\rho_3})\frac{1}{w_{c,p}} \tag{1}$$

where ρ_1, ρ_2, ρ_3 are the densities of PBT crystal (1.40 gram/cc), amorphous PBT (1.28 gram/cc) and PAr (1.21 gram/cc) [20], respectively. x is the mass fraction of PAr in the blends, and $w_{c,t}$, $w_{c,p}$ are the total and partial mass fraction of crystallinity, related by $w_{c,t} = w_{c,p}(1-x)$. We see from Figure 4, and Table 1 that the departure from interlamellar PAr placement begins at about 60/40 and as PAr fraction increases, the interlamellar placement becomes less probable. While we could not measure the (001) peak width for 20/80, on the assumption that 20/80 has the same c* as the other blends, we find a calculated L of 980Å. The large difference between the measured and calculated L makes an interlamellar structure for 20/80 unlikely.

CONCLUSIONS

Even at low compositions, PBT crystallizes in blends with PAr, forming phase-separated regions. The addition of the non-crystallizable component results in a decrease in crystallization kinetics and reduced partial degree of crystallinity for x>0.20. As a function of composition, differences in melting behavior and crystal perfection are seen from DSC and WAXS analyses, respectively. Dynamic mechanical analysis [10] also showed two glass transition relaxations. However, none of these methods can provide direct information about the location of the PAr fraction. Using small angle x-ray scattering, we find a large scattering intensity as s--->0 in blends with a large PAr component. We conclude that this scattering arises from large-scale structure interference, either spherulites or fibrillar bundles. In addition, we see strong coherent scattering from lamellar stacks. We calculated the long period that would result if the PAr placement were strictly interlamellar. This long period is only close to the measured long period for blends with low PAr fraction. As the PAr fraction increases, an interlamellar placement becomes less probable. For blends with PAr > 0.40 interfibrillar or interspherulitic placement is favored. In another similar system the segregation of amorphous polymer during crystallization from a miscible melt has been interpreted by considering the relationship between the linear growth rate of the lamellae and the mutual diffusion coefficient between the components [36]. A similar study of the PBT/PAr system is currently underway.

ACKNOWLEDGEMENTS

Research was supported by U. S. Army Contract DAAL03-91-G-0132. Research was carried out in part at the National Synchrotron Light Source, Brookhaven National Laboratory.

REFERENCES

1. T. Nishi, T. T. Wang. *Macromolecules*, **8**, 909 (1975).
2. T. T. Wang, T. Nishi. *Macromolecules*, **10**, 421 (1977).
3. P. B. Rim, J. P. Runt. *Macromolecules*, **17**, 1520 (1984).
4. B. Hahn, J. Wendorff, D. Y. Yoon. *Macromolecules*, **18**, 715 (1985).
5. B. R. Hahn, O. Herrman-Schonherr, J. H. Wendorff. *Polymer*, **28**, 201(1987).
6. J. P. Runt, C. A. Barron, X. Zhang, S. K. Kumar. *Macromolecules*, **24**, 3466 (1991).
7. R. S. Porter, L. H. Wang. *Polymer*, 33, 2019 (1992).
8. P. Tekely, F. Laupretre, L. Monnerie. *Polymer*, 26, 1081 (1985).
9. H. Ito, T. P. Russell, D. Y. Yoon. *Macromolecules*, **20**, 2213 (1987).
10. P. P. Huo, P. Cebe, M. Capel. *Macromolecules*, **26**, 4275 (1993).
11. P. P. Huo, P. Cebe. *Macromolecules*, **26**, 3127 (1993).
12. P. P. Huo, P. Cebe. *Macromolecules*, **26**, 5561 (1993).
13. Z. Mencik. *J. Polym. Sci., Polym. Phys.*, **13**, 2173 (1975).
14. I. H. Hall, M. G. Pass. *Polymer*, **17**, 807 (1976).
15. I. J. Desborough, I. H. Hall. *Polymer*, **17**, 807 (1976).
16. B. Stambaugh, J. L. Koenig, J. Lando. *J. Polym. Sci., Polym. Phys.*, **17**, 1053 (1979).
17. P. Peszkin, J. M. Schultz. *J. Polym. Sci., Polym. Phys.*, **24**, 2617 (1986).
18. J. Roebuck, R. Jakeways, I. M. Ward. *Polymer*, 33(2), 227 (1992).
19. K. Nakamae, M. Kameyama, M. Yoshikawa, T. Matsumoto. *J. Polym. Sci., Polym. Phys.*, **20**, 319 (1982).
20. M. Kimura, R. S. Porter, G. Salee. *J. Polym. Sci.: Polym. Phys. Ed.*, **21**, 367 (1983).
21. J. P. Runt, X. Zhang, D. Miley, K. Gallagher, A. Zhang. *Macromol.*, **25**, 3902 (1992).
22. J. P. Runt, D. M. Miley, X. Zhang, K. P. Gallagher, K. McFeaters, J. Fishburn. *Macromol.*, **25**, 1929 (1992).
23. J. Runt, K. P. Gallagher. *Polym. Comm.*, **32**, 181 (1991).
24. J. D. Hoffman, G. T. Davis, J. I. Lauritzen. In "Treatise on Solid State Chemistry", vol.3, Hanay, N. B., Ed. (Plenum Press: New York, 1975).
25. G. Grevecoeur, G. Groeninckx. *Macromolecules*, **24**, 1190 (1991).
26. K. H. Illers. *Colloid Polym. Sci.*, **258**, 117 (1980).
27. S. Z. D. Cheng, R. Pan, B. Wunderlich. *Makromol. Chem.*, **189**, 2443 (1988).
28. J. T. Yeh, J. P. Runt. *J. Polym. Sci., Polym, Phys. Ed.*, **27**, 1543 (1989).
29. M. E. Nichols, R. E. Robertson. *J. Polym. Sci., Polym. Phys.*, **30**, 305 (1992).
30. M. E. Nichols, R. E. Robertson. *J. Polym. Sci., Polym. Phys.*, **30**, 755 (1992).
31. P. Cebe, J. S. Chung. *Polymer Composites*, **11**, 265 (1990).
32. J. S. Chung, P. Cebe. *Polymer.*, 33, 2312 (1992).
33. J. S. Chung, P. Cebe. *J. Polym. Sci., Polym. Phys.*, **30**, 163 (1992).
34. P. P. Huo, P. Cebe, M. Capel. *J. Polym. Sci., Polym. Phys.*, **30**, 1459 (1992).
35. M. Kakudo, N. Kasai. *"X-ray Diffraction by Polymers"* (Kodansha Ltd., Tokyo, 1972).
36. B. Hsiao, B. Sauer. *J. Polym. Sci., Polym. Phys.*, in press.

THE EVOLUTION OF SOL-GEL FILMS IN THE ENVIRONMENTAL SCANNING ELECTRON MICROSCOPE.

STUART MCKERNAN*, JOHN WRIGHT** AND LORRAINE F. FRANCIS.**
* University of Minnesota, High-Resolution Microscopy Center, Minneapolis, MN 55455
** Department of Chemical Engineering and Materials Science, University of Minnesota, Minneapolis, MN 55455.

ABSTRACT

The ability to study the surface texture of a sol-gel film of lead titanate as it is heated *in-situ* in an environmental scanning electron microscope is demonstrated. Measurements of the shrinkage of the film at cracks and around delaminations of several microns have been made. Grain sizes of the final crystalline microstructure were determined to be ~ 0.2 μm. The formation of bumps in the film 1-5 μm in size was observed. The presence of these features may correlate with the onset of crystallization.

INTRODUCTION

The formation of thin films using sol-gel technology involves many different steps. After the preparation of the sols, the substrates must be coated, usually by repeated spin coating and drying to build a sufficiently thick layer without cracks, but dip-coating is also used. The dried sol-gel films are calcined to produce an amorphous film of the correct composition, and may also be fired to crystallize the thin film if necessary. Since, in the majority of these stages the films are either in a hydrated state or are non-conducting ceramics, it is difficult to follow the formation of these films in the scanning electron microscope. The advent of the Environmental Scanning Electron Microscope (ESEM) has created the possibility of following the evolution of thin films *in situ* without the necessity of modifying the specimen surface to prevent charging artifacts. Using a hot stage in a controlled environment, the different processes of drying, calcining and crystallization of the thin film can be followed and recorded on videotape.

For these studies, the ESEM has two advantages over the more conventional SEM[1,2]. The microscope is operated with the specimen chamber at a much higher pressure in the ESEM, typically 1-10 torr, than in the typical pressures in a conventional SEM(~10^{-5} torr). A variety of gases can be used in the specimen chamber; water vapor being the most commonly used. When the sample is cooled to a few degrees C the water vapor above the sample is enough to saturate the environment, and the evaporation of water from aqueous gels may be controlled. The second advantage of the ESEM is the mechanism for signal collection from the sample. Secondary and backscattered electrons which are emitted from the sample when it is struck by the electron probe collide with molecules of the residual gas in the chamber, ionizing it. The gas pressure in the chamber controls the magnitude of the resulting electron cascade. An electrode above the sample at a positive potential accelerates the electrons upwards towards the electrode, producing a signal current. The positive ions are attracted to areas on the specimen surface with a high negative potential, and thus neutralize any charge build-up on the specimen. The specimen surface therefore does not need to be coated with a conductive layer to prevent specimen charging artifacts.

The results of processing a ceramic oxide sol-gel films in-situ in the ESEM are reported, and compared with more traditional analyses. The microstructure of thin films is an important characteristic, particularly where the film is to be only one step of a multilayer structure - such as might be fabricated in the electronics industry. The growth of subsequent layers on a sol-gel derived film may well be influenced by the microstructure of the film. The microstructural evolution of the lead titanate films used in this study is fairly well understood[3] and the results presented here serve to illustrate the usefulness of this technique in evaluating the microstructure of the thin films. This technique should be valuable in assessing the influence of the substrate, for example, or of the processing conditions, on the quality of the films. Particular emphasis will be given to the advantages and disadvantages of the technique.

EXPERIMENTAL

Sol-gel films of lead titanate were prepared from 1 molar $PbTi(OR)_6$ solutions with R = $CH_2CH_2OCH_3$ (methoxyethoxide). Details of the method are as given in Budd and Payne[4]. The hydrolyzed solutions were aged overnight and spun coated onto cleaned silicon substrates at 2000 rpm for 60 seconds. The substrates were dried on a 200°C hotplate for 60 seconds between coatings. Most of the films examined were composed of 4 successive spin coatings. The films were aged at room temperature for several days prior to being examined in the environmental scanning electron microscope.

The Electroscan E3 ESEM was used for the heating study of the sol-gel films. The heating stage has a range from room temperature to ~1000°C. Cleaved fragments of the substrates (with air-dried sol-gel films) were placed in an alumina crucible in the hot stage. For temperatures of a few hundred degrees the heat shield was not installed to give the maximum field of view on the specimen. The actual temperature of the sol-gel film may have been slightly overestimated under these conditions. For higher temperature experiments the heat shield was installed to achieve the highest temperatures.

The films were imaged at 20 kV, which was found to give an acceptable compromise between surface sensitivity and spatial resolution, in an atmosphere of ~ 4 torr water vapor.

Drying experiments were attempted using a Peltier-cooled stage to minimize evaporation of the solvents from the sol. However the initial pump-down of the specimen chamber appears to be too hostile a change for the sol to survive. More informative results may be obtained from *in situ* deposition experiments, such as those described by Chiu et. al.[5]

RESULTS

The initial experiments were performed on relatively thick films prepared by depositing a drop of the sol on the substrate in the hot stage, and "vacuum-drying" it in the microscope. Under these circumstances the films were heavily cracked, and macroscopic features such as crack growth and the shrinkage of delaminated portions of the film were readily followed in the

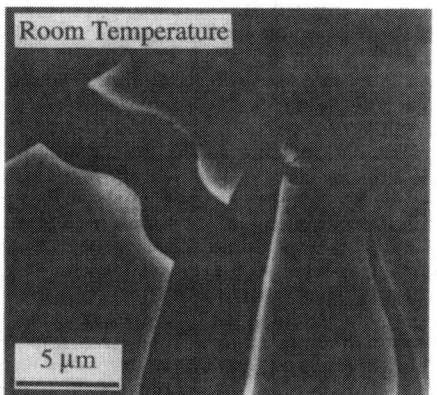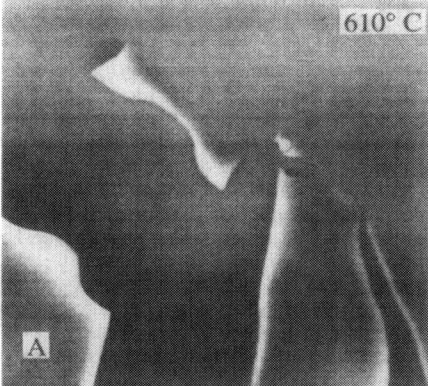

Figure 1.   Region of cracked and partly delaminated lead titanate film. The motion of the isolated island of film (A) and the width of the peninsula to the right was recorded as the temperature was raised from room temperature to 610°C.

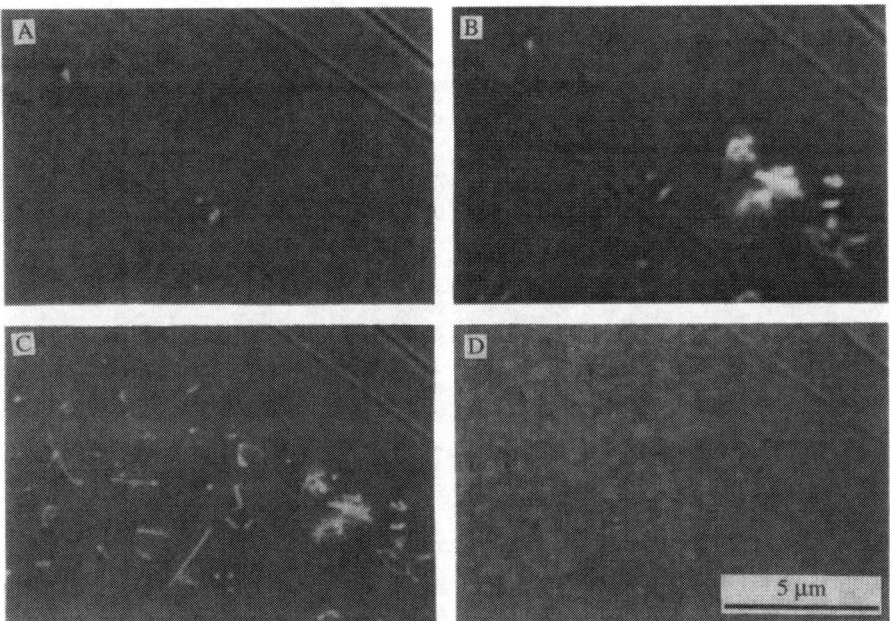

Figure 2.   a) Initial surface of aged PT sol-gel film, b) and c) same area as a) after increasing doses of electron beam bombardment. d) same area after heating to 1000°C. Note the mottled appearance of the film, and the presence of small dark pores.

microscope. Figure 1 shows a pair of images showing the gross change in morphology of the film between room temperature and ~600 °C. The gap between the bulk film and the separated fragment A has grown from 4.3 at room temperature to 5.7 μm at 80°C and 9.3 μm at 600°C. The shrinkage of isolated regions of the film, whilst smaller in magnitude, is also measurable. Small, dark spots seen at A appear at, higher magnifications, to be pores in the film approximately 0.2 μm in diameter.

Figure 2 is part of a sequence of images recorded as the films were heated to 1000°C showing the formation of similar pores. This region of the film was used for the initial alignment and focusing of the microscope (due to the presence of some surface contaminants). As a result some beam damage is evident in figure 2b and c, which were recorded at room temperature. Figure 2d shows the same area after heat treatment. The majority of the film now shows a speckled contrast on a scale of 0.1-0.2 μm, but the damaged area contains many small, dark spots which appear to be pores in the film. These pores were not observed in other areas of the film which had received a lower electron dosage. The appearance of the background contrast, which presumably results from the crystallization of the film, did not appear to be any different in the damaged region than elsewhere.

Noticeable differences in the surface texture of the film were observed by 700°C. The originally smooth surface was observed to develop the isolated contrast features approximately 5 μm in diameter shown in figure 3. Samples cooled from this temperature in the ESEM were carbon coated and examined in the SEM, the field of view is not the same in the two sets of images. The same morphological features were observed, although the directionality of the imaging system in the SEM allowed them to be more readily identified as raised bumps than was the case in the ESEM. Apart from the increase in surface contaminants in the SEM images, pores are visible in the heat-treated film which range in size from 1-3 μm.

ESEM image   10 μm

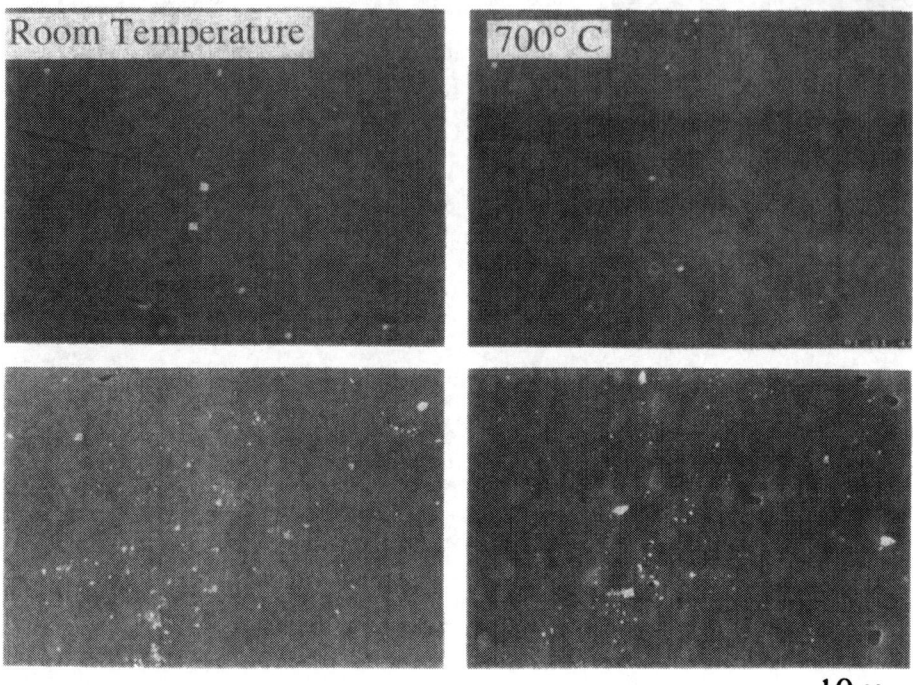

SEM image   10 μm

Figure 3.   Comparison of ESEM (top) and conventional SEM (bottom) images. The SEM images are from the same specimens as the ESEM images, which have been carbon coated prior to examination in the SEM. The surface morphology is very similar in both the room temperature (left) and 700°C (right) comparisons.

X-ray diffraction data taken from similar lead titanate films heat-treated in a furnace at 750°C for 15 minutes is shown in figure 4. The peaks indicate the presence of crystalline perovskite and pyrochlore in the films.

DISCUSSION

The ESEM enables aged sol-gel films to be studied as they are heated. Gross defects such as cracks and areas where the film has delaminated were easily studied, and their size and rate of growth were readily determined in real-time, particularly with the aid of a videotape recorder. Not surprisingly most of the growth occurred at lower temperatures, as most of the solvent evaporated from the film. Since the detection system of the ESEM depends on both the pressure and nature of the residual gas in the specimen chamber, sudden evolutions of material (gas) from the film result in a temporary change in contrast of the image. Changes in crack growth, for example, may then be correlated with any such events recorded on the videotape.

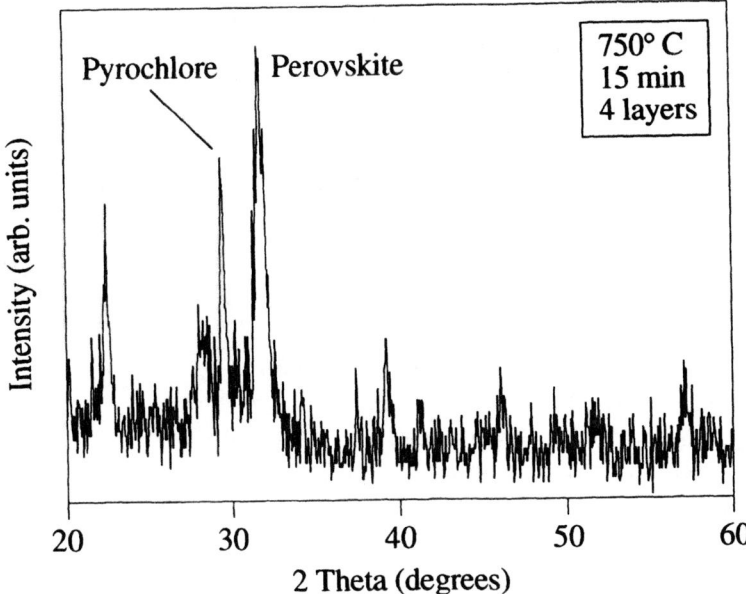

Figure 4.   XRD spectrum from lead titanate sol-gel derived film, heated at 750°C for 15 min. Peaks corresponding to the presence of crystalline perovskite and pyrochlore are indicated.

The lack of coating means that specimens can be imaged continuously while heating, and changes in surface textures examined as they occur, rather than making a sequence of films quenched from different temperatures. X-ray diffraction spectra obtained from similar lead titanate films heated to 750°C show that at least a portion of the film had crystallized forming both perovskite and pyrochlore structures. The surface texture shown in figure 3 may arise as a result of the crystallization of the film.

The 0.1-0.2 μm speckled contrast seen in figure 2 after the film was heated to 1000°C should represent the structure of the film in a nearly fully crystalline state. Images of the surface of similar films images at much higher spatial resolution in the atomic force microscope (AFM) show essentially the same morphology.

The comparison of data from the ESEM with information obtained by other techniques confirms that the surface microstructure apparent after heat-treatment in the ESEM is consistent with the microstructure produced in more conventional equipment, at least for this particular lead titanate system. In combination with these other techniques, particularly X-ray diffraction, a more complete understanding of the evolution of thin films, both in terms of morphology and phase composition, may be obtained using ESEM.

SUMMARY

Uncoated lead titanate sol-gel films have been successfully examined in the ESEM during *in-situ* heating experiments. Shrinkage of the film at cracks and delaminations have been measured as a function of temperature.

The surface structure of the films after heating in the ESEM correlates well with *ex-situ* results from conventional SEM and from AFM.

The ESEM provides microstructural data on the evolution of thin films as a function of temperature. This complements the crystallographic data obtained by XRD on the crystallinity of the films.

ACKNOWLEDGMENTS

Support was provided by the Center for Interfacial Engineering, an NSF Engineering Research Center.

REFERENCES

1. G. D. Danilatos, Journal of Microscopy, **160** (1), 9, (1990)

2. S. McKernan, Proc. **51**st MSA, 910, (1993)

3. J. S. Wright and L. F. Francis, J. Mater. Res., **8** (7), 1712, (1993)

4. K. D. Budd and D. A. Payne, Inst. Phys. Conf. Ser., **103** (1), 13, (1985)

5. R. C. Chiu, T. J. Garino and M. J. Cima, J. Am. Ceram. Soc., **76** (9), 2257 (1993)

IN-SITU TEM CRYSTALLIZATION OF ANORTHITE-GLASS FILMS ON α-Al$_2$O$_3$

MICHAEL P. MALLAMACI, JAMES BENTLEY* AND C. BARRY CARTER
Department of Chemical Engineering and Materials Science, University of Minnesota,
421 Washington Ave. SE, Minneapolis, MN 55455-0132
*Metals and Ceramics Division, Oak Ridge National Laboratory, P. O. Box 2008,
Oak Ridge, TN 37831-6376

ABSTRACT

Anorthite-glass films have been grown by pulsed-laser deposition on single-crystal α-Al$_2$O$_3$ substrates which were pre-thinned to electron transparency. The glass films were crystallized in the transmission electron microscope (TEM), which allowed direct observation and video-recording of the crystallization process. Crystallization of these films in the TEM resulted in the formation of hexagonal and orthorhombic anorthite. The orthorhombic phase was the predominant product of glass films grown at elevated substrate temperatures and displayed strong epitaxy with the underlying substrate. In contrast, the hexagonal phase was the major constituent of films grown at ambient substrate temperature and displayed no clear epitaxy with the substrate. The differences in degree of epitaxy and phase structure may be evidence of ordering at the original glass/oxide interface.

INTRODUCTION

Intergranular films are commonly observed in polycrystalline ceramics, and in many cases these films can control critical properties of the material.[1] The chemistry and crystallography of interfacial films may be manipulated by heat treatments to tailor materials properties. For example, alumina (α-Al$_2$O$_3$) and silicon-nitride ceramics often contain an amorphous intergranular phase which can be crystallized through post-sintering heat treatments in order to improve mechanical properties.[2,3] To investigate the role which α-Al$_2$O$_3$ grains play during the devitrification process, the crystallization of intergranular glass has been modeled by the deposition of thin glass films on single-crystal sapphire substrates followed by heat treatments to devitrify the film. Anorthite glass (CaO•Al$_2$O$_3$•2SiO$_2$) was chosen as the film composition because it commonly occurs in liquid-phase-sintered (LPS) alumina, and anorthite is a frequently observed phase in devitrified intergranular junctions. The use of sapphire substrates allows control of the grain crystallography; and basal sapphire was selected for these experiments since alumina grains in LPS material are often observed to facet preferentially parallel to the basal plane.

EXPERIMENTAL PROCEDURE

Commercially obtained basal sapphire pieces were cut with a low-speed saw equipped with a diamond wafering blade to a size suitable for examination in the TEM. The sapphire pieces were mechanically thinned to electron transparency by means of standard dimpling and ion-milling methods. Prior to film deposition, each pre-thinned substrate was carefully cleaned and annealed following a procedure detailed elsewhere.[4] A pulsed-laser-ablation system was used to deposit an anorthite-glass film on each substrate. Anorthite glass in bulk form served as the source material for film growth, and an oxygen partial pressure of 10 mTorr was maintained during deposition. Film-deposition conditions for each substrate were identical except for the nominal substrate temperature maintained during the deposition process, which ranged from ambient (~23°C) to 500°C.

The specimen geometry employed allowed observation of the film-substrate interface in plan-view, which was ideal for tracking the evolution of the film microstructure during

567

crystallization. The plan-view geometry complicated X-ray microanalysis in the TEM since aluminum was present in both the film and substrate.[5] As an independent confirmation of the film composition, glass was deposited on a pre-thinned MgO substrate in the same manner. X-ray microanalysis from this specimen showed that the Ca:Al:Si ratios for the deposited film were similar to the target material, which had a composition close to that of anorthite. The Ca/Si ratio of each alumina/glass specimen was checked prior to crystallization, and no significant deviation from the expected ratio for anorthite was observed.

Crystallization experiments were performed in a Philips CM12 analytical transmission electron microscope with a Gatan 628 single-tilt heating holder. Each specimen was heated in the microscope to approximately 1200°C and held at temperature for 20 minutes. The crystallization process was video-recorded during this time. Three glass films were crystallized in this study, corresponding to substrate temperatures during deposition of 23, 200 and 500°C. After crystallization, the specimens were placed in a standard double-tilt holder and electron-diffraction techniques were employed for microstructural analysis and phase identification. In some cases, the specimens were coated with amorphous carbon to alleviate charging problems.

RESULTS AND DISCUSSION

The anorthite-glass films crystallized in this study resulted in the formation of metastable hexagonal and orthorhombic polymorphs of the compound $CaAl_2Si_2O_8$, otherwise known in its natural triclinic state as anorthite . These metastable forms have been identified in anorthite glass crystallized in a nitrogen atmosphere, and are known to be structurally unrelated to the triclinic mineral.[6] These forms will be referred to as hexagonal anorthite and orthorhombic anorthite, since this is already the convention in the literature with regard to the hexagonal form.[7,8] The final crystallized microstructure was dependent on the substrate temperature maintained during deposition of the parent glass film. To distinguish between the different substrate temperatures, the three films will subsequently be referred to according to their deposition temperature, i.e., the "200°C-film." The reader is cautioned not to confuse this designation with the TEM-holder temperatures reported during the crystallization process.

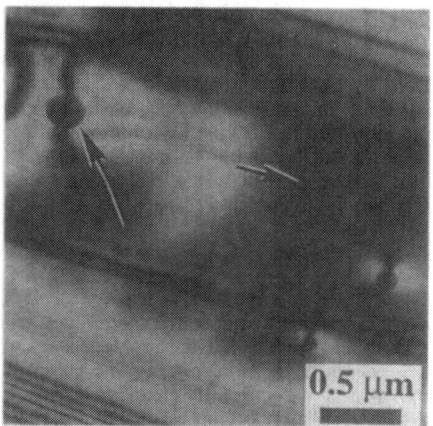

Figure 1: (0001) α-Al_2O_3 plan-view substrate after deposition of a glass film. Round glass particles and surface steps on the substrate are indicated.

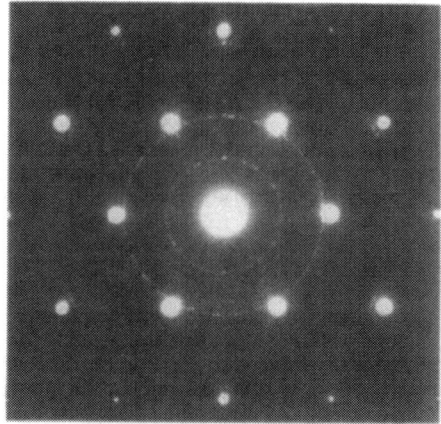

Figure 2: SAD pattern showing the rings from the polycrystalline hexagonal-anorthite film.

Observations During Crystallization

Figure 1 shows a bright-field image of an anorthite-glass film after deposition onto a (0001) α-Al$_2$O$_3$ substrate. Round glass particles seen in the image are characteristic of the laser-deposited glass films and are discussed elsewhere.[9] Other than these particles, the 30-nm-thick glass film covers the substrate surface in a uniform layer and does not contribute noticeable contrast to this image. Bend contours and surface steps on the substrate can be identified in the image; the formation of the 'step and terrace' structure on single-crystal α-Al$_2$O$_3$ surfaces after annealing at high temperatures has been well-documented.[4,10]

The round glass particles proved convenient for locating and monitoring specific regions during in-situ heating of the films. Each film was heated quickly to ~600°C, then heated in 100°C increments after observation periods of 4-5 minutes. No significant changes were ever noted until ~1000°C, at which point the formerly distinct edges of some glass particles became diffuse. All particles did not behave similarly; some completely diffused into the film, while others retained their shape until the onset of crystallization.

The onset of crystallization for each specimen was found to occur during heating from 1100°C to 1200°C. For the 23°C-film, crystallization occurred rapidly and did not appear to be associated with the round glass particles, i.e., crystals seemed to nucleate and grow in featureless regions of the film. The crystallized film was fine-grained in texture and a ring pattern, which is characteristic of polycrystalline materials, was obtained from electron diffraction. For the 200°C- and 500°C-films, larger crystals were obtained and the movement of crystallization fronts could be recognized by the development of moiré fringes in the film. For the 500°C-film, it was noted that some of the round glass particles crystallized before the rest of the film, and crystallization fronts in the film were observed moving away from the devitrified particles. It is likely that the round particles served as nuclei for film crystallization in this case.

Hexagonal Anorthite

The major phase which crystallized from the 23°C-film was identified as hexagonal anorthite using the selected-area-diffraction (SAD) pattern shown in Figure 2. This pattern was obtained from an area 24 µm in diameter while oriented at the [0001] pole of the substrate. The presence of the substrate reflections of known d-spacing allowed accurate calibration of the camera length for the pattern. The two most intense rings are due to the {10$\bar{1}$0} (d = 0.443 nm) and {11$\bar{2}$0} (d = 0.256 nm) planes of hexagonal anorthite, which has c = 1.4743 nm and a = 0.5113 nm; the grain size of this film was approximately 2.5 µm. Some grains of the hexagonal phase were also found in the 200°C-film, although it was not the major phase present in this film. When the hexagonal phase was observed in the crystallized films, it generally did not display epitaxy with the underlying substrate.

Orthorhombic Anorthite

The moiré fringes observed during crystallization of the 200°C and 500°C-films indicated the development of epitaxy between the crystallizing phase and (0001) α-Al$_2$O$_3$ substrate. The phase identified in this case was orthorhombic anorthite, which has a = 0.8224 nm, b = 0.8606 nm, and c = 0.4836 nm; this was the major phase for the 200°C- and 500°C-films after crystallization. The orthorhombic phase was also identified in isolated areas of the 23°C-film, an example of which is shown in Figure 3. Moiré fringes in orthorhombic grains are clearly visible in this image, as is a crystallized glass particle central to the grains. The round glass particle has crystallized as c-axis oriented hexagonal anorthite, although it is surrounded by orthorhombic grains. This general morphology for the orthorhombic phase is observed on a much coarser scale in the 200°C- and 500°C-films; an example from the 500°C-film is shown in Figure 4. It is clear that round glass particles (now crystalline) are observed in association with most orthorhombic anorthite grains, and hence may be involved in the nucleation of this phase. The epitactic relationships of each orthorhombic grain are [010]$_{anorthite}$‖[0001]$_{alumina}$ normal to the foil surface with

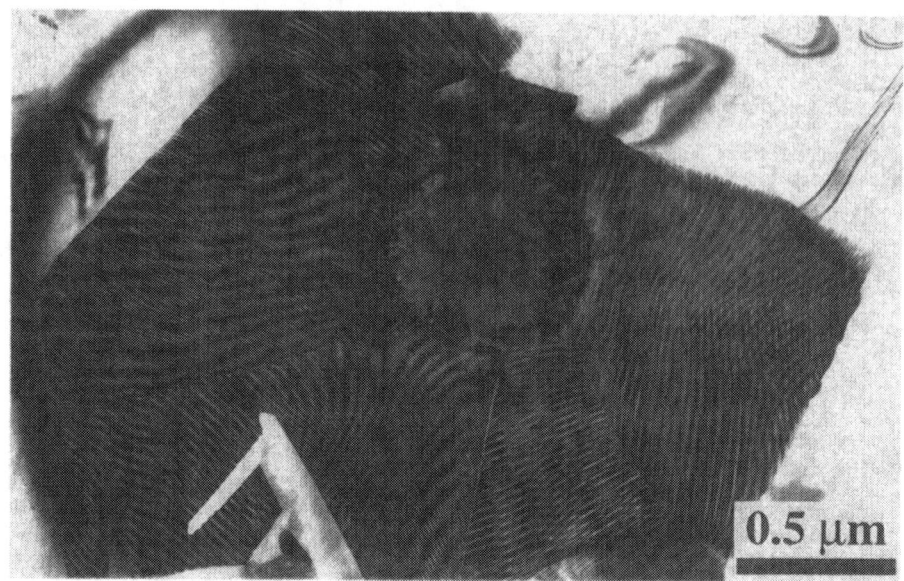

Figure 3: Bright-field image of orthorhombic-anorthite grains on the 23°C-film obtained at the [0001] α-Al$_2$O$_3$ zone axis.

Figure 4: Dark-field image from a 500°C-film using a diffraction spot generated by the orthorhombic-anorthite grains only. Specimen is oriented at the [0001] zone axis of the substrate. The three different variants for the orthorhombic grains are indicated by number.

[001]anorthite‖[11$\bar{2}$0]alumina and [100]anorthite‖[1$\bar{1}$00]alumina in the plane of the substrate. This type of bi-directional epitaxy parallel to the basal plane allows three different orthorhombic-anorthite variants on the surface, and all three variants can be seen in Figure 4. A SAD pattern of a single orthorhombic anorthite grain obtained when oriented at the [0001] substrate zone axis is shown in Figure 5, along with a schematic of this diffraction pattern for illustration of the epitactic relationships. The schematic shows direct overlap of orthorhombic anorthite and α-Al$_2$O$_3$ diffraction spots for clarity. These reflections are actually displaced due to the small lattice mismatch, which gives rise to the moiré fringes seen in the images. The orthorhombic grain selected for this SAD pattern was actually a thicker grain observed on a 500°C-film which was crystallized ex situ. Crystallization of a 500°C-film in air using a conventional furnace resulted in the formation of orthorhombic anorthite with the same epitaxy observed in the specimen crystallized in situ, the only notable difference being that extensive de-wetting of the film to form thicker crystallites occurred. The $h+k+l=odd$ reflections of the orthorhombic phase can be seen more readily in SAD patterns from thicker crystals, presumably due to double diffraction effects, whereas they are very weak in the (much thinner) films crystallized in situ. These reflections are useful for visualizing the 2-fold symmetries of the orthorhombic structure. Detailed discussion of air-crystallized films will be the subject of a future paper.

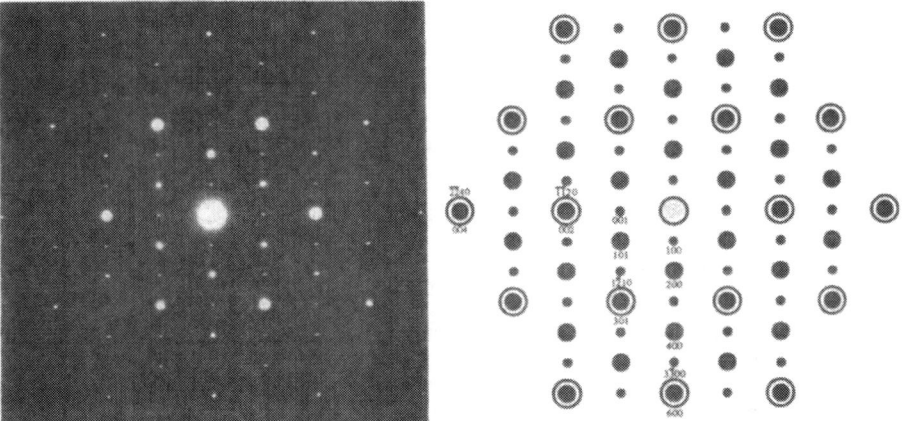

Figure 5: (a) SAD pattern from a single orthorhombic-anorthite grain oriented at the [0001] α-Al$_2$O$_3$ zone axis, which is also the [010] zone axis for the anorthite crystal. (b) Schematic representation of the diffraction pattern illustrating the observed epitaxy.

SUMMARY AND CONCLUSIONS

In-situ TEM crystallization of anorthite-glass films on basal alumina has resulted in the formation of hexagonal and orthorhombic anorthite. The crystallization behavior of the films was dependent upon the deposition conditions. The hexagonal phase was the major phase obtained when the original glass was deposited at 23°C, and no clear epitaxy with the substrate was observed. The orthorhombic phase showed strong epitaxy with the substrate and was the predominant phase obtained from glass films which were originally deposited at substrate temperatures of 200 and 500°C. It is possible that elevated surface temperatures during the deposition process allowed local ordering of the glass film at the interface; the existence of such ordering could promote epitaxy during crystallization. The epitaxy observed for the orthorhombic phase is directly attributable to nucleation and growth from the substrate surface and forbids the presence of a residual amorphous layer between the two oxides. This is in contrast with devitrified silicon-nitride materials, in which a residual amorphous layer between silicon nitride grains and crystallized intergranular glasses is commonly observed.[11,12] The close lattice match which orthorhombic anorthite shares with the basal plane of α-Al_2O_3 is remarkable. Since the same epitactic relationships for the orthorhombic phase are observed after both in-situ and ex-situ crystallization, it is proposed that epitaxy with the basal plane stabilizes the orthorhombic form of a mineral which is otherwise triclinic in its natural state. It is interesting to note that orthorhombic anorthite has not been reported in devitrified polycrystalline aluminas, despite the preference for alumina grains to facet parallel to the basal plane.[13,14]

ACKNOWLEDGMENTS

This research has been supported by the U. S. Department of Energy under Grant No. DE-FG02-92ER45465 and partially sponsored by the Division of Materials Sciences, U. S. Department of Energy, under contract DE-AC05-84OR21400 with Martin Marietta Energy Systems, Inc., and through the SHaRE Program under contract DE-AC05-76OR00033 with Oak Ridge Associated Universities. MPM is also affiliated with the Department of Materials Science and Engineering, Cornell University, Ithaca, NY. The authors would like to thank Mr. Mark Zika for his assistance with the photographic work. The TEM used at the University of Minnesota is part of the Center for Interfacial Engineering, an NSF Engineering Research Center.

REFERENCES

1. D. R. Clarke and M. L. Gee, in Materials Interfaces: Atomic-level structure and properties, edited by D. Wolf and S. Yip (Chapman & Hall, London, 1992), p. 255-72.
2. D. A. Bonnell, T.-Y. Tien and M. Ruhle, J. Am. Ceram. Soc. **70**, 460-65 (1987).
3. W. A. Zdaniewski and H. P. Kirchner, Adv. Ceram. Mat. **1**, 99-103 (1986).
4. D. W. Susnitzky and C. B. Carter, J. Am. Ceram. Soc. **69**, C-217-C-220 (1986).
5. M. P. Mallamaci, J. Bentley and C. B. Carter, in Laser Ablation in Materials Processing-Fundamentals and Applications, edited by B. Braren, J. Dubowski and D. Norton (Mat. Res. Soc. Proc. **285**, Boston, MA, 1992) pp. 433-38.
6. G. L. Davis and O. F. Tuttle, Amer. J. Sci. *Bowen Volume*, 107-14 (1952).
7. B. P. Borglum, J. M. Bukowski, J. F. Young and R. C. Buchanan, J. Am. Ceram. Soc. **76**, 1354-56 (1993).
8. C. Liu, S. Komarneni and R. Roy, J. Am. Ceram. Soc. **75**, 2665-70 (1992).
9. M. P. Mallamaci, J. Bentley and C. B. Carter, in Proc. 51st Annual Meeting of the Microscopy Society of America, edited by G. W. Bailey and C. L. Rieder (San Francisco Press, Inc., San Francisco, CA, 1993), p. 928-9.
10. G. C. Ndubuisi, J. Liu and J. M. Cowley, Micros. Res. and Techn. **21**, 10-22 (1992).
11. M. K. Cinibulk, G. Thomas and S. M. Johnson, J. Am. Ceram. Soc. **75**, 2037-43 (1992).
12. H.-J. Kleebe, M. K. Cinibulk, R. M. Cannon and M. Ruhle, J. Am. Ceram. Soc. **76**, 1969-77 (1993).
13. Y. K. Simpson and C. B. Carter, J. Am. Ceram. Soc. **73**, 2391-98 (1990).
14. W. A. Kaysser, M. Sprissler, C. A. Handwerker and J. E. Blendell, J. Am. Ceram. Soc. **70**, 339-43 (1987).

THE PYROLYTIC CONVERSION OF PERHYDROPOLYSILAZANE INTO Si₃N₄: X-RAY DIFFRACTION ANALYSIS

CHERYL R. BLANCHARD AND STUART T. SCHWAB
Southwest Research Institute, P.O. Drawer 28510, San Antonio, TX 78228

ABSTRACT

X-ray diffraction analysis was used to study the pyrolytic conversion of a preceramic polysilazane into Si_3N_4. Quantitative data for crystallite size, phase composition, and degree of crystallinity versus pyrolysis temperature and atmosphere (N_2 and NH_3) are presented. Pyrolytic products produced under N_2 and NH_3 atmospheres consist of microcrystals of silicon and α- and β-Si_3N_4. Under both atmospheres, a majority of the char is crystalline at ≈ 1270°C, and the entire char is crystallized at 1400°C. Pyrolysis under an NH_3 atmosphere produces near stoichiometric Si_3N_4, while pyrolysis under N_2 produces a silicon-rich material.

INTRODUCTION

The use of inorganic polymers as the precursor to ceramic materials has become popular in recent years because of the promise of low-cost fabrication of ceramic parts [1]. In addition to economics, processing with these so-called "preceramic polymers" offers a number of advantages over conventional powder-based methods including control of impurities, lower processing temperatures, and facile formation of complex shapes [1]. Preceramic polymers have been used as precursors to fibers, coatings, fine powders, and matrices in continuous fiber reinforced ceramic composites [1-3].

Silicon nitride (Si_3N_4) is considered to be one of the more technologically important advanced ceramics due to its high strength and hardness, and excellent resistance to creep, oxidation, and thermal shock [4]. Unfortunately, Si_3N_4 is difficult to sinter, and as such has become a popular candidate for processing through preceramic polymer routes. Polysilazanes (inorganic polymers with a Si-N backbone) have been shown to convert into either pure Si_3N_4 or a combination of Si_3N_4 and SiC, depending on the chemistry of the polymer and the pyrolysis conditions [1]. The chemistry of preceramic polysilazanes has recently been reviewed [5]. While most polysilazanes produce carbon containing ceramics, precursors to carbon-free Si_3N_4 have recently been reported by Arai *et al.* [6] and Schwab *et al.* [7].

Despite their potential importance, a full understanding of the polymer conversion process into a ceramic has yet to be developed. Other researchers [8,9] have used X-ray diffraction (XRD) to observe the polymer-to-ceramic conversion process and identify the final ceramic phases present, but ignored the critical quantitative information regarding crystallite size, degree of crystallinity and phase composition. This paper summarizes the results of an X-ray diffraction study of the pyrolytic products of a carbon-free perhydropolysilazane (PHPS). This PHPS was developed for use as a binder [10] for Si_3N_4 powder processing, but has been shown to be an effective matrix precursor for the processing of continuous fiber-reinforced Si_3N_4 composites [3] and to be an effective repair material for damaged oxidation protection coatings on carbon-carbon composites [2]. The phase composition, crystallite size, degree of crystallinity, and char density will be shown to be a function of both pyrolysis temperature and atmosphere.

EXPERIMENTAL PROCEDURE

Preceramic Polymers

Details of the polymer synthesis are provided elsewhere [10,11]. Briefly, dichlorosilane and trichlorosilane were reacted with ammonia (NH_3) in ether to yield PHPS and NH_4Cl as follows:

$$H_2SiCl_2 + HSiCl_3 + NH_3 \longrightarrow \left[\begin{array}{cc} \overset{\displaystyle H}{\underset{\displaystyle R}{\overset{|}{\underset{|}{Si}}}} & - & \overset{\displaystyle H}{\overset{|}{N}} \end{array} \right]_x + NH_4Cl \qquad (1)$$

$$R = N, H$$

All manipulations of PHPS were performed under anhydrous and anaerobic conditions using either an inert atmosphere/vacuum manifold system or a dry box.

To perform the pyrolysis studies, 5-10g samples of PHPS were weighed and placed in a tantalum crucible. The samples were pyrolyzed at temperatures ranging from 1000°C - 1700°C in a programmable, horizontal tube furnace (Model 54434 Tube Furnace with Eurotherm Controller 59256, Lindberg, Watertown, Wisconsin). All samples were heated at a rate of 10°C/min to the desired temperature, held at temperature for 15 minutes, and cooled to room temperature at an average rate of 15°C/min. The pyrolysis were performed under both nitrogen (N_2) and NH_3 atmospheres. The heat treatment yielded a porous, ceramic char, which was transferred to a dry box, ground with an Al_2O_3 mortar and pestle, and classified through a -325 mesh sieve. The samples were subsequently used for the XRD studies.

X-Ray Diffraction

The X-ray diffraction experiments were performed using a diffractometer (Phillips PW 1729 Diffractometer, Eindhoven, Netherlands) with a single crystal graphite monochromator. Each powder sample was poured and packed into a tape-backed aluminum holder to provide a randomly oriented, infinitely thick sample. Atmospheric exposure was minimized to prevent oxidation and water absorption. Step-scanned intensity data were generated using CuK_α radiation at a voltage of 40 kV and amperage of 40 mA. Data were gathered at room temperature over a 2θ range of 5° to 90° using a step size of 0.05° and a 3-second count time at each step, and were recorded directly into an ASCII text file for further manipulation.

Due to the significant amount of diffraction peak line-broadening observed on the XRD patterns from the lower temperature chars, crystallite sizes of each phase, α- and β-Si_3N_4 and silicon (Si), were calculated using the Scherrer equation [12]. Corrections were made for the inherent line broadening due to the X-ray source as described in detail elsewhere [11]. The crystallite sizes were determined for each phase by calculating the thickness of each crystal for a number of peaks and averaging.

To understand the evolution of the crystalline phases during pyrolysis, the relative amounts (based on weight fraction) of the three major phases present were calculated [11] with measured peak intensities using techniques based on the combined work of Gazzara and Messier [13], and Pigeon and Varma [14].

The chars pyrolyzed below a temperature of 1400°C exhibited broad maxima (characteristic of amorphous materials) that decreased with increasing pyrolysis temperature (see Figure 1). Although this phenomenon has been observed in glass-ceramics and some polymers, and used to quantify the degree of crystallinity, no calculations of this type had been previously attempted for a nitride ceramic. Therefore, a modification was developed based on the integrated area method described in the polymer literature [15], with a set of standards prepared for this material [11]. A detailed explanation of the technique used to calculate the volumetric degree of crystallinity, ϕ_c, and the standardization procedure will be found in reference [11].

RESULTS AND DISCUSSION

The XRD patterns for both the N_2 and NH_3 chars are shown in Figure 1. In general, as the pyrolysis temperature increases, the magnitude of the crystalline peaks increases indicating an increase in ϕ_c. In addition, line broadening decreases and the magnitudes of the Si peaks decrease with increasing temperature indicating an increase in crystallite size and a shift in the chemistry of the char, respectively. It should be noted that crystalline NH_4Cl is observed occasionally in the XRD patterns. NH_4Cl is a by-product of the ammonolysis reaction, and minor amounts often remain in the polymer. The final chars consist primarily of α- and β-Si_3N_4 and some Si.

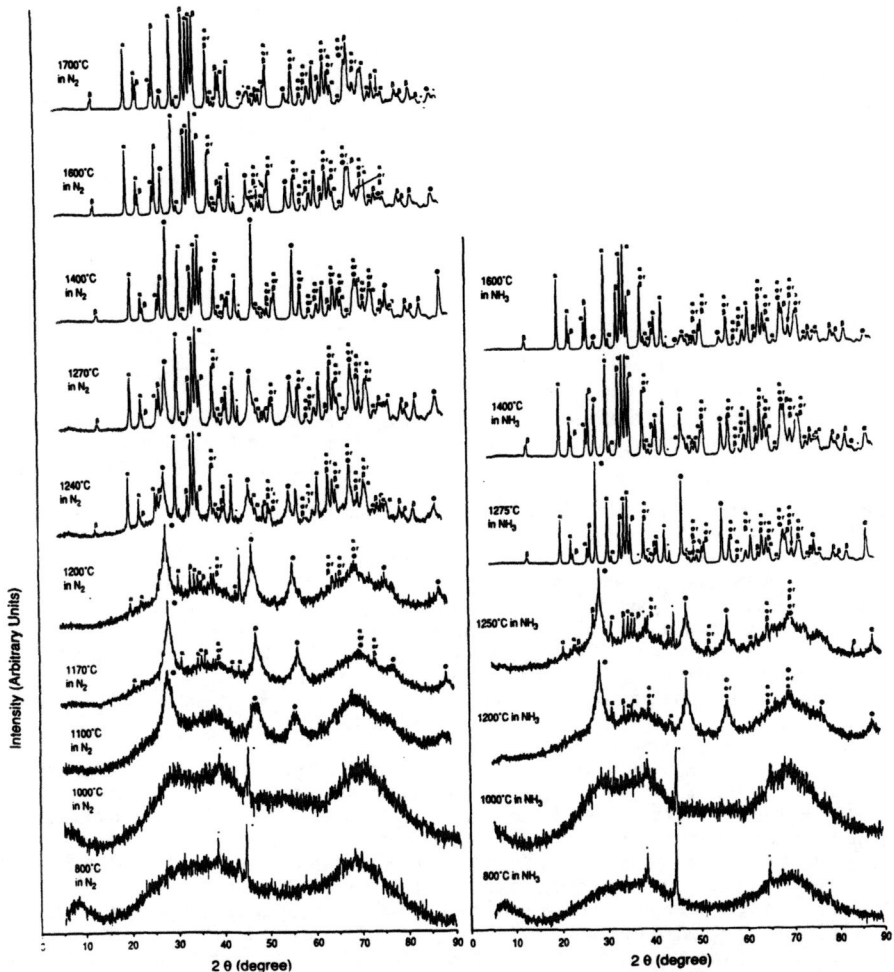

Figure 1. X-ray diffraction patterns of the polysilazane pyrolyzed at various temperatures in N_2 and NH_3. The peaks are labeled as follows: (⊗) Si, (α) α-Si_3N_4, (β) β-Si_3N_4, and (*) NH_4Cl.

Figure 2 presents a quantitative description of the volumetric degree of crystallinity, or the volume fraction of crystalline matter relative to the total matter, with respect to pyrolysis temperature. Above ≈ 1000°C, Si and some Si_3N_4 crystallized, resulting in an increase of ϕ_c from 0 to ≈ 0.09 at 1200°C. The degree of crystallinity increases to 1.0 between 1200°C and 1400°C, with most of the volume becoming crystalline between 1240°C and 1270°C for materials produced under both atmospheres. This increase in crystal fraction is consistent with the DTA crystallization exotherm for Si_3N_4 previously observed under both N_2 and NH_3 [11] at 1260°C for PHPS. Crystallization of PHPS pyrolyzed under NH_3 appears to take place within a more narrow temperature range than does that of the material produced under N_2. This is likely a result of the enhanced reactivity of NH_3 over N_2 [11]. The change in degree of crystallinity is also consistent with previous work reporting the density versus pyrolysis temperature of PHPS chars [11]. Specifically, the density increased from ≈ 1 g/cm³ at room temperature (representing

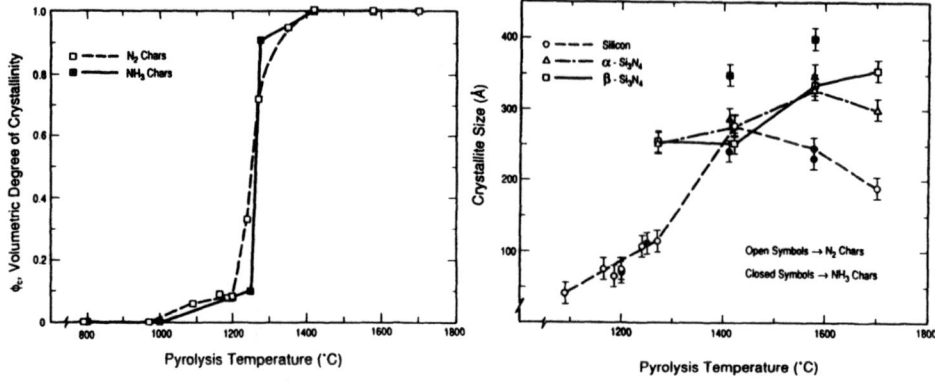

Figure 2. The effect of pyrolysis temperature and atmosphere on the volumetric degree of crystallinity.

Figure 3. Calculated crystallite size of each phase versus pyrolysis temperature for chars pyrolyzed under N_2 and NH_3 atmospheres.

the starting polymer) to ≈ 3 g/cm^3 at 1500°C. Also, previous thermogravimetric analysis revealed that PHPS has a ceramic yield of 80%.

The results of quantifying the diffraction line broadening are shown in Figure 3, which illustrates the relationship between the average crystallite size of each phase versus the pyrolysis temperature for both the N_2 and NH_3 chars. In both cases, the crystallites of Si are measured at ≈ 1100°C to be approximately 50Å in size. As the pyrolysis temperature increases to 1400°C, the Si crystallites increase in size to slightly larger than 250Å, the same size as the Si_3N_4 particles observed at that temperature. Above 1400°C, (the temperature regime at which Si was observed to nitride) the silicon particle size decreases as the α- and β-Si_3N_4 crystallites size increases. This inverse size relationship suggests that Si is nitriding to form Si_3N_4, or that Si_3N_4 is forming at the expense of the existing Si particles.

The evolution of the ceramic char was also found to be dependent on the relative amounts of each phase present. The phase composition versus pyrolysis temperature for the N_2 and NH_3 chars are presented in Figures 4 and 5, respectively. In Figure 4, the α-Si_3N_4 content is seen to increase from 5% to approximately 55% between 1200°C and 1270°C. The α-Si_3N_4 content increases and then decreases between 1270°C and 1700°C. In contrast, both the free Si and β-Si_3N_4 content increase slowly from 4% and 3%, respectively, to approximately 20% between 1200°C and 1350°C, and remain at that level up to 1420°C. Above that temperature to 1575°C, the β-Si_3N_4 content appears to increase at the expense of the free Si, again suggesting that the free Si is nitriding directly to β-Si_3N_4. Above 1575°C, the β-Si_3N_4 content increases at the expense of both free Si and α-Si_3N_4. At 1700°C, the material pyrolyzed under N_2 is 100% crystalline and contains approximately 62.5% α-Si_3N_4, 33.5% β-Si_3N_4, and 4% free Si by weight.

While the phase transformations taking place under NH_3 (Figure 5) are similar to those found for N_2, the changes generally occur at lower temperatures, in keeping with the greater reactivity of NH_3. The α-Si_3N_4 content increases sharply between 1250°C and 1575°C, reaching approximately 76% by weight. In contrast to the N_2 chars, the β-Si_3N_4 content of the chars produced under NH_3 rises from 4% to 20% between 1250°C and 1275°C, and remains essentially constant as the pyrolysis temperature increases. The free Si content also increases from 5% to 22% between 1250°C and 1275°C, and then decreases to $\approx 2\%$ as the pyrolysis temperature approaches 1600°C. The increase in α-Si_3N_4 content appears to correspond to the decrease in free Si content, suggesting that under NH_3, the free silicon nitrides directly to α-Si_3N_4. At 1575°C, the material pyrolyzed under NH_3 is 100% crystalline and contains approximately 76% α-Si_3N_4, 22% β-Si_3N_4 and 2% free Si by weight. The char pyrolyzed under NH_3 contained less Si and more α-Si_3N_4 at a given pyrolysis temperature compared to the N_2 char.

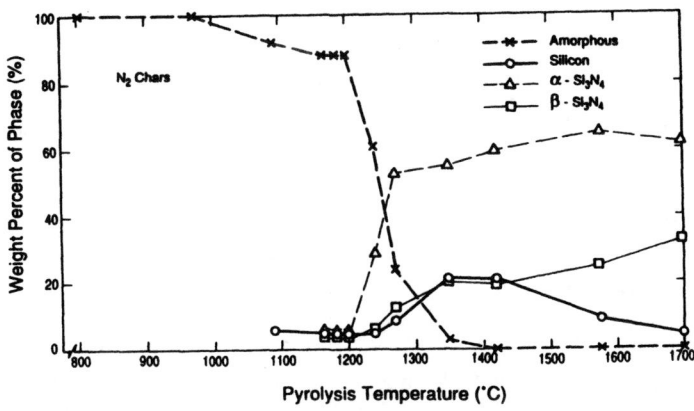

Figure 4. The effect of pyrolysis temperature on the phase composition of the N₂ chars.

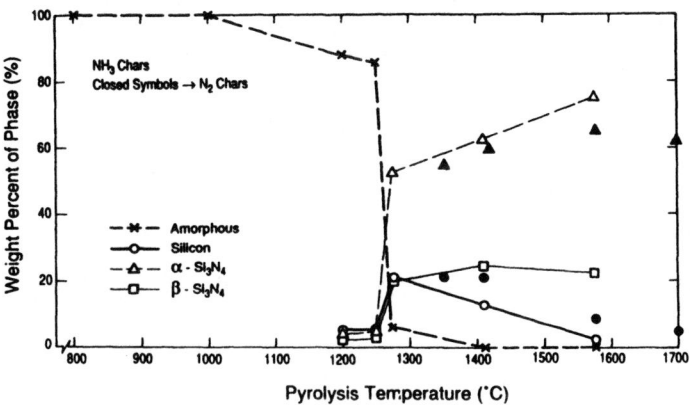

Figure 5. The effect of pyrolysis temperature on the phase composition of the NH₃ chars. (Data for the N₂ chars are included for comparison).

SUMMARY AND CONCLUSIONS

X-ray diffraction has been shown to be a powerful tool for the characterization of the pyrolysis products of preceramic polymers. Quantitative data on the crystallite size, degree of crystallinity, and phase composition of the char products were readily obtained. The PHPS-derived Si_3N_4 used in this study has been shown to crystallize at a temperature well below that of other polymer-derived ceramics, and to contain no detectable SiC or SiO_2 impurities. Specifically, a majority of the ceramic char was crystalline at a pyrolysis temperature of 1270°C, and by 1400°C had completely crystallized in both atmospheres studied. The pyrolysis products obtained under both N_2 and NH_3 atmospheres were microcrystalline chars consisting of Si, α-Si_3N_4, and β-Si_3N_4. In general, the amount of Si decreased and Si_3N_4 increased with increasing pyrolysis temperature under both N_2 and NH_3 atmospheres. While free silicon was present in the chars, its content could be minimized using NH_3 as the pyrolysis atmosphere. We believe this material may prove useful in the fabrication of fiber-reinforced composites, in which a crystalline matrix is desired, but the processing temperatures are limited by fiber degradation.

ACKNOWLEDGMENTS

The authors gratefully acknowledge the assistance of Messrs. J. L. Sievert, R. Railsback, S. Salazar, and W. Machowski, and Ms. L. Tufiño. The authors would also like to acknowledge the financial support of the Air Force Office of Scientific Research Under Contract No. F49620-91-C-0045.

REFERENCES

1. R. W. Rice, *Am. Cer. Soc. Bull.*, 62, 899 (1983).
2. S. T. Schwab and R. C. Graef, NASA CP-3133, Part 2, 781 (1991).
3. S. T. Schwab, R. C. Graef, Y. M. Pan, and D. L. Davidson, NASA CP-3175 Part 2, 721 (1992).
4. G. Ziegler, J. Heinrich, and G. Wötting, *J. Mater. Sci.*, 22, 3041 (1987).
5. D. Seyferth, G. H. Wiseman, J.M. Schwark, Y.-F. Yu, C. A. Poutasse, in Inorganic and Organometallic Polymers, Edited by M. Zeldin, K. J. Wynne, and H. R. Allcock, ACS Symposium Series 360 (American Chemical Society, Washington, DC, 1988), p. 143.
6. M. Arai, S. Sakurada, T. Isoda, and T. Tomizawa, *Polymer Preprints*, 28, 407 (1987).
7. S. T. Schwab, R. C. Graef, C. R. Blanchard, Y.-M. Pan, D. L. Davidson, G. E. Maciel, B. L. Hawkins, S. F. Dec, M. F. Davis, and R. Lewis, *Polymer Preprints*, 34, 286 (1993).
8. R.L.K. Matsumoto, *Mat. Res. Soc. Symp. Proc.*, Vol. 180, Materials Research Society, Pittsburgh, PA, pp. 797 (1990).
9. G. D. Soraru, F. Babonneau, and J. D. MacKenzie, *J. Mater. Sci.*, 25, 3886 (1990).
10. S. T. Schwab, C. R. Blanchard, and R. C. Graef, *J. Mater. Sci.* (in press).
11. C. R. Blanchard and S. T. Schwab, *J. Am. Cer. Soc.* (submitted, October 1993).
12. B. D. Cullity, Elements of X-Ray Diffraction, (Addison-Wesley, Redding, MA, 1956) p. 262.
13. C. Gazzara and D. Messier, *Am. Cer. Soc. Bull.*, 56, 777 (1977).
14. R. G. Pigeon and A. Varma, *J. Mater. Sci. Let.*, 11, 1370 (1992).
15. F. J. Baltá-Calleja and C. G. Vonk, X-Ray Scattering of Synthetic Polymers (Elsevier, New York, NY, 1989) p. 175.

HIGH RESOLUTION *IN SITU* TEM STUDIES OF SILICIDE-MEDIATED CRYSTALLIZATION OF AMORPHOUS SILICON

C. HAYZELDEN* and J.L. BATSTONE**
*Division of Applied Sciences, Harvard University, Cambridge, MA 02138
**IBM T.J.Watson Research Center, P.O. Box 704, Yorktown Heights, NY 10532

ABSTRACT

We report *in situ* high resolution transmission electron microscopy studies of $NiSi_2$-mediated crystallization of amorphous Si. Compared to conventional solid phase epitaxy of (111) Si, an enhancement of the growth rate by three orders of magnitude was observed and high quality twin-free needles of <111> Si were formed. Crystallization occurred via a ledge growth mechanism at the epitaxial Type A $NiSi_2$/crystalline Si (111) interface. A model for $NiSi_2$-mediated crystallization of amorphous Si involving the passage of kinks along <110> ledges at the $NiSi_2$/crystalline Si (111) interface is proposed.

INTRODUCTION

The transformation of amorphous Si (a-Si) to crystalline Si (c-Si) is an important aspect of semiconductor device processing. Pre-amorphization by ion implantation is frequently employed prior to doping, with subsequent annealing to induce recrystallization via solid phase epitaxy (SPE). As device dimensions shrink, the ability to form self-aligned metal contact regions by low temperature solid state reactions has stimulated great interest in the cubic metal silicides $NiSi_2$ and $CoSi_2$ (which have the CaF_2 structure and a lattice parameter mismatch with c-Si of 0.4-, and 1.2-%, respectively). Buried precipitates of $NiSi_2$ formed by Ni implantation of a-Si were reported by Cammarata et al.[1] to lead to low temperature crystallization of a-Si. Using *in situ* transmission electron microscopy (TEM), Hayzelden and Batstone,[2] and Hayzelden et al.[3] directly observed silicide-mediated crystallization of a-Si via migrating precipitates of $NiSi_2$. We have shown the rate of Si crystallization to be dependent on the thickness of the $NiSi_2$ in the growth direction, indicating diffusion controlled growth[3,4] and effective diffusivities through the $NiSi_2$ have been determined.[5] *Ex situ* high resolution electron microscopy (HREM) studies[3,4,6,7] by the present authors of the Type A $NiSi_2$/c-Si (111) interface have led to ball-and-stick models of the crystallization mechanism.[6,7] The phenomenon of silicide-mediated crystallization has recently received considerable attention.[8-12] We report here on dynamic *in situ* HREM studies of the atomistic mechanism of silicide-mediated crystallization of a-Si.

EXPERIMENTAL PROCEDURE

Amorphous Si thin films, 95nm thick, were deposited by low pressure chemical vapor deposition onto 100nm thick thermally grown SiO_2 layers on Si (100) wafers as described previously.[3] Ion implantation of Ni into the a-Si was performed at an energy of 55keV with a nominal dose of 5×10^{14} cm^{-2} to give a peak Ni concentration of about 2×10^{20} ions cm^{-3} at a depth of approximately 50nm. Prior to *in situ* TEM studies, the wafers were sealed in evacuated quartz backfilled with Ar to 380 Torr and annealed in a furnace at 400°C for 3hrs to form isolated $NiSi_2$ precipitates buried within the a-Si thin films. Specimens were prepared for *in situ* TEM experiments by removal of the c-Si substrate with chemical jet thinning in a 1:12 solution of $HF:HNO_3$. Initiation of Si crystallization within the a-Si films was studied *in situ* in a Philips EM430 high resolution electron microscope operated at an accelerating potential of 150keV and equipped with a Gatan single-tilt heating stage. The partially crystallized specimens were removed from the microscope and prepared for HREM by single sided ion-milling with Ar ions to remove the SiO_2 layer. *In situ* HREM of silicide-mediated growth of c-Si was performed at an accelerating voltage of 300keV with a point to point resolution of ~2.4Å.

579

RESULTS

Initiation of Crystallization

Annealing of the Ni implanted a-Si samples in a furnace at 400°C for 3hrs led to $NiSi_2$ precipitation close to the peak of the Ni concentration profile. Cross sectional TEM revealed the isolated precipitates to be completely buried within the a-Si film as demonstrated previously at higher implantation doses.[1] The $NiSi_2$ precipitates were regular octahedra bounded by eight {111} faces and exhibited edge lengths (which run along <110> directions) of ~30-35nm. Previous studies have revealed a suppression of the initiation of Si crystallization (though not of subsequent silicide-mediated growth) at accelerating voltages of 300keV, and this has been attributed to knock-on displacement damage of c-Si by electrons above the threshold energy of 185keV.[3]

Using an accelerating voltage of 150keV, a sample was heated *in situ* to 505°C to initiate crystallization. Fig. 1 shows plan-view TEM images of the same area of a furnace-annealed sample at different tilt angles after cooling to 452°C. Fig. 1(a) shows a region of silicide-mediated crystallization of a-Si together with two isolated precipitates. One of the precipitates, which is in a [110] orientation, has a step clearly visible on one of the {111} faces (arrowed). In the center of Fig. 1(a), a precipitate has given rise to silicide-mediated crystallization of a-Si, in which initiation of c-Si on a [110]-oriented $NiSi_2$ precipitate, has led to a series of growth fronts and to the breakup of the precipitate. At the leading edge of each of the growth fronts is a narrow band of $NiSi_2$. Fig. 1(b) shows the same region as Fig. 1(a) following tilting through an angle of ~42°. The triangular faces of the migrating $NiSi_2$ prisms appear dark against the Si single crystal and provide direct evidence of the morphology of the migrating prisms of $NiSi_2$.

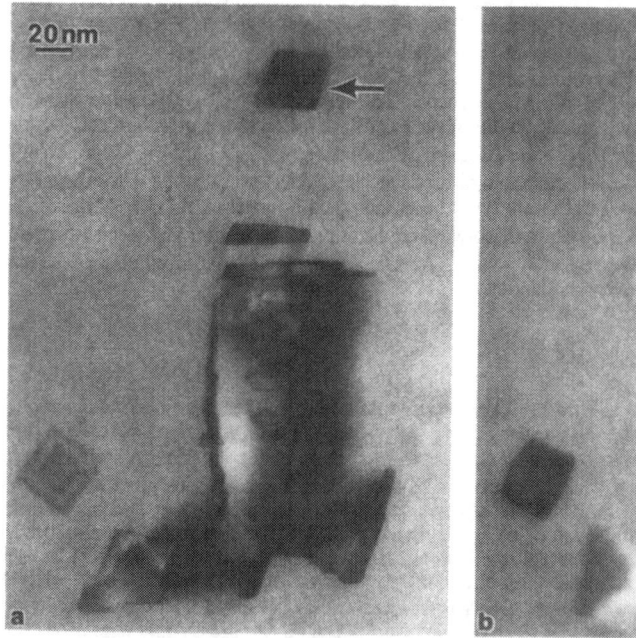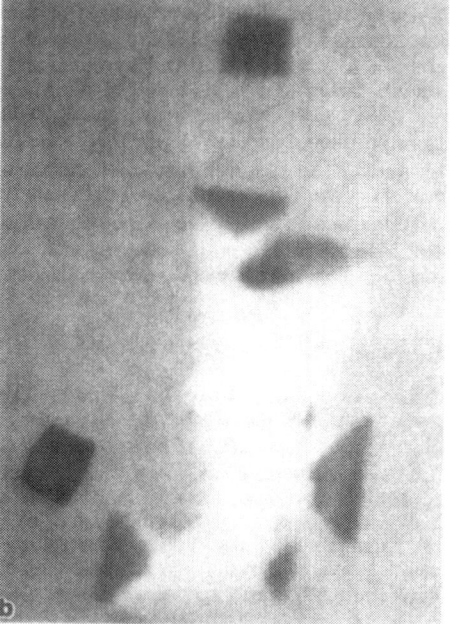

Fig. 1. TEM images of silicide-mediated crystallization of a-Si recorded *in situ* at 452°C. a) Showing dark bands of $NiSi_2$ at the growth fronts of the Si single crystal. A step appears on the [110]-oriented precipitate. b) Showing the triangular faces of the $NiSi_2$ prisms after tilting ~42°.

We have demonstrated previously that the initiation of silicide-mediated crystallization of a-Si occurs via the low-temperature formation of epitaxial Type A c-Si on one or more of the {111} NiSi₂ faces and that subsequent growth occurs via the migration of NiSi₂ precipitates in <111> directions trailing epitaxial c-Si.[3,5] Of the major zone axes, <100>, <110> and <111>, only the <110>-oriented precipitates provide {111} planes having normals within the plane of the thin film and are therefore suitably oriented for extensive <111> in-plane growth.[5] In Fig. 2(a), formation of c-Si is shown schematically on one face of a [110]-oriented precipitate. In Fig. 2(b), silicide-mediated growth of c-Si is shown by the subsequent migration of NiSi₂ *away* from this face, trailing epitaxial c-Si with a Type A interface behind it. The prism of NiSi₂ and trailing needle of c-Si are bounded by three {112} planes.

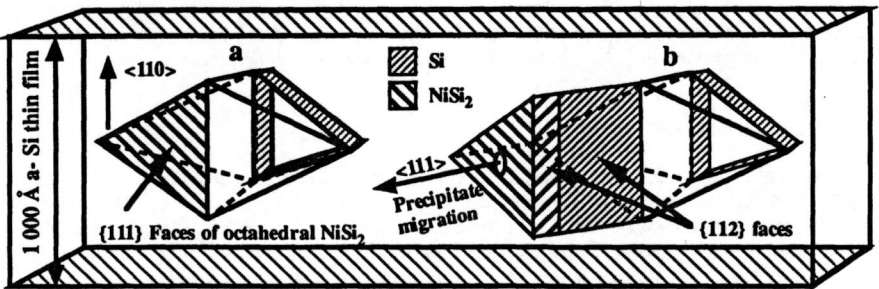

Fig. 2. a) Showing initiation of c-Si on {111} faces of buried precipitates of NiSi₂ in a-Si.
b) Showing silicide-mediated growth of c-Si at the trailing edge of a migrating NiSi₂ precipitate.

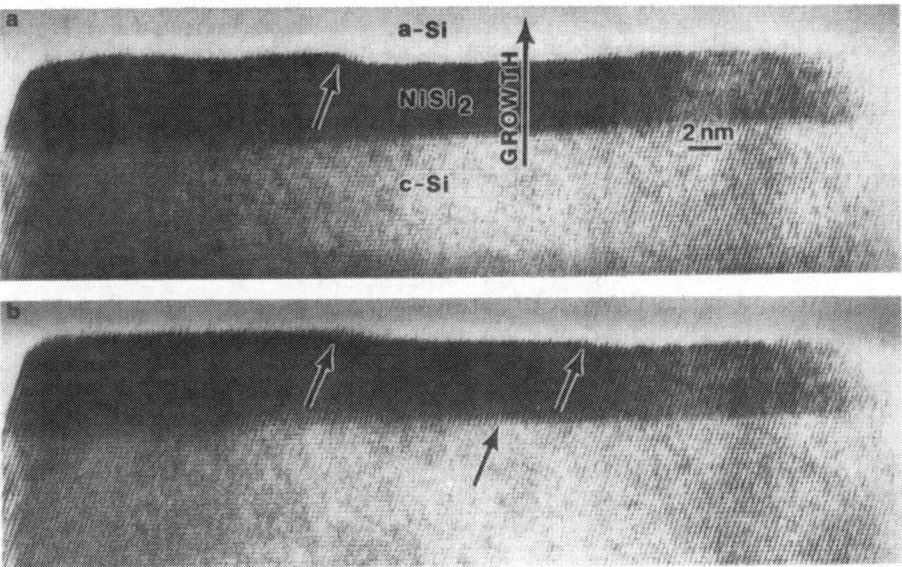

Fig. 3. HREM images of NiSi₂-mediated crystallization of a-Si recorded *in situ* at 457°C.
The images in a) and b) were taken ~20s apart and ledges may be seen traversing both interfaces.

Dynamic *in situ* HREM showed that c-Si growth proceeded via the passage of ledges at the leading (a-Si/NiSi$_2$) and trailing (NiSi$_2$/c-Si) interfaces as shown in [110] projection in Fig. 3. The NiSi$_2$/c-Si (111) interface is Type A and was inherited from the Type A orientation of the c-Si formed on {111} faces of the NiSi$_2$ precipitate. Ledges of varying heights were observed at both interfaces, although heights of d(111) and 3d(111) were most common. At 457°C an average growth rate of ~1Ås^{-1} was measured which is about three orders of magnitude greater than that expected for conventional SPE of (111) Si at this temperature.[13] The HREM images exhibit a distinctive "wedge" shaped appearance and are thicker on the left hand side than on the right hand side. No dislocations were observed at the NiSi$_2$/c-Si interface and none would be expected for NiSi$_2$ thicknesses less than ~40nm. Ledges were observed to run from left to right and from right to left (along <112> directions) at both the a-Si/NiSi$_2$ and NiSi$_2$/c-Si interfaces and their passage maintained a fairly constant NiSi$_2$ thickness of ~5nm in the growth direction.

DISCUSSION

We have shown that NiSi$_2$-mediated crystallization of a-Si results from the migration of precipitates of NiSi$_2$, which commonly exhibit a prism morphology, trailing epitaxial needles of <111> Si. Initiation of c-Si occurs on one or more of the {111} faces of the octahedral NiSi$_2$ precipitates, while growth occurs via migration of the NiSi$_2$ precipitate away from the region of c-Si initiation. Steps on {111} faces of the NiSi$_2$ precipitates, such as those observed in Fig. 1(a), may play an important role in selection of a NiSi$_2$ face for c-Si initiation and/or migration during c-Si growth. *In situ* HREM demonstrated that growth of c-Si occurred via the passage of ledges at the NiSi$_2$/c-Si interface. Fig. 4 presents a schematic model of the process of NiSi$_2$-mediated crystallization of a-Si, analogous to those of Spaepen and Turnbull,[14] and Williams and Elliman,[15] for conventional SPE of Si at an a-Si/c-Si interface. Fig. 4 shows an equilateral, triangular Type A NiSi$_2$/c-Si (111) interface at which transformation of NiSi$_2$ to c-Si is accomplished by the passage of ledges at the moving interface. A <110> TEM viewing direction is indicated on Fig. 4 and shows the origin of the appearance of a wedge shaped thickness in the HREM images of Fig. 3. We propose that NiSi$_2$ is transformed into c-Si by the passage of <110> ledges (or steps) across (111) terraces (where the NiSi$_2$/c-Si interfacial plane can be viewed as the (111) terrace). The <110> ledges traverse the (111) terraces to complete a new terrace by moving in <112> directions.

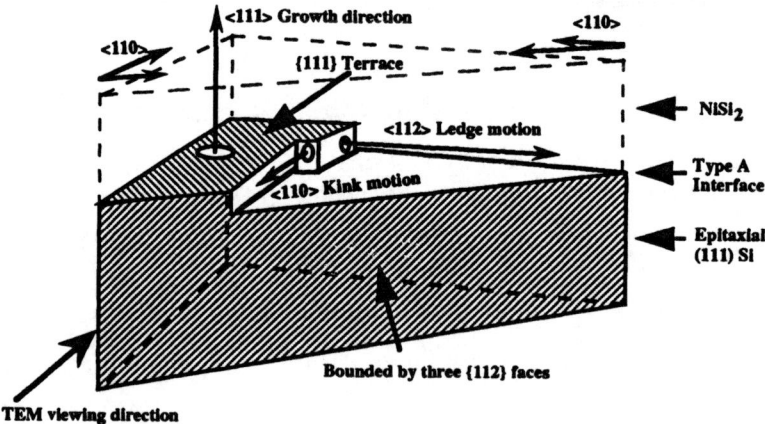

Fig. 4. Schematic representation of proposed ledge mechanism of NiSi$_2$-mediated crystallization of a-Si at the Type A NiSi$_2$/c-Si (111) interface.

The passage of ledges is accomplished via the motion of kinks along the <110> ledges. Fig. 5 illustrates this schematically. In Fig. 5(a) a triangular section of the NiSi$_2$/c-Si interface is viewed in a <111> direction to illustrate the passage of ledges across a terrace. The velocity of ledges across the terraces, V_L, moving in a <112> direction and the kink velocity, V_K, moving along a <110> ledge is indicated. Fig. 5(b) shows a section through the interface viewed in a <110> projection to correspond to the TEM viewing direction. Within the TEM in this orientation we would expect to observe V_L directly.

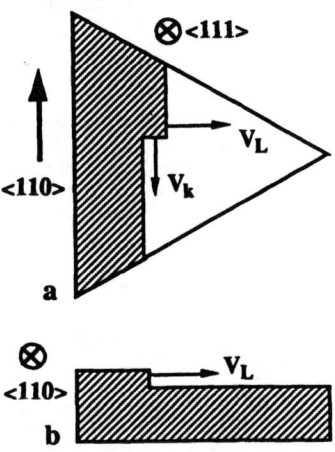

Fig. 5. a) Schematic representation of the NiSi$_2$/c-Si interface viewed in the <111> direction and showing a <110> ledge traversing a (111) terrace (shaded). The size of the ledge and kink (of the order of a few Ås) are shown greatly exaggerated for clarity. b) Schematic representation through the interface viewed in a <110> projection to correspond to the TEM viewing direction.

We have shown that NiSi$_2$-mediated crystallization of a-Si occurs via the motion of ledges at the a-Si/NiSi$_2$ and NiSi$_2$/c-Si interfaces, an observation indicating an interface-controlled growth process. Our earlier studies showed, however, that the rate of Si crystallization was dependent on the thickness of the NiSi$_2$ in the growth direction, and that the rate-controlling step for precipitate migration was diffusion-limited. These observations may be reconciled by consideration of two limiting cases that describe the diffusional process of NiSi$_2$-mediated crystallization of a-Si, namely, the non-dissociative and dissociative diffusion models. In the non-dissociative model, Si atoms would simply diffuse through the NiSi$_2$ layer and bond at the epitaxial c-Si. In the dissociative model, the NiSi$_2$ layer would dissociate to provide free Si for epitaxial growth at the trailing NiSi$_2$/c-Si interface, with new NiSi$_2$ formed at the leading a-Si/NiSi$_2$ interface. In the latter case, all of the Si atoms originally in the NiSi$_2$ layer would be incorporated in the epitaxially growing c-Si and replaced by Si atoms from the a-Si.

Our *in situ* HREM observations suggest that crystallization occurs via a dissociative-diffusion process in which diffusion occurs in a coordinated manner that involves both Si and Ni atom species. The passage of a <110> ledge at the NiSi$_2$/c-Si interface would thus involve the breaking of bonds around the Ni atoms via kink motion along <110> ledges, enabling Si atoms to hop into the vacated Ni lattice sites.[6,7] While long range diffusion of Si is not required, the Ni atoms would diffuse toward the a-Si (i.e. in the growth direction). The observation that NiSi$_2$ formation at the leading a-Si/NiSi$_2$ interface also occurs via a ledge mechanism suggests the coordinated diffusion of Ni atoms in the growth direction.

CONCLUSIONS

In conclusion, the atomistic nature of $NiSi_2$-mediated crystallization of a-Si has been investigated using the powerful technique of *in situ* HREM. The crystallization rate was enhanced by three orders of magnitude compared to conventional SPE of (111) Si and took place via a ledge growth mechanism at the epitaxial Type A $NiSi_2$/c-Si (111) interface. A model for $NiSi_2$-mediated crystallization of a-Si involving the passage of kinks along <110> ledges at the $NiSi_2$/c-Si (111) interface and the dissociative diffusion of Ni and Si atoms was proposed.

ACKNOWLEDGMENTS

This work was supported by the Harvard Materials Research Laboratory under NSF Grant DMR-89-20490 and by the David Wright Fund of Harvard University. The authors wish to thank D. Turnbull, F. Spaepen and D. A. Smith for their helpful comments and interest in this work.

REFERENCES

1. R. C. Cammarata, C. V. Thompson, C. Hayzelden and K. N. Tu, J. Mater. Res. **5**, 2133 (1990).
2. C. Hayzelden and J. L. Batstone, in Proc. 49th Annual Meeting of the Electron Microscopy Society of America, edited by. G. W. Bailey and E. L. Hall (San Francisco Press, California, 1991) pp. 826-827.
3. C. Hayzelden, J. L. Batstone and R. C. Cammarata, Appl. Phys. Lett., **60**, 225 (1992).
4. C. Hayzelden and J. L. Batstone, in Proc. 50th Annual Meeting of the Electron Microscopy Society of America, edited by G. W. Bailey, J. Bentley and J. A. Small (San Franscisco Press, California, 1992) pp. 1352-1353.
5. C. Hayzelden and J. L. Batstone, J. Appl. Phys. **73**, 8279 (1993).
6. J. L. Batstone and C. Hayzelden, in Proc. 8th Oxford Conf. Microsc. Semicond. Mater. edited by A. G. Cullis and A. Staton-Bevan, (Inst. Phys. Conf. Ser. **134**: Section 4 , London, 1993) pp. 165-172.
7. J. L. Batstone and C. Hayzelden, in Polycrystalline Semiconductors III - Physics and Technology, Solid State Phenomena, edited by H. P. Strunk, J. H. Werner, B. Fortin and O. Bonnaud, (Trans. Tech, Zurich), to be published.
8. B. Mohadjeri, J. Linnros, B. G. Svensson and M. Östling, Phys. Rev. Lett. **68**, 1872 (1992).
9. Y. N. Erokhin, R. Grötzschel, S. R. Oktyabrsky, S. Roorda, W. Sinke and A. F. Vyatkin, Mats. Sci. and Eng. **B12**, 103, (1992).
10 T. Hempel, O. Schoenfeld and P. Veit in Beam-Solid Interactions: Fundamentals and Applications, edited by M. Nastasi, L. R. Harriott, N. Herbots and R. S. Averback (Mater. Res. Soc. Proc. **279**, Pittsburgh, PA, 1993) pp. 267-272.
11 A. Yu. Kuznetsov, I. I. Khodos, V. N. Mordkovich and A. F. Vyatkin, Nucl. Instr. Meth. B80/81, 990 (1993).
12 A. Yu. Kuznetsov, I. I. Khodos, V. N. Mordkovich and A. F. Vyatkin, Appl. Surf. Sci. (to be published).
13. L. Csepregi, E. F. Kennedy, T. J. Gallagher and J. W. Mayer, J. Appl. Phys. **48**, 4234 (1977).
14. F. Spaepen and D. Turnbull, in Laser-Solid Interactions and Laser Processing, edited by S. D. Ferris, H. J. Leamy and J. M. Poate, (Am. Inst. Phys. Proc. **50**, New York, NY, 1979). pp.73-83.
15. J. S. Williams and R. G. Elliman, Phys. Rev. Lett. **51**, 1069 (1983).

SILICON CRYSTALLIZATION FROM ANNEALED Cu/a-Si:H BILAYERS: A MULTITECHNIQUE STUDY

C. A. Achete*, L. Bernardino*, F. L. Freire Jr.**,1, G. Mariotto** and H. Niehus***

* PEMM, COPPE, Universidade Federal do Rio de Janeiro, 21910, Rio de Janeiro, RJ, Brazil
** Dipartimento di Fisica, Università di Trento, 38050, Povo (TN), Italy
*** ICV/Forschungszentrum, Jülich, Federal Republic of Germany

ABSTRACT

Silicon crystallization has been observed to occur in copper/ a-Si:H thin film bilayers annealed at 280 °C. Copper-silicide formation was observed after annealing at 200 °C. Samples characterization was made by a combination of several analytical techniques: scanning electron microscopy, Raman spectroscopy through a microscope probe, Auger electron spectroscopy, elastic recoil detection analysis and Rutherford backscattering spectrometry. The possible role of hydrogen in this process is discussed.

1. INTRODUCTION

Interfacial reactions between metal and silicon are of great fundamental and technological interest since they involve processes such as interdiffusion, alloy formation, etc., which determine the physical properties of the contact. In particular, the study of Cu/Si interface has received increasing attention in order to develop copper-based metallization schemes for ULSI [1].

In the last years, metal mediated amorphous silicon crystallization has been intensively studied. The crystallization of amorphous silicon (a-Si) when it is in contact with noble metals occurs at surprisingly low temperatures. In fact, crystallization of a-Si has been observed to occur by random nucleation and growth at temperatures higher than 600 °C [2], and by solid phase epitaxy at temperatures beginning at 450 °C [3]. When is in contact with a gold film, a-Si crystallizes at about 180 °C [4]. The crystallization temperature is 200 °C in Al/Si multilayers [5]. Low temperature Si crystallization was also induced by silicide formation in Ni/a-Si:H bilayers annealed at 380 °C [6]. The fractal growth of silicon crystallites during Si crystallization in a Cu_3Si matrix was observed at the temperature of 485 °C [7].

In this work, we report the silicon crystallization in a Cu/a-Si:H thin film bilayer annealed at 280 °C. To probe the reactions at the interface and to characterize the Si crystallization, a multitechnique approach was followed. The possible role of hydrogen is also discussed.

---

1 - Permanent address: Departamento de Física, PUC-RIO
  22453, Rio de Janeiro, RJ, Brazil.

2. EXPERIMENTAL PROCEDURES

Films of intrinsic a-Si:H (~2 μm thick), were grown by conventional glow discharge decomposition of silane onto (100)-Si crystals. The substrates were kept at 250 °C during film deposition. Prior to the Cu deposition, a-Si:H were exposed to air for no longer than 5 min. Such air exposure is known to produce a thin oxide layer (~0.2 nm) [8]. The subsequent Cu deposition (80 nm thick) was done by e-beam evaporation in a vacuum chamber at a pressure lower than 10^{-4} Pa. Immediately after the Cu-film deposition, the samples were annealed at different temperatures between 150 and 380 °C, in a quartz tube furnace evacuated with a turbomolecular system at pressures of the order of 10^{-4} Pa. The annealing time was 30 minutes and no kind of sequential annealing was done.

The chemical analysis, as well as, the interface reaction analysis of as-deposited and of annealed samples, were carried out by both 2 MeV-He Rutherford backscatering spectrometry (RBS) and Auger electron spectroscopy (AES). Hydrogen profiles were measured by means of 2.2 MeV-He elastic recoil detection analysis (ERDA).

Surface characterization was done by scanning electron microscopy using a 30-kV Jeol microscope, equipped with energy dispersive X-ray analysis (EDX).

Raman scattering measurements were carried out using a microprobe apparatus. The samples were positioned under an optical microscope and viewed onto a monitor. Both excitation (λ=488.0 nm) and light collection occurred through the microscope objective (100X) in backscatering geometry. The beam spot was approximately 1 micron in diameter. The radiation was analyzed by a 1 m focal length Jobin-Yvon double-monochromator (Ramanor, mod. HG2 S), equipped with holographic gratings (2000 grooves/mm) and detected by a cooled GaAs photomultiplier operating in photon counting mode. All the measurements were performed with the samples at room temperature and the excitation power was kept below 2 mW at the samples to prevent laser annealing [9].

3. RESULTS

In fig. 1, we show the RBS spectra obtained from as-deposited and annealed samples (temperatures of 200 and 280 °C). The RBS spectrum for a sample annealed at 200 °C shows the formation of a metal-rich Cu-Si compound with a composition very close to Cu_3Si. The formation of the copper-silicide was confirmed by AES results: the spectra from treated samples present the characteristic splitting of the Si(LVV) Auger transition at 92 eV . This transition is sensitive to Si(3p) states and its splitting reflects the local environment of Si atoms when they were in a Cu matrix [10].

For annealed samples, RBS spectra also indicate the presence of oxygen at the surface, which can be attributed to the enhanced room temperature Si-oxide formation at copper-silicide surface [11].

RBS and AES analyses of samples treated at 150 °C do not indicate any kind of interfacial reaction.

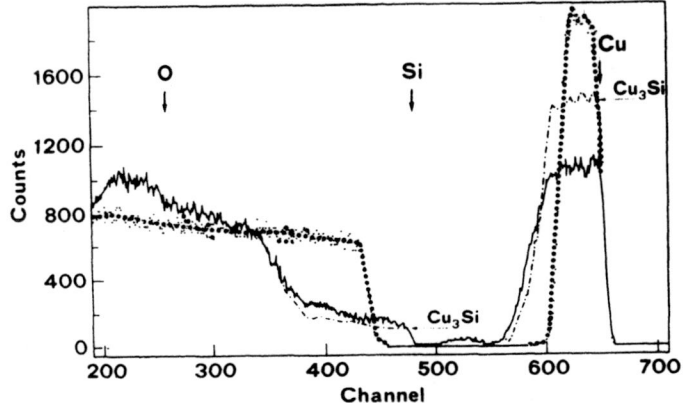

Figure 1- RBS spectra from the as-deposited Cu/a-Si:H bilayer (dotted line) and from samples annealed at temperatures of 200 °C (dash-dotted line) and 280 °C (full line).

For annealing temperatures higher than 200 °C, RBS spectra seems to indicate an increasing of the amount of silicon at the reacted layer. This result, however, is an artifact of the RBS technique. In fact, SEM micrographs presented in fig. 2a and fig. 2b, clearly show the inhomogeneous surface morphology of samples annealed at 280 and 380 °C, respectively. EDX analyses of samples treated at 280 °C indicate that the dark spots correspond to a Si-rich zone, whenever the bright regions contains both copper and silicon. EDX maps of samples annealed at 380 °C indicate that the dark areas are essentially silicon.

(a)   (b)

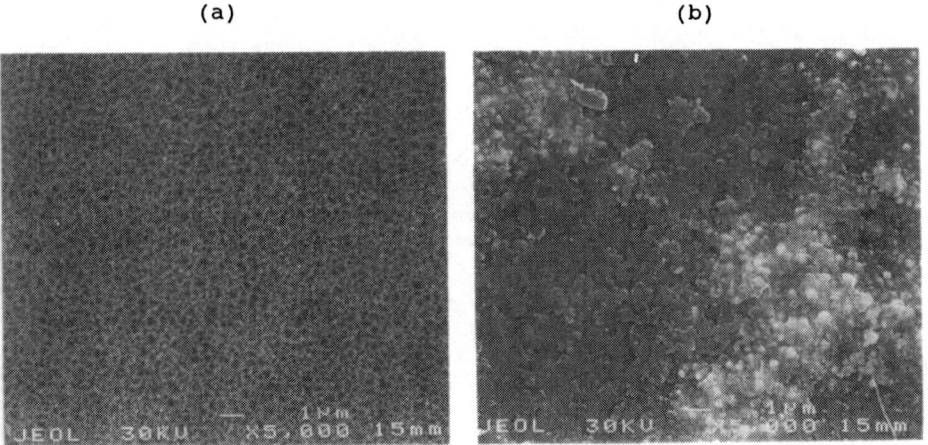

Figure 2- SEM pictures obtained from samples treated at: a) 280 °C (5000X) and b) 380 °C (5000X).

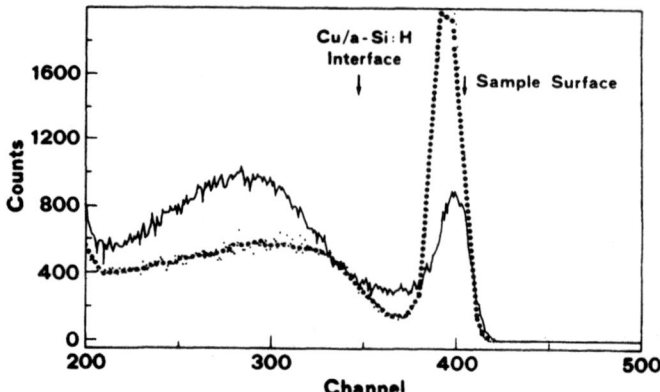

Figure 3- ERDA spectra measured on the as-deposited (dotted line) and on 280 °C annealed samples (full line).

Figure 3 presents ERDA spectra obtained for as-deposited Cu/a-Si:H bilayers and for samples annealed at 280 °C. The peak at about channel 400 is due to hydrogen-containing contamination adsorbed at the surface during the analysis. Inhomogeneous sample surfaces inhibit quantitative analysis by using this technique, however, the results obtained from samples annealed can be interpreted as follows: the spectrum of the sample annealed at 280 °C indicate the presence of hydrogen in the reacted layer, in the meanwhile the interface between this region and the underneath silicon layer acts as a trap for hydrogen out-diffusion.

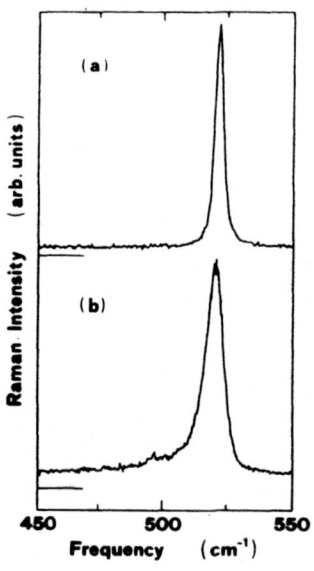

Figure 4- Raman spectra obtained from a Si single-crystal (a), and from a Cu/a-Si:H bilayer annealed at 280 °C (b).

Figure 4 presents a typical Raman spectrum, obtained using

the microscope probe, from a Si-rich zone of samples annealed at the temperature of 280 °C (dark region of the SEM pictures). For the sake of comparison, a Raman spectrum of a Si single-crystal obtained under identical experimental conditions is also shown. In both cases, the Raman spectra present a peak centered at 520 cm^{-1}, typical of crystalline Si, but with different linewidths: the peak-width measured from the sample treated at 280 °C (6.8 cm^{-1}) is twice the natural linewidth of Si-crystals [12]. Raman spectra obtained from the zones corresponding to the bright areas on the SEM micrographs, do not present any structure that can be associated either to amorphous or crystalline Si. A broad structure around 300 cm^{-1} was observed and can be attributed to the presence of Cu-Si alloy [13].

For silicon microcrystals with dimensions smaller than 30 nm, the 520 cm^{-1} peak broadens and shifts to lower frequencies [12]. The increase in the width values is associated with the disorder at the grain boundaries of the crystallites or to the average separation between defects or impurities in the crystals [14], which leads to a decrease of the phonon lifetime. The downshift is attributed to a size effect [12]. Recent calculations have shown that the asymmetric shape of the Raman peak can be attributed to a size grain distribution [15]. In our case, only peak broadening and peak asymmetry were observed. So, the absence of the downshift peak can be correlated to the fact that, the silicon grains were larger than the critical size to produce this effect, as can be easily seen in the SEM picture shown in fig. 2b.

4. DISCUSSION AND CONCLUSIONS

On the basis of the results presented in the previous section, some considerations concerning the interfacial reaction in Cu/a-Si:H bilayers and copper-silicide mediated silicon crystallization could be derived.

The first result concerns the Cu_3Si formation in those samples. Copper silicon reaction and subsequent alloy formation is expected to occur at clean interfaces even at room-temperature [16]. In our samples, silicide formation occurs only after annealing at 200 °C probably due to the presence of native oxides at the interface.

Silicon crystallization has been observed to occur in samples annealed at 280 °C. Raman results clearly indicate the presence of polycrystalline silicon in those samples. Up to now, the mechanism of metal mediated Si-crystallization is not well known, but it is suggested that distorted covalent structures can be easily disrupted by the presence of certain metals and can rearrange themselves to achieve a lower free-energy state [17]. However, the dynamic of such process is not known.

The role of hydrogen in that process is also unclear. In fact, Si crystallization in Cu/a-Si bilayers was observed at a higher annealing temperature, 485 °C, which suggests that the presence of hydrogen may lower the crystallization temperature in this system. In a recent publication [18], a similar behavior was observed for the system Ag/a-Si:H. This effect might be caused by the break of Si-H bonds, which may stimulate a preferential nucleation at the Cu-silicide interface resulting in a much lower crystallization temperature of the a-Si:H layer.

In summary, in this work we have studied the interfacial reaction in Cu/a-Si:H bilayers. Cu_3Si is formed after annealing at 200 °C. For annealing at temperatures higher than 280 °C, Raman measurements indicate the formation of polycrystalline silicon.

ACKNOWLEDGMENTS

We would like to thank Dr. E. Zanghellini (Università di Trento) for his skillful assistance during Raman analyses and R. Belli (Università di Trento) for her help during SEM measurements. One of the authors, F.L. Freire Jr., acknowledges the hospitality during his visit at the Università di Trento.

REFERENCES

1. J. Lian, Y. Shacham-Diamand and J.W. Mayer, Mat. Sci. Rep. **9**, 1 (1992).
2. J.L. Batstone, Phil. Mag. A **67** (1993) 51.
3. L. Csepregi,J.W. Mayer and T.W. Sigmon, Phys. Lett. A **54**, 157 (1975).
4. A.A. Pasa, M.B. Schubert, C.D. Adel, W. Beyer, W. Losch and G.H.Bauer in Amorphous Silicon Technology - 1992, edited by M.J. Thompson, Y. Hamakawa, P.G. LeComber, A. Madam and E.A. Schiff (Mater. Res. Soc. Proc. **258**, Pittsburgh, PA, 1992) pp. 129-134.
5. T.J. Konno and R. Sinclair, Phil. Mag. A **66**, 749 (1992).
6. Y. Kawazu, H. Kudo, S. Onari and T. Arai, Jpn. J. Appl. Phys. **29**, 2689 (1990).
7. S.W. Russel, J. Lian and J.W. Mayer, J. Appl. Phys. **70**, 5153 (1991).
8. C.C. Tsai, R.J. Nemanich and M. Thompson J. Vac. Sci. Technol. **21**, 632 (1982).
9. H.S. Mavi, A.K. Shukla, S.C. Abbi and K.P. Jain, J. Appl. Phys. **66**, 5322 (1989).
10. A. Cros, M. O. Aboelfotoh and K.N. Tu, J. Appl. Phys. **67**, 3328 (1990).
11. W.F. Banholzer and M.C. Burrel, Surf. Sci. **176**, 125 (1986).
12. I.H. Campbell and P.M. Fauchet, Solid State Commun. **58**, 739 (1986).
13. R. Vuppuladhadium, H.E. Jackson, J.T. Boyd, J. Appl. Phys. **73**, 4887 (1993).
14. J. Gonzalez-Hernandez, R.S. Tsu, G.H. Azarbayejani and F.H. Polak, Appl. Phys. Lett. **47**, 1350 (1985).
15. D.R. dos Santos and I.L. Torriani, Solid State Commun. **85**, 307 (1993).
16. H. Dallaporta and A. Cros, Surf. Sci. **178** (1986) 64.
17. A. Hiraki, Surf. Sci. Rep. **3**, 357 (1984).
18. B. Bian, J. Yie, B. Li and Z. Wu, J. Appl. Phys. **73**, 7402 (1993).

EFFECT OF Pb CONTENT ON THE TRANSFORMATION OF SPUTTERED PbxZr0.4Ti0.6O3 THIN FILM

JAE-HYUN JOO, DEOK-SIN KIL, SEUNG-KI JOO
Department of Metallurgical Engineering,
Seoul National University, Seoul Korea

ABSTRACT

PZT($Pb_xZr_{0.4}Ti_{0.6}O_3$) thin films were prepared by reactive co-sputtering and annealed by RTA(Rapid Thermal Annealing). Transformation kinetics and effect of Pb content on the transformation were intensively studied using EMA(Effective Medium Approximation). It has been found that depending on Pb content as well as RTA temperature, the crystal structure of PZT films changed greatly. It turned out that the transformation temperature for the perovskite phase can be lowered and the width of transition temperature region was reduced by increasing Pb content in the films. Dependence of transformation path on the Pb content has been studied.

INTRODUCTION

$Pb(Zr,Ti)O_3$ is famous for ferroelectric, piezoelectric and pyroelectric properties. It has been studied for many applications such as non-volatile memories, high density DRAM capacitors, SAW(surface acoustic wave devices), electro-optic switches etc.[1-3] For such application, PZT films must be perovskite phase. But the as-deposited phase is usually amorphous, so post-annealing process is necessary to crystallize the films into the perovskite phase. If the post-annealing process is improper, the centrosymmetric pyrochlore phase appears.[4]

To repress the formation of the pyrochlore phase, it is important to study the transformation kinetics from pyrochlore to perovskite during the post-annealing process. We have studied the transformation kinetics of sputter deposited PZT ($Pb_xZr_{0.4}Ti_{0.6}O_3$) films using an EMA(Effective Medium Approximation)[5] and have investigated the effect of Pb content on the transformation.

EXPERIMENT

PZT films 3500Å thick were deposited by co-sputtering method. The sputtering conditions used in the experiment are summarized in Table I. To crystallize the films, RTA(Rapid Thermal Annealing) was performed in air. The crystal structure of the films were characterized by XRD. To obtain the volume fraction of the perovskite phase, the refractive index was measured by an Ellipsometer.

591

Table. I Sputtering Conditions

Target	Pb, Zr, Ti
Substrate	Pt/SiO$_2$/Si
Gas flow(sccm)	Ar/O$_2$ (2/18)
Pressure	10 mtorr
Substrate temperature	350℃

RESULTS AND DISCUSSION

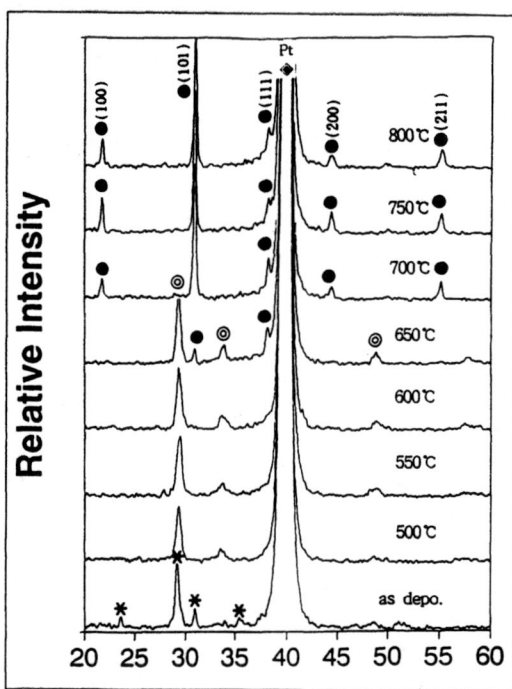

Fig.1 X-ray diffraction patterns of Pb$_{0.93}$Zr$_{0.4}$Ti$_{0.6}$O$_3$ films annealed for 30 sec with RTA temperature(*:as-deposited phase ◎:Pyrochlore ●:Perovskite).

Fig.1 shows the XRD patterns of PZT films annealed for 30sec. As RTA temperature increased, the as-deposited phase changed to the pyrochlore phase and then to the perovskite phase. The transition temperature of pyrochlore to perovskite was about 650℃~700℃. Fig.2(a) shows the change of refractive index with RTA temperature. There are three different regions. In the first region, the as-deposited phase changed to the pyrochlore phase. In the second region, the pyrochlore phase was stable. In the third region, the pyrochlore phase changed into the perovskite phase. Applying the EMA for the third region, we obtained the change of volume fraction of the perovskite phase with RTA temperature as in Fig.2(b). With increasing RTA temperature, the activation energy decreased from $\Delta E1$(~660kJ/mole) to $\Delta E2$(~71kJ/mole).

Fig.3 shows the XRD patterns of PZT films which were annealed at 650℃. As the RTA time increased, the pyrochlore phase changed to the perovskite phase. The transition time of the pyrochlore phase to the perovskite phase was 30sec~90sec. The change in the refractive index with RTA time is shown in Fig.4(a). The refractive index increased gradually with increasing RTA time.

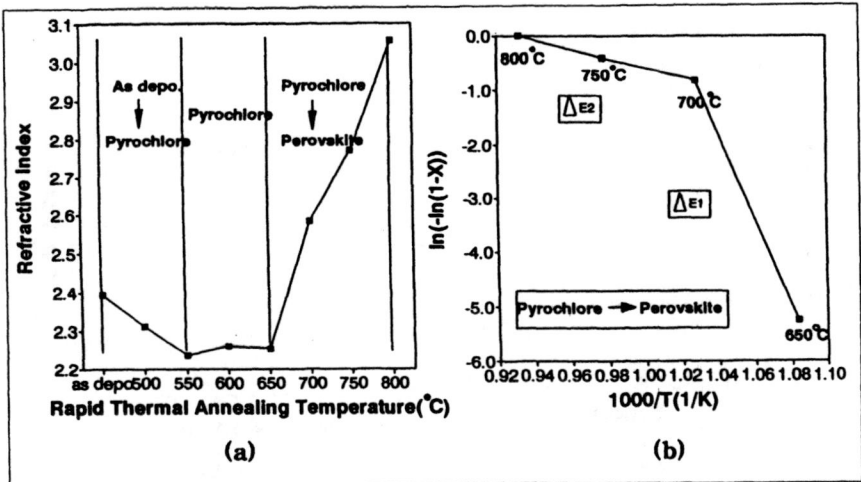

Fig.2 (a)Variation of refractive index and (b)change of volume fraction of the perovskite phase with RTA temperature of $Pb_{0.93}Zr_{0.4}Ti_{0.6}O_3$ films annealed for 30 sec.

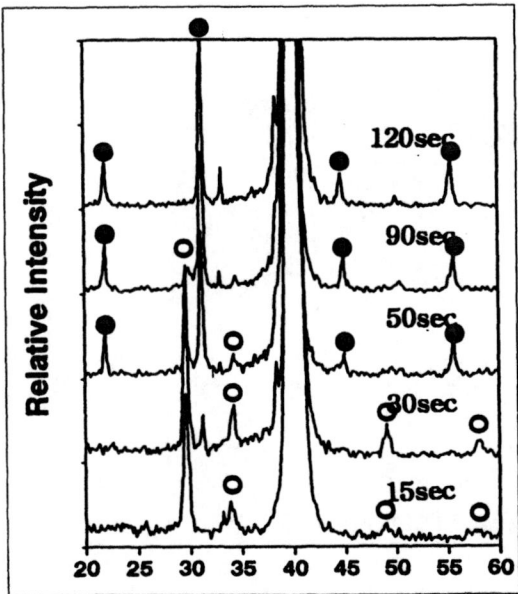

Fig.3 X-Ray diffraction patterns of $Pb_{0.93}Zr_{0.4}Ti_{0.6}O_3$ films annealed at 650℃ with RTA time.

Using EMA, the volume fraction of perovskite phase changed as in Fig.4(b). The calculated Avrami constant was about 3 in the constant nucleation region and Avrami constant about 2 in the depletive nucleation region. So we could find out that it is 2-dimensional interface controlled transformation.[6] Fig.5 is the schematic diagram of XRD patterns of PZT films with RTA temperature and Pb content. With increasing Pb content, the transition temperature decreased and the width of the transition region reduced.

Fig.6 shows the change of interplanar spacing ratio with Pb content for samples annealed at 800℃ for 30sec. With increasing Pb content, the interplanar spacing ratio increased for all perovskite peaks. From these results, we

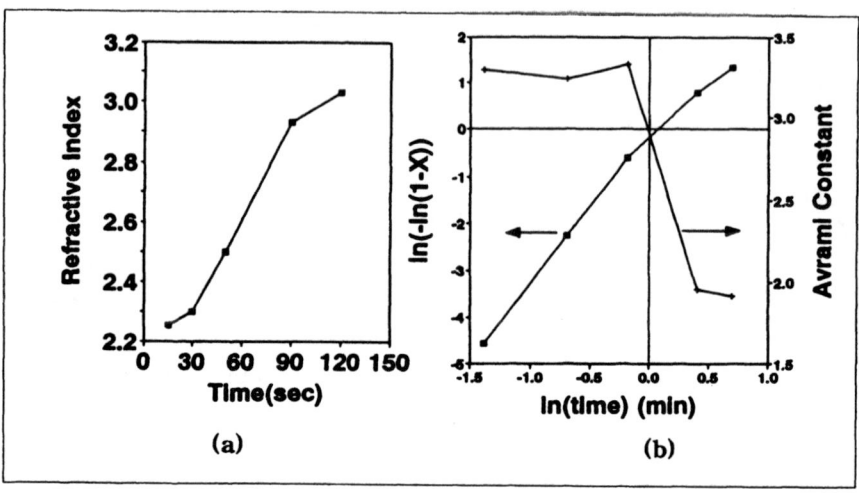

(a)   (b)

Fig.4 (a)Change of refractive index and (b)change of volume fraction of the perovskite phase and Avrami constant with RTA temperature of $Pb_{0.93}Zr_{0.4}Ti_{0.6}O_3$ films annealed at 650℃ ($X=1-exp(kt^n)$) : X=volume fraction of the perovskite phase , n=Avrami constant).

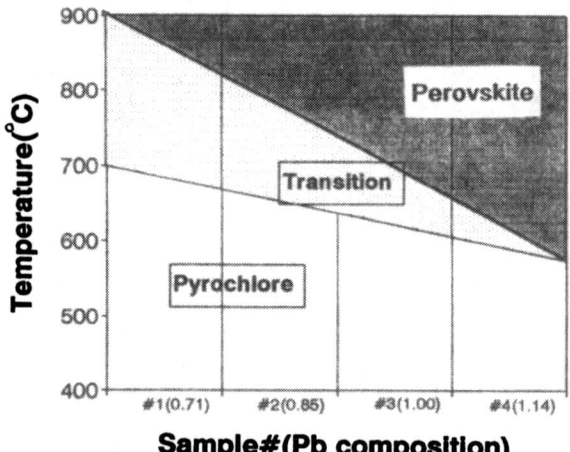

Fig.5 Schematic diagram of the phase transformation of $Pb_{0.93}Zr_{0.4}Ti_{0.6}O_3$ films with Pb content and RTA temperature.

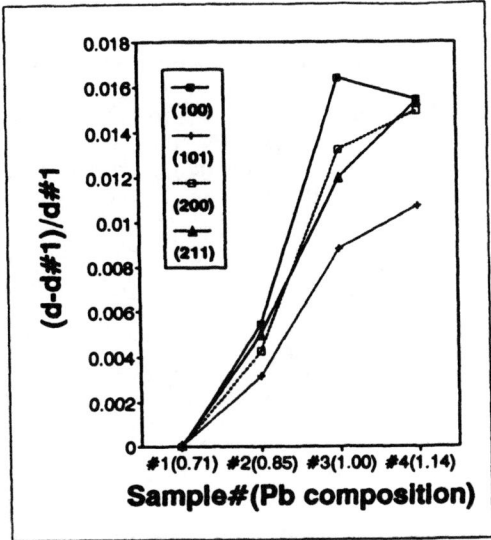

Fig.6 Interplanar spacing ratio of Pb$_x$Zr$_{0.4}$Ti$_{0.6}$O$_3$ films annealed at 800℃ for 30 sec with Pb content.

can conclude that as Pb content increased the lattice expanded in all directions.

From intensive investigation of XRD patterns, we found that three different paths exist depending on Pb content. If Pb content was sufficient in PZT films, the as-deposited phase changed to the perovskite phase directly. When the Pb content was about one, the as-deposited phase changed to the pyrochlore phase and then changed to the perovskite phase. If Pb content in PZT films was deficient, the as-deposited phase changed to perovskite phase through pyrochlore phase and PbTi$_3$O$_7$. Fig.7 shows the schematic diagram of these transformation paths.

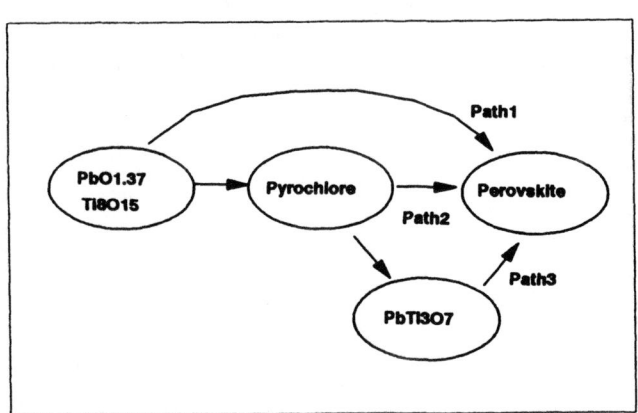

Fig.7 Schematic diagram of phase transformation path of Pb$_x$Zr$_{0.4}$Ti$_{0.6}$O$_3$ films with Pb content. (Path1 : Pb>1, Path2 : Pb~1, Path3 : Pb<1)

CONCLUSIONS

We have intensively studied the transformation kinetics of sputtered PZT films and effect of Pb content on the transformation. The transformation of the pyrochlore phase to the perovskite phase was 2-dimensional interface controlled. With increasing Pb content 1) the transformation temperature lowered 2) the width of transition region reduced 3) lattice expanded in all directions 4) transformation path changed depending on Pb content.

ACKNOWLEDGMENT

This work has been supported by KOSEF through RETCAM in Seoul National University.

REFERENCES

1. R.Srivastava and A.Mansingh, Ferroelectrics, **108**, 21-26 (1990)
2. S.Sinharoy, H Buhay, D.R.Lampe and M.H.Francombe, J.Vac.Sci.Tech.A, **10**(4), 1554-1561 (1992)
3. G.H.Haertling, J.Vac.Sci.Tech.A., **9**(3), 414-420(1991)
4. J.N.Reimers and J.E.Greedan, R.K.Kremer, E.Gmelin and M.A. Subramanian, Phys.Rev.B, **43**(4), 3387-3394(1991)
5. D.A.G.Bruggemam, Ann.Phys., **24**, 636 (1932)
6. R.H.Doremus, Rates of Phase Transformation (Academic Press,1985), p26

CRYSTALLIZATION DYNAMICS OF FERROELECTRIC PZT(52/48) THIN FILM PREPARED BY REACTIVE COSPUTTERING ON Pt ON Si(100)

WOONG KIL CHOO[*], HYO JIN KIM[**], KWANG YOUNG KIM[†], AND SUNG TAE KIM[†]
[*]Korea Advanced Institute of Science and Technology, Department of Electronic Materials Engineerig, Taejon 305-701, Korea
[**]Korea Advanced Institute of Science and Technology, Applied Science Research Institute, Taejon 305-701, Korea
[†]Gold Star Co., Inc., Central Research Laboratories, Lab 2, Seoul 137-140, Korea

ABSTRACT

The crystallization process and microstructural evolution of PZT(52/48) thin films deposited on Pt thin film electrode on Si(100) by reactive multitarget cosputtering technique have been studied as a function of post-annealing temperature and holding time. As annealing temperature increases, the amorphous PZT films as-deposited at low substrate temperature of 200 °C crystallize into pyrochlore at 450 °C and ferroelectric perovskite phase with pseudo-cubic structure at 550 °C in sequence. X-ray diffraction data show crystallization into perovskite phase to be complete in 30 minutes at 550 °C. Furthermore, the change of PZT/Pt/Ti/SiO$_2$/Si interfacial TEM morphology during heat-treatment has been closely scrutinized.

INTRODUCTION

The potential for their widespread practical applications in electronic and electro-optic devices has increased research activities on ferroelectric thin films. Much recent work has focused on lead zirconate titanate Pb(Zr$_x$Ti$_{1-x}$)O$_3$ (PZT) with a zirconium concentration x of about 0.5 as a radiation-hardened nonvolatile memory.[1-3] In PZT ceramic system, compositions with $x \approx 0.53$ crystallize in a two-phase system with equal content of tetragonal and rhombohedral perovskite phase (morphotrophic phase boundary–MPB). It is well known that the dielectric constant, remanent polarization, and piezoelectric coefficients of PZT ceramics located near the MPB show their maximum values.[4]

Experimental studies give evidence that material properties of ferroelectric thin films very often show pronounced differences with processing, which may be caused by a mere variation in the microstructure – grain size, surface or interfacial morphology – related to the crystallization process of the films. Processing parameters such as deposition technique, heat-treatment conditions and substrate are known to be contributing factors to the development of film crystallinity, grain size, crystallographic texture and growth morphology.

The fabrication of PZT thin films has been attempted by a number of techniques, including rf and dc magnetron sputtering, ion-beam sputtering, laser ablation, chemical vapor deposition (CVD), and sol-gel processing. We have recently applied a multitarget cosputtering technique, which is designed to solve the problems in the ceramic-target sputtering system, to the growth of PZT thin films.[5,6] In the present undertaking, we have explored the effect of post-annealing temperature and time on the crystallization process and microstructural evolution of PZT(52/48) thin films deposited on Pt/Ti/SiO$_2$/Si substrates by reactive dc magnetron cosputtering technique using Pb, Zr, and Ti metal-target. Furthermore, the change of PZT/Pt/Ti/SiO$_2$/Si interfacial morphology during heat-treatment has been closely investigated through cross-sectional transmission electron microscopy.

EXPERIMENT

The PZT thin films used in this study were of the morphotropic phase boundary composition, with a ratio of Zr/Ti = 52/48, and were prepared by a multitarget reactive dc magnetron

Mat. Res. Soc. Symp. Proc. Vol. 321. ©1994 Materials Research Society

cosputtering technique. The multitarget cosputtering system used here and optimized sputtering conditions for the growth of stoichiometric PZT(52/48) films are fully described elsewhere.[5,6] The films were annealed at the temperature range of 450–800 °C for 30 min to 3 h in oxygen atmosphere. The structural investigation was then performed using a Rigaku DMAX-3C glancing angle x-ray diffractometer. Changes in the surface morphology were observed using a Hitachi S-800 scanning electron microscope (SEM) equipped with a field emission gun.

Cross-sectional transmission electron microscopy (TEM) samples were prepared using a standard method. Sliced sections were mechanically polished, followed by dimple grinding. The samples were then thinned to electron transparency using ion milling with 4 kV Ar^+ ions in a Gatan Model 600 ion thinning instrument of dual type. Transmission electron microscopy was performed using a Philips CM20 scanning transmission electron microscope operating at 200 kV. Cross-sectional TEM micrographs and selected area electron diffraction patterns were recorded for each sample prepared under different post-annealing conditions.

RESULTS AND DISCUSSION

As-deposited PZT films were amorphous to x-rays and the films were subjected to an post-annealing treatment step to crystallize into the proper ferroelectric phase, i.e. perovskite phase. The film composition (Zr/Ti ratio) obtained by energy dispersive x-ray microanalysis was 52/48. As shown in Figures 1 and 2, the crystallization of as-deposited PZT film is a function of annealing temperature and holding time. Figure 1 illustrates that the amorphous PZT films were first crystallized into a pyrochlore phase of $Pb_2(Zr,Ti)_2O_{7-x}$ type for annealing temperature as low as 500 °C for 4 hours at substrate temperature of 200 °C. However, 30 minutes of annealing at 550 °C was required for the initial formation of perovskite phase (Figure 2). These results indicate that the transformation process from the amorphous to the crystalline state in the PZT films is very sluggish at lower temperatures and does not proceed at a measurable rate, as observed in most nonmetallic materials.[7] This implies that the growth rate strongly depends on

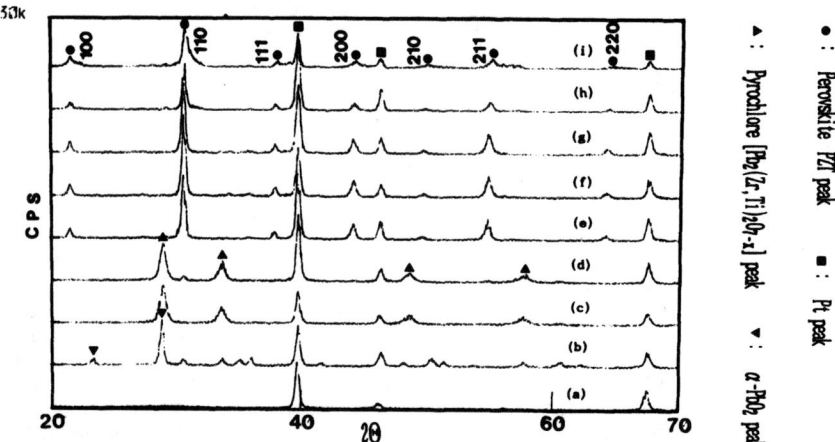

Figure 1. X-ray diffraction patterns of 0.72 μm-thick PZT(52/48) film deposited on Pt/Ti/SiO$_2$/Si at substrate temperature of 100 °C (a), 200 °C (b) and post-annealed at temeprature of 450 °C (c), 500 °C (d), 550 °C (e), 600 °C (f), 750 °C (h) and 800 °C (i) for 4 hours.

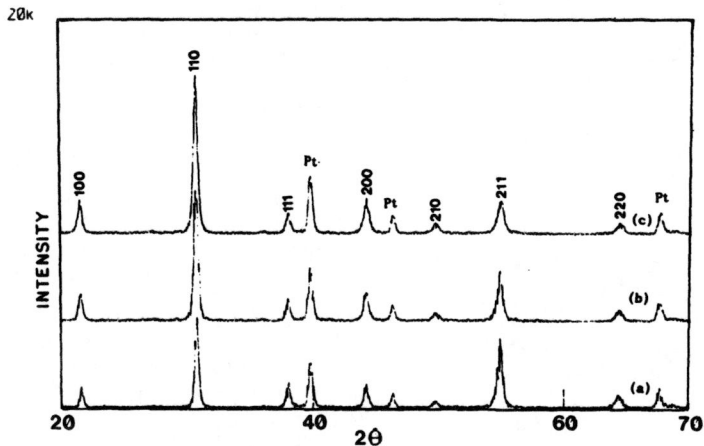

Figure 2. X-ray diffraction patterns of 2.1 μm-thick PZT(52/48) film annealed at 550 °C for 30 min (a), 1 hour (b), and 4 hours (c)

temperature with the dependence being of an Arrhenius type.

The continuous microstructural evolution of the 0.72 μm-thick PZT(52/48) film at different annealing temperatures is shown in Figure 3. The as-deposited film has a dense surface structure with grain sizes of 0.2–0.25 μm. It was revealed through the examination of film cross-section[5,6] that the individual grains are columnar with respect to the plane of the film. Heat treatments below 550 °C do not seem to alter the grain size significantly, whereas grains remarkably grow to

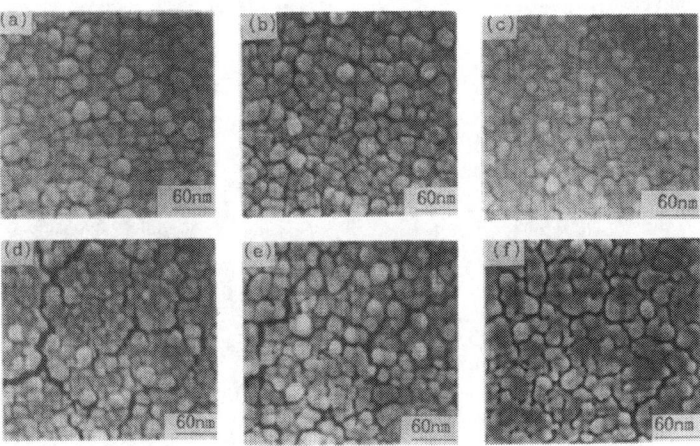

Figure 3. SEM micrographs of surface of PZT(52/48) thin films as-deposited (a), annealed at 450 °C (b), 500 °C (c), 550 °C (d), 600 °C (e), and 700 °C (f).

the size of 0.4 μm to 0.5 μm at 650 °C. The SEM micrographs clearly show significant grain boundary grooving above 550 °C, presumably due to the segregation and evaporation of lead atoms. Furthermore, at 650 °C, microvoids due to Pb evaporation during heat treatment are observed inside grains. Based on the microstructure observations including the results of x-ray diffraction, we conclude that 550 °C is the optimum annealing temperature for the PZT thin film.

The interfacial morphology or the presence of interfacial reaction between the film and the bottom electrode has significant effect on the measured electrical properties of the films.[8-10] Here, the investigation of interfacial morphologies as-deposited and after annealing in oxygen was performed using transmission electron microscopy.

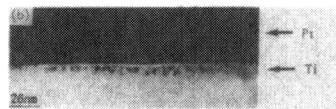

Figure 4. Cross-sectional TEM micrographs of as-deposited PZT, showing PZT/Pt bilayer (a) and Pt/Ti bilayer (b).

Figures 4–6 show the cross-sectional TEM images for the 0.72 μm-thick PZT(52/48) film sputtered on the Pt/Ti/SiO$_2$/Si substrate at several annealing temperatures. In Figure 4 for the as-deposited PZT film, sharp and discrete interfaces of PZT/Pt and Pt/Ti bilayer indicate that no significant interfacial reaction between the layers occurs. The electron diffraction pattern of ring shape[11] in the PZT layer implies that the as-deposited PZT film is amorphous, which agrees with the result of x-ray diffraction study (Figure 1).

For the multilayer film annealed at 500 °C, as can be seen from Figure 5(a), the PZT layer is divided into two regions: the upper part has a "rossette" structure, while the lower part is porous. It is well known that the rossette structure is typical of the grain structures observed in polycrystalline PZT thin films.[12] The porous region can be ascribed to damage during ion milling or microcavities formed during film deposition. For the Pt/Ti bilayer, a second phase layer of 400–500 Å thickness in TiO$_{2-x}$ layer is observed (Figure 5(b)). This isolated layer is considered to be caused by the

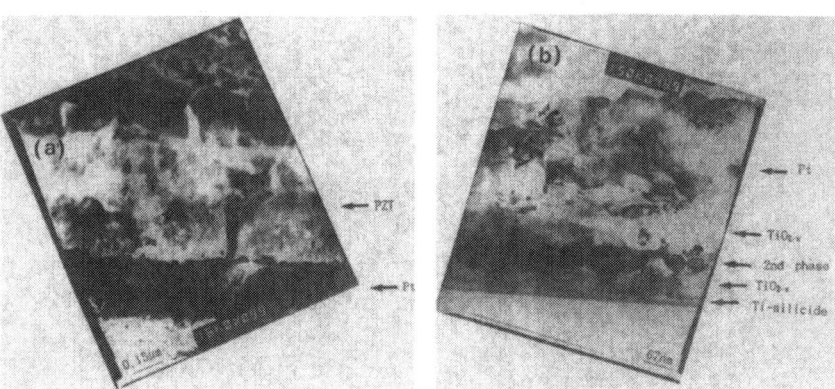

Figure 5. Cross-sectional TEM micrographs of PZT annealed at 500 °C, showing PZT/Pt bilayer (a), and Pt/Ti/SiO$_2$/Si multilayer (b).

diffusion of Pt atoms into TiO_{2-x} layer from Auger electron spectroscopy (AES) depth profile and is composed of Pt-Ti intermatallic compound, revealed by energy dispersive spectroscopy (EDS).[11] In addition, between TiO_{2-x} and SiO_2 layer, about 130 Å-thick reaction layer of Ti-silicide are observed.

At the annealing temperature of 650 °C, the interfacial morphology of PZT/Pt bilayer is presented in Figure 6. The PZT layer appears to have a columnar structure and boundaries of columns are porous. The surface region of columns coalesce and a continuous layer forms at the PZT/Pt interface, indicating that interdiffusion has taken place. These results are in basic agreement with our SEM observations.[5,6] Similarly as the sample annealed at 500 °C, a Ti-silicide reaction layer of about 130 Å thickness was found at TiO_{2-x}/SiO_2 interface. EDS analysis of the second phase inside TiO_{2-x} layer[11] showed that Pt_3Ti phase were formed.

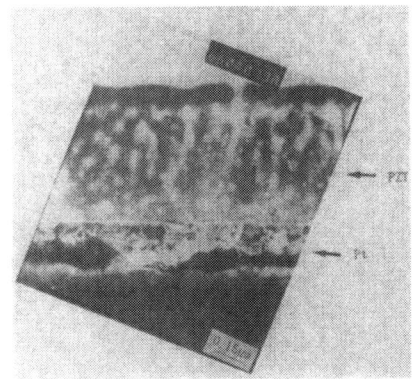

Figure 6. Cross-sectional TEM micrographs of PZT annealed at 650 °C, showing PZT/Pt bilayer.

An equivalent result was found by Olowolafe et al.[9] for $Pt/Ti/SiO_2/Si$ multilayer film annealed in oxygen ambient.

CONCLUSIONS

The crystallization process and the microstructure of the sputtered PZT(52/48) thin films strongly depend on the post-annealing temperature and holding time. A single perovskite phase with a pseudocubic structure has been obtained with annealing at 550 °C, which is the optimum annealing temperature for our PZT films. The kinetics of crystallization process is considered to be governed by an Arrhenius-type rule.

The interface analysis of $PZT/Pt/Ti/SiO_2/Si$ multilayer have revealed that the Pb-Pt reaction layer at PZT/Pt interface and Ti-silicide at Ti/SiO_2 interface are formed at annealing temperatures higher than 450 °C. Also, the intermetallic compound Pt_3Ti layer intervenes as isolated layer inside the TiO_{2-x} layer due to the interdiffusion of Pt/Ti and the barrier effects of TiO_{2-x} during heat-treatment.

REFERENCES

1. J. F. Scott and C. A. Araujo, Sciene **246**, 1400 (1989).
2. S. K. Dey and R. Zuleeg, Ferroelectrics **108**, 37 (1990).
3. L. H. Parker and A. R. Tarsch, IEEE Circuit and Device Magazine **6**, 17 (1990).
4. B. Jaffe, W. R. Cook, Jr., and H. Jaffe, *Piezoelectric Ceramics* (Academic, New York, 1971).
5. K. Y. Kim, S. T. Kim, and W. K. Choo, Jap. J. Appl. Phys. **32**, 1700 (1993).
6. W. K. Choo, K. Y. Kim, H. J. Kim, and S. T. Kim in *Evolution of Surface and Thin Film Microstructure*, edited by H. A. Atwater, E. Chason, M. Grabow, and M. Legally (Mater. Res. Soc. Proc. **280**, Pittsburg, PA, 1993).
7. E. I. Givargizov, *Oriented Crystallization on Amorphous Substrates* (Plenum Press, New York, 1991).
8. K. F. Etzold, R. A. Roy, K. L. Saenger, J.-W. Lee, and J. J. Cuomo in *Ferroelectric Films*, edited by A. S. Bhalla and K. M. Nair (Ceramic Transactions **25**, Amer. Ceram. Soc., Westerville, 1992), pp. 399–411.
9. J. O. Olowolafe, R. E. Jones, A. C. Campbell, P. D. Maniar, R. I. Hedge, and C. J. Mogab in

Ferroelectric Thin Films II, edited by A. I. Kingon, E. R. Myers, and B. Tuttle (Mater. Res. Soc. Proc. **243**, Pittsburg, PA, 1992), pp. 355–360.

10. R. A. Roy, K. F. Etzold, and J. J. Cuomo in *Ferroelectric Thin Films*, edited by E. R. Myers and A. I. Kingon (Mater. Res. Soc. Proc. **200**, Pittsburg, PA, 1990), pp. 77–82.

11. K. Y. Kim, PhD thesis, KAIST, 1993.

12. L. N. Chapin and S. A. Myers in *Ferroelectric Thin Films*, edited by E. R. Myers and A. I. Kingon (Mater. Res. Soc. Proc. **200**, Pittsburg, PA, 1990), pp. 153–158.

CHARACTERISATION OF FERROELECTRIC LITHIUM TANTALATE THIN FILMS PREPARED BY A SOL-GEL PROCESS

CHIANPING YE*, PAUL BAUDE, AND DENNIS L. POLLA,
E.E. Dept. University of Minnesota, Minneapolis, MN 55455.
* Present Address: Physics Department, Boston College, MA 02167

ABSTRACT

Thin $LiTaO_3$ films were prepared by spin coating of polymerized sol-gel precursor solution. Films have been deposited on single crystal silicon substrate, Ti/Pt or SiO_2 coated silicon substrate. Films were characterized by x-ray diffraction, dielectric and pyroelectric measurements. High Curie temperature (above 550 °C) was assumed for $LiTaO_3$ thin films from the temperature dependence of dielectric constant. Replacing 35% of tantalum by titanium atoms in the $LiTaO_3$ precursor solution has resulted the thin films with Curie temperature of 330 °C. The lower Curie temperature leads to the larger pyroelectric coefficient at room-temperature, which is more than double that of the undoped $LiTaO_3$ thin films. The dielectric, pyroelectric, and ferroelectric properties have been compared to the single crystal $LiTaO_3$ and ceramic $Li_{0.91}Ta_{0.73}Ti_{0.36}O_3$. $LiTaO_3$ thin films are available by sol-gel process at low temperature, and their properties may possibly be controlled by varying the composition of the sol-gel precursor solution.

Introduction

This work is the continuation of our efforts to develop new materials for integrated room temperature infrared detectors and for integrated electro-optical applications.

The change in polarization with temperature which occurs in any polar material is known as the pyroelectric effect. As a subgroup within polar materials, ferroelectric crystals generally exhibit the strongest pyroelectric effects. Our previous studies on infrared detectors based on $PbTiO_3$ and PLT have shown promising results[1,2]. For sensitive pyroelectric detectors, $LiTaO_3$ is also a good material due to excellent figure of merit [3]. Although $LiTaO_3$ does not have the perovskite structure, it also forms ABO_3 lattices with oxygen octahedra. On the other hand, $LiTaO_3$ pyroelectric sensors have a much lower microphonic sensitivity effect due to their relatively low piezoelectric coefficients (d_{33}=-8pC/N, d_{31}=-2pC/N). This provides better performance in a vibrational or intensely acoustic environment [4]. The very high Curie temperature (618 °C) provides a temperature independent sensitivity over a wide operational temperature range. Replacing 36% of tantalum by titanium atoms in $LiTaO_3$ ceramic has reduced the Curie temperature from 620 °C to 358 °C [5], which leads to better sensitivity as a room-temperature infrared detector.

On the other hand, $LiTaO_3$ has been reported as a better material for optical waveguide because it is highly resistive against photorefractive damage, and also because of its large nonlinearities, and short-wavelength transparency [6].

Integrated microsensor applications of pyroelectric materials require film deposition techniques to be compatible with conventional device processing. Sol-gel processes have been used to produce polycrystalline ceramic films at significantly lower temperatures than other more traditional ceramic fabrication methods [7]. Compared to the other deposition methods, such as rf sputtering, electron beam evaporation, and MOCVD, sol-gel processing also offers the possibility of independent fine control over both composition and crystalline structure.

Sol-gel method has been used for the deposition of $(Pb,La)(Zr,Ti)O_3$ thin films with various composition, and it has been combined with silicon micromachining for fabrication of microsensors in our laboratory [8, 9].

In the present work, both stoichiometric LiTaO3 and titanium doped LiTaO3 thin films were deposited on various substrates by sol-gel method. The pyroelectric, dielectric, and ferroelectric properties of the resultant films have been investigated, and the deviations from bulk material behavior were evaluated.

Precursor Solution Preparation and Thin Film Deposition

The solution preparation technique used in this work has been developed by Phule, et al. for preparation of ultrafine LiTaO3 powders at temperatures lower than 600 °C [9]. In the process of preparing LiTaO3 precursor solutions for thin films, organic precursors of the tantalum ethoxide type were reacted with Lithium acetate in an acidic environment. The resultant solutions are clear, colorless, and having equivalent LiTaO3 concentration of 4% or 5% by weight. The solutions with lower H_2O / alkoxides ratio used during preparation usually are stable for six to seven days; the gellation is usually completed in ten days. A hydrolyzed solution for spin coating was stable for two days, then the solution became cloudy and finally turned into white gel in four days. It was found that the hydrolysis condition and the equivalent concentration of LiTaO3 in the solution have strong influence on the stability of the precursor solution. The influence of these conditions on crystallization behavior, single layer thickness, and the surface roughness of PbTiO3 thin films has been investigated.

Spin coating was carried out at a speed of 2000 rpm for 30 seconds. Pyrolysis was carried out in air at 400 °C for 20 min after each deposition. Thick films were prepared by multiple spin depositions. A final annealing step was carried out to densify and crystallize the film in air at 500-600 °C for 30 min. Crack-free films with thicknesses ranging between 0.27 and 0.4 μm on crystalline silicon, platinum, silicon dioxide, and silicon nitride surfaces were obtained.

Thin Film Characterization

The film thickness resulting from four spin coatings from w.t. 5% solution was 2800Å before firing, 2350Å after firing. The refractive index is measured at wavelength of 6328Å using an ellipsometer. The refractive index is in the range of 1.8-2.0.

The crystalline phase formation and the preferred orientation of crystal growth were studied by X-ray diffraction analysis with Cukα radiation. Fig. 1 shows the XRD results of films deposited on (100) silicon, with lower firing temperature of 540 °C. The XRD result for the film fired at 590 °C is shown in Fig.2. The peak ratio is significantly different from Fig.1. The prominent peaks at (104) and (006) indicate the preferred c-axis orientation is perpendicular to the surface of the silicon wafer. In general, the Rhombohedral hexagonal phase is formed in films after firing at 540 - 600 °C for 30 minutes for both films on platinum covered and bare silicon surfaces. The resulting XRD peaks have very slight shifts from the corresponding powder XRD peak positions indicating that the internal stress is not significant in these films.

A simple capacitor structure was prepared for electrical measurement. The lower electrode is formed by sputtering of titanium and platinum layers onto a thermally oxidized silicon wafer, the top electrode is a sputtered platinum layer or a vacuum evaporated gold layer. The relative dielectric constant and the loss tangent of the sol-gel prepared LiTaO3 films were obtained from the measurement of capacitance and dissipation factor at 100kHz using a Hewlett Packard Model 4194A impedance / Gain-Phase Analyzer.

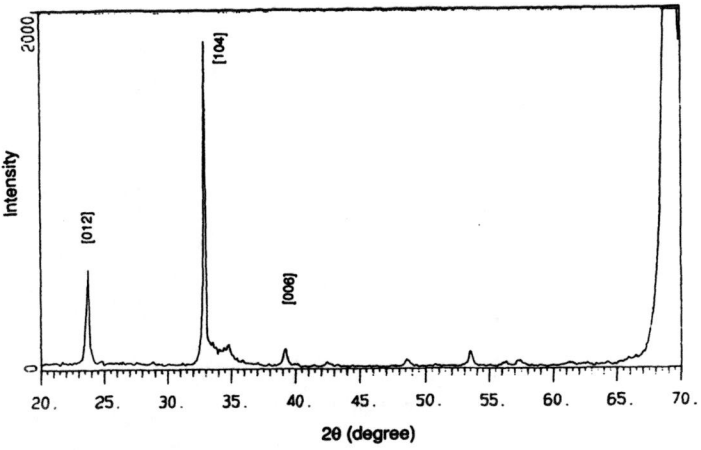

Fig. 1. X-ray diffraction spectrum of LiTaO3 polycrystalline thin film deposited on (100) silicon.
Annealing condition was 540 °C, 30 min.

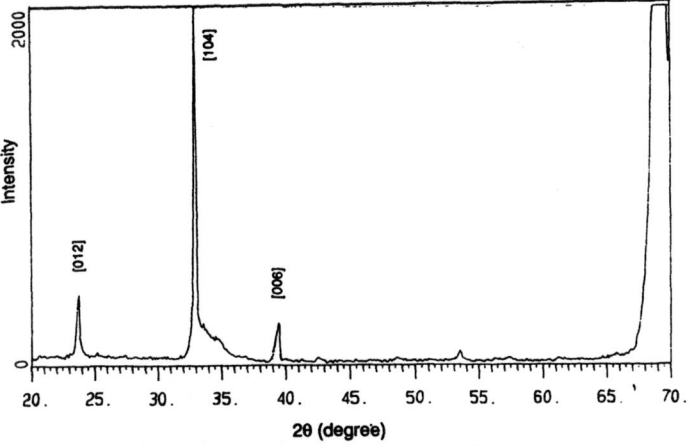

Fig. 2. X-ray diffraction spectrum of LiTaO3 polycrystalline thin film deposited on (100) silicon.
Annealing condition was 590 °C, 30 min.

The small-signal dielectric constant ε has been investigated in the temperature range of 25 °C to 550 °C. All data were collected by using a furnace with a stability of ±4 °C. The plots of the dielectric constant as a function of temperature are shown in Fig. 3 and Fig.4. The thickness of these films ranges from 3000Å to 4000Å. For LiTaO3, no maximum appeared up to 550 °C. The leaking current then became too large for the measurement at higher temperature. For Ti doped LiTaO3 the transition temperature is significantly lowered. It has shown a broad peak in the range of 300-360 °C. The maximum value around 330 °C is three times higher than the dielectric constant at room temperature. This temperature is assumed to be associated with the transition between ferroelectric and paraelectric phases.

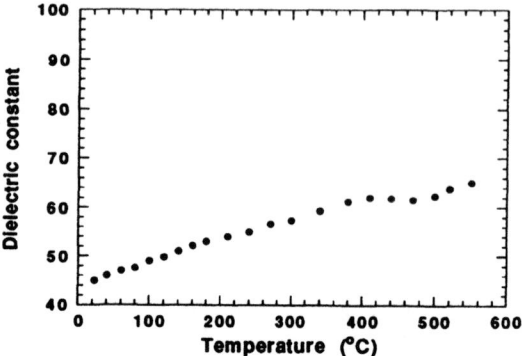

Fig.3 Temperature dependence of relative dielectric constant for LiTaO$_3$ thin films.

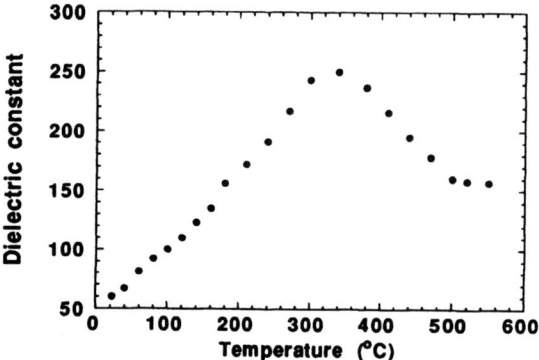

Fig.4 Temperature dependence of relative dielectric constant for Ti doped LiTaO3 thin films.

The pyroelectric coefficient normal to the electrode surfaces was measured by the *direct technique* initially used by R.L. Byer [11]. The variation of spontaneous polarization produces a displacement current I parallel to the polar axis given by

$$I = Ap^\sigma(T)\frac{dT}{dt} \tag{1}$$

where $p^\sigma(T)$ is the pyroelectric coefficient in the case of uniform heating, constant stress and low electric field in the sample, A is the active device area normal to the polar axis. In the present work the device was mounted on a copper block and heated (or cooled) at a constant rate. The pyroelectric current was measured using a Keithley Model 617 Electrometer. A temperature change rate in the range of 1.0-1.5 °C/min was applied using a Lake Shore Cryogenics temperature controller. The typical value of pyroelectric constant for $LiTaO_3$ samples is $7\pm2nC/cm^2$ K, for Ti-doped $LiTaO_3$ samples it is $15\pm2nC/cm^2$ K

Table 1 summarizes the results of measurement at room temperature for undoped $LiTaO_3$ and Ta modified $LiTaO_3$ thin films. The corresponding values of the undoped $LiTaO_3$ single crystals and the $Li_{0.91}Ta_{0.73}Ti_{0.36}O_3$ ceramics [5] are also presented in Table 1 for comparison.

Table 1. Dielectric constant ε and loss tangent tan δ at 100kHz

| | Present work-thin film | | Single crystal [5] | Ceramics [5] |
	$LiTaO_3$	Ti-$LiTaO_3$	$LiTaO_3$	$Li_{0.91}Ta_{0.73}Ti_{0.36}O_3$
Py nC/cm^2 K	7 ± 2	17 ± 2	23	-
ε (100kHz)	45 ± 2	60 ± 3	54	64.2
tan δ	$(1.5\pm0.5)\times10^{-2}$	$(1.0\pm0.5)\times10^{-2}$	6×10^{-3}	3×10^{-3}

The ferroelectric behavior at room temperature has been observed by a dynamic measurement with a Sawyer-Tower circuit and an oscilloscope readout . The polarization reversal in $LiTaO_3$ films were observed by applying 1000Hz sin-wave electrical field. The P-E hysteresis characteristics at 1000Hz are illustrated in Fig.5. The remanent polarization ranges from 17 to $19\mu C/cm^2$, and Coercive field is in the range of 60-80kV/cm. The polarization switching occurs over a field range which is relatively wide.

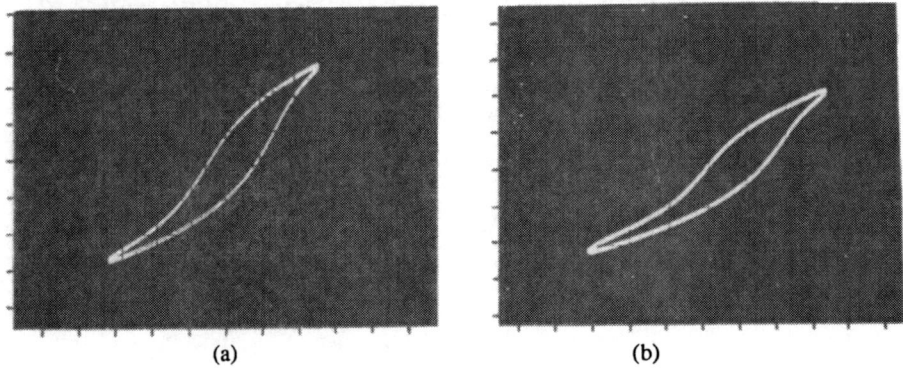

(a)   (b)

x-axis: 100kV/cm /div.,   y axis: $22\mu C/cm^2$ /div.

Fig. 5. P-E hysteresis loops observed at room temperature for (a) Pt -$LiTaO_3$ thin film -Pt capacitor, (b) Au - $LiTaO_3$ thin film -Pt capacitor.

Discussion and Conclusion

For $LiTaO_3$ crystal, the refractive index in the wavelength range of 0.60-0.70μm is reported as 2.1834 to 2.1652 [4]. The pyroelectric coefficient is $23nC/cm^2$ K [5]. The thin films deposited by the present sol-gel process exhibited lower refractive indices and pyroelectric constant. Another parameter which deviates significantly from bulk $LiTaO_3$ material is the spontaneous polarization.

A large room-temperature spontaneous polarization, $50\mu C/cm^2$, was reported on LiTaO$_3$ crystals[12]. The present data is substantially lower. The deviations from bulk crystal data may be related to the porosity of the films. Our experiments indicated that the higher level of hydrolysis water used in the solution preparation has resulted in an improved porosity. The precursor solution however, was stable for very short time. Improvements in the preparation of precursor solution to reduce the film porosity should result in improved material properties.

In conclusion, ferroelectric LiTaO$_3$ thin films have been deposited by sol-gel processing. Polycrystalline thin films having preferred c-axis orientation normal to the film have been obtained. The resultant films have shown their pyroelectric property and spontaneous polarization at room temperature. Most importantly, an enormous range of doped LiTaO$_3$ thin films can be made by this method, so that their electrical, electro-optical, and ferroelectric properties can be modified over a wide range.

References

[1]   R. Takayama, Y. Tomita, J. Asayama, K. Nomura and H. Ogawa, Sensors and Actuators, A31-A23, p508 (1990).
[2]   C. Ye, T. Tamagawa, P. Schiller, DL. Polla, Sensors and Actuators, 35, p77 (1992).
[3]   E. H. Puley, Infrared Physics, Vol. 20, p149 (1980).
[4]   LANDOLT-BORNSTEIN, Numerical Data and Functional relationships in Science and technology, Group III, Vol. 16.
[5]   K. K. Deb, IJ. Mater. Res. 2(5), p588, (1987).
[6]   K. Mizuuchi, K. Yamamoto, and T. Taniuchi, Appl. Phys. Lett. 58 (24), p2732 (1991).
[7]   K. D. Budd, S. K. Dey, and D. A. Payne, Br. Ceram. Proc., 36, 107 (1985).
[8]   Hsueh CC, Tamagawa T, Ye C, Helgeson A, Polla DL., Integrated Ferroelectrics 3, p21 (1993).
[9]   P. Baude , C. Ye, T. Tamagawa, DL. Polla, J. Applied Physics, 73 (11), p7960, (1993).
[10] P. P. Phule , T. A. Deis, and D. G. Dindiger, J. Mater. Res., 6 (7), p1567 (1991).
[11] R.L.Byer and C. B. Roundy, Ferroelectrics, 3, p333 (1972).
[12] R. C. Ruchanan, Ceramic Materials for Electronics, Ch.8, p 267 (1986).

ON PREPARATION OF CRYSTALLINE MgO-CERAMICS
FROM AMORPHOUS PRECURSORS

H. JOST*, CH. CARIUS* AND B. PEPLINSKI**
*KAI e.V., AG "Hochleistungskeramik", Rudower Chaussee 5, Gebäude 1.1, D-12489 Berlin, Germany
**Bundesanstalt für Materialforschung und Prüfung (BAM), Rudower Chaussee 5, Gebäude 1.5, D-12489 Berlin, Germany

ABSTRACT

Amorphous precursors are obtained as a result of the precipitation from magnesium acetate solutions with ammonia in the presence of citric acid. Compared with crystalline precursors they have a higher reactivity and a changed behaviour in calcination process as well as during sintering. Already at 1300 °C sintered pellets are achieved, whereas in technological process 1700 °C are needed.

INTRODUCTION

Sintered magnesia has a long-range application as a fireproof material in metallurgical furnaces. In large-scale technical process it is manufactured from seawater and natural magnesite or by precipitation from magnesium chloride solutions. A new wet chemical preparation technique is being used to improve MgO powders for ceramics.

EXPERIMENTAL

A 2M $Mg(CH_3COO)_2$ or $MgCl_2$ solution without (way A) or with (way B) an addition of citric acid is dropped into a solution of a 7.5-fold stoichiometric excess of an ammonia solution. The precipitated $Mg(OH)_2$ is filtrated, washed with water and isopropanol, calcinated at 900 °C for 2 hours, densified to green compacts at a pressure of 35 MPa and sintered at 1300 °C for 2 hours (see Fig.1).

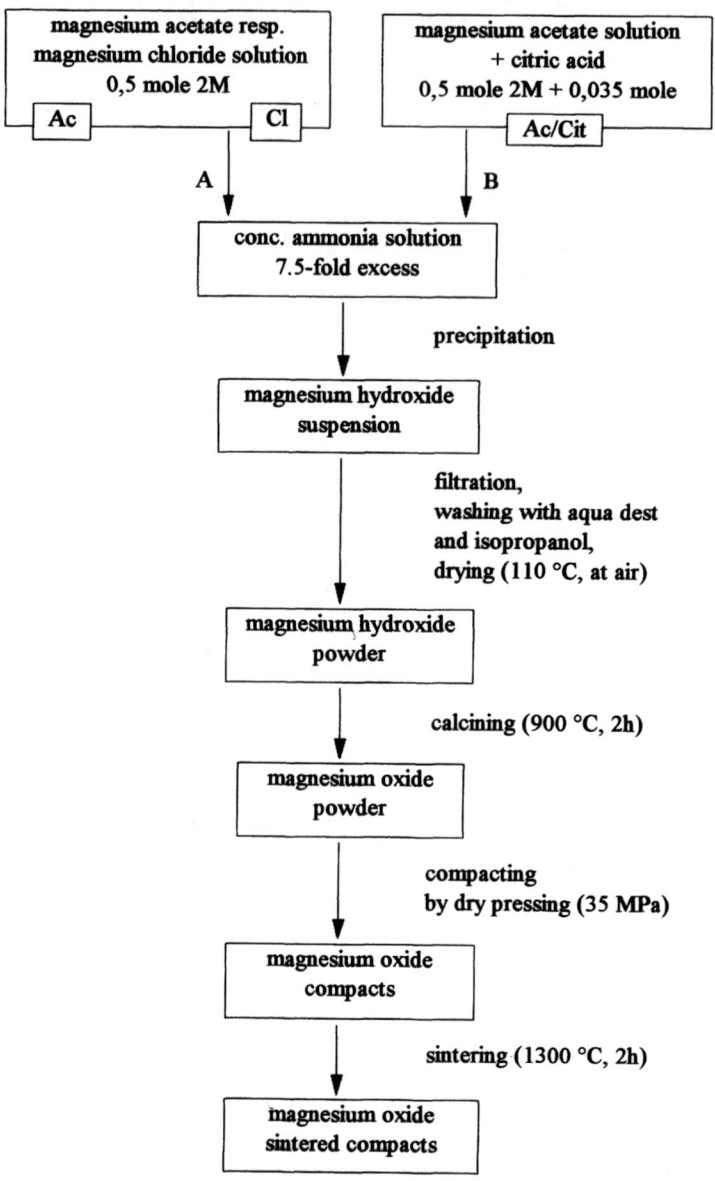

Fig. 1. Scheme of MgO preparation technique

RESULTS AND DISCUSSION

Resulting from the precipitation with an addition of citric acid the precipitate is an amorphous magnesium hydroxide powder *ac/cit* [1-3]. Refraining from adding citric acid leads to crystalline precipitates *ac*. When magnesium hydroxide is precipitated from a magnesium chloride solution with ammonia, so a crystalline material *cl* is also obtained (see Fig.2).

The specific surface area (BET) of $Mg(OH)_2$ depends on the concentration of citric acid (see Fig.3). An addition of 35 mmole citric acid leads to amorphous $Mg(OH)_2$ with surface areas of up to 650 m^2/g (m_1) and with theoretical particle sizes of 3 nm. The products of *ac* (m_2) and *cl* (m_3) have only surface areas of 80-100 resp. 60 m^2/g.

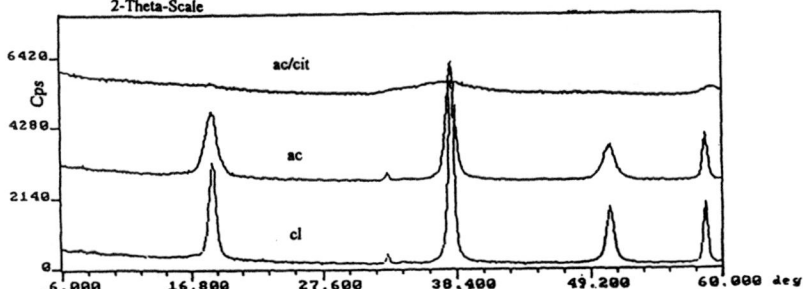

Fig.2. X-ray diffraction patterns of $Mg(OH)_2$ precipitates

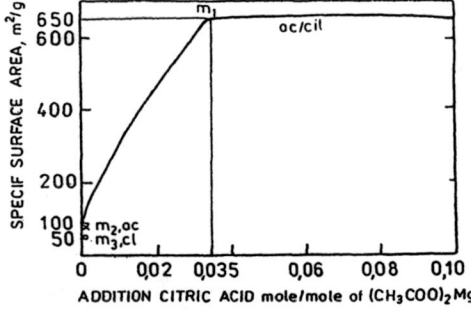

Fig.3.
Specific surface area (BET) of $Mg(OH)_2$ versus addition of citric acid to $(CH_3COO)_2Mg$

The amorphous $Mg(OH)_2$ powder (*ac/cit*) has been decomposed by heating from 20 °C to 1000 °C at air with a rate of 10 K/min (see Fig.4). The decomposition rate reaches its maximum at a temperature of 360 °C. X-ray diffraction patterns at several points before and after the maximum of decomposition (see arrows in Fig.4) have been measured (see Fig.5,6). At temperatures between 20 °C and 330 °C the precipitate is amorphous as the diffraction curves 1 and 2 show (see Fig.5). Reflections of crystalline MgO are detected in samples heated at 600 °C and 900 °C (see curves 3,4 in Fig.5). In the transition region (ranging from 250 °C to 480 °C) only amorphous pattern are observed up to 330 °C. Reflections of crystalline MgO appear above 360 °C (see Fig.6).

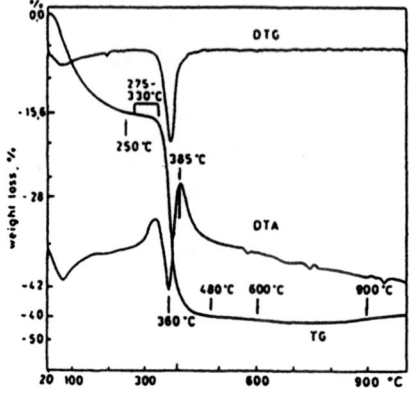

Fig. 4.
Decomposition of amorphous
Mg(OH)₂ precursors

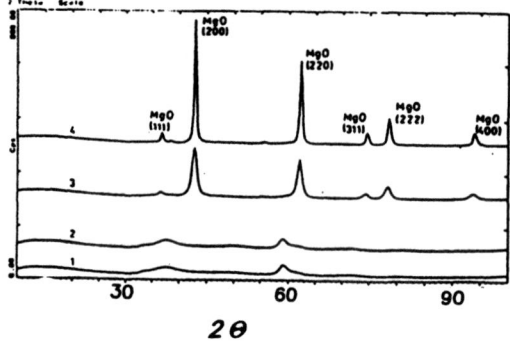

Fig. 5.
X-ray diffraction patterns of
precipitates
1   precipitate ac/cit at 20 °C
2   after thermoanalysis
     up to 230 °C
3   after thermoanalysis
     up to 600 °C
4   after thermoanalysis
     up to 900 °C

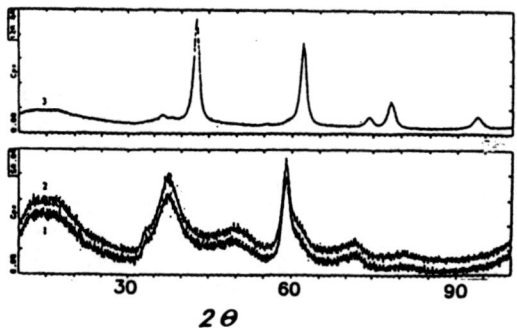

Fig. 6
X-ray diffraction patterns of
precipitates
1   after thermoanalysis
     up to 150 °C
2   after thermoanalysis
     up to 330 °C
3   after thermoanalysis
     up to 360 °C
1,2 amorphous products
     (zoomed diagr.)
3   crystalline MgO

Crystallite sizes of MgO powders were evaluated by means of the *Scherrer* method using the XRD-line broadening of the following MgO-reflections: (200) , (220) , (311) , (222) and (400). The crystallite sizes as determined from the reflex (200) are given in Table I.

Table I. Crystallite size of MgO, as determined from the reflex (200)

DECOMP. TEMP. [°C]	ac/cit [nm]	ac [nm]	cl or tech [nm]
360	9.6	23.4	—
385	9.9	—	—
480	9.9	24.2	—
600	10.6	—	—
900	29.4	42.6	48

The crystallite size of the crystalline product (*ac/cit*) is 9.6-10.6 nm and increases rapidly above 600 °C. After calcination at 900 °C the crystallite size is 29.4 nm. This value is essentially smaller compared with products, which are made without citric acid addition.

The shrinkage behaviour during sintering of green compacts is different (see Fig.7,8). *Ac/cit*-green bodies show their shrinkage maximum at a lower temperature (m_1: 1040 °C) than those from a *ac*- (m_2: 1260 °C) or *tech*-product (m_3: 1420 °C) respectively.

Some properties of MgO powder and compacts are summarized (see Table II).

Table II. Properties of MgO powders and compacts

PROPERTIES	MgO tech	ac	ac/cit
BET surface area of Mg(OH)$_2$ [m²/g]	60	80-100	550-700
BET surface area of MgO [m²/g]	12	40	50
crystallite size of MgO [nm]	48	42.6	29.4
green density at 35 MPa [g/cm³], (% th. d.)	1.67 (46.8)	1.42 (39.9)	1.39 (39.2)
sintering density at 1300 °C, 2 h [g/cm³], (% th. d.)	2.02 (56.8)	2.19 (61.5)	3.21 (90.2)
bending strength [MPa]	11 ± 1.8	17 ± 4.2	44 ± 10.0

Amorphous precursors lead to powders with advantageous properties:
* very high surface areas (BET) of *ac/cit*-Mg(OH)$_2$
* smaller crystallite sizes of *ac/cit*- than of *ac*-, *cl*- or *tech*-MgO
* low green densities of *ac/cit*- and *ac*-bodies
* high sintering densities of *ac/cit*-bodies at a temperature of 1300 °C and a duration of 2 h
* the best bending strength for sintered *ac/cit*-laboratory specimens

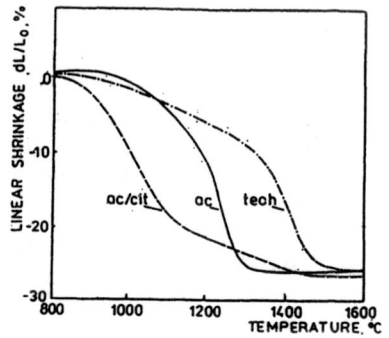

Fig. 7. Sintering behaviour of MgO compacts   Fig. 8. Sintering behaviour of MgO compacts

SUMMARY

Amorphous precursors are obtained as a result of precipitation from magnesium acetate solutions with ammonia in the presence of citric acid. Compared with crystalline precursors they have a higher reactivity and a changed behaviour in calcination process as well as during sintering. Already at 1300 °C sintered pellets with 90 % of the theoretical density are achieved, whereas in technological process 1700 °C are needed.

ACKNOWLEDGEMENT

This work was supported by the Deutsche Forschungsgemeinschaft.

REFERENCES

1. T. Shirasaki
   Denki Kagaku 29, 551 (1961)

2. P. Lessing
   Ceramic Bulletin 68, 1002 (1989) 5

3. P. Hidber, Th. Graule and L. Gauckler
   Jahrestagung der Deutschen Keramischen Gesellschaft vom 6. - 10. Okt. 1993 in Weimar
   Tagungsband, p. 260

NANOMETER-SCALE OXIDE PARTICLES IN GeSi FILMS GROWN BY WET OXIDATION

TAN-CHEN LEE*, ROBERT J. SOAVE*, YOSI Y. SHACHAM-DIAMAND** AND
JOHN SILCOX***
*Department of Materials Science and Engineering,
**Department of Electrical Engineering,
***School of Applied and Engineering Physics,
Cornell University, Ithaca, NY 14853

ABSTRACT

Amorphous GeSi films with different thicknesses and oxygen contents were electron beam evaporated onto Si(100) wafers and wet oxidized at 900 °C for 30 min. If there was no oxygen in the as-deposited film, an epitaxial GeSi film would be grown after wet oxidation. For the samples with oxygen, epitaxial growth broke down when the thickness of the epitaxy exceeded about 200 Å and polycrystalline GeSi films were formed. A dedicated STEM (scanning transmission electron microscope) was used to characterize the sample after oxidation. STEM BF (bright field), ADF (annular dark field), and energy filtered images revealed the presence of small oxide particles in the polycrystalline GeSi films. X-ray microprobe analysis with a windowless detector was employed to identify the oxide particles. The failure of the epitaxy is explained by the random nucleation and growth of GeSi grains on the oxide particles.

INTRODUCTION

Interfacial contamination is a very important consideration for the formation of epitaxial Si and Ge by SPE (solid phase epitaxy)[1]. Prokes et al.[2] reported the formation of epitaxial GeSi layers from amorphous GeSi films by wet oxidation at 900 °C. In contrast to the SPE growth of Ge and Si films, it was found that interfacial oxide contamination did not prevent epitaxy by this method. The formation of epitaxy was explained in terms of the motion of both Si and Ge during the oxidation process.[3] In the present work, GeSi films with various thicknesses and oxygen contents were deposited by electron beam evaporation. These samples were then wet oxidized at 900 °C. A VG HB501A STEM equipped with a windowless x-ray detector and an electron energy loss spectrometer (EELS) was employed to study the GeSi films grown by wet oxidation. Both EELS and the windowless x-ray detector are capable of detecting light elements such as oxygen. The probe size of the STEM can be 2 Å to 10 Å. High spatial resolution either in chemical analysis (e.g. local compositional or phase change) or imaging (e.g. lattice fringes) can thus be attained.

EXPERIMENTAL

Si (100) wafers were RCA cleaned[4] and then dipped in HF (10 %) before being loaded into an electron evaporator. GeSi films with different thicknesses and oxygen contents were deposited onto Si wafers by electron beam evaporation. The oxygen content was uncontrolled and might have resulted from the degassing of the inner wall of evaporator. The samples used in this study are listed in Table I.

Mat. Res. Soc. Symp. Proc. Vol. 321. ©1994 Materials Research Society

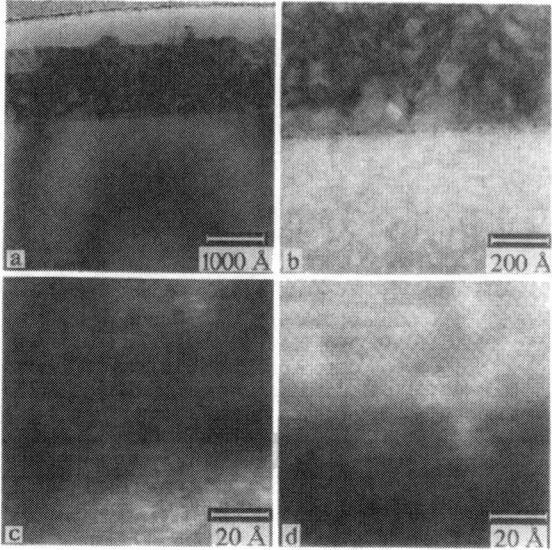

Fig. 1. STEM images of sample A1 after oxidation. (a) BF image shows strain contrast in the GeSi film. The GeSi/Si interface is sharp and well defined. (b) Higher mag. image of 1a) shows the stacking faults and twins. Mismatch dislocations are found at the GeSi/Si interface. The density of dislocations is higher than that in a totally relaxed GeSi film with the same Ge concentration. (c) High resolution BF image of the GeSi/Si interface. We speculate that the dark regions in the interface are dislocation cores. (d) High resolution ADF image of the same area as 1(c) displays both Z contrast and strain contrast from the lattice mismatch.

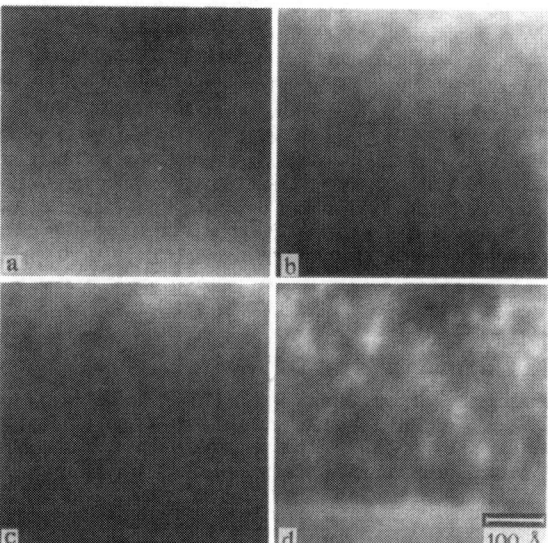

Fig. 2. STEM energy selected and ADF images of a GeSi film near the GeSi/oxide interface in sample A2 after wet oxidation.
(a) Zero loss image is similar to BF image. It is difficult to tell oxide particles from the GeSi matrix. (b) ADF image shows dark particles in the GeSi film. (c) Mass thickness mapping shows a possible atomic mass difference between the dark particles and the matrix. (d) Oxide plasmon mapping of the same area reveals the existence of oxides.

Table I

sample	Ge concentration (atomic %)	film thickness (Å)	oxygen content (atomic %)
A1	16	1800	0
A2	20	1800	16
B1	20	600	0
B2	18	600	18

The concentrations of Si, Ge and oxygen in A1 and B1 were measured by EDS (energy dispersive spectrometry) and those of A2 and B2 were measured by RBS (Rutherford backscattering spectrometry). After deposition, wafers were RCA cleaned, HF (10%) dipped then wet oxidized at 900 °C for 30 min. They were then made into cross-sectional TEM specimens and examined by using a STEM.

RESULTS AND DISCUSSION

a. As-deposited films with a thickness of 1800 Å

After oxidation, SAD(selected area diffraction) indicated that epitaxial GeSi films with twins were grown on Si substrates. In Fig. 1a, a STEM (BF) image, which is similar to a TEM BF image, shows the thickness of the GeSi film to be about 1300 Å. The variations of contrast in the GeSi films indicated strong strain fields in the epitaxy. The higher magnification picture (Fig. 1b) displays micro-twins and stacking faults. The interface was sharp and was decorated with dislocations. It is worth noticing the high density of mismatch dislocations that was observed at the GeSi /Si interface. The distance between the two nearest dislocations was about 30~50 Å. For a totally relaxed pure Ge film on Si, the distance between two nearest perfect dislocations would be 46 Å. However, the concentration profile (obtained from x-ray microprobe analysis) demonstrated that the concentration of Ge at the interface is only 10%. The highest Ge concentration in the GeSi film, which was found at the GeSi/oxide interface, was 30%. If the dissociation of partial dislocations was considered, this concentration is still lower than what is required for the formation of mismatch dislocations with such a high density. The high density of dislocations might be explained by the stacking faults observed in the GeSi layer. It is well-known that the oxidation of Si might induce extrinsic stacking faults in the Si substrate.[5,6] These interstitial planes in the GeSi thus increased the lattice mismatch between GeSi and Si. It apparently resulted in the strong strain contrast and high density of dislocations shown in Fig. 1a-b. High resolution images from STEM BF and ADF are shown in Fig. 1c and 1d, respectively. Lattice fringes are clearly seen in the ADF image and give further proof of the registry between the GeSi film and the Si substrate.

In contrast to the results from A1, polycrystalline GeSi films were found to grow in A2 under the oxide after wet oxidation. STEM Z-contrast ADF images (Fig. 2d) show the existence of 10-50 Å particles in the GeSi films. These particles were buried in the polycrystalline GeSi layer and are difficult to detect in conventional TEM. It is suggested that the contrast difference between these particles and GeSi matrix results from the variation of thickness or composition. Energy filtered images were used to characterize these particles. The energy filtered images were acquired with slit widths of 3 eV and 128 x 128 pixels. The collection time was 5 msec per pixel. Electrons that lost no energy were collected to form the zero loss image which is an energy filtered BF image. Electrons which lost 16.5 eV corresponding to the Si plasmon energy were collected to form the Si plasmon image. The mass thickness image (Fig. 2b) is a pixel by pixel ratio of the intensity in the Si first plasmon image divided by the zero loss image. Eq. 1 is the

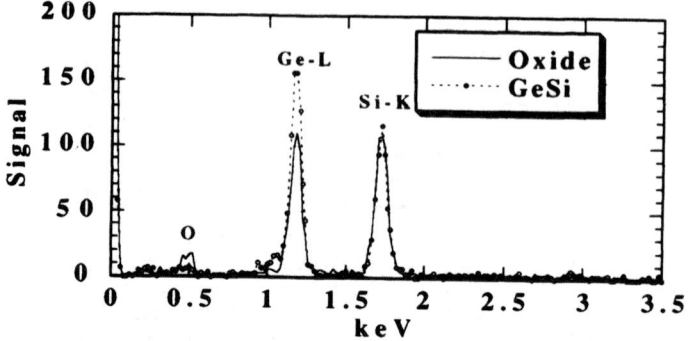

Fig. 3. X-ray spectra from GeSi and oxide particles
The oxygen signal is higher, while Ge signal is lower in the oxide particles than those in the GeSi matrix.

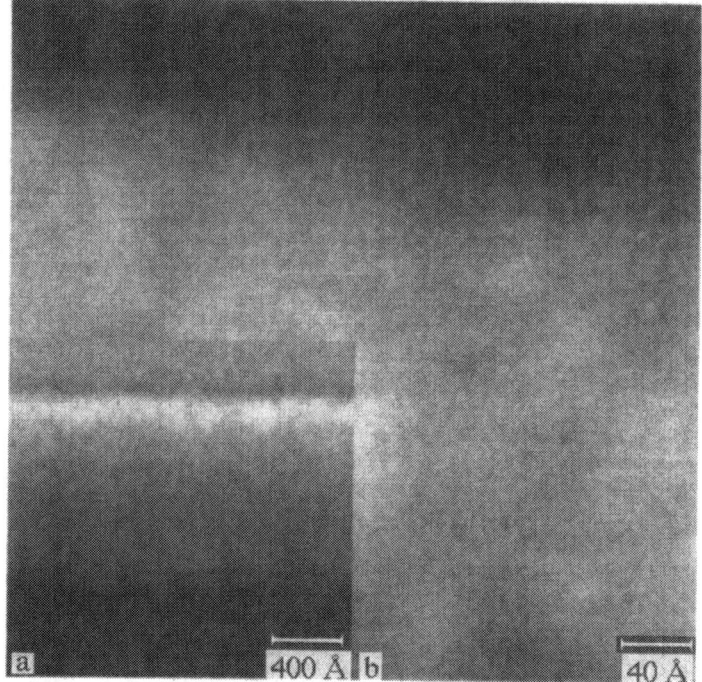

Fig. 4. ADF images of A2 after wet oxidation
(a) Higher intensity is observed in the GeSi film near the GeSi/interface. (b) Lattice fringes are found in the high resolution ADF image. It demonstrates the partial epitaxy of the GeSi film.

probability of the occurrence of each plasmon loss with a mean free path λ_p in a specimen of thickness t.

$$P_n(t)=(\frac{1}{\lambda_p})^n \frac{\exp(-t/\lambda_p)}{n!}$$ (1)

Since n equals zero for zero loss and n equals 1 for the first plasmon, Fig. 2b provides a map of t/λ_p. It was found that the thickness of the GeSi layer gradually changed. A difference in λ_p or the plasmon energy might cause the abrupt change of contrast in these particles. Silicon oxide mapping (Fig. 2c) by STEM energy filtered images revealed that these particles are oxides. X-ray microprobe analysis was performed with a probe size of 10 Å. The X-ray spectra from a similar sample are shown in Fig. 3. The oxygen signal was higher, while the Ge signal was lower in these particles than in those from the GeSi matrix. This provides evidence that these particles are silicon oxides. We believe that the existence of the oxide particles led to the failure of epitaxial growth. The abrupt contrast change in the GeSi/Si interface was unusual for a polycrystalline (Fig. 4a) film. A 200 Å-thick epitaxial region was found from high resolution ADF imaging (Fig. 4b)to form on Si substrate. Epitaxial growth broke down when the epitaxy grew thicker than 200 Å. The competition between polycrystalline and epitaxial growth was explained by the competition between random nucleation and growth and epitaxial growth. Since the small oxide particles can be treated as extra nucleation sites for random nucleation, we speculate that the delay time for random nucleation and growth was longer than that for epitaxial growth in this case. Thus, a GeSi film was grown epitaxially at the beginning even with the presence of oxide particles. When the epitaxial layer was grown to a certain thickness, strain induced by the lattice mismatch might then favor increased random nucleation and growth. Thus the polycrystalline GeSi was grown on top of the epitaxial film.

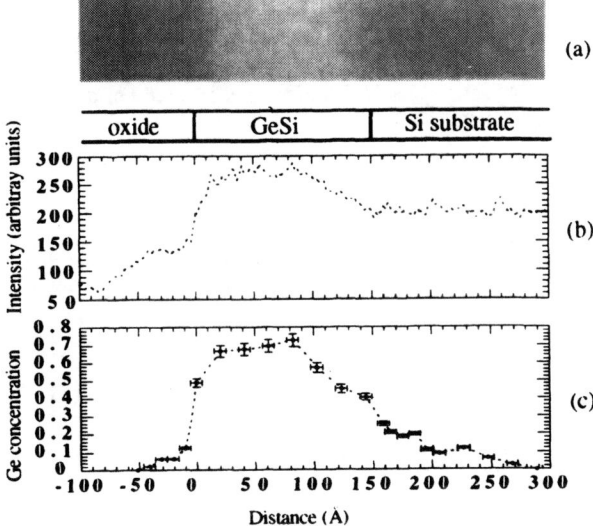

Fig. 5. (a) The Z-contrast ADF image is used to define the SiO_2/GeSi and the GeSi/Si interfaces in B2 after wet oxidation. (b) The profile of intensity versus distance was taken by a line scan across 5(a). (c) Ge concentration profile of corresponding area in 5(a) and (b). Ge atoms were rejected from the oxide and diffused into the Si substrate. This is consistent with the Z contrast image from 5(a) and (b).

b. As-deposited films with a thickness of 600 Å

For sample B1, the growth of an epitaxial GeSi film after wet oxidation was consistent with our previous experimental results from A1. Microtwins and stacking faults were also found in this film. STEM x-ray microprobe analysis was performed every 20-30Å to study the Ge distribution. The Ge concentration profile was for comparison with that from B2.

A 150 Å-thick GeSi film was formed after wet oxidation of sample B2. The GeSi film was epitaxial with the presence of oxygen in the as-deposited film. The interfaces of SiO_2/GeSi and GeSi/Si are well defined in the ADF image (Fig. 5a). The Ge concentration profile (Fig. 5c) of the corresponding area showed an abrupt change at the oxide/GeSi interface and a slow decrease of Ge concentration from the GeSi film toward Si substrate. This concentration profile was consistent with the ADF intensity profile(Fig. 5b) which was taken by a line scan across Fig. 5a. No detectable Ge was found in the oxide further than 50 Å away from the oxide/GeSi interface. Because the formation energy of germanium oxide is higher than that of silicon oxide, Ge atoms were rejected from the oxide to GeSi. This selective oxidation resulted in an increase of Ge concentration in the GeSi film thus increasing the concentration gradient toward the Si substrate. Ge thus diffused into the Si substrate to a distance about 130 Å from the GeSi/Si interface. The highest concentration in the GeSi film was 70%, which is the same as that of B1. The Ge concentration at the SiO_2/GeSi interface is 50 %. It was proposed[7] that if the Ge concentration was less than 50 % then Ge would not be oxidized. This is consistent with the Ge concentration profile obtained from the present study.

CONCLUSION

Wet oxidation of amorphous GeSi films with different thicknesses and oxygen contents was studied with a dedicated STEM. For samples with oxygen, oxide particles were found in the GeSi films after wet oxidation. It is not clear whether the oxide particles were formed during electron beam evaporation or during the wet oxidation process. However, the presence of oxygen in the as-deposited films does not influence the growth of epitaxy after wet oxidation if the thickness of the epitaxy is less than 200 Å. If there is no oxygen in the as-deposited films, the GeSi films can be grown epitaxially as thick as 1000 Å, although with the presence of high density of stacking defects. In this study, STEM with high spatial resolution is demonstrated to be very useful in the characterization of nanometer-scale oxides and the study of their effects on the epitaxial growth of GeSi films by wet oxidation .

ACKNOWLEDGMENTS

We are grateful to Earl Kirkland and Mick Thomas for their assistance on STEM. The GeSi films without oxygen were deposited by Ken Kramer from Prof. M.O. Thompson's group. This work is supported by the SRC. We also acknowledge the MSC and NNF (both supported by NSF) at Cornell for the use of facilities.

REFERENCE

1. M. Maenpaa, L.S. Hung, M.G. Grimaldi, I. Suni, J.W. Mayer, M.-A. Nicolet, and S.S. Lau, Thin Solid Films, **82**, 347 (1981).
2. S.M. Prokes, W.F. Tseng and A. Christou, Appl. Phys. Lett., **53**, 2483 (1988).
3. S.M. Prokes and A.K. Rai, Appl. Phys. Lett., **60**, 568 (1992).
4. W. Kern, and D. Puotinen, RCA Review **31**, 187 (1970).
5. G.R. Brooker and R. Stickler, Phil. Mag., **11**, 1303 (1965).
6. G.R. Brooker and W.J. Tunstall, Phil. Mag., **13**, 71 (1966).
7. H.K. Liou, P. Mei, U. Gennser, and E.S. Yang, Appl. Phys. Lett., **59**, 1200 (1991).

CRYSTALLIZATION OF AMORPHOUS METASTABLE CERAMIC PVD-COATINGS

O. KNOTEK*, F. LÖFFLER, L. WOLKERS
Materials Science Institute, Aachen University of Technology, D-52056 Aachen, Germany

ABSTRACT

Ceramic PVD coatings have got very often an X-ray amorphous structure. This is regarded for all classes of ceramics like covalent and ionic materials. These amorphous materials start to crystallize at temperatures about 800 °C, a quite low temperature for the practical use of such materials. This prevents the propagation of this materials in applications like high temperature wear protection, oxidation resistivity etc.

This paper describes several possible techniques to increase the thermal stability of the amorphous state of ceramic PVD-coatings like Al2O3 and SiC up to 1200 °C. The best way is the incoporation of an additional element like nitrogen into the thin filmsl. The change of the mechanical properties of alumina leads to new applications like diffusion barriers.

INTRODUCTION

Physical Vapor deposition (PVD) coatings are normally deposited in a metastable state, which is responsible for the mechanical and thermal behaviour of these materials. They form very often metastable phases, e.g. (Ti,Al)N, due to the deposition or have interstitials and vacancies which are not in the thermodynamic stable state. Ceramic PVD coatings like Al_2O_3. SiC and Si_3N_4 form X-Ray amorphous thin films with a very dense structure. These amorphous coatings can be superior in some applications to the comparable bulk materials. For example, the corrosion resistance of an amorphous Si-N thin film is higher than of the crystalline material. Reasoned by the metastable state of these thin films, they crystallize at higher temperatures, normally at 800 °C. This crystallization is accompanied by crack formation, the crystallization can also cause a coating failure. These effects prevent the practical use of such ceramic PVD-coatings at higher temperatures, as they are demanded for wear protection, oxidation resistance or diffusion barriers.

The aim of this study is to understand the way of crystallization of these thin films and to develop a way to prevent the crystallization by shifting it to higher temperatures. This paper investigates the systems Al-O and Si-C.

EXPERIMENTAL PROCEDURE

The structural and thermochemical stability of these coatings were investigated after deposition and annealing at various temperatures. The characterization techniques used were X-ray diffractometry and structure observation by SEM. Together with the measurement of mechanical properties like microhardness, adhesive strength and toughness, a full characterization is given. The mechanical properties were investigated on metal or cemented carbide substrates, so that all these properties are showing the behaviour and possible improvements of metal-ceramic composites. The coatings for the X-ray diffraction and the annealing tests were deposited on alumina. This substrate was chosen for the annealing tests, because it is known to have a extrem high thermal stability and nearly no reactivity with the coatings, so that mainly the coating properties and not the composite properties were examined. The annealings were carried out in a high vacuum furnace in 200 °C steps, with 3 hours annealing time.

Mat. Res. Soc. Symp. Proc. Vol. 321. ©1994 Materials Research Society

COATINGS IN THE AL-O SYSTEM

Due to the extremly high thermal gradient in the R.F. magnetron sputter ion plating process and the fact that ceramic materials need a long crystallization time combined with a high energy level [1], the sputtered Al-O coatings are deposited amorphous. This is in contrast to the crystalline CVD coatings [2,3]. In figure 1, which gives X-ray diffraction pattern measured with a Co-tube, is shown, that at higher temperatures, the coating starts to crystallize in several low temperature alumina modifications, e.g. γ-, δ- or θ-Al_2O_3 [4]. Above 800°C, the coating starts to form the α-phase. Due to this crystallization, the volume contraction is causing very high stresses in the coating so that cracks and finally peel off effects are destroying the substrate-coating composite.

Therefore a stabilization of the amorphous structure at higher temperatures is needed. The stabilization can be done in several ways. There is a number of variable technical and process parameters like power source, power, partial gas pressures, target materials, bias etc. which have an influence on the structure and morphology of the coatings. Most parameters are not suitable for a successful increase of the thermal stability, only small improvements were achieved.

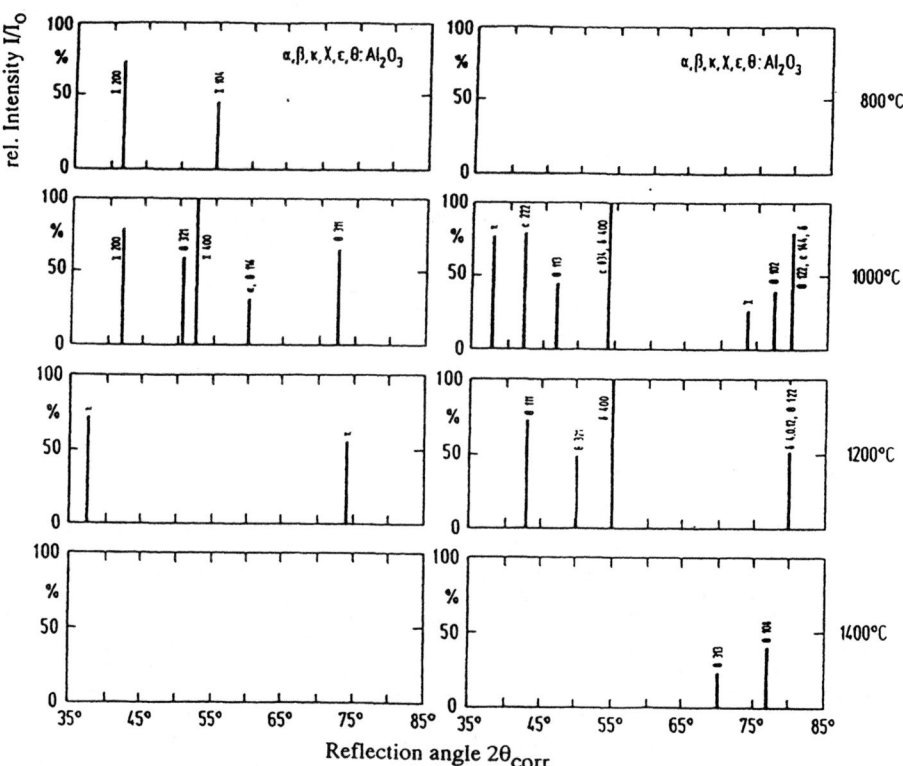

a)   b)
Figure 1: Crystallization of Al_2O_3 (a) and Al-O-N (b) coatings

Another way for the stabilization of the amorphous phase structure is a small change in the coating composition, for instance with an integration of a small element, like nitrogen, oxygen and carbon which can be solved at interstitial places. This is possible for the here used MSIP (Magnetron Sputter Ion Plating) process by the addition of a gas during the deposition. Remarkable improvements in the thermal stability are achieved with nitrogen, shown in figure 1. where the X-ray diffraction patterns for the systems, Al-O and Al-O-N, are given. The left side is showing the X-ray diffraction pattern of an Al_2O_3 coating, the right side shows an Al-O-N coating after several annealing steps. No crystalline phases are observed in Al_2O_3 below 800°C. Several Al_2O_3 phases are formed at 800 °C, 1000 °C and 1200 °C. At 1400 °C, no phases are given. This caused by the alumina substrate, which has got also the α-phase,. The proof is not directly possible. Examinations on other substrates like cemented carbide confirm this result. The Al-O-N coatings show a similar crystallization behaviour with one important difference: the starting temperature for the crystallization is 200 °C higher and at 1400 °C is not only the α-phase formed. Further investgations show that this temperature difference is a function of the nitrogen content in the coating [4]. With a higher nitrogen content the thermal stability increases.

	Microhardness HV0,05	Transition temperature [°C]	Deposition rate [μm/h]	Critical load [N]	Intrinsic stresses [N/mm²]	Growth stress [N/mm²]	Process	Target	Gas	Additional gas
Al_2O_3	2000	800	5	40	0		HF	Al_2O_3	-	
Al_2O_3 +N	1500	1100	3	45			HF	Al_2O_3	-	N_2

Figure 2: Properties of Al_2O_3 and Al-O-N Coatings

The change of the mechanical properties, shown in figure 2 through the built in of the nitrogen is essential. The microhardness decreases a lot. This effect increases with an increasing nitrogen content. For example. coatings stabilzed to 1300 °C have a microhardness of 500-800 HV0,05. This is too low new applications in the field of wear protection, especially abrasive wear. Potential other applications are corrosion or oxidation protection. A very promising application for this new material seems to be diffusion barriers for turbine blades [6]. The corrosion protection is not investigated, but can be assumed to be the simalar to amorphous alumina. The intrinsic and growth stresses, measured at a deposited bending strip, are so small that they are not measurable for both coating variations. The deposition rate for the stabilized coatings decreases, which can be explained by the higher pressure during deposition.

COATINGS IN THE SIC SYSTEM

Siliconcarbide is an interesting material for high temperature wear applications like alumina. It has an extremly high hardness and thermal stability. Like alumina it is deposited amorphous with an PVD [7] and PECVD [8] process. Crystalline SiC is only known to be deposited by CVD processes with temperatures around 1000 °C [9]. This is reasoned by the high energy and long time which is needed to form covalent bondings, which are the main bonding type in SiC. The covalent bonding is responsible for the properties of this material.

These thin films start to crystallize at 800 °C. The effect on the coatings are same as they are described for the alumina coatings. Therefore it is also necessary to stabilize the amorphous state, because the demanded temperatures for a crystalline deposition are much to high for a PVD-process. The variation of the technical parameters, as mentioned before, have a small influence on the thermal stability of the coatings, as it is regarded for the alumina. A succesful way for the stabilization is here again the build in of additional elements. Examined structure stabilizing elements are nitrogen, carbon and a combination of both. These are added to the plasma gas in small concentrations during the coating process.

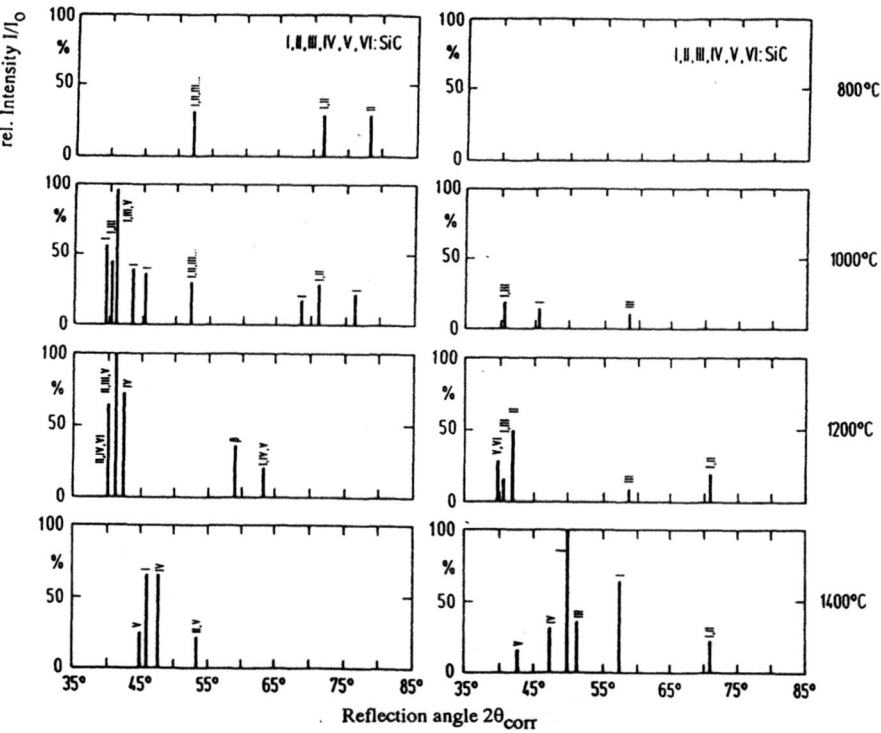

a)   b)
Figure 3: Crystallization of SiC (a) and Si(C,N) (b) coatings

The effect of the stabilization by nitrogen is given in figure 3, which shows the X-Ray diffraction of a normal SiC-coating and a nitrogen stabilized one. The nomenclature of the different α-modifications is done with the Thilbault model. SiC has a crystallization starting temperature of 800°C. At increasing temperatures an intensivated crystallization is seen. At 1400 °C a coating decomposition is observed, some coating material is evaporated. The crystallization behaviour of the stabilized SiC takes place in a different way. At 800°C there is no crystallization observed. The first SiC-crystalls are detected at 1000°C. The maximum of the crystallization is reached at 1400°C, a decomposition does not take place. Using carbon as an additional element an accelerated crystallization starts at 800°C. Using mixed reactive gases the crystallization takes place between these temperatures. It is remarkable that only α-SiC is examined. The expectable formation of β-SiC in the stabilized coating does not take place, although nitrogen is known to stabilize and catalyse the formation of β-SiC [9]. That means that only α-germs are formed through the depositio and first annealing steps. A prefered α-SiC modification and orientation can not be observed.

	Microhardness HV0,05	Transition temperature [°C]	Depositionrate [μm/h]	Critical load [N]	Intrinsic stresses [N/mm^2]	Growth stresses [N/mm^2]	Process	Target	Gas	Additional gas
SiC	4000-7000	800	8	25	-1000	800	DC	SiC	-	-
Si(C,N)	4000-7000	1000-1200	8	15	-750	1100	DC	SiC	-	N_2
SiC+C	5500	800	10	17	-700	1000	DC	SiC	-	CH_4
Si(C,N)+C	4300	1000	14	15	-730	1000	DC	SiC	-	N_2 + CH_4

Figure 4: Properties of SiC and stabilized SiC coatings

The mechanical properties of the stabilized coatings are of great interest, as shown in figure 4. The microhardness of the stabilized SiC coating shows a small decrease. But the hardness is quite high in comparison to other PVD-coatings especially the stabilized alumina. The critical load decreases a little bit, so do the intrinsic stresses. The deposition rate shows nearly no change. Great changes are seen at the transition temperature amorphous-crystalline, which increases a lot by the nitrogen stabilized coatings.

The other examined variations do not have such a effect. Using carbon the changes in the mechanical properties are small, an increase of deposition rate and a moderate decrease of the hardness and adhesion is investigated, the stresses do not change. Using the mixed atmosphere with a carbon and nitrogen carrier gas, a further increase of the deposition rate is given, the hardness decreases a lot, but is still high.

All examined variations of the SiC coatings have got the same problems in the practical use. The adhesion is quite low and deteriorated by the stabilization so that a bond layer is necessary for applications in the wear protection, for example indexible inserts. This problem has to be solved also for the stabilized Si-C PVD-coatings.

CONCLUSIONS

Amorphous deposited ceramic PVD-Coatings like Al_2O_3 and SiC crystallize at relatively low temperatures. A characteristic temperature is 800 °C, SiC forms only the α-SiC, while Al_2O_3 crystallize in several modifications like δ, ε -Alumina. This effect prevents the use of these materials in high temperature applications, because the crystallization is often accompanied by a crack formation in the thin films. The benefits of the materials like high hardness, high temperature wear resistance, oxidation resistance etc. can not be explored.

Because there is no chance to deposite crystalline ceramic coatings with a low temperature PVD-process, the amorphous state of the thin films has to be stabilized to higher temperatures. A useful tool is the incorporation of an additional element into the coating. This must be a very small element with no solubility in the crystalline phases of the coating material. This is fulfilled by nitrogen. A stabilization of the amorphous state up to 1000 °C and more is possible.

The changes in the mechanical properties through the incorporation are relativly small for SiC, while the changes for the alumina coating are drastically. The microhardness decreases here so much, that the coating is not useful for applications in the wear protection. New applications like diffusion barriers and corrosion protection at high temperatures up to 1000 °C and more are possible. SiC is still interesting for wear applications, but the adhesion of this coatings has to be improved.

REFERENCES

[1] H. Salmang, H. Scholze; Keramik; Springer Verlag, Berlin, Heidelberg; 6. Aufl. 1982
[2] E.A. Irene et. al.; J. of Electr. Mat. 4, 1975, 409 ff
[3] J.J. Sylvestri, et. al.; J. of Electr. Mat. 4, 1975, 429 ff
[4] T. Leyendecker; Doctoral Thesis, Aachen University of Technology, 1985
[5] A.v. Richthofen et. al., Proc. o. AMPT'93, Dublin, 1, 1993, 791
[6] P. Kettunen; Diffusion Barriers, Proc. Mat. for Power Eng. Components-Cost 501, Jülich, Germany, Oct. 13th - 14th 1992, 52
[7] O. Knotek et al, Amorphous SiC PVD coatings; Proc of Diamond Conf, Heidelberg, Germany, 31.8-4.9.1992
[8] V.A. Mernagh et al.; Surf. A. Coat. Techn, 49, 1991, 462
[9] A.R. Kieffer, P. Ettmayer, E. Gugel, A. Schmidt, Mat. Res. Bull. 4, 1969, 153

A COMPARATIVE STUDY BETWEEN HIGH AND LOW TEMPERATURE THERMALLY CONTROLLED CRYSTALLIZATION OF THIN FILMS

Richard D. Robinson and Ioannis N. Miaoulis*
Thermal Analysis of Materials Processing Laboratory, Mechanical Engineering Department,
Tufts University, Medford, MA 02155

ABSTRACT

Numerical simulation of zone-melting recrystallization (ZMR) was conducted to determine the heat transfer dynamics over a wide range of temperatures. ZMR is a thermal processing technique used to recrystallize materials. Therefore, the thermal effects induced by the ZMR process critically affect the crystallization dynamics. Parametric studies indicated that the conductive heat flux from the heat source through the gas accounted for at least 15% of the total energy heating the film for materials with melting points less than 800°C. The influence of this conductive heating has been neglected in past analyses. Also, materials with higher melting points are less sensitive to changes in the heat flux from the heat source. Slight variations of thermal gradients in the film can lead to different qualities of crystal, so care must be taken when processing materials with lower melting points, since they are more sensitive to temperature variation. This paper analyzes the dominant modes of heat transfer in ZMR over a wide range of temperatures that influence the recrystallization dynamics.

INTRODUCTION

Electronic materials used in thin-film microelectronic devices exhibit a wide range of melting points. These compounds include Ga (30°C), Se (217°C), Ge (937°C), Si (1412°C), and the new III-V electronic materials InSb (535 °C), GaSb (710°C), and InP (1070°C). By design or consequence, modern material processing techniques utilize heat transfer to refine and/or reorganize the material. Consequently, their thermal processing dynamics are dependent on the heat treatment process and the thermophysical properties of all the materials in the film structure. In particular, the modes of heat transfer (conduction, convection, and radiation) in the processing chamber have different levels of influence depending on the melting point of the material. In this paper, zone-melting recrystallization (ZMR) processing of different thin films (~1μm) on silicon substrates is investigated numerically. Moreover, the effect of the heat transfer modes on the crystallization dynamics for different thin film materials is examined.

ZMR is a directional solidification process used to recrystallize amorphous or polycrystalline films into a single crystal. Typically, a lower susceptor heats the film to a temperature near its melting point. A line heater situated above the film heats a narrow region, creating a molten zone. As the line heater is scanned across the film, the material in the wake of the moving molten zone recrystallizes in the form of a single crystal. A graphite strip with a square cross section was the line heat source used in this research.

The ZMR processing of silicon wafers and silicon-on-insulator (SOI) structures has been investigated by numerous individuals [1-10]. Analysis of the ZMR process revealed several different physical phenomena. In particular, the size of the molten zone and the crystallization dynamics at the solid/liquid interface were found to be affected strongly by the difference in radiative properties between solid and liquid phases [9,10], the radiative distribution from the line heater [10], the lower susceptor intensity [10], supercooling in the liquid phase [4,11], and scan speed [6,11]. Silicon ZMR technology has matured to a level where high quality material

---

* Author to whom correspondence should be addressed.

can be produced [12]. However, there has been little research done on processing materials other than silicon (T_m = 1412°C). Recently, research on gallium films (T_m = 30 °C) for low temperature ZMR processing has been reported [13].

The line heater temperature in silicon processing is above 2000°C, whereas the line heater temperature in gallium processing is around 250°C. In gallium processing, the conductive flux from the line heater through the gas is the dominant heating mechanism. This is in contrast to processing of silicon where the radiative flux is the dominant mode of heat transfer. The heat conducted from the line heater to the wafer travels through a stagnant region of gas below the strip heater [13,14]. Different ambient gases produce different temperature profiles in the film, even though heating of silicon is dominated by radiation [14]. Moreover, the stability of the interface and the quality of the resultant crystal are dependent on the type of heating (i.e. radiative or conductive) [6, 9, 13]. Clearly, the heat transfer issues must be addressed in more detail in order to apply the ZMR process to materials other than silicon.

In the ZMR process, the solid/liquid interface may be non-planar due to destabilizing microscale heat transfer effects [2,6,9,11,15]. Non-planar morphologies which may develop include repeating cellular, faceted, and dendritic structures. These characteristics are due to the modes of heating, the thermophysical properties, crystalline structure, and impurities.

This paper presents the results of a numerical investigation of the heat transfer effects in ZMR processing of materials which melt over a wide range of temperatures. Numerical models are used to calculate the heat transfer and fluid flow in the processing chamber, as well as the heat transfer and melting in the film. These models are described in the next section. General results of the parametric study are presented and the results for gallium and silicon are examined in detail.

MODELING

The ZMR process can be examined in two parts: the heat transfer in the chamber (external parameters) and the heat transfer in the film (internal parameters). The physical model and the numerical models are discussed in relation to these two reference points.

A. Physical Model

For these studies, the external parameters include the lower susceptor, the line heater, and the chamber walls. The size of the chamber itself is considered to be 10 cm wide and 6 cm tall. The ambient gas in the chamber is argon at atmospheric pressure. The walls of the chamber are considered to be water cooled to a temperature of 30°C. The film rests on a quartz sheet which is 3 mm thick. The quartz sheet rests atop the lower susceptor heater, which is typically a flat graphite sheet. Thus, heat from the susceptor arrives at the film in two ways: radiative heating passing through the semi-transparent quartz sheet and conductive heating through the quartz sheet. The line heater is a resistively heated graphite strip with square cross section, 3 x 3 mm. The spacing between the strip heater and the film is 1.5 mm. In the space between the heater and the film, a stagnant region of gas exists which allows heat to travel conductively between the two surfaces. Beyond the stagnant region, there is convective cooling of the film. Radiation from the graphite strip heater is distributed over the film with a distinct profile depending on the size and geometry of the heater [2].

The internal parameters of the film structure include the properties of the substrate and the recrystallizing thin film. In these studies, the substrate is always made of crystalline silicon. The substrate is considered not to melt or warp during processing. The thermophysical properties of the substrate are considered to be constant [2].

In these studies, a parametric range of different thin film materials are considered. Each thin film is characterized by its melting point. The reflectivity of the solid phase of the material is assumed to always be 0.5, while the liquid-phase reflectivity is 0.95. This represents the typical behavior of electronic materials which becomes more metallic with melting. Thermal conductivity assumes the same ratio; 0.5 W/cm°C in the solid phase and 0.95 in the liquid phase. The density of the material is 2.33 g/cm^3, the specific heat is 0.927 J/g°C, the solid state emissivity is 0.6 and the liquid state emissivity is 0.2. These values were chosen to be identical to silicon.

B. Numerical Model

The strip heater temperature, heat flux over the film, and the temperature distributions in the film are solved through an iterative procedure [14] using two numerical models. One model determines the heat transfer and fluid flow in the chamber, and the other finds the heat transfer dynamics and melting in the film.

The argon gas flow in the chamber is determined using FLOTHERM™ software [16]. The water-cooled wall temperature is fixed at 30°C. For these studies, the strip heater is considered to be stationary above the center of the film. This has been found to be a good approximation for high temperature processing [17] and low scan speeds (<300 μm/sec) [11]. However, at low processing temperatures, there is significant difference between stationary and moving results [13]. Stationary conditions are studied here to provide baseline results for all processing temperatures. The center of the film is set at its melting point and drops off to the susceptor temperature at its edges. The uniform susceptor temperature is set at 85% of the melting point of the thin film material. The strip heater temperature is also considered uniform and constant. The value of the strip heater temperature is determined iteratively between the chamber model and the film model.

The film is modeled with a two-dimensional control-volume formulation of the enthalpy approach. Details of the model are given elsewhere [2,18]. The temperature of the control volume is held at the melting temperature until the enthalpy of the control volume surpasses the latent heat of fusion (100 J/g). Thereafter, the temperature is allowed to increase normally. The change in properties with melting are modeled as changing linearly with the solid/liquid fraction. This allows for a smooth transition in modeling melting.

Often, incomplete melting of a region is calculated which can be interpreted as a non-planar interface. This partially melted region is known as the 'slush' region. Numerous factors, including the internal and external parameters, determine the size of the slush and complete liquid regions. The stability of these regions can be examined using the model. Explosive melting conditions and radiatively stable cases have been found in high temperature processing with silicon film structures [10]. In low temperature processing of gallium, slush regions cannot be created unless motion is considered in modeling. These phenomena will be discussed further in the next section.

RESULTS

Of specific interest to this study was the delivery of heat from the strip heater to the thin film. Depending on the strip temperature, the conductive heat flux through the gas and the radiation from the strip exhibited varying levels of influence. To investigate this, films with a broad range of melting temperatures were used. The region of critical importance in the ZMR process is the melting zone, and since this zone is very narrow it lies completely under the strip. Therefore, all heat flux measurements were exclusively from this region.

Two parametric tests were performed. In the first, the strip temperature was raised until a minimum melt width was achieved. This temperature is called the "breakthrough" point. Raising the strip temperature from this point resulted in more complete melting, while lowering it led to an unmelted film. The breakthrough point describes the minimum power required to induce melting. A melt width of 0.65 mm was used as the standardized breakthrough point. Heat transfer at the breakthrough point determines the minimum power requirements to induce melting, in the absence of phase change complications. The second parametric test evaluated the relative influence of the conductive heat flux and radiative heat flux by measuring the sensitivity of the melt zone width to increases in strip temperature.

A. Minimum Power Requirements

In this study the strip temperature was raised until the breakthrough point was reached in the thin film. The minimum power requirements to reach this point were then compared. The strip

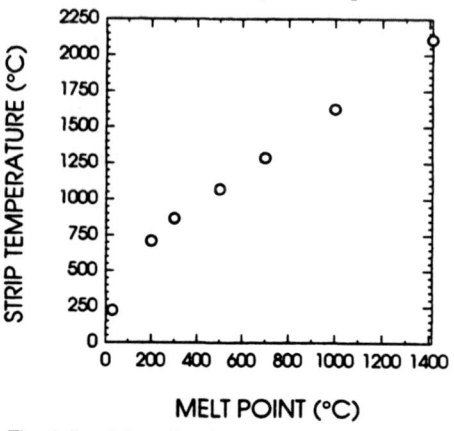

Fig. 1 Breakthrough strip temperature plotted as a function of melting temperature.

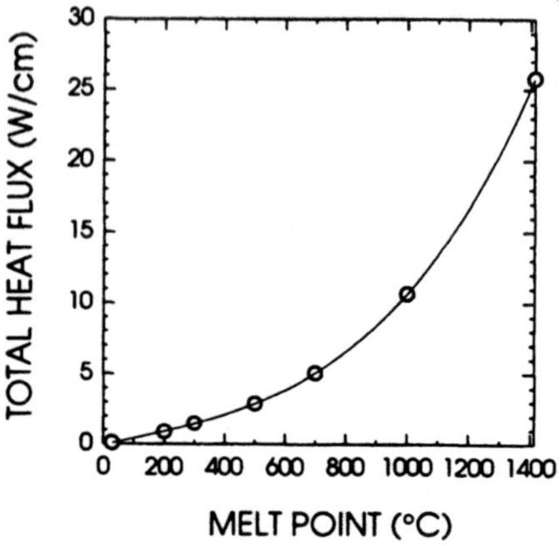

Fig. 2 Total heat flux at breakthrough point, plotted as a function of melting temperature.

temperature required to reach the breakthrough point is linearly related to melting point, except for the lowest melting point where it is parabolically related (Fig. 1). The total heat flux (conductive and radiative) from the strip is parabolically related to the melting temperature (Fig. 2). The minimum power required to reach the breakthrough point is higher for materials with higher melting points. As film temperatures rise so do the emissive losses, which are dominated by a quadratic term. At higher melting points, more strip heat is required to make up for these losses.

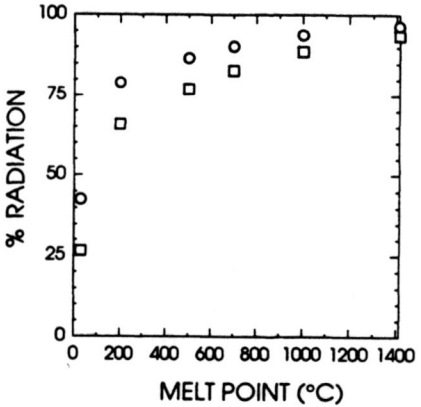

Fig. 3 Percentage of total energy from strip that is radiation, incident on film (O) and absorbed by film (□).

Figure 3 shows the percentage of total energy from the strip that is radiation, incident on the film (O) and absorbed by the film (□). Shown clearly in this figure is the early onset of radiation dominance. Once the melting point of a material is above about 100°C, radiation becomes the dominant source of heat from the strip. The conductive heat flux, however, remains a significant influence, accounting for at least 15% of the heat reaching the film from the strip for materials with melting points less than 800°C. Previous research on silicon and gallium films [13] corroborate the dominance of radiative and conductive heating, respectively.

B. Optical and Thermal Property Effects

Once melting passes the breakthrough point, the heat transfer dynamics of strip heating becomes more complicated. Material and optical properties such as emissivity, reflectivity and thermal conductivity are all affected by the change of phase from solid to liquid. In this study the heat transfer is examined in terms of the optical and thermal properties of the material. To do this, the strip temperature was raised to 40°C above the breakthrough point. The resulting melt width and heat fluxes from the strip were then recorded.

The sensitivity of the melt width to the change in total strip heat flux (radiative and conductive) is shown in figure 4. Films with low melting points show a greater amount of melt width sensitivity. As the melting point increases, the sensitivity of the melt zone decreases. We believe this is due to the increased reflectivity of the molten zone. The higher levels of reflectivity in the molten zone, upon melting, curb the amount of radiation absorbed by the film. Thus, when a molten zone is formed it acts to cut back the influence of the radiative heating. The cutback in radiation due to the increased reflectivity of the molten region has been noted in previous research on gallium (T_m = 30°C) [13] and silicon (T_m = 1412°C) [5,9,18].

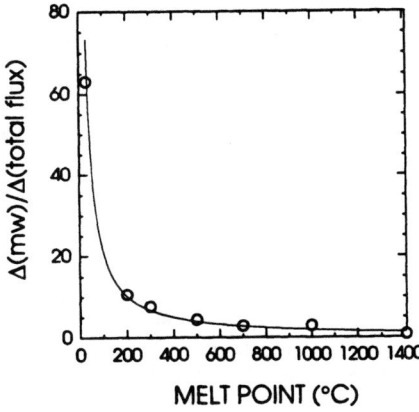

Fig. 4   Ratio of change in melt width to change in total flux, plotted as a function of melting point.

C. Material Processing Consequences

The conductive heat flux influences the melting process in a variety of ways. If the thermal conductivity of the material increases upon melting (as it does in silicon) the conductive heat flux will play a larger role in the heating once liquid regions are formed. This is in contrast to the radiative heating which decreases when liquid regions form (assuming that reflectivity increases upon melting, like silicon). The effects of increased thermal conductivity and reflectivity means that the conductive heat flux plays a larger role in heating the liquid region. Once a melt zone has been established, the temperature gradient in the film will be influenced by both the conductive heat flux and radiative heat flux: the liquid regions receiving more heating from the conductive heat flux than the solid material regions and vice versa for the radiation heat flux. To mitigate the effects from the conductive heat flux, ZMR can be performed in gases with lower thermal conductivity or in a vacuum.

CONCLUSIONS

A numerical simulation of graphite strip ZMR was developed to simulate the processing of films with a wide range of melting temperatures (i.e., T_m = 30 - 1412°C). Using two separate numerical models, the conductive heat flux from the strip through the gas and radiative heat flux from the strip were examined. Also examined were their combined effects on the melting dynamics of the films. A parametric study was conducted to evaluate the minimum power required to induce melting in the films, and to evaluate the response of the melt zone to increases in strip temperature. Results showed that the conductive heat flux was a significant form of heating. For materials with melting temperatures less than 800°C, the conductive heat flux accounts for at least 15% of the total energy heating the film to its breakthrough point. However, once a melt zone is established, the 15% threshold is raised by 100°C. For materials with melting points less than 900°C that have a melt zone, the conductive heat flux accounts for at least 15% of the total energy heating the film. Previous research has seriously neglected the influence of the conductive heat flux in ZMR processing.

At higher melting temperatures the melt zone width is less sensitive to changes in total heat flux (radiation and conduction) from the strip. More care must be taken when processing materials with lower melting points than higher, because slight variations in strip temperature can lead to large changes in melt width. This, in turn, alters the temperature gradients in the film which can lead to films with undesirable material quality.

ACKNOWLEDGMENTS

The authors would like to extend their gratitude to Peter Y. Wong for his careful and constructive advice in preparing this manuscript. And also to Harris J. Papamichael for his insight. This research was supported by the National Science Foundation under grants EID-9017557 and CTC-9157278.

REFERENCES

1. C. K. Chen , M. W. Geis, B-Y. Tsaur, R. L. Chapman, and J. C. C. Fan, J. Electrochem. Soc. **131**, 1707 (1984)
2. I. N. Miaoulis, P. Y. Wong, J. D. Lipman, and J. S. Im, J. Appl. Phys. **69**, 7273 (1991)
3. L. Pfeiffer, A. E. Gelman, K. A. Jackson, and K. W. West, Mat. Res. Soc. Proc. **74**, 543 (1987)
4. P. W. Mertens, Thesis, Katholieke Universiteit Leuven, Belgium 1991, (University Microfilms Incorporated dissertation services, Ann Arbor, MI, 1992)
5. R. D. Robinson and I. N. Miaoulis, J. Appl. Phys. **73**, 439 (1993)
6. J. S. Im, C. K. Chen, C. V. Thompson, M. W. Geis and H. Tomita, Mat. Res. Soc. Proc. **107**, 169 (1988)
7. D. Dutartre, Mat. Sci. Eng. **B4**, 211 (1989)
8. M. W. Geis, H. I. Smith, D. J. Silversmith, R. W. Mountain, and C. V. Thompson, J. Electrochem. Soc. **130**, 1178 (1983)
9. K. A. Jackson and D. A. Kurtze, J. Cryst. Growth **71**, 385 (1985)
10. J. S. Im, J. D. Lipman, I. N. Miaoulis, C. K. Chen and C. V. Thompson, Mater. Res. Soc. Proc. **157**, 455 (1990)
11. S. Y. Yoon and I. N. Miaoulis, J. Mater. Res. **7**, 124 (1992)
12. P. M. Zavracky, "Silicon-on-insulator Wafers by Zone Melting Recrystallization," *Solid State Technology* **34**, 55 (1991)
13. R. D. Robinson and I. N. Miaoulis, J. Appl. Phys. (to be published Feb. 1994)
14. B. D. Heilman, M. A. Marston, P. Y. Wong and I. N. Miaoulis, J. Mater. Res. **8**, 551 (1993)
15. W. W. Mullins and R. F. Sekerka, J. Appl. Phys. **35**, 444 (1964)
16. Flomerics Limited, *The Flotherm Reference Manual* (England, 1992)
17. J. D. Lipman, Thesis, Tufts University, 1989
18. J. D. Lipman, P. Y. Wong, I. N. Miaoulis, and J. S. Im, *HTD- Vol. 123, Collected Papers in Heat Transfer*, The American Society of Mechanical Engineers, pp. 211-217, Dec. 1989.

CRYSTALLIZATION OF GLASS FILMS ON SINGLE-CRYSTAL MgO SUBSTRATES

SUNDAR RAMAMURTHY, MICHAEL P. MALLAMACI AND C. BARRY CARTER
Department of Chemical Engineering and Materials Science, University of Minnesota,
421 Washington Ave. SE., Minneapolis, MN 55455

ABSTRACT

Thin, calcium magnesium silicate glass films have been deposited onto (001)-oriented single-crystal MgO substrates by pulsed-laser deposition (PLD). The substrates were thinned to electron transparency and characterized in the transmission electron microscope (TEM) before and after the deposition and following different heat treatments. Energy Dispersive X-ray Spectroscopy (EDS) was used in the TEM to characterize the chemistry of the films. The heat treatments were performed both in situ and ex situ. Direct evidence for crystallization was obtained by in-situ experiments at 950°C. Subsequent heat treatments were performed for longer times in air between 950°C and 1100°C. The crystallized phase was found to be monticellite ($CaMgSiO_4$) and the crystallites show moderate epitaxy with the underlying substrate. Films contaminated with aluminum resulted in the growth of magnesium aluminate spinel ($MgAl_2O_4$) epitaxially from the substrate. These observations of epitaxy indicated that crystallization initiated at the surface of MgO.

INTRODUCTION

Intergranular phases commonly observed in polycrystalline ceramics may affect the overall properties of the material.[1] By controlling the chemistry and crystallography of these phases, the properties may be suitably altered.[2-4] In view of the complexities associated with devitrification studies of glassy phases in polycrystalline materials, novel experiments may be performed on single-crystal substrates. The thin-film approach described in this paper has been previously employed to study glass-crystalline interactions in the alumina-anorthite system successfully.[5] In that system, epitactic crystallization of anorthite phases indicated that the surface of alumina plays a major role during the devitrification process.

Calcium magnesium silicate glass films are commonly observed in the intergranular regions of commercially prepared MgO compacts.[6,7] In the previous study involving interaction of liquid monticellite ($CaMgSiO_4$) at the grain boundaries of MgO, the intergranular regions were found to comprise of crystalline monticellite ($CaMgSiO_4$) and a glassy phase rich in the impurities present in MgO.[8] TEM studies revealed that crystalline monticellite is invariably in contact with the MgO grains, suggesting that the crystallization process initiated at the surfaces of the MgO grains. In order to understand the role played by MgO during the devitrification process, single-crystal substrates were chosen for this study.

EXPERIMENTAL

Substrates

Thin films of calcium magnesium silicate glass were grown by PLD onto specially prepared single-crystal (001) MgO substrates. The substrates were prepared from bulk by dimpling to about 15μm followed by ion milling from the dimpled side at 18° with 5 kV Ar ions. Prior to the deposition all substrates were acid cleaned and annealed at 1350°C for 0.5 hour. The acid cleaning procedure that was adopted is described elsewhere.[9] After annealing, all the substrates were characterized in the TEM. Such an approach of using TEM-ready foils as substrates for depositions ensures an unambiguous study of crystallization-phenomenon without introducing any specimen preparation artifacts.

<u>Glass-film deposition</u>

Glass films were deposited by PLD using an excimer laser operating with KrF (248 nm). The target material used was a calcium magnesium silicate pellet sintered to about 85% of the theoretical density. The experimental setup for PLD is described elsewhere.[10] Depositions were carried out with the laser operating at a pulse repetition rate varying from 10 Hz to 2 Hz, with an energy of about 200 mJ per pulse. The substrate temperature maintained during the deposition ranged between 200°C and 900°C. Prior to the deposition, the chamber was evacuated to about 1 μTorr, and then backfilled with 15-20 mTorr of oxygen. In all the depositions, the number of pulses was between 1000 and 5000; all yielding a continuous film. After deposition, the substrates were cooled to room temperature in the same oxygen pressure as used for deposition.

<u>Heat treatment and Characterization</u>

The films were characterized in a 300 kV Philips CM30 TEM equipped with an EDS system. A Gatan 628 single-tilt heating holder was used to perform the in-situ heating experiments. The glass films were also heat treated in air at different temperatures between 950°C and 1100°C for 0.5 to 6 hours in a conventional furnace. All the samples were characterized in plan-view. The chemistry of the film was characterized using EDS. The ratio of the peak intensities of Ca and Si was used to ascertain the composition of the film since the intensity in the Mg and O peaks arose from both the substrate and the film.

RESULTS AND DISCUSSION

Acid-cleaned and annealed substrates had surfaces with well-defined {001} steps and ledges. Glass films deposited onto the substrates were found to be continuous although the deposition parameters did have an influence on the morphology of the film. Figure 1(a) is a bright-field image of a 1000-pulse film grown at a nominal substrate-temperature of 900°C and a pulse-repetition rate of 2 Hz. The thickness of the film was estimated to be about 10 nm. The contrast in this image is indicative of de-wetting of the film on the surface of MgO. This was confirmed by imaging using the dark-field diffuse scattering technique. By this technique, the regions of glass film appear bright since the imaging conditions are such that the intensity from the diffuse scattering of the amorphous material is used to form the image. Figure 1(b) is a dark-field image formed under such conditions and the dark lines are regions where the glass film has de-wet the MgO surface. In the films grown at 600°C, however, de-wetting was not observed. Figure 2 is a dark-field image of a film grown at 600°C at a pulse-repetition rate of 5 Hz. The well defined grey levels of contrast in this image is characteristic of surface steps in annealed MgO. The glass films were continuous and generally found to be of uniform composition. Also seen are particles on a fine-scale which are presumably crystallites of monticellite. These particles have nucleated during the thin-film deposition in the chamber. The round particles (indicated by an arrow in the image) were found to be of a range of compositions, but generally enriched in calcium. The influence of the chemistry of the deposited film on the crystallization process and hence the products of crystallization is discussed later in this section.

<u>In-situ crystallization</u>

In-situ crystallization experiments were performed primarily to assess the approximate temperature at which the glass film began to crystallize. Crystallization was found to initiate at

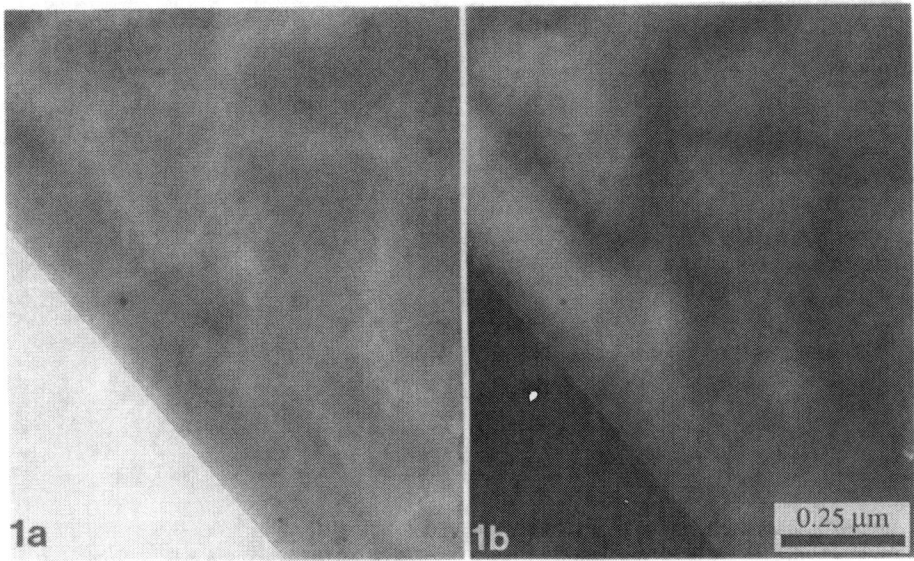

Figure 1(a). Plan-view bright -field image of 100 nm glass film deposited at 900°C, 2 Hz.
Figure 1(b). Dark-field image of the same area using the diffuse scattering technique.

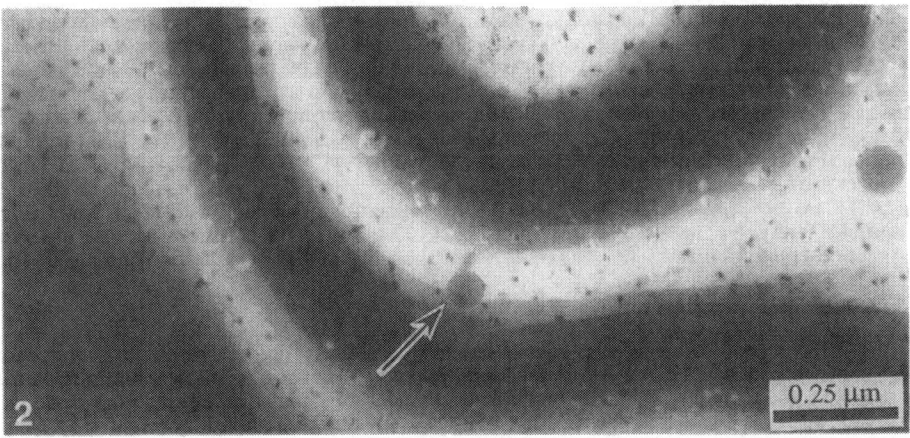

Figure 2. Dark-field image of a continuous film grown at 600°C, 5 Hz. The arrow indicates particulates which vary in composition from the film.

about 950°C on a film deposited at 200°C and a pulse-repetition rate of 10 Hz. Figure 3 is a dark-field image of an in-situ crystallized film at 950°C after 2 minutes. Epitactic crystallization is evidenced by the small faceted islands and fine moiré fringes that are visible in some regions of the film (indicated by an arrow).

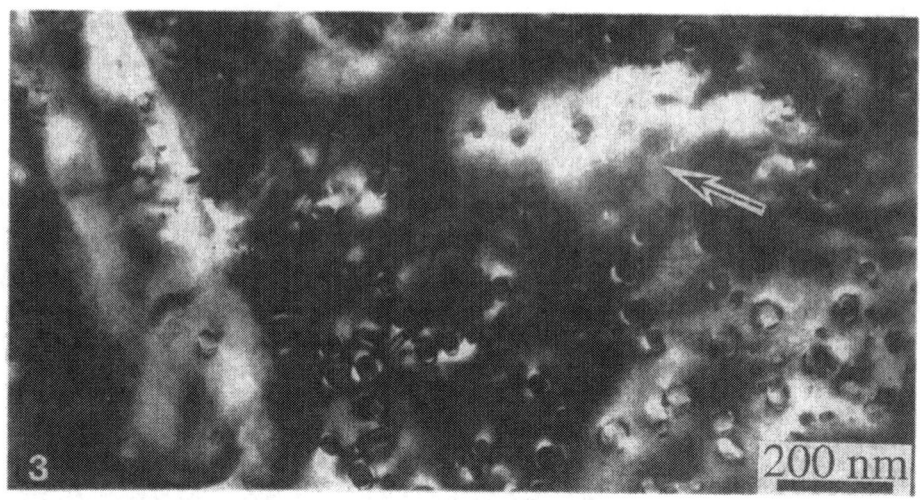

Figure 3. Dark-field image of a film heat treated at 950°C for 2 minutes in-situ the TEM. Fine moiré fringes (see arrow), and faceted islands are indicative of epitactic crystallization.

Heat treatments in air

All the glass films that were heat-treated in air between 950°C and 1100°C were found to result in similar morphologies of the crystallized film. The crystallites were found to be uniformly distributed and most of them showed moiré-fringe contrast when the electron beam direction was parallel to [001] direction of MgO (i.e., the foil normal). Selected-area-diffraction (SAD) patterns from some of these particles were analyzed and the majority crystallized phase was identified to be monticellite. In addition, phases found to be enriched in Si or Ca were also found. It is supposed that these crystallize due to possible variations in the local chemistry of the film. Figure 4(a) is a bright-field image showing a monticellite particle. The SAD from this region (Figure 4b) reveals an epitactic relationship of [001] $CaMgSiO_4$ ∥ [001] MgO. The moiré fringes seen in this image are due to the close matching of (230) $CaMgSiO_4$ with (200) MgO, (140) $CaMgSiO_4$ with (020) MgO, and the slight rotation associated with them. Monticellite particles with the epitactic relationship of [212] $CaMgSiO_4$ ∥ [001] MgO were also observed. These observations clearly show the influence of substrate on the crystallization of the glass film.

Compositions of the glass films

The crystallization of calcium magnesium silicate glass films can be markedly influenced by impurities such as Fe and Al. Initial depositions at 10 Hz were found to result in impurities such as Fe and Cr in the film. Such impurities, particularly Fe, may radically alter the products of crystallization as dictated by the Ca-Mg-Si-Fe-O system.[11] These impurities originated from the stainless steel target holder in the PLD chamber during deposition. To avoid Fe and Cr contamination, the target assembly was wrapped with an Al foil. As anticipated, EDS analyses of the glass films showed Al peak and the absence of Fe or Cr peaks. When these films were heat treated at 950°C in air for a few minutes, the resulting microstructure showed $MgAl_2O_4$ spinel particles uniformly distributed throughout the film with perfect epitaxy with the underlying MgO. Figure 5 is a bright-field image of a region close to the [001] pole of MgO showing the cube-on-cube relationship of the spinel particles. In the subsequent depositions, this problem of film-contamination was remedied significantly by lowering the pulse-repetition rate to 5 Hz or less.

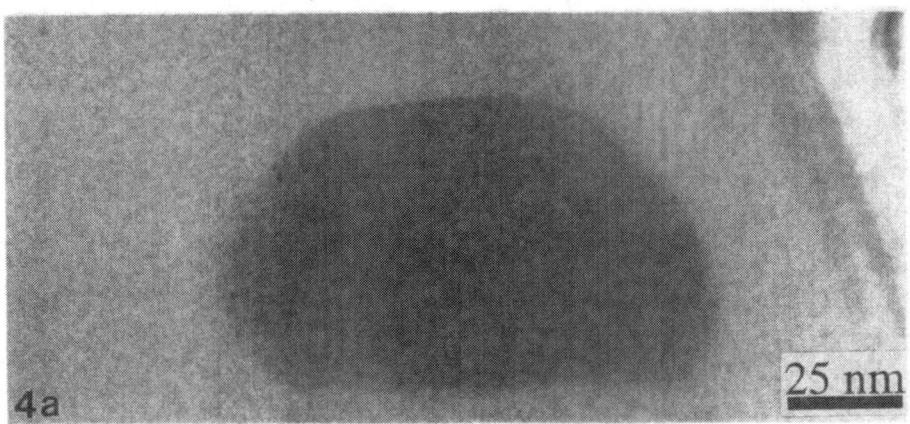

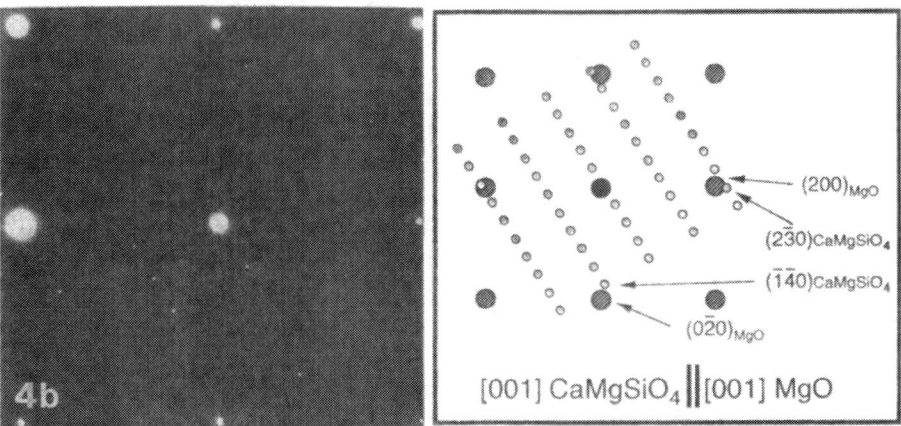

Figure 4(a). Bright-field image of a monticellite particle.
Figure 4(b). The SAD pattern from this particle and the accompanying schematic illustrates the orientation relationship.

CONCLUSIONS

Calcium magnesium silicate glass films have been deposited onto pre-characterized TEM-ready (001) MgO substrates and crystallized both in-situ and outside the TEM. Moiré fringe-contrast in these crystallized particles due to epitaxy with the underlying substrate indicate that the crystallization has initiated at the surface of MgO. Slight variations in the composition of the glass films can significantly alter the crystallization process and the chemistry of the products. The thin-film geometry adopted in this work could serve as a model for a fundamental understanding of glass-crystalline interactions.

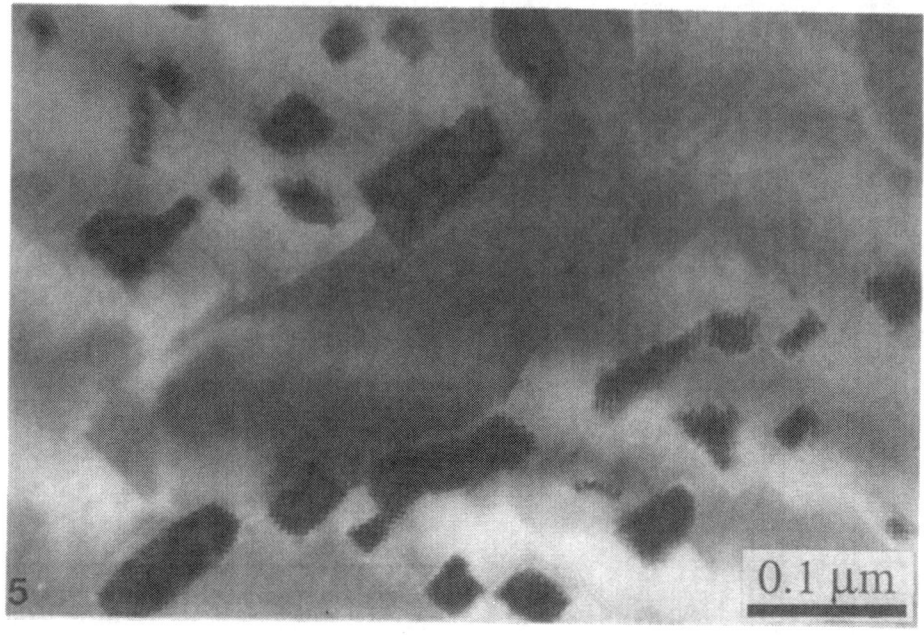

Figure 5. Bright-field image showing epitaxial growth of spinel particles from the MgO surface.

ACKNOWLEDGMENT

This research has been supported by the U. S. Department of Energy under Grant No. DE-FG02-92ER45465. The TEM is part of the Center for Interfacial Engineering, an NSF Engineering Research Center.

REFERENCES

1. D. R. Clarke, in *Surfaces and Interfaces of Ceramic Materials*, edited by L. C. Dufour (Kluwer Academic Publishers, 1989), p. 57-79.
2. D. A. Bonnell, T. Y. Tien and M. Ruhle, J. Am. Ceram. Soc. **70**, 460-65 (1987).
3. N. P. Padture and H. M. Chan, J. Am. Ceram. Soc. **75**, 1870-75 (1992).
4. M. K. Cinibulk and G. Thomas, J. Am. Ceram. Soc. **72**, 2037-43 (1992).
5. M. P. Mallamaci, J. Bentley and C. B. Carter, in edited by G. W. Bailey and C. L. Reider (51st Annual Meeting of the Microscopy Society of America 1993) pp. 928-929.
6. D. H. Hubble and N. B. Dodge, J. Am. Ceram. Soc. **43**, 343-347 (1960).
7. J.-J. Kim, B.-K. Kim, B.-M. Song and D.-Y. Kim, J. Am. Ceram. Soc. **70**, 734-37 (1987).
8. S. Ramamurthy, M. P. Mallamaci, C. B. Carter, P. R. Duncombe and T. M. Shaw, in edited by G. W. Bailey and C. L. Reider (51st Annual Meeting of the Microscopy Society of America 1993) pp. 948-949.
9. M. G. Norton, S. R. Summerfelt and C. B. Carter, Appl. Phys. Lett. **56**, 2246-2248 (1990).
10. P. G. Kotula and C. B. Carter, Mat. Res. Soc. Symp. **285**, 373 (1993).
11. W. A. Deer, R. A. Howie and J. Zussman, *Rock Forming Minerals: Orthosilicates,* John Wiley & Sons, New York, (1982)

$Au_{0.80}Cu_{0.10}Y_{0.10}$, A GOLD-RICH TERNARY ALLOY GLASS WITH T_c > 400°C AND ITS CRYSTALLIZATION KINETICS

SUNIL V. GOKHALE[a,c], KRASSIMIR G. MARCHEV[b,c], WELVILLE B. NOWAK[a] AND
BILL C. GIESSEN[a,b,c]
[a]Department of Mechanical Engineering, [b]Department of Chemistry, [c]Barnett Institute of
Chemical Analysis and Materials Science, Northeastern University, Boston, MA 02115

ABSTRACT

There are no reported gold-rich alloys that are both readily glass forming (RGF) upon rapid
solidification processing (RSP) and, in the glassy state, have crystallization temperatures T_c
sufficiently high to insure long metastable life times at room temperature. A representative of a
new family of ternary gold-based glasses is described that contain Cu and a rare earth (RE) metal
(or Y), with total addition element concentrations as low as 15 at. pct., and its crystallization
characteristics are reported. Under RSP processing by arc furnace hammer-and-anvil quenching,
the alloy $Au_{0.80}Cu_{0.10}Y_{0.10}$ readily forms a ductile glass, with T_c = 685 K, ΔH_c = 1.25 kJ/g-mole
and an activation energy of crystallization ΔE_a (cryst.) = 190 kJ/g-mole.

INTRODUCTION

Gold alloys have played a major role in the history of amorphous metals. $Au_{0.75}Si_{0.25}$ was
the first alloy glass to be prepared by rapid quenching from the melt [1]; also, the first definitive
thermal study demonstrating that an amorphous metal produced by this technique is indeed a
glass (by demonstrating the presence of a glass transition) was made on Au-Si [2] and Au-Si-Ge
[3]. Further, glass formation near the Au_4Si composition has been a major factor in formulating
theorical approaches to the glass forming ability of alloys based on structural [4] and electron
theoretical [5,6] criteria.

Despite their historical and theoretical importance, these Au-rich glasses are not conveniently
prepared or studied due to their low thermal stability (T_c = 307 K for $Au_{0.80}Ge_{0.12}Si_{0.08}$ [3]).
On the other hand, formation of less unstable Au glasses, e.g., in Au-lanthanide systems, requires
minimum concentrations of addition elements near 40 at. pct. [7]. Thus, it was of interest to
search for readily glass forming (RGF) compositions rich in gold (80-85 at. pct.) and with
relatively high T_c values; a study of promising ternary compositions was undertaken.

The rationale followed in this search is discussed more fully in a forthcoming note [8] on
silver-rich glasses prepared using the same concept; in either case, a smaller element (Cu) and a
larger element (Y or a rare earth) are added in amounts too small to produce ready glass
formation by themselves [9]. These additives satisfy the thermal criterion contributing to
compound formation [10] and hence, glass formation [11] (viz., there is an appropriately large
negative enthalpy of mixing) as well as a size ratio criterion (viz., addition element and majority
element sizes differ by ± 10 to 15%), as expressed for binary mixtures in ΔH_f - radius ratio maps
of RGF ability [11-13].

639

EXPERIMENTAL METHODS

Alloy buttons were arc-melted as usual and small alloy quantities were quenched by the arc furnace hammer - and - anvil device used and described previously [14]. The resulting foils were studied by X-ray diffraction (XRD, using Mo-Kα and Cu-Kα radiation on a GE-XRD6 diffractometer) and differential scanning calorimetry (DSC, using a Perkin-Elmer DSC-2 unit).

RESULTS AND DISCUSSION

Glass Formation: Glass formation was proven by both XRD and DSC, as documented here. The important low-k region (k = $4\pi\sin\Theta/\lambda$ = 2.1 to 3.8) of a representative XRD pattern (taken with Cu-Kα radiation) is shown in Fig. 1; this pattern, as well as others taken with Mo-Kα radiation up to larger k-values clearly indicates amorphization. The first broad peak presented here is exceptionally wide, consistent with the large range of interatomic distances expected to be present in the glass, in view of the large range of atomic sizes of the components [r_{12} (Au) = 1.44 Å; r_{12} (Cu) = 1.28Å; r_{12} (Y) = 1.80Å].

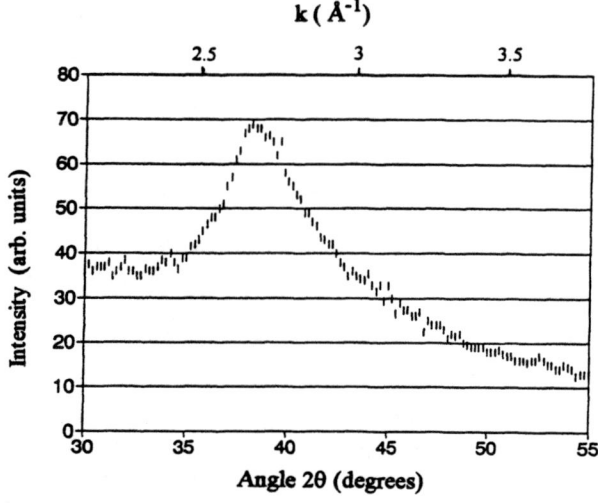

Fig. 1: X-ray diffraction pattern of amorphous $Au_{0.80}Cu_{0.10}Y_{0.10}$ taken with Cu-Kα radiation. Only first broad peak with maximum at $2\Theta \approx 38.3°$ is shown (see text).

Conversion of the peak maximum at k_m = 2.67 Å ($2\Theta_m \sim 38.3°$) into a principal interatomic distance r_m by use of the Ehrenfest relation $k_m \cdot r_m = \kappa$ with $\kappa \sim 7.70$ [15] shows r_m to be approx. 2.88 Å, equal to the twelve-coordinated elemental Au-Au distance which dominates the diffraction pattern; however, this close agreement between distances in the amorphous and crystalline states may be fortuitous because κ-values in metallic glasses can range from 7.6 to 8.0 [15].

As to elemental substitutions, it was noted that rare earth metals can substitute for Y and, to a more limited degree, that Si can be substituted for Cu. These results, together with a determination of the glass forming composition ranges in the gold-based systems will be published subsequently; a study on homologous silver-based glasses is being reported [8].

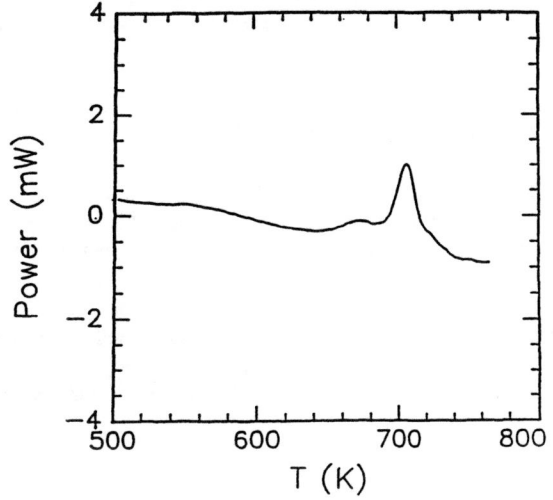

Fig. 2: DSC scan of amorphous $Au_{0.80}Cu_{0.10}Y_{0.10}$ taken at 40 K/min, showing crystallization exotherm at 685 K and pre-crystallization exothermic activity at ~ 650 K.

The presence of Si instead of Cu as the smaller-sized additive was found to reduce the RGF ability, contrary to what might be expected from the respective binary systems, where addition of 20 at. pct. Si has a high RGF effect [2] (although the resulting glass has extremely low resistance to crystallization, see above). By contrast, Cu, which is not expected to be an RGF additive in Au (with which it forms a continuous solid solution series), performs well as a co-glass-former in the present system, somewhat similar to its role in Pd-Si, where it substantially increases the RGF over that of the binary alloy [16].

Crystallization Data: A DSC scan taken at $\dot{T}$ = 40 K/min is shown in Fig. 2. Crystallization is often (but not always) preceded by unidentified exothermic activity beginning at ~650 K, followed by crystallization at T_c. The higher resistance to crystallization of the new glasses compared to the previous gold glasses is their most significant feature. The crystallization temperature T_c for $Au_{0.80}Cu_{0.10}Y_{0.10}$ is 685 K (412°C) (for $\dot{T}$ = 40 K/min), as compared to the low values of T_c = 307 K (34°C) for $Au_{0.80}Ge_{0.12}Si_{0.08}$ [3]. This strongly enhanced T_c indicates that the new alloys have sufficient long-term (meta)stability at room temperature and, hence, fulfill a necessary condition for potential usefulness in applications.

Next we compare the crystallization temperature T_c of the new glass to the solidus temperature of the corresponding multiphase crystalline alloy [15], i.e., in this case, the (estimated) eutectic temperature T_E in the pseudobinary system $(Au_{0.89}Cu_{0.11})_{0.9}Y_{0.1}$. In the absence of a ternary diagram, T_E was estimated from the corresponding binary phase diagrams [17] (using that of the crystal chemically related Au-Dy system in place of the undetermined Au-Y phase diagram), obtaining T_E = 1055 ±35 K. This yields T_c/T_E = 0.65 ± .02, to be compared to T_c/T_E = 0.48 for $Au_{0.80}Ge_{0.12}Si_{0.08}$, [3, 15], again showing the greater thermal stability of the new glass even after this normalization procedure. (Typical T_c/T_E values for metallic glasses lie between 0.44 and 0.68 [15]). It is reasonable to assume that the higher T_c values of the new gold glasses (as contrasted to those of the earlier gold glasses not containing transition metals) is correlated with stronger bonding (and hence, lower crystal nucleation and growth rates) in the new alloys; the stronger bonding is assumed to be due to their content of addition metals with d-electron bonding contributions, i.e., the Y (or RE) metal component and Cu, instead of Si or Ge.

The multiphasic product of crystallization consists primarily of an f.c.c. (Au,Cu) solid solution and an unidentified intermetallic phase (or phases) containing the Y component.

The observed $\Delta H_c = 1.25$ kJ/g-mole is relatively low, compared to $\Delta H_c = 6.2$ kJ/g-mole for $Au_{0.80}Ge_{0.12}Si_{0.08}$ [3], suggesting that, in the new glasses, strong metal-metal binding and short range order already exist in the glassy state, causing only a small amount of enthalpy to be released upon crystallization.

A Kissinger plot analysis [18] of the T_c maxima taken at different heating rates $\dot{T}$ is presented in Fig. 3. It shows good straight-line behavior between 10 and 80 K/min and yields an activation energy of crystallization, $E_a = 190$ kJ/g-mole. This may be compared to $E_a = 152$ kJ/g-mole for $Pd_{0.823}Si_{0.177}$ [19].

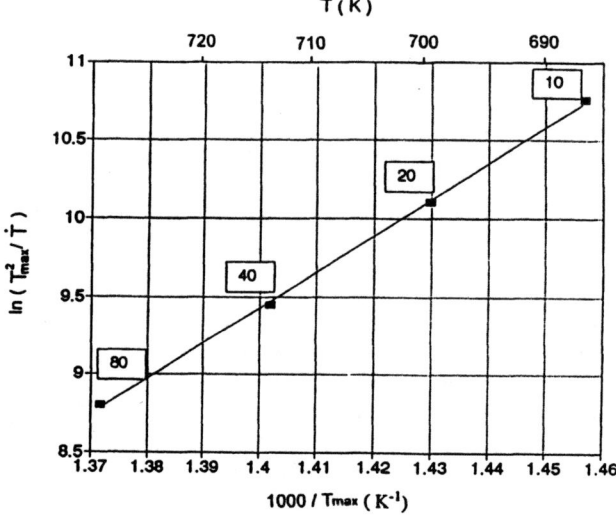

Fig. 3. Kissinger plot [17] analysis of DSC scans similar to the one shown in Fig. 2 taken at the cooling rates $\dot{T}$ (K/min) indicated in the figure. Plot shows ln $T^2_{max}/\dot{T}$ vs. $1000/T_{max}$ is the temperature at maximum of exotherm.

Mechanical Properties: As to their mechanical properties, the new gold glasses have fair bend ductility at RT (permitting a 180° bend test over a bend radius of 6 mm, although not over a razor edge) but they are not tough (showing fracture after re-bending).

Further Results: A larger range of RGF gold alloys related to $Au_{0.80}Cu_{0.10}Y_{0.10}$ is now being studied; the results will allow a comprehensive discussion of the thermal properties of these glasses as a function of composition. Initial experiments with Zr instead of Y have also shown glass formation, and suggest that the compositional pool of the new glass forming family may be even larger than suggested above.

We mention that some of the new alloy glasses have a potential for applications as electrical contact materials or as precursors for producing internally oxidized noble metal alloys.

ACKNOWLEDGMENTS

Support from the Fund for Materials Science at the Barnett Institute is gratefully acknowledged. Our special thanks for DSC measurements go to the Materials Research Laboratory at Harvard University, Cambridge, MA (Mr. Howard Zola and Mr. Frank Molea) and to Chomerics, Inc., Woburn, MA (Mrs. N. Rogan).

Communication No. 528 from the Barnett Institute.

REFERENCES

1. W. Klement, Jr., R.R. Willens and P. Duwez, Nature, 187, 869 (1960).
2. H.S. Chen and D. Turnbull, J. Appl. Phys., 38, 3646 (1967).
3. H.S. Chen and D. Turnbull, J. Chem. Phys., 48, 2560 (1968).
4. C.H. Bennett, D.E. Polk and D. Turnbull, Acta Met., 19, 1295 (1971).
5. S.R. Nagel and J. Tauc, Phys. Rev. Lett., 35, 380 (1975).
6. P. Häussler, in Proc. Fifth Internat. Conf. on Rapidly Quenched Metals, Würzburg (1984), eds. S. Steeb and H. Warlimont, Vol. 1, North-Holland, Amsterdam, 1985, p. 797.
7. T.R. McGuire and R.J. Gambino, J. Appl. Phys., 50, 7653 (1979).
8. K. G. Marchev, S. V. Gokhale, W. Nowak and B.C. Giessen, Mater. Lett., (1994).
9. D.E. Polk and B. C. Giessen, US Patent 4,116,682 (1978).
10. A.R. Miedema and D.F. de Chatel, in Theory of Alloy Phase Formation, ed. L.H. Bennett, TMS-AIME, Warrendale, PA, 1980, p. 344.
11. B.C. Giessen in Proc. Fourth Internat. Conf. on Rapidly Quenched Metals, Sendai (1981), eds. T. Masumoto and K. Suzuki, The Japan Institute of Metals, Sendai, Japan, Vol. I, 1982, p. 213.
12. B.C. Giessen and S.H. Whang, in Proc. Fourth Internat. Conf. on Liquid and Amorphous Metals, (LAM 4), Grenoble, J. de Physique, Colloque C-8, 41 (1980) p. C8-95.
13. J. A. Alonso, L.J. Gallego and J.M. Lopez, Phil. Mag., A 58, 79 (1988).
14. M. Fischer, D.E. Polk, and B.C. Giessen, in Rapid Solidification Processing, Principles and Technology, eds. R. Mehrabian, B.H. Kear, and M. Cohen, Claitor's Publ. Div., Baton Rouge, LA, 1978, p. 140.
15. B.C. Giessen and C.N.J. Wagner, in Liquid Metals, Chemistry and Physics, ed. S.Z. Beer, Marcel Dekker, New York, NY, 1972, p. 633.
16. H.S. Chen and K.A. Jackson, in Metallic Glasses, ASM, Metals Park, OH (now ASM International, Materials Park, OH), 1978, p. 74.
17. T. Massalski, ed. - in - chief, Binary Alloy Phase Diagrams, Second Edition, ASM International, Materials Park, OH, 1990.
18. H.E. Kissinger, Anal. Chem., 29, 1702 (1957).
19. N. Funakoshi, T. Kanamori, and T. Manabo, Jap. J. Appl. Phys., 17(1), 11-6 (1978).

CRYSTALLIZATION PROCESS AND CHEMICAL DISORDER IN FLASH EVAPORATED AMORPHOUS GALLIUM ANTIMONIDE FILMS

J. H. DIAS DA SILVA*, J.I. CISNEROS** AND L.P. CARDOSO**
*UNESP, FC, Depto. de Fisica. CEP 17033-060, Bauru, SP, Brazil, and also
**Universidade Estadual de Campinas, Instituto de Fisica. 13081-970.
Campinas, SP, Brazil.

ABSTRACT

In this work we describe a flash evaporation system specially built to produce amorphous films of III-V compounds and characterize GaSb films using optical, electrical and X-Ray diffraction measurements. Changes in the composition of the GaSb samples were obtained by the use of different crucible temperatures. In such samples, consequently, the optical absorption edge and the DC electrical conductivity were modified. The departure from stoichiometry in GaSb films is analyzed on the basis of these results which can be used as an evidence of the chemical disorder. This kind of disorder is represented by either wrong bonds or sites with different coordination.
Thermal annealing with a sequence of increasing temperatures first induced detectable variations in the optical absorption edge and in the vibrational properties of the amorphous GaSb. These variations are compatible with the GaSb local ordering and were observed by Raman scattering and infrared absorption spectra. The annealing at higher temperatures allowed the crystallization of the material confirmed by X-Ray diffraction. From these experimental results a crystallization mechanism based on the segregation of Sb excess coming from the crystallized regions toward the amorphous tissue is proposed.

INTRODUCTION

In the crystallization of bulk GaSb from the melt, when the thermodynamic equilibrium is practically attained, there is a strong tendency to the formation of stoichiometric material[1]. Differently, stoichiometry in amorphous GaSb films prepared by evaporation is a difficult task[2,3,4]. Even if the amorphous material is stoichiometric some degree of chemical disorder is present besides the structural disorder. The combined effects of these two kinds of disorder on the physical properties of the materials are not well understood at present. In this work we describe some experiments related to the chemical and structural disorder in the amorphous semiconductor GaSb, and the results are used to qualitatively check some recently developed models[5,6].

EXPERIMENTAL PROCEDURE

A set of a-Ga$_{1-x}$Sb$_x$ films with different concentrations $(0.50 \leq x \leq 0.59)$ and thicknesses ranging from 200 to 1200 nm were prepared on silica glass and crystalline silicon substrates using a specially designed flash evaporation system (fig.1). The evaporation assembly consists of a high vacuum chamber, with a feeder (F), a crucible (C) and a substrate holder (S).

The feeder is able to continuously supply the hot crucible with crystalline GaSb powder first kept in the reservoir (R) until the beginning of the evaporation process. At this time the powder is continuously supplied to the hot crucible by an externally rotated worm-gear(W-G). The rotation speed determines the film deposition rate. The gear which provides vibration (VG) to the feeder, and the tube (T) which mechanically guides the steel cable are important in order to get constant feeding rates.

As the vapor pressure of antimony is at least four orders of magnitude higher than that of gallium at common evaporation temperatures, the control of the film composition is not easy[2,3], even if the flash process is used. The relevant parameters in this respect are the crucible

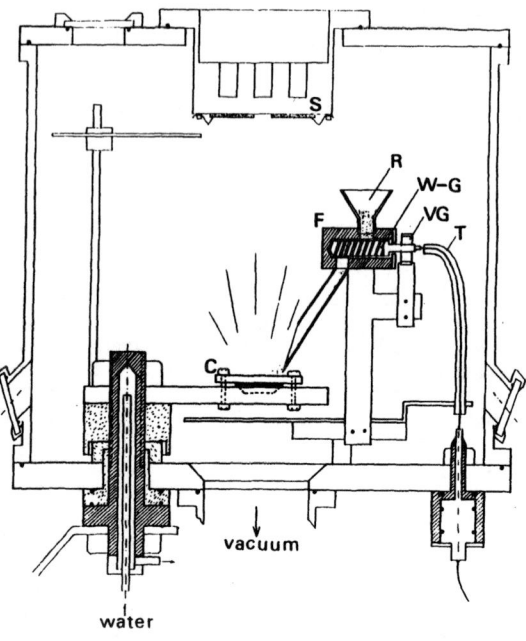

Fig.1 Cross section of the flash evaporation system described in the text.

temperature, the powder grain size, the deposition rate, and the substrate temperature. We were able to get materials with variable composition by controlling the crucible temperature and by keeping the other parameters constants. $a\text{-}Ga_{1-x}Sb_x$ films with x varying between 0.50 (stoichiometric) and 0.59 were obtained using crucible temperatures between 1070 and 990°C. The raw material was a 30μm powder obtained from a pure crystalline GaSb rod, with an acceptor concentration of $1.5 \times 10^{17} cm^{-3}$. A feeding rate of 0.3*grams/min.* resulted in deposition rates of approximately 3*nm/s*. The substrates (silicon (100) and silica glass) stayed at room temperature during all the depositions.

The compositions of the samples were measured[7] by a combination of XPS and AES techniques. No contaminant, specially oxygen, carbon, tungsten or iron were detected in the films after standard cleaning of the surface with argon ions.

The films were optically characterized using transmittance and reflectance measurements at normal incidence in a wide spectral region (50000 to 180cm^{-1}). These data were used to determine the optical constants by fitting and iterative[8] computer routines based in thin film optical theory.

Nearly stoichiometric films were submitted to thermal annealings at different temperatures(180, 210, 250 and 400°C). No detectable changes were observed below 180°C. These annealings were performed for 20 minutes in a pure nitrogen atmosphere in order to prevent surface contaminations. After each annealing, X-ray diffraction, UV-Vis., infrared absorption and Raman scattering measurements were performed in order to characterize the structural changes which had occurred.

RESULTS

Fig.2 shows the absorption coefficient vs. photon energies for the various compositions, calculated using the reflectance and transmittance data. In this figure it is included the absorption coefficient of a sample (7B) grown during the final stage of a standard evaporation (not flash evaporation). One can observe that the absorption edges for the materials with compositions nearly stoichiometric ($X = 0.50$, 0.51, 0.52) display similar behavior, while highly unbalanced samples ($X = 0.59$ and 7B) are different showing less steep absorption edges and higher absorption at low energies.

Fig.3 presents the absorption coefficient as a function of the photon energies, calculated for a nearly stoichiometric ($X = 0.51$) sample thermally annealed at various temperatures. One can verify that the annealing at $180°C$ causes a lowering of the absorption relative to the as grown condition, and a slight lowering of the Urbach[9] energy. These changes are compatible with a

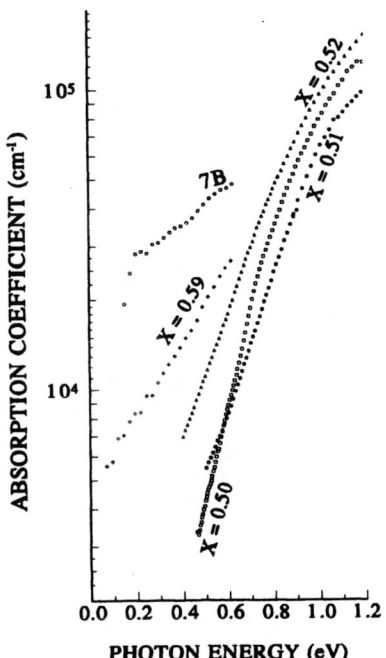

Fig.2 Absorption coefficient vs. photon energy of a–$Ga_{1-x}Sb_x$.

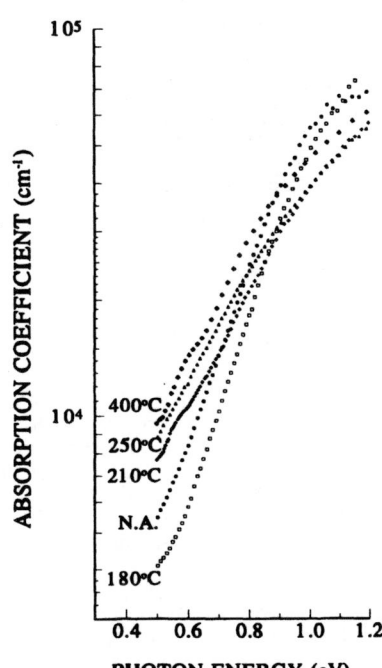

Fig.3 Absorption coefficient vs. photon energy of thermally annealed a–$Ga_{0.49}Sb_{0.51}$. The absorption edge for the not annealed (N.A.) condition was also included.

reduction of the potential fluctuations (in an atomic scale) and consequently with an increasing degree of the local order in the material[10,9]. A completely different feature occurs when the annealing temperature is $210^\circ C$ or higher. In these cases the absorption edges become less steep. These changes are associated with the increase of disorder as will be discussed later on.

Fig.4 shows the X-Ray pattern measurements of a nearly stoichiometric sample ($X = 0.51$) following the thermal annealing. It is clear from this figure, and from the UV-Vis. reflectance spectra[7] that the annealing at $180^\circ C$ did not induce appreciable long range ordering. On the other hand, the subsequent pattern ($210^\circ C$) displays a clear set of peaks, the positions and intensity ratios of which coincide with that of crystalline GaSb. The annealing at $250^\circ C$ is responsible for the raise of the crystallized material fraction, due to the increasing number of crystallites. In this case, a growth of the crystallite size

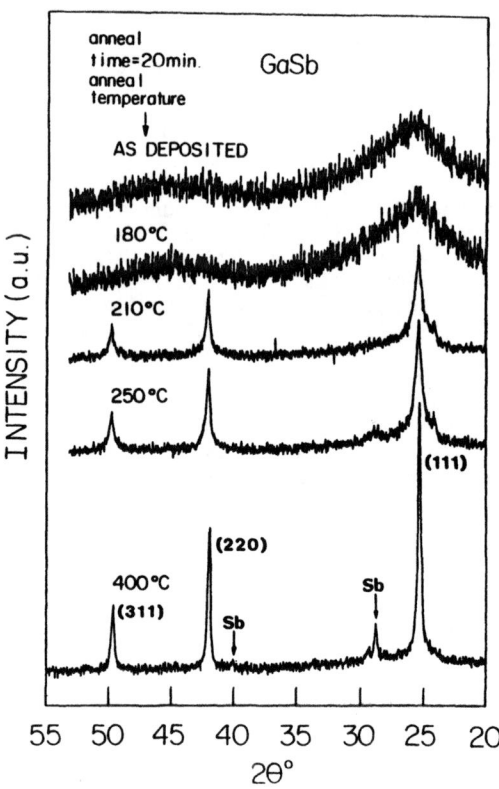

Fig.4 X-Ray pattern measuremets of a nearly stoichiometric material ($X = 0.51$) submitted to thermal annealings.

($\simeq$ 220 Å) is not expected since the diffraction peak widths are unchanged. A bump at $2\theta \simeq 29^\circ$ appears as a consequence of this annealing, and develops into two new peaks in the next annealing. At $400^\circ C$, the small peak observed at 29.26° corresponds to the planes (200) in crystalline GaSb, while the peak at 28.75 is related to the planes (102) of the crystalline Sb. The other small peak at $2\theta = 40.07^\circ$ is associated to the planes (014) of c-Sb. Therefore the annealing at $400^\circ C$ guarantees the formation of Sb crystallites inside the GaSb matrix.

DISCUSSION

In order to explain the previous results we will use two hypotheses. The first one is that the Sb excess is homogeneously distributed into the network of the amorphous material. The second hypothesis is that the main electronic effect of the excess atoms in the network is to increase the number of triply coordinated (C_3) atomic centers, which is also the main defect of the stoichiometric amorphous GaSb. The first hypothesis is based on experimental data[3,11,12], while the second is based on theoretical calculations of the electronic structure of amorphous III-V materials, specially a-GaAs[6,5].

Using these hypotheses we can explain the disorder effects observed in the unbalanced material as due to the bond angle fluctuations, that are enhanced with the Sb excess in the materials. According to the molecular dynamics calculations[5] the $Sb-C_3$ centers have trigonal symmetry, introducing large bond angle deviations in the material.

On the thermally annealed material ($180°C$) one can observe the effect of ordering in the absorption edge, also confirmed by infrared and Raman scattering results that will be published in a forthcoming paper. This ordering effect is probably due to small atomic arrangements, which in some way are able to reduce the bond angle distortions and eliminate part of the C_3 centers and part of the existent wrong bonds. On the other hand, annealings at higher temperatures, which are responsible for crystallite formation, may introduce a large amount of under coordinated centers (both $Sb-C_3$ and $Ga-C_3$) in the grain boundaries. Furthermore, the amorphous region that remains among the crystallites may receive the excess Sb migration from the crystallized region, which has the tendency of being stoichiometric. For this reason we assume that the main defects responsible for the disorder are the C_3 centers and the related bond angle distortion. Our measurements of the partially crystallized material reveal both the high degree of disorder (absorption edge) and the existence of the long range order in the crystallites (X-Ray and reflectance patterns).

As we are considering that the disorder in the unbalanced material and in the crystallized one are essentially due to the same cause, we could expect that the amplitude of variation of the disorder parameter E_o (Urbach energy) would be similar. Fig.5 shows E_o as a function of composition

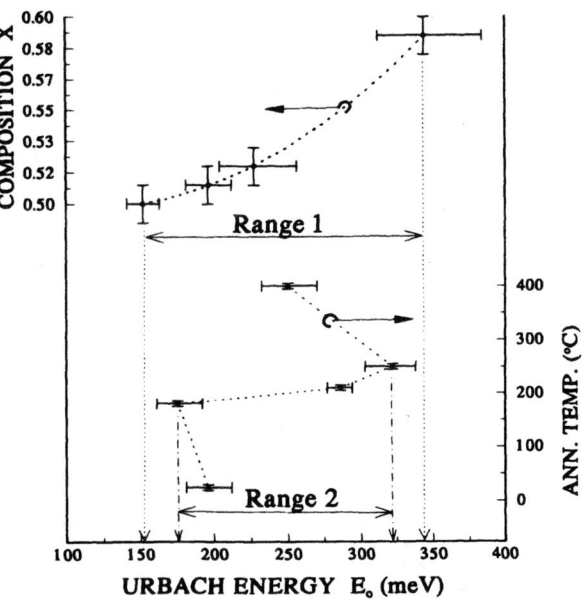

Fig.5 Urbach energy as a function of composition and of thermal annealings . The *Range* 1 corresponds to the composition variations, while the *Range* 2 corresponds to the annealing induced variations in a nearly stoichiometric ($X = 0.51$) sample.

for unbalanced samples and as a function of temperature for the annealed one. As expected the ranges of variation of E_o are similar.

CONCLUSIONS

From this work we could conclude that the thermal annealings at temperatures lower than that necessary to crystallize the a-GaSb, produced changes that are compatible with the local ordering or the elimination of point defects in the material. Higher temperature partially crystallize the material creating regions with a high density of defects and higher Sb concentration in the vicinity of the c-GaSb grains.

The absorption edges of both unbalanced and partially crystallized GaSb films, are less steep than stoichiometric a-GaSb. In both cases the effect can be interpreted by the formation of C_3 centers favored by the Sb excess.

ACKNOWLEDGEMENTS

The authors wish to thank G. Zampieri and M.M. Guraya for the surface analysis of the samples, and M.M.G. de Carvalho and C.E.M. de Oliveira for preparing the raw material. This work was financially supported by the brazilian agencies FAPESP and CNPq.

REFERENCES

1. D. Weiler and H. Mehrer, Philos. Mag. A, **49**, 309 (1984).

2. A. Gheorghiu, T. Rappeneau, J.P. Dupin and M.L. Theye, J. de Physique, **42** (C4), 881-889 (1981).

3. J. Dixmier, A. Gheorghiu and M.L. Theye, J. Phys. C: Solid State Phys., **17**, 2271 (1984).

4. J.E. Davey and T. Pankey. J. Appl. Phys., **49**, 212 (1969).

5. E. Fois, A. Selloni, G. Pastore, Q.M. Zhang, R. Car, Phys. Rev. B, **45**, 13378 (1992).

6. E.P. O'Reilly and J. Robertson, Phys. Rev. B., **34**, 8684 (1986).

7. J.H. Dias da Silva, J.I. Cisneros, C.E.M. de Oliveira, M.M. Guraya and G. Zampieri, J. Phys.: Condens. Matter 5, A343 (1993).

8. J.H. Dias da Silva, J.I. Cisneros, F.C. Marques and M.P. Cantão, in Current Topics on Semiconductor Physics, edited by O. Hipolito, A. Fazzio and G.E. Marques (World Scientific, Singapore, 1988), p.192.

9. See for instance N.F. Mott and E.A. Davis, Electronic Processes in Non-Crystalline Solids. (Pergamon, Oxford, 1971), p. 238.

10. J. Tauc, Mat. Res. Bull., **5**, 721 (1970).

11. N.J. Shevchik and W. Paul, J. Non-Cryst. Solids, **13**, 1 (1974).

12. N.J. Shevchik, J. Tejeda and M. Cardona, Phys. Rev. B, **9**, 2627 (1974).

EFFECTS OF GLASS AND CARBON FIBER ON NYLON 6,6 CRYSTALLIZATION

Krisda Siangchaew*, Theodore Davidson**, and Matthew Libera*
Stevens Institute of Technology, Castle Point on the Hudson, Hoboken, N.J. 07030
* Department of Materials Science and Engineering.
**Polymer Processing Institute (PPI) and Design and Manufacturing Institute (DMI).

ABSTRACT

The effects of addition of glass fiber and HMS4 carbon fiber on the crystallization of nylon 6,6 has been investigated using DSC and polarized optical microscopy (POM). DSC observations indicate that HMS4 fiber lowers the supercooling required to initiate crystallization. A transcrystalline layer is also observed in the near-fiber region of carbon fiber-reinforced composites after DSC. The presence of transcrystallinity and the earlier onset of crystallization for this composite are due to heterogeneous nucleation on carbon fiber surfaces. DSC measurements of nylon 6,6 with glass fibers show supercoolings similar to those of neat nylon 6,6. Transcrystallinity is also absent in the glass-reinforced composites. These glass fibers appear to be weak nucleation catalysts. Different transcrystalline layer thickness is found to be influenced by thermal processing condition. In addition to transcrystallinity, the HMS4 carbon fiber also influences the bulk morphology of the nylon 6,6.

INTRODUCTION

The presence of foreign matter within crystallizable materials can induce heterogeneous nucleation.[1] Thermoplastics, being semicrystalline polymers, have demonstrated heterogeneous crystallization in the form of substrate-induced morphology[2] and surface crystallization.[3] Such processes produce a distinguishable morphology known as transcrystallinity which originated from a crowding of primary nuclei along a catalytic interface. Kinetically, heterogeneous nucleation lowers the degree of supercooling needed to initiate the liquid-crystal transformation. The presence of a transcrystalline microstructure and a lower supercooling required to initiate crystallization serve as indications of heterogeneous nucleation. The intent of this paper is to investigate the effect of addition of fiber on the heterogeneous crystallization of both the bulk and the near-fiber region in nylon 6,6 thermoplastic polymer.

EXPERIMENTAL PROCEDURE

Neat nylon 6,6 pellets of a scientifically pure grade (Scientific Polymer Product Cat. No.033) and the glass fiber and HMS4 carbon fiber were used in this study. HMS4 (Hercules

651

Inc.) fiber was stored in air in a closed container after heat treatment in flowing ultrapure N_2 gas at 400 °C for three hours. The glass fiber was obtained by pyrolysis at 550 °C in air overnight of Verton long fiber pellets.

Neat and composite specimens were made by vacuum compression molding. During molding, nylon 6,6 pellets were held at 280 °C for 10 minutes before pressure was applied. Films were produced with thicknesses ranging from 60 to 100 μm. Samples were then cooled to room temperature under vacuum. To make composite specimens, the same procedure was followed except that fiber was sprinkled along with the neat Nylon 6,6 pellets onto the molding plate. Specimens were dried *in-vacuo* at 100 °C for 6 hours, cooled overnight, and desiccated prior to use. Different volume percent fiber was obtained by selectively cropping sections of nonuniformly distributed fibers in the molded composite film. Compression molded carbon fiber-reinforced nylon composite used in DSC and tube furnace is coded as CFN; glass composite, GFN.

DSC cooling scans were performed using a Perkin-Elmer DSC7. The pre-cool temperature profile is as follows: (1) heating from 50 to 250 °C at 50 °C/min; (2) heating from 250 to 280 °C at 10 °C/min; and (3) isothermal anneal at 280 °C for 10 min. The samples were cooled at -10 °C/min. To obtain the crystallization onset temperature from DSC traces, a baseline was drawn on each curve following the method described by Gray.[6] Composites underwent pyrolysis in air at 500 °C after DSC in order to obtain the weight percent fiber present in them.

Specimens for microstructural studies via polarized optical microscopy (POM) were made by annealing compression-molded neat, CFN, and GFN specimens under a nitrogen atmosphere. Static pressure was applied during the anneal in order to thin the specimens from ~60-100 μm to approximately 20 μm. Our previous investigations have found that the nylon 6,6 microstructure is more defined at 20 μm probably due to fewer spherulites in the through-thickness direction. 20 μm thick specimens of neat, CFN, and GFN were subjected to each of three thermal treatments: (1) continuous cooling at -10 °C/min from 280 °C; (2) continuous cooling at -10 °C/min from 280 °C and isothermal hold at 227 °C (neat-227, GFN-227, CFN-227); and (3) continuous cooling at -10 °C/min from 280 °C and isothermal hold at 232 °C (neat-232, GFN-232, CFN-232). Each isothermal anneal was done for 15 minutes. A fourth procedure was applied to CFN specimens where following treatment (3) above, they were rapidly heated to 250 °C after staying at 232 °C for approximately 2 minutes. These samples will be subsequently identified as CFN-250.

EXPERIMENTAL RESULTS

Confirmation of the catalytic potency of the carbon fiber in heterogeneously nucleating Nylon 6,6 is seen in Fig. 1. CFN samples with a higher volume percent of carbon fiber has a higher crystallization onset temperature. Lee and Porter[7] have also observed that increasing the carbon fiber in PEEK increased the crystallization onset of their composite. On the other hand, DSC traces (Fig.

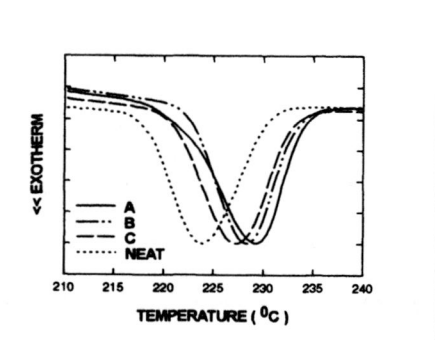

Fig. 1. DSC of CFN samples cooling from 280 °C at -10 °C/min showing an earlier onset of crystallization as compared to the neat. (A) 34.7, (B) 25.4 and (C) 23.3 wt %.

2) of glass fiber composite do not show any depression in the degree supercooling with an increase of fiber content and are similar to those of neat nylon. This points to the fact that HMS4 carbon fiber was more potent in heterogeneously inducing the transformation of the Nylon 6,6 than the glass fiber in agreement with other observations on different thermoplastics composite systems.[8,9]

The morphology characteristic of crystallization in neat, CFN, and GFN specimens was studied using transmitted polarized optical microscopy of 20 μm specimens. The results are shown in Fig. 3 and summarized in Table I. Following the DSC result, isothermal annealing temperature of 227 °C and 235 °C were chosen since these correspond to the temperatures for the onset of crystallization as measured by DSC for GFN and CFN composite, respectively.

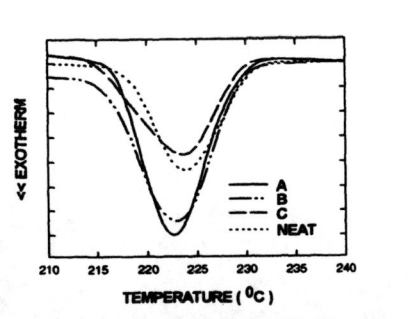

Fig. 2. DSC of GFN samples cooling from 280 °C at -10 °C/min showing an identical onset of crystallization as compared to the neat. (A) 53.6, (B) 45.4 and (C) 22.7 wt %.

Fig. 3 illustrates the variation of two morphological features: the presence/absence of transcrystallinity and the average spherulite diameter. Transcrystallinity is not present in the neat or the GFN specimens for any of the three thermal treatments (Fig. 3a-f). The inability for glass to induce any transcrystalline morphology was also reported by Hsiao et al[8] for the E-glass in PPS, PEEK, and PEKK. Transcrystallinity is clearly present in all three cases for the CFN (Fig. 3g-i). To ascertain that heterogeneous nucleation occurred along the HMS4 fiber/matrix interface, additional CFN samples (CFN-250) were quickly heated to 250 °C as soon as the sample reached its crystallization onset temperature (232 °C). As illustrated in Fig. 4 the width of the transcrystalline layer are largest in this case.

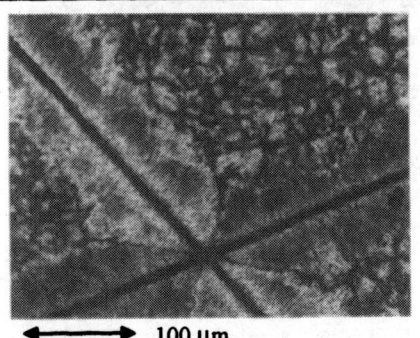

◄──► 100 μm

Fig. 4. Micrograph taken at room temperature of CFN-250 showing transcrystallinity and few granular bulk spherulites.

Observation of the bulk indicates coarser spherulites in the GFN and the neat when isothermally held at 232 °C (Fig. 3c and 3f) in contrast to those at 227 °C (Fig. 3b and 3e). This is consistent with the spherulitic growth observation made by McLaren[10] for the nylon 6,6. However, bulk spherulites in the CFN samples display a smaller influence of the processing temperature on their resultant morphology (Fig. 3h and 3i). In addition, except for the transcrystalline layer, bulk spherulites in the CFN (Fig. 3g) are finer than both the neat and the GFN (Fig. 3a and 3d) when continuously cooled at -10 °C/min. Since the controlling parameter in all three processing conditions is the presence (or absence) of different fiber types, current observations suggest that the influence of carbon fiber seem to extend beyond the fiber/matrix interface.

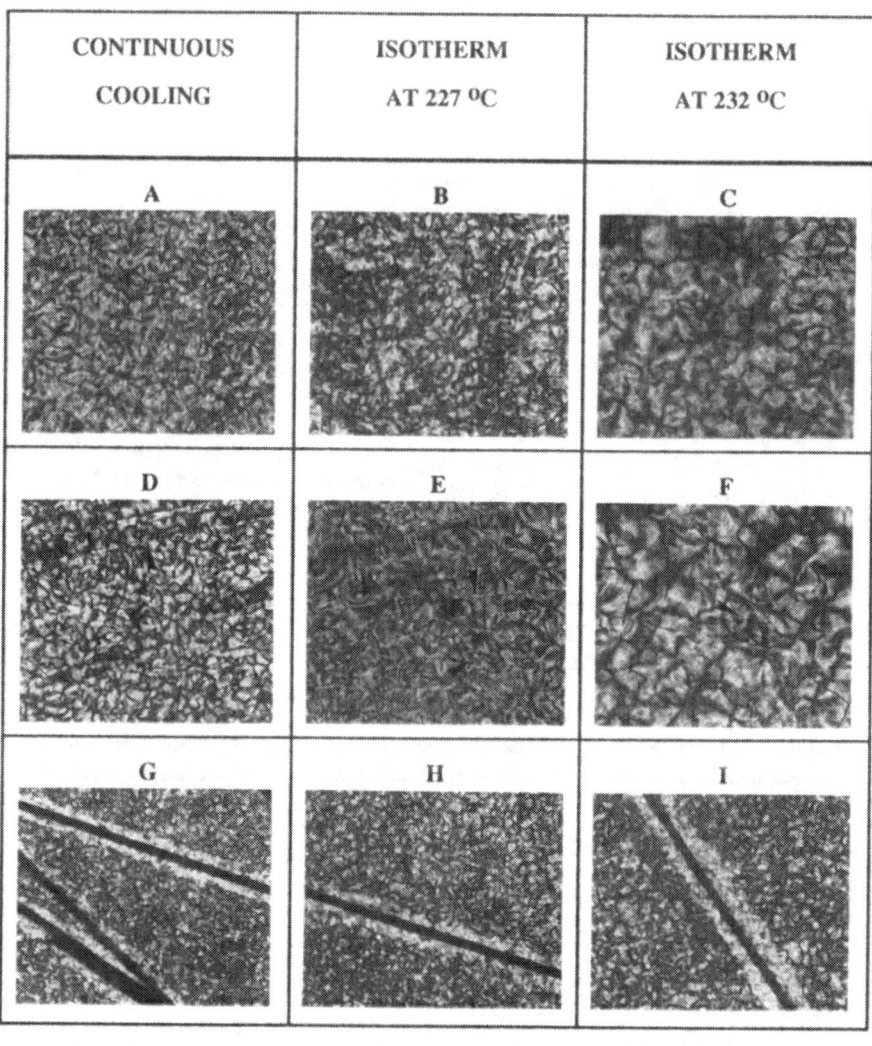

CONTINUOUS COOLING	ISOTHERM AT 227 °C	ISOTHERM AT 232 °C
A	B	C
D	E	F
G	H	I

◀———▶ 100 μm

Fig. 3. Micrographs of (a-c) neat, (d-f) GFN, (g-i) CFN. Specimens are subjected to the temperature profile described on the top of each column. Summary of the observation is in Table I. The arrow locates the glass fiber.

Table I. Summary of microstructures after prescribed heat treatments in the tube furnace.

Thermal treatment	Continuously cooled	Isotherm @ 227 °C	Isotherm @ 232 °C
Neat	Small spherulites.	Small spherulites.	Granular spherulites.
GFN	No transcrystalline. Medium spherulites in the bulk.	No transcrystalline. Small to medium spherulites in the bulk.	No transcrystalline. Granular spherulites in the bulk.
CFN	Transcrystalline. Fine spherulites in the bulk.	Transcrystalline. Fine spherulites in the bulk.	Transcrystalline. Small spherulites in the bulk.

DISCUSSION

DSC traces indicate that crystallization has significantly progressed in the CFN before the onset of neat crystallization. Intuitively, this initially crystallized portion, as determined by the area under the DSC curve, could be attributed to transcrystallinity. However, it is also observed that the termination of the crystallization in the composite itself is also earlier than that of the neat. Transcrystallization could not unambiguously account for this latter observation. Yet, this shift of the whole crystallization event must somehow be related to transcrystallinity since this shift was not observed in specimens without a transcrystalline layer, e.g., GFN.

Our baseline drawn on the continuously cooled DSC curves for the CFN indicates an onset of crystallization around 234 ± 1 °C. Since neither isothermal annealing at 227 nor 232 °C affected the transcrystalline width (Fig. 3h and 3i), nucleation and growth of the transcrystalline layer must be limited to a very early stage of crystallization. Furthermore, by heating to 250 °C after the sample reached 232 °C (CFN-250), not only did the transcrystalline layer grow, a few bulk spherulites also grew (Fig. 4). This suggests that bulk crystallization must have concurrently occurred with that heterogeneously nucleated by the HMS4 carbon fiber.

CONCLUSION

The present work has experimentally studied the effects of: (i) addition of glass and carbon fibers; and (ii) variations in thermal history on the crystallization of nylon 6,6. This investigation concludes:
1. DSC indicates that lower supercooling is necessary to initiate crystallization in carbon-fiber/nylon composite (CFN). The DSC traces characterizing the crystallization of glass-fiber/nylon composite (GFN) is similar to that of neat nylon 6,6.
2. A transcrystalline layer forms in the near-fiber region of carbon-fiber/nylon composite. Such transcrystallinity is not observed in glass-fiber/nylon composite.

3. The extent of transcrystallinity in carbon-fiber/nylon composite can be controlled by varying the thermal history during crystallization.
4. Comparison of the bulk morphologies in the neat, glass-fiber/nylon, and carbon-fiber/nylon composites suggests that HMS4 carbon fibers influence the morphology in the bulk polymer far from the fiber/matrix interface.

ACKNOWLEDGEMENTS

The authors wish to thank the Polymer Processing Institute (PPI) at Stevens for the use of their DSC system and Dr. Victor Tan and Matthew Blackburn for their assistance with the experiments. This project was supported, in whole or part, under Cooperative Agreement No. N00014-90-CA-0002 with the Office of the Chief of Naval Research and an agreement between Great Lakes Composites Consortium, Inc. (GLCC) and Stevens Institute of Technology. These agreements provide for a program entitled "Automated Design for Manufacturing of Polymer Composite Parts" as part of the Navy MANTECH efforts.

REFERENCES

1. D. C. Bassett, Principles of Polymer Morphology, (Cambridge Univ. Press, Cambridge, 1981).
2. A. M. Chatterjee and F. P. Price, J. Polym. Sci., 13, 2369 (1975).
3. D. R. Fitchmun and S. Newman, J. Polym. Sci., A-2, 1545 (1970).
4. S. Incardona, C. Migliaresi, H. D. Wagner, A. H. Gilbert, and G. Marom, Composites Science and Technology, 47, 43 (1993).
5. A. Lustiger, F. S. Uralil, and G. M. Newaz, Polymer Composites, 11, 65 (1990).
6. A. P. Gray, Thermochim. Acta, 1, 563 (1970).
7. Y. Lee and R. S. Porter, Polymer Eng. and Sci., 26, 633 (1986).
8. B. S. Hsiao and E. J. H. Chen, Interfaces in Composites, edited by C. G. Pantano and E. J. H. Chen (Mater. Res. Soc. Proc. 170, 1990) pp. 117.
9. E. J. H. Chen and B. S. Hsiao, Polym. Eng. and Sci., 32, 280 (1992).
10. J. V. McLaren, Polymer 3, 175 (1963).

LATERAL GRAIN GROWTH IN THE EXCIMER LASER CRYSTALLIZATION OF POLY-Si

H. Kuriyama*·***, K. Sano*, S. Ishida*, T. Nohda*, Y. Aya*, T. Kuwahara*, S. Noguchi*,
S. Kiyama*·***, S. Tsuda* and S. Nakano*
*Sanyo Electric Co., Ltd., 1-18-13, Hashiridani, Hirakata, Osaka 573
**Giant Electronics Technology Co., Ltd., 1-6-5 Higashinihonbashi, Chuo-ku, Tokyo 103

ABSTRACT

We have succeeded in obtaining nondoped, thin poly-Si film (thickness ~500Å) with excellent crystallinity and large grain size (maximum grain size ~4.5 μ m) by an excimer laser annealing method, which offers the features of low-temperature processing and a short processing time. The grain size distribution shrinks in the region around 1.5 μ m and this poly-Si film exhibits a strong (111) crystallographic orientation. Poly-Si thin film transistors using these films show quite a high field effect mobility of 440cm²/V · s below 600℃ process.

INTRODUCTION

Development of a polycrystalline silicon (poly-Si) thin film transistor (TFT) fabricated by a low-temperature process (<600℃) is expected as a solution to realize next generation giant-microelectronics such as large area, quick response, and high definition liquid crystal displays (LCDs). In particular, a low-temperature poly-Si fabrication technique is one of the required key technologies.

Among the various approaches being examined, we have focused on excimer laser annealing technology to obtain a high-performance poly-Si TFT on a large-area glass substrate. We selected this method because it causes little damage to the glass substrate thanks to its short and shallow melt-regrowth process. Moreover, due to the high power and large beam size of the laser, this is potentially a high throughput process. However, one drawback is that poly-Si films obtained by this method have a small grain size of less than 1000Å [1]. This is because the solidification velocity during excimer laser annealing is extremely high (~m/s)[1, 2].

We previously proposed, for the first time a new concept for the excimer laser annealing method to solve the grain size problem[2]. This concept is based on the intentional control of solidification velocity during laser irradiation. We were able to specifically clarify two parameters effective in reducing the solidification velocity of molten Si without adopting a special substrate structure. One was low-temperature ($\leq$400℃) substrate heating during excimer laser irradiation and the other was thinning of the a-Si thickness(~500Å) as a starting material. Using this method and controlling the above parameters, we have obtained a large grain size of over 5000 Å and a high field effect mobility of 280cm²/Vs[3].

In this study, we report dramatic lateral grain growth in poly-Si films using our new proposed method combined with multi-shot excimer laser annealing. We will also show the poly-Si film characteristics, grain growth mechanism and TFT characteristics using these high quality poly-Si films.

EXPERIMENTAL

Nondoped hydrogenated amorphous silicon (a-Si:H) film (500Å) was deposited on a quartz substrate by the plasma CVD method. The deposition was carried out using 100% SiH_4 decomposition at a substrate temperature of 170°C, an RF power density of 8mW/cm², a pressure of 26.6 Pa and a gas flow rate of 22 sccm. After dehydrogenation by furnace annealing at 550°C, the a-Si film was recrystallized by an ArF excimer laser in a vacuum chamber (<10^{-4} Pa) where the substrate temperature was set to 400°C. The laser beam was unified by beam-homogenized optics.

The crystallinity of the poly-Si films was evaluated by scanning electron microscopy (SEM), transmission electron microscopy (TEM), transmission electron diffraction (TED), and X-ray diffraction (XD).

RESULTS

Figure 1 shows SEM images of Secco-etched poly-Si films recrystallized using various numbers of shots for excimer laser annealing where the substrate temperature was 400°C and the laser energy density was 350mJ/cm².

In the case of two shots, the grain size was enlarged to about 5000Å due to the reduction in the solidification velocity of the molten-Si. However, as the number of laser shots was increased, dramatic lateral grain growth occurred, as the figure illustrates. The 64 shot-annealed poly-Si films exhibited a maximum grain size of about 3μ m, and note that the 128 shot-annealed poly-Si films have a maximum grain size of about 4.5μ m. This value is about ten times larger than that for the films we reported previously [3] and two orders larger than the film thickness.

Figure 2 shows oblique SEM images of poly-Si films with lateral grain growth. The grains are entirely of a large columnar structure that is many times larger than the film thickness. And the surface morphology of these poly-Si films is quite smooth rendering them well suited to thin film transistor applications.

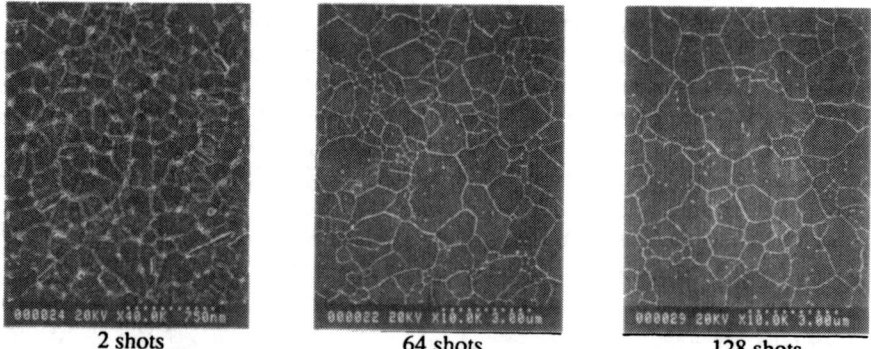

| 2 shots | 64 shots | 128 shots |

Fig. 1 SEM images of Secco-etched poly-Si films recrystallized using various numbers of shots for excimer laser annealing

Figure 3 shows the TEM lattice image and TED pattern of laterally grown grain. As this photograph illustrates, this large-grain poly-Si film has a low defect density within grains and no extra spot were observed in TED. This result indicates that these poly-Si films exhibited excellent crystallinity. Similar micron- size poly-Si films were reported using solid phase crystallization of a-Si in a furnace annealing at 600°C[4]. But the grains obtained by this technique were of a dendritic structure and have many types of internal defects, such as twins[5]. In contrast, the poly-Si films that we obtained exhibited excellent crystallinity compared with those by the solid phase crystallization method.

Figure 4 shows the grain size distribution undergoing lateral grain growth as revealed in Secco-etched SEM images. As shown in this figure, the two-shot-annealed poly-Si films exhibited almost monomodally distributed grains with an average grain size of about 2000 Å. The distribution of the 64-shot-annealed poly-Si films shifted to bimodal. This distribution seemed to occur during the process of lateral grain growth. Then the 128-shot-annealed poly-Si films were again almost monomodal with a narrow grain size distribution of about 1.5 μ m on average. In the case of solid phase crystallized poly-Si films, it is reported that the grain size was widely distributed and that the frequency decreased monotonically as the grain size became larger[6]. On the contrary, the poly-Si films which were obtained have a narrow distribution with micron size grain size. This result clearly shows that this method offers an additional advantage in terms of the distribution of grain size, and device characteristics using these poly-Si films are expected to become homogeneous compared with solid phase crystallized poly-Si films.

We evaluated these poly-Si films by X-ray diffraction spectra. As shown in Figure 5, both the two-shot and eight-shot-annealed poly-Si films appeared to have random polycrystalline

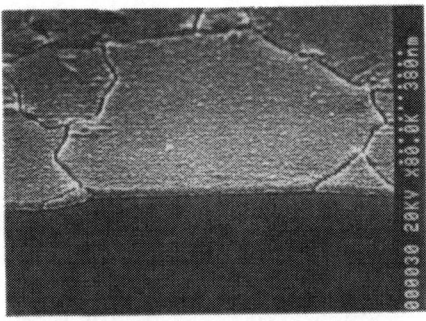

Fig. 2 Oblique SEM image of Secco-etched poly-Si film with lateral grain growth.

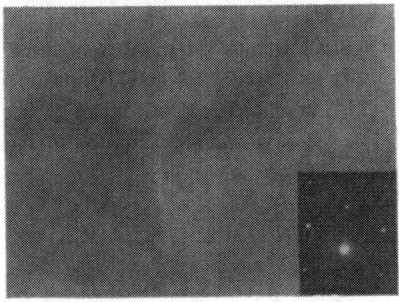

Fig. 3 TEM and TED pattern for laterally grown grain.

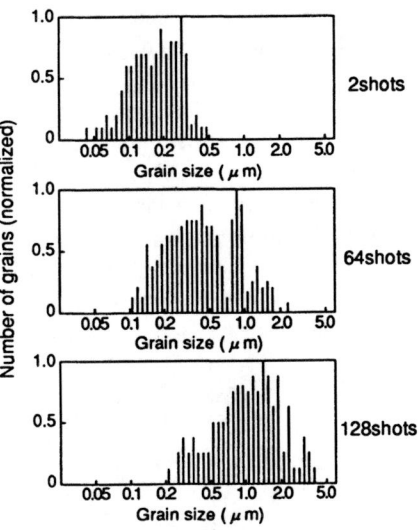

Fig. 4 Grain size distribution undergoing lateral grain growth.

659

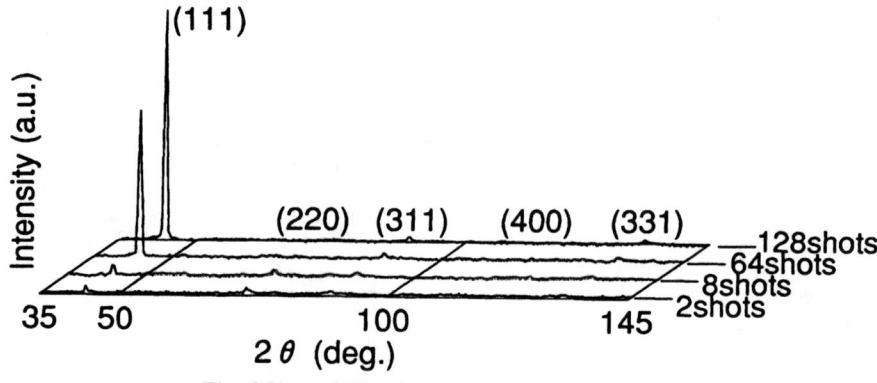

Fig. 5 X-ray diffraction spectra of poly-Si films.

orientation. But as the number of laser shots was increased, only the (111) peak intensity radically increased, and the (110) peak simultaneously disappeared. These results indicate that the (111) crystallographic orientation occurred with lateral grain growth. We also confirmed (111) crystallographic orientation of laterally grown grains by a selected-area transmission electron diffraction pattern.

DISCUSSION

The above experimental results clearly show that this lateral grain growth phenomenon by excimer laser annealing was caused by a certain anisotropy which exhibited a (111) preferential orientation. Figure 6 shows the schematic microstructure of the lateral grain growth that we observed in this study. A similar phenomenon was reported by C. V. Thompson et al. in the high-temperature (~1200°C) and long-time (~20 min) thermal annealing of doped-thin (< 1000 Å) poly-Si films[7]. They reported that grain growth in thin films was affected by anisotropy of the surface free energy, and that grains which exhibit lateral grain growth will have orientations which minimize surface free energy. In the case of Si, this is the (111) orientation. And their poly-Si films have similar characteristics to our poly-Si films, such as grains much larger than the film thickness with monomodally distributed micron size grains, uniform crystallographic (111) orientation and a low density of defects within grains. Considering the similarity of characteristics and conditions between the above lateral grain growth and the poly-Si films we obtained, we speculate that the basic driving force behind the lateral grain growth observed was also surface free energy anisotropy. But as for the difference

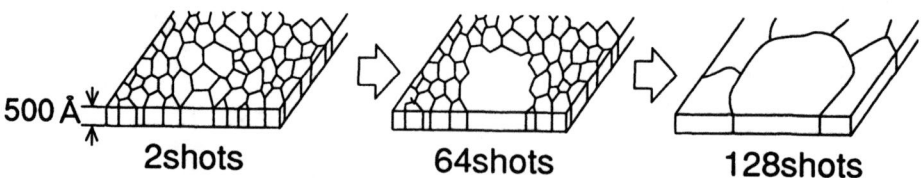

Fig. 6 Schematic microstructures for grain growth.

from the phenomenon which we observed, doping with donor impurities such as phosphorous or arsenic and long-time annealing (>20min) are indispensable conditions for realizing the lateral grain growth phenomenon by the high-temperature furnace annealing of thin poly-Si films. The reason for the necessity of the dopant for secondary grain growth was explained in terms of the increased atomic mobility of grain boundaries[8]. In our previous study, three-dimentional thermal analysis showed that the region of grain boundary melted completely and that the grain region was pinned with the Si melting tempetature (not fully molten) in the case of excimer laser irradiation to poly-Si film where the substrate temperature was 400℃. From result, and based on the phenomenon observed in the current study, it is possible to explain that this complete melting of the grain boundary enhances the atomic mobility of the grain boundary without doping, as well as the speed of the lateral grain growth phenomenon. To confirm this speculation, we examined as-deposited thin (500 Å) poly-Si films fabricated by the LPCVD method as a starting material instead of a-Si. This film showed a nearly columnar, fiberlike grain structure in which the average grain size in the plane of the film is smaller than the average grain size perpendicular to the plane of the film. In this film, the surface influence on grain growth is not so large compared with a-Si as a starting material. In this case, no dramatic lateral grain growth was observed even in multi-shot excimer laser annealing. This result supports the importance of the initial grain structure, in other words, the magnitude of the surface energy in the lateral grain growth phenomenon that we observed.

APPLICATION TO TFT

We applied these poly-Si films with lateral grain growth (128 shot-laser annealing) to the fabrication of poly-Si TFTs. The TFT was self-aligned coplanar structure with a channel dimension of 20/20 μ m. The gate oxide film (1500 Å) was deposited by the APCVD method at 430℃. As for the impurity doping for the source, drain and gate regions, phosphorus was introduced by an ion shower doping apparatus and activated by excimer laser irradiation at

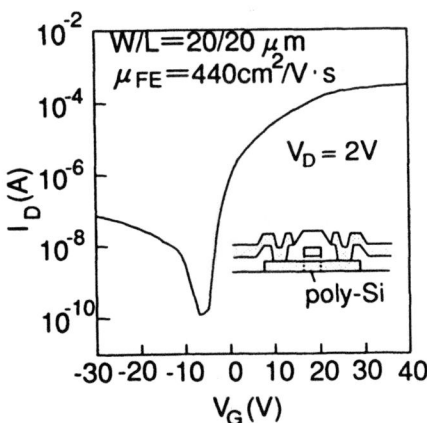

Fig. 7 Drain current versus gate voltage characteristics of poly-Si TFT.

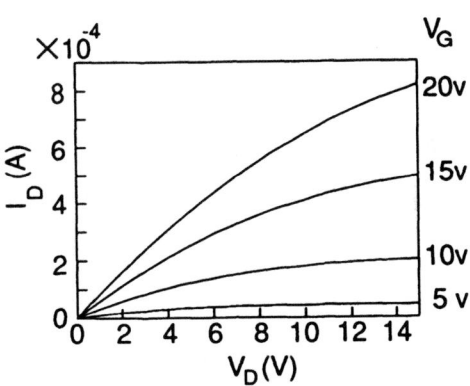

Fig. 8 Drain current versus drain voltage characteristics of poly-Si TFT.

room temperature. Including these process, all process temperatures were below 600°C.

Figure 7 shows the drain current versus gate voltage characteristics of maximum-mobility poly-Si TFT. This TFT provided a high field effect mobility of 440cm²/V·s in the linear region. This is the highest value ever reported for low-temperature processes below 600°C. And the ON/OFF ratio was greater than 10⁶.

Figure 8 shows the drain current versus drain voltage characteristics of this TFT. Well saturated characteristics were obtained.

CONCLUSION

We have succeeded in obtaining large-grain nondoped thin (500 Å) poly-Si films with excellent crystallinity, controlled orientation and a narrow grain size distribution by an excimer laser annealing method. This was achieved by using our new proposed method combined with multi-shot laser annealing. The films produced had a maximum grain size of 4.5 μ m with a strong (111) crystallographic orientation and a smooth surface. The crystallinity of these poly-Si films was also excellent with minimal internal defects.

On the analogy of a similar phenomenon observed in the high-temperature furnace annealing of doped poly-Si thin films, we speculate that the basic driving force behind the lateral grain growth which we observed was also surface free energy anisotropy.

Poly-Si TFT using these films showed quite a high field effect mobility (440cm²/V·s) for low temperature processing below 600°C.

REFERENCES

[1]T. Sameshima, M. Hara and S. Usui, Jpn. J. Appl. Phys. 28, 1789 (1989).
[2]H. Kuriyama, S. Kiyama, S. Noguchi, T. Kuwahara, S. Ishida, T. Nohda, K. Sano, H. Iwata, H. Kawata, M. Osumi, S. Tsuda, S. Nakano and Y. Kuwano, Jpn. J. Appl. Phys. 30, 3700 (1991).
[3]H. Kuriyama, S. Kiyama, S. Noguchi, T. Kuwahara, S. Ishida, T. Nohda, K. Sano, H. Iwata, S. Tsuda and S. Nakano, Int. Electron Devices Meet. Tech. Dig. p. 563 (1991).
[4]F. Emoto, K. Senda, E. Fujii, A. Nakamura, A. Yamamoto, Y. Uemoto and G. Kano, IEEE Trans. Electron Devices ED-37, 1462 (1989).
[5]T. Katoh, IEEE Trans. Electron Devices ED-35, 923 (1988).
[6]H. Kumomi and T. Yonehara, Appl. Phys. Lett. 44, 3565 (1991).
[7]C. V. Thompson and H. I. Smith, Appl. Phys. Lett. 44, 603 (1984).
[8]C. V. Thompson, J. Appl. Phys. 58, 763 (1985).

PART VIII

Crystallization in Amorphous Silicon

MULTIPLE PULSE IRRADIATION EFFECTS IN EXCIMER LASER-INDUCED CRYSTALLIZATION OF AMORPHOUS Si FILMS

H. J. Kim and James S. Im
Columbia University, Department of Chemical Engineering,
Materials Science and Mining Engineering, New York, NY

ABSTRACT

We have experimentally investigated the effects that are associated with multiple-pulse irradiation in the excimer laser processing of thin Si films on SiO_2. Double-pulse irradiation experiments revealed results, which are consistent with that which is known from single-pulse crystallization experiments, and these experiments confirm the applicability of the transformation scenarios, which were derived from single pulse-induced crystallization experiments [1, 2]. The results from the multiple-pulse irradiation experiments clearly show that gradual and substantial grain enlargement can occur — and *only* occurs — when the irradiation energy density is close to but less than the level that is required to melt the film completely. Based on these findings, we argue that the grain enlargement effect is a *near-complete melting phenomenon* that is associated with polycrystalline Si films, and we present a grain boundary melting model to account for this phenomenon. A brief discussion on the apparent similarities and physical differences between the observed phenomenon and the solid state grain growth processes is provided herein.

INTRODUCTION

It is becoming increasingly clear that the excimer laser-induced phase and structural modifications of thin Si films involve several melt-mediated and far-from equilibrium transformation processes [1, 2]. Figure 1 illustrates the categorization of various melt-mediated transformation mechanisms involved in single pulse-induced crystallization of thin LPCVD Si films, in accordance with the views expressed in recent investigations [1, 2].

SINGLE PULSE IRRADIATION OF LPCVD THIN Si FILMS ON SiO_2

I. *PARTIAL MELTING REGIME* (Low Energy Density Regime)
- Unmelted Si composed of continuous layer (i.e., Melt-depth < Film thickness)
- Explosive crystallization,
- Vertical regrowth and competitive occlusion of grains
- Fine-grained and small-grained poly-Si (Grain radius < Film thickness)

II. *NEAR-COMPLETE MELTING REGIME* (Super Lateral Growth Regime)
- Unmelted Si composed of discrete islands (i.e., Melt-depth ≈ Film thickness)
- Significant lateral growth can proceed before the impingement
- Up to ~ μm-sized grains observed

III. *COMPLETE MELTING REGIME* (High Energy Density Regime)
- No unmelted Si remains
- Deep supercooling followed by nucleation and growth of solids
- For low substrate temperatures, fine-grained and small-grained poly-Si observed (Grain radius < a few hundred Å)
- Amorphization observed for thinner films

Figure 1. Various melt-mediated transformation mechanisms involved in single pulse-induced crystallization of LPCVD a-Si films on SiO_2 [1, 2].

Mat. Res. Soc. Symp. Proc. Vol. 321. ©1994 Materials Research Society

In general, the excimer laser-induced crystallization of Si films can be divided into single and multiple pulse-based methods. In the single pulse-based method, much of the film (i.e., except the overlapping areas) is exposed to the beam only once, and in the multiple pulse-based method, the film is exposed to the beam numerous times. Since control and uniformity of the resulting microstructure *over the entire film* is critical for *both methods*, it is technologically imperative that one properly understands the effects that are associated with multiple-pulse irradiation.

To this end, it has recently become apparent that multiple-pulse irradiation of poly-Si films can lead to substantial enlargement of the grains [3]. The observed excimer laser-induced grain enlargement (ELGE) phenomenon has the potential to be scientifically noteworthy, as the responsible mechanism may represent a yet uncharacterized process for the microstructural evolution, since it is likely to be quite distinct from well known grain growth processes [4].

Here, we report on double- and multiple-irradiation experiments, which were designed in order to better characterize the ELGE process such that the results will help to unveil the precise transformation scenario through which grain enlargement occurs. Based on the resulting observations, it is suggested that the ELGE phenomenon corresponds to the near-complete melting and solidification process, which involves complete melting of the film occurring locally at and near the grain boundaries.

EXPERIMENTAL METHOD

Irradiation experiments were conducted with non-hydrogenated LPCVD a-Si films (1000Å-thick), which were deposited on oxidized Si substrates (with a 1000Å-thick thermal oxide layer). The samples were irradiated in atmosphere with 30-nanosecond XeCl excimer laser pulses (308 nm).

We have conducted two types — double- and multiple- pulse irradiation — experiments. Double-pulse irradiation experiments consisted of the irradiation and microstructural analysis of films, which were irradiated at various combinations of first- and second-pulse energy densities. This represents the most simple multiple-pulse irradiation scenario, and its main purpose was to investigate whether any dramatic results occur upon the second irradiation of first-pulse crystallized poly-Si films. We note that the second-pulse irradiation (and subsequent pulses in the multiple-pulse irradiation experiments) is differentiated from the first pulse in that the second pulse no longer "crystallizes" the a-Si film; instead, it melts and may modify the microstructure of already-crystallized poly-Si films.

Multiple-pulse irradiation experiments consisted of investigating the microstructure of the films after the film has been irradiated at a fixed energy density with increasing numbers of pulses. To simplify the subsequent interpretation of the results, irradiations were carried out (1) with a stationary beam and sample (i.e., no scanning) in order to avoid variations in the incident energy density, (2) using a low pulse frequency (<1 Hz) in order to avoid a rise in the substrate temperature — i.e., the beam heating effect, and (3) with low numbers of pulses in order to minimize the effects associated with the possible fluctuations in the incident energy densities (particularly the possible complete melting of the film, which can drastically alter the microstructure of the material [1]). After irradiation, a microstructural investigation of the irradiated samples was conducted using TEM analysis.

RESULTS

Figure 2 shows a particular set of double-pulse irradiation experiments in which the energy density of the first pulse was at 565 mJ/cm^2 and the energy density of the second pulse was varied from 300 mJ/cm^2 to 588 mJ/cm^2. It shows that the trend in the microstructure of the second-pulse irradiated films is identical to the trend observed in single-pulse crystallized Si films [1, 2]. A notable specific difference refers to the increases in the critical energy density, which is required in order to melt the film completely. Various combinations of energy densities were examined, and no

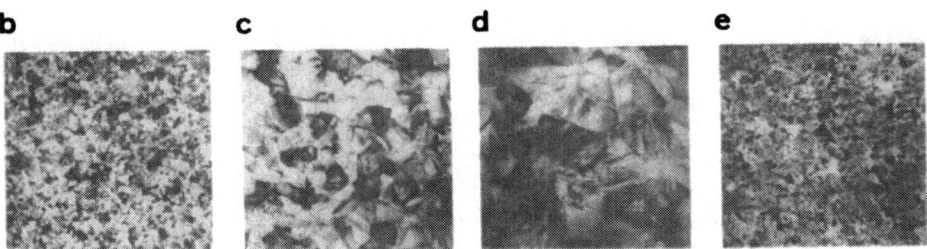

Figure 2. Planar view bright field TEM micrographs of laser crystallized Si films: (a) single pulse irradiation at 565 mJ/cm^2; (b) to (e) first pulse-crystallized at 565 mJ/cm^2 and second pulse-irradiated at 300 mJ/cm^2, 446 mJ/cm^2, 516 mJ/cm^2, and 588 mJ/cm^2, respectively.

major effect, which was introduced due to the second-pulse irradiation of the first-pulse crystallized Si films, was found.

Figure 3 illustrates the changes (or lack of changes), which accompany multiple-pulse irradiation, in the microstructure of Si films. The figure contains results from three sets of experiments in which originally identical films (i.e., first irradiation was conducted with an energy density of 594 mJ/cm^2) were irradiated multiple times (i.e., second and higher irradiation) at three different energy density levels. The energy densities of 318 mJ/cm^2, 500 mJ/cm^2, and 588 mJ/cm^2 correspond to partial melting, near-complete melting (but slightly less than the upper threshold of the near-complete melting energy density), and complete melting. Figures 3b, 3c and 3h, 3i show that multiple-pulse irradiation of the film at energy densities of 588 mJ/cm^2 and 318 mJ/cm^2 (which correspond to complete and partial melting, respectively) do not lead to noticeable further modifications in the microstructures other than that which was observed after the second irradiation.

Figures 3d to 3g, on the other hand, clearly show the appearance of large grains after being irradiated many times, where the size of the large grains increases with an increasing number of pulses. Microstructural inspection of the stacking faults, which are sometimes observed within these large grains, shows the angles, which suggest that these large grains are predominantly of (111) texture. It is also noteworthy that smaller grains appear to roughly preserve the original size distribution. However, more quantitative experiments must be conducted before details of the microstructural evolution can be properly assessed.

In Figure 4, the maximum width of the largest grains that are observed after being irradiated various times are plotted as a function of the number of pulses at the incident energy density of 500 mJ/cm^2. The plot projects the information pertaining to *the early stage* of the grain enlargement that accompanies multiple-pulse irradiation. In particular, the slope of the curve, which corresponds roughly to 400Å/pulse, represents a maximum increase — for the given incident energy density — in the diameter of the growing grains.

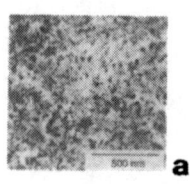

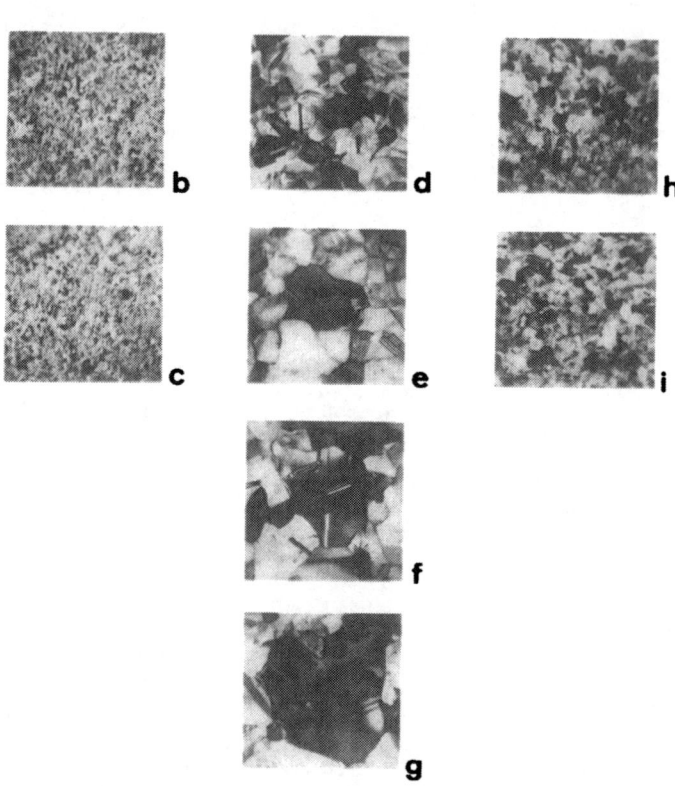

Figure 3. Planar view bright field TEM micrographs of laser crystallized Si films: (a) single pulse irradiated at 594 mJ/cm^2; (b) and (c) 2 and 20 times irradiated at 588 mJ/cm^2, respectively; (d) to (g) 2, 10, 18, and 25 times irradiated at 500 mJ/cm^2, respectively; (h) and (i) 2 and 20 times irradiated at 318 mJ/cm^2, respectively.

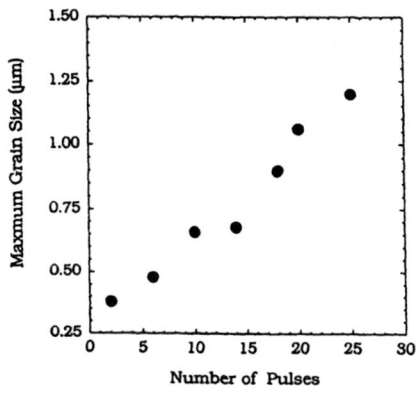

Figure 4. Maximum grain diameter observed as a function of number of pulses at an energy density level of 500 mJ/cm^2.

DISCUSSION

The above observations enable us to conclude the following points with respect to the ELGE phenomenon. First, as the time interval between consecutive pulses was sufficiently long and the number of pulses sufficiently low so as to not induce the self-heating of the substrates (i.e., gradual increase in the base temperature of the films), it can be concluded that ELGE is not associated with the substrate heating effect. Likewise, the stationary (both the beam and the sample) aspect of the experiments enables us to conclude that grain enlargement does not originate from nor require the scanning [3] of the beam (and the accompanying possible variations in the energy densities).

The absence of any dramatic or unusual results in the double-pulse irradiation experiments indicates that there are no unexpected microstructural modifications (other than that which have been already characterized based on single-pulse irradiation experiments [1, 2]) involved with irradiation of crystalline Si films. That is, three major solidification scenarios — (1) partial melting followed by vertical regrowth at low energy densities, (2) near-complete melting followed by lateral solidification at intermediate energy densities, and (3) complete melting followed by nucleation and growth of solids in supercooled liquids at high energy densities — which have been identified for single pulse-induced crystallization of LPCVD a-Si films [1, 2], are applicable to polycrystalline-Si films. Notable differences between the first- and second-pulse irradiation of Si films are that the transient reflectance traces do not reveal (as expected) oscillation signals associated with the explosive crystallization (which is possible only for amorphous films) and that the complete melting threshold of the first pulse-crystallized Si film associated with the second pulse irradiation is higher than that of the first irradiation. In general, higher threshold energy densities are needed for complete melting of the films with greater microstructural perfection.

These observations lead us to conclude that the ELGE must *first be a near-complete melting phenomenon and, more specifically, must involve complete melting of the film occurring at and near the grain boundaries.* The absence of changes in the microstructures of the films upon multiple-pulse irradiation at a low energy density level (in which a continuous layer of the original film remains on top of the oxide layer) and a high energy density level (in which the entire film is completely melted) bears to the fact that it must be a near-complete melting phenomenon. In addition, in order for incremental movement of the grain boundaries to proceed during each irradiation, the completely melted portion of the film (within the near-complete melting condition) must involve the grain boundaries. Plausibility of complete melting of the thin films occurring at and near the grain boundary is substantiated by the established knowledge regarding the lower melting point associated with the grain boundaries and the enhanced absorption of incident irradiation, which is expected to occur at the grain boundaries.

Based on the above deductions as a starting point, we propose the following grain boundary melting model (for which a more detailed discussion and specific implications of the model will be presented elsewhere [5]), which consists of three key components for the ELGE phenomenon: (1) a complete melt-through of the film occurring at and near the grain boundaries, likely as a consequence of the enhanced absorption and the lower melting point associated with the grain boundaries, (2) laterally asymmetric complete melting of grains near the grain boundaries, arising presumably from differences between the adjacent grains in the interfacial free energies between the Si-oxide interface (i.e., lateral melt distance of the higher interfacial energy grain is greater than that of the more favorably-oriented grain), and (3) subsequent solidification (with comparable solidification rates) and collision of the solidification fronts that lead to the net displacement of the grain boundaries.

The above scenario predicts that while all grains may participate and evolve, the most dramatic growth would be experienced by the grains, which have interfacial energy-minimizing orientations (such as (111) textured grains), and that the grains with high interfacial energy orientations on average would shrink and would eventually disappear. Such an evolution should eventually lead to development of polycrystalline films with restricted crystallographic textures. The model also indicates that the net displacement of the grain boundary can be much less than the lateral melt distances which grains undergo and further that complete melting of grains is *not a necessary requirement* for the ELGE to occur. (Indeed, it is a phenomena, which is intimately related to the Super Lateral Growth (SLG) phenomenon observed in single-pulse crystallization of LPCVD a-Si films [1, 2].)

Finally, we note that there exists a rather remarkable apparent similarities (as far as the microstructures are concerned) between the observed ELGE and grain growth processes in general (and the surface energy-driven secondary grain growth (SEDSGG) in particular [6]). However, it must be emphasized that these apparent similarities, striking though they may be, do not necessarily imply a common mechanism (or origin) of these phenomena.

In fact, it is clear that these processes are physically quite distinct from each other for the following reasons. Specifically, SEDSGG is a solid phase process that is directly driven by the differences that exist in the interfacial energies between the adjacent grains; the movement of the grain boundary is thermodynamically driven and proceeds *in order to lower the total free energy of the closed system.* ELGE of thin Si films, on the other hand, occurs in an open system that never reaches steady state, in which melting and solidification kinetics govern and subsequently lead to the evolution in the initial and irradiated microstructures; it is a melting and solidification-induced phenomenon. According to the model presented herein, relocation of the original grain boundary (and therefore the *apparent* "growth of the grain") occurs upon impingement of the resolidifying interfaces. Finally, it is emphasized that the model presented in this paper represents and should be considered a preliminary working model; it would require further experimentation in order to refine the details or for that matter to evaluate critically the basis of the model.

The authors would like to acknowledge Prof. H. J. Frost of Dartmouth University for helpful discussions.

REFERENCES
1. J. S. Im, H. J. Kim, and M. O. Thompson, *Appl. Phys. Lett.* **63**, 1969 (1993). H. J. Kim, J. S. Im, and M. O. Thompson, *MRS Symp. Proc.* **283**, 703 (1993).
2. J. S. Im and H. J. Kim, to be published in *Appl. Phys. Lett.* (1994).
3. R. I. Johnson, G. B. Anderson, S. E. Ready, and J. B. Boyce, presented at the Spring 1993 Mat. Res. Soc. Symp. and to be published in *MRS Symp. Proc.*
4. R. W. Cahn, in *Physical Metallurgy* edited by R. W. Cahn (North Holland, Amsterdam, 1970), Chapter 19.
5. J. S. Im and H. J. Kim, to be published.
6. C.V. Thompson, *J. Appl. Phys.* **58**, 763-772 (1985).

LARGE GRAIN CREATION AND DESTRUCTION IN EXCIMER LASER CRYSTALLIZED AMORPHOUS SILICON

J. B. BOYCE, G. B. ANDERSON, D. K. FORK, R. I. JOHNSON, P. MEI, and S. E. READY

Xerox Palo Alto Research Center, 3333 Coyote Hill Road, Palo Alto, CA 94304

ABSTRACT

For fast-pulse laser-crystallized thin-film Si on non-crystalline substrates, the average grain size exhibits a peak as a function excimer laser energy density at a characteristic laser fluence F_m. The average grain size increases with increasing laser fluence and can reach a maximum value on the order of 10 μm or about 100 times the film thickness. The grain size then decreases with further increases in fluence. This peak in grain size is accompanied by a similar peak in the Hall electron mobility and x-ray scattering intensity. Our experiments have investigated as-deposited and ion-implanted samples, using a double-scan laser crystallization process. Devices have also been fabricated and studied. The results are consistent with the increase in grain size occurring because of the destruction of nucleation sites with increasing laser fluence (i.e., increased heating and complete melting). But substrate damage occurs in the vicinity of F_m, creating nucleation sites which give rise to small grain sizes in the solidified film. The disruption of the interface causes substantial current leakage through the dielectric of bottom-gate transistors, implying that devices should be laser fabricated below F_m.

INTRODUCTION

Fast-pulse excimer-laser crystallization of amorphous silicon on non-crystalline substrates is an important processing technique for large-area electronic devices [1-11]. Due to its short pulse length (~ 10 ns) and small optical absorption length (≈ 7 nm for 308 nm laser wavelength), the silicon is melted and solidified in times of the order of 100 ns. This fast heating into the melt and cooling has the advantage that it produces polycrystalline silicon material of quality comparable to that produced by standard techniques at high temperatures (> 600°C) while keeping the average substrate temperature low (< 600°C). As a result, the process is compatible with low-temperature glass substrates as well as other temperature sensitive materials and can be used to produce hybrid amorphous and polycrystalline silicon material and/or devices in neighboring regions of the same substrate [12-13].

For this laser processing, the sought-after beam profile is a top-hat or a mesa with steep slopes. But for currently available excimer lasers and optical systems, the beam is inhomogeneous to varying degrees. One method used to diminish the inhomogeneity is to expose each point of the amorphous silicon to multiple shots while the beam is scanned over the film in steps that are smaller than the beam size. This multiple-shot crystallization provides a partial averaging of the beam inhomogeneities, but the final pulse at each point might be expected to imprint any beam inhomogeneities into the polysilicon. The step-size-to-beam-size ratio is adjusted to arrive at a compromise between averaging and throughput.

Under scanned, multiple-shot conditions, we have observed that a narrow peak exists in the Si (111) x-ray peak intensity, the average grain size, and the electron mobilities at a particular laser fluence, F_m, for a given substrate temperature and film thickness [8, 9]. While an increase in melt depth and melt time with increasing laser fluence may explain the growth of larger crystals with improved electron mobilities, it fails to elucidate the drop in the grain size and mobility as the energy density continues to increase above F_m. To interpret these observations, we have previously proposed a model involving the heterogeneous nucleation from nucleation sites created at the silicon-substrate interface at the higher laser fluences [9]. Here we describe experiments that support this model and shed further light on the large grain growth and destruction with laser fluence. The experiments also detail the implications for devices.

EXPERIMENTS

Two sets of experiments have been performed: (1) A structural study of LPCVD a-Si on fused silica, laser crystallized at varying laser energy densities; (2) Device measurements on bottom-gate thin film transistors (TFT's), fabricated from PECVD a-Si:H and laser dehydrogenated/crystallized at varying laser energy densities.

For the first set of experiments, 100 mm silica wafers were coated with 100 nm of intrinsic LPCVD a–Si. Both as-deposited and Si-ion-implanted samples were laser crystallized with an XMR excimer laser operating at 308 nm with a 50 ns pulse length. The beam was homogenized and focussed down to a beam spot size of approximately 4×15 mm. The laser beam was scanned in one direction with a laser pulse rate of 20 Hz and a pulse overlap of 99%, i.e., a shot density of 100 laser exposures per unit area and a step size between laser shots of 40 microns. All laser crystallizations were performed in a vacuum of about 10^{-6} Torr.

For the second set of experiments, 70 nm, undoped, hydrogenated amorphous silicon films were deposited by PECVD on Corning 7059 glass substrates containing patterned metal gate electrodes covered with a dual dielectric insulating layer. The dual dielectric consisted of 10 nm of plasma-deposited silicon nitride over the gate electrode followed by 150 nm of plasma-deposited silicon dioxide. The a-Si:H was laser-dehydrogenated and crystallized in a three-step process: a first scan at 140 mJ/cm^2, a second scan at 230 mJ/cm^2, and a third and final scan at varying laser fluence [12, 13]. The first two scans gradually remove most of the H so that the film does not ablate during crystallization. The final scan, being the highest in laser energy, removes the remainder of the H and determines the ultimate grain size. For the device experiments, the crystallized silicon was further processed into n-channel TFT's with the deposition of phosphorous-doped a-Si:H source and drain contacts and a capping layer.

STRUCTURAL RESULTS FROM MATERIALS EXPERIMENTS

The crystallinity of the laser crystallized LPCVD Si was investigated using x-ray diffraction and TEM. Fig. 1 shows the Si (111) x-ray peak intensity of a 100 nm thick sample crystallized under the conditions described above with varying laser fluence. The peak observed in the vicinity of $F_m \approx 500$ mJ/cm^2 is also evident in the average grain size extracted from these x-ray data and in Hall mobility measurements. The peak in grain size at F_m is also seen directly in TEM plane view images. This increase in grain size with fluence can be understood as follows: As the laser energy increases, the depth and the duration of the melt both increase, destroying more nucleation sites at the film/substrate interface. The decrease in the nucleation site density reduces the number of competing crystals and thereby allows for larger grain growth during solidification. At $F_m \approx 500$ mJ/cm^2 the areal density of nucleation sites has reached a minimum. If these nucleation sites are associated with regions of crystalline order or with actual crystallites at the interface, then Si-ion implantation to amorphize these regions before crystallization would be expected

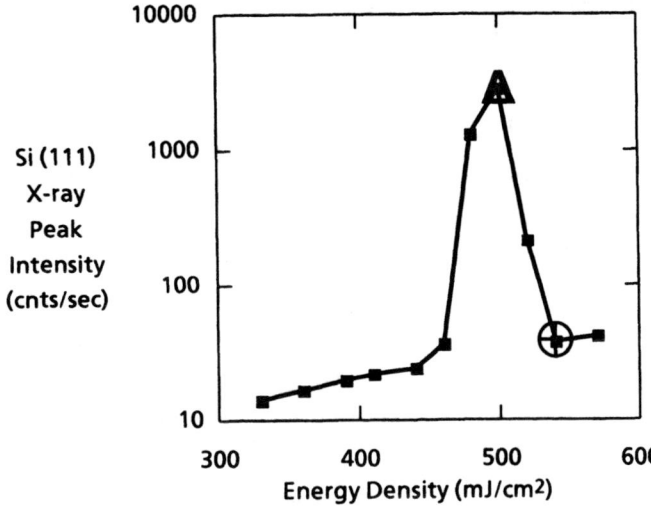

Fig. 1.

Si (111) x-ray peak intensity as a function of laser fluence for 100 nm thick LPCVD Si on fused silica substrates. The symbols, also shown in Fig. 3, mark the energy density of the samples scanned a second time.

to lower the laser energy density required to achieve the large grain material. This is indeed the case as shown in Fig. 2, which provides a comparison of the Si (111) x-ray peak intensity of another set of LPCVD a-Si samples crystallized under similar conditions. The mean grain size exhibits a similar variation with laser fluence and a shift in the position of the peak. One of the samples was Si implanted at 100 keV and a dose of $2 \times 10^{15}/cm^2$, the standard conditions used to amorphize any crystalline grains at the a-Si/silica interface that form during the deposition of the a-Si film. These serve as nucleation sites during thermal crystallization, giving rise to smaller grain material than with fewer crystallites. For laser crystallization, these small crystallites can survive the melting of the film and foster the growth of small grains

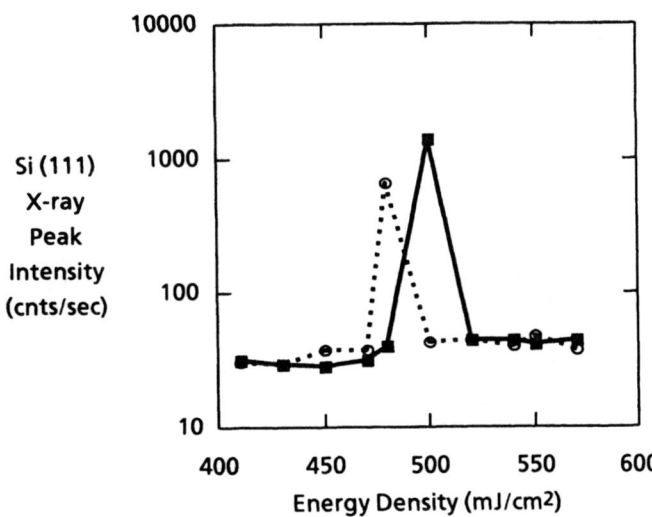

Fig. 2.

Si (111) x-ray peak intensity as a function of laser fluence for LPCVD Si on silica substrates: as-deposited (solid curve) and Si-implanted (dotted curve).

in the bulk of the film. The fact that the peak shifts to lower laser energy density with Si implantation, although only a small shift, supports this notion.

Beyond F_m, the decrease in grain size at the higher energies can be attributed to a dramatic increase in the number of nucleation sites. Our previous studies have shown evidence for substrate damage as a likely nucleation source [9]. A roughening of the interface is readily observed in TEM cross sections (see Fig. 4 below and Fig. 3 in Ref. 9). That substrate damage can occur is not unreasonable since heat flow calculations indicate that the silicon/substrate interface can exceed 2000°C at these laser energy densities. An alternate proposal is that the drop in grain size is caused by homogeneous nucleation from a supercooled melt [10].

To further test these ideas, we have performed a two-scan experiment. That is, we have rescanned films first crystallized at and above the fluence of the peak, $F_m \approx$ 500 mJ/cm2. The results, shown in Fig. 3, reveal that samples crystallized above F_m

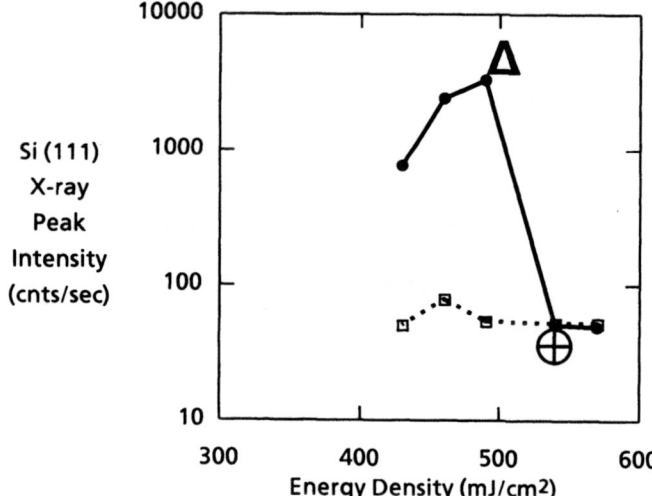

Si (111) X-ray Peak Intensity (cnts/sec)

Energy Density (mJ/cm2)

Fig. 3.

Si (111) x-ray peak intensity as a function of laser fluence for 100 nm thick LPCVD Si laser crystallized a second time. The first crystallization was at the peak (solid curve) and above the peak (dotted curve). The symbols, also shown in Fig. 1, mark the energy density of the first scan.

have undergone an irreversible change: the large-grain material cannot be reproduced. Similar variations are also observed in the average grain size from x-ray diffraction and from TEM. The peak in the grain size cannot be be reproduced for first-scan samples exceeding F_m, while it is recreated for first-scan samples exposed at or below F_m. Fig. 4 shows a cross-sectional TEM image of two of these samples: one crystallized twice near F_m and a second crystallized above F_m and then at F_m. A smooth interface characteristic of the starting conditions before crystallization is seen for the first sample but a roughened one for the second sample. The irreversible change can thereby be attributed to the observed interface roughening, producing an enhanced density of nucleation sites and an augmented heterogeneous nucleation rate. This enhanced heterogeneous nucleation results in small grain poly-Si in the solidified films.

DEVICE RESULTS

The structures that were laser crystallized for the devices consisted of a glass substrate/10 nm silicon nitride/150 nm silicon dioxide/70 nm PECVD a-Si:H. This sandwich differs from the previous materials in three ways that are relevant for grain growth: (1) the silicon thickness is 70 nm rather than 100 nm, (2) the a-Si

Fig. 4.
Cross-sectional TEM images of Si on fused silica where the first laser scan was at F_m (490 mJ/cm^2, upper) and above F_m (540 mJ/cm^2, lower) and the second scan was at 490 mJ/cm^2 ($\approx F_m$) in both cases.

contains hydrogen which is removed by a three-scan process [12, 13], as mentioned above, and (3) the interface consists of Si on a deposited oxide rather than on a fused silica wafer. These differences generate quantitative differences in the size and structure of the resulting laser crystallized grains as is evident in Fig. 5. A peak in x-ray intensity (and grain size) occurs as before but at a different F_m and of smaller magnitude, all associated with the differences enumerated.

The crystallized silicon was further processed into bottom-gate TFT's, and the electrical characteristics were measured. The resulting I-V characteristics were of good quality, but the gate insulator became leaky as the laser fluence went through $F_m \approx 300$ mJ/cm^2. Twenty two identical TFT's (45 microns wide by 15 microns long)

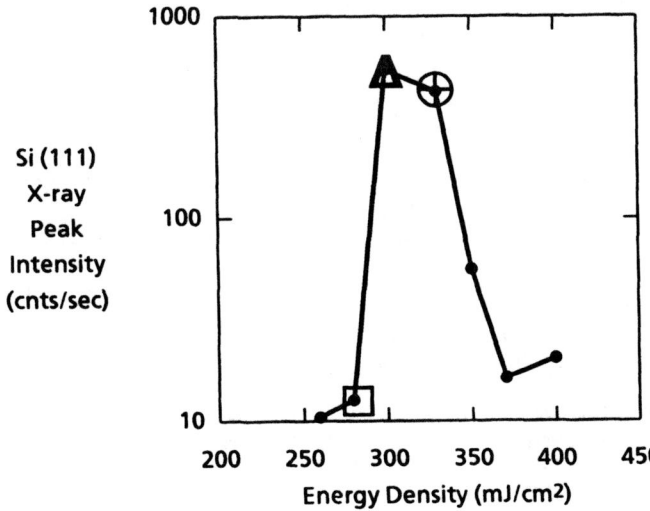

Fig. 5.
Si (111) x-ray peak intensity as a function of laser fluence for the final scan in the three-scan laser dehydrogenation/ crystallization process for 70 nm of PECVD Si on dielectric layers on glass substrates. The symbols mark the corresponding energy density of the device results in Fig. 6.

Si (111) X-ray Peak Intensity (cnts/sec)

Energy Density (mJ/cm^2)

675

were tested at each value of laser fluence, and the number exhibiting dielectric leakage determined. The fraction is displayed in Fig. 6. The number of leaky devices increases as F_m is reached and exceeded; all suffer from severe leakage above 350 mJ/cm^2. These results corroborate the explanation arrived at above that interface damage occurs above F_m and correlates with the decrease in grain size. The device-processing implications of the damaged substrate or compromised bottom dielectric layer are clear: fabricate devices below F_m.

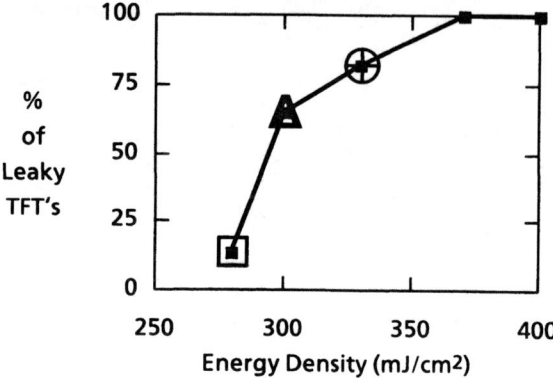

Fig. 6.

Fraction of TFT's exhibiting leakage current through the dielectric as a function of laser fluence for a set of 22 identical bottom-gate devices. The symbols mark the corresponding energy density of the x-ray results in Fig. 5.

ACKNOWLEDGMENT

We acknowledge the assistance of S. Chen and S. Jones at XMR, Inc, Santa Clara, CA, where the laser crystallizations were performed. A part of this work was funded under DARPA Contract No. F33615-92-C-5811.

REFERENCES

1. T. Samashima and S. Usui, Mat. Res. Soc. Symp. Proc. **71**, 435 (1986).
2. S. E. Ready, J. B. Boyce, R. Z. Bachrach, R. I. Johnson, K. Winer, G. B. Anderson, and C. C. Tsai, Mat. Res. Soc. Proc. **149**, 345 (1989).
3. K. Sera, F. Okumura, H. Uchida, S. Itoh, S. Kaneko, and K. Hotta, IEEE Trans. Electron Devices **36**, 2868, 1989.
4. K. Winer, G. B. Anderson, S. E. Ready, R. Z. Bachrach, R. I. Johnson, F.A. Ponce, and J. B. Boyce, Appl. Phys. Lett. **57**, 2222 (1990).
5. R. Z. Bachrach, K. Winer, J. B. Boyce, S. E. Ready, R. I. Johnson, and G. B. Anderson, J. Electron. Mat. **19**, 241 (1990).
6. K. Shimizu, O. Sugiura, and M. Matsumura, Jpn. J. Appl. Phys., 29, L1775, 1990.
7. T. Samashima and S. Usui, J. Appl. Phys. **70**, 1281 (1991).
8. R. I. Johnson, G. B. Anderson, S. E. Ready, D. K. Fork, and J. B. Boyce, Mat.Res. Soc. Proc. **258**, 123 (1992).
9. R. I. Johnson, G. B. Anderson, J. B. Boyce, D. K. Fork, P. Mei, S. E. Ready, and S. Chen, Mat.Res. Soc. Proc. **297**, 533 (1993).
10. J. S. Im, H. J.Kim, and M. O. Thompson, Appl. Phys. Lett. **63**, 1969 (1993).
11. S. D. Brotherton, D. J. McCulloch, J. B. Clegg, and J. P. Growers, IEEE Trans. Electron Devices **40**, 407, 1993.
12. P. Mei, J. B. Boyce, M. Hack, R. A. Lujan, R. I. Johnson, G. B. Anderson, S. E. Ready, D. K. Fork, and D. L. Smith, Mat.Res. Soc. Proc. **297**, 151 (1993).
13. P. Mei, J. B. Boyce, M. Hack, R. A. Lujan, R. I. Johnson, G. B. Anderson, D. K. Fork, and S. E. Ready, to be published.

EXCIMER LASER CRYSTALLIZATION OF AMORPHOUS SILICON FILMS FOR POLY-SI TFT FABRICATION

HIROSHI TANABE, KENJI SERA, KEN-ICHI NAKAMURA,
KAZUMI HIRATA, KATSUHISA YUDA AND FUJIO OKUMURA
Functional Devices Research Laboratories, NEC Corporation,
4-1-1 Miyazaki Miyamae-ku Kawasaki 216, Japan

ABSTRACT

Excimer laser crystallized silicon films have been studied as a function of the number of laser shots, and the influence of the use of such polycrystalline films in thin film transistors (TFTs) has also been investigated. It is found that electron mobility, one of the most important of all TFT characteristics, increases monotonically with the number of irradiations, with maximum mobility being obtained at about 20 shots. This result is not due to grain size, since transmission electron microscopy indicates that the number of laser shots does not affect grain size in polycrystalline silicon films. Raman studies and TFT carrier transport analysis, on the other hand, suggest that this increase in electron mobility may be explained by the decrease in grain boundary defects and defects inside grains.

INTRODUCTION

Polycrystalline silicon thin film transistors (poly-Si TFTs) have attracted much attention in such large-area microelectronics devices as liquid crystal displays[1)2)3)] and contact image sensors[4)]. Excimer laser crystallized poly-Si film is a key material for fabricating these high-mobility TFTs on a glass substrate[5)6)]. Film quality is strongly influenced by such crystallization conditions as energy density[7)8)] and the number of laser shots[9)]. While the effects of energy density on the melt-recrystallization process, a process of temperature rise and fall, have been studied to a significant degree, it remains to be determined precisely what effects might be produced by a series of laser shots, i.e. what film properties might change, and just how they might change, as a function of the number of shots.

This paper reports detailed results concerning the effects, as a function of number of laser shots, on poly-Si TFT characteristics and on the properties of poly-Si film. The main estimated TFT characteristics studied were field effect mobility, threshold voltage, and grain boundary trap state density. Measurements were made by means of transmission electron microscopy (TEM), Raman spectroscopy, and secondary ion mass spectroscopy (SIMS).

EXPERIMENTAL PROCEDURE

Excimer Laser Crystallized Poly-Si Films

Low pressure chemical vapor deposited (LPCVD) amorphous silicon (a-Si) films on quartz substrates were laser crystallized by the NEC excimer laser annealing system XL-561, which was equipped with the NEC XL-120B XeCl excimer laser with an energy output of 200 mJ at a wavelength of 308 nm. The laser beam was focused on the Si surface through a beam homogenizer. The laser beam was 5 x 5 mm^2 on the surface of the sample and was clocked at 10 Hz. The films were crystallized in an argon atmosphere at room temperature.

<u>TFT Fabrication</u>

Figure 1 shows the fabrication process for an n-channel poly-Si TFT. The TFT was fabricated on a quartz substrate at low temperatures, below 600℃. A sputtered tungsten-silicide (WSi_2) layer was employed for source/drain metallization. N^+ poly-Si layers were prepared as source/drain regions by an ion implantation method. For the active layer, an a-Si layer was deposited by LPCVD, and it was crystallized by the XeCl excimer laser mentioned above. After annealing, the poly-Si film was plasma-etched to define the island pattern. A SiO_2 film was deposited as the gate insulator by LPCVD. An aluminum layer was formed for gate metalization. Grain boundary passivation was accomplished by using hydrogen plasma.

RESULTS AND DISCUSSION

<u>TFT Characteristics</u>

Figure 2 shows the I_d-V_g characteristics of poly-Si TFTs fabricated, with the number of laser shots ranging from 1 to 5. TFT characteristics vary dramatically between 1 and 2 shots, and only slightly after that. This is due to the fact that the first shot produces an amorphous-crystalline transition, while successive shots produce crystalline-crystalline transformations. Figure 3 shows mobilities and threshold voltages in terms of the number of shots. Mobility increases monotonically with the number of irradiations, with maximum mobility being obtained at about 20 shots. Our results indicate that film crystallinity varies with each individual irradiation step.

In order better to understand the influences being exerted on TFT characteristics, we estimated the grain boundary trap densities of effective channel regions. Such density is known to be a major factor in determining those characteristics. We conducted our estimations using the theory proposed by Proano et al.'s[10]

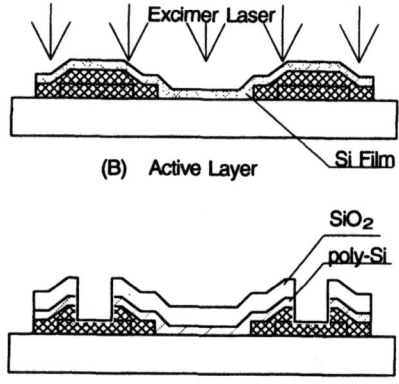

(A)   Source-Drain Region

(B)   Active Layer

(C)   Gate Insulator / Contact Hole

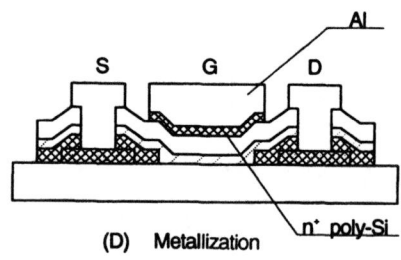

(D)   Metallization

Figure 1.
Process sequence for n-channel poly-Si TFT fabrication by using excimer laser annealing.

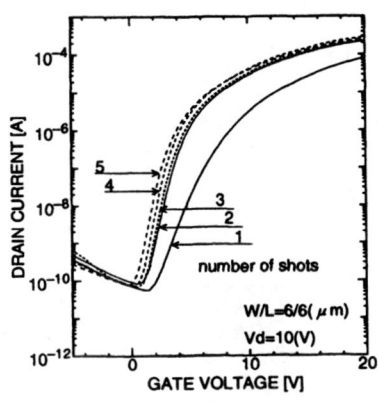

Figure 2.
Typical Id-Vg characteristics for poly-Si TFTs. Annealing energy density is 406 mJ/cm^2 and number of shots is ranging from 1 to 5 shots.

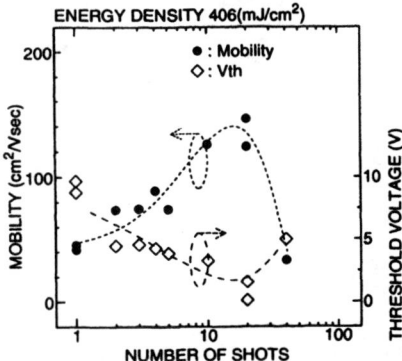

Figure 3.
Electron mobility and threshold voltage for n-channel TFTs dependence on number of shots.

Figure 4 shows a plot of grain boundary trap densities in terms of the number of shots. Trap density decreases to approximately 1×10^{12} cm^{-2} at 20 shots, which is less than a third of that for film crystallized with only one shot. This resembles closely observed changes in mobility. The number of shots that minimizes trap density is the same as that which creates maximum mobility, i.e. mobility seems to be correlated with the grain boundary conditions.

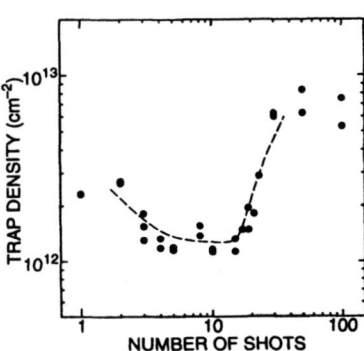

Figure 4.
Grain boundary trap density dependence on number of shots.

TEM Observations

It is known that the electrical properties of poly-Si films are determined by grain size and by conditions in disordered regions, such as defects in grains and grain boundaries. Figure 5 contains TEM microphotographs of films crystallized by 1, 2, 20 and 100 shots, respectively, at the same energy density of 380 mJ/cm^2. They show that for these films the number of shots has no effect on the grain size, which is about 150 nm diameter. This suggests that mobility depends on the defects in grain and the grain boundaries conditions, rather than on the grain size.

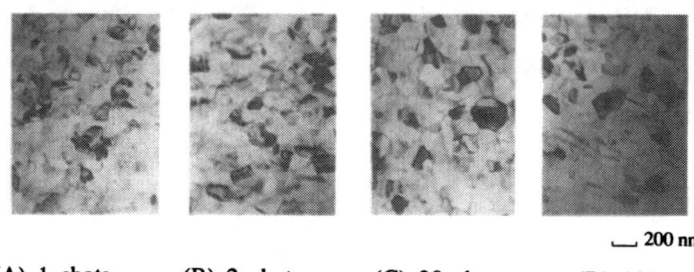

⌐⌐ 200 nm

(A) 1 shots   (B) 2 shots   (C) 20 shots   (D) 100 shots

Figure 5.
TEM micrographs of poly-Si films. Shot densities are 1, 2, 20 and 100 shots, respectively.

Raman Spectroscopy

A Raman scattering spectrum shows a clear distinction between ordered and disordered states[11]. And the observation of a sharp 520 cm^{-1} line of crystalline silicon provides a unique tool for detecting traces of crystallinity[12)13)]. Figure 6 shows the Raman spectra of both irradiated (380 mJ/cm^2, 1 shot) and non-irradiated films. A broad band near 510 cm^{-1}, corresponding to an amorphous-like structure[13)], is seen in the non-irradiated film. A sharp peak near 520 cm^{-1} is observed in the first-irradiated film. This symmetrical peak shape suggests the formation of a fine polycrystalline structure, although the peak is wider than single crystalline silicon and its peak position shifts to a lower frequency. These results indicate that the crystallization has already been completed in the one shot process. The above-mentioned TEM observations support these results.

Figure 7 shows plots of both peak positions and peak widths in terms of the number of shots. The peak position of laser crystallized silicon varies with the number of shots and is about 4 cm^{-1} lower than that of single crystalline. The shifts to lower frequency can be explained by the residual strains inside the grains and the crystalline size[14)15)]. However, a size effect on Raman shift can be ruled out here because of the results of TEM observations. Thus, the shifted peak may be considered to be due to residual strains inside the grains. Such strain is caused by tensile stress. A molten silicon-quartz interface will be fixed during the melting process induced by the laser, and the expanded volume of the molten silicon will contract during the recrystallization[16)]. The variety of positions arises from formations of buffered layers at silicon-silicon dioxide interfaces by successive laser annealing. We suspect that the buffered layers are formed by a diffusion of oxygen atoms or a transformation of dioxide surfaces. Oxygen diffusion is observed in a SIMS measurement of impurity doped silicon-quartz interface processed by multiple laser shots. The mobility decrease resulting from an excessive number of shots can also be explained by this oxygen diffusion.

Peak-width decreased with the number of shots, and the minimum width has been shown to occur at about 7 shots (see figure 7). Decreasing peak width can either be due to the enlargement of grains, or to the decrease of silicon disorders, such as defects in grain and grain boundaries. Since the TEM observations showed that no enlargement of grains occurred, it appears to have been caused by reduced disorder. However, the shot number that provides the maximum mobility is different from the one that provides the minimum Raman peak width. This inconsistency may be considered to be caused by the difference between the channel region and the Raman measurement region. The TFT channel region that contributes to electron conduction is located in a part of poly-Si film near poly-Si/SiO$_2$ interface. By contrast, the Raman measurement region involves the whole poly-Si film in the direction for incidence of an Ar ion laser.

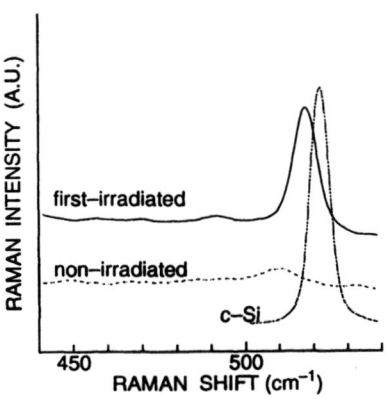

Figure 6.
Raman scattering spectrum for starting film, poly-Si film crystallized by one laser shot and silicon wafer, respectively.

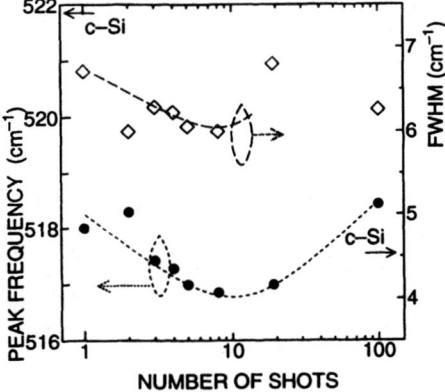

Figure 7.
Plots of both Raman peak positions and peak widths in terms of the number of shots

681

CONCLUSIONS

Investigated here are the effects of excimer laser crystallization on the properties of a number of different polycrystalline films. These effects were studied as a function of the number of laser shots, and the influence of the use of such films in thin film transistors (TFTs) was also investigated. While it was discovered that electron mobility, one of the most important of all TFT characteristics, increased monotonically with the number of irradiations, with maximum mobility being obtained at about 20 shots. Grain boundary trap density was reduced to 1×10^{12} cm^{-2}, which is less than a third of that for film crystallized with a single shot, under the same condition. TEM observations showed that for these poly-Si films the number of shots had no effect on grain size, which was about 150 nm diameter. Raman study indicated that the disorders of those films including defects in grain and grain boundaries, were improved by several successive shots and then degraded by excessive shots. These results indicate that controlling the number of shots is important for poly-Si TFT fabrication.

ACKNOWLEDGMENT

The authors would like to thank Dr. S. Esho, Mr. C. Tani, Dr. T. Saitoh, and Mr. S. Kaneko for their invaluable support, as well as members of thin film semiconductor devices research group for device fabrication and members of the Display Device Research Laboratory for their helpful advice.

REFERENCES

1. Y. Takafuji, T. Yamashita, T. Shimada, Y. Akebi, T. Matsumoto, K. Tsubota, K. Fujioka and K. Awane, SID 93 Digest of Technical Papers, 383 (1993)
2. H. Ohshima, T. Hashizume, M. Matsuo, S. Inoue and T. Nakazawa, SID 93 Digest of Technical Papers, 387 (1993)
3. S.A. Haws, S. Fluxman and P. Rundle, SID 93 Digest of Technical Papers, 895 (1993)
4. K. Sera, H. Asada, F. Okumura, H. Tanabe, K. Nakamura, H. Sekine, I. Fujieda, S. Tujumura and S. Kaneko, SID 93 Digest of Technical Papers, 356 (1993)
5. K. Sera, F. Okumura, H. Uchida, S. Itoh, S. Kaneko and K. Hotta, IEEE Trans. Electron Devices, 36, 2868 (1989)
6. T. Sameshima, M. Hara and S. Usui, Jpn.J.Appl.Phys. 28, 1789 (1989)
7. R.Z. Bachrach, K. Winer, J.B. Boyce, S.E. Ready, R.I. Johnson and G.B. Anderson, J.Electron.Materials 19, 241 (1990)
8. J.S. Im, H.J. Kim and M.O. Thompson, Appl.Phys.Lett. 63, 1969 (1993)
9. R.I. Johnson, G.B. Anderson, S.E. Ready and J.B. Boyce, Mat.Res.Soc.Proc. 219, 407 (1991)
10. R.E. Proano, R.S. Misage and D.G. Ast, IEEE Trans.Electron Devices, 36, 1915 (1989)
11. R. Tsu, M. Izu, S.R. Ovshinsky and F.H. Pollak, Solid State Communs., 36, 817 (1980)
12. S. Nakashima, S. Oima, A. Mitsuishi, T. Nishimura, T. Fukumoto and Y. Akasaka, Solid State Communs., 40, 765 (1981)
13. Z. Iqbal and S. Veprek, J.Phys.C;Solid State Phys., 15, 377 (1982)
14. Y.M. Cheong, H.L. Marcus and F. Adar, J.Mater.Res., 2, 902 (1987)
15. S. Nakashima and M. Hangyo, IEEE J.Quantum Electronics, 25, 965 (1989)

SELECTIVE CRYSTALLIZATION OF A-SI:H FILMS ON GLASS

AIGUO YIN AND STEPHEN J. FONASH
Electronic Materials and Processing Research Laboratory , The Pennsylvania State University, University Park, PA 16802.

ABSTRACT

We have found that an oxygen plasma exposure of a-Si:H films can cause these films to crystallize with a much lower thermal-budget than that required for the same a-Si:H films without this plasma exposure. Based on this unique finding, a selective area crystallization process has been developed to successfully form patterned polycrystalline Si films. In this study, 1500 Å PECVD a-Si:H films were first covered by 500 Å sputtered SiO_2 which was then patterned by lithography to form various islands covered by the SiO_2. These patterned films were exposed to an oxygen plasma and were then thermally annealed in a furnace at 600 °C for 6 hours. After this annealing it was found that the islands covered by SiO_2 during the oxygen plasma treatment remained a-Si while the oxygen plasma exposed regions were crystallized completely. Both XRD and TEM were used to establish the existance of these controlled regions of a-Si and poly-Si in these films.

INTRODUCTION

It has been established that the amorphous phase of silicon (a-Si) is an excellent precursor material for forming polysilicon films that can be far superior to as-deposited polysilicon. This has been demonstrated for both low pressure vapor deposited (LPCVD) precursor films deposited at T ~ 550 °C [1] [2] and plasma enhanced chemical vapor deposited (PECVD) precursor films deposited at T ~ 250 °C [3]. In the case of these PECVD precursor films we have previously shown that a-Si films deposited on Corning 7059 glass at 250 °C can be crystallized at thermal budgets as low as 700°C/4 minutes using rapid thermal annealing [4]. We have also shown that this low crystallization thermal budget can be reduced further to 650°C/ 4 minutes with the aid of an ultra-thin Pd surface treatment [5]. Since we can conceive of applications where it could be advantageous to crystallize areas for drive circuitry while leaving amorphous regions for TFTs or sensors, it could be advantageous to be able to selectively crystallize amorphous silicon. This would be especially advantageous if it did not require laser sintering for the selective crystallization. We have achieved a selective crystallization process based on depositing Pd ultra-thin films in the regions to be crystallized before rapid thermal annealing [6].

We have recently demonstrated that an oxygen plasma exposure of PECVD a-Si:H films can cause these films to crystallize with a much lower thermal-budget than that required for the same a-Si:H films without this plasma exposure [7]. We further report here this approach also can be developed into selective crystallization. We show that by selective masking a-Si:H with SiO_2 during the oxygen plasma exposure, films can be selectively crystallized after furnace annealing at 600 °C for 6 hours.

EXPERIMENTAL PROCEDURE

The hydrogenated amorphous silicon used in this study was deposited by plasma-enhanced chemical vapor deposition on 7059 glass substrate at a substrate temperature of 250 °C. The thickness of the amorphous silicon was 1500 Å. The films were first covered by 500 Å SiO_2 deposited by magnetron sputtering and were then patterned by lithography to form various islands covered by the SiO_2; that is, some regions of amorphous Si samples were covered by SiO_2 (region B) and some regions were not (region A). These patterned films were exposed to an oxygen plasma in an electron cyclotron resonance (ECR) reactor. The samples were placed in the ECR plasma chamber in the same configuration as previously reported in our studies of ECR hydrogenation of thin film transistors. In this selective crystallization study, a 60 minute ECR plasma treatment was used with the microwave power at 700 watts, oxygen pressure at 4×10^{-4} Torr. and substrate temperature at 400 °C.

The 500 Å patterned SiO_2 was removed by etching in a buffered HF solution after oxygen plasma treatment. After this step, these plasma processed a-Si films were thermally annealed in a furnace at 600 °C for 6 hours. The annealed films were characterized using UV reflectance measurement, X-ray diffraction (XRD), and transmission electron microscopy (TEM). The electron transparent samples needed for TEM were prepared by using dilute HF to etch the film away from the substrate [6].

To explore the thermal budget space further, plasma processed a-Si films were also thermally annealed at 600 °C for other periods ranging from 2 hours to 20 hours. UV reflectance measurements were taken on all samples to determine the different crystallization times of the patterned regions.

EXPERIMENTAL RESULTS

In Figure 1, the UV reflectance spectra of a silicon film (region A of the sample) that was not covered by SiO_2 during the oxygen plasma exposure prior to 6 hours annealing at 600 °C is compared to that of a silicon film (region B of the sample) which received the same furnace annealing but was covered by the 500 Å SiO_2 during oxygen plasma treatment. As a reference, the UV reflectance spectrum of a Si wafer is also shown (see Fig. 1(a)). This single crystal Si spectrum shows two reflectance peaks at 276 nm and 365 nm which are due to optical transitions at the X point and along Γ-L axis of the Brillouin zone [8]. As shown in Fig. 1(b), these characteristic reflectance peaks are also observed in region A of the sample. This indicates that region A film has been crystallized. In Fig. 1(c), no reflectance peak is discernable for the region B film. This demonstrates that region B of the sample remains amorphous. Both region A and region B of the annealed film were also characterized using x-ray diffraction. As shown in Figure 2, region A of the sample is crystallized while region B of the sample is still amorphous.

To monitor the selective crystallization process and determine the crystallization thermal budget of different regions in an oxygen plasma treated sample, a reflectance peak height ΔR was defined as the difference between the minimum reflectance at around 240 nm and the maximum reflectance at 276 nm (4.5 ev). In Figure 3, this reflectance peak height ΔR is shown as a function of annealing time. As demonstrated by curve (a), region A film that exposed to oxygen plasma directly can be converted to polycrystalline material in only 5 hours at 600 °C, while region B film which was covered by only 500 Å of SiO_2 during oxygen plasma exposure requires a 13 hour anneal at 600 °C to become polycrystalline silicon (curve (b)). Therefore, after 6 hours annealing at 600 °C, the oxygen plasma treated a-Si film is selectively crystallized, as demonstrated by XRD (Fig. 2) and UV reflectance spectra (Fig. 1).

A Phillips EM-420ST scanning transmission electron microscope (TEM) was used for the TEM investigation of the amorphous and crystallized regions of the selectively crystallized silicon

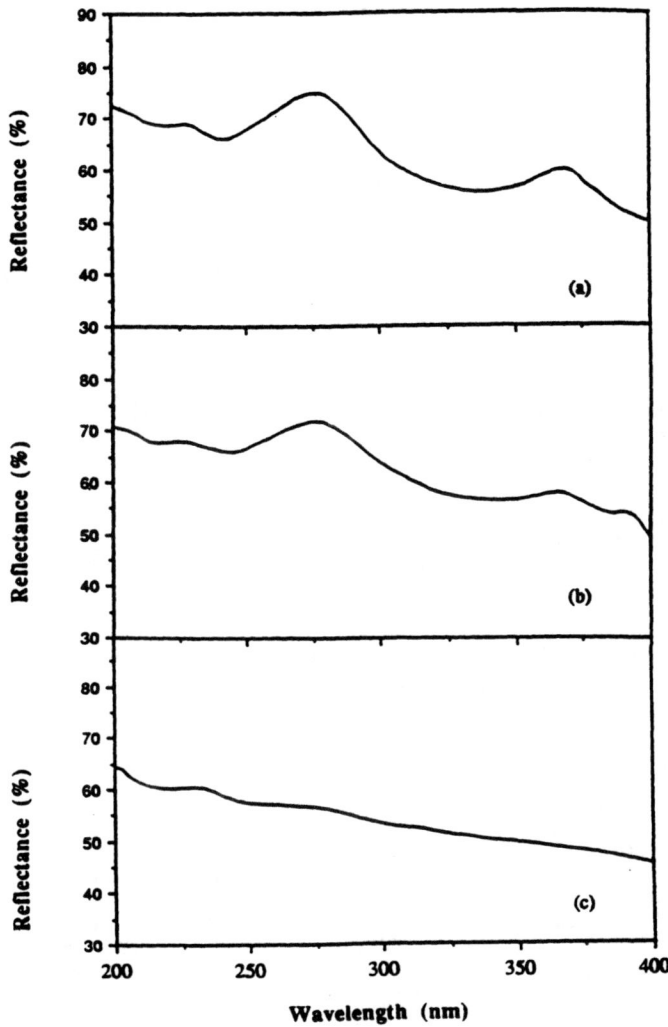

FIG. 1. UV reflectance spectra: (a) reference sample (Si wafer); (b) region A of the oxygen plasma treated sample, 6 hours furnace annealing at 600 °C; (c) region B of the oxygen plasma treated sample, 6 hours furnace annealing at 600 °C.

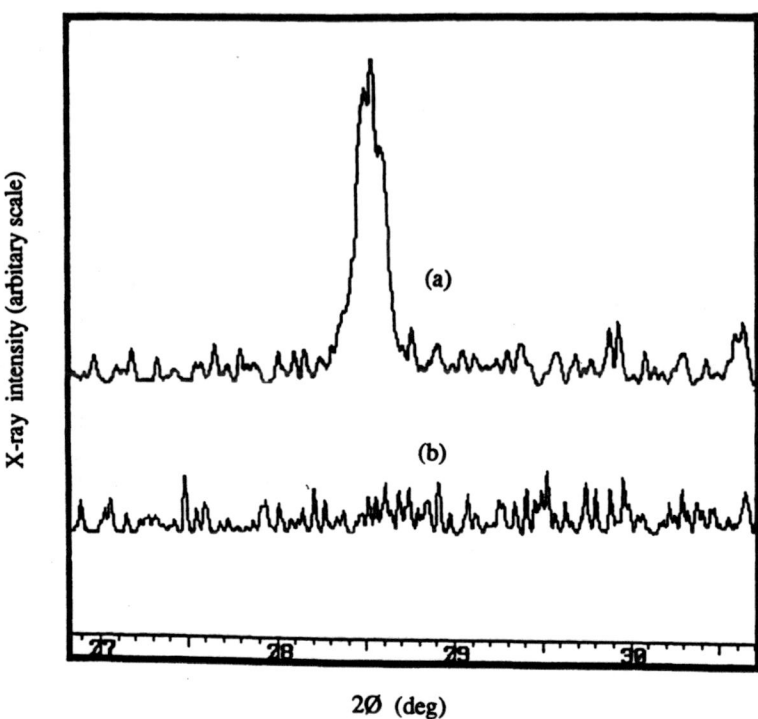

FIG. 2. XRD spectra of the oxygen plasma treated sample with
6 hours furnace annealing at 600 °C: (a) crystallized region A;
(b) amorphous region B.

films. Figure 4 (a) and (c) show the bright field TEM micrographs of region A film that exposed
to oxygen plasma directly and of region B film which was covered by 500 Å SiO_2 during
oxygen plasma exposure but did have the same thermal annealing at 600 °C for 6 hours as region
A film. Figure 4 (b) and (d) show the corresponding selective area diffraction patterns (SADP).
As clearly demonstrated by these TEM pictures, a selectively crystallized silicon film was
obtained in which one region is amorphous while another region is completely polycrystallized
with typical grain sizes up to 1 μm. We note that, in general, the boundary on a sample between
the amorphous and crystallized region is quite sharp after the selective crystallization process.

DISCUSSION

We have proposed a model for this crystallization enhancement provided by our ECR oxygen
plasma treatment. This model is based on the picture that the bombardment from a high density
plasma source such as ECR can cause a chemical reaction to release excess silicon atoms or cause
a physical action to release the excess silicon atoms. These "extra" silicon atoms act to enhance
silicon diffusion and enhance poly-Si grain growth during the subsequent crystallization process
and, therefore, enhance this crystallization process and reduce the crystallization thermal budget
dramatically.

In this report, we show that the crystallization thermal budget is not reduced if the a-Si film is covered by a SiO_2 layer during the oxygen plasma exposure. We suggest that this SiO_2 coating is sufficient to protect the a-Si film surface from both chemical and physical interaction with the incoming radical and ion flux. The result is that no "extra" silicon atoms are produced in this coated a-Si film region during the oxygen plasma treatment; hence, there is no enhanced subsequent crystallization process in this region. Therefore, while region A a-Si film regions which are exposed to oxygen plasma directly can be crystallized by 6 hours annealing at 600 °C due to the enhanced crystallization process, the region B a-Si films that are covered by SiO_2 during oxygen plasma exposure remain amorphous after the same 6 hour annealing at 600 °C. As shown in this study, this can be expected to give an easily controlable selective crystallization process.

CONCLUSION

We have demonstrated that an ECR oxygen plasma exposure of PECVD a-Si:H will reduce its crystallization thermal budget dramatically. We have also shown that with a thin SiO_2 cover layer on such PECVD a-Si:H films during the oxygen plasma exposure, the ECR oxygen plasma exposure will not reduce crystallization thermal budget of the covered region. Thus, with a selective SiO_2 cover layer on an a-Si film during our oxygen plasma exposure, the a-Si film can be selectively crystallized by subsequent low thermal budget furnace annealing at 600 °C for 6 hours. High quality polysilicon films with grain sizes up to 1 μm have been obtained in the crystallized regions while remaining film remains amorphous.

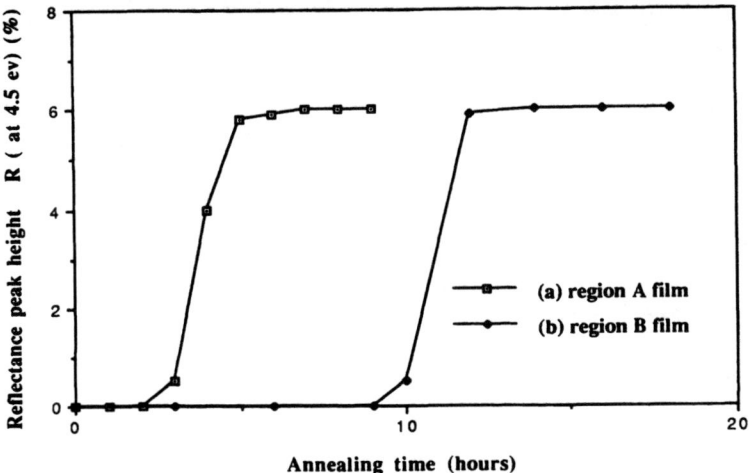

FIG. 3. UV reflectance peak height ΔR at 276 nm (4.5 ev) as a function of 600 °C furnace annealing time: (a) region A of the oxygen plasma treated sample; (b) region B of the oxygen plasma treated sample.

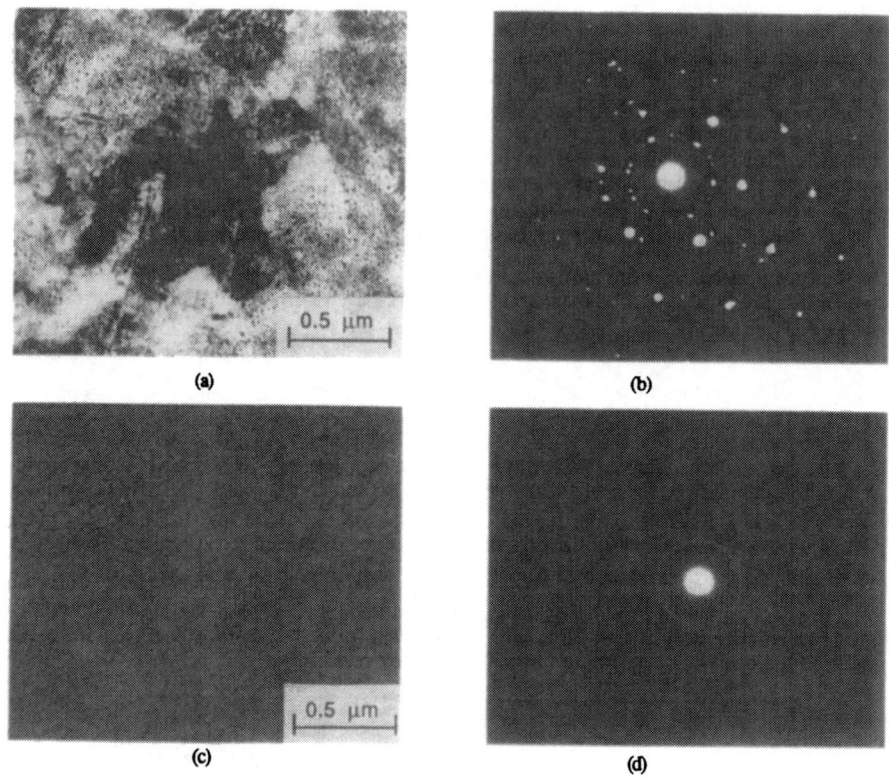

FIG.4. TEM of the oxygen plasma treated sample with 6 hours annealing at 600 °C:
(a) (b) TEM image and TEM SADP of reqion A film of the sample; (c) (d) TEM image and TEM SADP of region B film of the sample.

ACKNOWLEDGMENTS

The TEM work was performed in the electron microscopy facility of Materials Characterization Laboratory at Penn State University.

REFERENCES

1. R. A. Ditizio, G. Liu, and S. J. Fonash, Appl. Phys. lett. 56(2), 1140(1990)
2. K. H. Milliadis, J. Appl. Phys. 63, 2260 (1988)
3. R. Kakkad, J. Smith, W. S. Lau, and S. J. Fonash, J. Appl. Phys. 65(5), 2069(1989),
4. Gang Liu and S. J. Fonash, Jap. J. Appl. Phys. 30 (2B), L269 (1991)
5. Gang Liu and S. J. Fonash, Appl. Phys. Lett. 62(20), 2554(1993),
6. R. Kakkad, Gang Liu and S. J. Fonash, J. of Non-Crystalline Solids 115 (1989),
7. Aiguo Yin and S. J. Fonash, presented at the 40th AVS conference (unpublished),
8. C. Harbeke, Polycrystalline Semiconductors. Physical Properties and Applications (Spinger, Berlin, 1985)

Excimer Laser Induced Crystallization of Amorphous Silicon-Germanium Films

A. Slaoui[1]*, C. Deng[1], S. Talwar[2], J.K. Kramer[2], B. Prevot[3] and T.W. Sigmon[1]

[1] Electrical Engineering and Applied Physics Dept. ,Oregon Graduate Institute 20000 N.W. Walker Road, Portland, OR-97291-1000

[2] Stanford Electronics Laboratories, Stanford University, Stanford,CA-94305

[3] Laboratoire GRPM-PHASE ,23 rue du Loess,F-67037 Strasbourg, France

ABSTRACT

Application of excimer laser crystallization of amorphous silicon (a-Si) has introduced a new, interesting potential technology for the fabrication of polycrystalline (poly-Si) thin film transistors. We are currently studying polycrystalline $Si_{1-x}Ge_x$ thin films in order to determine whether this material can lead to improved electrical properties or to better processing requirements when compared with polycrystalline Si films.
In this work we analyze by RBS , TEM , Raman spectroscopy and surface reflectance, the structure of thin amorphous $Si_{1-x}Ge_x$ films after irradiation with a XeCl excimer laser. The amorphous SiGe films were prepared by evaporation of Si and Ge onto oxidized Si substrates using an electron gun in vaccum .The effects of laser energy fluence during irradiation are investigated. The amorphous to crystalline transition is followed by in-situ measurement of time-resolved reflectivity .

INTRODUCTION

The use of lasers for the crystallization of amorphous silicon thin films has been investigated extensively in order to fabricate polycrystalline silicon thin film transistors (poly-Si TFTs) [1,6]. Due to the high absorption of UV-light in amorphous silicon, it is possible to melt the silicon at a very short time (nanoseconds) using a pulsed laser. The crystallization of the molten layer starts from the underlying substrate with a very high solid-liquid interface velocity typically of several meters per second. Since the laser energy is confined within a very small layer, this process can be executed without damage to the underlying glass substrate thereby allowing the use of low cost substrate materials.This process presents the unique advantage of permiting selective local crystallization methods to integrate , on the same glass substrate, both the high mobility poly-Si TFTs for the peripheral circuits and the low-mobility pixel addressing a-Si TFTs [7]. To date, poly-Si TFTs having mobilities higher than 300 cm^2/V.s were recently obtained by Zhang et al. [8] using a XeCl excimer laser. They were obtained by controlling the amount of impurities in the as-deposited films, by optimizing the laser parameters (laser energy density, number of shots and substrate temperature) and by using thin a-Si films of 50nm in thickness.

On the other hand, King. et al. [9] have recently shown that polycrystalline silicon-germanium (poly-SiGe) films with Ge fractions up to 0.6 are completely compatible with standard VLSI fabrication processes. They also observe an enhancement in the carrier mobility as a function of Ge content in the film. In addition, poly-$Si_{1-x}Ge_x$ is reported to be an attractive alternative to poly-Si in technologies like TFTs with limited thermal budget allowances [10,11].

This low-temperature processing capability, along with potential enhancement in carrier mobilities [9] have motivated us to investigate the formation and properties of poly-$Si_{1-x}Ge_x$ films using excimer laser to induce the crystallization of deposited amorphous silicon-germanium.

*Presently at Laboratoire PHASE, C.R.N,23 rue du Loess, F-67037 Strasbourg cedex, France

EXPERIMENTAL PROCEDURE

Multilayers of Ge/Si (MLS) are deposited at room temperature by alternate electron beam evaporation of Ge and Si. The background vacuum before avaporation was 3×10^{-9} Torr , during deposition, the recorded pressure is about two orders of magnitude higher. Oxidized <100> silicon wafers with a 500nm thick SiO_2 layer are used as substrates. The deposition rate was 0.1nm/s for both elements. The films consist of twelve alternating layers of the same thickness (10nm), the surface layer chosen to be Ge. As the thickness of this layer is greater than the penetration depth of the irradiating laser beam (7nm), the total laser energy is mainly absorbed in this upper layer.

The crystallization step is performed in a He ambient (500Torr) by means of single pulse (FWHM 27ns) from a XeCl (308nm) excimer laser (Questek). The laser energy density is varied between 100-700 mJ/cm^2. Irradiations are made using a homogeneizd beam with a spot size of typically 4x4 mm^2. To monitor the melt and recrystallization processes, a polarized HeNe laser is normally incident on the sample. Reflected light is monitored by a fast response avalanche photodetector (1ns rise time) which provides a measurement of the high reflectivity (liquid) phase.

The amorphous to polycrystalline transition and film crystalline quality are characterized by ultraviolet-visible reflectance (UVR) , Raman spectroscopy (RS), and transmission electron microscopy (TEM). The structure and intermixing of the layers are measured by Rutherford backscattering spectroscopy (RBS) using a 2.2 MeV ^{4}He$^+$ beam.

RESULTS AND DISCUSSION

Measurements of the melt duration as a function of incident energy fluence for theGe/Si MLS are shown in Fig.1. The melt duration of the surface was determined by measuring the duration of the high-reflectivity phase associated with surface melting [12] using the TRR system. The duration of the irradiation induced process increases sharply at a certain energy fluence (0.17J/cm^2) suggesting the existence of a transformation threshold associated with the onset of mixing. This threshold is probably related to surface melting.

The melt duration at a given value of irradiation laser fluence is found to be higher in our films compared to that reported in pure amorphous silicon of comparable thickness [13]. Amorphous silicon has both higher thermal conductivity and melting point [14]. As a consequence, shorter melt-time durations and melt-depths are expected for the process when compared to a-Si$_{1-x}$Ge$_x$ films irradiated at the same energy fluence. We see from Fig.1 that the melt duration increases with increasing laser fluence. Also, the MLS surface became rough above an energy fluence of 0.45J/cm2.

Typical RBS spectra obtained in the as-deposited and irradiated Ge/Si MLS films are shown in Fig.2. The spectrum of the as-deposited MLS film presents six well separated peaks corresponding to the Ge and Si layers. After a single pulse irradiation at the fluence of 0.17 J/cm^2 , it is observed that the height of the first five peaks decreases, indicating substantial interdiffusion of Si and Ge over the100nm thick layer. Following laser irradiation at an energy fluence of 0.22 J/cm^2 ,the multilayers are completely mixed with a smooth concentration gradient found throughout the film depth. At much higher laser fluences (> 0.5 J/cm^2), the films are heavily damaged even by a single laser pulse with ablation observed.

690

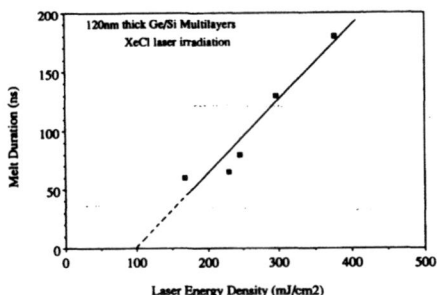

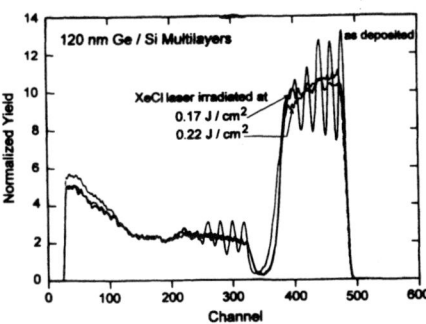

Fig.1.Measurements of melt duration vs energy density in 120nm thick SiGe multilayers.

Fig.2. RBS spectra of as-deposited and after laser melt SiGe multilayersamples.

As noticed before, the intermixing procedure requires melting of the amorphous germanium and silicon layers. As the beam energy fluence increases, the melti-time increases and the molten zone becomes deeper [15]. Consequently, more Ge/Si layers are mixed, allowing the formation of thicker $Si_{1-x}Ge_x$ films.

The crystallization of the $Si_{1-x}Ge_x$ layers after laser irradiation is characterized by the ultraviolet-visible reflectance (UVR) method. Figure 3 shows optical reflectivity spectra of Ge/Si MLS films before and after laser processing with a single pulse at various laser energy fluences. The reflectivity spectra of Si and Ge are also shown in Fig.3 for comparison. The as-deposited Ge/Si MLS reflectance curve does not display any structure in the 250-650nm range. Following laser irradiation, however, the appearance of the peak E_2 around 275nm characterizes the amorphous-to-polycrystalline transition. On the other hand,the position of the E_1 peak characterizes the degree of mixing as this peak can shift from 365 to 560nm as the Ge concentration in the Si decreases from 1 to 0 [16].

At the lowest energy fluence shown here, the spectrum exhibits a broad peak around 460nm and another at about 275nm, these are characteristic of the formation of a crystallized silicon-germanium layer in the surface region.The position of the first peak indicates that mixing of the Si and Ge is not complete. For laser fluences aroud 0.22J/cm2, the broad peak is observed to shift towards lower wavelengths, this means that the $Si_{1-x}Ge_x$ alloy-layer film is more homogeneously crystallized in respect to the distribution of Si and Ge atoms [17]. The lowering of the reflectance signal and the smoothness of the peaks after laser illumination for fluences higher than O.45J/cm^2 are due to an overall signal loss induced by surface degradation process and was confirmed by profilometry and RBS measurements.

Figure 4 compares Raman spectrum of the as-deposited MLS film with that obtained after a single XeCl laser shot at an energy fluence of 0.22J/cm^2. The control trace is typical of amorphous germanium with a broad structure located at around 275 cm^{-1}. No Raman peaks typical of amorphous silicon are observed because of the strong absorption coefficient of a-Ge (which is on the top of the MLS structure) at the Ar probe laser wavelength used [17]. After laser irradiation, the amorphous band totally disappears and the Raman signature is dominated by very narrow lines peaking at 290 , 405 and 485 cm^{-1}. These phonons peaks are usually attributed to Ge-Ge, Ge-Si and Si-Si vibration modes. The positions and half widths of these peaks are characteristics of polycrystalline $Si_{1-x}Ge_x$ films with x=0.5 [18].

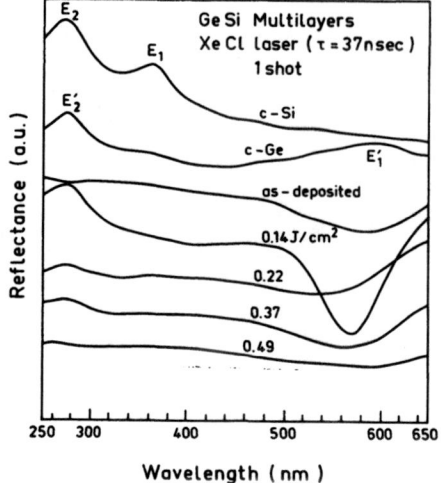

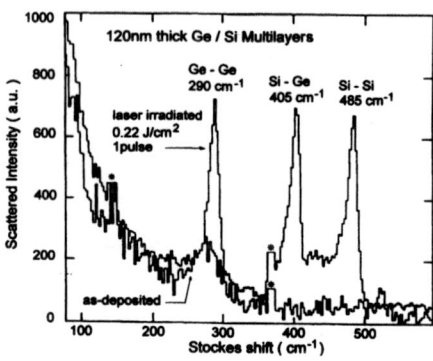

Fig.3. Optical reflectivity spectra of 120nm SiGe multilayers before and after a single shot laser irradiation at various fluences.

Fig.4. Raman stokes TO-phonons spectra of 120nm-thick SiGe multilayer before and after laser irradiation at 0.22J/cm2.

The crystallinity and morphology of the laser crystallized $Si_{1-x}Ge_x$ MLS films were also evaluated by Transmission Electron Microscopy (TEM) and Transmission Electron Diffraction (TED). Figure 5 shows plan-view TEM and TED images of a 120 nm thick poly-$Si_{1-x}Ge_x$ film fabricated by laser irradiation at a fluence of 0.25J/cm2. The resulting films have much in common with the polycrystalline silicon obtained by excimer laser processing of amorphous silicon [4-6,13] . Diffraction patterns taken from a single poly-$Si_{1-x}Ge_x$ grain show only one set of rings characteristic of the diamond crystal structure (Figure 5b). Thus the TED and TEM analysis confirm that the laser crystallized films are $Si_{1-x}Ge_x$ alloys, rather than clusters of Ge within a silicon matrix.TEM examination of the crystallized $Si_{1-x}Ge_x$ film shows that the material consists of relatively large grain up to about 200nm in size. Thus the grain structure of the poly-$Si_{1-x}Ge_x$ films is essentially the same as that of the underlying poly-Si. However, the average grain size in the laser crystallized poly- $Si_{1-x}Ge_x$ films appears to be larger than that in poly-Si crystallized by excimer laser at room temperature [4-6,13].

On the other hand, larger grain sizes (up to 600nm) are usually obtained by classical thermal crystallization [9,19]. This fact does not embarace the laser crystallized material because it is shown that despite of the smaller grains in the laser crystallized poly-Si, higher mobilities are found [4-6,13]. Among the possible explanations,we suggest that some types of purification phenomena is occuring in the crystallized layer due to the simultaneous removal of surface contaminants and segregated impurities through the laser melting process. The segregation coefficients of metal impurities (Cu, Fe, In, Bi) are known to be substantially less than unity [20] and should be segregated out of the active regions to the surface.While the behavior of such impurities in $Si_{1-x}Ge_x$ is unknown, similar results are expected.

In conclusion, melting and subsequent crystallization of electron beam deposited Ge/Si multilayers on oxidized Si has been shown to present an interesting approach for the fabrication of poly-$Si_{1-x}Ge_x$ alloys. Good quality mixing and crystallization has been observed by RBS and TEM for samples consisting of 50% Ge at 120nm thickness films. Concurrent work includes in-situ laser doping of the laser crystallized poly-$Si_{1-x}Ge_x$ alloy films.

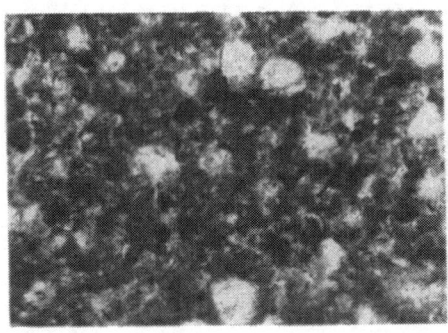 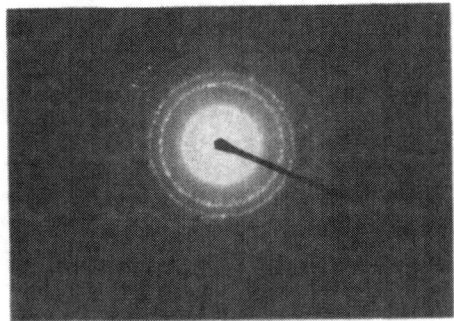

Fig.5. Plan-view TEM (a) and Electron Diffraction (b) micrographs of a 120nm-thick SiGe multilayer laser irradiated at 0.22J/cm2 and 1 shot. 1cm on the TEM micrograph corresponds to 200nm.

Acknowledgments:
The authors would like to thank Dr. T. Derry from Swansea University,U.K, for the TEM measurements.

REFERENCES

1. T. Serikawa, S. Shirai, A. Okamoto and S. Suyama, IEEE Tans. Electron Dev. ED-36, 1929 (1989)
2. T. Sameshima, M. Hara and S. Usui, Jpn. J. Appl. Phys. 28, 1789 (1989)
3. K. Sera, F. Okumura, H. Uschida, S. Itoh, S. Kaneko and K. Hotta, IEEE Tans. Electron Dev. ED-36, 2868 (1989)
4. K. Winer, G.B. Anderson, S.E. Ready, R.Z. Bachrach, R.I. Johnson, F.A. Ponce and J.B.Boyce, Appl. Phys. Lett. 57, 2222 (1990)
5. E. Fogarassy, H. Pattyn, M. Elliq, A. Slaoui, B. Prevot, R. Stuck, S. Unamuno and M.L. Mathé ,in Materials Surface Processing , E-MRS Symp. Proc. V32, p.231, ed. by M. Stucke, E. E. Marinero and I. Wishiyama, North-Holland, Amsterdam, 1993
6. S.D. Brotherton, D.J. McCulloch, J.B. Clegg and J.P. Gowers , IEEE Tans. Electron Dev. ED-40, 407 (1993)
7. A. Chiang , M.W. Geis and L. Pfeifer, MRS Symp. Proceeding V.53, (ed. MRS, Pittsburg, PA, 1986)
8. H. Zhang, N. Kusumato, T. Inushima and S. Yamazaki, IEEE Tans. Electron Dev. Lett. EDL-13, 297 (1992)
9. T.J. King, J.R. Pfister, J.D. Shott, J.P. McVittie and K. C. Saraswat, IEDM Tech. Dig. 1990, p.253
10. T.J. King, K.C. Saraswat and J.R. Pfister, IEEE Elect. Dev. Lett. V12,584 (1991)
11. E.F. Crabbé, J.H. Comfort, W. Lee, J.D. Cressler, B.S. Meyerson, A.C. Megadanis, J.Y.C. Sun and J.M.C. Stork, IEEE Elect. Dev. Lett. 13, 259 (1992)
12. P.S. Peercy, M.O. Thompsonand J.Y. Tsai, Mater. Res. Soc. Symp. Proc., V74, 15 (1987)
    J. Solis and C.N. Afonso, J. Appl. Phys. ,69, 2105 (1992)
13. H.J. Kim, J.S. Im and M.O. Thompson, Mater. Res. Soc. Symp. Proc. V.283, p.703
14. M.O. Thompson, P.S. Peercy, T.Y. Tsao and M.J. Aziz, Appl. Phys. Lett, 49, 558 (1986)
15. T. Sameshima , S. Usui and M. Sekiya, , J. Appl. Phys. 62, 711 (1987)
16. J.C. Phillips, Bonds and Bands in Semiconductors, Academic Press, New York, 1973

17. J. Humlicek, M. Garriga, M.I. Alonso and M. Cardona, J. Appl. Phys. 65, 2827 (1989)
18. N. Nakano, L. Marville, S.M. Jang, H. Liao, C. Tsai, J. Tsai, H.W. Kim and R. Reif, J. Appl. Phys. 73,414 (1993)

19. H. Pattyn, J. Poortmans, P. Debenest, C?. Caymax, P. Vetter, M. Elliq, E. Fogarassy, Z. Nenyer, J. Nijs and R. Mertens, ESSDERC'92, Leuven (Belgium) 1992
20. P. Baeri, S.U. Campisano, G. Foti and E. Rimini, Phys. Rev. Lett. 41, 1246 (1979)

Acknowledgments:
The authors would like to thank Dr. T. Derry from Swansea University,U.K, for the TEM measurements.

694

CONDUCTION MECHANISMS IN CRYSTALLIZED SILICON FILMS ON MOLYBDENUM SUBSTRATES

J. PALMER, J. YI, R. WALLACE, AND W. ANDERSON
State University of New York at Buffalo, Department of Electrical and Computer Engineering, Center for Electronic and Electro-Optic Materials, 215 Bonner Hall, Buffalo, NY 14260

ABSTRACT

Hydrogenated amorphous silicon films deposited on molybdenum sheet metal substrates have been crystallized by thermal annealing at 850°C for 4 hours in a nitrogen atmosphere. X-ray diffraction and scanning electron microscopy results indicated that the average grain size in the crystallized films was approximately 200Å. Palladium contacts were fabricated and the resulting Pd/Si/Mo structures were electrically characterized. Current-voltage-temperature measurements for phosphorus doped and undoped annealed samples resulted in a $J \propto V^2$ characteristic consistent with space-charge-limited current. Using this data, mobility as a function of temperature from 100K-300K was obtained. In phosphorus doped samples, the mobility appeared to be limited by energy barriers at the grain boundaries. In undoped samples, a $\sigma \propto T^{1/2} \exp(-b/T^{1/4})$ temperature dependence consistent with variable-range hopping conduction was observed.

INTRODUCTION

For solar energy to become practical in terrestrial applications, a method of fabricating inexpensive large-area solar cells needs to be developed. Conversion efficiencies of approximately 23% are possible using single crystal silicon cells [1], but such cells are too expensive to be made into large area panels. Thin-film cells made of hydrogenated amorphous silicon (a-Si:H) can be fabricated inexpensively with efficiencies over 12% [2]. However, a-Si:H exhibits the Staebler-Wronski Effect [3] whereby the photovoltaic response of the material degrades as a function of light exposure. An alternate method for producing solar cells would involve crystallizing an a-Si:H layer deposited on a sheet metal substrate into a polycrystalline film [4]. Such a procedure would combine the cost-effectiveness of a-Si:H cells with the stability of single crystal cells. This paper presents the conduction mechanisms found in silicon films prepared in this manner.

EXPERIMENT

The starting material consisted of a-Si:H films deposited on molybdenum sheet metal substrates using a dc glow discharge in silane gas. Two different types of samples were compared in this study. One used undoped a-Si:H, 5μm thick, and the other used phosphorus doped ($N_D \cong 10^{17} cm^{-3}$) a-Si:H, 4μm thick. Both a-Si:H films had a thin n+ layer adjacent to the molybdenum to insure that the back contacts were ohmic. The samples were annealed at 850°C for 4 hours in a tube furnace using a nitrogen gas flow of 2 liters/minute to crystallize the silicon films.

Mat. Res. Soc. Symp. Proc. Vol. 321. ©1994 Materials Research Society

Molybdenum was chosen as the substrate metal because of its relatively low thermal expansion coefficient compared to other metals. Also, an Auger electron spectroscopy depth profile on annealed samples was unable to detect any Mo in the silicon, indicating that almost none of the metal had diffused into the silicon during annealing.

X-ray diffraction and scanning electron microscopy were used to obtain the average grain size in the annealed films. Samples were then cleaned using acetone, methanol, and de-ionized water and etched in a buffer hydrofluoric acid solution to remove the surface oxide. Circular palladium contacts 1mm in diameter were evaporated onto the silicon films using a shadow mask in a vacuum of 10^{-6} Torr. The resulting Pd/Si/Mo structure is shown in Fig 1. Current-voltage-temperature (I-V-T) measurements were performed on these samples using a Keithley programmable voltage source and picoammeter. For this measurement, the sample was mounted on a stage that could be cooled with liquid nitrogen. All measurements were performed over a temperature range of 100-300K.

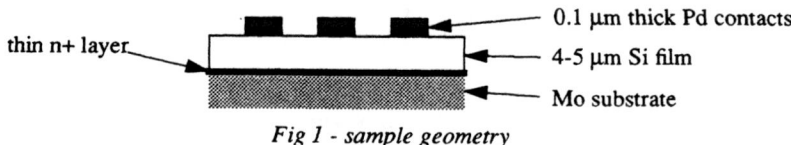

thin n+ layer — 0.1 μm thick Pd contacts
— 4-5 μm Si film
— Mo substrate

Fig 1 - sample geometry

RESULTS AND DISCUSSION

Grain Size

Fig 2 is a sketch of relative peak intensities and widths measured by X-ray diffraction (XRD) from annealed silicon films. The average grain size in the film was obtained using the Scherrer formula [5].

$$t = \frac{0.9\lambda}{B\cos\theta_B} \qquad (1)$$

$$B^2 = B_M^2 - B_I^2 \qquad (2)$$

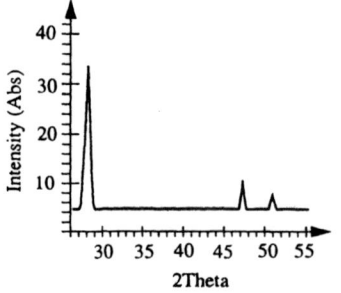

Fig 2 - Sketch of XRD data obtained from annealed films. Peaks shown are (111), (220), and (311) Si in order of increasing 2Theta.

In Eqs 1 and 2, t is the average grain size, λ is the wavelength of the X-rays, θ_B is the Bragg angle, B_M is the full-width-half-maximum of the measured peak, and B_I is instrumental broadening. Using Fig. 2, an average grain size of about 200Å was obtained. This result was confirmed by scanning electron microscopy which detected a few grains as large as 2000Å randomly spaced in a fine grain background. These results are similar to those reported by others [6, 7], who crystallized silicon films deposited on oxidized silicon wafer substrates.

Space-Charge-Limited Current

Current-voltage-temperature measurements on the annealed films resulted in a $J \propto V^2$ characteristic for both undoped and phosphorus doped films at all temperatures, as shown in Figs 3 and 4. This relationship indicated the presence of space-charge-limited current (SCLC). SCLC occurs when the cathode, the Mo substrate, injects an uncompensated density of electronic charge which extends through the silicon thickness to the Pd contact. The standard expression for SCLC is given by Eq 3 [8, 9].

$$J_{SCLC} = \frac{9\varepsilon\mu V^2}{8(4\pi)L^3} \tag{3}$$

Here, ε is the permittivity of silicon, μ is the electron mobility, and L is the thickness of the silicon film. Mobility as a function of temperature was obtained from the slopes of the lines in Figs 3 and 4, assuming that the permittivity was approximately constant as temperature varied. The room temperature mobility values of the two samples are similar to those measured by others [10-12].

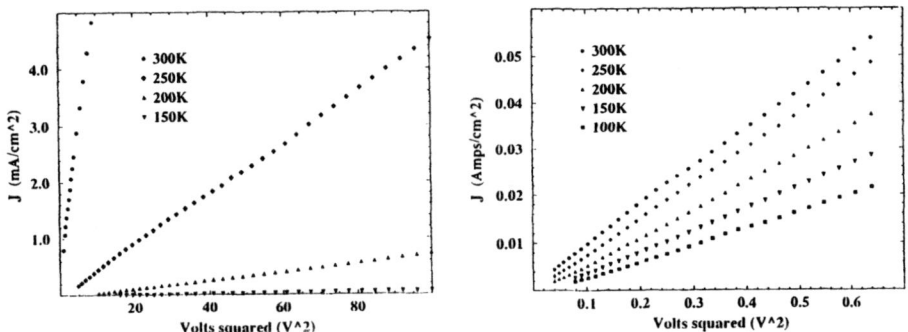

Fig 3 - SCLC data from undoped samples.
$\mu(300K)=0.7\ cm^2/Vsec$
$\mu(250K)=0.06\ cm^2/Vsec$
$\mu(200K)=0.01\ cm^2/Vsec$
$\mu(150K)=0.001\ cm^2/Vsec$

Fig 4 - SCLC data from doped samples.
$\mu(300K)=55\ cm^2/Vsec$
$\mu(250K)=52cm^2/Vsec$
$\mu(200K)=40cm^2/Vsec$
$\mu(150K)=30cm^2/Vsec$
$\mu(100K)=23cm^2/Vsec$

Grain Boundary Barrier Model

The Seto grain boundary model [13] is commonly used in characterizing doped polycrystalline materials. This model states that free electrons become trapped in gap states at the grain boundaries, resulting in an electric field between trapped carriers and donor ions in the grains. The resulting potential barriers are shown in Fig 5. The mobility as a function of temperature according to the Seto model is given by Eq 4.

$$\mu = tq(2\pi mkT)^{-1/2} exp(-\frac{E_B}{kT}) \qquad (4)$$

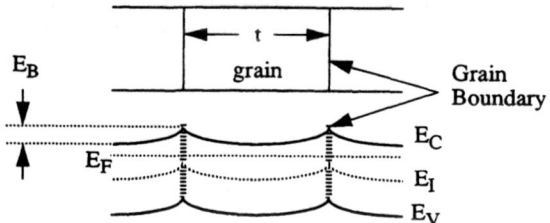

Fig 5 - Grain boundary barrier with continuous distribution of traps.

In Eq 4, t is the grain size, m is the effective electron mass, and E_B is the energy barrier present at grain boundaries. Fig 6 shows the mobility values obtained from the SCLC equation plotted to show agreement with Eq 4. As expected, mobility values from the doped sample agree with the Seto model, while those from the undoped sample do not. An E_B value of 18meV was obtained for the doped sample, in approximate agreement with the work of others [10,13].

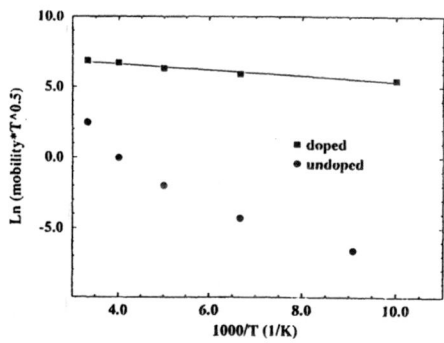

Fig 6 - Mobility values obtained from SCLC data plotted to show agreement with the Seto grain boundary model.

Variable-Range Hopping

Variable range hopping conduction [14] is commonly used in characterizing undoped polycrystalline silicon [15,16] at temperatures well below 300K. This model assumes that most carriers become trapped in gap states at the grain boundaries and conduction occurs as carriers "hop" from state to state. Variable range hopping has a $\sigma(T)$ dependence given by Eq 5.

$$\sigma\sqrt{T} = \sigma_o exp(\frac{-b}{T^{1/4}}) \qquad (5)$$

The factors b and σ_0' are complex parameters involving the density of localized states at the Fermi level, the decay constant of the wave function associated with the localized states, and the phonon frequency. Fig 7 shows $\sigma=(L/A)(\Delta I/\Delta V)$ values from the undoped sample plotted to show agreement with Eq 5.

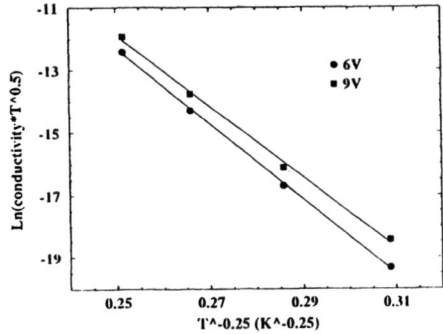

Fig 7 - *Conductivity data from undoped samples plotted to show agreement with variable range hopping conduction.*

The conductivity was a function of voltage because of the non-linear I-V characteristic. However, the slope and intercept changed by less than 8% as the applied voltage changed from 6 to 9V, and values of $b \cong 115K^{1/4}$ and $\sigma_0' \cong 10^7 K^{1/2}/\Omega$-cm are valid for all voltages. These values are in approximate agreement with results obtained from evaporated silicon films [17], but they do not agree with the results of Refs 15 and 16. One reason for this is that the presence of SCLC in our films probably affected the hopping conduction.

CONCLUSIONS

In this work, a-Si:H films deposited on molybdenum substrates were crystallized by thermal annealing. The electrical properties of the resulting polycrystalline films were characterized. The main conclusions were the following.
• The average grain size in both doped and undoped crystallized films was approximately 200Å.
• Doped and undoped films both exhibited a $J \propto V^2$ characteristic consistent with SCLC at temperatures from 100-300K.
• Mobility values obtained from the SCLC data of doped films were consistent with the Seto grain boundary model and an E_B value of 18meV was obtained.
• Conductivity values of the undoped films had a temperature dependence consistent with variable-range hopping conduction.

ACKNOWLEDGEMENTS

The a-Si:H films were deposited at the Solar Physics Corporation, Long Island, NY. Prof. D.D.L. Chung gave helpful comments in data interpretation. The financial support of the New York State Energy Research and Development Authority is acknowledged.

REFERENCES

1. M. Green, A. Blakers, J. Zhao, A. Milne, A. Wang, and X. Dai, IEEE Trans. Elec. Dev. **37** (4), 331-336 (1990).
2. Y. Arai, M. Ishii, H. Shinohara, and S. Yamazaki, IEEE Elec. Dev. Let. **12** (8), 460-461 (1991).
3. D. Staebler and C. Wronski, Appl. Phys. Let. **31** (4), 292-294 (1977).
4. J. Yi, R. Wallace, N. Sridhar, Z. Wang, K. Xie, D. Chung, C. Wie, K. Etemadi, W. Anderson, M. Periard, R. Cochrane, Y. Diawara, J. Currie, and J. Coleman, Solar Cells **30** (1-4), 403-413 (1991).
5. B. Cullity, Elements of X-Ray Diffraction, 2nd ed. (Addison-Wesley Publishing, 1978), p. 102.
6. A. Sakai, H. Ono, K. Ishida, T. Niino, and T. Tatsumi, Jap. J. Appl. Phys. **30** (6A) L941-L943 (1991).
7. E. Korin, R. Reif, and B. Mikic, Thin Solid Films **167**, 101-106 (1988).
8. N. Mott and R. Gurney, Electronic Processes in Ionic Crystals, 2nd ed. (Dover Publications, 1964), pp. 168-173.
9. A. Grinberg, S. Luryi, M. Pinto, and N. Schryer, IEEE Trans. Elec. Dev. **36** (6), 1162-1170 (1989).
10. P. Jeanjean, P. Sellitto, J. Sicart, J. Robert, G. Chaussemy, and A. Laugier, Semicond. Sci. and Tech. **6**, 1130-1134 (1991).
11. S. Solmi, M. Severi, R. Angelucci, L. Baldi, and R. Bilenchi, J. Electrochem. Soc. **129** (8), 1811-1818 (1982).
12. G. Queirolo, E. Servida, L. Baldi, G. Pignatel, A. Armigliato, S. Frabboni, and F. Corticelli, J. Electrochem. Soc. **137** (3), 967-971 (1990).
13. J. Seto, J. Appl. Phys. **46** (12), 5247-5254 (1975).
14. N. Mott and E. Davis, Electronic Processes in Non-Crystalline Materials, (Clarendon Press, 1979), pp. 32-37.
15. R. Sharma, A. Shukla, A. Kapoor, R. Srivastava, and P. Mathur, J. Appl. Phys. **57** (6), 2026-2029 (1985).
16. C. Dimitriadis and P. Coxon, J. Appl. Phys. **64** (3), 1601-1604 (1988).
17. M. Knotek, Solid State Com. **17** (11), 1431-1433 (1975).

AMORPHOUS SILICON CRYSTALLIZATION FOR TFT APPLICATIONS

J. Yi, R. Wallace, N. Sridhar*, D. D. L. Chung*, and W. A. Anderson
Department of Electrical and Computer Engineering, Center for Electronic and Electro-optic Materials, State University of New York at Buffalo, 215 Bonner Hall, Buffalo, N.Y. 14260

* Department of Mechanical and Aerospace Engineering

ABSTRACT

Thin film hydrogenated amorphous silicon (a-Si:H) was deposited on molybdenum (Mo) substrates by d.c. glow discharge. We investigated the a-Si:H crystallization using four anneal techniques; nitrogen atmosphere furnace, vacuum, rapid thermal anneal (RTA), and excimer laser anneal. Anneal temperature ranged from 100 to 1200 °C. Excimer laser energy per pulse ranged from 90 to 340 mJ. Transmission electron microscopy (TEM) revealed microstructure of crystallized Si film with grain size over 0.5 μm. X-ray diffraction (XRD) and Raman spectroscopy were employed to determine the degree of crystallization. The a-Si:H started to crystallize at temperatures over 600 °C. An 850 °C anneal reduced film resistivity to 10^5 (Ω-cm) for intrinsic and 1 (Ω-cm) for n-type. Coplanar type thin film transistors (TFT) with gate channel length of 25 μm and width of 220 μm were fabricated with various insulating layers; rf sputtered SiO_2, Si_3N_4, $BaTiO_3$, MgO, and evaporated SiO. The first two exhibited the least leakage current. The as-grown intrinsic a-Si:H field effect mobility was around 0.03 (cm²/V.s) and delay time was 5×10^{-7} s. The solid phase crystallized silicon film exhibited high leakage current. The delay time of an excimer laser anneal treated TFT was reduced to 2.5×10^{-7} s. Crystallized Si film mobility was improved to 15 (cm² /V.s).

INTRODUCTION

In the early 1970's, Spear and Lecomber showed that a-Si:H could be doped and demonstrated MOS transistor-like field effect devices [1]. Probably the most promising application of a-Si:H thin film transistors (TFTs) is their use in addressable liquid crystal displays. Intrinsic a-Si:H shows weak n-type conductivity, hence TFTs operate in the electron accumulation mode. TFTs for flat panel displays requires the following properties: high I_{on}/I_{off} ratio of 10^5; high off resistance $\geq 10^{11}$ (Ω); and a low, stable threshold voltage (V_T) ~ 5 (V) for SiO_2 dielectric; high field effect mobility ⟩ 1 (cm²/V.s). Achieving high ON current depends mainly on obtaining high field effect mobility by optimizing the a-Si:H layer, the gate dielectric layer, and the interface between them. Achieving high OFF resistance can be achieved using a very thin semiconductor layer to increase the OFF resistance. Finally, threshold voltage is a strong function of the fixed charge in the gate and passivation of dielectric and is strongly influenced by the deposition process. The low field effect mobility in a-Si:H (0.3-1.5 cm²/V.s) has long been an obstacle to many practical applications. Recently, various technologies which improve the field effect mobility of poly-Si TFTs have been reported such as low pressure chemical vapor deposited (LPCVD) poly-Si, solid phase crystallization, laser crystallization, and hydrogenation [2-5]. This paper reports the mobility improvement using various crystallization techniques. Increased mobility can contribute to increase the ON current and the switching speed of TFT's.

Mat. Res. Soc. Symp. Proc. Vol. 321. ©1994 Materials Research Society

EXPERIMENTAL

Due to the importance of the poly-Si applications for TFTs, substantial experimental effort has been directed towards different crystallization methods and subsequent analysis of the crystallized material. The a-Si:H films were deposited onto 4"x8" Mo metal substrates by d.c glow discharge. Deposition temperature varied from 225 to 380 °C giving a deposition rate of around 0.5 μm/min. The a-Si:H films were cut into 1/2"x1/4" using a shear press. These small samples were then subjected to different anneal techniques. The employed crystallization techniques were furnace anneal in nitrogen atmosphere, anneal in a vacuum, rapid thermal anneal (RTA), and excimer laser anneal. A solid phase crystallization was done by Lindberg quartz tube furnace annealing. Nitrogen gas flowing at 2 (lpm) created a nitrogen dominating environment which prevented oxidation during the anneal process. Anneal temperature ranged from 100 °C to 1200 °C with 4 hour duration. The furnace was cooled at a rate of 2.2 °C/min. When the furnace temperature reached 200 °C, the crystallized samples were unloaded for the next procedure. RTA was done using an AG Associates Heat Pulse 210 in nitrogen atmosphere by a proximity capping method. Isothermal vacuum annealing employed a Vycor tube and turbomolecular pump to create a vacuum. A sample was inserted in a Vycor tube (or a quartz tube for an annealing higher than 1000 °C). Then, the turbomolecular pump was employed to create a high vacuum (low 10^{-6} torr). An acetylene and oxygen gas torch was used to encapsulate the Vycor tube with steps of softening, twisting, and pulling. This ampoule was then heated isothermally in the Lindberg furnace as in nitrogen atmosphere annealing with temperature ranging from 100 to 1100 °C for times ranging from 1 h to 40 h. An explosive crystallization was carried out using a argon-fluorine (ArF) excimer laser at the pulse repetition frequencies of a few hertz. The excimer laser specifications are beam spot size of 4x4 mm^2 and wavelength of 193 (nm). The excimer laser anneal was done in Ar atmosphere and in air. The laser beam was perpendicular to the a-Si:H sample with energy of 90 to 340 mJ. The degree of crystallization was studied with x-ray diffraction (XRD), transmission electron microscopy (TEM), and Raman scattering spectroscopy. Partial or the whole Mo layer was removed to characterize the bare thin film Si. The Mo etching removed the interference effects of the Mo metal substrate. TEM and Van der Pauw resistivity were measured after the Mo substrate removal. Finally, TFTs were fabricated to examine applicability of the poly-Si. Al was evaporated for source and drain electrodes. The deposited insulators were evaporated SiO, r.f deposited SiO_2, r.f. magnetron sputter deposited amorphous $BaTiO_3$, Si_3N_4 and MgO. Mainly, SiO, SiO_2, and Si_3N_4 insulators were investigated because other insulators exhibited high leakage current characteristics. Thickness of the insulator was varied from 80 (nm) to 1000 (nm). Al or Pd gate metallization was the final step for the TFT fabrication. TFT gate dimension was 220 μm for gate width (W) and 25 μm for gate length (L). Optical microscope and scanning electron microscope (SEM) were employed after the TFT fabrication to examine gate alignment. Current-voltage characteristics were measured by using an HP4145B semiconductor parameter analyzer. To optimize TFT insulator quality, metal-insulator-semiconductor (MIS) and metal-insulator-metal (MIM) capacitors were investigated. The capacitance-voltage (C-V) data of the capacitors were obtained using an HP4280A 1 MHz meter/plotter.

RESULTS

The influencing factors for the crystallization were doping, temperature, film thickness, and annealing time. Doped amorphous silicon showed better crystallization than did intrinsic Si. Doping induced defect states may have assisted crystallization at lower temperature. Anneal time duration has only a small effect on crystallization. However, anneal temperature greatly affects crystallization. As anneal temperature increased, the Si (111) peak increased, and FWHM was decreased from 0.583 at 600 °C to 0.343 at 1000 °C for n-type silicon. This temperature effect was generally observed for various anneal techniques such as RTA, furnace anneal, and vacuum anneal. Increasing anneal temperature led to increased grain size. TEM examination on a preferentially etched 850 °C, 4 h anneal treated sample showed etch pits of around 0.5 μm indicating that grain size may be greater than 0.5 μm. Triangular type etch pits indicated (111) direction preferrential crystallization. Excimer laser anneal treatment caused an explosive crystallization. Cross-sectional SEM study of the laser annealed sample indicated that only the top thin layer was crystallized at a laser energy of 310 mJ. This result indicates that the excimer laser anneal is useful for only thin film Si crystallization. Amorphous and microcrystalline mixed phases formed after the laser energy of 130 mJ. As laser energy increased, the film crystallinity was improved. A strong surface degradation was observed for a laser energy above 270 mJ and anneal temperature above 1000 °C .

The anneal treatment changes structural properties as well as electrical and optical characteristics of the film. Resistance (R), capacitance (C), dielectric constant (ε), refractive index (n), and light absorption coefficient (α) are known to be reduced with crystallization of the a-Si:H. Mobility (μ) and transmittance (T) are reported to be increased after crystallization. C-V and ellipsometry measurement indicated that the dielectric constant changed from 10 for as-grown a-Si:H to 4 after 850 °C anneal treatment. Van der Pauw resistivity (ρ) measurement was employed to find resistivity and activation energy (E_a) of the thin film Si. Resistivity was reduced from 10^8 (Ω-cm) for as-grown intrinsic a-Si:H to 10^5 (Ω-cm) after 850 °C, 4 h anneal treatment. By neglecting the grain boundary effect on resistivity, activation energy of the intrinsic poly-Si was approximately half of the energy gap ($E_g/2$). In single crystal Si, the energy gap is 1.12 eV. For an intrinsic sample, therefore, E_a is predicted to be 0.56 eV which is in good agreement with the experimental values of 0.51 eV. Calculated activation energy showed lower values (0.04 eV) in n-type poly-Si than for the intrinsic sample because of the reduced energy gap from the donor level to the conduction band

The carrier mobility is an important device parameter. It influences the device behavior through its frequency response or time response. The carrier velocity is proportional to the mobility for low electric fields. The device current depends on the mobility and higher mobility materials have higher current. Higher currents charge capacitances more rapidly, resulting in a higher frequency response. Figure 1 summarizes the measured drift mobilities as a function of anneal temperature. Mobility increased from 0.1 for a-Si:H to 15 (cm²/V.S) for the 850 °C annealed poly-Si sample. Mobility was reduced after 400 °C due to increased defect density after weakly bonded hydrogen evolution. Mobility increased with high temperature anneal because of the improvement in crystallinity. Decreased mobility after 1000 °C anneal may be related to the structural degradation of the Si film. Increased mobility can contribute to increase ON current and switching speed of the TFT.

The MIS structure capacitor C-V study showed the existence of positive fixed charge and the effect of relatively high mobile charge on SiO and Si_3N_4. For r.f sputter grown SiO_2, we observed negative fixed charge for high r.f power and positive fixed charge for low r.f power. Figure 2 shows that fixed charges are compensated by using double layers of SiO_2 deposited at low and high r.f power. Flatband voltage shift indicated oxide trapped charges to be greatly reduced.

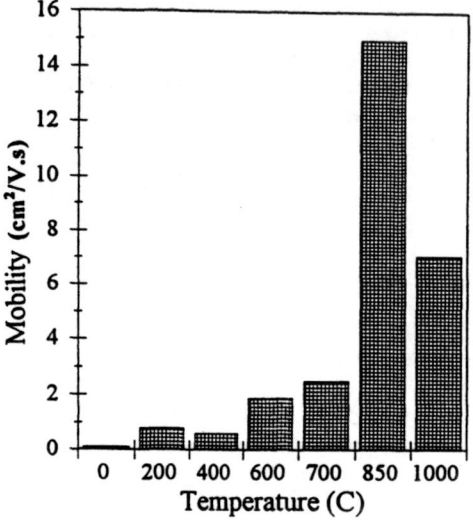

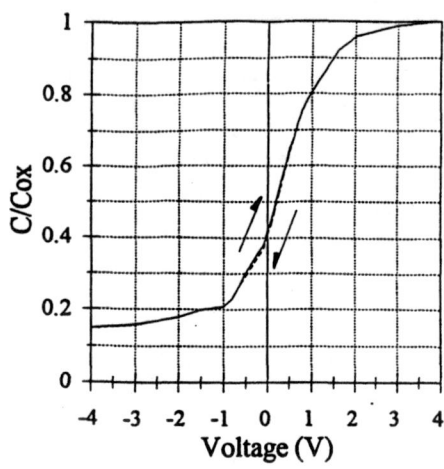

Figure 1. Drift mobility versus anneal temperature for a-Si:H and films after anneal

Figure 2. C-V (1 MHz) on a MIS capacitor using a double layer oxide on 50 Ω-cm crystalline n-Si.

Figure 3 shows that a double layer of SiO_2 grown by low and high r.f power exhibits leakage current almost comparable to a thermally grown oxide. A TFT should be fabricated after the Si film crystallization, otherwise any crystallization anneal step will degrade the insulator quality and the interface properties. A coplanar structure TFT was fabricated before and after anneal treatment. Figure 4 shows the cross-sectional view of the fabricated TFTs with gate width of 220 μm and length of 25 μm. A TFT on as-grown a-Si:H exhibited I_{on}/I_{off} ratio of about 10^4.

A TFT on as-grown a-Si:H exhibited a field effect mobility of 0.03 ($cm^2/V.s$) with evaporated SiO gate insulator layer. Low field effect mobility may come from imperfect oxide quality. Figure 5 shows the I_d-V_{ds} characteristics of the TFT on as-grown a-Si:H. The TFT on as-grown intrinsic a-Si:H shows a relatively high turn-on voltage of 12 (V).

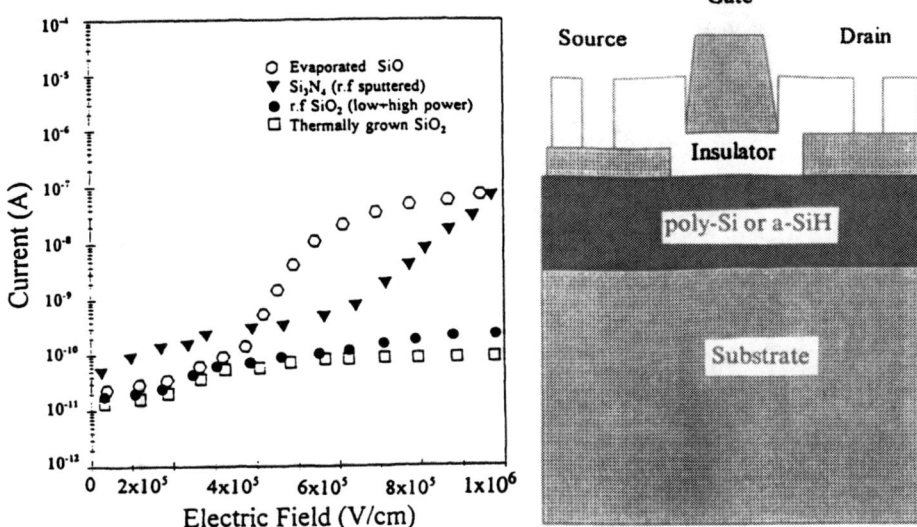

Figure 3. I-E characteristics for various insulators on single crystal Si.

Figure 4. Cross-sectional view of a TFT.

A TFT on a thermally annealed intrinsic Si film showed high ON current, however OFF current also increased due to large leakage current. The I_{on}/I_{off} ratio was less than 10^4 due to large leakage current. TFTs on anneal treated intrinsic Si film showed poor drain saturation characteristics because of defect states at the grain boundaries. The defect states at grain boundaries of the crystallized Si film can be reduced by hydrogen passivation. The TFT on doped polycrystalline film exhibited a large drain to source current and a small gate to source field effect current. The anneal treatment on n and p-type Si could enhance the dopant migration along the grain boundaries. The carriers can then easily tunnel through the grain boundaries which causes the drain to source current to be dominated by the grain boundary current contributions. Laser anneal treated poly-Si also exhibited poor drain current saturation characteristics. Fortunately, dopant migration along the grain boundaries after the laser anneal was not as severe as for thermally annealed samples. Figure 6 shows the reduced turn on voltage and increased ON current using laser anneal treated n-type poly-Si. The TFT delay time was measured by applying a square pulse voltage on the gate and by monitoring the drain current response across the drain resistor. The delay time was improved by 2.5×10^{-7} second on the 230 mJ laser anneal treated sample.

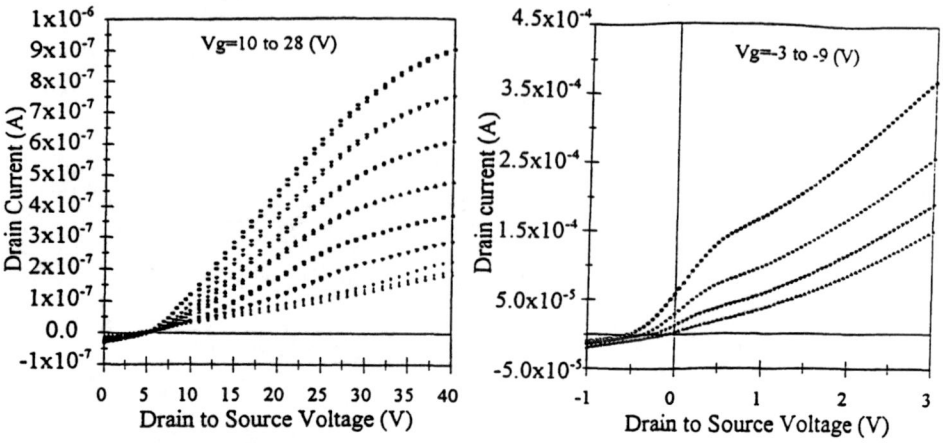

Figure 5. I_d-V_{ds} on as-grown a-Si:H.   Figure 6. I_d-V_{ds} on excimer laser annealed Si (E=230 mJ).

CONCLUSIONS

The a-Si:H started to crystallize at a temperature of 600 °C and excimer laser energy of 130 mJ. Although doped samples exhibited better crystallization, electrical conduction mechanism was dominated by the heavy doped grain boundaries giving poor field effect characteristics. The 850 °C anneal treatment on intrinsic a-Si:H improved drift mobility to 15 cm²/V.s. The double layer of low and high r.f power grown oxide exhibited the most promising insulator properties. TFTs on poly-Si gave data which indicates that grain boundary passivation must be done to reduce defect states. The most promising electrical properties were achieved by the laser anneal with anneal energy less than 270 mJ. Laser anneal treatment improved TFT delay time response by 2.5×10^{-7} s.

ACKNOWLEDGMENT

We recognize financial support from the New York State Energy Research and Development Authority and National Renewable Energy Laboratory. Mr. John Coleman of Plasma Physics Corp. provided samples of a-Si:H.

REFERENCES

1. W. E. Spear and P. G. Lecomber, J. Non-cryst. Solids, Vol. 8-10, 727, (1972).
2. M. Miyasaka, T. Nakazawa, and H. Ohshima, J. Appl. Phys. Vol. 74, 2870, (1993).
3. S. Brotherton, D. McCulloch, and J. Gowers, IEEE Trans. on Elect. Dev., Vol. 40, 407, (1993).
4. C. Liu and K. Lee, IEEE Elect. Dev. Lett., Vol. 14, 382, (1993).
5. T. Chikmura, S. Hotta, and S. Nagata, MRS Proc., Vol. 95, 421, (1987).

CRYSTALLIZATION OF SILICON THIN FILMS PREPARED BY ECR PECVD

Seok-Woon Lee, Yoo-Chan Jeon, and Seung-Ki Joo
Dept. of Metallurgical Eng., Seoul Nat'l Univ., Seoul 151-742 Korea

ABSTRACT

Silicon thin films were deposited by Electron Cyclotron Resonance PECVD using silane as a source gas at room temperature. Deposited films were crystallized either by conventional furnace annealing(FA) or by rapid thermal annealing(RTA) process. The films deposited on SiO₂/Si wafer substrates were amorphous or microcrystalline depending on the microwave power. Deposited films were annealed at 600℃ in a furnace. As expected, higher crystallinity was obtained in the case of the amorphous films than the microcrystalline films after 7.5 hours annealing. It took 15 hours at 600℃ for the amorphous films to reach their maximum crystallinity in case of FA, but it only took 1 second at 900℃ for RTA. In addition, it was shown that RTA can be applied to the rapid crystallization of amorphous silicon thin films deposited on a fused quartz substrate utilizing a new film structure.

INTRODUCTION

In the field of Active Matrix Liquid Crystal Displays, recently, poly-Si thin film transistors(TFTs) have received considerable attention because of their many advantages over a-Si TFTs such as a higher field effect mobility and a lower photoconductivity. Poly-Si TFTs have been fabricated by various methods, among which the crystallization of amorphous silicon films deposited by Low Pressure Chemical Vapor Deposition or Plasma Enhanced Chemical Vapor Deposition(PECVD) and the deposition of poly crystalline silicon films are typical methods[1,2]. The former has been preferred because the usage of inexpensive glass limits the process temperature, and solid phase crystallization(SPC) has been widely studied rather than the other methods[3]. One of the weaknesses of SPC, however, is that it takes a long time. Rapid Thermal Annealing(RTA) process has made a great progress in Si processing field because of its high throughput and ability to preferentially raise the surface temperature, but few reports its usage[4] in the field of poly-Si TFTs due to the low absorption coefficient of silicon in visible region. Electron Cyclotron Resonance(ECR) PECVD is currently attracting much attention because of its lack of electrode and its ability to create high densities of charged and excited species at low pressure(<10⁻⁴ Torr)[5]. Despite the many benefits over the other PECVD processes, not much has been reported on the deposition of silicon films[6] and their crystallization behavior.

In this paper, we demonstrated that we could deposit the amorphous or microcrystalline silicon films at room temperature simply by adjusting microwave power and that the higher crystallinity could be obtained for the amorphous films than for the microcrystalline films after furnace annealing. And we also showed that RTA can be introduced to rapidly crystallize the amorphous films on SiO₂/Si substrate and thus can be applied to the rapid crystallization of amorphous silicon films on a fused quartz substrate.

EXPERIMENTAL

Silicon films were deposited on SiO₂/Si (100) wafer substrates using silane(SiH₄) as a source gas by ECR PECVD. Details in our ECR system is described elsewhere[7]. Silane gas was introduced into chamber and its flowrate was 4 sccm, which resulted in 0.3 and 0.8 mTorr chamber pressure before and after plasma generation, respectively. Microwave power was varied from 100W to 800W and the sustrates were not heated during the deposition process. Most of the experiments were performed with the silicon films of 1000Å thickness, and thickness was varied to 500, and 1500Å when the thickness effects on crystallization process were investigated.

Deposited films were annealed either in a conventional furnace or in a rapid thermal annealing system using tungsten-halogen lamps, and annealing time as well as annealing temperature was varied. The crystallinity of the films was measured by UV reflectance spectroscopy and Raman spectroscopy. For UV reflectance spectroscopy, the peak around 280nm is caused by optical interband transition at X point in Brillouine zone and its peak area between 238nm and 318nm is known to be closely related to the crystallinity of silicon films[8]. The crystalline fractions(crystallinity) were determined as the ratio of the peak area of the films to that of the single crystalline silicon wafer.

In addition, for the purpose of investigating the possibility of RTA application to the crystallization of amorphous silicon films in a very short time, amorphous silicon films of 1000 Å were deposited on a fused quartz substrate and annealed at 900℃ in the RTA system.

RESULTS AND DISCUSSIONS

Figure 1 shows the UV reflectance spectra of 1000Å thick films deposited at microwave power of 300W and 600W. As can be seen in the figure, the films deposited at 300W were amorphous while the films at 600W showed crystallinity. Reduced peak intensity of the films at 600W can be explained with the roughness increase of as deposited microcrystalline films, but that at 300W was to be explained with the native reflectance characteristics of hydrogenated silicon films. That is, the background reflectance of the amorphous films became higher to the level of single crystalline silicon wafer right after the hydrogen evolution during the post annealing processes, even though the films still remained amorphous. Corresponding Raman spectra are shown in figure 2. It can be readily confirmed that the films deposited at

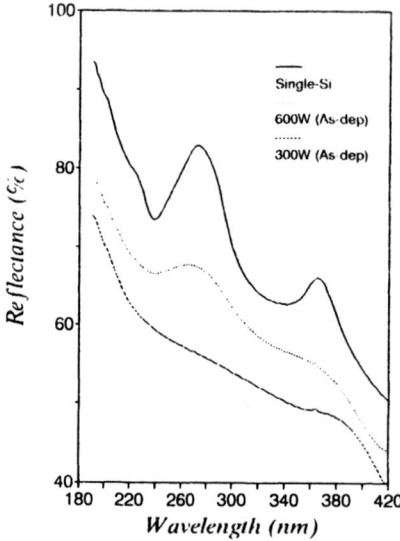

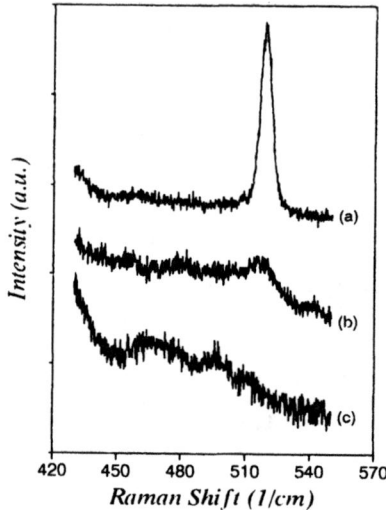

Figure 1 UV reflectance spectra of 1000Å silicon films deposited at 300W and 600W. Single-crystalline silicon is compared.

Figure 2 Raman spectra of 1000Å silicon films and single crystalline silicon wafer
(a) single Si (b) 300W (c) 600W

300W is amorphous but that the films deposited at 600W is microcrystalline.

Figure 3 shows the crystallinity of as deposited films and the deposition rate with the variation of applied microwave power. Deposited films showed no crystallinity until the microwave power reached 400W, and the crystallinity increased with increasing microwave power thereafter. This can be attributed to the high electron energy caused by high microwave power during plasma deposition[9].

Figure 4 shows the crystallinity change of the films deposited at different microwave power with the annealing time. The annealing was performed in a furnace at 600°C. After 3.75 hours' annealing, amorphous films deposited at 300W were crystallized while films at 200W and 100W were not. For all the films deposited at the higher power than 400W, the crystallinity was slightly increased from the original value to a similar extent, so that the trend of the crystallinty with microwave power continued. After 7.5 hours' annealing, crystallinity of 300W continued to increase to around 65% and the films of 200W also began to crystallize, but the films of 100W was not yet crystallized. The crystallinity of the films of the higher power than 400W, however, hardly increased. Therefore, the films of 300W showed highest crystallinity after 7.5 hours' annealing. After prolonged annealing of 15 hours, highest crystallinity of around 80% was obtained for the films of 300W and 200W(the films of 100W has not been measured), and the films of 600W showed the crystallinity of around 50%, only 15% higher than that of as deposited films. Therefore, it can be concluded that the higher crystallinity is obtained for amorphous films after annealing than for microcrystalline films. Figure 5 shows the corresponding UV reflectance spectra of the films annealed for 15 hours at 600°C.

It seems that a very small number of crystallites exist in the amorphous matrix for 300W, and that less or none exist for 200W. This can be inferred by the fact that the crystallinity appeared earlier for 300W than for 200W, and that the higher crystallinity was obtained for 200W than for 300W. In the case where microwave power was higher than 400W, the microcrystalline films seem to have amorphous phase as well as crystalline phase and the size of crystallites appears to increase with increasing microwave power. This is why the crystallinity slightly increased by a similar amount and the initial trend continued even after the annealing.

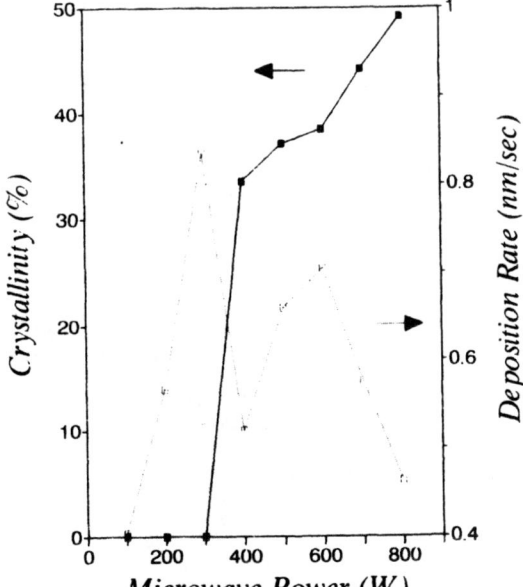

Figure 3 Crystallinity of 1000Å silicon films and their deposition rate with microwave power.

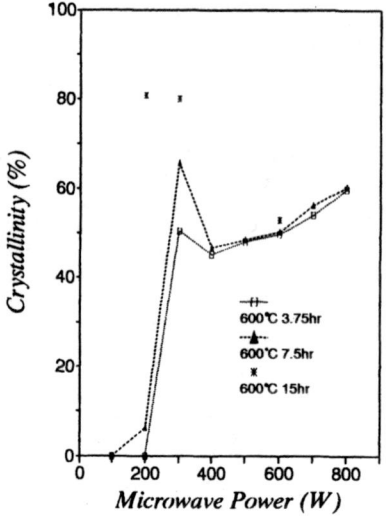

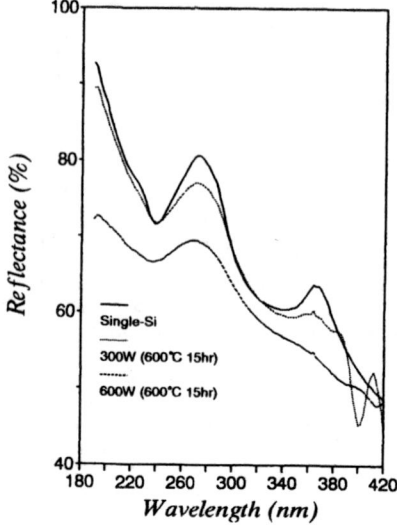

Figure 4 Crystallinity change with microwave power when annealed at 600℃

Figure 5 UV Reflectance Spectra of the films at 300W and 600W after annealing at 600℃ for 15 hours

In order to investigate the effect of film thickness on crystallization, the thickness was varied as 500, 1000, and 1500Å and the microwave power was set at 300W. When furnace annealing was performed at 600℃, the film thickness less than 1000Å turned out to retard the crystallization, but the film thickness more than 1000Å had no effect on the crystallization as can be seen in figure 6. The maximum crystallinity of around 80% was obtained in 15 hours for 1000Å and 1500Å, but it took 45 hours for 500Å to reach the same value. The thickness effect also turned up in the case of RTA at 800℃, but it was not observed at 900℃. Figure 7 shows the change of the crystallinity with RTA time. Despite the crystallinity began to appear from 5 and 15 seconds for 1000Å and 500Å, respectively, the same maximum crystallinity of around 65% was attained after all. At 900℃, however, 1 second's annealing was enough to attain the maximum crystallinity of around 70% for both films. It seems that thickness effect may still occur at 900℃, but the crystallization rate is so fast at 900℃ that it is not observable. It can be readily seen from the figure that the ultimate crystallinity is determined by RTA temperature.

One of the problems of the application of RTA to the crystallization of amorphous silicon thin films on a glass or a quartz substrate is that both the silicon films and the substrate is almost transparent to the radiation of tungsten-halogen lamps. It is well known that most of tungsten halogen lamps' radiation lied in visible wavelength region[9]. In order to solve this problem, we proposed a new film structure. Figure 8 (a) shows a simple structure(type A), silicon films on a fused quartz substrate which hardly absorb lamp radiation and (b) shows a new structure(type B) that will absorb much of lamp radiation. In this new structure, cobalt metal films were introduced on the top of the films for the purpose of absorbing radiation and silicon nitride films(Si_3N_4) were intervened between cobalt and silicon films to block the silicide formation reaction. During RTA process, temperature was monitored by a thermocouple placed on the top of the films(silicon for a type A structure and cobalt for a type B structure). In fact, the thermocouple cannot be used for the correct estimation of the real temperature of the films because it receives radiation directly in itself, particularly in the case as type A, where silicon films are not effective to absorb radiation. Therefore, it requires much time for the films to reach the temperature of thermocouple for type A. But for type B, the real temperature is very close to the reading temperature because the metal films absorb much

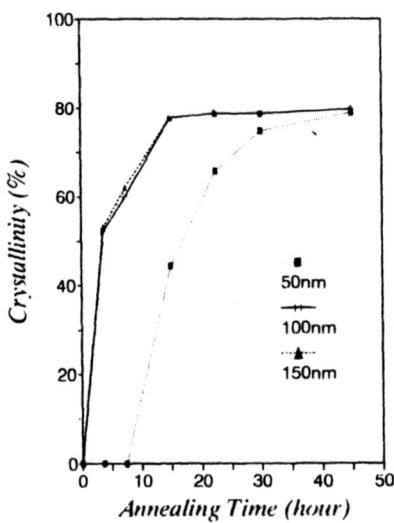

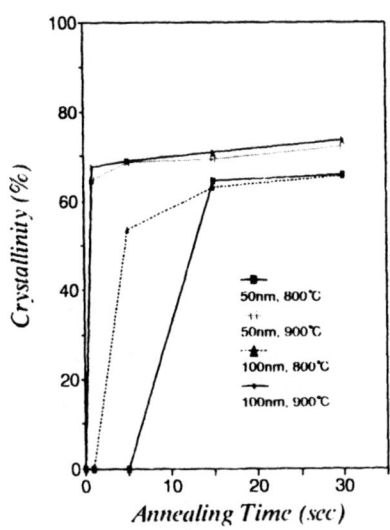

Figure 6 Effect of film thickness on the crystallization as a function of furnace annealing time at 600℃

Figure 7 Effect of film thickness on the crystallization as a function of rapid thermal annealing time at 800℃ and 900℃

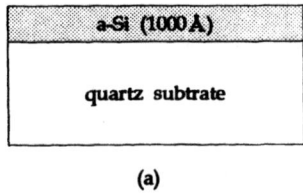

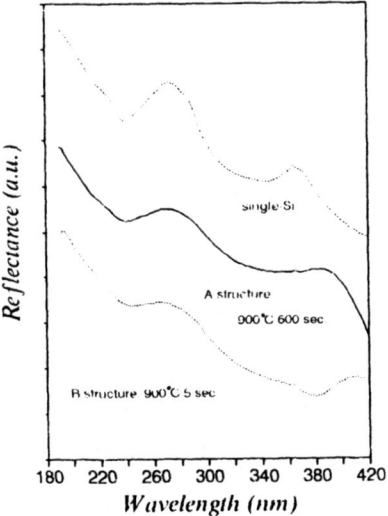

Figure 8 Sample structures for RTA
(a) Simple structure, type A
(b) New structure, type B

Figure 9 UV Reflectance Spectra of silicon films on a quartz substrate after RTA annealing at 900℃

of radiation in the same way as thermocouple does, so the films are to reach their real temperature in a very short time. Figure 9 shows the UV reflectance spectra of type A and B after RTA at 900℃. The silicon films of type B showed crystallinity only after 5 seconds' annealing, it took a long time as long as 600 seconds for type A samples to reach the similar crystallinity. These results provide us with many possibilities, one of which is an application of this process to a glass substrate, and it is currently being studied.

CONCLUSIONS

Silicon thin films were deposited on SiO₂/Si wafer substrates by ECR PECVD, and the deposited films turned out to be amorphous or crystalline depending on the applied microwave power. When the power was lower than 300W, amorphous films were deposited. But above 300W, crystalline phase showed up and the crystallinity increased with increasing power up to 800W. As a result of furnace annealing at 600℃, higher crystallinity was obtained for the amorphous films than for the microcrystalline films and the crystallinity of the microcrystalline films only slightly increased after a prolonged annealing. Among the amorphous films, the films deposited at 300W crystallized most rapidly and reached highest ultimate crystallinity of around 80%. Moreover, they showed the highest deposition rate.

In the other experiments where RTA was applied to the crystallization of amorphous films on SiO₂/Si substrates, it was found out that it requires only 1 second to reach the crystallinity of 70% at 900℃, which value is slightly lower than that value obtained from 15 hours' annealing at 600℃. It was also shown that RTA process can be successfully applied to the rapid crystallization of amorphous films on a fused quartz substrate utilizing metal films for absorbing lamps' radiation.

ACKNOWLEDGEMENTS

This work has been supported by KOSEF through RETCAM in Seoul National University.

REFERENCES

1. Y. Masaki, P. G. Lecomber, and A. G. Fitzerald, J. Appl. Phys., 74, 129 (1993)
2. S. Hasegawa, M. Morita, and Y. Kurata, J. Appl. Phys., 64, 4154 (1988)
3. A. T. Voutsas and M. K. Hatalis, Appl. Phys. Lett., 63, 1546 (1993)
4. M. Bonnel, N. Duhamel, M. Guendouz, L. Haji, B. Loisel, and P. Ruault, Jpn. J. Appl. Phys., 30, L1924(1991)
5. J. Asmussen, in Handbook of Plasma Processing Technology, edited by S. M. Rosnagel, J. J. Cuomo, and W. D. Westwood(Noyes Publications, New Jersey, 1990), p. 285.
6. T. Matsuura, T. Ohmi, J. Murota, and S. Ono, Appl. Phys. Lett., 61, 2908 (1992)
7. Y.-C. Jeon, H.-Y. Lee, and S.-K. Joo, J. Electron. Mat., 21, 1119 (1992)
8. G. Habeke and L. Jastrzebski, J. Electochem. Soc, 137, 696 (1990)
9. Yoo-Chan Jeon, Seok-Woon Lee, and Seung-Ki Joo, Proc. of 1993 MRS FALL Meeting, to be published
10. C. M. Kyung, IEEE Trans. Electon Dev., 31, 1845 (1984)

KINETICS OF HYDROGEN EVOLUTION AND CRYSTALLIZATION IN HYDROGENATED AMORPHOUS SILICON FILMS STUDIED BY THERMAL ANALYSIS AND RAMAN SCATTERING

Nagarajan Sridhar[*], D. D. L. Chung[*], W. A. Anderson[**], W. Y. Yu[+], L. P. Fu[+] and A. Petrou[+], Center for Electronic and Electro-Optic Materials, State University of New York at Buffalo, NY 14260-4400, and J. Coleman, Plasma Physics Corp., P. O. Box 548, Locust Valley, NY 11650.

[*] Also with Department of Mechanical and Aerospace Engineering
[**] Also with Department of Electrical and Computer Engineering
[+] Also with Department of Physics and Astronomy

ABSTRACT

We observed the processes of hydrogen evolution and crystallization in hydrogenated amorphous silicon 0.5-7 μm thick films (deposited by dc glow discharge on molybdenum) by differential scanning calorimetry (DSC), Raman scattering and thermogravimetric analysis (TGA). Investigation was made as a function of doping, deposition temperature and film thickness. For all the films, an endothermic DSC peak was observed at 694 $^{\circ}$C (onset). That this peak was at least partly due to hydrogen evolution was shown by TGA, which showed weight loss beginning at 694 $^{\circ}$C, and by evolved gas analysis, which showed hydrogen evolution at 694 $^{\circ}$C. This temperature (658-704 $^{\circ}$C) increased with increasing heating rate (5-30 $^{\circ}$C/min). Doping reduced this temperature from 694 to 625 $^{\circ}$C for boron doping and to 675 $^{\circ}$C for phosphorous doping. Hydrogen evolution kinetics and FTIR results suggest that the silicon-hydrogen bonding in the intrinsic film was a mixture of SiH and SiH_2, and was predominantly SiH in the phosphorous doped films and SiH_2 in the boron doped films. Crystallization was independent of silicon-hydrogen bonding in the as-deposited amorphous silicon film. It was bulk (not interface) induced. No exothermic DSC peak accompanied the crystallization. The film deposition temperature had little effect on the DSC result, but crystallization was enhanced by a higher deposition temperature.

INTRODUCTION

Amorphous silicon is a potential candidate for solar cell applications [1, 2]. It is well known that the presence of hydrogen in amorphous silicon, known as hydrogenated amorphous silicon (a-Si:H), is important to achieve a material with a low density of electron states within the energy band gap. This is due to the saturation of dangling bonds in the presence of hydrogen. However, a-Si:H solar cells suffer from degradation of solar cell efficiency due to light exposure [3]. On the other hand, such a stability problem does not occur in polycrystalline silicon solar cells.

It has been reported that polycrystalline silicon obtained by the crystallization of amorphous silicon has better structural and electrical properties than that produced by direct deposition [4, 5]. The structure of amorphous silicon is greatly dependent on the deposition technique and conditions [6-8]. Among these techniques, a-Si:H produced by the glow discharge technique shows better electrical characteristics and hence this is the technique which can produce device quality films. To our knowledge, there has been no systematic work done on the influence of hydrogen in glow discharge deposited a-Si:H films on the crystallization behavior. It is known that hydrogen evolution takes place on heating a-Si:H, as detected by mass spectrometry [9, 10]. However, this technique cannot detect physical and chemical processes taking place during the thermal treatment of a-Si:H, such as bond breaking and crystallization. Differential scanning

713

calorimetry (DSC) is a technique that can do so. The objective of this work is to understand the nature of hydrogen and its influence on the crystallization behavior of a-Si:H thick films deposited by glow discharge on molybdenum through DSC, Raman scattering and thermogravimetric analysis (TGA). This investigation was conducted as a function of doping, deposition temperature and film thickness.

EXPERIMENTAL

The a-Si:H films were deposited at Plasma Physics Corp. on molybdenum substrates by dc-glow discharge in a gradient field with cathodic deposition from silane with deposition rates up to 1 μm min^{-1}. The deposition temperatures were 225 and 300 ^{o}C; the pressure was 250-300 mtorr; the SiH$_4$ flow was 10 cm^{3}min^{-1}. The film thickness ranged from 0.5 to 7 μm.

For DSC and TGA measurements, the films were detached from the substrates by etching the molybdenum substrate in an H$_2$O:HNO$_3$:H$_2$SO$_4$ (2:1:1) solution. After washing and drying, the film (in the form of flakes after detaching from the substrate) was put in an alumina DSC/TGA sample pan. For DSC measurements, the weight of the film ranged from 0.3 to 2 mg; for TGA measurements, it ranged from 2 to 10 mg.

DSC was performed using a power compensated Perkin-Elmer 7 series instrument. In a power compensated DSC, the sample and reference pans are placed in two distinct cells (made of platinum). Each cell is heated by a furnace. The DSC measures the increase or decrease in power provided to the sample furnace in order to keep the temperature the same as that of the reference furnace. All the runs were made at a constant heating rate of 5-30 ^{o}C/min from room temperature to 725 ^{o}C (the maximum temperature of this instrument) and then rapidly cooled to room temperature. A few runs were performed by cooling the sample at a constant rate to check the effect of the cooling rate. The baseline of each run was obtained by using the same heating and cooling program as that of the corresponding sample run, except that, during the baseline run, the sample pan was empty. The baseline was then substracted from the sample run to obtain the DSC curve of the sample. All the runs were performed in nitrogen.

Weight loss measurements were performed using a Perkin-Elmer 7 series TGA. In all TGA runs, the films were heated from room temperature to 800 ^{o}C at a rate of 20 ^{o}C/min in helium.

For crystallization measurements, the sample was heated in the DSC at a constant rate to 725 ^{o}C and then cooled rapidly to room temperature. The sample was then removed from the DSC and Raman scattering was performed on the sample at room temperature to determine the crystallinity of the film. Raman scattering was performed in the backscattered mode using a Spex 1877 triplemate Raman spectrometer equipped with a CCD detector. The wavelength of the Ar laser was 4880 A and the laser power was 100 mW. For all the crystallization experiments, the molybdenum substrate was not removed from the film. The presence of molybdenum did not alter the DSC response. Evolved gas analysis (EGA) was done by heating the film in an evacuated Vycor tube (10^{-6} torr) at a constant rate of 20 ^{o}C/min to 700 ^{o}C and simultaneously measuring the partial pressure of the H$_2^+$ ions using a quadrupole mass spectrometer (VG Instruments). Fourier transform infrared spectroscopy (FTIR, Mattson Cygnus100, NU30000) was performed in the reflection mode from 400 to 4000 cm^{-1} under a nitrogen atmosphere.

RESULTS AND DISCUSSION

Fig. 1 shows a DSC heating curve of a 7 μm thick intrinsic a-Si:H film heated to 725 ^{o}C at a rate of 20 ^{o}C/min. The film deposition temperature was 225 ^{o}C. An endothermic peak was observed at an onset temperature of 694 ^{o}C. Fig. 2 shows a TGA curve of the same sample heated at 20 ^{o}C/min to 800 ^{o}C. Weight loss was observed at 694 ^{o}C also. That this weight loss was due to hydrogen evolution was confirmed by the following two tests. The first test was done by heating the film at 20 °C/min to 700 °C while performing mass spectrometry, which showed hydrogen evolving from the film both at 365 and 694 °C. However, the amount of hydrogen evolved at 365 °C (less than 1 % of total hydrogen evolved from the film) was so little that no DSC peak was observed at that temperature. The hydrogen evolved at 365 °C has been attributed to the release of molecular hydrogen residing in the voids of the a-Si:H film [10]. Thus, the hydrogen content of the film was calculated to be around 20 at % from the weight loss occurring at 694 °C. The second confirmatory test was the absence of the endothermic peak in the DSC curve of a sputtered film (which had no hydrogen) having the same thickness as the glow discharge film of Fig. 1 and 2. Since gas evolution is accompanied by weight loss, which corresponds to an apparent heat content loss, gas evolution appears as an exothermic process. In contrast, an endothermic reaction was observed in this work. Previous workers [11, 12] performed DSC of glow discharge amorphous silicon alloys (1-3 μm thick) using two different types of DSC instruments, namely power compensated and heat flux instruments The former is more quantitative in the heat change measurement. They found that at temperatures greater than 300 °C, there was an endothermic peak for the power compensated DSC, whereas for the heat flux type it was exothermic. They attributed this peak to be due to hydrogen evolution and they attributed this discrepancy to the difference in instrumental response.

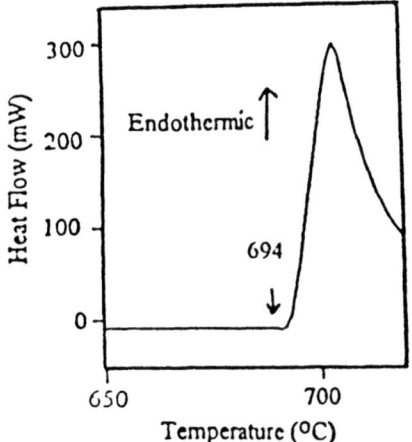 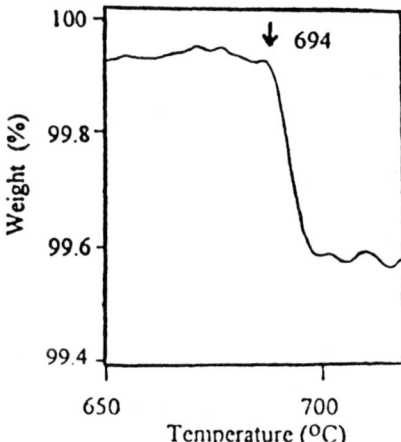

Fig. 1   DSC heating curve of 7 μm thick intrinsic a-Si:H film.

Fig. 2   TGA heating curve of 7 μm thick intrinsic a-Si:H film.

It was found that the onset temperature of the endothermic peak in the DSC heating curve of the 7 μm thick film of Fig. 1 and 2 increased with increasing heating rate. The onset temperature was 658, 672, 694 and 704 °C at heating rates of 5, 10, 20 and 30°C/min respectively. This trend further confirms that the endothermic peak was due to hydrogen evolution (probably diffusion controlled [13, 14]). The activation energy for hydrogen evolution was determined from the variation of the hydrogen evolution temperature with the heating rate, using the equation [13]

$$\ln[d^2h/(\pi^2 k T_H^2)] = \ln(D_0/E_D) - E_D/k T_H, \qquad (1)$$

where d is the film thickness, h is the heating rate, D_0 is the diffusion prefactor, E_D is the activation energy, T_H is the onset temperature for hydrogen evolution and k is the Boltzmann's constant. The slope of the Arrhenius plot of $\ln(D/E_D)$ vs. $1/T$ for this film gave $E_D = 2.77$ eV and the intercept gave $D_0 = 1.5 \times 10^5$ cm^2/s. The value of E_D suggests the type of bonding between silicon and hydrogen in the film. Comparison of this value to the bond energies of SiH and SiH_2 (3.5 and 1.5 eV respectively [15]) suggests that the bonding in the film is a mixture of SiH and SiH_2. This estimate is consistent with FTIR results on this film, which showed peaks at around 630, 880 and 2000-2100 cm^{-1}. The presence of SiH bonding is indicated by the occurrence of peaks at 630 and 2000 cm^{-1} and for the SiH_2 bonding at 630, 880 and 2100 cm^{-1}. Our E_D and D_0 values were considerably higher than other reports on glow discharge films [13]. On the other hand, our values were similar to that of Ref. 16 where hydrogen diffusion was studied in self implanted amorphous silicon of low hydrogen concentration. Ref. 16 explained this difference by proposing that in the low H concentration regime, the motion of hydrogen occurred not through the breaking of silicon-hydrogen bonds followed by interstitial transport (as proposed by Ref. 13), but through a bond-switching process facilitated by mobile dangling bonds that are thermally generated and migrating readily through the a-Si network. However, due to the high hydrogen content in our film and a reasonable consistency between the hydrogen evolution kinetics and FTIR results, it is not clear at this moment if the hydrogen diffusion model proposed by Ref. 16 is plausible in our films.

Raman scattering results obtained at room temperature after heating at a constant rate (in the range from 5 to 30 ^oC/min) to 725 ^oC showed that decreasing the heating rate enhanced the crystallization of the intrinsic film. After heating at 5 ^oC/min, a sharp Raman peak was observed at 520 cm^{-1}, which corresponds to crystalline silicon. However, there was no exothermic peak seen in the DSC curve of this film. This indicates that either the heat of crystallization of this film was too small to be detected by DSC or the strong endothermic peak due to hydrogen evolution masked the crystallization peak. To check if crystallization occurred at a temperature above 725 ^oC, the film was heated from room temperature up to the melting point of silicon in a Shimadzu differential thermal analyser (DTA). No exothermic peak was oberved in the DTA curve at any temperature. In contrast, Ref. 11 reported a strong exothermic peak (due to crystallization) at around 700 ^oC in both power compensated and heat flux DSC instruments for a-Si:H films.

As hydrogen evolution and crystallization is inhibited by increasing heating rate for the intrinsic film and that our hydrogen evolution activation energy value (2.77 eV) matches that of solid phase (epitaxial) crystallization of Si [17], it might appear that the two processes might be related to each other. However, the onset temperature of crystallization of this film, as determined by Raman scattering performed after heating at 5 ^oC/min to different temperatures and subsequent rapid cooling to room temperature was 37 ^oC higher than the onset temperature of hydrogen evolution. This indicated the existence of an intermediate amorphous state (pure a-Si) between the initial state (a-Si:H) and the final state (c-Si), in agreement with Ref. 18 where solid phase crystallization kinetics of glow discharge a-Si:H thin films were studied. This suggests that there might not be a relation between hydrogen evolution and crystallization in glow discharge a-Si:H films. In contrast, from the equality of the activation energies and the H diffusion mechanism, Ref. 16 suggested that the sum of the formation and migration energies of dangling bonds are identical in both processes and hence hydrogen diffusion and solid epitaxial crystallization of a-Si are related to each other for a-Si films of lower H content.

The effect of boron (p-type) and phosphorous (n-type) doping on the DSC curve was also investigated. The intrinsic and doped films were the same in thickness (7 μm) and deposition temperature (225 ^oC). At a heating rate of 20 ^oC/min, the hydrogen evolution temperatures were 694, 675 and 625 ^oC for the intrinsic, phosphorous doped and boron doped films respectively.

This reduction was particularly significant for the boron doped film, in agreement with Ref. 9. The E_D values obtained from hydrogen evolution kinetics in our work suggest that the bonding was predominantly SiH_2 in the boron doped film, and predominantly SiH in the phosphorous doped film. This was consistent with FTIR results.

Doping not only reduced the hydrogen evolution temperature, but also enhanced the crystallization at high heating rates (particularly for boron). However, it is known that crystallization was enhanced even for pure amorphous silicon films prepared by ion implantation [19] and by chemical vapor deposition [20], on phosphorous and boron doping. This enhancement has been suggested to be due to the variation of the charge state of defects such as vacancies and dangling bonds, on doping [21]. This suggests that the enhanced crystallization in doped films might not be due to the lowering of hydrogen evolution temperature. This enhancement occurred despite the fact that the silicon-hydrogen bonding was stronger for the phosphorous doped films (predominantly SiH) than the intrinsic film (SiH and SiH_2). This indicated that silicon-hydrogen bonding in the as-deposited a-Si:H film does not play any role in its crystallization process, in agreement with Ref. 22.

To check the influence of the presence of hydrogen in the as-deposited film on the crystallization behavior, the onset temperatures of crystallization of two different amorphous silicon films having the same thickness (7 μm) were determined by the same procedure as described before; an intrinsic glow-discharge film and a sputtered film (which had no hydrogen). The former film began to crystallize at 695 °C, whereas for the latter film it began at 750-800 °C. This suggests that the absence of hydrogen in the as-deposited film probably results in an amorphous silicon network that does not crystallize easily. This might agree with solid phase crystallization of rf sputtered pure a-Si on fused silica which occurred at 741 °C heated at 10 °C/min [23], whereas solid phase epitaxial growth of pure as-implanted a-Si on single crystal Si (100) occurred at around 690 °C [24]. These differences may be due to the method of obtaining amorphous silicon, the method of crystallization and the presence of impurities in the film.

There was no significant change in the DSC curve of intrinsic glow discharge films deposited at two different temperatures, namely 225 and 300 °C. Hydrogen evolution activation energy and FTIR measurements showed similar silicon-hydrogen bonding in these films. However, crystallization occurred even at high heating rates for the film deposited at 300 °C, indicating that crystallization was enhanced for the film deposited at a higher temperature. This was in agreement with Ref. 25 for LPCVD a-Si films deposited on fused quartz substrates.

Films of two different thicknesses (namely 0.5 and 7 μm) and deposited at 225 °C were heated to 725 °C at different heating rates and then cooled rapidly to room temperature. Subsequent Raman scattering showed no difference between the Raman spectra of the two films for the same heating rate. This indicates that the two films, though of different thicknesses, exhibited the same ease of crystallization. This means that the nucleation of crystallization was bulk induced and not surface (or interface) induced. The penetration depth of the Ar laser for Raman scattering was about 1 μm. Thus, for the 0.5 μm film, Raman scattering could detect the presence of any nucleation site at the film/substrate interface, whereas, for the 7 μm thick film, the detection was confined only to the top 1 μm thick layer. Therefore, if nucleation were to occur at the film/substrate interface, the Raman spectrum would have shown a crystalline silicon peak for the 0.5 μm thick film and an amorphous peak for the 7 μm thick film. This result was in agreement with Ref. 26.

CONCLUSION

This work was focused on hydrogenated amorphous silicon thick films deposited from silane by dc glow discharge. The nature of hydrogen and the influence of hydrogen on the crystallization behavior were studied by DSC and TGA, supplemented by Raman scattering, EGA and FTIR. The following findings were obtained. An endothermic peak was observed upon heating at a temperature (658-704 °C) which increased with increasing heating rate. This peak was due to hydrogen evolution, as confirmed by TGA and EGA. From the hydrogen evolution kinetics, the silicon-hydrogen bonding for the intrinsic film was estimated to be a mixture of SiH and SiH_2, whereas for the boron doped film, it was predominantly SiH_2, and, for the phosphorous doped film, it was predominantly SiH. These estimates are consistent with FTIR results. Doping lowered the hydrogen evolution temperature. The crystallization process was independent of the silicon-hydrogen bonding in the as-deposited a-Si:H film. Crystallization was enhanced for a film deposited at a higher temperature and it was induced in the bulk of the film.

ACKNOWLEDGEMENT

The authors acknowledge financial support from National Renewable Energy Laboratory.

REFERENCES

1. D. E. Carlson and C. R. Wronski, Appl. Phys. Lett. 28, 671 (1976).
2. D. E. Carlson, J. Non-Cryst. Solids 35-36, 707 (1980).
3. D. L. Staebler and C. R. Wronski, Appl. Phys. Lett. 39, 292 (1977).
4. H. S. Yoon, C. S. Park and Sin-Chong Park, J. Vac. Sci. Technol. A 4 (6), 3095 (1986).
5. M. K. Hatalis and D. Greve, J. Appl. Phys. 63, 2260 (1988).
6. R. A. Street, Hydrogenated Amorphous silicon (Cambridge University Press, Cambridge, 1991) and references cited in.
7. W. C. Sinke, T. Warabisako, M. Miyao, T. Tokuyama, S. Roorda and F. W. Saris, J. Non-Cryst. Solids 99, 308 (1988).
8. W. Paul, A. J. Lewis, G. A. N. Connell and T. D. Moustakas, Solid State Comm. 20, 969 (1976).
9. W. Beyer, H. Wagner and H. Mell, Solid State Comm. 39, 375 (1981).
10. M. Kumeda, H. Komatsu and T. Shimizu, Thin Solid Films 129, 227 (1985).
11. L. Battezzatti, F. Demichelis, C. F. Pirri, A. Tagliaferro and E. Tresso, J. Non-Cryst. Solids 137 & 138, 87 (1991).
12. L. Battezzatti, F. Demichelis, C. F. Pirri and E. Tresso, Physica B 176, 73 (1992).
13. W. Beyer and H. Wagner, J .Appl. Phys. 53, 8745 (1982).
14. D. K. Beigelson, R. A. Street, C. C. Tsai and J. C. Knights, Phys. Rev. B 20, 4839 (1979).
15. CRC Handbook of Chemistry and Physics (CRC, West Palm Beach, FL, 1977).
16. J. A. Roth, G. L. Olson, D. C. Jacobson and J. M. Poate, Mat. Res. Symp. Proc. 297 (1993).
17. G. L. Olson and J. A. Roth, Mater. Sci. Rep. 3, 1 (1988).
18. Y. Masaki, P. G. LeComber and A. G. Fitzgerald, J. Appl. Phys. 74, 129 (1993).
19. C. Licoppe and Y. I. Nissim, J. Appl. Phys. 59, 432 (1986).
20. K. Zellama, S. Squelard, J. Magarino and D. Kaplan, J. Non-Cryst. Solids 59 & 60, 807 (1983).
21. J. Magarino, D. Kaplan, A. Friederich and A. Deneuville, Phil. Mag. 45, 285 (1980).
22. J. C. Chou, S. K. Hsiung and C. Y. Lu, Jpn. J. Appl. Phys. 26, 1971 (1989).
23. J. C. C. Fan and C. H. Anderson, Jr., J. Appl. Phys. 52, 4003 (1981).
24. E. P. Donovan, F. Spaepen, D. Turnbull, J. M. Poate and D. C. Jacobson, J. Appl. Phys. 57, 1795 (1985).
25. S. Hasegawa, S. Sakamoto, T. Inokuma and Y. Kurata, Appl. Phys. Lett. 62, 1218 (1993).
26. S. Roorda, D. Kammann, W. C. Sinke, G. F. A. Van de Walle and A. A. Van Gorkum, Mater. Lett. 9, 259 (1990).

LASER ANNEALING OF
HYDROGENATED AMORPHOUS SILICON THICK FILMS

Nagarajan Sridhar*, D. D. L. Chung*, W. A. Anderson**, W. Y. Yu+, L. P. Fu+ and A. Petrou+, Center for Electronic and Electro-Optic Materials, State University of New York at Buffalo, NY 14260-4400, and J. Coleman, Plasma Physics Corp., P. O. Box 548, Locust Valley, NY 11650.

* Also with Department of Mechanical and Aerospace Engineering
** Also with Department of Electrical and Computer Engineering
+ Also with Department of Physics and Astronomy

ABSTRACT

Hydrogenated amorphous silicon thick films deposited by dc glow discharge on molybdenum substrates were annealed by a pulsed Nd:glass laser. Mass spectrometry showed hydrogen remaining in all the laser annealed films. The amount of hydrogen remaining decreased with decreasing scan rate. The hydrogen evolved upon heating at 365 °C and mainly at 658 °C before laser annealing, but at 365, 575 (mainly) and 645 °C after laser annealing, indicating weakening of the silicon-hydrogen bonding after laser annealing. The presence of hydrogen inhibited crystallization, as indicated by Raman scattering. The photo and dark conductivity of the film increased by one and three orders of magnitude respectively with increasing laser energy density up to 12 J/cm^2 at a fixed scan rate. This means that the photoresponse was decreased with laser annealing, in spite of the associated increase in crystallinity. This photoresponse decrease is attributed to the hydrogen evolution.

INTRODUCTION

Polycrystalline silicon film supported on a substrate is a promising candidate for the development of large area low cost solar cells [1]. This can be achieved by first depositing hydrogenated amorphous silicon (a-Si:H) at low temperatures where unwanted impurity diffusion and thermal stress problems can be eliminated and then crystallizing the a-Si:H to polycrystalline silicon by a thermal treatment. However a major limitation in polycrystalline silicon is the high density of recombination centers at the grain boundaries and this degrades the electrical properties of the device. The role of hydrogen in polycrystalline silicon is likely to be similar to that in amorphous silicon, which is the passivation of the dangling bonds [2]. However, thermal treatment of a-Si:H by furnace annealing results in the evolution of hydrogen from the film. Subsequent hydrogenation of polycrystalline silicon by a plasma treatment results in the passivation of the grain boundaries [3]. An alternate way is to crystallize the a-Si:H film in a short duration such that the out diffusion of hydrogen can be minimized. Therefore, it has been suggested that laser crystallization of a-Si:H could serve this purpose [4]. Previous work in this area [5] have shown improved electrical properties. Prior studies had focussed only on laser annealing of glow discharge a-Si:H thin films deposited on insulating substrates (c-Si or glass) [6-8], where the film thickness was not greater than 2 μm. However, the minimum thickness necessary for a polycrystalline silicon film solar cell is 5 μm. To our knowledge there has been no work done on laser

719

crystallization of thick a-Si:H films. For such a purpose, only a laser in the infrared region (such as a Nd:YAG, Nd:glass or a CO_2 laser) could be used. The objective of this work is to study the crystallization and detect the presence and form of hydrogen in the dc glow discharge deposited intrinsic a-Si:H thick film on a molybdenum substrate after being annealed by a Nd:glass laser. The study was performed as a function of laser energy density and scan rate. These results are then correlated with the electrical properties of the film.

EXPERIMENTAL

Intrinsic a-Si:H films were deposited at Plasma Physics Corp. on molybdenum substrates by dc glow discharge in a gradient field with cathodic deposition from silane with deposition rates up to 1 µm/min and using conditions shown in Table 1. The film thickness was 7 µm. All the samples had an n^+-type a-Si:H thin layer (formed by the addition of 1 % PH_3 to H_2) deposited on molybdenum prior to the deposition of the intrinsic a-Si:H film, in order to achieve a good ohmic contact between molybdenum and the intrinsic film. The thickness of the n^+ layer was about 500 Å.

The a-Si:H films were annealed by a Nd:glass pulsed laser having an output emission wavelength of 1060 nm. The maximum output energy of the laser was 60 Joules at 1 ms. The laser energy density was altered by employing beam splitters and changing the beam size (by means of a convex lens) and the supply voltage. The beam size was altered by employing a convex lens. The pulse width of the laser was 0.2 ms and the pulse frequency was 1 pulse/s. During annealing, the sample stage was translated in a direction perpendicular to the laser beam by a DC step motor in order to obtain a quasicontinuous laser scan.

The crystallinity in the film was studied by Raman scattering in the backscattered mode. The Raman spectrometer was a Spex 1877 triplemate equipped with a CCD detector. The wavelength of the Ar laser was 4880 Å and the laser power was 100 mW during the Raman measurement. The crystallinity in the annealed films was studied from the front and back sides.

Measuring the crystallinity from the back side of the silicon film gave information on the minimum laser energy density necessary to completely crystallize the film. Raman scattering on only the front side would be insufficient, because the laser beam could penetrate only a depth of 1 µm. The back side of the film was exposed by attaching the front side of the film to a ceramic substrate (by means of an adhesive) and then etching away the Mo substrate.

The presence of hydrogen and its forms were studied in the following manner. The laser annealed sample was inserted in a Vycor glass tube and evacuated by a turbomolecular pump to a base pressure of 10^{-6} torr. The sample was then heated to 700 °C at a constant rate. The

Table 1 Parameters for a-Si:H film deposition from silane by dc glow discharge

Deposition temperature (°C)	Pressure (mtorr)	Voltage (V)	SiH_4 flow (standard $cm^3 min^{-1}$)	Current (mA)
225	250-300	500	10	50

partial pressure of hydrogen was measured during the heating by monitoring the H_2^+ ions evolved during heating using a quadrupole mass spectrometer (VG Instruments).

The electrical properties were determined by measuring the dark conductivity and photoconductivity. The ohmic contacts were fabricated by patterning and etching the molybdenum after attaching the front side of the film to a ceramic substrate. This technique does not involve any heating, which is usually required during the evaporation of contacts.

RESULTS AND DISCUSSION

The 7 μm thick intrinsic a-Si:H film was laser annealed at various energy densities. At each laser energy density, the film was scanned at different rates. For all energy densities and scan rates, a sharp and strong peak at 520 cm^{-1} was obtained from Raman scattering performed on the front side of the film, indicating that the laser energy density employed in this study were above the threshold value necessary for crystallization (at least partial) to occur.

More complete information on the crystallinity of these films was obtained from Raman scattering performed on the back side of the film. Table 2 shows the data of such a study as a function of energy density and scan rate. Complete crystallization was achieved at a high laser energy density and/or a low scan rate. Though the film crystallized completely at 15 J/cm^2 (the highest energy density employed), the problem was that of laser induced film damage. On the other hand, for a lower energy density, such as 8 J/cm^2, a low scan rate, such as 0.25 mm/s, had to be employed to achieve maximum crystallinity. However, a low scan rate had an adverse effect on the surface roughness. During laser annealing, the energy density was high enough to melt the silicon film at least partly. Due to the short pulse duration, there was not enough time for the melt to spread uniformly in the lateral direction. On cooling the

Table 2 Effect of laser energy density and scan rate on the crystallinity, as studied by Raman scattering performed on the back side of each film.

Scan rate (mm/s)	Laser energy density (J/cm^2)			
	5	8	12	15
0.25	A	C	C	-
0.45	A	C+A	C	C
0.55	A	A	C+A	C
0.9	A	A	A	C+A
1.2	A	A	A	C+A

Note: A = amorphous; C = crystalline

melt, the solid phase formed hillocks, thus resulting in a rough surface. Surface roughness could also be due to hydrogen evolution from the molten film which might cause blister formation. Surface roughness was reported in Ref. 8, which was concerned with crystallizing glow discharge a-Si:H thin films on glass substrates, using an ArF excimer laser. Ref. 8 attributed the surface roughness to hydrogen evolution.

Table 3 shows the mass spectrometric hydrogen evolution results of laser annealed a-Si:H films scanned at different rates for the same laser energy density (8 J/cm^2). Two hydrogen evolution peaks (at 365 and 658 °C) were observed for the as-received film (before laser annealing), whereas three peaks (at 365, 575 and 645 °C) were observed for the laser annealed films. The $[H_2^+]$ partial pressure obtained by integrating the curve of the hydrogen evolution rate vs. temperature for each peak gave in qualitative terms the amount of hydrogen evolved, which is related to the amount of hydrogen in the film prior to the heating during which mass spectrometry was performed. The $[H_2^+]$ partial pressure divided by the film thickness and surface area of the film at these three peak temperatures (T_1, T_2 and T_3) are c_1, c_2 and c_3 respectively in Table 3, where $c_T = c_1 + c_2 + c_3$ corresponds to the total amount of hydrogen evolved. Note that c_1 is smaller than c_2 or c_3 at all conditions. Before laser annealing, c_3 dominates; after laser annealing, c_2 dominates. Thus the main form of hydrogen before laser annealing was more strongly held to the silicon than that after laser annealing. All of c_1, c_3, c_3 and c_T decreased with decreasing scan rate. Comparison of Tables 2 and 3 indicates that c_1, c_3, c_3 and c_T decreased with increasing crystallinity. Hence, the presence of hydrogen inhibited the crystallization process. This is in close agreement with Ref. 8. This trend is opposite to what was reported in Ref. 9 where crystallization was enhanced by the presence of hydrogen in the as-deposited film. The difference is because, in Ref. 9, the crystallization process took place in the solid phase by slow heating, whereas in this work crystallization occurred from the liquid phase and took place in a short duration. An additional hydrogen evolution peak occurred at 575 °C (T_2) for the laser annealed films, suggesting that the silicon-hydrogen bonding may be changed by the laser annealing. This was confirmed by the absence of the SiH vibrational mode (2000

Table 3 Effect of scan rate on hydrogen in the laser annealed film, as studied by mass spectrometry

Scan rate (mm/s)	Peak temperature (°C)			$[H_2]^+$ partial pressure (torr/µm.cm^2)			$c_T = c_1 + c_2 + c_3$ (torr/µm.cm^2)
	T_1	T_2	T_3	c_1	c_2	c_3	c_T
As-received	365	-	658	8×10^{-10}	-	8.8×10^{-9}	8.8×10^{-9}
0.18	365	575	645	2.83×10^{-10}	2.7×10^{-9}	3.55×10^{-10}	3.3×10^{-9}
0.45	365	575	645	4.5×10^{-10}	3.76×10^{-9}	1.22×10^{-9}	5.4×10^{-9}
0.9	365	575	645	5×10^{-10}	6×10^{-9}	1.52×10^{-9}	8×10^{-9}
1.2	365	575	645	5.3×10^{-10}	6.7×10^{-9}	1.18×10^{-9}	8.4×10^{-9}

Table 4 Effect of laser annealing on the dark conductivity (σ_d) and photoconductivity (σ_p), and the photoresponse (σ_p/σ_d).

Sample condition	σ_d (ohm.cm)$^{-1}$	σ_p (ohm.cm)$^{-1}$	σ_p/σ_d
As-received	7.5×10^{-6}	1.83×10^{-4}	24
Laser annealed at 5 J/cm^2	2.15×10^{-4}	3.0×10^{-4}	1.39
Laser annealed at 8 J/cm^2	8.2×10^{-4}	1.16×10^{-3}	1.41
Laser annealed at 12 J/cm^2	1.25×10^{-3}	1.8×10^{-3}	1.44
Laser annealed at 15 J/cm^2	8×10^{-4}	9.5×10^{-4}	1.18

cm^{-1}) in Raman spectra obtained from the back side of the films after laser annealing and the presence of this mode before laser annealing. Ref. 4 reported the infrared absorption spectra of glow discharge deposited a-Si:H thin films on crystalline silicon substrates crystallized by a Q-switched ruby laser. Consistent with this work, they found that the intensity of the SiH vibrational modes at 630 and 2000 cm^{-1} decreased with increasing laser energy density. They attributed this decrease either to the loss of hydrogen or to the fact that SiH bonds were broken and the hydrogen resided in a different bonding configuration.

Table 4 shows the dark conductivity σ_d and photoconductivity σ_p measured as a function of the laser energy density (with the scan rate fixed at 0.45 mm/s). It was observed that the dark conductivity increased by 2-3 orders of magnitude as the laser energy density increased, in agreement with Ref. 5 and 6, while the photoconductivity increased only by an order of magnitude. The drop in photoconductivity observed from 12 to 15 J/cm^2 may be due to the laser induced film damage. The photoresponse (σ_p/σ_d) was thus greatly decreased by the laser annealing, probably due to the formation of dangling bonds resulting from the absence of SiH bonding (as observed by Raman scattering).

CONCLUSION

Intrinsic a-Si:H thick films deposited by dc-glow discharge on molybdenum were annealed by a Nd:glass pulsed laser resulting in at least partial crystallization. Complete crystallization of the film was achieved by increasing the laser energy density and decreasing the scan rate. The laser annealed films showed a rough surface morphology. The amount of hydrogen after laser annealing decreased with decreasing scan rate for a given laser energy density. The presence of hydrogen inhibited crystallization. After laser annealing, an additional peak appeared at 575 $^\circ$C in the curve of the hydrogen evolution rate vs. temperature, suggesting that the silicon-hydrogen bonding was altered by the laser annealing. This was also suggested by the vanishing of the SiH Raman peak after laser annealing. The dark and photoconductivity increased with increasing laser energy density up to 12 J/cm^2. However, the photoresponse of all the laser annealed films was much lower when compared to the as-received film, indicating the formation of dangling bonds due to hydrogen evolution during laser annealing.

ACKNOWLEDGEMENT

The authors acknowledge financial support from National Renewable Energy Laboratory. The technical assistance of J. Yi of State University of New York at Buffalo in photolithography is greatly appreciated.

REFERENCES

1. T. L. Chu, J. Cryst. Growth 39, 45 (1977).
2. D. E. Carlson, RCA Internal Report, SAN1286-8 (1978).
3. C. H. Seager and D. S. Ginley, Appl. Phys. Lett. 24, 337 (1979).
4. R. S. Sussmann, A. J. Harris and R. Ogden, J. Non-Cryst. Solids 35, 249 (1980).
5. Y. K. Koh, H. Okamoto, K. Murakami, K. Gamo, Y. Hamakawa and S. Namba, Jap. J. Appl. Phys. 19, 849 (1980).
6. I. D. Calder, K. L. Kavanaugh, H. M. Naguib, C. Brassard, J. F. Currie, P. Depelsenaire and R. Groleau, J. Electron. Mater. 11, 303 (1982).
7. J. I. Pankove, C. P. Wu, C. W. Magee and J. T. McGinn, J. Electron. Mater. 9, 905 (1980).
8. E. L. Mathe, A. Naudon, M. Elliq, E. Fogarassy and S. de Unamuno, Appl. Surf. Sci. 54, 392 (1992).
9. N. Sridhar, D. D. L. Chung, W. A. Anderson, W. Y. Yu, L. P. Fu, A. Petrou and J. Coleman, MRS Fall Proceedings, Symposium E (1993).

THE MECHANISM OF EXCIMER LASER-INDUCED AMORPHIZATION
OF ULTRA-THIN Si FILMS

T. Eiumchotchawalit and James S. Im
Columbia University, Department of Chemical Engineering,
Materials Science and Mining Engineering, New York, NY

ABSTRACT

To better understand the involved phase transformation mechanism, we are studying the excimer laser-induced amorphization (ELA) of ultra-thin Si films on oxidized Si substrates. In this paper, we show that the onset of amorphization of hydrogen-free Si films on SiO_2 substrates upon increases in the energy density is associated with the onset of complete melting of the film. Once complete melting occurs, further increases in the incident energy density and/or increases in the substrate temperature can lead to incomplete amorphization of the film. Planar view TEM analysis of nearly-amorphized Si films reveals a heterogeneous microstructure, which consists of a mixture of densely dispersed amorphous-like annular regions (~20 to 40 μm^{-2}), embedded within and typically separated by a region containing fine-grained small crystals. Such a cellular microstructure strongly suggests that amorphization occurred not via a homogeneous but via a heterogeneous transformation. In particular, the microstructure paints a scenario in which amorphization proceeded via nucleation of solids, which is then followed by interfacial amorphization. The experimental results unambiguously reveal (1) that the previously proposed criteria of the melt duration and the vertical temperature gradient are irrelevant in determining amorphization of supercooled liquid Si films and (2) that the quenching rate, not surprisingly, is the important parameter.

INTRODUCTION

It was discovered in recent years that excimer laser irradiation of a thin Si film (a-Si:H) on an oxide surface at high energy densities can lead to complete amorphization (i.e., conversion to four-fold coordinated amorphous Si) of the film [1, 2]. The observed phenomenon is remarkable in that, due to the nonparticipating and inert nature of the oxide interface [3, 4] (i.e., it does not initiate growth of solid Si nor does it well catalyze the heterogeneous nucleation of crystals), the thermal and kinetic path through which formation of amorphous Si proceeds in these experiments is expected to be *clearly distinct from the much-investigated interfacial amorphization of the surface region of Si wafers observed in pulsed laser-induced melting and regrowth experiments* [5]. The situation, on the other hand, is physically more similar to rapid quenching and supercooling of isolated liquid droplets [6, 7] and is intimately related to deep supercooling of liquid Si films on SiO_2 [3, 4].

It is now possible to envision a number of transformation scenarios, ranging from schemes based on discontinuous first-order transitions to those which approach continuous transitions. Fundamentally, the situation warrants our attention because the involved cooling rates of l-Si represent the highest quenching rates (>10^{10} K/sec) that are encountered in rapid solidification experiments to date. When coupled with the beneficial simplicity of the system being elemental and the experimental flexibility inherent in thin film methods, the situation presents materials scientists with a unique opportunity to probe experimentally into the previously uncharted kinetic territory of phase transformations in a rigorous manner.

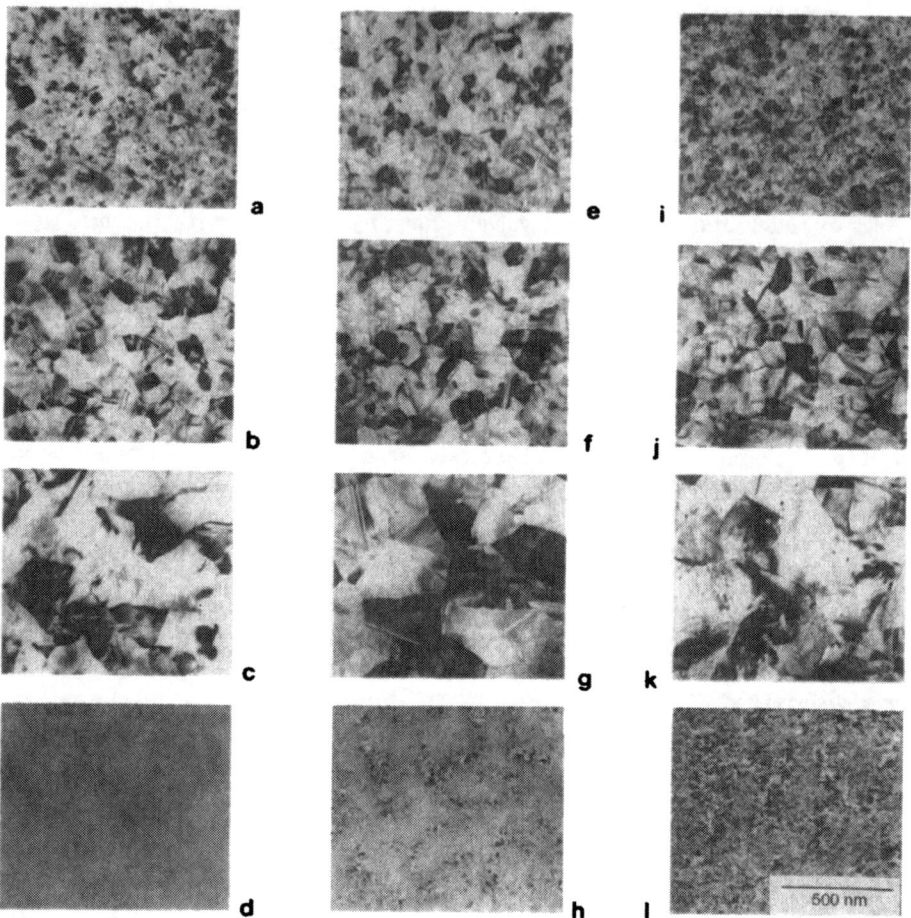

Figure 1. Planar view bright field TEM images of laser irradiated Si films at various energy densities. The first sequence ((a) to (d)), the second sequence ((e) to (h)), and the third sequence ((i) to (l)) correspond to the series in which the temperature of the substrates were at room temperature, 320°C, and 425°C, respectively. The following are the specific energy densities: (a) 150 mJ/cm^2, (b) 164 mJ/cm^2, (c) 181 mJ/cm^2, (d) 192 mJ/cm^2, (e) 71 mJ/cm^2, (f) 119 mJ/cm^2, (g) 132 mJ/cm^2, (h) 138 mJ/cm^2, (i) 119 mJ/cm^2, (j) 125 mJ/cm^2, (k) 139 mJ/cm^2, and (l) 143 mJ/cm^2.

In this paper, we report on an experimental investigation, which is conducted in order to first characterize the conditions that lead to amorphization of thin Si films and second to probe into the details of the phase transformation mechanism behind ELA of Si films on SiO$_2$. In particular, we elaborate on the microstructures of nearly-amorphized Si films, and show that the observed heterogeneous microstructure suggests that amorphization occurs via discontinuous first-order transitions in which the interfacial amorphization is triggered by solids that are nucleated within deeply supercooled liquid Si.

EXPERIMENTAL METHOD

Single-pulse irradiation experiments were conducted on non-hydrogenated LPCVD a-Si films (500Å-thick), which were deposited on oxidized Si substrates (with 1000Å-thick thermal oxide) and capped with a 500Å-thick SiO_2 layer. The oxide capping layer was utilized in order to ensure that the completely molten Si film encounters identical physical boundaries at the top and bottom. The samples were irradiated inside the vacuum chamber (10^{-6} torr) with a 30-nanosecond XeCl excimer laser pulse (308 nm) at various substrate (i.e., preheating) temperatures and various energy densities. The energy density of the incident beam was controlled by varying the distance between the sample and the focusing lens and the substrate temperature of the samples was controlled by controlling the temperature of the hot stage. After irradiation, TEM samples were prepared using the back-etch technique, and were analyzed via bright-field image and electron diffraction analyses.

RESULTS

Figure 1 shows the bright-field TEM micrographs of Si films, which were irradiated at various energy densities with the temperature of the substrate held at three different settings. Figures 1a to 1d (which correspond to a room temperature irradiated series) show that the microstructures of the irradiated film follows a recently reported trend for thicker Si films [8] with one important difference; when the energy density of the incident pulse is sufficiently high (i.e., the high energy density regime [8]), amorphous-Si film, instead of a fine-grained microstructure, is obtained [8]. Figures 1i to 1l, which correspond to a higher substrate temperature of 425°C, show the appearance of fine-grained microstructures (Figure 1l) upon entering the higher energy density regime.

At an intermediate substrate temperature of 320°C, a remarkable heterogeneous microstructure (shown in Figure 1h) is obtained at the onset of the high energy density regime. The microstructure can be best described as being composed of a mixture of amorphous-like annular regions, (with an approximate density of 20 to 40 μm^{-2}) which appear to be embedded within and often separated by a narrow region containing fine-grained small crystals.

Indeed, for a given substrate temperature at 275°C (within the high energy density regime), TEM micrographs and the corresponding diffraction patterns in Figures 2a to 2c show that it is possible to go from obtaining amorphous Si film at an energy density of 307 mJ/cm^2 to mixed microstructure at an energy density of 467 mJ/cm^2 to fine-grained microstructure at 600 mJ/cm^2. Figures 2d to 2f show that an identical trend is obtained when the substrate temperature is increased for a fixed energy density of 378 mJ/cm^2.

DISCUSSION

A first glimpse into the origin of amorphization is obtained through the examination of the overall microstructural trend shown in Figures 1a to 1d and the comparison of these results to those which were observed for the thicker (1000Å-thick) Si films [8]. The observed trend in microstructures of the room temperature irradiated

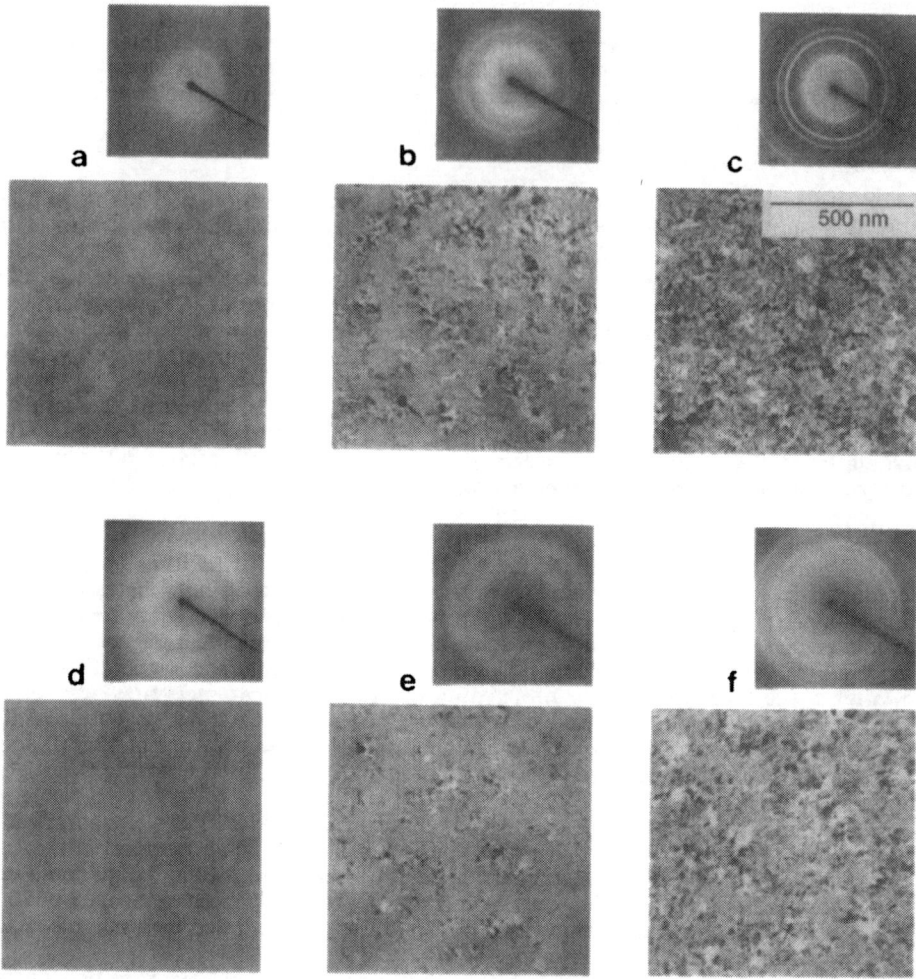

Figure 2. Planar view bright-field TEM micrographs and corresponding electron diffraction pattern of Si films, which were completely melted and solidified. Figures (a) to (c) correspond to a series (substrate temperature = 275°C) in which the incident energy density was increased from (a) 307 mJ/cm², (b) 467 mJ/cm², and (c) 600 mJ/cm²; figures (d) to (f) correspond to a series in which the substrate temperature was increased from (d) 225°C, (e) 320°C, and (f) 385°C with the constant incident energy density at 378 mJ/cm².

films as a function of the incident energy density is initially identical to the thicker film (i.e., within what has been identified as the low energy density regime), but a completely amorphous film — as opposed to the fine-grained Si film — is obtained as the film enters the high energy density regime. The transition accompanying the collapse of large-grained Si films (obtained within the narrow SLG regime) upon increases in the energy density has been previously identified as the transition from near-complete to complete melting of the Si film [8]. As such, it follows that the *necessary but clearly not sufficient condition for the amorphization of thin Si films on SiO_2 is the complete melting of the film.*

Additionally, we note that the observation of incomplete amorphization (i.e., partial amorphization) of Si films upon further increases in the incident energy density (Figure 2b) and the substrate temperature (Figure 2e) — all within the complete melting regime — directly contradicts the explanation previously suggested for the ELA of Si films on SiO_2 [1, 2]. The scenario advocated by Sameshima and Usui suggests that amorphization of the Si film occurs when the vertical temperature gradient is lower than and the melt duration of Si film is longer than certain critical values, which were derived from one-dimensional numerical simulations. Since the effect of increases in either the incident energy density (for a given substrate preheating temperature) or increases in the substrate temperature (for a given incident energy density) is *to increase the melt duration* and *to decrease the vertical temperature gradient for a given temperature at the interface,* the observations (Figure 2) make it clear that the assignment of critical values for the vertical temperature gradient and/or the melt duration represent neither physically meaningful nor relevant criteria for ELA of Si films.

The above results, on the other hand, unambiguously identify and clearly attest to the significance of the quenching rate of completely molten Si films; the cooling rate, which decreases with increasing substrate (i.e., preheating) temperatures and/or increasing energy densities (and which *actually* is a function of liquid temperatures for a given irradiation condition), of molten Si films must be high enough during the supercooling of the molten Si in order to amorphize the Si film. Indeed, the increase in the quenching rate (which accompanies the reduction in the Si film thickness) accounts for the fact that amorphization can be achieved with 500Å-thick Si films, while it is not possible to attain the amorphization upon complete melting of the film, in the case of 1000Å-thick Si films [8]. Likewise, we attribute the incomplete amorphization of 500Å-thick Si films on the quartz substrates [1, 2] to the reduction in the quenching rate that results from lower conductivity associated with the quartz substrate (as opposed to Si substrates with a 1000Å-thick oxide layer).

The most important outcome of this experiment pertains to conclusions, which can be derived from analysis of the heterogeneous microstructures of the partially amorphized Si films. It can be seen in Figures 1h, 2b, and 2e that the microstructures of these nearly-amorphized films can be described as non-homogeneous, cellular-like structures, which consist of a mixture of two types of regions: densely dispersed amorphous-like annular regions and a partially connected network of narrow crystalline regions. Such morphological details suggest strongly that the conversion of highly supercooled liquid Si to amorphous Si occurs not continuously or homogeneously but via an interface-led first-order (i.e., discontinuous) transformation. The cellular structure of the type shown in Figures 1h, 2b, and 2e is well known to be associated with the microstructures, which result from nucleation and growth [9]; each cell corresponds to a unit, which is formed via interfacially-controlled growth that originates from the nucleation site at the center of the cell.

Here, the temperature of the liquid — at the onset of the transformation — is expected to be sufficiently low (i.e., estimated to be around or less than 500°C below the equilibrium point [4, 10]). Subsequently, once the solid has nucleated within the supercooled liquid, it is the a-Si, which is expected to outpace crystals and dominate the growth [11], provided that the ensuing recalescence does not push the interfacial temperature and/or the temperature of the liquid high enough so as to cause the appearance of crystals. In other words, when interpreted in the context of the above transformation scenario, the origin of the heterogeneous structure observed for partially amorphized Si films can be readily accounted for by noting that crystalline structures can and will appear when the temperatures of the interface and the remaining liquid rise rapidly due to recalescence as the solidification fronts originating from adjacent nucleation centers impinge on each other. A corollary to the above statement is that *complete amorphization of the film takes place when the temperatures of the interface and the liquid fail to rise sufficiently during the transformation to cause the appearance of crystallites.*

We also note that, of all the possible scenarios, the transformation of highly supercooled liquid Si to the amorphous phase occurring via nucleation of solids followed by amorphization in the form of interfacial growth (in contrast either to a glass transition or well defined second order transition) is most consistent with the established view that considers a-Si to be a metastable phase, which is structurally and thermodynamically distinct from the supercooled liquid Si [12, 13, 14].

Finally, it is noted that we have not addressed (as the above observations in and of themselves are insufficient to permit us to decisively conclude) the issues regarding the details of the nucleation which, in fact, triggers subsequent amorphous growth. Specifically, the central question regarding the nature of nucleation (i.e., a-Si and/or x-Si, steady state and/or transient, homogeneous and/or heterogeneous) remains to be examined.

The authors would like to acknowledge Dr. A. L. Greer of Cambridge University and Dr. P. S. Peercy of Sandia National Laboratories for their helpful discussions.

REFERENCES

1. T. Sameshima and S. Usui, *Appl. Phys. Lett.* **59** 2724 (1991).
2. T. Sameshima and S. Usui, *J. Appl. Phys.* **74**, 6592 (1993).
3. D. Dutartre, *J. Appl. Phys.* **59**, 1977 (1986).
4. S. R. Stiffler, M. O. Thompson, P. S. Peercy, *Phys. Rev. Lett.* **60**, 2519 (1988).
5. M. O. Thompson, J. W. Mayer, A. G. Cullis, H. C. Webber, N. G. Chew, J. M. Poate and D. C. Jacobson, *Phys. Rev. Lett.* **50**, 896, (1983).
6. Yeon-Wook Kim, Hong-Ming Lin and T. F. Kelly, *Acta Metall.* **37**, 247 (1989).
7. P. V. Evans, G. Devaud, T. F. Kelly, and Yeon-Wook Kim, *Acta Metall.* **38**, 719 (1990).
8. James S. Im, H. J. Kim, and M. O. Thompson, *Appl. Phys. Lett.* **63**, 1969 (1993). H. J. Kim, James S. Im, and M. O. Thompson, *Mat. Res. Soc. Symp. Proc.* **283**, 703 (1993).
9. A. Gettis and B. Boots, *Models of Spatial Processes* (Cambridge, UK: Cambridge University Press, 1979).
10. S. R. Stiffler, M. O. Thompson, P. S. Peercy, *Phys. Rev. B* **43**, 9851 (1991).
11. P. A. Stolk, A. Polman, and W. C. Sinke, *Phys. Rev. B* **47**, 5 (1993).
12. D. Turnbull, *Mat. Res. Soc. Symp. Proc.* **51**, 71 (1985).
13. J. M. Poate, *Mat. Res. Soc. Symp. Proc.* **13**, 263 (1983).
14. J. Q. Broughton and X. P. Li, *Phys. Rev. B* **35**, 9120 (1987).

COMPARISON IN THE GROWTH AND PROPERTIES OF rf SPUTTERED μc-Si:H AND GLOW DISCHARGE-CHEMICAL VAPOR DEPOSITED μc-Si:H FILMS

J.Y.LIN,* B.H.TSENG* K.C.HSU**, and H.L.HWANG**
*Instituteof Materials Science and Engineering, National Sun Yat-Sen University, Kao-Shiung, Taiwan, R.O.C.
**Department of Electrical Engineering, National Tsing-Hua University, Hsin-Chu, Taiwan, R.O.C.

ABSTRACT

Properties of μc-Si:H films grown by rf sputtering and by glow discharge-chemical vapor deposition (GD-CVD) using diluted-hydrogen and hydrogen-atom-treatment method were compared employing TEM, X-ray diffraction, Raman scattering and FT-IR. The films deposited by both methods all exhibited comparable grain sizes in the range of 10-18 nm. and showed the same tendency in almost all the measurements.

INTRODUCTION

It is noted that μc-Si:H films can be grown by GD-CVD with diluted hydrogen by high hydrogen gas dilution ratio $(H_2/(H_2+SiH_4))$[1,2] and with alternative hydrogen atom treatment[3,4]. Recently, microcrystalline Si film with grain sizes less than 3nm was deposited at liquid-nitrogen substrate temperature by rf sputtering in a 100% hydrogen plasma[5]. The deposition rate is low because of the low sputtering efficiency of hydrogen ions. In this work, an argon-hydrogen gas mixture and a 24.2 MHz rf frequency were used to improve the deposition rate. It would be interesting to compare the properties of films deposited by these two methods. TEM, X-ray diffraction, Raman scattering, and FTIR were used for the characterization, in which the bonding configuration, grain size, energy band gap and the hydrogen content were correlated.

EXPERIMENTAL METHODS

The sputtering system consists of a vacuum chamber equipped with a diffusion pumping system, a substrate heater, a magnetron gun and a rf power supply with variable rf frequency settings. The gas pressure was controlled by a high-precision leak valve. The argon pressure was kept at 4.65×10^{-3} torr while the hydrogen pressure was varing from 3.5×10^{-4} torr to 1.5×10^{-2} torr. The background vacuum was 2×10^{-6} torr. A boron-doped silicon wafer with resistivity about 5 Ω-cm was used as the sputtered target. The target was presputtered for 20 min before the film deposition. The substrate was not intentionally heated. However, the radiation heat coming from plasma heats the substrate to 180°C. Two substrates, one was the Corning 7059 glass plate and the other was the phosphorous-doped (100) Si wafer, were deposited at the same time. The film grown on the glass substrate was characterized by four-point probe, X-ray diffractometry, Raman spectroscopy and absorption spectroscopy. The film grown on Si wafer was characterized by FTIR.

The GD-CVD method involved two deposition processes at a substrate temperature(Ts) of 250°C. a) diluted-hydrogen method employed a 13.56 MHz rf glow discharge with the fraction of hydrogen flow rate, $Xc=H_2/(SiH_4+H_2)$, between 40 to 98% with a vacuum pressure of 0.5 torr, and a rf forward power density of 0.1 W/cm². b) hydrogen-atom-treatment (HAT) method by alternately exposing the growing film surface to hydrogen atoms which are generated by an rf plasma in the same deposition chamber. The deposition time was kept 10s with Xc of 90% and a vacuum pressure of 1.1 torr, rf power density of 0.1 W/cm^2. The duration of the HAT step was varied from 20s to 50s with a vacuum pressure of 1.0 torr in hydrogen gas

The hydrogen contents were determined from the infrared absorption of the wagging band [6]. The Raman scattering spectra were measured from an excitation of a 150 mW, 488 nm argon ion laser. The average grain sizes of the μc-Si phase were determined from TEM and the full width at half maximum (FWHM) of the X-ray diffraction peak. The optical bandgaps were determined from the Tauc's plot using UV-VIS absorption spectrophotometry.

RESULTS AND DISCUSSION

The Si:H films deposited by the sputtering method showed no μc-Si:H phase but amorphous phase when the hydrogen pressure was vared from 3.5×10^{-4} to 2.0×10^{-2} torr with a rf power of 300 W and a substrate temperature(Ts) of 177 °C. When the hydrogen pressure was increased beyond 4×10^{-2} torr, the Raman peak position of the a-Si band was shifted to the higher energy with the FWHM narrowed, and a sharp TO band of crystalline silicon at 511 cm^{-1} was present. This indicates that the μc-Si phase was created. The grain size of the μc-Si:H flims with hydrogen pressure higher than 4×10^{-2} torr are around 13-18 nm as-identified from the X-ray diffraction pattern, which is almost independent of the hydrogen pressure up to 1.15×10^{-1} torr.as shown in Fig. 1. The typical TEM bright field photograph and the TED photograph are shown in Fig. 2, in which the grain size was estimated to be 20nm, the result is very close to that estimated from the X-ray data.

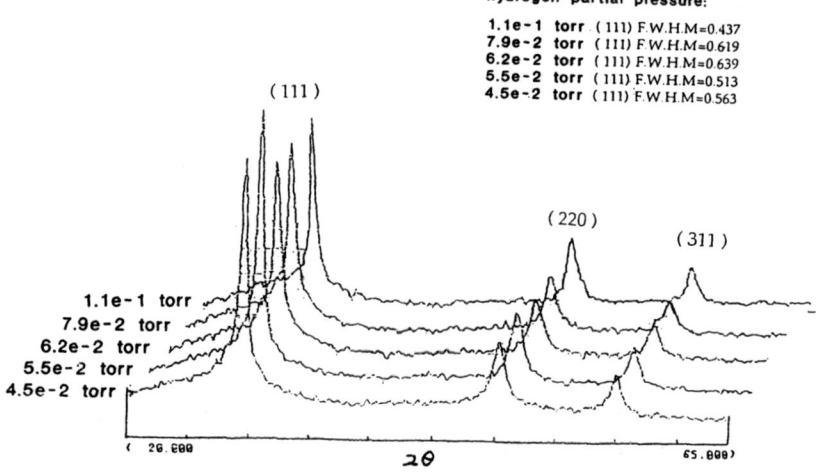

hydrogen partial pressure:

1.1e-1 torr (111) F.W.H.M=0.437
7.9e-2 torr (111) F.W.H.M=0.619
6.2e-2 torr (111) F.W.H.M=0.639
5.5e-2 torr (111) F.W.H.M=0.513
4.5e-2 torr (111) F.W.H.M=0.563

Fig. 1 The X-ray diffraction pattern of Si:H samples deposited by sputtering method under different hydrogen pressure.

It was found that when the hydrogen gas pressure ratio, $P(H_2)/(P(H_2)+P(Ar))$, was smaller than 0.9, no μc-Si:H films was formed. But when the hydrogen gas pressure ratio was higher than 0.9, μc-Si:H films started to be formed. Although the grain size was almost indenpendent of the hydrogen gas ratio whenever μc-Si:H films began to to formed. High hydrogen ratio

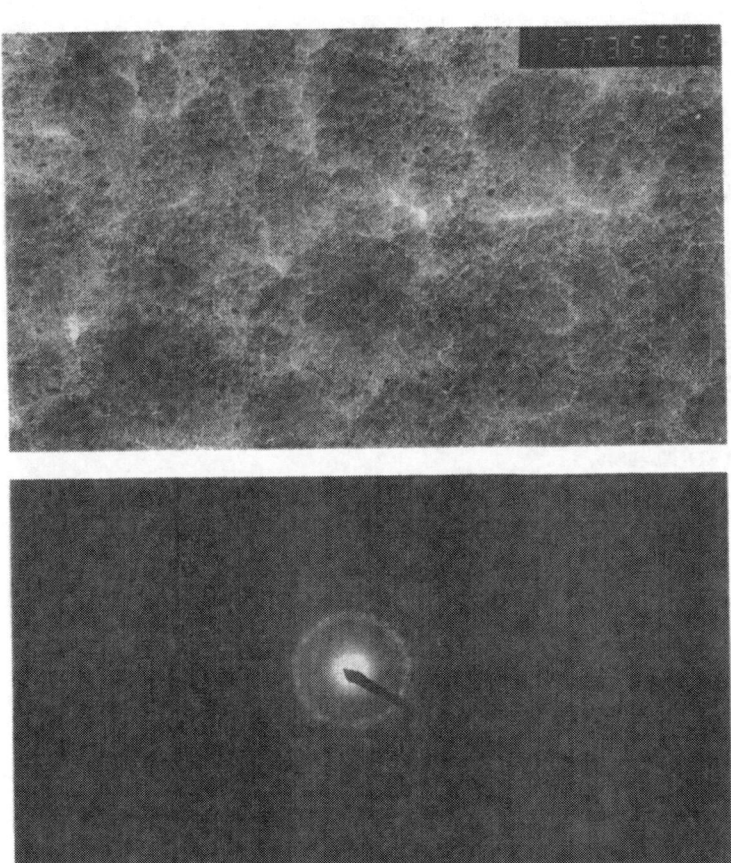

Fig. 2 The TEM bright field photograph and TED photograph of Si:H samples deposited by sputtering method under hydrogen pressure of 1.1×10^{-1} torr.

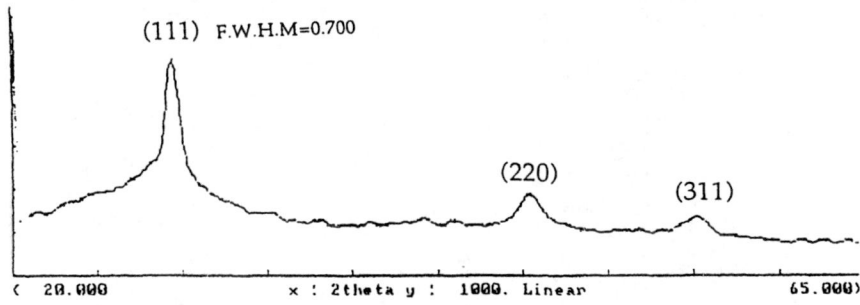

Fig. 3 The X-ray diffraction pattern of Si:H samples deposited by sputtering method at liquid nitrigon substrate temperature.

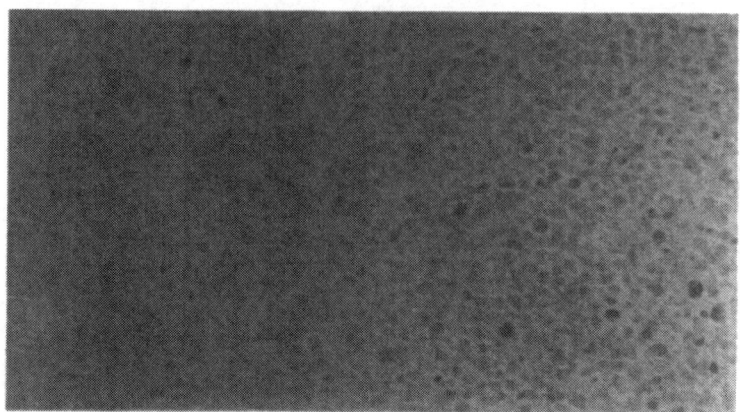

Fig. 4 The TEM bright field photograph of Si:H samples deposited by GD-CVD diluted-hydrogen
method with Xc=98%.

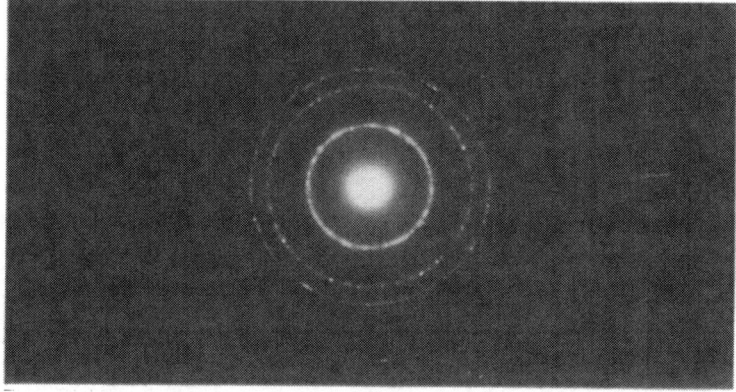

Fig. 5 The TEM bright field photograph and TED photograph of Si:H samples deposited by GD-CVD
hydrogen-atom-treatment method with HAT time 40s.

enhance the formation of μc-Si:H phase. μc-Si:H films with good crystallinity can be deposited by rf sputtering with a hydrogen gas pressure ratio above 0.96.

When the Ts was cooled down to liquid nitrigen temperature, the μc-Si:H flims can also be deposited with a hydrogen pressure of 6.0×10^{-2} torr and 300W. The X-ray diffraction pattern is shown in Fig. 3. In this figure, it was found that the lattice constant of the silicon bonding was changed resulting in a longer d spacing that might be due to stress change in the low temperature deposition. It is assumed that the crystal structure of silicon was changed from cubic system to orthorhombic system. The sample deposited with 250W at LN$_2$ cooled Ts showed amorphous phase only.

For Si:H samples deposited by GD-CVD with diluted hydrogen method, a hydrogen gas dilution ratio (Xc=H$_2$/(H$_2$+SiH$_4$)), of 0.95-0.98, and a Ts of 250°C was used for the deposition of μc-Si:H films with good crystallinity as identified from the X-ray and Raman scattering data The average grain size was calculated to be about 13 nm at Xc=98%. The grain size can also be estimated from the TEM bright field photograph as shown in Fig 4. to be around 15nm. The sample deposited at Xc=93.5% showed amorphous phase and there are microcracks presented as those usually occured in the sputtering prepared samples.

Also, the HAT method with a 20s HAT step produced amorphous silicon phase as-identified from the TEM photograph and the Raman spectra. When the HAT time was increased to 40s, μc-Si:H films can also be produced with the average grain size of 5nm. The TEM bright field and TED photographs are shown in Fig 5, which identifies the formation of μ c-Si:H but the grain size was smaller than the 98% diluted samples(10 nm). When a longer HAT time of 50s was applied, it resulted in a film that was characterized to contain only amorphous phase as identified by the Raman and X-ray measurement. But the TEM bright field photograph showed that it contains very small grains as well as lattice distortion as that observed in the cooled sample.

The hydrogen content calculated from the wagging mode of the Si-H bonding in the films prepared by both methods show the same tendency that as the μc-Si:H film was formed the hydrogen content was decreased to less than 10 atomic percent. The optical bandgap also showed the same tendency that it was narrowed down to less than 1.5eV as the μc-Si:H was formed.

The comparison of properties characterized by different methods are summarized in Table I. However, there are distinct differences in the properties such as Raman shift between the two category of samples that it showed a better bond structure quality in the GD-CVD sample but larger grain size was obtained in the sputtered samples. It may be indicative of higher degree of crystallinity and possibly less plasma damage results that could be obtained from the method of GD-CVD.

From the above results, it is concluded that the appropriate hydrogen incorporation during deposition is a necessary condition for the formation of μc-Si:H film either in GD-CVD of rf sputtering with hydrogen gas ratio higher than 90%. The most possible mechanism describing the formation of μc-Si:H film might be the "etching" effect of hydrogen atom during film deposition. The etching mechanism selectively eliminates the energically less stable site which is responsible for the formation of μc-Si:H films.

CONCLUSIONS

μc-Si:H films can be grown by RF sputtering in a hydrogen-argon mixture and by GD-CVD using hydrogen-dilution or hydrogen atom treatment. For RF sputtering, a gas pressure ratio, P(H$_2$)/(P(H$_2$)+ P(Ar)), above 0.96, a RF power of 300 W and a substrate temperature of 177°C

Table I Comparison of measurement data of some typical Si:H samples deposited by sputtering and GD-CVD.

METHOD/CONDITIONS				RAMAN Peak	XRD L	Hydrogen Content	Bandgap
Method	Conditions	rf Power	Ts (C)	(cm-1)	(nm)	(%)	(eV)
Sputter	83 %	300 W	180	--	--	21.7	1.74
Sputter	91 %	300 W	180	511	14.5		1.34
Sputter	94 %	300 W	180	511	13.2	7.9	1.45
Sputter	96 %	300 W	180	511	18.7	8.4	1.34
Sputter	93 %	250 W	180	511	11.7		1.36
Sputter	93 %	300 W	100	511	10.4	20.1	1.78
CVD-D	93.5 %	0.1 W/cm2	250	500	--	14.8	1.70
CVD-D	95 %	0.1 W/cm2	250	500	7.5	8.3	1.63
CVD-D	98 %	0.1 W/cm2	250	501	13.0	6.3	1.48
CVD-H	20 s	0.1 W/cm2	250	--	--	8.5	1.65
CVD-H	40 s	0.1 W/cm2	250	516	5.0	6.9	1.57
CVD-H	50 s	0.1 W/cm2	250	--	--	7.9	1.66

CVD-D: diluted-hydrogen method
CVD-H: hydrogen-atom-treatment method
Conditions : hydrogen pressure ratio for sputtering,
             hyd. gas ratio for CVD-D, HAT time for CVD-H
Ts : substrate temperature
L : average grain size calculated from FWHM of XRD peak

was used for the deposition of μ-Si:H films with good crystallinity.. For hydrogen diluted GD-CVD, a hydrogen gas dilution ratio Xc=95%-98%, and a substrate temperature of 250 °C was used for the deposition For films deposited by both methods, their properties were correlated using TEM, X-ray diffraction, and Raman scattering data. Films deposited by the two methods all exhibited comparable grain sizes in the range of 10-18 nm. The reduction in the optical bandgap and the hydrogen content in the films prepared by both methods show that they have the same tendency and they are some examples of the many similarities. However, there still exist distinct differences in the properties (e.g. Raman shift), which is the indicative that higher degree of crystallinity and possibly less plasma damage results were obtained by the method of GD-CVD.

REFERENCES

[1] K. Kumeda, Y. Yonezawa, A. Morimoto, S. Suda and T. Shimizu, J. Non. Cryst. Solid, 59&60, 775 (1983).
[2] C. C. Tsai, G. B. Anderson, R. Thompson, J. Non. Cryst. Solid, 137, 673 (1991).
[3] M. Fang, J. B. Chevrier and B. Drevillon, J. Non. Cryst. Solid, 137&138, 791 (1991).
[4] Y. J. Lim, J. Jang, C. Lee, J. Non-Cryst. Solids, 137&138, 705, (1991).
[5] M. Jonouchi, F. Moriyama and T. Miyasato, Jpn, J. Appl. Phys., 29, L3556 (1990).
[6] M. Cardona, Phys. Stat. Sol., B118, 463 (1983)

THIN FILM TRANSISTORS MADE FROM HYDROGENATED MICROCRYSTALLINE SILICON

K.C. WANG*, B.Y. CHEN**, K.C. HSU*, T.R. YEW***, AND H.L. HWANG*
*Department of Electrical Engineering, National Tsing-Hua University,
**Department of Materials Science and Engineering, National Tsing-Hua University,
***Materials Science Center, National Tsing-Hua University, Hsinchu, Taiwan, 30043, ROC.

ABSTRACT

Microcrystalline silicon films were deposited by diluted-hydrogen method and hydrogen-atom-treatment method at 250°C in a plasma enhanced chemical vapor deposition system and they were characterized by nuclear magnetic resonance, Raman spectroscopy, and optical bandgap measurements. One-mask a-Si:H thin film transistors (TFT's) were fabricated with those microcrystalline materials as the channel layer. The highest electron mobilities of the TFT's fabricated by diluted-hydrogen method and hydrogen-atom-treatment method were 1.23 and 1.04 cm^2/V•s, respectively without any thermal treatment steps.

INTRODUCTION

Hydrogenated amorphous silicon (a-Si:H) films are indispensable for many large area devices such as solar cells, contact image sensors, and thin film transistors (TFT's). The factors limiting their applications are the low electron mobilities and light induced degradation of electrical conductivity of a-Si:H [1], which are generally related to the intrinsic network structure [2-4]. Changing instrinsic network of a-Si:H into microcrystalline phase could be the most effective way to improve the characteristics of devices mentioned above. Microcrystalline silicon films could be deposited by diluted-hydrogen method and hydrogen-atom-treatment method at 250°C. In this paper we report Si:H thin film transistors fabricated with the microcrystalline silicon as the channel layer. These materials were characterized by nuclear magnetic resonance (NMR), Raman spectroscopy, optical bandgap measurements. The I_{ds}-V_{ds} characteristics of TFT's were determined by HP4145 meter.

EXPERIMENTAL METHODS

The conventional (Conv.) method and hydrogen diluted (Dilu.) method employed a $H_2/(SiH_4+H_2)$ fraction of 40% and 90-98%, respectively. The operation pressure was 0.5 Torr, and the RF power density was 0.1 W/cm^2. The hydrogen-atom-treatment (Htr) method involved alternately exposing the growing film surface to hydrogen atoms which are generated by a RF plasma in the same deposition chamber. The growth time was 10 s, the other growth conditions were the same as the diluted-hydrogen method, and the hydrogen treatment time was 40 s. The $H_2/(SiH_4+H_2)$ fraction in the deposition step of Htr method was 90%, the operation pressure was 1.1 Torr, and the RF power density was 0.1 W/cm^2, while in hydrogen treatment step the SiH_4 flow was turned off. The deposition temperature was fixed at 250°C.

The nuclear magnetic resonance (NMR) studies [5] were employed to analyze the silicon-hydrogen bonding configurations. The ^1H NMR spectra of hydrogenated silicon films were measured with Bruker MSL-300 FT-NMR at a Larmor frequency of 300.13 MHz. The Raman scattering spectra were measured from the right angle scattered light with an excitation of a 150mW, 488nm argon ion laser. The optical bandgap were determind from Tauc's plot using UV-VIS absorption spectrophotometry.

For convenience, one-mask structure was adopted to fabricate TFT, and it is shown in Fig. 4. In order to confirm the effect of structure change of the materials studied above, thin film transistors based on hydrogenated silicon were fabricated. The Si:H materials made with hydrogen dilution and atomic hydrogen treatment were used as the channel layer of the TFT's. Inverted staggered structures of TFT's with PECVD deposited silicon nitride (SiN$_x$:H) as the gate insulator and n$^+$ a-Si:H for ohmic contact with aluminum source/drain metal were fabricated. The W/L was 2500 µm/ 50 µm. The deposition conditions and characterization of the layers used for the TFT's are summarized in Table I.

RESULTS AND DISSCUSION

Fig. 1 shows the results of NMR spectra. Fig. 1 (a) shows the NMR spectrum for the Si film deposited by the convetional method (H$_2$/(SiH$_4$+H$_2$) : 40%). The spectrum includes a narrow Lorentzian-shape spectrum with full-width half-maximum (FWHM) of about 3.4 KHz and a broad spectrum with FWHM of about 23.5 KHz. The narrow line-shape corresponds to the randomly distributed Si-H structure in the a-Si:H network, and the broad spectrum corresponds to the region associated with clustered hydrogen which may contain internal surface, poly-hydride groups, and poly-silane chains. Fig. 1 (b) shows the NMR spectrum of the Si film deposited by Dilu. method in which both the Lorentzian-shape spectrum and the Gaussian-shape spectrum were narrower than those of Fig. 1 (a). Fig. 1 (c) shows the NMR spectrum of the Si film deposited by Htr. method. Sharp line-shape spectrum were obtained from the samples shown in Figs. 1 (c) and (d) .The sharp line-shape should come from molecular hydrogen.

The Raman spectra were investigated to study the microstructure of Si films deposited by the methods mentioned above. The typical Raman shift is peaked at 480 cm^{-1} for amorphous Si and peaked at 520 cm^{-1} for crystalline Si. Fig. 2 shows the Raman spectra of the Si films deposited by (a) Conv. method, (b) 90% Dilu. method, (c) 98% Dilu. method, and (d) 90% Htr. method. Figs. 2 (a) and (b) show the broad spectra with a peak at around 480 cm^{-1} and having a FWHM of 60 cm^{-1}. The result means that the silicon films deposited by Conv. method and 90% Dilu. method were mostly amorphous phase and possibly with some small fraction of microcrystalline phase. Figs.2 (c) and (d) both show the sharp spectrum with a peak centered at 520 cm^{-1} having a FWHM close to 15 cm^{-1}. The fraction of microcrystalline phase is much enhanced in the silicon film deposited by 98% Dilu. and 90% Htr. method.

The room temperature optical (Tauc's) bandgap of amorphous silicon is about 1.7 eV, and the optical bandgap of crystalline silicon is 1.1 eV. Fig. 3 shows the optical bandgap of silicon films deposited by different methods. The optical bandgap of sample made by the Conv. method (40% Dilu.) is 1.69 eV, and this value increased a little for the sample made by 90% Dilu. method. As the H$_2$/(SiH$_4$+H$_2$) fraction was further increased, the optical bandgap descreased. 1.64 eV bandgap was obtained when the H$_2$/(SiH$_4$+H$_2$) fraction was 95%. When H$_2$/(SiH$_4$+H$_2$) fraction was increased to 98%, the bandgap was reduced to 1.48 eV. This result suggests that it is easier to obtain microsrystalline silicon with higher H$_2$/(SiH$_4$+H$_2$) fraction in the Dilu. method. With the specified conditions, hydrogen atom treatment method produced sample with a bandgap of 1.56 eV. The decrease of bandgap with increasing hydrogen dilution during deposition suggests that the microstructure was changed with increasing either the volume fraction of the c-Si or the crystallite size. It is also noted that the bandgap of sample deposited by combining the 40s hydrogen plasma treatment with the 90% hydrogen dilution is much lower than those of the 90% to 95% hydrogen-diluted samples, except it was higher than that of the 98% diluted sample. These results show that bandgap reduction of the diluted hydrogen samples

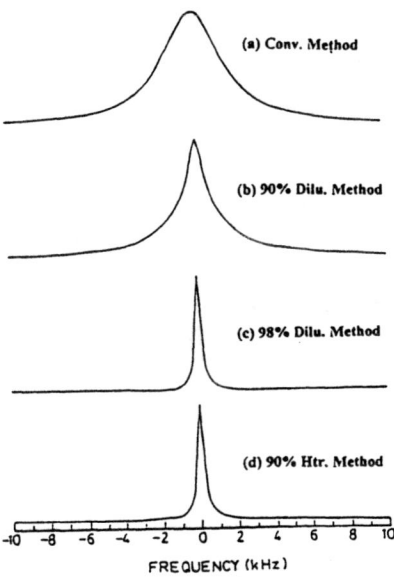

Fig. 1 1H NMR spectra of Si:H samples prepared by (a) Conv., (b) 90% Dilu., (c) 98% Dilu., and (d) Htr. methods.

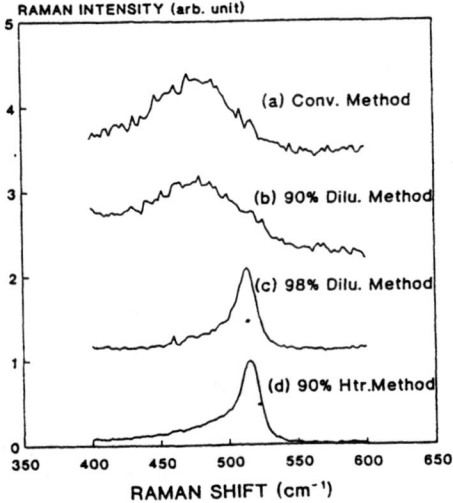

Fig. 2 Raman scattering spectra of Si:H samples prepared by (a) Conv., (b) 90% Dilu., (c) 98% Dilu., and (d) Htr. methods.

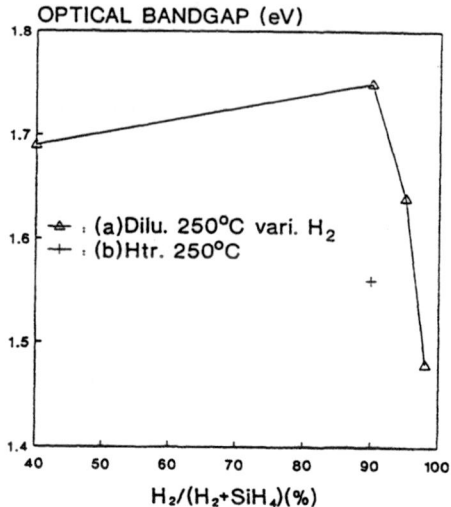

Fig. 3 The optical bandgap of Si:H samples prepared by (a) hydrogen dilution with different dilution ratios and (b) 90% Htr. methods

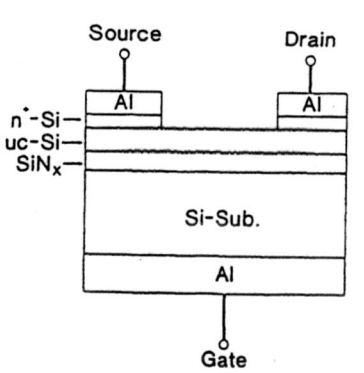

Fig. 4 Structure of thin film transistors.

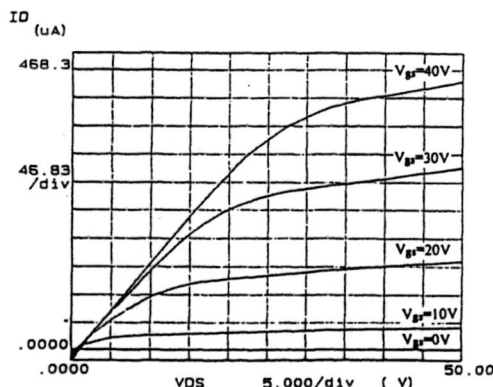

Fig. 5 I_{ds}-V_{ds} characteristics of TFT made by 98% Dilu. method.

is effective in the range between 95% to 98%, and the Htr method is also effective for further bandgap reduction.

Samples prepared by the above methods showing distinct microcrystalline characteristics like the sharp line-shape NMR spectrum, Raman shift, optical bandgap reduction, etc., suggest that the treated films should possess a more compact structure than the conventional ones. These results suggest that with appropriate hydrogen incorporation or atomic hydrogen treatment during the deposition, the degree of crystallinity of hydrogenated silicon films can be systematically adjusted.

Table II shows the electron mobility of TFT's made by Conv. method, Dilu. method, and Htr. method. The mobility of the TFT made by 98% Dilu. method is the highest one. The I_{ds}-V_{ds} characteristics of TFT with the highest electron mobility is shown in Fig. 5.

When the V_{gs} was 40V, the mobilities were 0.53, 0.828, 1.227, and 1.036 $cm^2/V \bullet s$, respectively, for TFT's made by Conv. method, 90% Dilu. method, 98% Dilu. method, and 90% Htr method without any thermal treatment steps. The results of I-V characteristics and electron mobility of the TFT's suggest that both the Dilu. and Htr. methods are very effective to improve the performance of TFT's. So far, TFT made by the 98% Dilu. method gave the highest electron mobility.

The reasons for such a mobility improvement might be due to the formation of microcrystalline phase in the Si:H film since the electron mobility inside the crystalline grain is much higher than that in the amorphous networks. A "transition" amorphous silicon layer of 100-300 Å thick might be grown between the amorphous substrate (SiN_x) and the μc-Si, which is called as a "incubation period" [6], during the film deposition of μc-Si. The existence of the transition layer together with the interface scattering between the channel layer and the gate insulator may reduce the electron mobility of the μc-Si channel layer. The relationship between the film structure and the field effect mobility are being under investigation in our laboratory. More accurate estimation of the electron mobility should be done in our future work, however, the electron drift mobilities in these Si:H films had been proved to be enhanced compared to the conventional ones. Wang et.al. had observed that μc-Si film could be grown on SiO_2 without transition to amorphous layer when a surface sulfurization process of the substrate was employed before the deposition [7]. This might be a solution for further study in the application of Si:H film to TFT's.

The typical on/off current ratio of TFT's is 10^3. The use of microcrystalline material for TFT's active layer might induce a degradation of on/off current ratio. Improvements on the on/off current ratio of TFT's are under way by certain post thermal annealing steps, and the results will be published in the near future.

CONCLUSIONS

μc-Si:H phase was formed as the hydrogen dilution was higher than 95% or atomic hydrogen treatment was employed with sufficient hydrogen atom treatment time. These properties can be characterized by NMR and Raman spectra.

Si:H samples showed μc-Si characteristics when incorporating a large amount of hydrogen, this suggests that these films possess a more compact structure than the conventional one's. These results also suggest that with appropriate hydrogen incorporation or with the combination of atomic hydrogen treatment during the deposition, the degree of crystallinity of hydrogenated silicon films can be systematically adjusted.

The electron drift mobility was found to be improved for TFT's using channel layer deposited by dilution-hydrogen method and hydrogen-atom-treatment methods. In our current

Table I. Deposition conditions of TFT Layers.

LAYERS	GAS	FLOW RATE RATIO	PRES. (Torr)	RF POWER (W/cm^2)	DEP. RATE (Å/s)	THK. (Å)
SiN$_x$	SiH$_4$/NH$_3$	1 : 4	0.5	0.15	0.1	2750
Conv.	SiH$_4$/H$_2$	1.6 : 1	0.5	0.1	1.2	3000
Dilu. 90%	SiH$_4$/H$_2$	1 : 9	0.5	0.1	0.76	3000
Dilu. 40%	SiH$_4$/H$_2$	1 : 49	0.5	0.1	0.17	3000
Htr. 40s	SiH$_4$/H$_2$	1 : 9	1.1	0.1	0.16	3000
n$^+$ a-Si:H	SiH$_4$/PH$_3$*/H$_2$	4 : 20 : 1	1.0	0.12	1.6	500

Conv. : convention method
Dilu. : hydrogen-dilution method
Htr. : hydrogen-atom-treatment method

Pres. : deposition pressure
PH$_3$* : 1% phosphine diluted in helium

Table II. Electron mobility of TFT's made by several methods.

Method	Electron Mobility (cm^2/V$\bullet$s)
Conv.	0.53
90% Dilu.	0.83
98% Dilu.	1.23
90% Htr.	1.04

work, further studies on the standard staggered TFT's are being carried out to investigate the effect of the microcrystalline silicon on the on/off current ratio, threshold voltage, and leakage current of TFT's.

ACKNOWLEDGEMENTS

This work was supported by the Republic of China National Science Council under a Contract No. of NSC82-0404-E-007-230.

REFERENCES

[1]. D.L. Staebler and C. R. Wronski, J. Appl. Phys. 51(1980) 3262.
[2]. J.I. Pankove and J.E. Berkeyheiser, Appl. Phys. Lett. 37(1980) 705.
[3]. H. Dersch, J. Stuke,and J. Beicher, Appl. Phys. Lett. 38 (1981) 456.
[4]. S.R. Elliolt, Philos. Mag. B39(1979) 349.
[5]. S. Hayashi, J. Non. Cryst. Solid 59&60(1983) 779.
[6]. C.C. Tsai, B. Anderson, R. Thompson, MRS Symp. Proc., 192(1990) 475.
[7]. K.C. Wang, T.R. Yew, H.L. Hwang, MRS Symp. Proc., 1993 impress.

TRANSIENT PHASES AND THEIR TRANSITION TEMPERATURES OF a-Si IN NON-ISOTHERMAL PROCESSES

MASAKUNI SUZUKI AND AKIO KITAGAWA
Kanazawa University, Department of Electrical & Computer Engineering,
Faculty of Technology, Kodatsuno 2-40-20, Kanazawa 920, Japan

ABSTRACT

The heating rate dependence of the phase transition temperature was formulated based on the temperature dependence of nucleation of a new phase. The glass transition temperature of a-Si was explained in terms of van der Waals fluid of a-Si pseudo-molecules which are produced by the fragmentation of continuous random networks of Si atoms. Transient phases and their transition temperatures as a function of the heating rate are summarized in the phase diagram.

1. INTRODUCTION

Tetrahedrally bonded amorphous solids such as amorphous silicon (a-Si) exhibit peculiar behavior in rapid thermal processes. Of particular interest are the reduced melting temperature,[1-3] direct amorphization from the supercooled melt,[4-9] and unique recrystallization behavior.[3,10-13] Those phenomena are observed almost exclusively in rapid thermal processes such as laser or electron beam irradiation, where the heating and cooling rates reach 10^6 to 10^{12} K/s. On the other hand, a-Si usually exhibits no other phenomenon than crystallization in non-isothermal differential scanning calorimetric (DSC) measurements, where the heating rates are lower than ~1 K/s. A variety of a-Si observed in heating and cooling processes must be controlled by the heating or cooling rate. We have, therefore, developed a procedure to estimate the heating rate dependent transition temperature from the transient time of the phase transition measured in a series of isothermal anneals.

In this paper, various features of a-Si observed in non-isothermal ramp annealings are arranged from the view point of the heating rate, and a model for the glass transition in a-Si is presented based on van der Waals fluid of a-Si pseudo-molecules. Results are summarized in a phase diagram of a-Si giving transient phases and their transition temperatures as a function of the heating rate.

2. Reduced melting temperature and the glass transition of a-Si

The melting temperature of a-Si has been estimated to be ~1400 K based on DSC measurements of the heat of crystallization.[14] The transition from the solid Si to the liquid Si involves a fundamental change in bonding from the covalent 4-fold coordination to the metallic ~6-fold coordination,[15,16] so the transition is discontinuous and first order.[1] The supercooled metallic (mt) liquid Si has been experimentally detected above 1400 K by transient reflectance and transient conductance measurements.[3,15-18] It should be pointed out that the supercooled mt-liquid Si appears in very rapid heating. The lowest heating rate to produce supercooled mt-liquid Si is inferred to be ~10^8 K/s from the data obtained by Olson et al. using pulsed dye laser irradiation.[18]

Mat. Res. Soc. Symp. Proc. Vol. 321. ©1994 Materials Research Society

The glass transition in a-Si

When the heating rate is not high enough to produce the supercooled mt-liquid state, a-Si crystallizes at some temperature below 1400 K.[18,19] There are, however, some indications that the softening of a-Si occurs at ~1200 K when the heating rate is higher than ~10^2 K/s. In fact, further low tempera-ture melting of a-Si down to ~1170 K is reported to occur under pulsed elec-tron beam irradiation of ~10^6 K/s.[2] Lower temperature melting of a-Si is further supported by the fact that enhanced diffusion of impurity atoms occurs at ~1200 K in ramp anneals of ~10^2 K/s. Such a deep supercooled liquid state is not metallic, since its conductance and reflectance are almost the same as those of the solid state a-Si. The deep supercooled liquid is strongly suggested to be ~4-fold coordination from the fact that unique microstruc-tures such as disk-shaped Si flakes having amorphous cores were observed in fine-grained poly-Si produced by explosive crystallization.[13] Thus, the behav-ior of a-Si in the heating rate between ~10^2 to ~10^8 K/s is summarized as follows: a-Si turns to supercooled non-metallic liquid at around 1200 K and then crystallizes at a certain temperature below 1400 K. This behavior of a-Si is quite similar to the glass transition followed by crystallization observed for glasses in the heating processes, thus the temperature at which the deep supercooled liquid state appears could be the glass transition temperature.

The glass transition temperature T_g of a-Si has been calculated by Turnbull[23] and Stiffler et al.[24] The former can not be applied to high heating rate processes as 10^2 K/s, since continuous random networks formed by tetrahe-drally bonded atoms are too rigid to exhibit fluidity in high heating rate processes.[23] The latter is also not to be applied to the glass transition of a-Si, since metallic glass has not yet been observed for Si under any quench-ing conditions.[25] There is no satisfactory model for the glass transition in a-Si so far.

The a-Si involves a high density of dangling bonds[26] and weak bonds,[27] thus a-Si is expected to fragment with increasing temperature if crystalliza-tion does not intervene. The fragmentation would at last produce relatively stable clusters, which may be called pseudo-molecules. Polk's continuous random network (CRN) model for a-Si and a-Ge,[28] having no dangling bond in-side, could be applied to pseudo-molecules of interest. Then, such pseudo-molecules are presumed to consist of, at most, a few hundred Si atoms. The solid consisting of pseudo-molecules is a van der Waals (VDW) solid, which would turn to VDW fluid by further heating. VDW fluid of pseudo-molecules must be semiconductive (sc) liquid, since constituents of the fluid are ~4-fold coordination. These processes are capable of providing the transition from solid a-Si to supercooled sc-liquid Si.

3. THE PHASE TRANSITION TEMPERATURE AS A FUNCTION OF THE HEATING RATE

The transient behavior of phase transitions in condensed systems have been treated by many authors.[29-32] Phase transitions under the isothermal condition take place by nucleation of the new phase followed by successive growth of nuclei. In the classical theory, nucleation is assumed to occur by the formation of clusters of the new phase. The free energy of formation of a cluster goes through a maximum and then decreases with the cluster size be-cause of their large surface-to-volume ratio. As a result, an incubation period or induction time for nucleation t_{in} is required for the cluster to surpass a critical size.[29-36] The phase transition begins when a nucleus is produced in the initial phase where there are no preexisting clusters. Then, the system passes over the point of no-return in the course of the phase transition. Although it is not easy to detect experimentally such a strict

744

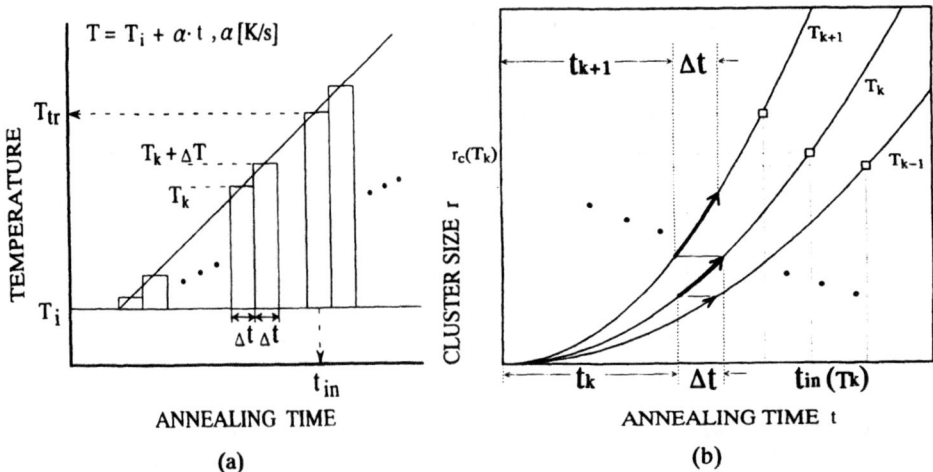

$$T = T_i + \alpha \cdot t \,, a\,[K/s]$$

TEMPERATURE

T_{tr}

$T_k + \Delta T$

T_k

T_i

$\Delta t\, \Delta t$

t_{in}

ANNEALING TIME

(a)

$r_c(T_k)$

CLUSTER SIZE r

T_{k+1}

T_k

T_{k-1}

t_{k+1} Δt

t_k Δt $t_{in}(T_k)$

ANNEALING TIME t

(b)

Fig.1   Non-isothermal rate-scan annealing and nucleation processes. (a) A medley of ascending isothermal annealings. (b) Cluster size in successive isothermal annealings is inherited from previous annealings.

onset of the transition, t_{in} could be approximated by extrapolating, for example, the plot of the crystalline fraction vs annealing time to a zero level fraction.[32,36]

If t_{in} is known as a function of temperature, the heating rate dependent transition temperature T_{tr} in the rate scan annealing can be estimated phenomenologically as follows: A rate scan annealing is regarded as a medley of ascending isothermal anneals as shown in Fig.1 (a). The nucleation processes at each new temperature are assumed to be inherited from the previous anneals as shown in Fig.1 (b). Thus, the inherited time $t_k^{(e)}$ from the previous annealings plus the annealing time Δt at temperature T_k is reduced to $t_{k+1}^{(e)}$ which is succeeded to the next annealing at T_{k+1} ($= T_k + \Delta T$) as given below:

$$t_{k+1}^{(e)} = \gamma(T_{k+1})\,(t_k^{(e)} + \Delta t)\,\frac{t_{in}(T_k + \Delta T)}{t_{in}(T_k)} \qquad (1)$$

where $\gamma(T_{k+1})$ is a parameter involving nonlinear characters of the nucleation process. $\gamma(T_k)$ approaches unity when Δt approaches zero.

The system reaches the transition temperature T_{tr} in the non-isothermal rate scan anneal when the sum of the reduced annealing time reaches $t_{in}(T_{tr})$. This situation is expressed by the following equation :

$$\int_{T_i}^{T_{tr}} \frac{1}{t_{in}(T)}\frac{dT}{\alpha} = 1 \qquad (2)$$

where T_i is the initial temperature at which the rate scan anneal starts and α is the heating rate. This equation provides the relation between the transition temperature T_{tr} and the heating rate α.

Solid Phase Crystallization of a-Si

At heating rates lower than ~10 K/s, only solid phase crystallization is observed for a-Si. The crystallization temperature T_c of amorphous solids can not be determined uniquely, since it depends on the heating rate. T_c of a-Si is, in fact, a function of the heating rate.

In the case of the phase transition from a-Si to crystalline Si, the transient time t_{tr} has been measured using X-ray diffraction, Raman scattering, transmission electron microscopes, conductance changes etc.,[32-36] and the experimental data of t_{tr} obtained from those measurements in a series of isothermal anneals fit well the following Arrhenius form with the activation energy E_{in}.[32-26]

$$t_{tr}(T) = \tau_0 \, exp(\frac{E_{in}}{kT})$$ (3)

where T is the isothermal annealing temperature, and τ_0 is the pre-exponential factor. Experimental results of t_{tr} involve, more or less, growth processes of crystallites after nucleation. Data of $t_{tr}(T)$ in ref.35 contain minimal traces of the growth process of crystallites after nucleation, since the detection level of crystallites is reduced as low as possible. Those data could be approximated to be $t_{in}(T)$, so eq.(2) can be calculated in the case of solid phase crystallization.

Transient phase diagram

Various features of a-Si observed in ramp annealings are arranged in a phase diagram, in which the phase transitions are shown as a function of the heating rate.

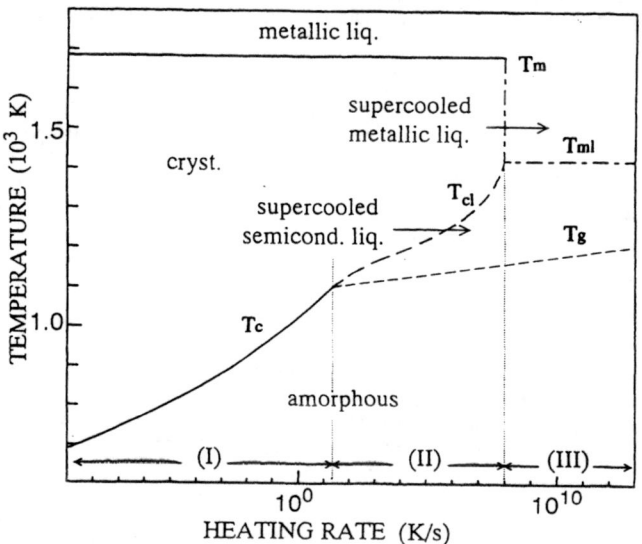

Fig.2 Transient phase diagram for a-Si. T_c, T_{cl}: crystallization temperatures in the liquid state and in supercooled liquid state respectively. T_g: the glass transition temperature. T_{ml}: the melting temperature (metallic liquid).

A solid curve T_c in Fig.2 is the calculated crystallization temperature from eq.(2) using data of $t_{in}(T)$ in ref.35. A measured T_c at 40 K/min obtained from DSC[37] coincides well with the calculated T_c shown in Fig.2. The calculated T_c is invalid above T_g. The supercooled liquid phase may appear in advance of crystallization in some heating conditions, as mentioned in the previous section. At present we do not have any data about $t_{in}(T)$ for nucleation of the supercooled sc-liquid phase, so numerical calculation of eq.(2) for T_g can

746

not be performed. T_g, however, could be inferred from experimental data reported so far. T_g at $\sim10^6$ K/s is taken to ~1170 K from the experimental results reported by Baeri et al.[2] T_g at 40 K/min is higher than at least ~1020 K, since T_c at 40 min is reported to be 1014 K.[37] Taking into account the weak heating rate dependence of T_g, it could be taken to around 1100 K at $\sim10^2$ K/s. Then a dotted curve for T_g is drawn as shown in Fig.2, where T_g intersects T_c at the point ($\sim2\times10^1$ K/s, ~1100 K). In this scheme, the supercooled sc-liquid state appears in the ramp anneals with α higher than $2\sim3\times10^1$ K/s.

Numerical calculation of the heating rate dependent crystallization temperature, T_{cl}, above T_g is once again not possible because of the lack of data about $t_{in}(T)$ for crystallization in the sc-liquid phase. T_{cl}, however, can be directly inferred from experimental results under some heating rates. For example, the transient reflectance measurements using cw Ar$^+$ ion laser irradiation showed that heating around 10^6 K/s produces not supercooled mt-liquid Si but poly-Si,[19] so that T_{cl} at $\sim10^6$ K/s is estimated to be lower than 1300 K.

Pseudo-molecules are expected to dissociate into atoms at ~1400 K, since the heat of crystallization calculated for CRN model of a-Si[38] agrees roughly with that of a-Si which gives ~1420 K as T_{ml}.[14] Thus, T_{cl} must terminate at ~1400 K on the line of α at 10^8 K/s, so a broken curve for T_{cl} could be drawn from ($\sim2\times10$ K/s, ~1100 K) to ($\sim10^8$ K/s, ~1400 K) as shown in Fig.2. An almost vertical line on 10^8 K/s in Fig.2, which separates the supercooled mt-phase from the crystalline phase, should be drawn from ~1400 K to the melting temperature T_m (1685 K) of crystalline Si. T_{ml} of a-Si is defined only for α larger than $\sim10^8$ K/s.

The behavior of a-Si in the rate scan anneal is, thus, categorized in three characteristic regions as shown in Fig.2. In the region (I) of low heating rates, solid-phase crystallization exclusively takes place at T_c. In the region (II), a-Si turns to the supercooled sc-liquid state at T_g and then crystallizes at T_{cl}. On the other hand, in the region (III) of high heating rates, crystallization does not occur. The a-Si goes into the supercooled mt-liquid phase through the supercooled sc-liquid phase. Similar phase diagrams for other amorphous materials will be obtained applying similar procedures. Most glasses do not have the supercooled mt-liquid phase, so rather simple diagrams will be obtained if they do not exhibit phase separation. Phase diagrams for the cooling processes could also be developed.

4. CONCLUSION

Transient phases and their transition temperatures as a function of the heating rate are summarized in the phase diagram. The heating rate dependence of the transition in the rate scan annealing is controlled by a competition between clusters of different new phases in which one exceeds first the critical size for stable growth. The phase transition temperature in a rate scan annealing was, thus, formulated as a function of the heating rate based on the temperature dependence of a new phase. The glass transition in a-Si was explained in terms of van der Waals fluid of a-Si pseudo-molecules which are produced by the fragmentation of continuous random networks of Si atoms.

ACKNOWLEDGEMENTS

The authors would like to express their thanks to Dr.S.Usui and Dr.D.P.Gosain of Sony Central Research lab., to Dr.S.Tsuda and Mr. S.Noguchi of Sanyo Functional Mater.Res.Center and to Prof. M.Kitao of Shizuoka University for providing useful data and helpful discussion. The authors also wish to thank Mr.Dean Schimpf for useful suggestions and encouragement. This work is supported in part by Betsukawa Foundation and Iketani Science Foundation.

REFERENCES

1) F.Spaepen and D.Turnbull, AIP Conf.Proc. 50,73(1979); B.G.Bagley and H.S.Chen, ibid., 50, 97 (1979).
2) P.Baeri, G.Fotti, J.M.Poate and A.G.Cullis, Phys.Rev.Lett.45,2036 (1980).
3) M.O.Thompson, G.J.Galvin, J.W.Mayer, P.S.Peercy, J.M.Poate, D.C.Jacobson, A.G.Cullis and N.G.Chew, Phys.Rev.Lett. 52, 2360 (1984).
4) R.Tsu, R.T.Hodgson, T.Y.Tan J.E.Baglin, Phys.Rev.Lett.42,1356 (1979).
5) P.L.Liu,R.Yen,N.Bloembergen and R.T.Hodgson,Appl.Phys.Lett.34,864(1979).
6) A.G.Cullis, H.C.Webber and N.G.Chew, Phys.Rev.Lett.49,219 (1982).
7) M.O.Thompson, J.W.Mayer, A.G.Cullis, H.C.Webber, N.G.Chew, J.M.Poate and D.C.Jacobson, Phys. Rev. Lett. 50, 896 (1983).
8) P.V.Evans,G.Devaud,T.F.Kelly and Y-W.Kim,Acta.Metall.Mater.53,719(1990).
9) T.Sameshima and S.Usui, J.Appl.Phys.70,1281 (1991).
10) J.Narayan,C.W.White,O.W.Holland and M.J.Aziz,J.Appl.Phys.56,1821(1984).
11) W.Sinke, F.W.Saris, J.C.Barbour and J.W.Mayer, J.Mater.Res. 1,155 (1986).
12) P.S.Peercy,M.O.Thompson and J.Y.Tsao,Mater.Res.Soc.Symp.Proc.74,15(1987).
13) D.H.Lowndes, S.J.Pennycook, G.E.Jellison,Jr., S.P.Withrow and D.N.Mashburn, J. Mater. Res. 2, 648 (1987).
14) E.P.Donovan, F.Spaepen, D.Turnbull,J.M.Poate and D.C.Jacobson, Appl. Phys. Lett. 42, 698 (1983); J. Appl. Phys. 57, 1795 (1985).
15) K.Murakami, H.C.Gerritsen, H.van Brug, F.Bijkerk, F.W.Saris and M.J.van der Wiel, Phys. Rev. Lett. 56, 655 (1986).
16) H.C.Gerritsen, H.van Brug, F.Bijkerk, K.Murakami and M.J.van der Wiel, J. Appl. Phys. 60, 1774 (1986).
17) D.H.Auston, C.M.Surko, T.N.C.Vekatesan, R.E.Slusher and J.A.Golovchenko, Appl. Phys. Lett. 33, 437 (1978).
18) G.L.Olson, J.A.Roth, E.Nygren, A.P.Pogany and J.S.Williams, Mat. Res. Soc. Symp. Proc. 74, 109 (1987).
19) S.A.Kokorowski,G.L.Olson,J.A.Roth and L.D.Hess,Phys.Rev.Lett.48,498(1982).
20) J.Narayan, O.W.Holland, R.E.Eby, J.J.Wortman, V.Ozguz and G.A.Rozgony, Appl. Phys. Lett. 43, 957 (1983).
21) R.Kalish, T.O.Sedgwick and S.Mader, Appl. Phys. Lett. 44, 107 (1984).
22) F.F.Morehead and R.T.Hodgson, Mat. Res. Soc. Symp. Proc. 35, 341 (1985).
23) D.Turnbull, Mater. Res. Soc. Symp. Proc. 7, 103 (1982).
24) S.R.Stiffler, P.V.Evans and A.L.Greer, Acta.Metall.Mater. 40,1617 (1992).
25) S.R.Stiffler, M.O.Thompson and P.S.Peercy, Phys. Rev. B, 43, 9851 (1991).
26) M.H.Brodsky,R.S.Title,K.Weiser and G.D.Pettit, Phys.Rev. B,1, 2632(1970).
27) T.Motooka and O.H.Holland, Appl. Phys. Lett. 61, 3005 (1992).
28) D.E.Polk, J.Non-cryst.Solids, 5, 365 (1971).
29) B.K.Chakraverty, Surf.Sci. 4, 205 (1966).
30) D.K.Kashchiev, Surf. Sci. 14, 209 (1969).
31) K.F.Kelton,A.L.Greer and C.V.Thompson, J. Chem. Phys. 79, 6261 (1983).
32) R.B.Iverson and R.Reif, J. Appl. Phys. 62, 1675 (1987).
33) N.Blum and C.Feldman, J.Non-cryst. Solids. 11, 242 (1972).
34) U.Koester, Phys. stat. sol. (a) 48,313 (1978).
35) R.Bisaro, J.Magarino, K.Zellama, S.Squelard, P.Germain and J.F.Morhange, Phys. Rev. B, 31,3568 (1985).
36) M.Suzuki, M.Hiramoto, M.Oguiura, W.Kamisaka and S.Hasegawa, Jpn. J. Appl. Phys. 27, L1380 (1988).
37) J.C.C.Fan and C.H.Anderson,Jr., J. Appl. Phys. 52, 4003 (1981).
38) G.A.N.Connel and W.Paul, J.Non-cryst.Solids, 8-10, 381 (1972).

Author Index

751

Subject Index

Langer-Schwartz model, 338
laser irradiation
    amorphization by, 429, 725
    annealing by, 719
    crystallization by, 393, 429, 671, 677, 689
    multiple-pulse, 665
lateral grain growth, 657
lead
    pyrophosphate, 19
    titanate, 561
lithium
    disilicate, 239
    tantalate, 603
luminescence, 363

magic-angle spinning, 123
mechanics, molecular, 71
melting, 3, 725
metals, 99
metamict, 25
metastable, 461
MgO, 609, 633
microcrystalline, 707
microdiffraction, electron, 117
microscopy, in situ, 271, 283, 579
microwave, 707
mirrors, x-ray, 215
modeling, numerical 223
modulus, mechanical, 203
molecular structures, 113
monticellite, 633
multilayers, 331

nanocrystals, 251, 331
    Si and Ge, 363
newberyite, 16
Ni-B, 65
Ni-Zr, 289, 331
nonequilibrium, 53
nucleation, 143, 223, 271, 355, 671
    classical theory of, 236
    heterogeneous, 349
    homogeneous, 233, 337
    internal, 239
    kinetics of, 283
    rate, time-dependent, 307
    transient, 223
nylon, 651

optical fiber, 209
order
    medium-range, 143
    short-range, 289
oxidation, wet, 615

phase
    diagram, 87, 743
    separation, 3, 531
phenolformaldehyde, 441
plastic yield, 161
polycarbonate, 543, 549
poly(e-caprolactone), 549
polyesters, 71

poly(ether ether ketone), 301
poly(ether imide), 301
polyethylene, terephthalate, 543
polymerization, 239
polymers, 99, 113
    blends, 47, 301, 531, 543, 555
    crystalline, 87
    dynamics of, 155
    glassy, 161
    liquid crystalline, 81
    morphology of, 511
    preceramic, 573
    semicrystalline, 511, 523
polysilazane, 573
polysulfone, 113
polytetrafluoroethylene, 155
positron annihilation, 47, 245
precipitates, 251, 263
precipitation, 467
pressure, hydrostatic, 59
pulsed-laser deposition, 567
pyroelectric, 603
pyrolysis, 123
pyrolytic conversion, 573
PZT, 591, 597

radial distribution function, 25, 39
Raman spectroscopy, 33, 350, 363, 393, 585, 713
reflectivity, 399
    time-resolved, 503
regrowth dynamics, 503
relaxation, 179, 197
    network, 203
    strain, 215, 375
    stress, 99
    structural, 161, 167
    time, 162
relocation, grain boundary, 668
roughening, 375, 491
roughness, 503
Rutherford backscattering, 585

Sb-Se, 429
scanning tunneling microscopy, 497
scattering
    neutron, 549
    small-angle, 549, 555
    x-ray, 549
shear, 523
Si, 167, 233, 417, 479, 497, 579, 585, 677, 695, 701, 707, 713, 719, 743
    microcrystalline, 731, 737
    polycrystalline, 657
Si-C, 387, 467, 473, 621
Si-Ge, 59, 307, 375, 461, 467, 485, 491, 615
    amorphous, 689
    polycrystalline, 689
silicide, cobalt, 271
    copper, 585
silicon nitride, 573
simulation, 295
    molecular dynamics, 65, 99, 135, 423
    reverse Monte Carlo, 65

CPSIA information can be obtained at www.ICGtesting.com
Printed in the USA
LVOW06s0750210514

386631LV00009B/307/P